The World of Mineral Deposits

Florian Neukirchen • Gunnar Ries

The World of Mineral Deposits

A Beginner's Guide to Economic Geology

 Springer

Florian Neukirchen
Berlin, Germany

Gunnar Ries
Marxen, Germany

ISBN 978-3-030-34345-3   ISBN 978-3-030-34346-0   (eBook)
https://doi.org/10.1007/978-3-030-34346-0

This Springer imprint is published by the registered company Springer Nature Switzerland AG
The registered company address is: Gewerbestrasse 11, 6330 Cham, Switzerland

Preface

Interest in where ore and energy resources can be found is not only relevant to geologists. This book is primarily intended for engineers, geographers, economists, chemists, mineral collectors, investors, and everyone else interested in the topic. Nevertheless, we are sure that geologists and geology students can also benefit from this book. A basic knowledge of geology is certainly helpful, although we have avoided or explained technical terms as much as possible. Background information is provided in introductory sections and boxes. Additional help is provided by a geological timescale in Sect. 1.5 and a glossary at the back of the book. Important ore minerals that are mentioned time and again throughout the book are presented in detail in Sect. 1.2, while others can be found in Chap. 2.

Our unconventional concept—a mixture of popular science, university textbook, and reference work—has obviously succeeded. The first edition of the German title quickly sold out and resulted in the authors consistently receiving very positive reviews and letters. We are pleased that the book has been translated into English and will be available to a larger audience. The translation is based on the second German edition and has been updated in a few places.

We attach particular importance to the processes that have led to the enrichment of corresponding metals. Our aim is to present them according to the current state of research in such a way that even complicated interrelationships can be understood. We show the interplay of different factors contributing to ore-forming processes. Deposits of high-tech metals such as tantalum and rare-earth elements are also explained in detail. While the main part of the book is sorted according to relevant processes, the second chapter offers an alternative starting point with an overview of individual metals. We have also explained modern applications of metals such as platinum in catalysts, which many may not think of at first glance. The chapter on fossil fuels also describes in detail increasingly important unconventional deposits including associated problems. Last but not least, non-metallic industrial rocks and minerals such as sand, gravel, limestone, and clay have not been ignored.

Some of the German deposits featured may seem exotic to readers from other continents because mining is currently pretty insignificant in Europe. However, they are well researched and serve as good examples for important deposit types found worldwide.

We thank Susanne Herting-Agthe from the Mineralogical Collections of the Technical University of Berlin for patiently digging out countless specimens from cabinets and showcases and providing us with photos from the archive. Photos of first-class mineral specimens from the Black Forest were taken with permission from a multivolume opus written by Gregor Markl. We are very happy to have been able to use them. We would like to thank Thorsten Eckardt for helping with the literature research and Lars Fischer, Walter A. Franke, Detlev Seibert, Eberhard Strehl, and Markus Hauser for helpful comments. We would also like to thank everyone who provided us with pictures, answered our questions, or otherwise supported us.

Berlin, Germany
Marxen, Germany
June 2019

Florian Neukirchen
Gunnar Ries

Contents

About the Authors

Florian Neukirchen is a mineralogist and non-fiction author. After his studies in Freiburg he worked on alkaline rocks at the University of Tübingen. Research trips led him to Oldoinyo Lengai in Tanzania, through the High Atlas in Morocco, and to the Ilimaussaq intrusion in Greenland. He currently lives in Berlin. He has written several books on geology in German.

Gunnar Ries studied mineralogy in Hamburg. His interest in the weathering behavior of East African carbonatites led him to Uganda and Tanzania. He works at CRB Analyse Service GmbH in Hardegsen (Germany) and in his spare time blogs about geoscientific topics at Scilogs.com.

Introduction

The consumption of primary raw materials has increased ever faster in recent decades, not least due to rapid economic growth in countries such as China, India, and Brazil. Petroleum and natural gas continue to be the most important energy sources driving engines and turbines worldwide, while petroleum is also the most important raw material for the production of plastics, various medicines, and other products of the chemical industry. Metals are an integral part of our world. Whether in the form of steel girders or copper wires, gold bars enclosed in safes or proudly presented silver jewelry, platinum in catalytic converters, or tantalum in electronic capacitors we use metals everywhere and in many different ways. At the same time, thanks to new high-tech applications the names of elements that until recently led a shadowy existence at the edge of the periodic table are on everyone's lips.

However, resources are limited and the effort to find and develop new mineral deposits is increasing. Questions as to how deposits formed and where they can be found are thus becoming increasingly important. Contrary to some pessimistic forecasts no commodities have yet been exhausted, although with some critical resources it may well become necessary to look for alternatives.

Fortunately, the substances that make up Earth are not homogeneously distributed. The formation of deposits involves processes that have led to economically viable enrichment of the elements concerned. The processes of fractionation are just as different as the metals being sought.

A particularly simple, yet effective means of fractionation is the deposition of gold nuggets and gold flakes by a river. In the course of a river the flow velocity changes and particles transported in the water are deposited and sorted according to size and density. Gold is preferentially deposited in certain places where it forms so-called gold placers (Sect. 5.9). With a washing pan or with chemicals we can finally separate it from the sand. Another sedimentary process is evaporation that leads to the deposition of evaporites (Sect. 5.7) such as salt and gypsum. These can also be enriched in lithium, an alkali metal that is particularly soluble as a chloride and is used in rechargeable batteries. Not only sedimentation but also weathering (Sect. 5.10) can lead to the formation of deposits. In this case we are mainly interested in those elements that are left in place by chemical weathering, while all soluble substances are leached out.

Many deposits are the result of hydrothermal processes (Chap. 4) in which certain substances are dissolved in hot water and precipitated elsewhere. This can be at a hot spring on the seabed, within the pores of a rock, in fine cracks, or along a fault. In other cases the space is created by dissolving the rock at the same time or by an exchange of substances between water and rock. In hydrothermal systems very different chemical reactions can take place and very different minerals can be formed. Hydrothermal deposits are correspondingly diverse.

Another important fractionation process is magmatism (Chap. 3) because melting and subsequent crystallization lead to strong fractionation between melt and rock. The number of different types of magmatic deposits (also called igneous-related deposits) is low, but there are some giants among them. Magmatism should also not be underestimated as the first fractionation step for the subsequent formation of hydrothermal or sedimentary deposits. This already shows that the classification of deposits is not always easy (Box 1.1).

The origin of petroleum, natural gas, and coal, on the other hand, goes back originally to living beings followed by a series of geological processes that are discussed in Chap. 6.

Of course, new deposits are still being formed today, but this is simply negligible compared with our consumption. As a rule, geology is about periods of time that are beyond human imagination as we are talking about millions of years (Fig. 1.21). That is an impossibly long period to wait for a

© Springer Nature Switzerland AG 2020
F. Neukirchen and G. Ries, *The World of Mineral Deposits*,
https://doi.org/10.1007/978-3-030-34346-0_1

new deposit, so we are dependent on the finite deposits that have formed over the course of the Earth's history—without disappearing again as a result of erosion.

Box 1.1 The problem with the classification of deposits

No two deposits are the same. In each case different processes took place one after the other and this happened under changing conditions. This makes classification difficult no matter how many "drawers" are used for classification. There are always some examples that don't fit or that belong best in between two drawers. Even the coarsest classification (i.e., magmatic, hydrothermal, and sedimentary deposits) is problematic since there are also transitions that flow here. Accordingly, there are countless approaches to classification and it should come as no surprise that an occurrence is pushed into a different "drawer" by different authors.

From an economic point of view it may be obvious to group deposits according to their metals. From a geological point of view, however, this has not proved successful. At most, it makes sense to indicate the economically most important metals as an additional detail. We then speak of lead–zinc veins or high-sulfidation epithermal gold–silver veins.

Of course, it is much more attractive to group deposits according to the most important processes involved in their formation. However, there are a couple of disadvantages: on the one hand, interpretation is always involved and the classification could change with further research and, on the other hand, we have to balance the importance of the processes involved.

In addition, very different types of deposits can often arise from one and the same magma or hydrothermal water. These types therefore often occur together, although they belong in different drawers. Although genetically related occurrences could be summarized, there will be some types that need to be mentioned several times. The same result would be obtained by trying to sort deposits according to how they relate to plate tectonics.

Some authors attach great importance to rock in which the ores occur (e.g., sediment bound), others more to the shape of the ore bodies (e.g., veins, breccias, disseminated in the rock), or the immediate process of precipitation (e.g., impregnation, replacement).

Another possibility is to classify on the basis of important prototypes such as Cyprus type, Carlin type, or Mississippi Valley type. Quite apart from the fact that we can define any number of prototypes it is often difficult in individual cases to decide whether a deposit corresponds more to one or the other type. The advantage is that a single keyword readily describes a complex system including the genetic model.

Dill (2010) proposes an interesting 2D scheme combining the economic enrichment of each metal with the most important processes, but of course he has to discuss each type of deposit with more than one metal correspondingly frequently.

No classification system has really established itself. Usually a mix of everything is used as in this book. In individual cases one should not worry so much about distinguishing different types, but rather about which processes have taken place during their development. A look at the similarities between different types of deposits can be helpful.

1.1 What Is an Ore?

Ore is a mineral aggregate or rock that can be mined out of economic interest—usually to extract metals. The term therefore does not only cover the ore minerals themselves such as chalcopyrite, sphalerite, and magnetite, but also the respective rocks that contain more or less high proportions of these minerals (Fig. 1.1). What is profitable naturally depends on the current situation of markets and on technological developments.

The required concentration of metal to call a rock "ore" is very different depending on the metal. Iron ore often contains more than 50% iron, while in the case of zinc and lead it is only a few percent of the respective metal. Copper and nickel contents start at 0.5% and gold at 0.0001%. If several metals are extracted from the ore the values may be correspondingly lower. The concentration of economically interesting metal in the ore is called ore grade. However, metal content is not the only consideration. Equally important is how easily the metal can be extracted from the ore by mineral processing and smelting (see Sects. 1.15 and 1.16). The size of a deposit (those with large tonnage may be worthwhile even having a low ore grade) and the cost of mining (e.g., through low-cost open-pit mining or expensive deep shafts) are also important. Cutoff grade is the minimum grade at which it is worthwhile to mine a certain deposit. Everything below it (low-grade ore or poor ore) is strictly speaking no longer ore even if metals can be extracted from it. The useless side rock is called sterile or barren.

Fig. 1.1 Massive ore with galena (*gray*) and chalcopyrite (*brass*) from the Friedrich-Christian Mine in Wildschapbach, Black Forest (Germany) (*photo* © F. Neukirchen, Markl Collection/Tübingen)

Ore grade (i.e., the concentration of economically interesting metal) is expressed as a percentage by weight, with very small contents usually as grams metal per metric ton rock. The following rule applies: 1 g/t = 1 ppm = 0.0001%.

Only a few metals and metalloids occur in nature in elementary form (i.e., native) such as platinum, gold, silver, copper, mercury, antimony, arsenic, and very rarely a few others. The precious metals gold and platinum occur preferentially in elementary form, while the others usually occur in the form of compounds. Therefore, metals have to be extracted from other minerals. The most important ore minerals are sulfides for copper, zinc, lead, silver, nickel, and molybdenum and oxides or hydroxides for iron, manganese, aluminum, chromium, titanium, tin, uranium, niobium, and tantalum. Silicate minerals are, with a few exceptions (beryl, spodumene, zircon, eudialyte, garnierite, and so on), not suitable as ore because aggressive chemicals have to be used to dissolve them, while for most metals minerals that are easier to process are available. In contrast,

carbonates (malachite, azurite, siderite, cerussite, smithsonite, and magnesite) are easy to process. However, they usually only occur in limited quantities and have a lower metal content than corresponding oxides or sulfides. Since carbon dioxide (CO_2) released from the ore during smelting impedes the reduction process, carbonates are usually first converted into oxides by heating in the presence of oxygen (calcination).

Ores from other mineral groups such as arsenides, arsenates, tungstates, or vanadates generally occur only in minor quantities, but can be of local importance. Rare-earth elements (REEs) represent a special case. They are mainly extracted from the minerals monazite (REE phosphate) and bastnäsite (REE fluorocarbonate). Lithium and zinc can even be obtained from specific brines, with such salt waters being regarded as liquid ore (see Sect. 5.7.2 and Box 4.29).

Ore usually contains other minerals such as quartz, feldspar, calcite, baryte, and fluorite that are separated if possible before smelting from interesting ore minerals.

Miners call them gangue. Nowadays some gangue minerals are in high demand themselves (Sect. 7.14).

Often several metals can be extracted profitably from a single ore. The economically most important metal is rarely the one with the highest content. In many sulfide deposits (e.g., pyrite, pyrrhotite, arsenopyrite, and chalcopyrite) iron is the most common metal, but it ends up in waste.

Many metals are only obtained as by-products. This applies above all to rare metals or to metals consumed in small quantities despite there being important high-tech applications. Supply and demand often vary widely resulting in highly volatile and sometimes extremely high prices.

Any occurrence of ore is called an ore deposit or a mineral deposit. Whether a deposit can be mined or not depends not only on the prices that can be achieved but also of course on the costs of mining, the development costs (infrastructure), and the size of the deposit. Box 1.2 introduces some important terms.

> **Box 1.2 Stratiform, syngenetic, epigenetic, stratabound …**
>
> A deposit is **syngenetic** when ores are formed simultaneously with the rock that surrounds them such as placer deposits (Sect. 5.9), banded iron formation (BIF) deposits (Sect. 5.2), and layered mafic intrusions (LMIs) (Sect. 3.3). Although sedimentary exhalative (SEDEX) deposits (Sect. 4.17) and volcanogenic massive sulfide (VMS) deposits (Sect. 4.16) are largely syngenetic, they include replacements in older rocks.
>
> A **diagenetic** deposit is formed when unconsolidated sediment is turned into solid rock. It's almost syngenetic, so to speak. For example, copper-bearing shale (Sect. 5.1) and certain zones in SEDEX.
>
> A deposit is **epigenetic** when ores are formed within older surrounding rocks. This applies to most hydrothermal deposits such as veins, replacements, impregnation, and metasomatism (see Chap. 4).
>
> A **stratiform** deposit has the form of a layer that lies parallel to the rock layers (i.e., concordant). It can be syngenetic or epigenetic. For example, copper-bearing shale (Sect. 5.1), coal seams (Sect. 6.1), and massive ores in SEDEX deposits (Sect. 4.17).
>
> A **stratabound** deposit can only be found in a certain rock layer. It can be stratiform (e.g., SEDEX; Sect. 4.17), irregular (e.g., many deposits in sandstones; Sect. 4.13), or discordant (e.g., chimneys; Sect. 4.8). It can be syngenetic, diagenetic, or epigenetic.
>
> A deposit is **discordant** when it cuts through rock layers (in particular, veins and volcanic breccias).

1.2 Selected Ore Minerals

1.2.1 Sulfides

Chalcopyrite

$CuFeS_2$ (Fig. 1.2).

Crystal system	Tetragonal
Color	Golden to brass with a green tinge
Luster	Metallic
Streak color	Black, greenish black

The most important copper ore has a copper content of 35%. It usually forms massive aggregates and occurs in many types of sulfide deposits. In hydrothermal deposits it is often found together with bornite and pyrite (Box 1.3) and in magmatic sulfide deposits with pentlandite and pyrrhotite (FeS).

Bornite

Cu_5FeS_4 (Fig. 1.3).

Crystal system	Orthorhombic
Color	Bronze to copper, tarnishes colorfully
Luster	Metallic
Streak color	Grayish black

Fig. 1.2 Chalcopyrite on quartz. Silberwiese Mine, Oberlahr, Westerwald (Germany) (*photo* © Monika Günther/Archive Mineralogical Collections of the TU Berlin)

Fig. 1.3 Bornite with typical tarnish colors from Mexico (*photo* ©
Géry Parent, CC-BY-SA, Wikimedia Commons)

The copper content is 63%. Frequent copper ore that
forms massive aggregates. Often occurs together with chalcopyrite as primary ore.

Covellite (covelline)

CuS (Fig. 1.4).

Crystal system	Hexagonal
Color	Dark blue
Luster	Semimetallic
Streak color	Bluish black

Often present in small quantities as a secondary mineral.
It has a copper content of 66%.

Fig. 1.4 Covellite and pyrite from Cuka Dulkan, near Bor (Serbia)
(*photo* © F. Neukirchen/Mineralogical Collections of the TU Berlin)

Fig. 1.5 Chalcocite from Mammoth Mine, Mount Isa-Cloncurry
Territory   (Queensland,   Australia)   (*photo*   ©   Rob
Lavinsky/iRocks.com, CC-BY-SA, Wikimedia Commons)

Chalcocite

Cu_2S (Fig. 1.5).

Crystal system	Monoclinic
Color	Lead gray, steel gray, dull and dark tarnishing
Luster	Metallic on fresh fracture, dull tarnishing
Streak color	Grayish black

Important copper ore with a copper content of 80%.
Mainly occurs in the cementation zone (Box 4.16) of copper
deposits.

Tennantite and tetrahedrite (fahlore or fahlerz)

Tennantite ($Cu_{12}As_4S_{13}$) and tetrahedrite ($Cu_{12}Sb_4S_{13}$)
(Fig. 1.6).

Crystal system	Cubic
Color	Steel gray, greenish to bluish
Luster	Metallic
Streak color	Gray–black, reddish gray

There is a complete series of compositions (solid solution) between antimony and arsenic endmembers. Can have
high contents of Fe, Zn, Ag, and Hg. Silver content often in
the percentage range. In freibergite (Ag-rich tetrahedrite) up
to 18%. Important copper and silver ore.

Fig. 1.6 Tetrahedrite from Wenzel Mine in Oberwolfach, Black Forest (Germany) (*photo* © F. Neukirchen, Markl Collection/Tübingen)

Enargite

Cu_3AsS_4.

Crystal system	Orthorhombic
Color	Gray, gray–black
Luster	Metallic to dull
Streak color	Black

In hydrothermal veins (especially, in the Andes region).

Pentlandite

$(Ni,Fe)_9S_8$.

Crystal system	Cubic
Color	Bronze yellow
Luster	Metallic
Streak color	Black

Most important nickel ore. Often as segregation lamellae in pyrrhotite (FeS) or in aggregates with pyrrhotite. Often together with chalcopyrite.

Fig. 1.7 Galena on siderite and quartz from Neudorf, near Harzgerode, Harz (Germany) (*photo* © Rob Lavinsky/irocks.com, CC-BY-SA, Wikimedia Commons)

Galena

PbS (Fig. 1.7).

Crystal system	Cubic
Color	Lead gray
Luster	Metallic
Streak color	Grayish black

Most important lead ore. Usually has a silver content of 0.01–0.3%—sometimes even more—making it an important silver ore due to its frequency. Often in hydrothermal sulfide deposits together with sphalerite and often as well-formed crystals such as cubes and combinations of cubes, octahedra, and rhombic dodecahedra.

Fig. 1.8 Honey-colored sphalerite from Santander (Spain) (*photo* © F. Neukirchen/Mineralogical Collections of the TU Berlin)

Sphalerite (zincblende)

ZnS (Fig. 1.8).

Crystal system	Cubic
Color	Yellow (honey), brown, red, oil green, black
Luster	Adamantine
Streak color	Yellowish to dark brown

The edges of zinc sulfide are transparent to translucent. It often occurs in hydrothermal deposits together with galena. Crystals especially as tetrahedra and rhombic dodecahedra. Often high iron content and then dark-colored and opaque. Iron-free sphalerite is yellow and transparent (honey colored). Also brown concentric layered aggregates (colloform or botryoidal sphalerite or Schalenblende; Box 4.21) of sphalerite and wurtzite (also ZnS). Sphalerite usually also contains manganese, cadmium, and traces of indium, gallium, tellurium, and germanium.

1.2.2 Oxides and Hydroxides

Chromite

$FeCr_2O_4$ (Fig. 1.9).

Crystal system	Cubic
Color	Black, brownish black
Luster	Metallic to semimetallic
Streak color	Dark brown

Mineral of the spinel group. Most important chromium ore; see Box 3.8.

Fig. 1.9 Chromite (cockade ore) from a podiform chromium deposit (location unknown) (*photo* © Andrew Silver, USGS)

Magnetite

Fe_3O_4 (Fig. 1.10).

Crystal system	Cubic
Color	Black, gray
Luster	Metallic
Streak color	Black

Very common mineral, contained in many rocks. Important iron ore, sometimes important titanium ore due to its high titanium content. Mineral of the spinel group with both divalent and trivalent iron: $Fe^{2+}(Fe^{3+})_2O_4$. Magnetic.

Fig. 1.10 Beautiful magnetite crystals octahedral in form are found less in iron deposits than in metamorphic rocks. These crystals were collected from a serpentinite near Zermatt (Switzerland) (*photo* © F. Neukirchen)

Fig. 1.11 Hematite. Fibbia, Gotthard (Switzerland) (*photo* © Rob Lavinsky/iRocks.com, CC-BY-SA, Wikimedia Commons)

Fig. 1.12 Ilmenite from Løvjomås, Froland, near Arendal (Norway) (*photo* © F. Neukirchen/Mineralogical Collections of the TU Berlin)

Hematite

Fe_2O_3 (Fig. 1.11).

Crystal system	Trigonal
Color	Reddish gray to black
Luster	Metallic
Streak color	Cherry red

Crystals often flaky to tabular. Frequently, there are kidney-shaped or cauliflower-shaped aggregates with a very shiny surface called kidney iron ore, fibrous red iron ore, or colloform or botryoidal hematite (Box 4.21). Important iron ore.

Ilmenite

$FeTiO_3$ (Fig. 1.12).

Crystal system	Trigonal
Color	Black, brown–black, steel gray
Luster	On fresh breaks metallic, otherwise dull
Streak color	Black, finely ground dark brown

Important titanium ore.

Pyrolusite

β-MnO_2 (Fig. 1.13).

Crystal system	Tetragonal
Color	Dark gray
Luster	Metallic
Streak color	Black

Fig. 1.13 Pyrolusite from Gremmelsbach, near Triberg, Black Forest (Germany) (*photo* © F. Neukirchen, Markl Collection/Tübingen)

Most important manganese ore. Also as colloform (botryoidal) aggregates (fibrous manganese oxide; Box 4.21).

Cassiterite

SnO_2 (Fig. 1.14).

Crystal system	Tetragonal
Color	Black, black–brown, yellow–brown
Luster	adamantine to adamantine–metallic, greasy
Streak color	Yellow to almost colorless

Fig. 1.14 Cassiterite (twin crystal) from Cínovec, Erzgebirge, Ore Mountains (Czech Republic) (*photo* © Monika Günther/Archive Mineralogical Collections of the TU Berlin)

Most important tin ore. Crystals tabular, needle shaped (needle–tin ore), bipyramids. Brown colloform aggregates (wood tin). In tin granites, hydrothermal deposits, and placers.

Uraninite (pitchblende)

UO_2 to U_3O_8.

Crystal system	Cubic, crystal lattice often destroyed by radiation
Color	Black
Luster	Greasy to metallic
Streak color	Brownish black

Highly radioactive. The uranium content decreases with increasing age due to radioactive decay.

Goethite

α-FeOOH (Fig. 1.15).

Crystal system	Orthorhombic
Color	Black–brown to light yellow
Luster	Metallic, silky, earthy
Streak color	Brown, yellow–brown

Fig. 1.15 Colloform goethite from Rossbach Mine, Siegerland (Germany) (*photo* © Bernd Kleeberg/Archive Mineralogical Collections of the TU Berlin)

Often formed by the weathering of sulfide deposits. Also found as colloform (botryoidal) aggregates (fibrous brown iron ore). Earthy yellow masses consisting of different iron hydroxides are called limonite.

Gibbsite

γ-Al(OH)$_3$.

Crystal system	Monoclinic
Color	Colorless, white, gray
Luster	Vitreous, pearly
Streak color	Whitely

Together with other minerals as a component of bauxite (aluminum ore). Also included in laterite.

Diasporite (diaspore)

α-AlOOH.

Crystal system	Orthorhombic
Color	Colorless, white, gray, greenish, reddish
Luster	Vitreous, pearly
Streak color	Whitely

Together with other minerals a component of bauxite (aluminum ore). Also found in laterite.

1.2.3   Carbonates

Malachite

$Cu_2CO_3(OH)_2$ (Fig. 1.16).

Crystal system	Monoclinic
Color	Green
Luster	Vitreous, silky
Streak color	Light green

Very common in the oxidation zone (Box 4.16) of copper deposits.

Azurite

$Cu_3(CO_3)_2(OH)_2$ (Fig. 1.17).

Fig. 1.16 Malachite from Gottesehre Mine, Urberg, Black Forest (Germany) (*photo* © F. Neukirchen, Markl Collection/Tübingen)

Fig. 1.17 Azurite on dolomite from Tsumeb (Namibia) (*photo* © Bernd Kleeberg/Archive Mineralogical Collections of the TU Berlin)

Crystal system	Monoclinic
Color	Azure blue
Luster	Vitreous
Streak color	Light blue

In the oxidation zone of copper deposits.

Siderite

$FeCO_3$ (Fig. 1.18).

Crystal system	Trigonal
Color	Yellow, brown
Luster	Vitreous, pearly
Streak color	Yellowish white

Fig. 1.18 Siderite from Hoffnung und Segen Gottes Mine in Stollberg, Harz (Germany) (*photo* © Monika Günther/Archive Mineralogical Collections of the TU Berlin)

Box 1.3 Pyrite

Pyrite, the gold-colored iron sulfide with metallic luster (FeS_2; Fig. 1.19), is by far the most abundant and widespread sulfide mineral. Museums often show perfect cubes or pentagonal dodecahedra, sometimes also pyritized fossils or shales with sun-shaped aggregates. Pyrite is colloquially called fool's gold.

This mineral can be formed under very variable conditions in a variety of environments. During the solidification (diagenesis) of sediments (especially, in shales) it is formed by the action of sulfate-reducing bacteria. Under these conditions it often occurs together with marcasite that has the same composition in a different crystal lattice (orthorhombic).

Pyrite is also found in almost all hydrothermal deposit types and is often the dominant mineral (e.g., in some SEDEX deposits; Sect. 4.17 and in certain zones of VMS deposits; Sect. 4.16). Finally, pyrite can also be formed magmatically—especially, in mafic intrusions where it can occur subordinate to low-sulfur pyrrhotite (FeS).

In fact, pyrite can occur when there is sufficient Fe^{2+} and S^{2-} regardless of temperature. Accordingly, reduced conditions are the most important prerequisite. In addition, a hydrothermal fluid must be at least slightly acidic but preferably strongly acidic.

However, a fluid can be supersaturated with several metals at the same time—for example, on copper or arsenic in addition to iron so that chalcopyrite ($CuFeS_2$), arsenopyrite (FeAsS), or other minerals are formed. Depending on the ratios of the metals precipitated they occur together with pyrite or other sulfides. The result is that pyrite is often formed both very early and very late in hydrothermal systems interrupted by the crystallization of other sulfides.

Sometimes pyrite contains tiny gold grains where fool's gold in this case is actually gold ore. Pyrite is unsuitable as iron ore from today's point of view. It can only be smelted if it is first converted into iron oxide by roasting that releases large amounts of environmentally harmful SO_2. It is easier to use oxides (magnetite, hematite). Nevertheless, pyrite is mined and roasted primarily to produce sulfuric acid. However, this is no longer very important as sulfuric acid can also be produced as a by-product in the smelting of other sulfides (in particular, in copper production).

Historically, pyrite played a role in early iron production at least in some regions. Pyrite was even more important in early history because of another characteristic. If you strike another stone (e.g., a flint stone) against it, sparks are produced and a fire can be ignited. Pyrite is also to blame for acid rain. Coal and lignite contain some pyrite and other sulfur compounds that are oxidized during combustion. The SO_2 released dissolves in condensed water droplets to sulfuric acid. The weathering of pyrite and other sulfide minerals is very similar. Therefore, extremely acid mine water often accumulates in abandoned mines. In such water only specialist microorganisms can survive such as *Acidithiobacillus ferrooxidans* whose metabolism is based on the oxidation of sulfides, which accelerates weathering.

At first glance it may seem confusing that the sulfur in pyrite appears to have an oxidation state of −1 and not −2 as it should be. This is because two S^{2-} form a covalent bond. In the crystal lattice Fe^{2+} alternates with dumbbell-shaped $(S-S)^{2-}$.

1.3 Resources, Reserves, and Consumption

Although it is well known that ores, fossil fuels, and other resources are only available in limited quantities, it is almost impossible to indicate when a particular resource might be depleted (e.g., Stürmer and Schwerhoff 2013). New deposits are discovered time and again, while in the case of known deposits estimates of the quantity of existing ore are regularly corrected upward or downward. Rising prices or new technologies can also make deposits profitable that were previously considered uneconomic. In the long term, deposits in remote areas and low-grade deposits will become increasingly important.

Well-known deposits that can be exploited profitably with current technology for the given market situation are referred to as reserves. The term resources, on the other hand, also covers deposits that are only suspected and those currently

Fig. 1.19 Pyrite (FeS_2) (*photo* © F. Neukirchen)

not worth exploiting. It is almost always the rule that the global reserves of a certain material correspond to the demand of only a few decades. This is no cause for concern as resources turn into reserves through rising prices, better technology, and continued exploration. Supply bottlenecks can nevertheless occur such as when sudden outages occur in politically unstable regions.

Another unknown factor is consumption. The consumption of raw materials is increasing particularly massively in emerging countries such as Brazil, India, and above all China. If a country's infrastructure is expanded quickly, this means high consumption of steel, copper, and so on. Moreover, expansion occurs in industries that consume metal and that not only supply their own market but export all over the world. China has for years been the largest consumer of many metals with still rapid growth rates. Metal consumption can be extrapolated to a certain extent into the future, but there are always unforeseeable deviations. Economic crises such as happened in the years following the 2007 financial crisis lead to a collapse in demand, falling prices, and as a result the closure of mines. Technological developments can have an equally abrupt effect (especially, if widespread high-tech components are replaced by those with a different composition).

The quantity of metal extracted (Fig. 1.20) adapts to demand as much as possible. One problem is that production reacts only very slowly to a changed market situation. A mine can be shut down relatively quickly if it is no longer profitable, but it usually takes a decade or more to open up a new mine if prices rise. Similarly, the capacity of existing plants can only be expanded with large investments and structural changes. When prices are low there is hardly any investment in the search for new deposits. If prices rise suddenly the opposite is the case. After about a decade so many new mines may go into operation at the same time that the price crashes again and some have to be closed once more. As a result reserves have increased for the time being.

Contrary to some pessimistic forecasts no resources have yet run out. The frequently quoted study titled *The Limits to Growth,* which was presented by the Club of Rome in 1972, assumed that many resources would be exhausted after a few decades at the exponential growth assumed at that time, which as is well known did not happen. However, it is relatively likely that certain materials will become significantly more expensive in the future when easily accessible deposits are depleted. The cost of production will therefore increase considerably. This can be partly offset by improvements in mining, processing, and smelting technology.

In general, it can be assumed that a certain commodity will not suddenly be completely exhausted after constant or rising production, but that from a certain point in time annual production will no longer be increased and then slowly begin to decline. Petroleum is a commodity that is already in short supply today and "peak oil" (Sect. 6.6) is already in sight. Some believe it has already been exceeded, while others are still giving us a few decades before this happens. Even the rather optimistic BGR (German Geological Survey) assumed in its 2009 energy study that no increase in oil production (including unconventional oil) will be possible after 2035. In the follow-up study of 2012 this statement is no longer to be found. This is not only because of new oil discoveries, but also because far more unconventional oil than assumed has already been produced. Nevertheless, it can be assumed that almost all oil deposits that can be extracted at favorable prices have been known for a long time and that new discoveries can only be extracted at high cost. We really are going to run out of cheap oil soon. Other fossil fuels such as natural gas and coal are available in much larger quantities. Phosphate is another scarce resource that could soon reach its peak (Sect. 5.8).

Also problematic are elements that do not form their own deposits and are only extracted as by-products from mining other metals. These include important high-tech metals such as germanium, indium, and gallium. Germanium and indium are obtained almost exclusively from zinc ore, while gallium comes almost exclusively from aluminum ore. Their production therefore depends on the production rate of the respective main metal; otherwise higher capacities can only be achieved by improving processing and refining. They are among resources currently considered critical (Box 1.4).

If ever-lower ore grades are exploited for the production of metals, then larger open-pit mines will be required to produce the same quantity of metal and at the same time larger quantities of material will have to be stored in overburden stockpiles and sludge ponds. The impact on the environment will therefore increase.

Box 1.4 Critical resources
A number of mineral resources are considered strategically important to industry because they are indispensable for high-tech applications (Elsner et al. 2010), even though the amount consumed is very small compared with the bulk metals. In addition, some are produced in only a few countries, including politically unstable regions, which can lead to unpredictable outages and supply shortages. Which resources are considered critical therefore depends not only on available deposits, but also on political developments and technological innovations. A publication by the European Union in 2010 lists 14 critical resources: antimony, beryllium, fluorite, gallium, germanium, graphite, indium, cobalt, magnesium, niobium, platinum-group elements, REEs, tantalum, and tungsten.

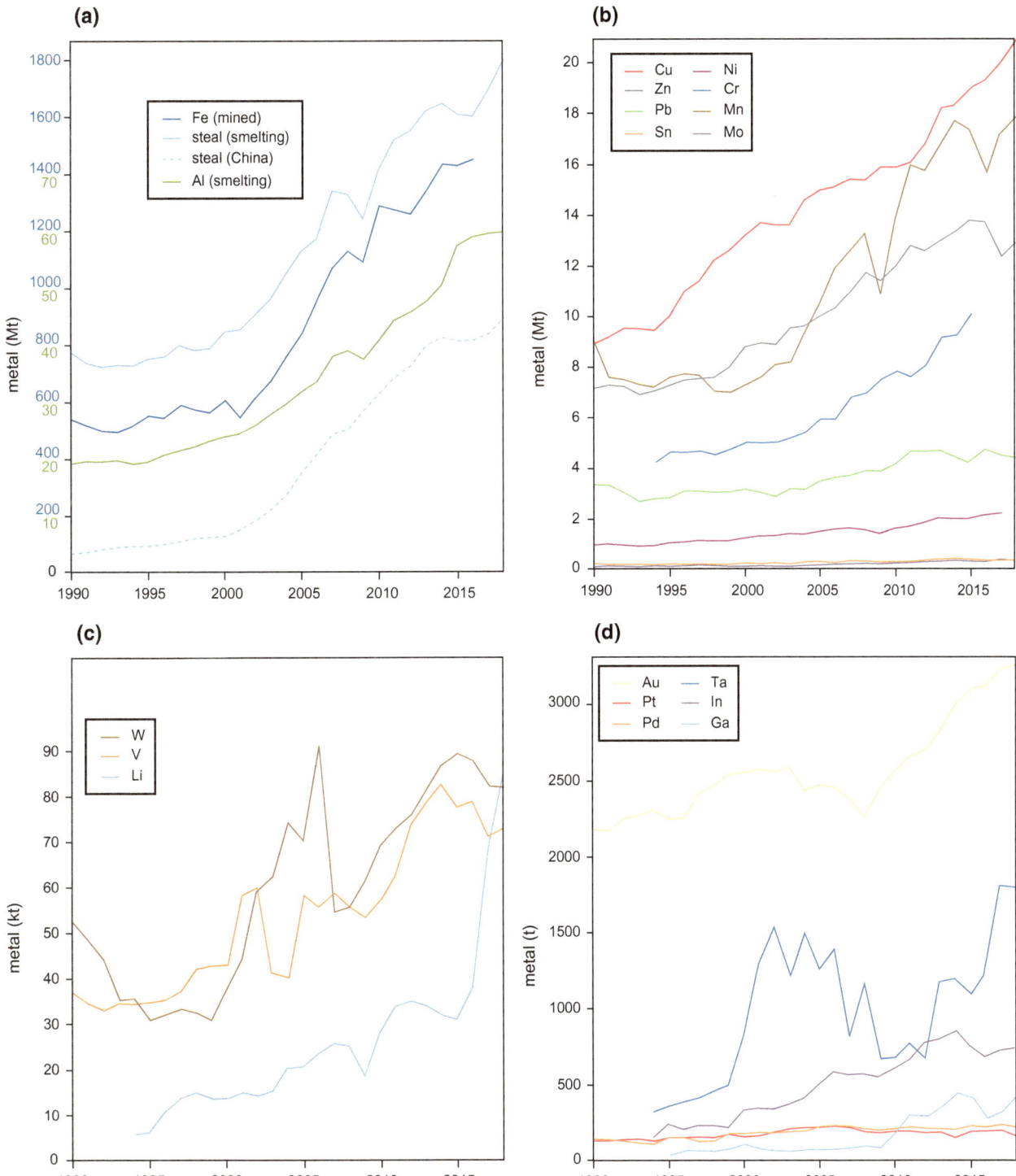

Fig. 1.20 Development of global metal 1990–2018 (note different scales). **a** Iron in mined ore, global steel production, steel production in China, and primary aluminum from smelting. Recycling is important for both metals. **b** and **c** Metal content in ore mined worldwide for copper, zinc, lead, tin, nickel, chromium, manganese, molybdenum, tungsten, vanadium, and lithium. Only part of the ore is smelted to pure metal, while other parts go into steel production or directly to the chemical industry. **d** Production of selected precious and high-tech metals. Gold, platinum, palladium, and tantalum through mine production; indium and gallium as by-products (refining). Data from *USGS Minerals Yearbooks*, USGS Commodity Summaries

1.4 Markets

A selection of commodities are traded on exchanges such as the New York Mercantile Exchange (NYMEX), the London Metal Exchange (LME), and the London ICE Futures. These commodities include petroleum, the most important industrial metals such as aluminum, copper, nickel, zinc, and tin, and precious metals such as gold, silver, platinum, and palladium. In addition to the spot market, futures are of great importance as a hedge against undesirable price developments.

Petroleum (i.e., crude oil) can be of different quality and composition, so different types of oil are traded on the exchanges. The most important varieties are Brent (from the North Sea) and WTI (West Texas Intermediate) whose prices also determine the price of other types of petroleum. While the price of oil is almost the same worldwide (apart from transport costs and subsidies), natural gas has regional markets with large differences. In some countries the price of gas is linked to the price of oil.

The most important trading places for precious metals are not the stock exchanges, but the London Bullion Market and the London Platinum and Palladium Market where such metals are traded as bars over the counter between producers, banks, traders, and consumers from all over the world. A fixed price is agreed once or twice a day at which as many transactions as possible are processed.

A large part of commodity trading takes place off-exchange through direct contracts between producers and consumers and via intermediaries with the price being based on stock exchanges. The five largest commodity traders, all based in Switzerland, have a combined turnover similar to Switzerland's national product. The largest, Glencore, is repeatedly criticized for human rights violations and corruption.

The remaining metals are not freely traded but delivered through long-term supply contracts directly from the producer or through intermediaries. Some of these are not mined in special mines but are by-products of the mining of a particular metal. The supply side is thus linked to the production conditions of another metal and changes in demand can only be responded to by optimizing the respective processing and smelting with corresponding effects on the price. Other metals such as tantalum and REEs are produced and consumed in such small quantities that just a few mines account for much of world production. Bottlenecks or overcapacities can therefore easily occur, which are reflected in sharply fluctuating prices.

The United States Geological Survey (USGS) concludes in its analysis (Papp et al. 2008) that metal prices are driven by a variety of factors such as wars and recessions, the collapse of the Soviet Union, China's economic rise, fiscal policies, exchange rates, technological developments, strikes, mine closures, and newly opened facilities. Looking at the long-term trends of different commodity prices the first thing to be noticed is that they are subject to enormous fluctuations. From time to time there are peaks brought about by a rapid rise in price followed by a rapid crash when the peak price is often a multiple of the initial price. In the middle of the twentieth century fluctuations were significantly lower than at the beginning and end of the century. It is also noticeable that the prices of most raw materials have risen in the long term. The most extreme increase was in the years before 2008 when almost all metals reached their historic price peak. Then as a result of the financial crisis the recession led to a sharp decline. Since then most prices have stabilized at a relatively high level.

The figures look somewhat different when price movements are adjusted for inflation. Although curves continue to oscillate strongly, they do so mostly on an almost constant level. Adjusted for inflation aluminum was many times more expensive at the beginning of the twentieth century than it is today; when it comes to other metals the extreme increase before 2008 remained within historical fluctuations. In interpreting these figures it is important of course to be cautious because rising prices of raw materials are passed on directly to the consumer and thus have an effect on inflation whose effect is then deducted from the nominal price. It is therefore hardly possible to tell from the prices (be they nominal or inflation adjusted) whether a commodity is becoming scarcer and how the level of necessary investment in the mining sector develops in comparison with the economy as a whole.

Gold by the way should be regarded as a normal commodity. It differs from other commodities only in that speculation has a greater impact on price fluctuations, which is mainly due to the myth that it is a safe investment.

While strongly fluctuating prices of all raw materials offer the possibility of speculation, the mining industry and the metal-processing industry are interested in price development that can be planned as far as possible. Even when there is a monopolist in the game, the monopolist usually strives to ensure that supply and demand match each other as closely as possible. Stocks can at least temporarily compensate for deviations between production and consumption. If growing demand leads to rising prices, ever-more deposits can be exploited that were previously not worthwhile. Although, as already mentioned, it takes some time for new mines to go into production, which can lead to bottlenecks.

Monopolists cannot necessarily exploit their position to their advantage. Although they may have enormous production capacity that lowers the cost of production, it may lead to oversupply on the market and thus depress prices. While booting out the competition, this does not ensure high profits either. In the case of tantalum the Wodgina Mine (Australia), which previously supplied 50% of world production, was no longer profitable in 2009 because the price of the metal crashed during the economic recession due to a lack of demand. In the following years the loss of the mine's enormous capacity naturally led to price rises; hence the mine was reopened, closed, and reopened again (the reserves are now exhausted). For more on the Chinese monopoly on REEs see Box 1.5.

An important point is that the costs involved in exploiting different deposits are very different. With some deposits high profits can be expected, but with many others the running costs can just be covered. This is of course due to geological conditions such as ore grade, size of the deposit, and depth of mining. However, there are also transport costs, wages, environmental regulations, exchange rates, and so on. In this sense price is linked not only to supply and demand, but also to the average production costs of all active mines, which together allow worldwide capacity to meet demand.

Box 1.5 Rare-earth elements and China's monopoly

With around 95% of world production China virtually has a monopoly on REEs (Braune 2008; Margonelli 2009). Apart from the huge Bayan Obo deposit (Box 3.16) other REE carbonatites (Sect. 3.10) and lateritic REE deposits (Sect. 5.11.4) are mined there. The monopoly must be regarded as problematic because these elements are just as essential for various products from entertainment to telecommunications technology as they are for the use of renewable energy sources and so-called energy transition (Sect. 2.5). Since REEs are not as rare as their name might suggest, how is it that the world has become dependent on China's monopoly for this group of commodities? Despite the minerals of REEs being not so rare there is still a problem with them because economically exploitable deposits are rare. Rare-earth minerals very often contain uranium and thorium and these radioactive ingredients cause problems. They must be separated and then stored in dumps and tailings. Moreover, there can be major problems as shown by the example of Bukit Merah (Malaysia) where 20 years ago a rare-earth refinery contaminated the soil and water. The consequences were significantly increased cancer rates in the affected region and ongoing clean-up efforts (Consumer Association of Penang 2013; Bradsher 2011).

Ores from the large Mountain Pass Mine (USA) (Sect. 3.10) are moderately radioactive. The deposit was discovered in 1949 when two geologists were searching for uranium ore with a borrowed Geiger counter (Anonymous 2013). Until 1989 this mine was the main producer of REEs in the world (Margonelli 2009). However, Deng Xiaoping had already recognized the importance of these elements in 1992 and compared China's position in REEs with that of the Middle East in terms of petroleum (Anonymous n.d.). In the 1990s China began supplying the world with cheap REE causing prices to fall sharply from USD11,700 per metric ton in 1992 to USD7430 in 1996. This led to the United States and other countries increasingly relying on imports from China and neglecting their own reserves. For mines like Mountain Pass that could not compete with the prices of Chinese suppliers such a price war would have been ruinous were it not for political support. Added to this were the rising costs of environmental regulations that were much stricter in the United States than in China. The mine was closed in 2002 after a series of accidents in 1998 involving large amounts of radioactive effluent entering Ivanpah Dry Lake.

Rare-earth prices started to rise in 2005 when China began to limit exports. In the following year and increasingly from 2010 Chinese exports were again limited (Jacoby and Jiang 2010; Anonymous 2012a, b). Other nations saw themselves cut off from the supply of important raw materials and complained to the World Trade Organization.

Reducing dependencies and resulting economic constraints resulted in frenetic activity. The Mountain Pass Mine was acquired by a newly formed consortium in 2008 under the old name of Molycorp and started production again in 2013 (Molycorp 2013). Various REE projects were also launched in other countries (particularly, those with a corresponding industry). Examples were Germany (DRAG 2013; Nestler 2013) with the deposit in Storkwitz (Box 1.8), Australia where the deposit on Mount Weld (Sect. 5.11.4) known since 1988 is probably one of the richest REE deposits in the world (Utter 2010) and where mining is already beginning, and Japan that is searching for such metals in its territorial waters (Germis and Nestler 2013).

1.5 Where to Search and How?

Every now and then geologists find a new deposit by chance often while looking for something else. The most spectacular example is the huge Olympic Dam copper–uranium deposit (Box 4.20) discovered in Australia in 1975 in an area not previously known as a mining district. However, these are exceptions and prospecting (the search for unknown deposits) is generally very expensive when it comes to replacing depleted mines and meeting increasing demand. This begs the question as to whether it is better to search in an area where many deposits are already known or where nobody has searched before.

Grassroots exploration (the second option) has a very low success rate similar to that of lotteries where there is a winner every now and then. Such exploration involves geologists and adventurers systematically roaming a large area. Some are looking for something specific such as gold grains in a river, others have their eyes open to anything conspicuous that could indicate deposits of any kind. As a rule larger companies only appear on the scene when ores have already been found and when it is a question of assessing the quantity and profitability of the ore available. At that point it is still not known whether it will actually be worth exploiting the deposit or not.

The first option is more promising. Once it is realized that respective ore-forming processes have taken place in a particular region it is reasonably safe to assume there will be more in the vicinity of a known deposit. This is the reason intensive searches are carried out in mining areas and geophysical methods are used to discover deposits that are not visible on the surface. Since such regions often already have the necessary infrastructure in place exploration and subsequent mining and processing become cheaper.

The worldwide search for areas with similar geology to corresponding known mining districts is also very promising since it is quite possible that deposits of the same type can be found there.

Whereas the Earth's surface in the past could simply be searched for ore minerals or alteration zones such easily detectable deposits hardly exist today. This obliges us to use methods that indicate hidden deposits in the depths termed blind deposits.

Successful prospecting depends on comprehensive knowledge of the type of deposit sought and use of a genetic model. This means any exploration model developed must include not just stratigraphy (Fig. 1.21) and tectonic structures but also important rock types and corresponding geophysical and geochemical parameters. Which of the methods presented in the following sections are used in the search depends not only on the type of deposit sought but also on local conditions.

During exploration—which includes prospecting and more detailed investigation of potential deposits—enormous amounts of data accumulate. Searching for the most important information in the data is so time-consuming that many geoscientists are primarily concerned with data mining: trying to obtain as much information as possible using statistical methods and sophisticated algorithms. Geoinformation systems (GIS) are used for the spatial allocation and interpretation of data and special software is used in many areas.

Remote sensing can be used in large areas to search for suitable sites (prospects) where relevant parameters apply and where more detailed investigation is worthwhile. Geophysical methods, more accurate mapping, rock sampling (see Box 1.6), and initial drilling will then be used to determine whether a site might be economically interesting (prefeasibility study). If this is the case, then more detailed investigations (evaluation) are carried out with the aim of determining as precisely as possible the spatial size and content of the ore as a planning basis for mining. The rock's mechanical properties must also be known. It is also important to decide whether only particularly rich and accordingly lucrative zones should be mined or also zones with a low ore grade, even if their exploitation is actually not profitable. In addition, experiments often have to be conducted to find out how best to process the ore. All such results are incorporated into feasibility studies together with other factors such as the market price of the commodity concerned, investments in infrastructure, and environmental requirements. Extraction will only take place if the study comes to the conclusion that it is profitable. It typically takes more than a decade from discovery of a deposit to start of mining. As to who owns the deposit see Box 1.7.

Box 1.6 Ore microscopy

Examining ores with a microscope is becoming increasingly rare, but it can be very helpful in interpreting the texture and determining the temporal sequence of mineral paragenesis. This knowledge is important not only to understand the formation of the deposit but also to find a suitable method to process it. Since almost all ores are opaque reflected light microscopes have to be used—not transmitted light microscopes. Unfortunately, most ore minerals look very similar and attention has to be paid to minimal color differences in light pastel shades.

Box 1.7 Mining law

Who owns a deposit? There are different traditions and legislations in different countries. Current German mining law goes back to the tradition of *Bergfreiheit* (freedom of mining) according to which all underground resources are unowned (i.e., belong neither to the State nor to the landowner). The State, however, regulates who may bring these resources to the surface and collects appropriate taxes that in earlier times were called tithes.

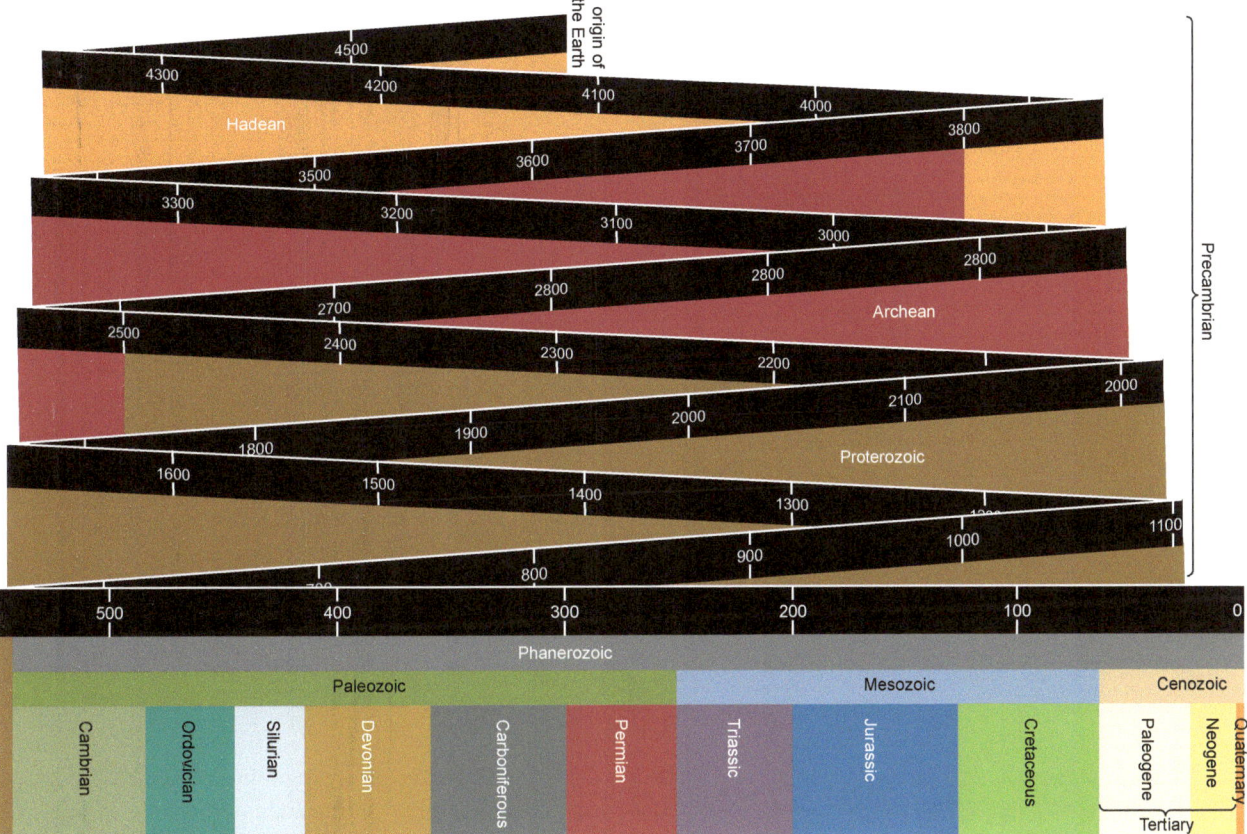

Fig. 1.21 Geological timescale in millions of years

1.6   Remote Sensing

According to a common form of mining law in the Middle Ages termed bergregal (from the German *Bergregal*) all resources were the property of the emperor, king, or sovereign. The latter was able to lend the mining rights and received tithes in return. Even today there are countries whose resources are in state ownership, which is particularly common in the case of petroleum.

In the Commonwealth countries and the United States all resources located under a piece of land are the property of the landowner with the exception of gold and silver in the Commonwealth, which are considered state property. In the United States the State reserves the right to strategically important resources. In the case of deposits on public land an annual rent can be paid for a claim.

In France and some other countries near-surface deposits belong to the landowner, while deeper ones belong to the State.

Satellites and aircraft are used to collect images and geophysical data that can be used to search vast areas for potential deposits. This works best in arid and semiarid areas, while in areas with dense vegetation the possibilities are limited. Aircraft or helicopters equipped with the appropriate equipment allow detailed investigation that is not only specially tailored to respective requirements, but is also associated with corresponding costs. Accordingly, satellites have the advantage that a large amount of existing data can already be accessed.

Tectonic structures can be mapped by simple interpretation of aerial and satellite images. Since faults and especially their intersections can serve as ascent routes for magmas or hydrothermal solutions, deposits are usually lined up along tectonic structures.

Multispectral remote sensing offers much more (Van der Meer et al. 2012; Brandmeier 2010; Rowan and Mars 2003; Yamaguchi and Naito 2003; Sabins 1999). Multispectral

Fig. 1.22 The Terra satellite launched by NASA in late 1999 carries the Japanese instrument ASTER (Advanced Spaceborne Thermal Emission and Reflection Radiometer) whose data are frequently used for exploration purposes (picture © NASA)

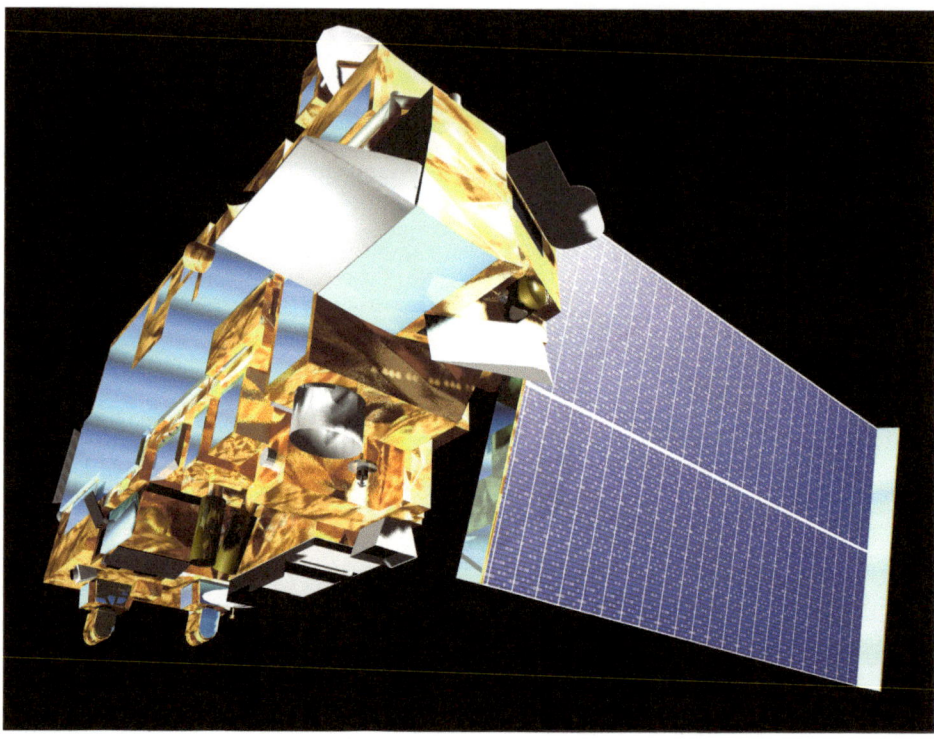

remote sensing concerns radiation emitted by the Earth's surface made up of a mixture of reflected light and light that has penetrated a little farther into matter and lost parts of its spectrum through absorption. The Landsat satellites (mainly the Landsat Thematic Mapper) provided the corresponding data for decades, but since the launch of the Terra satellite at the end of 1999 the data of the Japan-built ASTER instrument has been used preferentially (Fig. 1.22). The way it works is in principle somewhat similar to a photo. In a digital photo each pixel consists of three channels—red, green, and blue (RGB)—whose brightness values determine the color. The three channels correspond to three wavelengths that represent only small sections of the light spectrum, but are sufficient for our eyes to perceive them as color. However, electromagnetic radiation in multispectral exploration is registered in a whole series of channels with specific wavelengths not only in the visible light range but also for infrared. ASTER has three visible and near-infrared (VNIR) channels, six shortwave infrared (SWIR) channels, and five longwave infrared (TIR) channels, the wavelengths of the channels being well chosen for our purposes. There are also devices that cover the interesting range of the spectrum more or less continuously with a large number of channels (hyperspectral). Although these have so far been mainly used from aircraft because the large amount of data and noise are still problems with satellites, they are likely to become more important in the future.

Light is absorbed differently by different minerals. In particular, transition metals such as iron and hydroxyl groups (OH⁻) have an effect on this. Therefore, we can get an impression of which minerals or rocks lie on the Earth's surface from false-color images (Fig. 1.23) composed of suitable channels. The simplest possibility would be to assign one channel of the sensor to each of the RGB channels of a monitor, but better results are obtained using ratios of suitable channels (ratio images). In combination with corresponding on-site fieldwork detailed geological maps can be produced quickly.

Using mathematical methods such as principal component analysis (PCA) or the calculation of so-called mineral indices (MIs) data can even be used to create maps highlighting areas with a high content of certain minerals. The ASTER data can for example be used to distinguish pixel by pixel (with a certain degree of uncertainty) clay minerals (e.g., kaolinite and illite); mica, chlorite, various iron oxides, carbonates (e.g., calcite, dolomite); and sulfates (e.g., alunite, baryte). However, the result for each individual case should be checked on-site.

Using even more complex algorithms and corresponding calibration of the instrument the spectrum measured for each pixel can be compared with an entire database of spectra of different minerals, which in some cases even allows quantitative estimation of the mineral content. The more channels measured the better this works.

In areas covered by dense vegetation this method is unfortunately very limited. However, it is possible to map the damage to vegetation caused by heavy metals, which in turn can give indications of deposits.

Fig. 1.23 False-color image of the edge of Saline Valley (California, USA) by ASTER. Channels 4, 6, 8 from the shortwave infrared are shown as RGB. Colors depend on contents of clay, carbonate, and sulfate minerals (*photo* © NASA, GSFC, MITI, ERSDAC, JAROS, and the US/Japanese ASTER team)

Multispectral remote sensing is particularly helpful in mapping hydrothermal alteration zones (Box 4.14) in dry regions because different types of alteration can be distinguished. Although not every alteration zone belongs to a deposit, good mapping makes it possible to decide where to look more closely. A frequently cited example is the Collahuasi copper deposit (Chile). Copper had long been mined there from hydrothermal veins. Analysis of the alteration zone at that time with Landsat data revealed that there were systematically arranged anomalies in a large area. Geophysical methods then confirmed that a huge porphyry copper deposit (Sect. 4.4) was hidden at a greater depth, now one of the most important copper mines in the world.

There are also examples where very special applications of multispectral remote sensing have been used to find and interpret natural gas and oil leaks and to map the borate content in the Salar de Uyuni (Sect. 5.7.2).

In addition to such passive systems that use natural electromagnetic waves there are active systems, which are discussed in the following section. Some geophysical methods can be performed from an aircraft or helicopter and are therefore part of remote sensing.

1.7 Geophysical Exploration

When searching for deposits a number of physical parameters can be exploited such as the magnetic field, the gravitational field, electrical conductivity, and gamma radiation. However, which methods are likely to lead to success depend on the type of deposit and various local conditions. Geophysical methods not only provide information about rocks on the surface but also about rocks at a certain depth. The search is mainly for anomalies or strong deviations from normal values. Normal values can be very different depending on the type of rocks. Usually several parameters are measured simultaneously and evaluated together. Some parameters are measured from an aircraft or helicopter (airborne geophysics) and are therefore part of remote sensing, while others use geophysical methods from ships such as the search for offshore oil and gas.

Remote sensing is usually followed by detailed measurements on the Earth's surface. Raw data can only be interpreted directly to a certain extent because the depth of an ore body, its geometry, and the rocks in between have a large impact on what is measured at the surface. Complicated algorithms can be used to create a 3D model of the subsurface capable of explaining the measured data as well as possible. This step is called inversion. A computer has to be fed with numerous other assumptions and the degree to which the local geology is more or less well known is accordingly helpful.

Finally, sensors can be placed into boreholes to measure parameters in a vertical profile and explore the surroundings of the borehole. This enables detailed 3D models that can be used to estimate the size of the deposit. There are cases where deep drilling has discovered deposits that could not be detected from data obtained on the surface.

Gravimetry is the measurement of the strength of a gravitational field. The strength of the Earth's gravitational field (Fig. 1.24a) and hence gravitational acceleration vary not only with latitude and absolute altitude but also with the relief and density of the subsurface. Since we are only interested in the latter other factors can be ignored in the data. Anomalies in corrected data (Bouguer anomalies) are due to variations in the density of the subsurface and of different rocks and their geometry in the Earth's crust and mantle. Data may indicate high-density ore or a low-density salt dome. Measurements are taken from the air, on the ground, at sea, or underground in a mine. Special satellites are also used to take such measurements.

Magnetics or magnetometry is an important method used to measure the direction, strength, or relative change of a magnetic field (Fig. 1.24b). The strength of the Earth's magnetic field varies according to latitude (31,000 nT at the equator, 63,000 nT at the poles) and fluctuates daily. It is also influenced by magnetizable minerals underground. Such minerals develop their own magnetic fields overlaying that of the Earth. This effect is strongest with magnetite (Fig. 1.25). Even a rock with a low magnetite content can have a strong influence on the magnetic field measured making a magnetite-rich iron deposit all the more noticeable (Box 5.3). Although ilmenite, hematite, pyrrhotite, and other minerals have an effect but to a lesser extent, they can also lead to strong anomalies if they are present in sufficient quantities. Magnetics is not only helpful in finding iron deposits but also in the search for other metals as long as their deposits contain magnetite or one of the other minerals

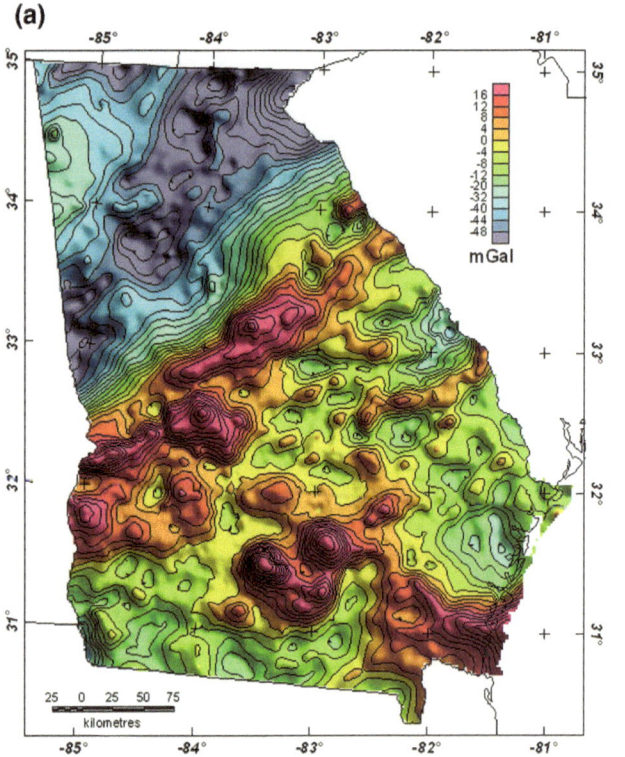
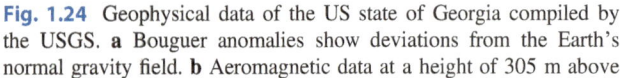
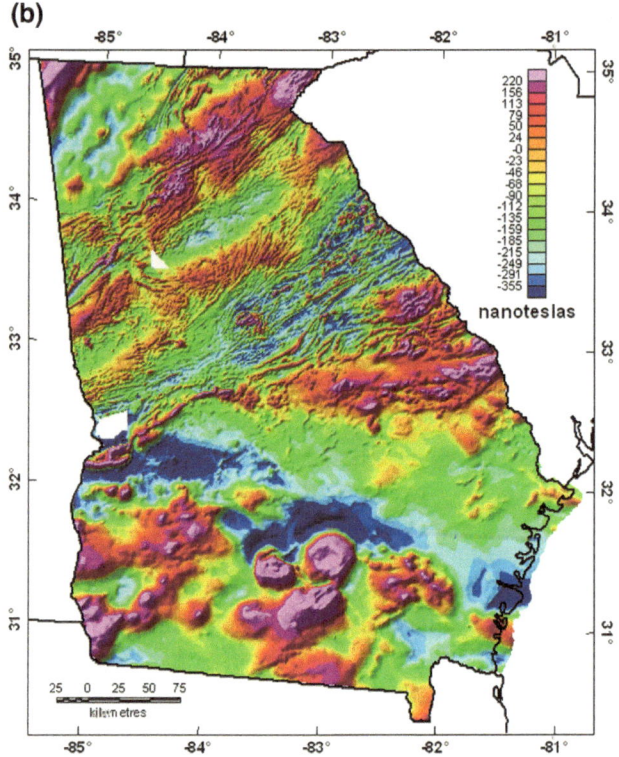

Fig. 1.24 Geophysical data of the US state of Georgia compiled by the USGS. **a** Bouguer anomalies show deviations from the Earth's normal gravity field. **b** Aeromagnetic data at a height of 305 m above the Earth's surface. The actual flight altitude was 150 m. The data were subsequently adapted to the national standard (© USGS)

Fig. 1.25 Magnetite grains from a banded iron formation (Sect. 5.2) in Mauritania. They can be pulled out of the desert sand using a magnet as shown (*photo* © Thomas Finkenbein)

mentioned. This is the case with many sulfide deposits, iron oxide copper gold (IOCG) deposits (Sect. 4.7), and skarns (Sect. 4.9).

Magnetic fields can be measured from the air (aeromagnetics), at sea, on the ground, and in boreholes making diverse interpretations possible. Usually, just the strength of the field is measured, but sometimes the direction is too. Measurements from high altitudes are particularly useful for geological mapping, while measurements on the ground or at low altitudes can also be used to find near-surface

magnetized bodies of small size. Differences calculated from data measured at different levels are particularly helpful. During evaluation additional stationary ground stations can be operated to deduce temporal fluctuations in the Earth's magnetic field.

Electrical methods are used to measure the conductivity of the subsurface or electrical potentials. Such methods are more complex because electrodes (e.g., steel rods) have to be placed in the ground and arranged in different ways depending on the application. The distance between electrodes determines the penetration range of the method; hence vertical profiles or depth measurements are possible by systematically arranging such electrodes.

Resistivity is used to measure the conductivity or rather electrical resistance of the subsurface. Electrical current is conducted via two electrodes through the ground and measured at two further electrodes. Every rock conducts electricity to a certain degree. This is mainly due to pore water containing a more or less high content of dissolved ions. Sulfides and graphite also have very high conductivity that makes them noticeable. However, resistivity is used above all for the investigation of groundwater and for the delimitation of rocks with different water contents.

Induced polarization (IP), on the other hand, is very helpful when exploring for ores (Fig. 1.26). When electric current is conducted into the ground certain substances such as sulfides, native metals, graphite, and to a lesser extent clay minerals are electrically charged. Discharge can be measured via two further electrodes when the current is switched off. Resistance can also be measured during excitation. This method is used not only at the Earth's surface but also in boreholes.

Compared to systematic placing of electrodes, **electromagnetic methods** (EM) have the advantage that they can be carried out quickly from an aircraft or helicopter, from a

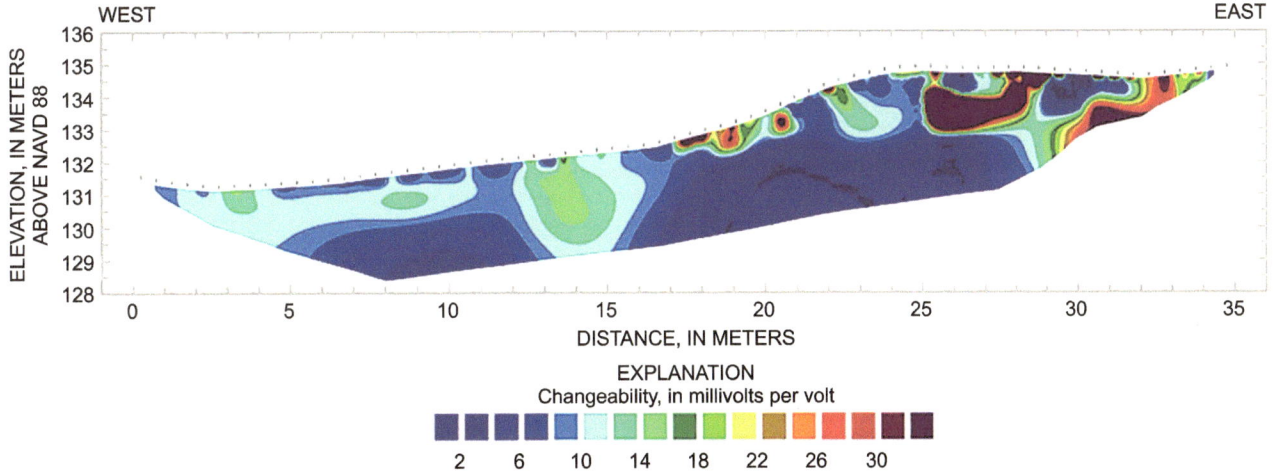

Fig. 1.26 Example of a profile created using induced polarization (© USGS)

moving car, using a portable device, from a ship, or even in a borehole. A transmitter loop generates an electromagnetic field that induces electrical currents in conductive bodies in the subsurface such as sulfides, graphite, or water that in turn generate a measurable field that is registered by the receiver loop. The depth at which the waves penetrate the subsoil depends on the frequency and the conductivity of the rock and is typically several hundred meters. Low-frequency waves and longwave radio waves are mainly used. There are two different procedures. Either the field is continuously generated using one or more frequencies whose wave is filtered from the signal measured by the receiver (frequency domain, FEM) or the field is generated in pulses of 20–40 ms and in the following pause the response from the induced field is measured (time domain, TEM). In the second case the transmitter can also be used as the receiver.

If the underground environment has low conductivity, the reflection of radio waves from near-surface inhomogeneities can be used. Although such a **georadar** (or ground-penetrating radar) is most often used in archeology, it is sometimes also used in underground mining. Radio waves can also be used to obtain morphological data from the air, which can be helpful in forests.

Magnetotellurics is sometimes used in the search for oil and gas in addition to seismics (see below) and sometimes for geothermal energy and mining. Similar to the above-mentioned electromagnetic methods it deals with the interplay of induced electric and magnetic fields such that this method can be used to investigate conductivity at a depth of hundreds to thousands of meters depending on the frequency measured. Using electrodes and magnetometers placed in the ground both the electric and perpendicular magnetic components of the electromagnetic field can be measured for a few hours. In classical magnetotellurics (MT) natural currents in the ionosphere and magnetosphere created by the

solar wind interacting with the Earth's magnetic field serve as excitation. Audiomagnetotellurics (AMT) uses higher frequencies and therefore reaches greater depths. Either natural or (especially, at sea) artificial waves—controlled source audiomagnetotellurics (CSAMT), a.k.a. controlled source electromagnetic (CSEM) method—are used for excitation.

Radiometry is frequently used and refers to the measurement of gamma radiation using a spectrometer. Such radiation is produced not only by the decay of certain daughter nuclides of uranium and thorium but also by the decay of the radioactive potassium isotope ^{40}K. Since each of these decay processes releases a gamma quantum of a certain energy the measured spectrum can be used to determine the contents of uranium, thorium, and potassium in the ground. Not only can uranium deposits be mapped and discovered in this way but also potassium salts, potassium-rich rocks, and corresponding alteration zones.

Seismics is the investigation of the subsurface using artificially generated seismic waves. It takes advantage of the fact that the waves are reflected and refracted at certain boundary surfaces. This happens on contact with rocks of different densities or even with different pore fluids such as oil and water. Reflection seismics (Fig. 1.27) is the most important method used in the search for oil and gas. It can be used to create profiles or 3D models of the subsurface reaching depths of several kilometers. This enables the discovery of structures such as folds, faults, and salt domes where oil may have accumulated (Sect. 6.3) and the location of important sediment layers to be tracked. It is not always easy to see which reflectors are oil or gas deposits. Geologists investigate the profiles for direct indicators of hydrocarbons such as bright spots (high amplitude), dim spots (reduced amplitude typical of sand with lower pore volume), phase changes, and shadow zones. If successful, then 3D

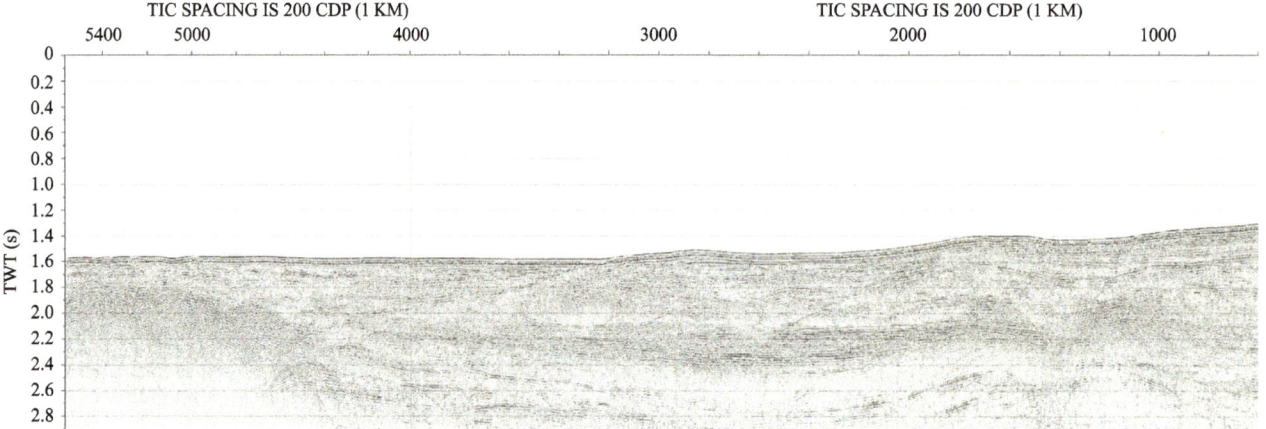

Fig. 1.27 Seismic profile from the Gulf of Mexico. CDP refers to the process of data processing. TWT is the time required by the seismic wave to reach the reflective layer and then return to the surface. This corresponds to the depth, but the relationship is not linear. *CDP*, Common depth point; *TWT*, two-way time (© USGS)

Fig. 1.28 At sea a seismic airgun serves as the seismic source. The waves reflected in the subsurface (schematically in *blue*) are received by hydrophones attached to streamers several kilometers long

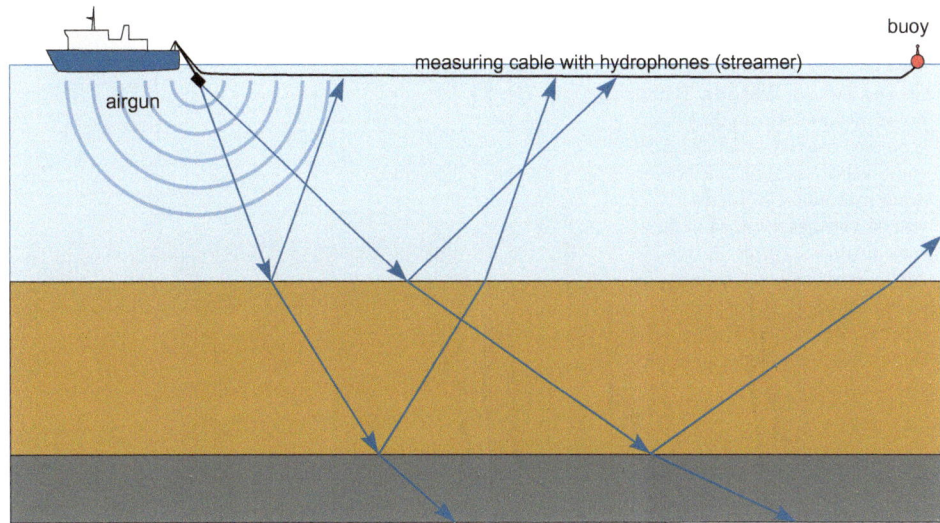

models can be used to decide exactly where to drill. The method can also be used to continuously measure the remaining amount of hydrocarbons and their position in producing fields (4D seismics).

Investigations at sea are carried out onboard specially equipped ships (Fig. 1.28). As a rule an airgun hanging in the water serves as the seismic source. To measure the source the ship tows a streamer with cables several kilometers long equipped with hundreds of hydrophones. It used to be the case in the past for only one streamer to be used, but today the standard is to use several (especially, to create 3D models).

Sometimes measurement systems are distributed on the ocean floor instead (especially, when a producing oil field is to be continuously monitored).

On land either blasting or the so-called vibroseis technique (Fig. 1.29) can be used as the seismic source. The vibroseis technique involves special heavy vehicles that press a plate onto the ground and trigger vibrations using hydraulics and applying their own weight. Normally, several vehicles are used at the same time driving in a coordinated way from point to point in the measuring area. Beforehand geophones are laid out over an area of several square kilometers. These convert ground movements into electrical signals that in turn are transmitted via cables to a receiving vehicle. Typically, this arrangement is systematically repeated over a larger area during a measurement campaign lasting several months.

Creating a profile from the travel times of seismic waves requires some computing effort. In addition to various corrections such as for topography and to reduce background noise common depth point (CDP) stacking or common midpoint (CMP) stacking is an important step. Since each point reflecting a wave is targeted from different directions

during the various measurements the data are searched for these common depth points and their position is calculated. The depth is specified in the resulting profile as the travel time of the wave. To transform the timescale into a real depth scale the data still have to be migrated. The problem here is that the wave velocity depends of the density of the respective rocks so there is no linear correlation. This also means that a non-migrated profile reproduces reality in a distorted way. However, unlike the case with earthquakes we only have P waves and therefore cannot calculate the velocity from the transit time difference between P and S waves. Therefore, a model of the existing rocks (and their density) must first be assumed for data migration. The geology should already be more or less well known (e.g., by drilling).

1.8 Geochemical Exploration

Certain substances are often enriched in the wider surroundings of a deposit by alteration processes during the formation of the deposit (primary halo) or subsequently by weathering (secondary halo). Systematic sampling of soil, rocks, plants, water, recent sediments, or gases can detect geochemical anomalies. It is usually not the metal enriched in the deposit but the more mobile elements that show us the way as pathfinder elements. For example, around a SEDEX deposit elevated manganese concentrations are likely to be found at great distances. Zinc is comparatively soluble in water and can therefore be detected in the sediments of streams. On the other hand, lead will only be found in the immediate vicinity of the deposit. Corresponding knowledge of the solubility of metals (which depends on the properties of water such as pH and Eh; see also Chap. 4) is

Fig. 1.29 The vibroseis technique involves special vehicles used in land-based seismic surveys. The plate visible in the picture between the wheels is pressed onto the ground and the entire vehicle is set into vibration using hydraulics. Normally, several vehicles are used at the same time. The waves reflected in the underground environment can be registered using geophones (*photo* © Interfase, CC-BY, Wikimedia Commons)

indispensable. In addition to dissolved substances there are also particles transported by the flow (in particular, heavy minerals) in rivers.

Large-scale geochemical exploration is carried out to delineate potentially interesting areas and to look in detail at smaller areas. The sediments of streams and rivers are most frequently sampled because they give an impression of the entire upstream catchment area. Moss growing directly along streams is also well suited to creating such an impression. When it comes to soil samples it is important to ensure that the correct soil horizon is sampled and that the soil has not been transported from elsewhere. Some plants prefer to be enriched by certain metals and have the advantage that they can be collected easily. Normally, the first step in exploration is the orientation phase used to identify how the best contrast can be achieved.

Samples are normally analyzed more or less automatically in the laboratory and evaluated using statistical methods. Moreover, there are devices capable of making measurements in the field such as portable X-ray fluorescence devices.

1.9 Drilling

Drilling plays an important role in the production of oil, gas, and water, as well as for the utilization of geothermal energy. But it also is used for the exploration and evaluation of ore deposits. Samples can easily be taken with a soil auger in loose rock. Simple percussion drills and rotary drills are used in solid rock to drill blast holes such that dirt (dry) or drilling mud (wet) can be interpreted. To reach greater depths percussion drills or rotary drills are used. They have an extendable hollow drilling rod through which compressed air is blown in with the aim of cooling and blowing out cuttings; in other cases water is pumped in. Since the dust released does not necessarily originate from the tip and may contain material from the remaining hole the process of reverse circulation (RC) drilling was developed. It uses an inner pipe within the drill pipe. Air or water is pressed down through the outer pipe and rises again inside the inner pipe together with the cuttings. Drilling rigs have a small swiveling drilling tower usually mounted on a vehicle.

However, the main purpose of drilling a borehole is to explore the geology and extract rock samples. There are special mobile rigs for this purpose called **diamond drilling rigs** (Fig. 1.30). A core drill bit is attached to the rotating drill pipe. Diamonds embedded in the steel of the drill bit or other hard materials such as tungsten carbide cut into the rock in such a way that a cylindrical drill core (Fig. 1.31) remains in the inner tube. Once this has reached a certain length the inner tube together with the drill core is pulled upward with a rope. This means that drilling can continue without removing and reinstalling the drill pipe. Drill pipes can be extended by bolting on further tubes (typically, reaching depths of 1000–1500 m). Water is pumped through the pipe to cool it, reduce friction, and transport the cuttings. Drill cores provide a direct way of capturing rock layers in

Fig. 1.30 Backlit diamond drill rig used to explore a banded iron formation in Mauritania (*photo* © Thomas Finkenbein)

Fig. 1.31 Drill cores from a banded iron formation in Mauritania (*photo* © Thomas Finkenbein)

the underground environment to determine the ore grade of different areas of a deposit or to measure porosity. However, pulling cores is much slower and more expensive than normal drilling.

Many parameters can be measured directly in situ by inserting a probe in a borehole: such as magnetic field, temperature, gamma radiation, electrical conductivity and induced polarization (see also Sect. 1.7), density, porosity, water content, pH, and Eh. It is also possible to use optical scanners to record a digital drill core; there are even instruments to measure the chemical composition of rocks in situ (e.g., by X-ray fluorescence, XRF). Boreholes can also be used to "X-ray" a rock volume three-dimensionally using seismics and electromagnetic methods.

Today, the **rotary drilling method** uses special drilling fluid called drilling mud. The method is used almost exclusively for deeper drillings (Fig. 1.32). This involves working with a rotating drill pipe that can be extended as the depth increases by bolting on pipes. It is driven by a motor suspended from a pulley in the drilling rig (power head or top drive). Older machines have a so-called rotary table directly above the borehole that transfers the rotation of a motor to the drill pipe. The drill bit is attached to the lower end of the drill pipe and grinds the rock into cuttings. There are various variants such as roller bits that have three rollers fitted with hard metal teeth and fixed heads fitted with diamonds or other hard materials (Fig. 1.33).

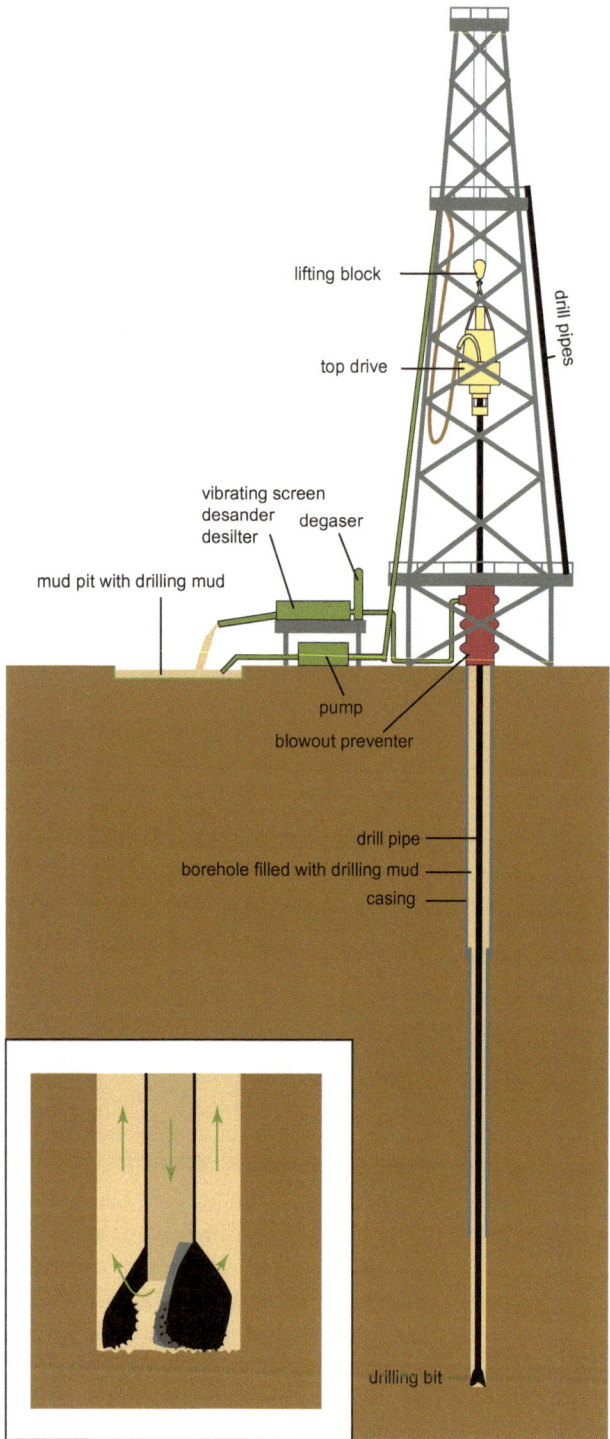

Fig. 1.32 The rotary drilling method uses a rotating drill pipe that is nowadays driven by a top drive suspended in the drilling rig (derrick). The drilling fluid (drilling mud) is pumped through the drill pipe, exits through the borehole, and is used again after cleaning. A blowout preventer is a valve system that prevents uncontrolled leakage of pressurized oil or gas

Fig. 1.33 Diamond drill head for oil and gas drilling (*photo* © Hhakim/iStockphoto)

Drilling fluid (drilling mud) is pumped through the drill pipe, flows upward laterally, and removes cuttings from the hole. The fluid is a muddy bentonite–water mixture. Bentonite is a swellable clay frequently used in construction engineering. Baryte (Sect. 7.14) is added for deeper drilling to increase density. Mud is used to remove the cuttings and indirectly monitor them by examining them. It is also used to cool the drill bit, to reduce the friction of the rotating drill pipe in the hole, to aid abrasion of the rock by the drill bit, and to stabilize the borehole. The pressure of the mud against the borehole's side walls prevents it from collapsing. When drilling a reservoir with pressurized oil or water the weight of the mud prevents uncontrolled discharge (blowout). Accordingly, the mud must always be adjusted to the correct density, manually done in simple systems and automatically in large modern drilling rigs. The mud emerging from the borehole is cleaned using such equipment as vibrating screens, desanders, and desilters and used again.

Systems of different sizes are used depending on the application. For deep oil wells drilling starts with borehole diameters up to 1 m and ends with smaller diameters that may not exceed 1 dm. In geothermal drilling the borehole diameter is often in the centimeter range.

The borehole is lined with a steel pipe (casing) to permanently stabilize it. After such a pipe has been lowered into the hole it is fixed with cement and then drilling continues inside the casing with, of course, a smaller diameter. A plastic tube is inserted for geothermal drilling.

Drill cores can also be taken using the rotary method with corresponding core drill bits. However, this method is much more time consuming and correspondingly more expensive than normal drilling.

It is not a trivial task to drill vertically downward with simple systems since the drill bit is often deflected by structures in the rock. In the past the borehole had to be surveyed regularly, which meant the drill pipe had to be removed. If it was necessary to correct the direction, then a special drill head with an additional rotating part was used driven by the mud flow (mud motor). As long as the drill pipe rotated this had no effect, but when the drill pipe stopped the head would mill sideways.

The drill bits of modern equipment can be deflected in any direction by remote control. To do this either the drill head is bent slightly or there are so-called control ribs directly behind the bit. They can be moved out and pressed against the borehole wall. In addition, ever-more sensors are being installed in drill heads not only to measure the position during drilling (with inclinometers and gyro compasses) but also to collect geophysical parameters via a process called measurement while drilling (MWD). For data transmission so-called mud pulses are usually sent through the drilling fluid. More recently so-called wired drill pipes have also been used. Directional drilling allows developing a larger oil field from a single platform. The target can be drilled at any angle from different directions and problematic zones can be avoided by taking detours. It is possible to reach targets at a depth of several kilometers that are not easily accessible. MWD also reduces the risk involved in drilling oil under salt horizons because it is now possible to adjust the density of the drilling fluid at exactly the right moment. Somewhat simpler are drill heads that specialize in automatic drilling (vertical or horizontal).

1.10 Open-Pit Mining

Open-pit (or open-cast) mining is most effective for deposits near the surface. This can only be done by removing the overlying rock (overburden). Nevertheless, it is often much more advantageous to exploit a large-volume deposit with a low ore grade by open-pit mining than a small, high-quality deposit by underground mining. The exact procedure depends on the topography, the shape of the deposit, and the hardness of the rock.

Lignite is often found in shallow seams within soft sedimentary rocks. Both the rock and lignite can be removed relatively quickly with large bucket wheel excavators or bucket excavators (Fig. 1.34). Conveyor belts are usually used for transport. The normal procedure is for mining to take place strip by strip, while the other side of the mine is refilled with overburden.

Mining a flat seam in hilly areas follows topographical contours. A variant of coal mining practiced mainly in the Appalachian Mountains (USA) is so-called mountain top removal mining in which mountain peaks are completely removed to expose a seam and the overburden is mostly dumped into the valleys. The method is so cheap that coal can not only be economically produced but also compete on the world market. However, the method is also very controversial since it concerns not only the loss of mountain landscapes and forests but also the pollution of rivers with sulfuric acid from sulfide weathering (Palmer et al. 2010).

Since flat ore bodies are rarely found and open-cast mines (Figs. 1.35 and 1.36) reach great depths it is not possible to dump the overburden of ore deposits in other parts of an open-cast mine; hence it must be dumped on an overburden stockpile (dump, spoil heap). The largest open-cast mines such as Chuquicamata (Chile) and Bingham (USA) have diameters of several kilometers and depths of more than 1000 m. A ramp-like road is created for the transport of the ore and the hard barren rock after blasting. They are loaded onto trucks using excavators. A whole fleet of large dump trucks (Fig. 1.37) normally commute continuously between the mining site and the edge of the pit. The steepness of a mine's flanks depends on the rock's mechanical properties.

Fig. 1.34 Extraction of lignite in Garzweiler (Germany) (*photo* © Raimond Spekking, CC-BY-SA, cropped image, Wikimedia Commons)

Fig. 1.35 Blasting in an open-pit mine. The Twin Creeks gold mine (Nevada, USA) is a Carlin-type deposit (Sect. 4.11) (*photo* © Geomartin, CC-BY-SA, Wikimedia Commons)

Fig. 1.37 A dump truck in Chuquicamata (Chile) (*photo* © James Byrum, CC-BY-ND, flickr.com)

Fig. 1.36 The Bingham Canyon Mine (Utah, USA) exploits a porphyry copper deposit (Sect. 4.4). The open-pit mine is about 1000 m deep, 4 km wide, and is one of the largest man-made holes in the world (*photo* © Tim Jarett, CC-BY-SA, Wikimedia Commons)

The normal procedure is to lay out terraces that not only allow excavation at different levels but also reduce the risk of rockfall. However, that does not always help. On April 10, 2013 Bingham suffered a massive landslide. Part of the northwestern flank set in motion more than 150 million tonnes of rock burying the deepest point of the mine under a 100-m-thick layer of broken side rock. Fortunately, the mine was evacuated in time since the flanks are monitored with sensors, but the financial damage was enormous. It will take some time for the operation to reach full capacity again.

Highwall mining is intermediate between surface and underground mining. It is increasingly used in coal production. The process involves milling galleries several hundred meters deep into the coal seam of an open-cast mine using a remote-controlled machine. Walls remain between the galleries for stabilization purposes making it possible to continue mining when an open-cast mine cannot be extended.

1.11 Underground Mining

When it is not possible or economical to remove the rock above a deposit, mining takes place underground. This is also the case with open-cast mines that can no longer be expanded. Underground mining is sometimes used to extract deeper ores.

This starts by sinking a vertical shaft (Fig. 1.38) or milling a horizontal adit into the mountain. Often a spiral-shaped decline (underground access ramp) is built ready for use by trucks. Smaller galleries branching from the main shaft or gallery lead to the ore body or coal. Different levels of the mine can be connected via blind shafts. While part of the deposit is being mined, other parts of the mine are being prepared for future mining and further deposits are being investigated in exploration galleries. Underground filling stations, magazines, jaw crushers, and so on can also be installed.

The larger and deeper a mine the more important the supply of fresh air (ventilation). Ventilation also removes exhaust gases, fine dust, and pit gas released from the rock and provides cooling in deep mines. At least two openings are required at the surface to create an air flow through the mine; special ventilation shafts are usually built. In mines it may be necessary to direct the air flow using doors, airlocks, and pipes (air ducts). Water flowing into the pit must be drained off via tunnels or lifted to the surface with pumps.

Fig. 1.38 Winding tower of a coal mine. Shaft 3 of Sophia-Jacoba Mine in Hückelhoven (Germany) (*photo* © Sven Teschke, CC-BY-SA, Wikimedia)

The shape and arrangement of chambers depend on the type of deposit to be mined. How large such chambers can be without collapsing depends on the rock's mechanical properties. Supporting mine props, wire meshes, bolts, and rock anchors driven into the rock are used to stabilize the surrounding rock. Miners have a number of options available to them when it comes to the size of chambers. They can make the chambers so small that they remain stable, but this usually means that some of the ore is left behind. They can fill chambers with backfill (e.g., overburden, sand, concrete) and then the remaining part of the deposit can be mined (cut and fill). They can allow cavities to collapse on purpose (caving in). The latter is a cheap option often used for the extraction of low-grade ores. However, it means that crater-shaped collapse structures may form at the surface.

Flat or slightly inclined coal seams and similarly shaped ore deposits are usually mined by **longwall mining** (Fig. 1.39). Chambers here are usually 100–300 m long and only a few meters wide. A coal shearer moves along the longwall and cuts coal over its entire length and a conveyor belt transports the coal. In former times the cavity was secured with mine props. However, today it is supported by self-advancing hydraulic roof supports. The excavation cavity always moves in the same direction through the seam. The remaining cavity beyond the roof supports is either filled with backfill or is allowed to cave in. Although caving in is cheaper, it can lead to subsidence at the surface and thus to mining damage such as cracks in buildings and disturbances in river courses. The ground surface in the Ruhr area of Germany has lowered (in some cases up to 40 m). This is the reason the Emscher River now flows between dams a few meters above the level of the surrounding area and streams that once flowed into the river now have to be pumped upward. The Ruhr area would become a lake if the pumps were switched off. Cost savings underground have to be weighed against so-called eternity costs.

Room-and-pillar mining is used for massive seam-like ore deposits, salt deposits, and some coal deposits (Fig. 1.40). In the process systematically arranged pillars are left standing between the mining chambers. In thick deposits mining takes place in chambers either upward by blasting the ceiling (overhand stoping) or downward along staircase-shaped steps (underhand stoping). Stoping can be repeated at several levels in such deposits. If the pillars themselves are to be extracted afterwards, then either they

Fig. 1.39 In longwall mining a seam is mined via a long excavation cavity just a few meters wide. The cavity moves into the seam in the direction of mining. The other side collapses or is filled with backfill

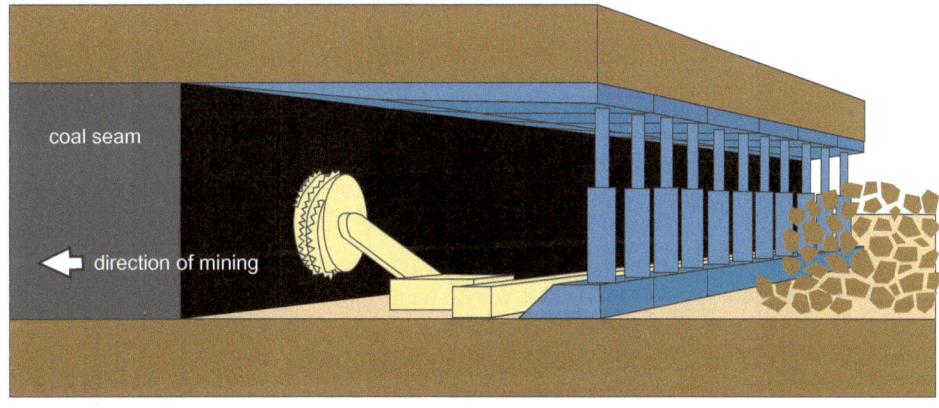

coal seam

direction of mining

coal shearer and conveyor belt

self-advancing hydraulic roof supports

backfill or collaps (caving-in)

Fig. 1.40 Room-and-pillar mining with pillars remaining to support the overburden

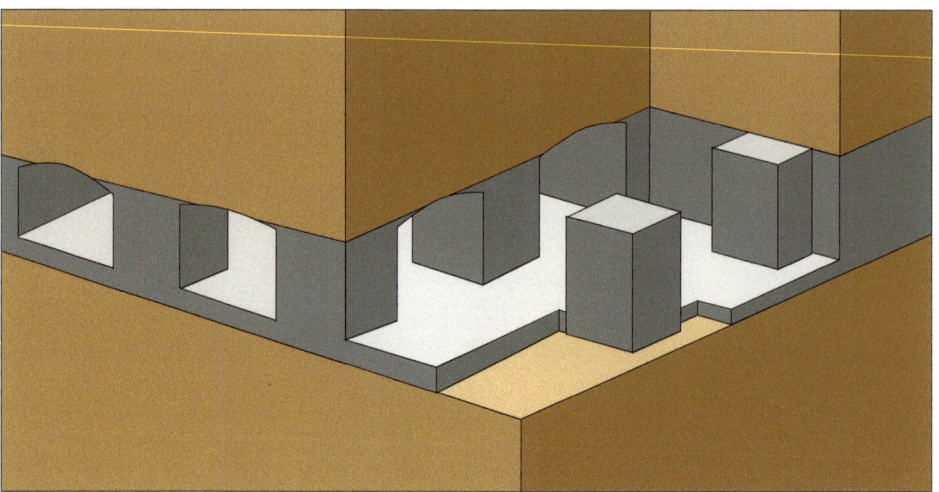

are systematically removed causing the roof to collapse or the chambers are filled with backfill first. In the case of smaller, irregularly arranged ore deposits mining takes place in chambers arranged accordingly (possibly without pillars).

One variant of room-and-pillar mining used in steep deposits is called **cavity working** (a.k.a. roomwork). Such deposits are accessed by a system of working levels where the ore between the levels is blown up pillar by pillar into the ever-increasing room. The result is a high chamber bounded laterally by the surrounding rock—not by supporting pillars.

Vein deposits and steep coal seams are usually mined by **shrinkage stoping** in which mining takes place from bottom to top (overhand). This involves creating a gallery following the vein at a deep level. Immediately above this level a second level is laid out for the purpose of mining. Ore is blasted from the ceiling and falls on the floor. Only a small part is cleared because the blocks lying on the ground serve as a working platform for further blasting operations. Only when the next higher level has been reached is the ore removed via the lower level. Depending on the procedure miners work their way straight up or diagonally and additional boxes, filling stations, and working platforms may be installed. Sometimes part of the cavity has to be filled with backfill before the rest can be removed. Although the process is relatively complex and expensive, historically it was very common.

A cost-effective and universally applicable method used for massive or steep deposits is **vertical retreat mining**. The ore body is accessed by a system of superimposed levels (Fig. 1.41). Ore located between these levels is divided into pillars that are blown up one after the other. Ore plunging to the bottom in blocks is loaded into wheel loaders. By retreating on every level miners work on ever-deeper levels. Subsequent crumbling of the overburden leads to the formation of large collapse craters on the surface.

Block caving is used for low-stability ore. Filling stations are installed below the ore body in preparation for mining. The lowest part of the ore is then removed. This leads to the entire ore body breaking and falling in the form of blocks into the filling stations.

Ore can be transported from the mining site to the surface or to a shaft by trucks, special articulated vehicles, mine railways, or conveyor belts. Special shafts are often constructed through which the ore falls directly into a filling station located near the main shaft.

Although mining at a depth of dozens or hundreds of meters is normal, some mines reach a depth of more than 1000 m. In exceptional cases depths greater than 2000 m have been reached. Gold mines in the Witwatersrand (South Africa) are by far the deepest mines reaching a depth of almost 4 km. In addition to transport being more expensive and time consuming in deep mining, heat also becomes problematic. Temperatures of up to 45 °C prevail in the gold mines in the Witwatersrand resulting in mining sites having to be cooled down to a tolerable temperature (see also Box 1.8 on current mining in Germany).

Box 1.8 Renaissance of mining in Germany?

In Germany hardly any active mines are left. However, while the last mines in the Ruhr Valley have been closed, new ones might be opened in Saxony where some deposits have been reassessed in recent years. Many still contain significant quantities of ore and some may even be worth mining at least if commodity prices remain high. However, mining costs are likely to be very high by international standards as a result of problems with processing.

A first step in this direction was the excavation of a new adit at Niederschlag, a small village on the Czech border between Annaberg-Buchholz and

Fig. 1.41 Vertical retreat mining in which ore above the working level is blown up, plunges to the bottom of the level, and is then cleared out. Breaking of the overburden leads to the formation of collapse craters (modified from Wirtschaftsvereinigung Bergbau 1994)

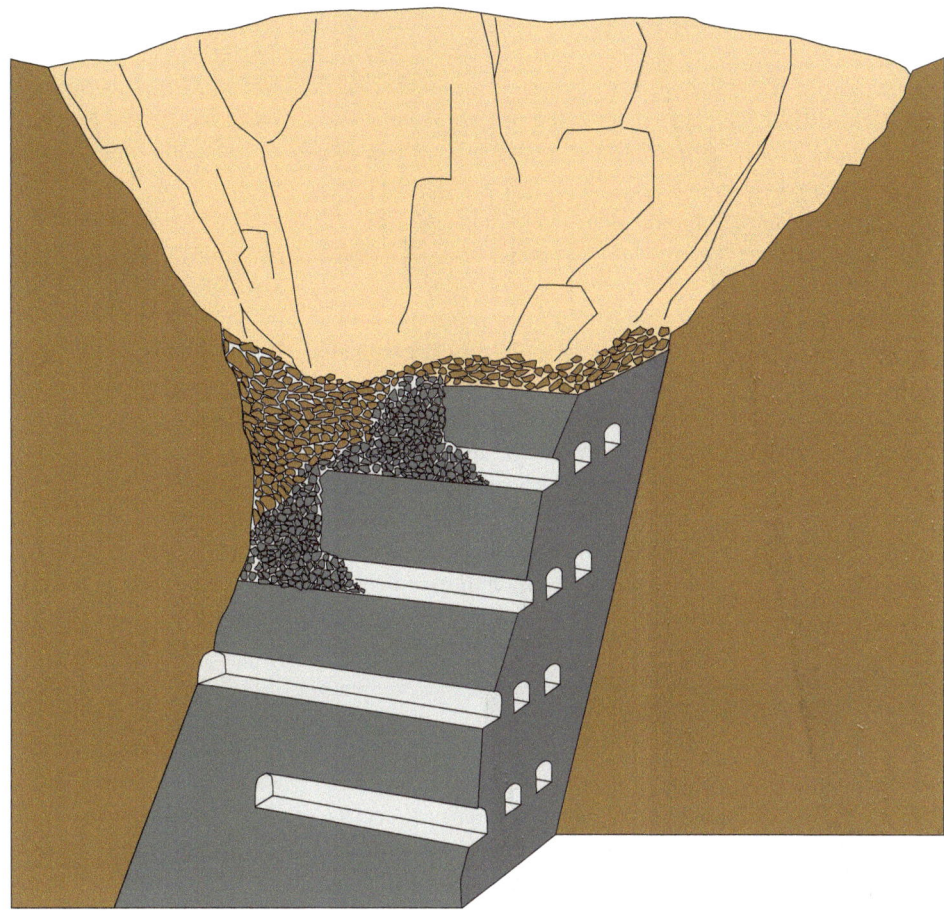

Oberwiesenthal. The intention is to reopen a hydrothermal deposit in which uranium has already been mined, but this time it will be about baryte and fluorite (Sect. 7.14). Further mines are expected to be opened soon in the Ore Mountains (Erzgebirge) and the Thuringian Forest to extract these minerals. In the Black Forest such minerals have been mined in the Clara Mine in Oberwolfach for quite some time.

However, metals could soon be mined in Germany again. Several prospects for lithium in the Ore Mountains are promising. Notably, a major occurrence of zinnwaldite (lithium mica) holding at least 96,000 t of lithium is believed to exist underneath the village of Zinnwald on the border with the Czech Republic.

Deutsche Rohstoff AG is exploring tin deposits in Gottesberg (Vogtland) and Geyer (Erzgebirge). It describes the greisen (Sect. 4.6) of Gottesberg as the largest tin deposit in Europe where 114,000 t of tin with an ore grade of 0.43% Sn are believed to exist; copper, tungsten, and gallium could be mined as by-products. Most of the deposit is located at a depth of 100–400 m with individual peaks of the ore body rising to the surface where tin and tungsten were

previously mined. Geyer is a skarn deposit containing 43,500 t of tin; zinc and indium could be mined as by-products. Further deposits are currently being investigated in the Ore Mountains. The reprocessing of old heaps and tailings is also believed to have great potential (Box 1.11).

Another example is interest shown in copper deposits around Spremberg in Lusatia (Germany) that belong to the copper-bearing shale called Kupfer-schiefer (Sect. 5.1). It lies here at depths of 800–1500 m and has a relatively high grade. It is estimated that 150 Mt of ore contain 1.5 Mt of copper.

The countryside around Delitzsch, a small town situated between Leipzig and Bitterfeld, is also being explored. Buried under 100 m of sediments there are several genetically unrelated deposits discovered during exploration by East German geologists looking for uranium. However, these sediments were never mined. Carbonatite (Sect. 3.10) is being explored for REEs under the village of Storkwitz. It is expected to contain 38,000 t of REE_2O_3 and 78,000 t of Nb_2O_3 down to the explored depth of 600 m, but the ore grade is very low with an average of 0.48% REE_2O_3.

1.12 In Situ Leaching

Sometimes it is possible to leach a desired substance out of the rock without bringing the ore to the surface. Prerequisites are that the rock is permeable and that a solvent can flow from one borehole to the next. Although fracking (Sect. 6.7) can assist if necessary, it needs to be ensured that the chemicals used cannot leak into an aquifer used as a source of drinking water.

Water can be used as a solvent in salt mining and is used accordingly in many salt mines. Uranium is occasionally extracted from uraniferous sandstones by in situ leaching when sulfuric acid, ammonium carbonate, or sodium hydrogen carbonate dissolved in water are used as solvents. An example is Königstein, Saxony where in situ leaching with sulfuric acid was introduced after a few years of conventional mining. Occasionally, other metals such as copper are extracted in this way (particularly, from crushed rock in disused and active mines).

1.13 Deep-Sea Mining

Extracting metals from the deep sea was first proposed in the 1960s. It mainly concerned manganese nodules (Sect. 5.6) that lie on the bottom in large parts of the deep sea and contain nickel, copper, cobalt, other metals, and manganese. Implementation was not easy since equipment had to be developed to withstand water pressure at a depth of several thousand meters. Such equipment had to be light enough not to get stuck in the mud. Caterpillar vehicles were developed to collect the nodules in many imaginative ways. Another problem involved transporting the material to the water surface. Bucket chains proved to be impractical. A better option would be to pump the material through pipes. Several projects at the end of the 1970s actually produced several hundred tonnes of nodules. Compared with mining on land this was a negligible amount and the costs were so high that it was not economical. It would only be worth it if huge areas could be collected relatively cheaply. Mining in the deep sea remains science fiction for the time being.

Another problem is that little is known about the effects large-scale mining on the sea floor would have on the ecosystem. Organisms living directly on the seabed would be affected and the oxidation of ploughed sediments would remove oxygen from the water. Sludge is disturbed by nodule collectors in the deep sea and sludge clouds are added at shallow depths when worthless material separated onboard the mining ship is returned to the sea. Sludge drifting in ocean currents can affect the gills of fish and any reduction in the amount of light worsens the conditions for

photosynthesis. In 1989 a German research project ploughed the seabed over a small area. It took a decade for the same species to resettle the area. The affected area would be many times larger in a real mining project.

Since the turn of the millennium high commodity prices have meant that the deep sea has moved back into the limelight despite all reservations. Claims have been staked in the Pacific and the Indian Ocean and countries such as Germany, France, China, Japan, Korea, and Russia have acquired exploration licenses. At 75,000 m^2 the area claimed by Germany alone is larger than Lower Saxony and Schleswig-Holstein combined. Claims are awarded by the International Seabed Authority, a UN agency based in Jamaica. The agency also imposes requirements for the protection of the environment. It is based on the 1994 Convention on the Law of the Sea that has not been ratified by some states including the United States. The German Geological Survey (BGR) has the primary objective of investigating how mining can be carried out in an environmentally friendly way. It remains to be seen whether the next few decades will see commercial deep-sea mining on a grand scale.

However, massive sulfides that form at hot springs such as black smokers (Sect. 4.15.1) are more rewarding than scattered manganese nodules. There are great differences in the depth of the water, the size of the deposits, and the content of worthwhile metals. Some deposits are near volcanic islands and not in international waters. Exploration licenses have also been granted for these massive sulfides. In the Bismarck Sea near the islands of New Britain and New Ireland, which belong to Papua New Guinea, the first commercial mining is already being prepared. The deposit is located at a depth of 1600 m and has a high grade of copper, gold, and silver. However, the tonnage is relatively small with the total quantity of gold contained of about 15 t corresponding to the annual production of a single South African gold mine. Therefore, the deep sea is not expected to make a significant contribution to the supply of metals in the foreseeable future.

REEs could also be extracted from the deep sea one day. A Japanese research team (Kato et al. 2011) measured REE content in more than 2000 samples and found that the mud at the bottom of the eastern Pacific has consistently high ore grades in the uppermost meters—sometimes higher than onshore deposits. These appear to be bound to iron hydroxides and zeolites. It would be relatively easy to suck off the sludge and leach out the metals with acid. Sludge lying on 1 km^2 could cover about one fifth of current annual demand. However, the consequences for the ecosystem would be even more devastating than would occur with the collection of manganese nodules.

1.14 Environmental Degradation, Land Use, and Social Responsibility

There is no denying that mining can cause significant environmental and health problems. Moreover, in many cases the living conditions of the population are made worse (see also Box 1.9). The problem is particularly acute in countries where environmental standards and social plans do not exist or are not enforced and in countries where corrupt politicians renounce these rules for their private advantage. An appropriate sense of responsibility is necessary to minimize consequences.

Box 1.9 The Marikana massacre

South Africa is currently regarded as the country with the greatest social inequality and the situation has even worsened since the end of apartheid. Inequality is rife in the mining industry where high profits are brought about by keeping wages low and turning the cheek to frequent accidents at work. In a wave of strikes in 2012 the struggle for higher wages got out of hand. The fact that the government-related trade union called National Union of Mineworkers (NUM) tried to suppress wage disputes was probably due to the fact that the leadership of the ruling African National Congress (ANC) had developed into an oligarchy who benefited directly from mining. By contrast, the smaller Association of Mineworkers and Construction Union (AMCU) organized wildcat strikes that resulted in a wage increase of 125% being achieved at the Rustenburg platinum mine. A few months later miners at the Marikana platinum mine went on strike, which the NUM called a criminal act. The founder and former chairman of this trade union was a delegate member of the supervisory board of the operating company who in emails demanded that the police take tough action against the "criminals." Within a few days 10 people had died in the clashes including miners, trade unionists, and policemen.

On August 16, 2012 the police surrounded the strikers who had gathered on a hill. According to the official version the police opened fire with automatic weapons after being attacked with spears, sticks, and machetes resulting in 34 workers being killed and 78 injured. This version of events was later questioned. According to evidence uncovered by an investigative journalist (Marinovich 2012) only a dozen of the victims died in the place filmed by the police. Most of the victims had tried to escape from the hill in another direction. They were run over by jeeps or shot in the back from a short distance. While ministers and trade union officials defended police action, to many observers this brought back bad memories of the apartheid regime.

The workers finally accepted a wage increase of 11–22% and a one-off payment of ZAR2000 each. As Marikana resumed operations the wave of strikes spread to many more mines. Around 75,000 miners were on strike at platinum, gold, iron, and coal mines as of October 2012.

The enormous land consumption that results from open-pit mines, spoil heaps, and tailings is a socially explosive subject. A new copper mine in the Atacama desert hardly disturbs anyone but in densely populated or agriculturally used areas it is something else. A German example is the villages that fall victim to lignite mines and whose inhabitants are resettled. In poor countries mining rights are often granted for areas where ownership is not regulated, where land is simply expropriated, or where land is already officially state owned—areas that have always been used by the local population for subsistence (e.g., as pasture). In this way the population is deprived of its living conditions without receiving any compensation. The effects on the environment are similarly problematic when forests are cleared, rivers diverted, and entire mountains removed. This applies in particular to ecologically sensitive regions (especially, when nature reserves are sometimes reduced in size for mining purposes). Since the demand for metals and ores of ever-lower grade is increasing, land consumption will continue to increase in the future. Although land consumption is lower in underground mining, there is always the risk for subsidence at the surface and damage to buildings and roads. Rehabilitation or renaturation of abandoned mines is of course possible to a certain extent, but in countries where there are no regulations this is not even attempted.

Mining can have a serious impact on the water supply. Water that has infiltrated a mine is inevitably pumped out with the result that the groundwater level in the surrounding area drops significantly. This can cause springs and wells to dry out and the drinking water supply downstream can be impaired by contamination.

The release of toxic and possibly also radioactive substances into the environment is probably the biggest problem with mining. In addition to various toxic heavy metals many ores also contain substances that are toxic in traces such as mercury, arsenic, or cadmium. Low-grade ores end up on heaps (dumps) and materials that cannot be extracted profitably end up in residues from processing (tailings) that are sedimented in sludge ponds and as residue of smelting in slags. Exhaust gases from smelting are also problematic when they are not adequately filtered. Some of these substances can be dissolved in water or transported as suspended particles by water, while others are transported as dust particles by the wind. For example, the Chilean mining town of Chuquicamata is so heavily polluted with arsenic and other toxic substances that its inhabitants had to be relocated to Calama in 2004. However, there is a second reason for this. Despite the inhabitants being obliged to move, the mining of copper ore continues below the town allowing the now empty ghost town to one day give way to open-cast mining.

Consequences can be reduced by taking such measures as sealing the tailings and using appropriate filters. The same applies to chemicals such as cyanides that are used for leaching. Although these are recovered whenever possible, small quantities are also discharged into the tailings. This is particularly dangerous when the dam of a tailings pond breaks and large quantities of contaminated water are suddenly released or sludge contaminated with heavy metals flows out as a cement-like mudflow. On April 25, 1998 between 5 and 7 million m^3 of toxic mud poured out of the tailings pond at the Los Frailes Mine in Aznalcóllar (Spain). Even more catastrophic was the dam burst in Baia Mare (Romania), where large quantities of cyanide were released (see Box 1.10).

Fig. 1.42 The Río Tinto (Spain) is strongly acidic (pH at about 2) and contains dissolved metals in such high concentrations that they can even be extracted from the water. Under these conditions only special microorganisms can survive in the water. Moreover, soils in the immediate environment are contaminated with heavy metals (*photo* © Riotinto2006, public domain, Wikimedia Commons)

Box 1.10 Cyanide disaster in Romania
On January 30, 2000 the dam of the tailings pond in the gold mine in Baia Mare (Romania) broke late in the evening during a thaw and after heavy rain. In the process 100,000 m^3 of water contaminated with heavy metals and about 100 t of cyanide poured over the adjacent area to the Someş River and reached the Danube via the Tisza. Several thousand tonnes of dead fish floated on the rivers and in some sections of the Tisza almost all life-forms died abruptly. Some wells were contaminated and in several cities there was no drinking water supply for days. Only years later did the river ecosystem begin to recover. Moreover, large quantities of heavy metals have been found in the soil since then.

Water escaping from mines receives less attention. The weathering of sulfides (especially, in sulfide deposits and coal mines) produces strongly acidic and toxic mine water with high metal concentrations (acid mine drainage). In some cases mine water also contains dangerous substances such as arsenic and cadmium in high concentrations. Such solutions are also caused by microorganisms that oxidize iron(II) and sulfur, dissolve sulfides, and produce sulfuric acid. These solutions can seep from heaps or tailings. When it comes to the pits themselves the problem arises mainly when they are shut down and filled with water. Acid mine water can escape through old galleries or seep through the subsoil. Acidic and poisonous lakes have been known to

form in old open-cast mines. One example is the Berkeley Pit of Butte (Montana, USA) where in 1995 an entire flock of snow geese died from poisoning and burns. There are also extremely acid lakes in the Iberian pyrite belt (Sect. 4.16.4) (Sánchez España et al. 2008). The Río Tinto is well known there (Fig. 1.42) for being acidic and contaminated with heavy metals. The river's cargo comes from the VMS deposit of the same name that has been mined since the Bronze Age. The bed of such watercourses usually has a striking red or yellow color because dilution with freshwater or neutralization by carbonates leads to the precipitation of iron hydroxides that of course also disturb the ecosystem. A sample from a mine in Iron Mountain (California) with a negative pH of −3.6 and a metal content of 200 g/L (Nordstrom et al. 2000) shows how extreme the composition of mine water can become. It is possible to remedy the situation and neutralize acid mine water by adding CaO or other substances, resulting in precipitation of the metals. There have also been attempts to produce economically interesting metals profitably.

The working conditions in mines are another important consideration. Safety and health standards differ widely from country to country so much so that in many countries miners have a significantly shorter life expectancy than in others.

Commodities originating from war zones are ethically questionable. In particular, coltan (Box 3.14), gold, and diamonds enabled the various warring factions to finance the civil war in the Congo and other crisis areas. This led to the waging of fierce battles for profitable mines and exploitation of people through forced labor. However, attempts have been made to exclude such minerals from markets by means of a system of certificates. The Dodd Frank Act passed in the United States in 2010 stipulates that companies must prove the "clean" origin of their raw materials.

1.15 Mineral Processing

Only rarely is the quality of the ore so high that it can be smelted directly. This is the reason attempts are first made to separate unusable parts using suitable methods and to produce an ore concentrate. In the case of polymetallic ores the respective metals can only be enriched in the various ore concentrates. Processing usually takes place near the mine, while subsequent smelting is often carried out elsewhere by other companies.

The ore must first be crushed to a suitable particle size. Jaw crushers do the rough work here causing larger pieces to break between the V-shaped, reciprocating jaws until they are small enough to fall through the slot between the jaws. The material is then ground to powder in a **mill**. Ball mills are normally used because they can process large quantities: the ore is placed in a rotating drum together with balls made of a hard material. If it is decided to follow this up with wet processing, water is also added to the mill producing an ore-containing slurry. If grain size is important, then appropriate sieves are used.

A **magnetic separator** is the simplest way to separate magnetite and other magnetizable minerals as well as screws or tools that have accidentally ended up in the ore. Mechanical methods that separate minerals according to their density are mainly used for placer deposits. Gold, cassiterite, and other heavy minerals can be enriched using washing pans or washing troughs.

Flotation is an effective and very frequently used process for mineral separation. It involves the use of a foam bath through which gas bubbles rise (Fig. 1.43). If the surface of a mineral is hydrophobic, then gas bubbles rising through the separation medium adhere to it. A foam loaded with ore particles accumulates on the surface and is skimmed off. Which minerals accumulate in the foam and which remain behind can be influenced by the addition of certain chemicals. Combining different flotation cells allows not only the separation of useless minerals such as sterile rock, gangue, and pyrite, but also the enrichment of the respective ore minerals of different metals. The separation of sulfides is particularly effective using flotation. The method is also sometimes used to enrich iron ore or gold and to purify phosphates and coal. A distinction needs to be made between the three groups of chemicals used in flotation. Frothers are responsible for foam formation. Collectors adhere to the surface of certain minerals and coat them with

Fig. 1.43 Foam loaded with sulfides during the flotation of copper ore at Jameson Cell, Prominent Hill (Australia) (*photo* © Geomartin, CC-BY-SA, Wikimedia)

a water-repellent film that strengthens or enables their flotation. Depressants have the opposite effect in that they improve the wettability of certain minerals and prevent their flotation. The pH value also has a great influence and must be adjusted to the appropriate value. Other substances such as flocculants can further improve the process.

Thickeners, filters, and drying ovens are used one after the other to dry the sludge obtained. Ore concentrates are then delivered to a smelter. Unusable residues of the treatment (tailings) are fed into sludge ponds (tailings ponds) (Fig. 1.44) and left to sediment there. Such heavy metal–rich deposits are not without problems since dust that has dried out can be transported by the wind into the environment or

Fig. 1.44 Tailings (i.e., sludge with unusable ore components resulting from mineral processing) are sedimented in sludge ponds. The figure shows tailings from the Antamina Mine (Peru). A copper–zinc skarn formed in connection with porphyry copper is mined here. Molybdenum and silver are also produced as by-products (*photo* © Xtremizta, CC-BY-SA, Wikimedia Commons)

dissolved materials can seep out of ponds that have not been sealed adequately.

Another treatment method is **leaching** in which acids or other chemicals are used followed by hydrometallurgical processing of the solution. This is often done when processing oxidic copper ores. Sulfide ores can only be dissolved after roasting (see next section). The ores are spread out on a prepared surface and sprinkled with sulfuric acid. Copper and other metals leach into the solution and the "pregnant" acid is then collected and transported on with pipes.

The next step in the separation of dissolved ions is usually solvent extraction in which the acid is mixed with a special organic solvent containing a chelating agent. While copper gets incorporated into the chelate complexes, other substances remain in the aqueous solution. The mixture then flows into a quiet (i.e., kept perfectly still) tank in which the organic solution separates from the acid aqueous solution (a. k.a. raffinate). The latter can be used for further leaching. The organic solution is mixed with a strong acid in another mixer where the chelate complexes become unstable and copper dissolves in the acid. The two solvents are then separated in another quiet tank and the organic solvent can then be used for the first step. Finally, high-purity "cathode copper" is obtained from the acid by deposition at a cathode (electroplating, electrowinning). The entire process is also known as **solvent extraction/electrowinning (SX/EW)**. However, in SX/EW the transition between processing and smelting is fluid allowing both to take place in the immediate vicinity of the mine. What is attractive about this process is that the sulfuric acid required is sometimes produced in the same mining area during the smelting of sulfide ores (see below and Fig. 1.45).

Gold usually occurs in the form of tiny inclusions in other minerals. After the minerals have been finely ground the gold can be extracted by two processes. The traditional method is the amalgam process that is mainly used in small mines. The method uses mercury to form a liquid alloy (liquid as long as the mercury content is high) with gold (namely, amalgam). Mercury evaporates when heated leaving the gold behind. The other process is cyanide leaching that exploits the solubility of gold as a cyanide complex. The complex is sprinkled with sodium cyanide solution, the leachate is filtered, and then the gold is precipitated by adding zinc dust. Although the cyanide solution can be reused for further leaching, traces of the highly toxic substance remain in the tailings. Both processes can also be used for silver production in a similar way. However, cyanide leaching only works if the minerals in which gold is trapped are not dissolved in the process. Because sulfides would be dissolved, cyanide leaching only works when sulfides are roasted to oxides beforehand.

REEs are also processed by leaching, but the details depend on the type of ore. A mineral concentrate is often first produced by flotation or other methods and then dissolved in a suitable chemical. Some REE minerals such as bastnäsite and eudialyte can be dissolved in hydrochloric acid. Monazite can be dissolved in sulfuric acid (acidic path) or in caustic soda (alkaline path). To separate REEs from other elements such as uranium, thorium, and iron they are usually precipitated as sodium REE double sulfates and then dissolved again. Different processes such as ion exchange and solvent extraction are used to separate individual elements despite the difficulty of doing so as a result of their very similar chemical properties. The only REEs that can easily be separated by redox reactions are cerium and europium since they can occur not only as trivalent ions but also as Eu^{2+} and Ce^{4+}.

Uranium ores are also processed by acidic or alkaline leaching (depending on the type of ore). Sometimes an oxidizing agent is added to oxidize the hardly soluble U^{4+} to UO_2^{2+}. The solution is purified by solvent extraction or ion exchange in which precipitation produces uranium concentrate called yellow cake.

Microorganisms are increasingly being used to process low-grade ores and tailings (Box 1.11). This is particularly the case in copper production where the process is referred to as bioleaching or biomining (Schippers et al. 2011). Bacteria such as *A. ferrooxidans* and *Acidithiobacillus thiooxidans* oxidize sulfides to sulfate and iron(II) to iron(III), respectively. These organisms also live in acid mine waters and are responsible for their low pH and high heavy metal content. Under their influence sulfide minerals weather, water develops into sulfuric acid, and copper, zinc, nickel, and other metals dissolve. Iron(III) acts as an oxidizing agent and thus helps to decompose sulfide minerals.

Biomining can be exploited by filling special tanks with crushed ore and bacterial strains (tank leaching) or by dumping the ore and bacteria as a heap on a prepared surface (heap leaching). Sometimes biomining is also carried out directly on a mine dump (dump leaching). To supply the bacteria with nutrients the material is irrigated with a suitable solution (e.g., ammonium, phosphate, etc.). The metal-rich solution is then collected at the bottom of the tank, heap, or dump. Depending on the type of ore (especially, chalcopyrite) it may be necessary for leaching to be carried out at elevated temperatures with suitable bacterial strains.

When it comes to copper recovery the process takes a few weeks (in tanks) or almost a year (in heaps). Nevertheless, it is very cost-effective and allows a higher yield than conventional processes. Copper is then extracted from the solution using the SX/EW process (see above). About 20% of the copper produced worldwide currently comes from bioleaching. Moreover, there are mines that work entirely

Fig. 1.45 In the Chuquicamata copper mine (Chile) (Sect. 4.4) both methods of processing and smelting are used. Sulfide ores are mined in the main mine where they are enriched with flotation and then pyrometallurgically smelted. Impure "refined copper" is refined into copper cathodes by electrolysis. By-products include molybdenum concentrate, sulfuric acid, and anode sludge (from which selenium and precious metals can be extracted). In the Mina Sur (the other mine) near-surface oxidic ore is mined from which copper is extracted by leaching with sulfuric acid and SX/EW. It remains to be shown whether other metals can be extracted from waste (poor ores, tailings, metallurgical dust) by leaching or bioleaching. It is expected that Mo concentrate will also be smelted onsite in the future

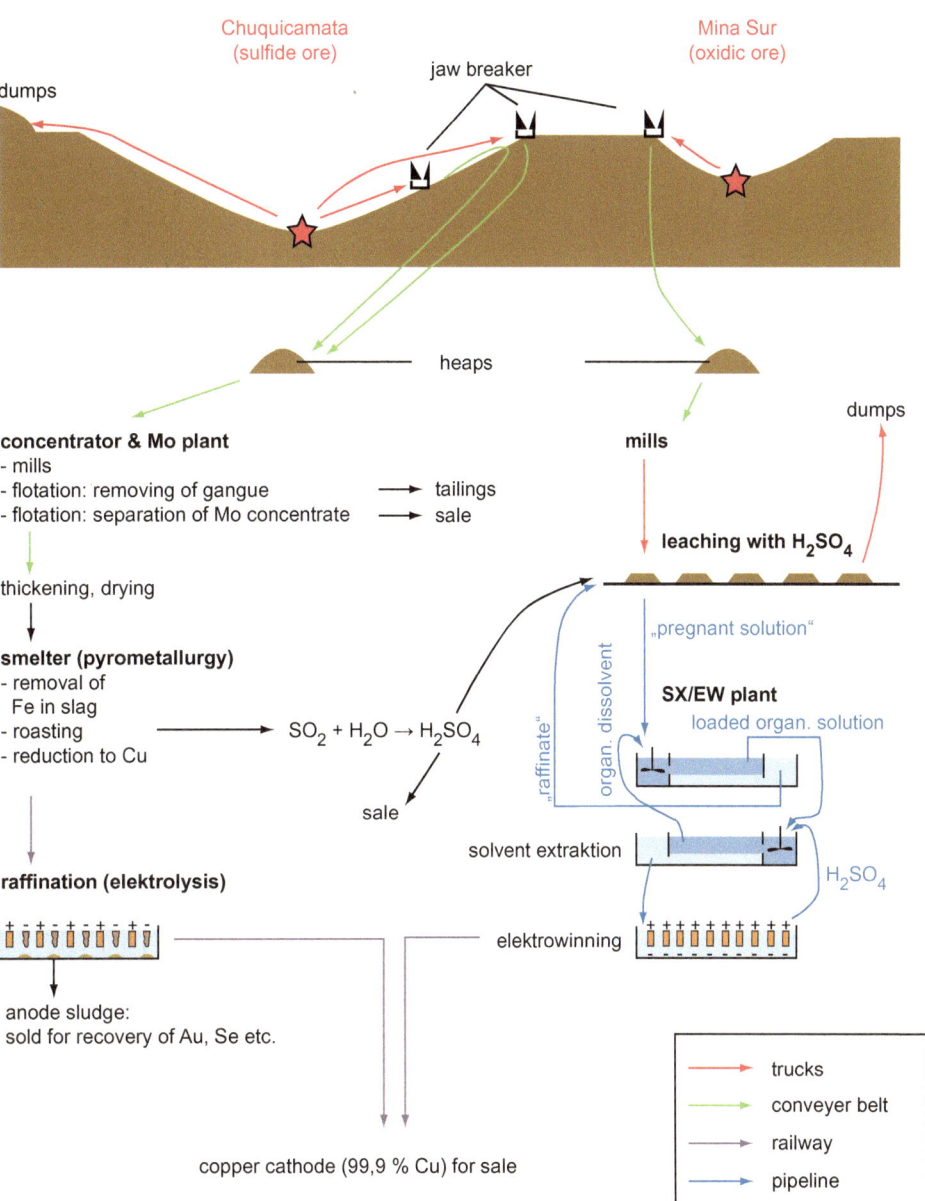

with bioleaching instead of conventional smelting. Similarly, cobalt, nickel, and zinc can also be extracted. The process is particularly worthwhile for polymetallic ores that are difficult to separate. The BGR has indicated that indium, gallium, and germanium can also be separated in this way (as can uranium).

Biomining is also widely used in gold mining when it is then referred to as **biooxidation** because the gold itself does not dissolve. Often tiny gold grains are enclosed in sulfides such as pyrite or arsenopyrite thus preventing direct extraction by cyanide leaching. Sulfides are removed with the help of microorganisms and then the gold is leached out of the residue with cyanides.

Sometimes biomining can be carried out in situ without bringing the ore to the surface. The caving method (Sect. 1.11) leaves a cavity filled with rock fragments that according to the BGR are well suited for leaching processes after the necessary infrastructure has been installed.

A variant of biomining is phytomining. It can also be used for the remediation of soils contaminated with heavy metals. This is done by taking advantage of the fact that some plants absorb and accumulate certain metals. Such plants can be cultivated, harvested, and incinerated and then the ash can be smelted recouping some of the decontamination costs. In this way nickel, gold, and thallium can be successfully extracted (Anderson et al. 1999).

Box 1.11 Processing old dumps and tailings

Material dumped on heaps as waste during earlier mining or left behind as sludge in tailings ponds may contain substances that are in demand today or others that methods of the time could not extract. Such material often also contains originally produced metals in low concentrations. Therefore, investigations are being stepped up into how this waste can be used for the economic production of metals using new processing methods. There are a number of advantages here such as no mining costs are involved and it is an opportunity to rehabilitate old dumps and tailings that often have not been sealed allowing the escape of acidic and heavy metal–rich seepage water into the environment.

Bioleaching is normally used when low concentrations of metals are involved possibly in combination with mechanical separation and methods such as acid and cyanide leaching. A good example are tailings from the Ticapampa copper mine (Peru) that consist of 1.6 million tonnes of material including 4.6 t of gold and 135 t of silver. Biooxidation experiments have shown that 97% of the gold and 50% of the silver can be recovered (Nagy 2008).

The Helmholtz Institute for Resource Technology (2012) in Freiberg (Germany) is developing methods to process overburden stockpiles and tailings in the Ore Mountains (Erzgebirge). Up to 50% of the original tin remains in the tailings of Altenberg. Moreover, in dumps in the Freiberg district significant amounts of zinc, lead, germanium, and indium are suspected. The latter two are important high-tech metals and are considered critical resources (Box 1.4). The institute is particularly interested in looking into the possibility of developing both localities for the production of indium tin oxide used in flatscreens and touchscreens.

1.16 Smelting

Reducing ore to metal and cleaning metal from unwanted components is in most cases carried out in special furnaces (pyrometallurgical) or by electrolysis in a solution or melt (electrometallurgical). In both cases the process is energy intensive in that it consumes large amounts of coal, gas, or electricity.

So-called oxidic ores such as oxides, hydroxides, carbonates, sulfates, and arsenates (see also Box 4.16) can in some cases be reduced relatively easily by carbon monoxide produced in the furnace during the combustion of charcoal or coke at a correspondingly high temperature. For some

metals a simple **shaft furnace** is still used where the furnace merges directly with the chimney. It is filled with charcoal (or coke) and ore. After reduction the molten slag and metal are tapped through an opening. Finally, the shaft furnace is emptied and only then refilled.

When it comes to reducing iron ore in a **blast furnace** (Figs. 1.46 and 1.47) the furnace is charged from above alternately with ore and coke, while the liquid pig iron (crude iron) and molten slag are regularly tapped at the bottom. This means that the blast furnace can be kept in continuous operation for 10–20 years without ever cooling down. Modern blast furnaces are about 25–35 m high and have a capacity of 5000–10,000 t pig iron per day.

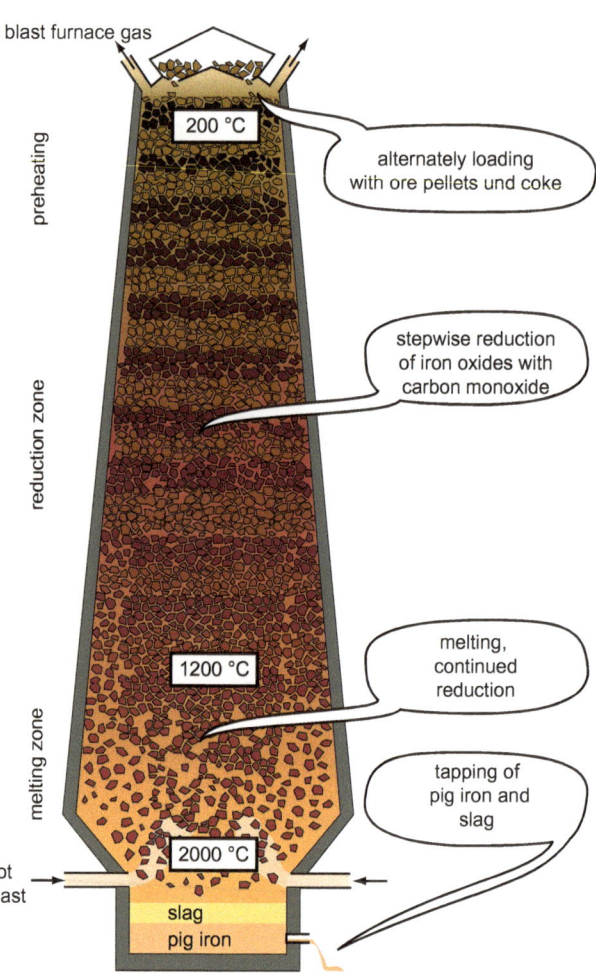

blast furnace gas

preheating

200 °C

alternately loading with ore pellets und coke

reduction zone

stepwise reduction of iron oxides with carbon monoxide

1200 °C

melting, continued reduction

melting zone

tapping of pig iron and slag

2000 °C

hot blast

slag
pig iron

Fig. 1.46 Iron ore is reduced to iron when exposed to carbon monoxide in a blast furnace. The blast furnace is in continuous operation and charged alternately with ore and coke from above. Molten pig iron and liquid slag are tapped regularly from below. Inside the furnace the material sinks into areas of increasing temperature. Oxide reduction takes place in step with the formation of molten slag from silicate ore components and additives. The air required to burn the coke (hot blast) is preheated with gas escaping from the blast furnace as fuel

Fig. 1.47 Tapping a blast furnace (*photo* © Getty Images/iStockphoto)

The process is optimized by rolling the ore and additives together to form pellets with a diameter of 1–2 cm before being fed to the furnace. The furnace is fed via "sluices" called charging bells to minimize heat loss. The exhaust gas escaping from the top of the blast furnace (blast furnace gas) consists of carbon dioxide, carbon monoxide, nitrogen, and a little hydrogen and methane. Hot air is blown into the furnace from below preheated using blast furnace gas as fuel. It allows the coke to burn to carbon monoxide providing the required heat and reducing agent.

Material inside the blast furnace forms layers that slowly sink downward into ever hotter zones. Reduction takes place in several steps as temperature increases from iron(III) oxide (hematite, Fe_2O_3) via iron(II)–iron(III) oxide (magnetite, Fe_3O_4) to iron(II) oxide (wüstite, FeO) and finally to elemental iron in the lower part of the furnace. This is helped by the fact that CO_2 is reduced to CO once again when it comes into contact with coke at more than 1000 °C. Other substances such as manganese oxides and phosphates are also reduced.

Limestone added as flux decomposes into CaO and CO_2 as it makes its way down toward the bottom of the furnace. CaO is used to form slags that remove undesirable substances such as silicates. This melt (i.e., liquid slag) collects in the lower part of the blast furnace and is regularly tapped. It solidifies into calcium silicates and other minerals.

In the melting zone at the bottom of the furnace solid iron melts and the remaining iron oxide reduces. Iron melt accumulates at the bottom of the blast furnace and is regularly tapped. Such melt is called pig iron and due to its high carbon content (3–4%) is brittle and not forgeable.

To convert pig iron into steel the carbon content must be reduced in a further step called refining. Oxygen is blown through the lance of a **converter** into the molten pig iron (Linz–Donawitz process). The carbon oxidizes to CO_2 thus providing the necessary heat to keep the metal molten. Sulfur is also oxidized and escapes as SO_2. By adding limestone and other substances a slag can also be produced here to remove undesirable elements.

An alternative method of producing steel mainly used for the production of particularly high–quality steels is reduction in an **electric arc**. A gas discharge occurs between two electrodes mounted at a small distance from each other to which a very high voltage is applied. In this discharge the gas between the electrodes transforms into electrically conductive plasma at temperatures of several thousand degrees. The electrodes are attached to a cover that swivels; the ore and scrap are then melted in the trough-shaped furnace and finally reduced. The metal and slag are then tapped. Since no coke is used an alloy with a low carbon content can be produced immediately. Other metals used in small quantities as steel refiners can be added right from the beginning or added later.

Sulfide minerals are the most important ore for many metals. Sulfide ores cannot be reduced directly with carbon monoxide. Heating produces a molten sulfide that solidifies into an artificial sulfide during cooling called matte. The ores or the matte must be **roasted** in an intermediate step. When heated under oxidizing conditions sulfides convert into oxides releasing sulfur dioxide. This environmentally harmful gas can be converted into sulfuric acid, an important by-product of copper production. Often further steps need to be taken to separate other elements.

The iron contained in the most important copper ore chalcopyrite ($CuFeS_2$) must be removed in a further intermediate step. The ore (or matte in which some of the iron has already been oxidized by roasting) is heated to 1100 °C in a **converter** together with sand and limestone and then oxygen is added. Fe^{2+} is oxidized to Fe^{3+} that flows off with the silicate melt (slag). Some of the sulfur is oxidized to sulfur dioxide (SO_2). Cu^{2+} is reduced to Cu^{+} and flows off as molten copper sulfide that solidifies to Cu_2S (matte). The matte is then roasted to produce Cu_2O and SO_2. In the next step Cu_2O reacts in a converter with the remaining Cu_2S to produce metallic copper and SO_2. Such raw copper can still contain impurities of a few percent. It then has to be oxidized once again to Cu^{+} by adding oxygen and slag-forming substances. Zinc, antimony, and arsenic escape as gaseous oxides, while iron, cobalt, nickel, and other elements flow off with the slag. Cu^{+} can be reduced back to Cu when exposed to natural gas.

Refined copper produced in this way still contains impurities of about 0.3%. By **electrolysis** (electrolytic refining) refined copper can be converted to copper cathodes that have a purity of 99.9%. This is done by pouring refined copper into large anodes hundreds of which are hung in a basin filled with copper sulfate solution. Pure copper is deposited at the cathode. The voltage applied is such that copper is electrically oxidized, passes through the solution as Cu^{+}, and is electrically reduced again at the cathode, while other metals remain in the solution. More noble metals are not oxidized and accumulate under the anode as anode sludge; hence gold, silver, and platinum-group elements as well as selenium and tellurium accumulate in the anode sludge. Such metals are often obtained as by-products. Suitable copper ores processed by leaching and solvent extraction (Sect. 1.15) often undergo electrolysis as the last step. Nickel, lead, and silver are also purified by electrolytic refining.

Zinc is special in that its boiling point of 907 °C is so low that it evaporates in a normal furnace. It can be distilled off in a retort, leached out with sulfuric acid, and obtained by electrolysis.

In aluminum production (Fig. 1.48) electrolysis is not carried out in an aqueous solution but in a melt at a temperature of around 950 °C (Hall–Héroult process). The aluminum ore bauxite is first dissolved in hot concentrated caustic soda (Bayer process) leaving behind iron oxides and other impurities. Cooling and diluting the lye results in the precipitation of pure aluminum hydroxide that is then fired in a rotary kiln at more than 1200 °C to form aluminum oxide (alumina). For electrolysis this is mixed with cryolite (Na_3AlF_6) whose proportion is 80–90%. Only synthetically produced cryolite is used today, whereas it used to originate from Ivigtut (Greenland) (Sect. 3.12). Cryolite is used to reduce the extremely high melting point of aluminum oxide

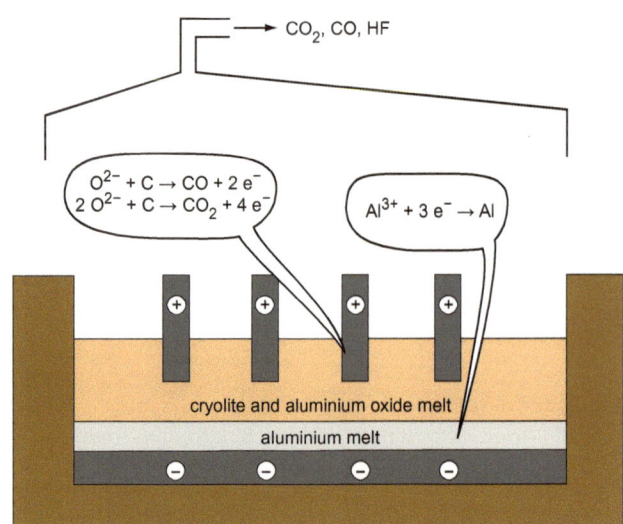

Fig. 1.48 Aluminum is produced by electrolysis in a cryolite–alumina melt at 950 °C where cryolite (Na_3AlF_6) is used to reduce the melting point. The electrodes are made of graphite. At the cathode Al^{3+} absorbs electrons and is reduced to aluminum that accumulates as a melt at the bottom. Oxygen reacts with the carbon of the anodes to form CO and CO_2

(2050 °C) to the eutectic temperature (cf. Box 3.3) of the mixture, which is only 950 °C. Graphite electrodes are lowered into the melt held in a tub lined with graphite as soon as the material has melted. Graphite electrodes serve as the anode in such a tub and the bottom of the tub serves as the cathode. The aluminum oxide of the melt absorbs electrons at the cathode and is thus reduced to liquid aluminum that accumulates on the bottom of the tank. The oxygen released at the anode reacts with graphite to form carbon monoxide and carbon dioxide. The exhaust gas also contains hydrogen fluoride (HF) released by the cryolite. The energy requirements for aluminum production are so high that aluminum smelters are mainly located in regions where electricity prices are low. Magnesium is produced in a very similar way from magnesium chloride.

Titanium is a good example of a metal that can be produced using even more complex processes in which rutile (TiO_2) or ilmenite ($FeTiO_3$) are used as ore. Ilmenite is first converted to TiO_2 either in an arc furnace using carbon where liquid iron accumulates on the ground and CO is released, or it is dissolved in acid and titanium oxide is precipitated. TiO_2 is converted into titanium tetrachloride using chlorine and coke at temperatures just below 1000 °C. This is purified by fractional distillation. Provided with a protective gas atmosphere it can be reduced to titanium using magnesium at temperatures just below 1000 °C resulting in molten magnesium chloride. The titanium is still porous and brittle (sponge titanium) and must be forged by constant turning or other processes.

1.17 Recycling

Since it is becoming ever-more difficult to find new deposits the need to develop effective recycling methods to obtain secondary raw materials from waste grows ever-more important. Scrap is routinely processed into steel, while aluminum, copper, lead, nickel, zinc, cadmium, and cobalt are increasingly recycled. In some cases recycling even plays a significant role in the production of these metals. This is particularly the case with closed circuits (e.g., by recycling nickel–cadmium batteries). Recycling ranges from the use of special sorting machines to people disassembling old equipment by hand.

The recycling of precious metals and high-tech metals has as yet not been given the importance it deserves. Only a fraction of discarded electrical equipment has so far been recycled despite often containing a significantly higher content of gold, palladium, copper, tantalum, indium, europium, neodymium, and other sought-after high-tech metals than is the case with corresponding mineral deposits. Separation is not made any easier by the fact that the amount of metal used in electrical appliances continues to rise and an increasing number of different metals are used in each product. However, if only equipment whose metal content is precisely known is recycled, it is even less profitable due to the small quantity. The preparation and separation processes used in normal metal production cannot simply be transferred directly. They must be adapted to the special metal mix. There is still a considerable need for research in this area. Much electronic scrap has so far been "exported" to developing countries like Ghana where it ends up in waste dumps searched by local people looking for aluminum parts and copper cables.

1.18 Cast, Forged, and Chased

Metals are special in that they can be machined and shaped by very different methods. During processing the structure of crystallites changes (see Box 2.1) thus influencing hardness, elasticity, and other properties. Crystallites deform when metal is cold-worked leading to stresses; therefore cold-working is only possible to a certain degree. However, these new structures can have greater strength. The microstructure recrystallizes as a result of heat treatment or hot-working. Forming at high temperature also has the advantage that less force needs to be applied.

The first step involves casting the molten metal to obtain either the final shape such as machine parts, rims, bells, and manhole covers or an intermediate product for further processing such as ingots, bars, and forgings.

Simple forms can be produced by casting in permanent molds usually made of cast iron, steel, or a copper alloy.

A simple mold is sufficient when casting blocks, ingots, and slabs. Other molds may be composed of two or more parts with an appropriately shaped cavity and feed channels between them. A variation is slush casting in which the melt remaining after a metal layer has solidified along the walls of the mold is poured out again. In low-pressure die casting the melt is sucked into the mold under negative pressure.

Continuous casting is a process used for the production of bars, tubes, and slabs. Metal melt is poured through a water-cooled mold that is open at the bottom. The continuously emerging strand is either further processed as an endless strand or sawn into smaller strands.

Although more complicated shapes are possible with lost mold casting, each mold can only be used once. Sand casting is a widespread method in which the cavity is formed between an upper box and a lower box filled with sand (mixed with clay, oil, or resin and possibly graphite powder). Cavities can be created by inserting models called patterns. For series production a permanent pattern is used and pressed into the sand. In the lost wax process even the model is lost. It is formed from wax (or plastic) and then coated with a heat-resistant material like clay. Liquid metal fills the cavity left behind after the wax has melted.

Which metals and alloys are used for casting depends on their suitability. Any decrease in volume during cooling varies according to the metal or alloy used and may have to be compensated for by larger molds. The temperature of the melt also varies according to the metal or alloy used with some alloys taking a long time to solidify. The cooling rate affects the structure of the metal and thus the properties of the material.

Rolling (Fig. 1.49) involves intermediate products such as blocks, slabs, or bars being rolled out into plates, sheets, foils, strips, rails, and coarse wires. A blank is moved through the gap between two rotating drums that can be cylindrical (for sheets and foils) or have a profile (for wires, rails, and so on). Rolling mills comprise a number of rotating drums arranged in a specific way for a certain product. As a rule processing takes place above the recrystallization temperature making the metal not only softer but also significantly more deformable thanks to relaxation of the microstructure. Thin sheets can also be processed to a certain degree by cold rolling. Rings, wheel tires, and tubes can also be produced using specially arranged drums.

Forging is the free forming of (mostly hot) metal by the application of pressure using a hammer and anvil, a machine hammer, or a machine press. Chasing and repoussé are mostly cold forming processes in which metal is plastically deformed with a hammer in such a way that it is locally stretched or compressed (e.g., reliefs or uncreased bent sheets). Subject to requirements the workpiece is placed on a sandbag, on an anvil, on a specially shaped punch, over a hollow, or embedded in putty. Workpieces can be decorated

Fig. 1.49 Rolling mill for steel processing (*photo* © Industrieblick/Fotolia)

by chasing reliefs into the surface using a liner. This is different from engraving where metal chips are removed using a graver.

Various mechanical forming processes are used in industry depending on the purpose and the properties of the metal. Wire drawing involves a coarse hot-rolled wire being drawn through a smaller diameter ring at room temperature. Extrusion involves forming a hot bar or block by pressing it through a die. Aluminum can be used to form complicated profiles such as heatsinks in this way. This is different from impact extrusion in which the workpiece is pressed into the die at high pressure by a punch allowing cold or hot metal to flow as a result of the pressure. Deep drawing involves sheet metal being pressed by a punch into a cavity (drawing ring, die) and repeated (e.g., can production) using different dies. Sheet metal can be bent using a so-called brake that clamps one side and bends the other by folding the front plate of the machine. Sheet metal can also be bent by die bending where a punch shaped like a ruler presses the sheet metal into a die causing both sides of the sheet metal to fold upward like wings. Coins and medals are minted by pressing two stamps, between which a cold blank is clamped, with force into the metal. Blanks were previously punched out of sheet metal.

Several metal parts can be joined together by generating a liquid metal layer on the surface through heat. Welding involves melting the workpieces themselves (if necessary with an additional welding wire). Soldering involves using a special alloy (solder) with a low melting point while the workpieces remain solid. The fusion film in welding can be produced with a flame, an electric arc, electric current (leading to heat generation at the point with the greatest resistance), friction, a laser, a strongly exothermic redox reaction (thermite welding), and other methods. All that is needed for soldering is an electrically heated soldering iron.

Finally, a workpiece can be coated with a thin layer of another metal to protect it from corrosion or for aesthetic reasons such as zinc plating, chromium plating, tinning, gold plating, and silver plating. The most common process is electroplating (galvanization) in which the workpiece hangs in an electrolyte solution and serves as cathode. The metal of the anode is electrochemically oxidized, migrates in the form of ions through the solution, and is deposited by electrochemical reduction at the cathode. This contrasts with hot-dip galvanizing and hot-dip tinning in which the workpiece is immersed in molten zinc (melting point 419 °C) or tin (melting point 232 °C).

1.19   The Composition of the Earth

Deposit formation involves enriching certain elements already present in the Earth system (Box 1.12); hence the composition of the Earth is important. Carbonaceous chondrite is a type of meteorite that can best model the average composition of the whole Earth (Box 1.13).

Box 1.12 How the elements were created

Shortly after the Big Bang the matter comprising the universe consisted mainly of hydrogen (three quarters) and helium (one quarter). Tiny traces of unstable lithium and beryllium isotopes were also present. Formation of the first elements is called primordial nucleosynthesis and was completed about 3 min after the Big Bang because the density and temperature of the universe were already too low for nuclear fusion. The remaining elements—which still make up only a tiny part of the universe's matter—were therefore created later.

Nuclear fusion in stars makes an important contribution to the formation of heavy elements. If temperature and density are sufficiently high, then protons (i.e., hydrogen nuclei) fuse to form helium releasing a great deal of energy. Several intermediate steps are needed for fusion to take place in which various possibilities occur with a certain probability. The most important chain of reactions is the so-called proton–proton chain: in a first step two protons merge to form a nucleus consisting of a proton and a neutron (deuterium, heavy hydrogen). When a proton is converted to a neutron a neutrino and a positron are released (the positron is destroyed as soon as it hits an electron). The second step involves heavy hydrogen fusing with another proton to form helium-3 emitting a gamma quantum. The third step involves two helium-3 reacting to form one helium-4 and two hydrogen nuclei.

The fusion of hydrogen to helium takes place in a star like the Sun and lasts for several billion years. The nucleus of the star in which the fusion takes place is therefore becoming ever richer in helium while the outer shell does not change. When all the hydrogen in the core is used up there is a material exchange between the core and outer shell. Since thermal energy is lowered, the equilibrium pressure is higer and the core collapses. The temperature in the center increases and triggers the fusion of helium to carbon and oxygen. This process is only possible at very high temperatures because three helium nuclei have to fuse at the same time. This releases significantly less energy than the fusion of hydrogen to helium. Hydrogen can be burned again in a thin shell around the helium core, while the cool outer shell expands to 50 times its volume. The star has a lower luminosity and has become a red giant.

When the helium is used up a star the mass of our Sun will only glow as a white dwarf and finally cool down to a black dwarf. However, more massive stars can still experience several upheavals that bring material from the outer shells into the core and set fusion in motion again for a while. Although heavy elements can also be built within very massive stars, little energy is released. Carbon produces neon, magnesium, and sodium; oxygen produces silicon, sulfur, and phosphorus. In extreme cases iron would be produced. However, fusion to heavier elements would consume energy rather than releasing it.

Lithium, beryllium, and boron are three light elements that are skipped in stellar fusion processes. This is due to their nuclear binding energy being low making them unstable at temperatures prevailing in a star and they immediately decay again. This is also the reason these elements are comparatively rare. One way in which these elements can nevertheless develop is through spallation in which cosmic rays interact with matter and heavy elements lose some of their elementary particles.

All the other elements in the periodic table formed as a result of neutron capture and the radioactive decay of unstable isotopes.

Free neutrons do not survive long. Their capture depends on a neutron source being present. Certain fusion processes taking place in red giants may be one possibility. The capture of a neutron by an atomic nucleus produces a heavier isotope of the same element. Often this isotope is not stable and decays. On the assumption this is beta decay a neutron transforms into a proton emitting an electron (beta radiation) and an antineutrino. The atom has thus slipped one field to the right in the periodic table. However, despite losing an electron it still has approximately the same mass as before decay. Neutron capture is so slow in stars that unstable isotopes decay before the next neutron can be captured. This is referred to as the s-process (for "slow"). In this way elements up to lead can be created. The processes that take place in a supernova are of a different order. During the explosion of a dying star so many neutrons are released that capture can take place faster than beta decay. This is referred to as the r-process (for "rapid"). The heaviest elements and exotic isotopes can be created in this way as of course can light elements. It is believed that bismuth, thorium, uranium, and so on were all formed as a result of a supernova (Fig. 1.50).

Fig. 1.50 Around the year 1680 a red giant 10,000 light years away exploded in the constellation Cassiopeia. This false-color image composed of data of different wavelengths shows the remains of this supernova. In the center there is a neutron star surrounded by a shell of gas and dust (*photo* © NASA)

Box 1.13 Chondrites, birth of the Earth, and a core of iron

It is widely assumed that the composition of certain kinds of meteorites called carbonaceous chondrites corresponds almost exactly to the average total composition of the Earth. These meteorites are lumps of the original material that condensed during the birth of the solar system—the building material from which the inner planets originated.

Such meteorites consist largely of small globules of glass containing fine silicate minerals called chondrules. These are condensed melt droplets baked by dust grains comprising tiny crystals such as corundum, perovskite, and spinel. Meteorite researchers refer to them as Ca–Al-rich inclusions. Other dust grains such as diamond, graphite, silicon carbide (SiC, a.k.a. carborundum), and various oxide and silicate minerals are also present.

The solar system originated from a spherical cloud of gas and dust. Presumably triggered by a supernova not too far away the cloud collapsed into a rotating accretion disk. Around 99% of the mass collected in the center of the disk where temperatures were so high that the nuclear fusion of hydrogen to helium began. The Sun was born. At some distance from the Sun the disk cooled rapidly and condensed about 4.57 billion years ago forming successively Ca–Al-rich inclusions, melt droplets, and further dust particles. Chondritic meteorites are pieces of the first small celestial bodies called planetesimals that formed from these particles.

At that time the solar system thronged with dust and planetesimals. Gravity ensured rapid growth of larger bodies by sucking in all the material in the immediate environment. In this way planets would have grown the size of Mars in a relatively short time. Collisions between such planets eventually led to the formation of larger planets like Earth. In contrast to the large outer planets that could accumulate large amounts of gas and volatile elements the Earth is comparatively depleted in such elements.

The growth of Earth was completed about 100 million years after the birth of the solar system. It had reached today's mass little more than 10 million years later (Canup and Agnor 2000) when accretion became ever slower. This was an astonishingly short period of time in the context of the Earth's history. An important event was the formation of the Moon about 30 million years after the birth of the solar system. According to current theory a planet about the size of Mars collided with the Earth. Our planet was partially melted and it incorporated the core of the impactor. Debris orbiting the Earth accreted to form the Moon that has been orbiting Earth ever since.

Another important event of early Earth was the fractionation of originally homogeneous material into two differently composed shells: the core and the mantle. The Earth was so hot that an iron melt either seeped into the depth due to its high density or sank diapir-like and accumulated in the center as a core. It is no longer assumed that such an "iron catastrophe" took place at a given time; rather it was a process that took place over and over again during the rapid growth of the early Earth. A critical question here is how much of the Earth had to be melted to enable effective separation because the mobility of molten iron in a solid silicate rock is very limited. Probably early Earth was partially or even largely molten several times as a result of collisions. It takes some time for such deep magma oceans to solidify. According to a model proposed by Solomatov (2000) an iron-rich layer accumulated at the bottom of such magma oceans from which diapirs of molten iron sank into the depths. If the collision leading to the formation of the Moon was one of many such collisions, then this would explain the comparatively low iron content of the Moon.

Perhaps core creation began much earlier. Experiments have shown that even small amounts of iron melt can migrate through rock and that core formation could already have begun in planetesimals with diameters of about 30 km (Yoshino et al. 2003). Therefore, it is possible that Earth was composed of several planetesimals that already had a core.

The Earth's crust developed later through magmatism from the early primitive mantle. However, since none of the oldest crustal rocks have survived, the present crust has to be the result of long-term magmatism, on the one hand, and repeated recycling in the rock cycle that takes place within the framework of plate tectonics, on the other hand.

Since chondrites are believed to be the best model for the bulk composition of the Earth, element contents are often given relative to their chondritic composition (especially, in magmatic systems). This standardization makes it easy to notice fractionation processes. Of course, there are different chondritic meteorites originating from different depths in planetesimals of different sizes that have already undergone fractionation. At present carbonaceous chondrites of the CV3 type are considered the best model for the Earth. The Allende meteorite that fell in Mexico in 1969 belongs to this type.

The Earth is made up of several differently composed shells (Figs. 1.51 and 1.52). Only the outermost of these shells (i.e., the Earth's thin crust) can be reached by humans. Figure 1.53 gives an overview of the average composition of the upper part of Earth's crust. Since the Earth's crust is the most heterogeneous of these shells it facilitates finding deposits that are little more than geological anomalies unusually highly enriched in certain elements. The emergence of shells during the history of early Earth was an extreme form of fractionation affecting the entire Earth (see Box 1.14). The composition of individual shells is the starting point when it comes to understanding ore-forming processes.

Continental crust is a wild mix of different rocks including sediments such as limestone and sandstone, metamorphic rocks such as schist and gneiss, and igneous (i.e., magmatic) rocks such as granite and gabbro. On average the composition roughly corresponds to tonalite, an igneous rock that lies roughly between granite and gabbro (Box 3.2). The composition of the upper crust deviates more from the average in the direction of granite and the composition of the lower crust more from the average in the direction of gabbro. In general, elements enriched in the continental crust do not fit well into minerals of the other shells. Typically, continental crust has a thickness of 35–40 km, but where there are high mountain ranges it has a thickness of up to 70 km.

Oceanic crust is about 5 km thick and consists largely of basalt and gabbro. Both are igneous rocks whose composition is identical. The only difference being that basalt has flowed out at the Earth's surface or on the seabed and solidified as fine-grained rock, while gabbro has solidified as

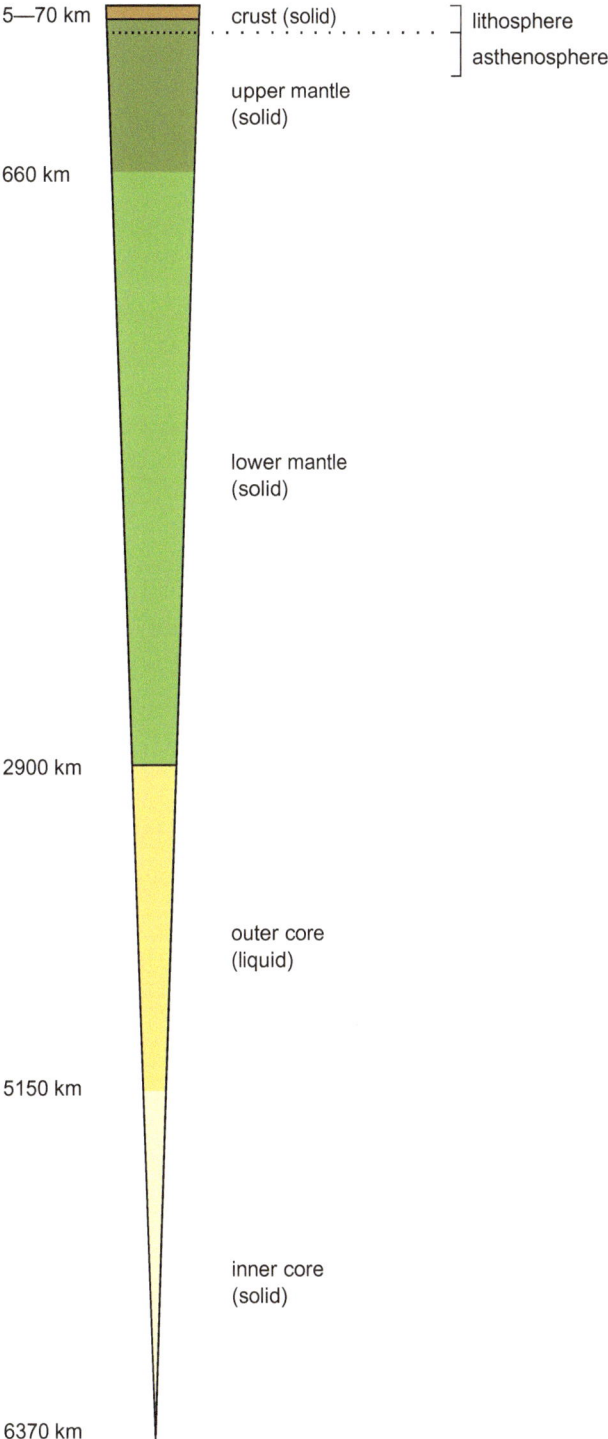

Fig. 1.51 The shells of the Earth. Although the crust, mantle, and core have different compositions, only the uppermost "bark" of the Earth's crust is accessible to humans

intrusions deep in the surface and crystallized to a coarse-grained rock. Magma comes directly from the Earth's mantle. Compared with continental crust, oceanic crust is characterized by a lower content of SiO_2 and alkalis and a

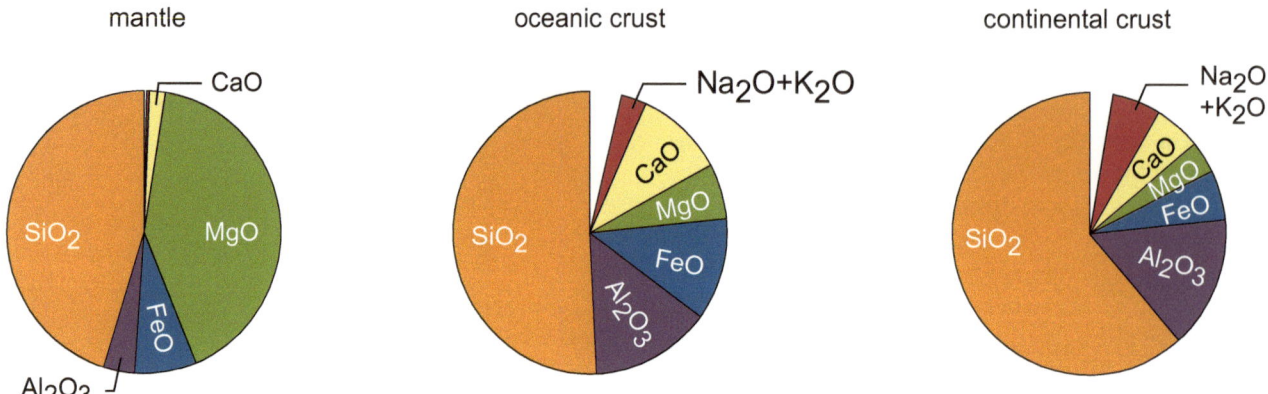

Fig. 1.52 Average chemical composition of mantle, oceanic crust, and continental crust (all iron as FeO). The large gap in basalt is mainly TiO_2. Representing compositions as oxides makes comparisons easier

Fig. 1.53 Abundance of elements in the upper part of continental crust. Abundance is given relative to 10^6 atoms Si and logarithmically plotted, while elements are sorted by atomic number (data from USGS)

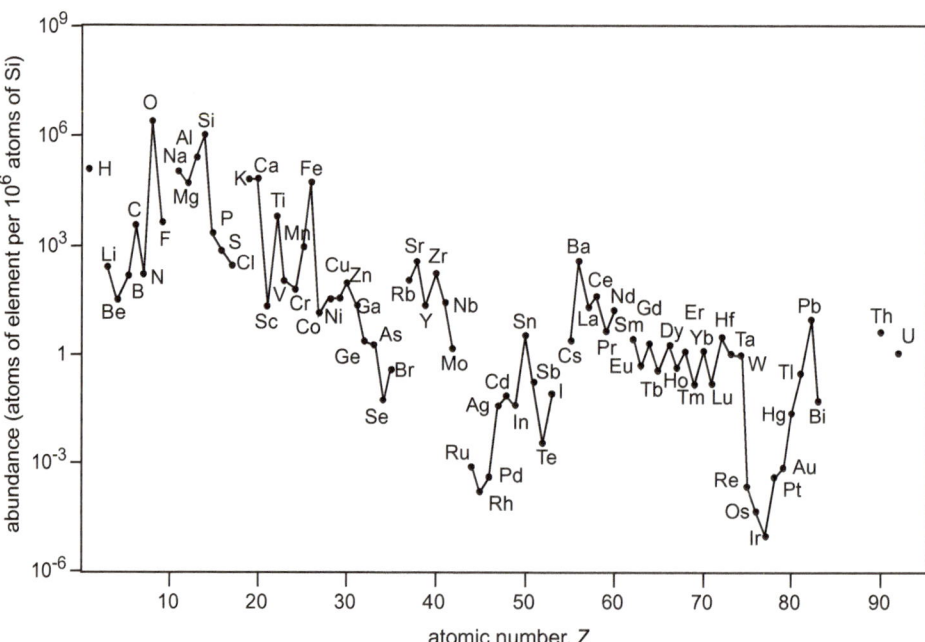

significantly higher content of FeO, TiO_2, MgO, and CaO. Important trace elements are chromium and nickel.

The Earth's **mantle** makes up a much larger proportion of the total Earth. It reaches a depth of 2900 km and is surprisingly homogeneous. Compared with the crust it has a very low SiO_2 content and a very high MgO content. Although Al_2O_3, FeO, and CaO are present, all other elements are present only in traces. Nevertheless, the content of chromium and nickel is significantly higher than in the crust. The rock of the Earth's mantle is peridotite consisting mainly of the minerals olivine ($MgSiO_4$), diopside (clinopyroxene, $CaMgSi_2O_6$), and enstatite (orthopyroxene, $Mg_2Si_2O_6$). Some of the Mg^{2+} is replaced by Fe^{2+} in all three minerals. Also present is an aluminum-bearing mineral that is plagioclase at very low pressure, spinel at low

pressure, and garnet at high pressure. This mineralogical composition only applies to the upper mantle. At even higher pressures several reactions produce other minerals that have even more densely packed structures despite the chemical composition of the rock remaining the same.

The uppermost part of the upper mantle is rigid and "sticks" to the underside of the Earth's crust. Together with the Earth's crust it forms the more or less rigid lithosphere, forming the migrating plates of plate tectonics. The lithosphere floats on the comparatively soft asthenosphere, the part of the Earth's mantle that can be easily deformed due to its heat and in places even contains small amounts of melt.

The Earth's **core** has a diameter of 3470 km and consists of an alloy of iron and some nickel. A small percentage of the core comprises silicon and other elements. The outer

Table 1.1 Average abundance of selected metals in the Earth's crust, typical minimum ore grade for a deposit (1 g/t = 1 ppm), and corresponding enrichment factor (from Robb 2005)

	Average abundance in the crust	Typical minimum exploitable grade	Approximate enrichment factor
Al	8.2%	30%	×4
Fe	5.6%	50%	×9
Cu	55 ppm	1%	×180
Ni	75 ppm	1%	×130
Zn	70 ppm	5%	×700
Sn	2 ppm	0.5%	×2500
Au	4 ppb	5 g/t	×1250
Pt	5 ppb	5 g/t	×1000

core is liquid, while the inner core is solid due to its extreme density. Apart from the Earth's magnetic field the core has hardly any effect on the rest of the Earth. The largest metal reservoir on Earth ironically has no significance when it comes to deposit formation (see also Boxes 1.13 and 1.14).

It is interesting to compare the metal content of a deposit with the average composition of the Earth's crust. Table 1.1 shows the enrichment factors necessary to form a useful deposit.

Box 1.14 The late veneer theory

Formation of the Earth's core has to have been the most comprehensive fractionation in the history of the Earth; after all, the volume of the Earth's crust compared with that of the entire Earth is negligibly small. The Earth's core consists largely of iron and some nickel, but also contains small amounts of silicon and other elements. In particular, the iron-loving (siderophile) elements (Sect. 1.20) were transported by iron melt migrating down into the core during core formation. This must have been particularly effective for some rare elements such as gold, platinum, palladium, and iridium.

However, these elements now occur in the crust and mantle too. Although average contents in the mantle are low, they are still significantly higher than they should be in equilibrium with the core. The simplest explanation is that this increased content is due to meteorites that fell to the Earth only after formation of the Earth's core. This is the late veneer theory in which such economically important commodities should be regarded as a greeting from space like crumb coating on a finished cake.

This theory is hotly debated. In particular, relative conditions of the platinum-group elements raise questions about the composition of these meteorites and about the extent processes in the mantle can change corresponding signatures (Lorand et al. 2008).

Experiments have shown that some of the relevant elements like palladium do not fractionate in the core as effectively as had been assumed under the extreme conditions prevailing in the interior of the early Earth (Righter et al. 2008). When it comes to osmium and iridium this is still the case (Brenan and McDonough 2009). It is therefore still too early to quantitatively assess the significance of "crumb coating".

1.20 Geochemical Classification of Elements

The behavior of elements during the geological process of fractionation depends above all on how well they fit into the respective crystal lattice of the minerals involved. To understand the behavior of elements during fractionation in the Earth system the founder of geochemistry Victor Moritz Goldschmidt conducted investigations into meteorites and iron smelting. His classification is particularly helpful as a rough overview regarding ore-forming processes or the composition of Earth's shells. Goldschmidt divided elements with similar geochemical behavior into four groups (Fig. 1.54) depending on their tendency to preferentially form an alloy with metallic iron, to form sulfides, to form silicates, or to accumulate as gas in the atmosphere.

Siderophile elements (iron-loving) preferably form an alloy with iron. They are preferentially found in the Earth's core (see also Box 1.14) and in the metal phase of meteorites (Box 1.15). In the blast furnace they are found in the molten metal. Examples are nickel, cobalt, gold, and platinum.

Chalcophile elements (copper-loving or ore-loving) preferably form sulfides. They are found in sulfide deposits and as scattered mineral grains in the Earth's crust and mantle. In the blast furnace they are found in the matte (synthetic copper sulfide). Examples are silver, mercury, lead, zinc, and arsenic.

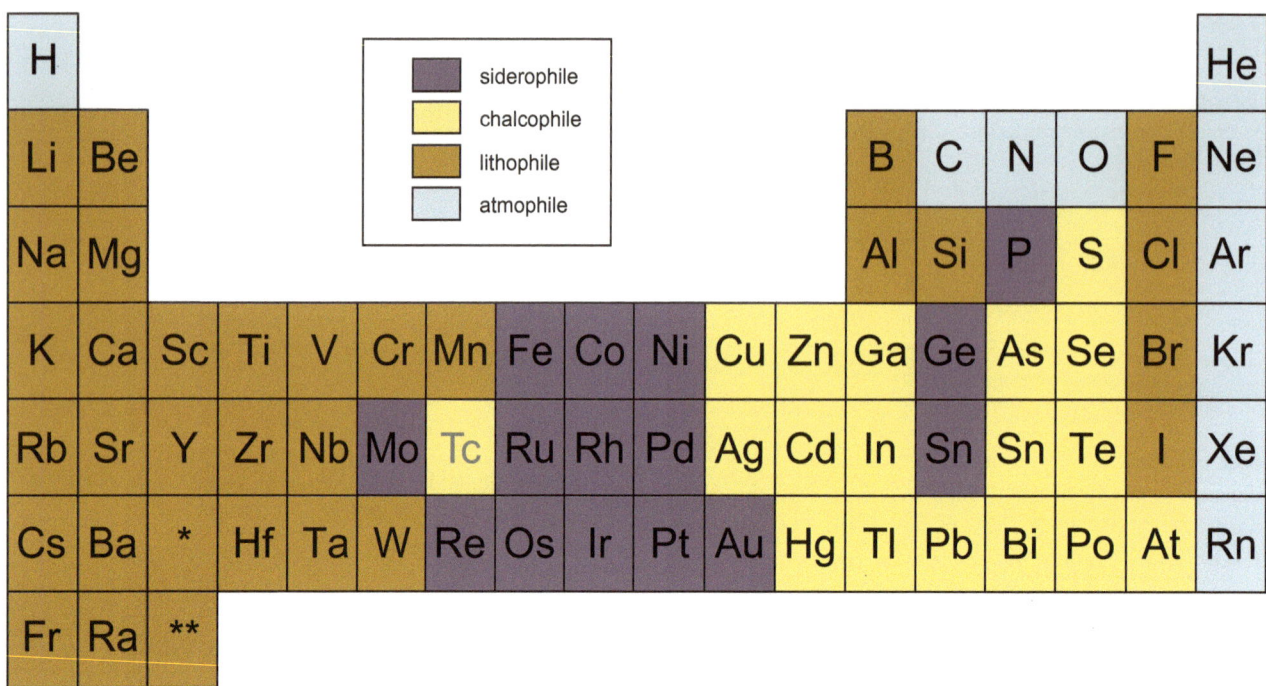

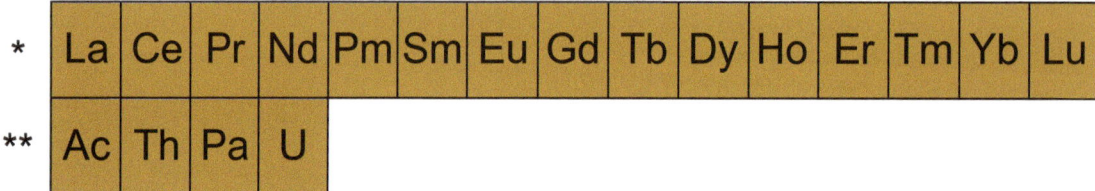

Fig. 1.54 Classification of elements according to their geochemical properties. Just the main affinities are indicated, but they are not always clear in individual cases (from Markl 2008)

Lithophile elements (rock-loving) preferably form in silicate minerals and thus in the Earth's mantle and crust. In the blast furnace they are found in the slag. Examples are sodium, potassium, magnesium, and aluminum.

Atmophile elements (gas-loving) are highly volatile and escape in gases from the blast furnace. They are mainly found in the atmosphere and the oceans. Examples are hydrogen and nitrogen.

Many elements cannot be assigned unambiguously because they fit into two or even three of these groups at one and the same time. Examples include iron, cobalt, copper, and nickel which can be found in an iron alloy, in sulfide minerals, and in silicates and could therefore be described as chalcophile, siderophile, and lithophile; and oxygen that can be either atmophile or lithophile; and so on. Although only the main affinity is generally given, it is not always the same in different publications.

This classification ultimately goes back to the electronegativity of the respective elements (i.e., to their ability to attract electrons and form anions). The difference between the electronegativity of two elements determines whether the bond is ionic or covalent. In a metallic bond the difference is almost zero, in a covalent bond it is small, and in a pure ionic bond it is very large. The bond in sulfides has a stronger covalent character than the bond in silicates.

In addition to classifying elements in such a rough way it helps to group together certain elements that have very similar geochemical behavior. Ions with the same charge and similar radius generally behave very similarly and typically occur together. They can replace each other in a crystal lattice and form series of compostions (solid solution) between the theoretical pure endmembers of different minerals. In other cases the elements a group have in common determine whether they fit with certain minerals or not.

Platinum-group elements (PGEs a.k.a. platinum metals or platinoids) are platinum, iridium, osmium, ruthenium, rhodium, and palladium. They have very similar properties and are among the rarest elements in the Earth's crust.

Rare-earth elements (REEs) are a group of elements that behave very similarly. The group includes lanthanum and lanthanides (in the periodic table cerium to lutetium) as well as yttrium and scandium. They play an important role in a wide variety of high-tech applications. In the Earth's crust they are more frequent than their name suggests.

Large-ion lithophile elements (LILEs) have a very large ionic radius such as potassium, rubidium, barium, cesium, and strontium. There is no precise definition of the term. In magmatic processes they act like incompatible elements (Box 4.4). LILEs are especially enriched in the Earth's crust.

High-field-strength elements (HFSEs) have a high ionic potential (i.e., a high ionic charge in relation to the ionic radius). HFSEs include uranium, thorium, zirconium, niobium, tantalum, and REEs. With magmatic processes they act like incompatible elements (Sect. 3.4). They are mainly enriched in the Earth's crust.

Box 1.15 Iron meteorites: metal from space

Iron meteorites represent a small percentage of all meteorites. They are fragments of celestial bodies that were so large that they had an iron-rich core. Their mother bodies had already probably been destroyed by collisions in the early days of the solar system. On the basis of their composition most iron meteorites can be assigned to one of 13 groups that obviously originate from 13 different mother bodies. However, there are also some ungrouped iron meteorites of unknown origin. There is evidence of people extracting metal from these meteorites even before the beginning of the Iron Age.

Iron meteorites consist largely of an iron–nickel alloy with 4–30% Ni. Minerals such as troilite (FeS), graphite (C), cohenite ((Fe,Ni,Co)$_3$C), and schreibersite ((Fe,Ni,Cr)$_3$P) also occur in small quantities; they are mainly enriched in aggregates.

Iron–nickel alloy in meteorites consists of two phases: kamacite (cubic inner-centered crystal lattice with maximum content of 7.5% Ni) and taenite (cubic surface-centered crystal lattice where a significantly higher nickel content is possible). The fine pearlitic structure (Box 2.1) of both phases is called plessite. It is found in the spaces between larger crystallites. A corresponding melt solidifies to form taenite from which bar-shaped kamacite with a lower nickel content is separated during cooling.

On the basis of their composition and texture iron meteorites are subdivided into hexahedrites, octahedrites, and ataxites. Octahedrites are the most common (approx. 7–15% Ni) where nickel segregation starts at high temperatures and continues to relatively low temperatures. At high temperatures the diffusion of nickel is still relatively rapid and slow cooling (periods of millions of years) causes the kamacite to grow to considerable size. The resulting so-called Widmanstätten patterns can be made visible by etching a polished cut surface.

Ataxites have a higher nickel content. Nickel segregation can only begin at a temperature less than 600 °C, which means very slow diffusion. The result is taenite and kamacite with very fine-grained structures.

Hexahedrites contain less than 6% nickel. Taenite is completely transformed into kamacite even at high temperatures. Such meteorites are very homogeneous. During etching parallel lines called Neumann lines can become visible; these are twin lamellae that have been formed by deformation during impact.

Literature

Anderson, C.W.N., R.R. Brooks, A. Chiarucci, C.J. LaCoste, M. Leblanc, B.H. Robinson, R. Simcock, and R.B. Stewart. 1999. Phytomining for nickel, thallium and gold. *Journal of Geochemical Exploration* 67: 407–415.

Anonymous. n.d. Baotou national rare earth hi-tech industrial development zone. Rare earth: An introduction. http://www.rev.cn/en/int.htm .

Anonymous. 2012a. China erlaubt höheren Export von Seltenen Erden. Die Welt, 22.8.2012. http://www.welt.de/newsticker/dpa_nt/infoline_nt/wirtschaft_nt/article108740465/China-erlaubt-hoeheren-Export-von-Seltenen-Erden.html.

Anonymous. 2012b. Umweltschutz als Ausrede: China hält Seltene Erden knapp. Financial Times Deutschland, 25.7.2012.

Anonymous. 2013. Mountain Pass Mine. Cal Poly Pomona. Acessed June 2013. http://geology.csupomona.edu/drjessey/fieldtrips/mtp/mtnpass.htm.

Bradsher, K. 2011. Mitsubishi is quietly cleaning up a former rare earth refinery. NYTimes.com. http://www.nytimes.com/2011/03/09/business/energy-environment/09rareside.html?_r=2&.

Brandmeier, M. 2010. Remote sensing of Carhuarazo volcanic complex using ASTER imagery in Southern Peru to detect alteration zones and volcanic structures—A combined approach of image processing in ENVI and ArcGIS/ArcScene. *Geocarto International* 25: 629–648.

Braune, G. 2008. Seltene Erden: begehrte Spezialmetalle. Handelsblatt. 8.12.2008. http://www.handelsblatt.com/finanzen/rohstoffe-devisen/rohstoffe/seltene-erden-begehrte-spezialmetalle/3002682.html.

Brenan, J.M., and W.F. McDonough. 2009. Core formation and metal-silicate fractionation of osmium and iridium from gold. *Nature Geoscience* 2: 798–801.

Canup, R., and C. Agnor. 2000. Accretion of the terrestrial planets and the Earth-Moon system. In *Origin of the Earth and Moon*, ed. R. Canup and K. Richter. Tucson: The University of Arizona Press.

Consumer Association of Penang. 2013. Chronology of events in the Bukit Merah Asian rare earth development. Accessed June 27. http://www.consumer.org.my/index.php/health/454-chronology-of-events-in-the-bukit-merah-asian-rare-earth-developments.

Dill, H.G. 2010. The „chessboard" classification scheme of mineral deposits: Mineralogy and geology from aluminium to zirconium. *Earth-Science Reviews* 100: 1–420.

DRAG. 2013. Deutsche Rohstoff AG: Seltenerden Storkwitz. Accessed June 28. http://www.rohstoff.de/geschaftsbereiche/hightech-metalle-seltene-erden-zinn-wolfram/storkwitz/.

Elsner, H., F. Melcher, U. Schwarz-Schampera, and P. Buchholz. 2010. Elektronikmetalle – zukünftig steigender Bedarf bei unzureichender Versorgungslage? BGR Commodity Top News 33.

Germis, C., and F. Nestler. 2013. Industriemetalle: Japan entdeckt Seltene Erden in seinen Gewässern. Frankfurter Allgemeine Zeitung, 28.3.2013. http://www.faz.net/aktuell/finanzen/devisen-rohstoffe/industriemetalle-japan-entdeckt-seltene-erden-in-seinen-gewaessern-12124757.html.

Helmholtz Institute for Resource Technology. 2012. Spurenelemente für eine gesunde Wirtschaft. Broschüre.

Jacoby, M., and J. Jiang. 2010. Securing the supply of rare earths. Chemical Engineering News 88: 9–12.

Kato, Y., K. Fujinaga, K. Nakamura, Y. Takaya, K. Kitamura, J. Ohta, R. Toda, T. Nakashima, and H. Iwamori. 2011. Deep-sea mud in the Pacific Ocean as a potential resource for rare-earth elements. Nature Geoscience 4: 535–539.

Lorand, J.P., A. Luguet, and O. Alard. 2008. Platinum-group elements: a new set of key tracers for the Earth's interior. Elements 4: 247–252.

Margonelli, L. 2009. Clean energy's dirty little secret. The Atlantic 05/2009. http://www.theatlantic.com/magazine/archive/2009/05/clean-energys-dirty-little-secret/307377/.

Marinovich, G. 2012. The murder fields of Marikana. The cold murder fields of Marikana. Daily Maverick, http://www.dailymaverick.co.za/article/2012-08-30-the-murder-fields-of-marikana-the-cold-murder-fields-of-marikana.

Markl, G. 2008. Minerale und Gesteine: Mineralogie – Petrologie – Geochemie, 2nd ed. Heidelberg: Spektrum Akademischer Verlag.

Molycorp. 2013. Molycorp announces new rare earth complex is operational and ramping up toward full-scale production. Accessed June 28. http://www.molycorp.com/molycorp-announces-new-rare-earth-complex-is-operational-and-ramping-up-toward-full-scale-production.

Nagy, A.A. 2008. Edelmetallrecycling beim Rückbau sulfidhaltiger Erzabgänge. Dissertation, TU Clausthal.

Nestler, F. 2013. Rohstoffe: Seltene Erden erstmals in Deutschland bestätigt. FAZ. http://www.faz.net/aktuell/wirtschaft/rohstoffe-seltene-erden-erstmals-in-deutschland-bestaetigt-12046040.html.

Nordstrom, D.K., C.N. Alpers, C.J. Ptacek, and D.W. Blowes. 2000. Negative pH and extremely acidic mine waters from Iron Mountain, California. Environmental Science and Technology 34: 254–258.

Palmer, M.A., E.S. Bernhardt, W.H. Schlesinger, K.N. Eshleman, E. Foufoula-Georgiou, M.S. Hendryx, A.D. Lemly, G.E. Likens, O.L. Loucks, M.E. Power, P.S. White, and P.R. Wilcock. 2010. Mountaintop mining consequences. Science 327: 148–149.

Papp, J.F., E.L. Bray, D.L. Edelstein, M.D. Fenton, D.E. Guberman, J. B. Hedrick, J.D. Jorgenson, P.H. Kuck, K.B. Shedd, and A.C. Tolcin. 2008. Factors that influence the price of Al, Cd, Co, Cu, Fe, Ni, Pb, Rare Earth Elements, and Zn. USGS Open File Report 2008-1356.

Righter, K., M. Humayun, and L. Danielson. 2008. Partitioning of palladium at high pressures and temperatures during core formation. Nature Geoscience 1: 321–323.

Robb, L. 2005. Introduction to Ore-Forming Processes. Malden: Blackwell Science.

Rowan, L.C., and J.C. Mars. 2003. Lithologic mapping in the Mountain Pass, California area using advanced spaceborne thermal emission and reflection radiometer (ASTER) data. Remote Sensing of Environment 84: 350–366.

Sabins, F.F. 1999. Remote sensing for mineral exploration. Ore Geology Reviews 14: 157–183.

Sánchez España, J., E. López Pamo, E. Santorimia Pastor, and M. Diez Ercilla. 2008. The acidic mine pit lakes of the Iberian Pyrite Belt: An approach to their physical limnology and hydrogeochemistry. Applied Geochemistry 23: 1260–1287.

Schippers, A., J. Vasters, and M. Drobe. 2011. Biomining – Entwicklung der Metallgewinnung mittels Mikroorganismen im Bergbau. BGR Commodity Top News 39.

Solomatov, V.S. 2000. Fluid dynamics of a Terrestrial Magma Ocean. In Origin of the Earth and Moon, ed. R. Canup and K. Richter. Tucson: The University of Arizona Press.

Stürmer, M., and G. Schwerhoff. 2013. Non-renewable but Inexhaustible—Resources in an Endogenous Growth Model. MPI Collective Goods Preprint.

USGS. Mineral Yearbooks. https://www.usgs.gov/centers/nmic/minerals-yearbook-metals-and-minerals.

USGS. Mineral Commodity Summaries. https://www.usgs.gov/centers/nmic/mineral-commodity-summaries.

Utter, T. 2010. Expertenkommentar: Seltene Erden Metalle: zwischen Panik und Realität. http://www.goldinvest.de/index.php/seltene-erden-metalle-zwischen-panik-und-realitaet-von-prof-dr-thomas-utter-19207.

Van der Meer, F.D., H.M.A. van der Werff, F.J.A. van Ruitenbeek, C. A. Hecker, W.H. Bakker, M.F. Noomen, M. van der Meijde, E.J.M. Carranza, J.B. de Smeth, and T. Woldai. 2012. Multi- and hyperspectral geologic remote sensing: A review. International Journal of Applied Earth Observation and Geoinformation 14: 112–128.

Wirtschaftsvereinigung Bergbau. (ed.). 1994. Das Bergbau-Handbuch (5th ed). Essen: Verlag Glückauf.

Yamaguchi, Y., and C. Naito. 2003. Spectral indices for lithologic discrimination and mapping by using the ASTER SWIR bands. International Journal of Remote Sensing 20: 4311–4323.

Yoshino, T., M.J. Walter, and T. Katsura. 2003. Core formation in planetesimals triggered by permeable flow. Nature 422: 154–157.

Further Reading

Lohmann, D., and N. Podbregar. Im Fokus: Bodenschätze. Auf der Suche nach Rohstoffen. Heidelberg: Springer.

Misra, K.C. 2000. Understanding Mineral Deposits. Dordrecht, Niederlande: Kluwer Academic Publishers.

Nesse, W.D. 2016. Introduction to Mineralogy, 3rd ed. Oxford: Oxford University Press.

Okrusch, M., and S. Matthes. 2009. Mineralogie: Eine Einführung in die spezielle Mineralogie, Petrologie und Lagerstättenkunde, 8th ed. Heidelberg: Springer.

Pohl, W.L. 2011. Economic Geology. Chichester: Wiley-Blackwell.

Rothe, P. 2010. Schätze der Erde. Darmstadt: Primus Verlag.

Seidler, C. 2012. Deutschlands verborgene Rohstoffe: Kupfer. Hanser, München: Gold und seltene Erden.

Strunz, H. 2001. Strunz Mineralogical Tables: Chemical-Structural Mineral Classification System, 9th ed. Stuttgart: Schweizerbart.

The World of Metals

This chapter is intended as a reference and is therefore kept short. It may be used as an alternative way of accessing the book by listing important application areas and ore minerals and referring to respective text passages in the following chapters. Geologically, it makes little sense to describe deposits by sorting the individual metals they contain since every single metal occurs in a multitude of different types of deposits and occurs there together with other metals (e.g., Dill 2010).

The most important metals when it comes to demand for quantities produced (sorted by tonnage) are iron, aluminum, copper, and zinc (data from *USGS Minerals Yearbooks*). Precious metals such as gold and platinum are produced in comparatively small quantities, but due to their high price they have a certain economic weight. Many of the so-called high-tech metals are only consumed in small quantities because they are used in tiny electronic components or as traces in special alloys. However, they are indispensable to the corresponding applications (Elsner et al. 2010). In a suitable analogy they are sometimes called "spice metals." Of course, ores are not only needed for metal production but also as a raw material for the chemical industry.

The classification of metals into steel refiners, non-ferrous metals, precious metals, and so on is partly arbitrary. Nickel, for example, is usually classified as a non-ferrous metal but is also an important steel refiner.

2.1 Iron and Steel Refiners

2.1.1 Iron (Fe)

Iron or rather steel is by far the most widely used metal accounting for about 95% of total metal production. Although it is readily available and correspondingly inexpensive, it also has excellent properties (Fig. 2.1). Pure iron is relatively soft and very susceptible to corrosion, but a certain amount of carbon and alloyed steel refiners lead to materials whose properties can be adapted to respective applications.

Iron with a carbon content below 0.01% is called wrought iron. Although it is relatively soft and rusts quickly, it is easy to forge. Steel (Box 2.1) has a carbon content between 0.01 and 2%. Although it is much harder, it is still plastically deformable. Various added steel refiners reduce susceptibility to corrosion (stainless steel), increase the hardness, and so on. With a carbon content above 2% the metal is very hard, brittle, and can no longer be forged. However, it can be easily cast into molds and is correspondingly called cast iron. Depending on the cooling rate the carbon is present as iron carbide (white cast iron) or as graphite (black cast iron). Pig iron flowing out of a blast furnace has a carbon content of 4–5%. It can be used directly as cast iron or can be decarburized into steel.

Steel is mainly used in construction (reinforced concrete and steel structures); for machines and tools; engines; and in the manufacture of automobiles, railways, and ships. For example, machine parts, manhole covers, and ladles are cast from cast iron. Stove plates, railings, street lamps, columns, and bridges made of cast iron were popular in the nineteenth century. Even earlier cast iron was used for cannons and cannon balls.

The ferromagnetic property of iron is also important. Electromagnets, generators, electric motors, and transformers are based on the induction of magnetic fields into the ferromagnetic or ferrimagnetic core wound around a coil. This core consists of iron or an alloy of iron with, say, aluminum, cobalt, or nickel. Sometimes ferrites are used in which case the term refers to ferrimagnetic but non-conductive ceramics with iron oxide or other metal oxides. The same compositions are also used to shield electromagnetic waves.

Iron oxides are used in relatively large quantities as color pigments.

World steel production in 2010 amounted to 1.42 billion tonnes, China alone accounted for 46% of this. In addition, demand has to be met for all the other iron applications

© Springer Nature Switzerland AG 2020
F. Neukirchen and G. Ries, *The World of Mineral Deposits*,
https://doi.org/10.1007/978-3-030-34346-0_2

Fig. 2.1 Iron can be used in many ways. The Könneritz Bridge (Leipzig) completed in 1899 (*photo* © F. Neukirchen)

mentioned. The metal content of iron ore mined in the same year was 1.28 billion tonnes. Some of the demand is met by recycling.

Continental crust consists on average of 5.6% iron, the fourth most common element after oxygen, silicon, and aluminum. However, ferrous silicates cannot be used. The most important ore minerals (Table 2.1) are magnetite (Fig. 2.2) and hematite. A product of frequent weathering is the brownish–yellowish earthy limonite, an aggregate of different iron hydroxides.

Typical ore grades are 55–60% Fe with a minimum of about 32%. Ores with a high content of phosphorus, sulfur,

and other metals may require special processing. Iron deposits are widespread and very diverse, with BIFs (banded iron formations; Sect. 5.2) being the most important. Other examples are layered mafic intrusions (LMIs; Sect. 3.3), Kiruna type (Sect. 3.6), iron skarns (Sect. 4.9), and metasomatic siderite deposits (Sect. 4.10). Since magnetite can

Table 2.1 Important iron minerals and their metal content

Magnetite (Fig. 2.2)	$Fe^{2+}Fe_2^{3+}O_4$ (often with Ti, Mn, Mg, Al, V)	72% Fe
Hematite	$Fe_2^{3+}O_3$	70% Fe
Goethite	α-$Fe^{3+}OOH$	62% Fe
Lepidocrocite	γ-$Fe^{3+}OOH$	62% Fe
Siderite	$FeCO_3$	48% Fe

Fig. 2.2 Magnetite from Saint Paul, Banat (Romania) (*photo* © F. Neukirchen/Mineralogical Collections of the TU Berlin)

be separated very easily using magnets, magnetite concentrate is also a by-product of some other types of deposits. Historically important were hydrothermal veins (Sect. 4.1), Lahn-Dill type (Sect. 4.18), iron oolites (Sect. 5.3), residual ore (Sect. 5.3), bean ore (Sect. 5.3), iron concretions (Box 5.19), and bog iron (Box 5.19).

Box 2.1 Steel

Iron alloys with a carbon content of 0.01% up to a maximum of 2% are referred to as steel. By varying the contents of carbon and added metals (so-called steel refiners) steels with very different properties (hardness, elasticity and tensile strength, susceptibility to corrosion, heat resistance, weldability, magnetizability, and so on) can be produced. Important steel refiners are chromium, manganese, cobalt, molybdenum, vanadium, nickel, titanium, niobium, and tungsten. Stainless steel contains at least 11% chromium, which also improves hardness. Manganese binds unwanted sulfur, reduces wear, and increases hardenability, while tungsten improves heat resistance and so on. Steel also contains other components such as sulfur and phosphorus that are generally undesirable but can only be reduced with considerable effort.

Iron and other elements are not evenly distributed in steel. Instead, steel consists of microscopically small grains called crystallites each of which has a very specific composition in a very specific crystal lattice such as ferrite, pearlite, cementite, martensite, and austenite (Fig. 2.3). The texture of these different crystallites can be changed by forging, rolling, drawing, hardening, annealing, and so on (Fig. 2.4). The properties of steel thus depend, on the one hand, on the phases present and their proportions and, on the other hand, on the microstructure. Similarly, other alloys such as bronze and brass consist of different crystallites.

Ferrite, the normal modification of pure iron at room temperature, has a cubic room-centered crystal lattice. It is ferromagnetic (i.e., magnetizable), soft, and susceptible to corrosion. Carbon is only contained in traces. Chromium can be mixed as a solid solution in ferrite, which significantly improves its properties. With very high chromium content an intermetallic phase FeCr is formed as well.

However, at higher temperatures iron is present in the cubic surface-centered crystal lattice as austenite. This phase can be stabilized at room temperature by adding nickel, cobalt, carbon, manganese, or nitrogen. Austenite can contain significantly more carbon than ferrite. The contents of other elements make it somewhat harder than ferrite and less susceptible to corrosion. It is not ferromagnetic.

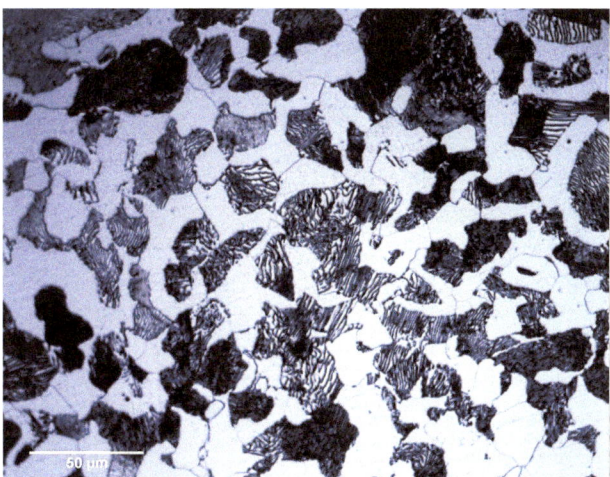

Fig. 2.3 Steel with ferritic–pearlitic microstructure. Ferrite is light-colored and pearlite is dark-striped. The carbon content of the steel is 0.45% (*photo* © Samson00, CC-BY-SA, Wikimedia Commons)

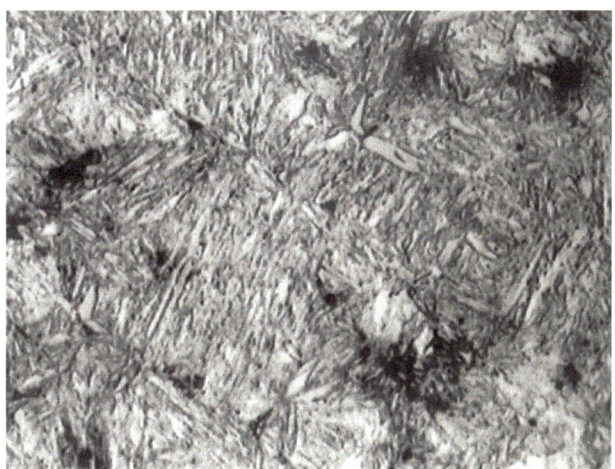

Fig. 2.4 Hardened steel with 0.35% carbon quenched from 870 °C water (*photo* © Unbound, public domain, Wikimedia Commons)

Cementite is an iron carbide with the composition Fe_3C. It is very hard, but brittle. White cast iron contains a lot of cementite, which is why it cannot be forged. Cementite only occurs in steel in small quantities in the form of very small crystallites.

Pearlite grains consist of closely intergrown, randomly arranged ferrite and cementite lamellae (pearlitic). It has a eutectic microstructure (Box 3.3) with a eutectic point of 0.8% carbon at 723 °C. Pearlite is an important component in steel and has very good properties. The closer the carbon content of the steel to the eutectic point the higher the pearlite content.

Martensite is a metastable phase produced by shear motion. In steel it has a similar composition to austenite, but in a tetragonal crystal lattice. Although martensite grains make steel much harder, they are relatively brittle. They can be produced by heating the steel to a certain temperature and then quenching it, a process known as hardening.

Steel refiners and carbon content have an effect on which phases are formed, in which proportions, and how exactly they are composed. Grain size and structure also change in detail. The best way to determine the effect of a particular mixture on the properties of the alloy is to do it empirically. There are thousands of standardized steel grades ranging from mass-produced steel for construction to high-quality alloys for special applications.

The term quality steel refers primarily to steel whose sulfur and phosphorus content is low. Unalloyed steel (mainly used in construction) does not contain any steel refiners, while micro-alloyed steel contains traces of these. In low-alloy steels the respective contents of added metals are less than 5%, while in high-alloy steels they are higher.

High-speed steel (HSS) contains up to 2.06% carbon and up to 30% steel refiners (tungsten, molybdenum, vanadium, cobalt, nickel, and titanium). During heat treatment (tempering) secondary carbides that provide extreme hardness are formed. HSS is wear resistant, heat resistant, and mainly used in tools for drilling and cutting.

Fig. 2.5 Rhodochrosite from N'Chwaning Mine, Kalahari manganese field (South Africa) (*photo* © Rob Lavinsky/iRocks.com, CC-BY-SA, Wikimedia Commons)

2.1.2 Manganese (Mn)

Manganese is primarily used as a steel refiner. It binds oxygen and sulfur and ensures hardenability and reduced wear. Manganese is also used in aluminum and copper alloys. Other applications include alkaline manganese batteries ("alkalines") and zinc manganese ferrites (ferrimagnetic ceramics, see iron).

Manganese ores are often earthy masses (umber) or solid crusts (psilomelane) composed of various manganese oxides and hydroxides (Table 2.2). There are also beautiful crystals. The most important deposits are found in marine sediments (Sect. 5.5). Hydrothermal exhalative occurrences are formed in the deep sea in the distal area of hot springs, partly also in

connection with BIFs. Oolithic manganese ores such as those found at Chiatura (Georgia), Nikopol (the Ukraine), and Groote Eyland (Australia) originate near the coast. There are also hydrothermal (iron–)manganese veins (Sect. 4.1) and metasomatic replacements in carbonates. Weathered manganese-rich rocks can produce residues that consist mainly of manganese oxides such as found in Nsuta (Ghana) and Orissa (India). Manganese nodules (Sect. 5.6) in the deep sea should also be mentioned. Ore grades range from 10 to 50% Mn. Ore concentrates are traded.

2.1.3 Chromium (Cr)

Chromium (Kropschot and Doebrich 2010) is used in large quantities to produce stainless steel (with 13–25% Cr plus Ni), which also hardens the steel. Various metals can also be chrome plated, the thin layer serving as corrosion protection and decoration. Chromium is also a component in superalloys (Box 2.2) such as used for jet engines of aircraft and for rockets. Chrome compounds are used as color pigments and tanning agents. Chromite with a low chromium content is used in refractory materials.

The only ore mineral is chromite ($(Fe,Mg)(Cr,Al,Fe)_2O_4$), a mineral of the spinel group. Most production comes from LMIs (Sect. 3.3) with the Bushveld Igneous Complex (South Africa) alone accounting for about half of world production. Podiform chromium deposits (Sect. 3.2) are common but mostly small. Chromium is mainly traded as an iron–chromium alloy (ferrochromium).

Table 2.2 Important manganese minerals and their metal content

Pyrolusite	MnO_2	63% Mn
Braunite	$3Mn_2O_3 \cdot MnSiO_3$	64% Mn
Hausmannite	Mn_3O_4	72% Mn
Rhodochrosite (Fig. 2.5)	$MnCO_3$	49% Mn
Manganite (Fig. 2.6)	$MnOOH$	62% Mn

Box 2.2 Superalloys

To cope with extreme conditions such as found in jet engines, in gas turbines, on oil rigs, and in reactors of chemical plants expensive superalloys are used that are made entirely or largely of metals such as nickel, chromium, cobalt, vanadium, and molybdenum. There are many such alloys most of which are traded under brand names. Some of them are also used in catalysts.

2.1.4 Nickel (Ni)

Nickel is mainly used in the manufacture of stainless steel (Boland 2012b). Metal objects can also be nickel plated for corrosion protection. In addition, various nickel alloys that can withstand extreme conditions are used in jet engines, oil refineries, chemical plants, pipes for drilling platforms, and ship hulls. Depending on prices on the commodity exchange the reserves that are available worldwide and worth exploiting are estimated at around 70–170 million tonnes according to various estimates (Claasen 2007). In 2006 some 1.34 million tonnes were produced.

The most important ore mineral in magmatic deposits is pentlandite that is mostly intergrown with pyrrhotite (FeS) (Table 2.3). Garnierite is a mixture of nickel-rich silicates (Ni-serpentine, Ni-talc, Ni-smectite, Ni-chlorite, and others) found in laterites. Nickeline is widespread in hydrothermal deposits (Fig. 2.7). The most important deposits are LMIs (Sect. 3.3), komatiites (Sect. 3.4) and laterites (Sect. 5.11). Nickeline also occurs in anorthosites (Sect. 3.5; Voisey's Bay, Canada) and other mafic/ultramafic igneous rocks. Hydrothermal deposits such as polymetallic veins (Sect. 4.1) and SEDEX (sedimentary exhalative deposits; Sect. 4.17) are of minor importance.

2.1.5 Cobalt (Co)

Cobalt (Boland 2011) is used as a steel refiner (heat-resistant permanent magnets, high-speed steel) and for corrosion-resistant and heat-resistant superalloys (especially, for jet

Fig. 2.6 Manganite from Ilfeld, Nordhausen in the Harz Mountains (Germany) (*photo* Rob Lavinsky/iRocks.com, CC-BY-SA, Wikimedia Commons)

engines). Lithium cobalt oxide batteries are increasingly being replaced by lithium–ion batteries. Cobalt blue is an important color pigment for glass, glazing, and lacquers. Geochemically, it behaves in a similar way to nickel. It is extracted as a by-product of polymetallic deposits (in particular, from Ni ores such as LMIs; Sect. 3.3 and Ni laterite; Sect. 5.11.2) and from stratiform Cu deposits such as the Central African Copper Belt (Sect. 5.1.2). It also occurs in polymetallic vein deposits (Sect. 4.1). The name goes back to miners in the German Ore Mountains (Erzgebirge) who found cobalt instead of silver in deeper parts of the veins, which they attributed to the activity of goblins (*Kobold*). Important cobalt minerals are shown in Table 2.4.

2.1.6 Molybdenum (Mo)

Molybdenum (Kropschot 2010) is an important steel refiner for high-strength steel grades and a component of superalloys. Electrical conductors made of molybdenum are used in thin-film transistors (TFTs) (e.g., for flat screens), thin-film solar cells, and halogen lamps. The mineral molybdenite is used as a solid lubricant and sometimes suspended in lubricating oils.

The most important ore mineral is molybdenite (Fig. 2.11, Table 2.5). Deposits are porphyry copper (Sect. 4.4) and porphyry molybdenum (Sect. 4.4.1). It is also found in skarns (Sect. 4.9) and some tin–tungsten deposits (Sect. 4.5) such as greisen (Sect. 4.6).

Table 2.3 Important nickel minerals and their metal content

Pentlandite	$(Ni,Fe)_9S_8$	Up to 35% Ni
Nickeline (Fig. 2.7)	NiAs	44%
Gersdorffite	NiAsS	35%
Rammelsbergite	$NiAs_2$	Up to 28% Ni
Nickelskutterudite	$(Co,Ni)As_3$	Up to 28% Ni
Garnierite	Mineral mixture	25–30% Ni

Table 2.4 Important cobalt minerals and their metal content

Cobaltite (Fig. 2.10)	CoAsS	35% Co
Skutterudite (Fig. 2.8)	$(Co,Ni)As_3$	Up to 24% Co
Linneite	$(Co,Ni)_3S_4$	Up to 58% Co
Carrollite	$CuCo_2S_4$	29% Co
Erythrite (Fig. 2.9)	$Co_3(AsO_4)\cdot2-8H_2O$	37% Co

Fig. 2.9 Erythrite from Neustädtel, Schneeberg, Ore Mountains (Germany) (*photo* © F. Neukirchen/Mineralogical Collections of the TU Berlin)

Fig. 2.7 Nickeline (NiAs) from Gottesehre Mine in Urberg, Black Forest (Germany) (*photo* © F. Neukirchen, Markl Collection/Tübingen)

Fig. 2.8 Skutterudite from Neuglück Mine in Wittichen, Black Forest (Germany) (*photo* © F. Neukirchen, Markl Collection/Tübingen)

Fig. 2.10 Cobaltite from Håkansboda (Sweden) (*photo* © Rob Lavinsky/iRocks.com, CC-BY-SA, Wikimedia Commons)

Fig. 2.11 Molybdenite from New South Wales (Australia) (*photo* © F. Neukirchen/Mineralogical Collections of the TU Berlin)

Table 2.5 Common molybdenum minerals and their metal content

Molybdenite (Fig. 2.11)	MoS_2	60% Mo
Wulfenite	$PbMoO_4$	26% Mo
Powellite	$CaMoO_4$	48% Mo

2.1.7 Vanadium (V)

Vanadium is an important steel refiner for hard, wear-resistant steel grades due to the formation of vanadium carbide. It is also used in superalloys. Vanadium oxide is used as a catalyst in sulfuric acid production.

The most important ore (Table 2.6) is magnetite. It sometimes contains about 1% vanadium together with titanium (titanium–vanadium magnetite). The most important deposits are LMIs (Sect. 3.3) (in particular, those found in the Bushveld Igneous Complex; Sect. 3.3.3). Vanadium is an important trace element in petroleum (Sect. 6.2). Ash produced during petroleum combustion is also used since it contains up to 20% V. Oil shale (Sect. 6.8) and black shale (Sect. 5.1) also have high contents.

Sandstone-bound uranium deposits (Sect. 4.14) in the Uravan Belt (Colorado, USA) contain vanadium minerals such as montroseite, carnotite (Fig. 2.12), and tujamunite. Vanadium is extracted here as a by-product. Sometimes

Table 2.6 Vanadium minerals and their metal content

Coulsonite component in magnetite	FeV_2O_4 in Fe_3O_4	Up to 2.8% V
Montroseite	VOOH	42% V
Carnotite (Fig. 2.12)	$K_2(UO_2)_2(VO_4)_2 \cdot 3H_2O$	20% V
Tujamunite	$Ca(UO_2)_2(VO_4)_2 \cdot 5–8H_2O$	20% V
Vanadinite	$Pb_5(VO_4)_3Cl$	19% V

Fig. 2.12 Carnotite from the Colorado Plateau (USA) (*photo* © F. Neukirchen/Mineralogical Collections of the TU Berlin)

vanadium minerals such as vanadinite occur in hydrothermal lead–zinc–copper veins (Sect. 4.1), but this holds greater interest for mineral collectors.

2.1.8 Tungsten (W)

Tungsten (a.k.a. wolfram) has a very high melting point and high electrical resistance. This is the reason the filaments of classic light bulbs and the electrodes in electron tubes and gas discharge lamps are made of this metal. It is used as a steel refiner for hard and heat-resistant steels and in superalloys. Tungsten carbide is extremely hard and pressure resistant. It is used for grinding and cutting tools and in high-pressure cells. The high density is exploited in armor-piercing ammunition and in sports (e.g., arrowheads in archery).

The ore minerals are wolframite and scheelite (Table 2.7). Tungsten almost always occurs together with tin. This is especially the case in skarns (Sect. 4.9), greisen (Sect. 4.6), and porphyry tin deposits (Sect. 4.5). It is also found in polymetallic veins (Sect. 4.1). Very many tungsten deposits are located in Asia in a belt extending from Sumatra to Kamchatka.

2.1.9 Tantalum (Ta) and Niobium (Nb)

Tantalum and niobium have extremely high melting points, are hardly ever attacked by acids, and scarcely ever susceptible to corrosion.

Table 2.7 Most important tungsten minerals and their metal content

| Wolframite | $(Fe,Mn)WO_4$ | Approx. 61% W |
| Scheelite (Fig. 2.13) | $CaWO_4$ | Approx. 64% W |

Table 2.8 Important niobium and tantalum minerals and their metal content

Pyrochlore	$(Na,Ca)_2(Nb,Ti,Ta)_2O_6(OH,F,O)$	Up to 52% Nb
Columbite	$(Fe,Mn)(Nb,Ta)_2O_6$	Up to 55% Nb
Tantalite	$(Fe,Mn)(Ta,Nb)_2O_6$	Up to 70% Ta
Wodginite	$(Mn,Sn,Ti)Ta_2O_6$	Approx. 57% Ta

Fig. 2.13 Scheelite from Gharmung, Skardu (Pakistan) (*photo* © Rob Lavinsky/iRocks.com, CC-BY-SA, Wikimedia Commons)

Fig. 2.14 Tantalite from Minas Gerais (Brazil) (*photo* © F. Neukirchen/Mineralogical Collections of the TU Berlin)

Niobium is mainly used for the production of particularly hard, corrosion-resistant, and heat-resistant steel grades from which aircraft engines and oil and gas pipelines are made. Superalloys such as niobium–titanium are used in spacecraft and among other things as superconductors in the electromagnets of particle accelerators.

Tantalum is mainly used in the manufacture of high-performance capacitors widely used in mobile phones, computers, and electrical components of cars and aircraft. Its exceptional suitability for this purpose is not only due to having very good electrical properties, but also to the fact that extremely thin foils can be produced that are protected from deeper oxidation by a wafer-thin oxide layer.

Other applications include bone nails, jaw screws, and prostheses in medical technology. Superalloys containing tantalum are used as linings in reactors in the chemical industry and in aircraft engines.

Tantalum and niobium are always found together in nature, although one of them usually predominates. The most important ore minerals (Table 2.8) are pyrochlore for niobium and the mixture series columbite–tantalite (Fig. 2.14) (coltan) with tantalum and niobium. Separating the two metals is a complex procedure usually done by solvent extraction.

Niobium is obtained almost exclusively from pyrochlore from carbonatites (Sect. 3.10). About 75% of the world production of niobium comes from the Araxa carbonatite in Minas Gerais (Brazil), the remainder is currently largely mined in Brazil and Canada.

Minimum ore grades are about 0.3% for niobium and 0.03% for tantalum. Global tantalum resources are estimated at 153,000 t (Burt 2010) 44% of which are in South America, 27% in Australia, 13% in Asia, and only 12% in Africa. So far tantalum has been extracted almost exclusively from pegmatites (Sect. 3.8) and secondary placer deposits formed from them (Sect. 5.9). Tantalum is also a by-product of niobium and tin production and is increasingly obtained through recycling. Tin granites (Sect. 3.7.1) and greisen (Sect. 4.6) account for 61% of known tantalum resources but are of low grade. Agpaitic nepheline syenites (Sect. 3.11) can also be important deposits of niobium and tantalum.

It has long been the case that more than half the world's tantalum production came from the Wodgina Mine in Australia despite having a low ore grade and a high cost of production. Production in other modern mines in Australia, Brazil, Canada, China and some African countries varies widely. Often this is only worthwhile if high prices can be achieved. A growing proportion of tantalum comes from artisanal mining in Africa (especially, Central Africa). Due to extremely low wages, often inhuman working conditions, and even forced labor in countries undergoing civil wars the coltan extracted there is often cheaper than production in large mines. So-called blood coltan from the Congo largely financed the country's bloody civil war (Box 3.14). Attempts are being made to prevent any trade in blood coltan.

The consumption of both metals has increased steadily apart from a drop in the course of the economic crisis around 2009. Although tantalum and niobium are not openly traded on the market, they are distributed through long-term contracts. So it is difficult to quote a price. Nevertheless, the prices for tantalum are very volatile; even the opening or closing of a large mine can make itself felt. With the upturn in the high-tech industry at the turn of the millennium tantalum was even more expensive than silver.

2.2 Non-Ferrous Metals

2.2.1 Copper (Cu)

Copper (Fig. 2.15) has excellent electrical conductivity (only silver conducts better), is a good heat conductor, and is relatively corrosion resistant. It can also be processed easily. Today it is mainly used (Doebrich 2009) as an electrical conductor. A single automobile contains an average of 1.5 km of copper wires weighing between 20 and 45 kg of copper depending on the model.

Copper alloys such as brass (copper and zinc) and bronze (copper and tin) are also widely used. Other alloys such as copper–nickel or copper–aluminum (aluminum bronze) are used for special applications. Brass is very easy to forge and is used to make musical instruments and machine parts. The antiseptic effect of copper is the primary reason it is used to make brass door handles. Bronze is historically very important since it was long the most important metal of humankind. Copper sulfate is used as a pesticide and fungicide.

Copper consumption has risen sharply in recent decades. China, in particular, has more than doubled its demand since 2000 and currently consumes almost one third of world production. In 2010 some 16 million tonnes of copper were produced worldwide. Copper is the third most consumed metal.

Although chalcopyrite is the most common and important copper ore, some other copper minerals are equally important (Table 2.9). About three quarters of the copper produced worldwide comes from porphyry copper deposits (Sect. 4.4). Other important deposit types are volcanogenic massive sulfide (VMS) deposits (Sect. 4.16), SEDEX deposits (Sect. 4.17), iron oxide copper gold (IOCG) deposits (Sect. 4.7), and stratiform sediment-hosted deposits such as copper-bearing shale like Kupferschiefer in Europe (Sect. 5.1.1) and the Central African Copper Belt (Sect. 5.1.2). There are also

Fig. 2.15 Chile has erected a monument to copper, its most important resource, in Antofagasta (*photo* © F. Neukirchen/Blickwinkel)

Table 2.9 Important copper minerals and their metal content

Native copper	Cu	100% Cu
Chalcopyrite	$CuFeS_2$	34% Cu
Bornite	Cu_5FeS_4	63% Cu
Covellite	CuS	66% Cu
Chalcocite	Cu_2S	80% Cu
Digenite	Cu_9S_5	79% Cu
Enargite	Cu_3AsS_4	48% Cu
Tetrahedrite-tennantite	$Cu_{12}(As,Sb)_4S_{13}$ with Fe, Zn, Ag, Hg	35–50% Cu
Azurite	$Cu_3(CO_3)_2(OH)_2$	55% Cu
Malachite	$Cu_2CO_3(OH)_2$	57% Cu
Cuprite	Cu_2O	88% Cu

occurrences in skarns (Sect. 4.9), various vein deposits (Sect. 4.1) such as copper–tin veins (e.g., in Cornwall, UK; Box 4.19), chalcopyrite siderite veins, and cordillera-type veins (Box 4.9). There are also impregnations in sandstone

(Sect. 4.13). Near Lake Superior hydrothermal impregnation of native copper occurs in basalts (Box 4.11). Magmatic deposits are found together with nickel in LMIs (Sect. 3.3) and the Phalaborwa carbonatite (Box 3.15). Secondary enrichment in the oxidation zone plays an important role in many copper deposits (Box 4.16). Typical ore grades are 0.5–1.5% for very large deposits, while the minimum grade for smaller deposits is several percent.

2.2.2 Lead (Pb)

Lead is soft, has a high density, is corrosion resistant, has a low melting point (327 °C), and is easy to cast. Most lead production is used (Kropschot and Doebrich 2011a) for automobile batteries (lead accumulators). These contain a lead and a lead(IV) oxide electrode in diluted sulfuric acid. Other applications include protection against radiation, as a weight for balancing automobile wheels, as immersion weights, and in ammunition. Babbitt or bearing metal is an alloy with lead, tin, antimony, and copper used to build plain bearings. Solder often contains lead. Lead oxide is used for glazing.

Lead is toxic, and soluble lead compounds and vapors are particularly problematic. Therefore, lead is no longer used in many areas. Historically important applications were leaded petrol (which contained tetraethyl lead as an antiknock agent), lead for letters in letterpress printing, lead-covered church roofs, and water pipes. Lead compounds have also been used as color pigments (lead white, red lead) and rust inhibitors.

Today, recycling fulfills more than half the demand. In 2010 some 4.1 million tonnes of lead were produced from mines worldwide, while consumption in the same year was 9.4 million tonnes.

The most important lead ore (Table 2.10) is galena (Fig. 2.16). It often contains some silver and is also the most important silver ore due to its abundance. Common secondary lead minerals are cerussite (Fig. 2.17), anglesite, pyromorphite, and mimetesite.

Lead occurs in hydrothermal deposits together with zinc. The most important deposits are SEDEX (Sect. 4.17) with about half the world's resources, Mississippi Valley type (MVT; Sect. 4.12), and VMS (Sect. 4.16). Although lead–zinc veins are widespread, they are of little economic importance today (Sect. 4.1). Lead is also found in chimneys

Fig. 2.16 Galena from Joplin (Missouri, USA) (*photo* © F. Neukirchen/Mineralogical Collections of the TU Berlin)

Fig. 2.17 Cerussite from Tsumeb (Namibia) (*photo* © Didier Descouens, CC-BY-SA, Wikimedia Commons)

and mantos (Sect. 4.8), skarns (Sect. 4.9), and as impregnations in sandstones (Sect. 4.13). Minimum ore grades are between 3 and 10% lead plus zinc.

2.2.3 Zinc (Zn)

Zinc as a pure metal was unknown in Europe until the middle of the eighteenth century. This was because of its low boiling point. It was known and produced much earlier in India and China. Today it ranks fourth after iron, aluminum, and copper in terms of the quantity produced. In 2010 production was 12 million tonnes worldwide.

Table 2.10 Important lead minerals and their metal contents

Galena (Fig. 2.16)	PbS (often some Ag)	86% Pb
Cerussite (Fig. 2.17)	$PbCO_3$	77% Pb
Anglesite	$PbSO_4$	68% Pb

Zinc is brittle and hard, has a low melting point (420 °C) and boiling point (907 °C), oxidizes easily, and reacts well with other substances. Although these may not sound like ideal properties for a metal, it compensates by having a major property. When mixed with air a thin layer of zinc oxide and zinc carbonate (patina) forms on the surface of a metal protecting deeper areas from corrosion.

The most important application (Kropschot and Doebrich 2011b) consuming about half of zinc production is the galvanization of iron and steel parts. A thin layer of zinc protects iron against corrosion (rust). The zinc layer can be applied electrochemically by electroplating or by immersing the workpiece in a zinc melt (hot-dip galvanizing). Molten zinc also lends itself to being sprayed using compressed air.

In alloys with other metals zinc can be used to protect against corrosion. Brass is an alloy of copper and zinc that is easily processed and shaped. It is widely used in mechanical engineering and for musical instruments. Zinc has many uses such as the manufacture of vehicle parts from alloys of zinc with aluminum and magnesium, zinc sheet (typically, an alloy of zinc and some titanium) used in roofs and rain gutters, and the anode in many batteries. In chemistry it is used as a reducing agent.

The production of zinc oxide is important because of its antiseptic effect. It is a component of ointment widely used for the treatment of wounds, rashes, and burns. It protects against sunburn and is contained in some sun creams. Zinc oxide can protect plastics and rubber polymers against UV radiation. It serves as a catalyst in the production of synthetic rubber and in vulcanization. Zinc oxide is also used as a white pigment (zinc white).

Smaller quantities of other zinc compounds are also produced such as zinc chloride used as a fire retardant and wood preservative and calcium zinc compounds used as stabilizers in plastics.

Zinc is an essential trace element in the diet absorbed through meat, cheese, lentils, nuts, and seafood. In the body it is a component of many enzymes and plays an important role in metabolism, the formation of hormones, cell growth, and the immune system.

In hydrothermal deposits lead and zinc almost always occur together and often with copper. The most important ore mineral (Table 2.11) is sphalerite (Fig. 2.18), which often contains a few percent cadmium. Wurtzite has the same composition as sphalerite but in another structure. Both occur in colloform sphalerite (Box 4.21). Common secondary zinc minerals are smithsonite and hemimorphite (a.k. a. calamine).

Zincblende is the old name given to sphalerite. The historical term **blende** (zincblende, pitchblende) was used by German miners for "deceptive" minerals that looked like ores, but from which no metals could be extracted at least with methods available at that time. The German word *blenden* means "to dazzle."

More than half the world's resources of zinc are accounted for by SEDEX deposits (Sect. 4.17). The remainder is mainly found in MVT deposits (Sect. 4.12) and VMS deposits (Sect. 4.16). Lead–zinc veins are also widespread (Sect. 4.1), but they are of little importance compared with the large deposits of massive ores. Zinc is also found in chimneys and mantos (Sect. 4.8), skarns (Sect. 4.9), and as impregnations in sandstones (Sect. 4.13). Minimum ore grades are between 3 and 10% lead plus zinc.

2.2.4 Cadmium (Cd)

Cadmium coatings are used to protect iron against rust. The metal is also used in nickel–cadmium batteries. Cadmium yellow and cadmium red are color pigments consisting of cadmium sulfide, cadmium selenide, and other compounds. Cadmium is very toxic and when ingested accumulates in the body over a long period of time.

Although there are very rare cadmium minerals such as greenockite (CdS) and otavite ($CdCO_3$), the metal is obtained almost exclusively from the zinc minerals sphalerite and smithsonite, which often contain a few percent cadmium. This makes it a by-product of zinc mining.

2.2.5 Tin (Sn)

Tin is a soft, corrosion-resistant metal with a very low melting point (232 °C). Pewter figures and dishes are made from it. However, it is mainly used in alloys such as bronze, bearing metal (see lead), and solder. Solder is an alloy with a very low melting point (e.g., the eutectic composition of tin–lead). Organ pipes used to be made from a tin–lead alloy. Tin-coating of other metals is also important (especially, for tableware and food containers). Tinplate is tinned iron sheet (it is no coincidence that cans are called tins in Britain) that can be rolled out as wafer-thin foils (tin foil). Tin foil has now been replaced by aluminum foil. Tin compounds are used in fungicides, dentistry (tin fluoride, amalgam), plastics, wood preservatives, and many other areas. Indium tin oxide is used in touchscreens.

Table 2.11 Important zinc minerals and their zinc content

Sphalerite (zincblende) (Fig. 2.18)	ZnS (with Fe, Mn, Cd, Ga, In, Ge, Tl, As, Se, Hg)	38–66% Zn
Wurtzite	ZnS (with Fe, Mn, Cd …)	
Smithsonite	$ZnCO_3$ (often with Cd)	52%
Hemimorphite	$Zn_4Si_2O_7(OH)_2 \cdot H_2O$	54%

Fig. 2.18 Sphalerite (*dark gray*) with fluorite (*cubes*) and baryte (*white*) from Gottes Segen Mine in Schnellingen, Black Forest (Germany) (*photo* © F. Neukirchen, Markl Collection/Tübingen)

Table 2.12 Important tin minerals and their metal content

Cassiterite	SnO_2	78% Sn
Stannite	Cu_2FeSnS_4	27% Sn

Tin pest describes silver–white β-tin (stable between 16 and 181 °C, tetragonal) that has transformed into gray–black α-tin (cubic). The transformation starts with dark spots that slowly spread and then increase in volume causing infested objects to decompose into tin powder. Such a conversion can only be triggered in extreme cold, but alloys with antimony or bismuth prevent this from happening.

The most important deposits are tin placers (Sect. 5.9), greisen (Sect. 4.6), pegmatite (Sect. 3.8), skarns (Sect. 4.9), porphyry tin deposits (Sect. 4.5), polymetallic veins (e.g., Ore Mountains; Sect. 4.1.2 and Cornwall; Box 4.19), and tin granite (Sect. 3.7.1). The most important ore mineral (Table 2.12) is cassiterite.

2.3 Precious Metals

Precious metals in the colloquial sense are particularly corrosion-resistant metals that are both rare and expensive (especially, gold, silver, and the platinum-group elements). Defined from the chemist's point of view precious metals are not easily oxidized, have positive electrode potential (referring to their behavior relative to a standard hydrogen electrode), and hence are not attacked by simple acids. According to this definition mercury and copper can also be deemed precious metals in addition to the elements already mentioned.

2.3.1 Gold (Au)

Gold has always been used for jewelry and as a status symbol. It lost most of its significance as money when gold exchange for the dollar ceased. Nevertheless, central banks still hold large quantities of gold. The largest gold reserves are in the Federal Reserve Bank of New York where many foreign banks also store their bars (Fig. 2.19) and in Fort Knox (Kentucky, USA). Gold is particularly popular in times of crisis as an investment despite not yielding interest and the possibility of its price falling.

Gold is also frequently used in technology because it has good electrical conductivity and does not corrode. Gold wires and gold-plated contacts are used to connect computer

Fig. 2.19 A couple of gold bars (*photo* Apollo 2005, CC-BY-SA, Wikimedia Commons)

processors. Gold-plated mirrors are used in optics to reflect infrared light, which they do particularly well. Dentists use gold as a filling and dental prosthesis. Gold can be hammered to wafer-thin foils (gold leaf) that can be applied to roofs, icons, and other objects. Galvanization is mainly used today for gold plating.

The total quantity of gold mined so far is estimated at 160,000 metric tons (tonnes or t), which corresponds to a cube with an edge length of just over 20 m. About 80% have been produced since 1900 and more than 2500 t are added annually.

The purity of gold (fineness) is expressed in carats (24 carats = 100% gold). This unit should not be confused with the weight unit of the same name used for gemstones.

In nature gold occurs primarily as native gold in the form of tiny flakes or larger nuggets containing 2–20% silver and small amounts of other metals. A natural alloy with a higher silver content is called electrum. Gold is often enclosed as tiny particles in sulfides such as pyrite or arsenopyrite. Gold nuggets with several percent palladium are also known. Gold tellurides such as sylvanite ((Au,Ag)Te$_4$), calaverite (AuTe$_2$), and krennerite ((Au,Ag)Te$_2$) are rare minerals.

One third of the gold found today comes from placer deposits (Sect. 5.9) and another third from orogenic gold veins (Sect. 4.2) including intrusion-related gold (Sect. 4.4.2) since the respective classification of veins is not always possible. Approx. 13% come from epithermal deposits (Sect. 4.3), 10% are extracted as by-products from porphyry copper (–gold) deposits (Sect. 4.4) and skarns (Sect. 4.9), 8% from Carlin-type deposits (Sect. 4.11), and only 1–2% each from VMS (Sect. 4.16), IOCG (Sect. 4.7), and various magmatic deposits (Frimmel 2008). Au–laterite should also be mentioned (Sect. 5.11.3). The proportion of placers was greater historically with the Witwatersrand in South Africa alone (Box 5.18) being responsible for around 40% of the total quantity ever produced.

Typical grades are 1–10 g/t. Combining large open-pit mines with heap leaching can even make significantly lower ore grades economical. Gold from placer deposits can be extracted mechanically with a washing pan or washing trough. Most gold mines use the amalgam process or leaching with cyanides. In copper production gold is being enriched in anode sludge.

2.3.2 Silver (Ag)

Silver is used for jewelry, table silver, sacred objects in churches, transverse flutes, and still partly as coin. It has the highest electrical conductivity of all metals. Silver and silver alloys are used for special applications in electronics and optics. Pure silver is rarely used since silver is relatively soft. Sterling silver (named for its use in British coinage) is an

Fig. 2.20 Native silver from the Sophia Mine, Wittichen, Black Forest (Germany) (*photo* © F. Neukirchen, Markl Collection/Tübingen)

alloy with 92.5% silver. Silver halides were the basis of analogue photography. The antiseptic effect of silver is used in medicine in coatings for devices, in creams, and in wound dressings. Silver particles and coatings have a similar function in water filters and textiles. It is also used in certain catalysts.

The most important silver ore is galena (PbS), which is very common and contains a few percent of silver. Tetrahedrite is relatively common and can contain up to 18% silver (freibergite). Native silver usually only occurs in small quantities (Fig. 2.20). A special case is Kongsberg (Norway) where large quantities of native silver occur in hydrothermal veins often in the form of wires that can be as thick as arms. Although silver minerals usually occur only in small quantities, in some individual deposits they can be important (Table 2.13).

Silver is extracted from a wide variety of hydrothermal deposits. Silver veins (Sect. 4.1) include lead–zinc veins such as those in the Black Forest and the Harz (Germany), various polymetallic veins such as the silver–cobalt–nickel–arsenic–bismuth–uranium veins in the Ore Mountains (Sect. 4.1.2) and Kongsberg, the tin–silver–bismuth–

Table 2.13 Selected silver minerals and their silver content

Native silver (Fig. 2.20)	Ag (with Au, Cu, Hg, As, Sb, Bi)	Up to 100% Ag
Acanthite	Ag$_2$S	87% Ag
Proustite	Ag$_3$AsS$_3$	65% Ag
Pyrargyrite	Ag$_3$SbS$_3$	60% Ag
Stephanite	Ag$_5$SbS$_4$	68% Ag
Polybasite	(Ag,Cu)$_{16}$Sb$_2$S$_{11}$	Up to 74% Ag
Chlorargyrite	AgCl	75% Ag
Tetrahedrite (freibergite)	(Cu,Ag)$_{12}$(SbS$_3$)$_4$ (with Fe, Zn, Ag, Pb, Hg)	Up to 18% Ag

tungsten veins of Potosi (Box 4.17), and polymetallic veins of the Cordillera type (Box 4.9). There are epithermal deposits (Sect. 4.3) of gold and silver or predominantly silver. Chimneys and mantos (Sect. 4.8), VMS (Sect. 4.16), and SEDEX (Sect. 4.17) contain silver in massive sulfides. Silver also occurs as a by-product of porphyry copper deposits (Sect. 4.4) and gold mining. Copper-bearing shale like Kupferschiefer (Sect. 5.1.1) and sandstone (Sect. 4.13) may also contain some silver. Since silver is commonly associated with non-ferrous metals the minimum grade of ore is highly dependent on the grades of such metals. Typical ore grades range from 20 to 700 g/t.

2.3.3 Platinum-Group Elements

Platinum is more expensive than gold and is becoming increasingly popular for jewelry. The precious metal is temperature resistant, not susceptible to corrosion, and is only attacked by a few chemicals. Therefore, it is often used for laboratory equipment. Many other devices also contain platinum or platinum-containing alloys. The most important use of platinum and other platinum-group elements is in catalytic converters (Box 2.3) such as the three-way catalytic converters of vehicles.

Box 2.3 Catalysts

A catalyst accelerates the speed of a chemical reaction by enabling an energetically more favorable reaction path. The three-way catalytic converter of automobiles converts dangerous components of exhaust gases such as carbon monoxide, hydrocarbon compounds, and nitrogen oxides into almost harmless carbon dioxide, water, and nitrous oxide. However, catalysts are also used on a large scale in the chemical industry enabling the synthesis of a wide variety of substances such as ammonia, sulfuric acid, nitric acid, and methanol.

Every application needs its own special catalyst. Some are in a solution together with the reactants and products of the reaction. More common are solids (especially, metals and metal oxides) that drive reactions in a solution or in gases. In principle, such solids do not react with the educts and products involved (apart from atoms at the surface). Although the valence electrons on the inside of each atom are involved in a bond with neighboring atoms, a neighboring atom is missing on the surface and the corresponding valence electron can stretch out a helping hand, so to speak. This makes it possible to first degrade the starting materials into intermediate products that attach themselves to the surface of the catalyst. The various intermediate products are then combined to form the reaction products.

It is therefore important that the surface area of the catalyst is as large as possible. The simplest variant is a powder that is then filtered out of the solution. More frequently, nanopowder is solidified by sintering to form a sponge-like structure that coats a honeycomb-shaped ceramic. Some zeolites (Sect. 7.15) activated by chemical reactions are also used as catalysts. The reaction here takes place within a wide-meshed crystal lattice. A good example is the cracking of heavy crude oil when some of the aluminum in the zeolite structure is replaced with lanthanum and other rare-earth elements.

Every catalyst needs suitable conditions (temperature, pressure, etc.) to function optimally. The active substance is exactly matched to the respective reaction. Expensive precious metals such as platinum, rhodium, and palladium are used in vehicles as is cerium(IV) oxide. Platinum-group elements are also used as catalysts in the synthesis of nitric acid in the Oswald process. When the Haber–Bosch process is used for ammonia synthesis the catalyst consists of magnetite (reduced to iron) together with K_2O, Al_2O_3, and SiO_2. Copper, copper oxide, zinc oxide, and chromium oxide are used in the production of methanol. There are many other combinations, the most noteworthy being lanthanum perovskites such as $LaCoO_3$ and $LaMnO_3$.

It is no coincidence that combinations of several metals and metal oxides are often used. Such phases give different intermediate products a "helping hand" enabling reactions to take place mainly at respective phase boundaries. Other components cause defects in the crystal lattice or build up a wafer-thin coating as the active surface. Since it is seldom known why the respective mixture works and what exactly happens during the reaction the catalysts are mainly developed by trial and error.

Platinum (Pt), palladium (Pd), rhodium (Rh), iridium (Ir), osmium (Os), and ruthenium (Ru) have very similar chemical properties and always occur together. Platinum and palladium make up the largest part with platinum usually predominating. Platinum occurs primarily as native platinum (usually tiny particles but sometimes nuggets on placers), as an alloy with other precious metals, and as traces in sulfides or chromite. There are also exotic platinum minerals (Table 2.14).

Rhenium (Re) is used in superalloys and catalysts and has some similarity with PGEs. It is extracted almost exclusively

Table 2.14 Selected platinum minerals and their respective metal content

Native platinum	Pt	Up to 100% Pt
Cooperite	PtS	Approx. 82% Pt
Braggite	(Pt,Pd,Ni)S	Approx. 59% Pt, 20% Pd
Sperrylite	PtAs$_2$	Up to 54% Pt
Moncheite	(Pt,Pd)(Te$_2$,Bi)	Approx. 37% PGEs, approx. 40% Te
Laurite	RuS$_2$	Approx. 61% Ru

from porphyry copper deposits (Sect. 4.4) where it is found in molybdenite. It also occurs as traces in some other types of deposits.

Apart from gold and rhenium the PGEs are the rarest stable elements in the Earth's crust. Iridium, rhenium, rhodium, and ruthenium are the rarest (the exact order is given differently in the literature). Annual production of platinum and palladium is around 200 t. Typical ore grades are between 3 and 10 g/t, even lower if nickel or chromium can be mined at the same time. PGEs are obtained almost exclusively from LMIs (Sect. 3.3). The Bushveld Complex (South Africa) alone (Sect. 3.3.3) contains 75% of the world's platinum, 54% of palladium, and 82% of rhenium resources (Naldrett et al. 2008). Apart from a few percent the remaining known resources are accounted for by the Great Dyke (Zimbabwe), Norilsk (Siberia), Sudbury (Canada), and Stillwater (USA).

There are also some podiform chromium deposits (Sect. 3.2) with a relatively high platinum content. Small ultramafic intrusions (Alaska–Urals type) usually contain platinum only in the subeconomic range. Placer deposits with platinum nuggets (Sect. 5.9) such as those in the Urals are not very important today, but PGEs (in particular, osmium and iridium) are by-products of gold purification in the Witwatersrand (Box 5.18). From time to time platinum is found in black shale (Sect. 5.1.1).

2.4 Light Metals

2.4.1 Aluminum (Al)

Aluminum and aluminum alloys are used wherever low weight is important such as aircraft and vehicle construction. It is also widely used as packaging material (aluminum foil, Tetra Pak™). Since aluminum is a very good electrical conductor it is sometimes used instead of copper. Aluminum's highly exothermic thermite reaction is used for welding in the construction of rails when Fe$_2$O$_3$ powder is mixed with Al. The two components then react to Al$_2$O$_3$ and molten iron. Aluminum dust can even ignite explosively

when exposed to air. The metal is also used in solid-fueled rockets, fireworks, as a mirror coating, and as a pigment.

Aluminum is the second most consumed metal with 41 million tonnes produced worldwide in 2010. In the Earth's crust it is the third most common element after oxygen and silicon. It is largely found in silicate minerals. Bauxite (Sect. 5.11.1) is an ore comprising a mixture of different aluminum hydroxides. Nepheline is an ore from alkaline rocks (Sect. 3.9) but is used in insignificant quantities. The Ivigtut cryolite deposit used to be historically important (Sect. 3.12).

2.4.2 Titanium (Ti)

Titanium is light, strong, ductile, corrosion resistant, and heat resistant. What makes the metal relatively expensive is not the lack of ore but the costly smelting process. Titanium and titanium alloys are used in the aerospace industry, in high-quality sports equipment, as housings for watches and laptops, in medicine as implants, and for other special applications. It is also an important steel refiner. Titanium oxide is used in large quantities as a white pigment (titanium white).

The most important ore minerals are rutile (Fig. 2.21) and ilmenite. Perovskite is also usable (Table 2.15). The very common mineral titanite (CaTiSiO$_5$) is not used. Leukoxene is a mixture of different titanium and iron oxides. The most important deposits are titanium placers (Sect. 5.9) (especially, sandy beaches where heavy minerals such as rutile, ilmenite, zircon, garnet, monazite, and xenotime are present). Primary deposits are located in anorthosites (Sect. 3.5), in LMIs (Sect. 3.3), and in ultrabasic cumulates of alkaline complexes (Sect. 3.9).

Fig. 2.21 Rutile from the Binn Valley, Valais (Switzerland) (*photo* © F. Neukirchen)

Table 2.15 Titanium ore minerals and their metal content

Rutile (Fig. 2.21)	TiO_2	60% Ti
Anatase	TiO_2	60% Ti
Ilmenite	$FeTiO_3$	32% Ti
Perovskite	$CaTiO_3$	35% Ti

2.4.3 Magnesium (Mg)

Magnesium is a light metal used as an alloying component in aircraft and vehicle construction. Pure magnesium oxidizes very easily, is used in incendiary devices, tracer ammunition, and as a reducing agent. It can be obtained from seawater, salt lakes (Sect. 5.7.2), and potash salts (Sect. 5.7; with the mineral carnallite ($KMgCl_3 \cdot 6H_2O$)); from ultrabasic rocks with minerals such as brucite ($Mg(OH)_2$), magnesite ($MgCO_3$; Sect. 7.10), serpentine (($Mg,Fe)_6Si_4O_{10}(OH)_8$); even from the common mineral dolomite ($CaMg(CO_3)_2$); and from the fly ash of lignite-fired power stations. Production is very energy intensive in every case.

2.5 Rare-Earth Elements

Box 2.4 Discovery of rare-earth elements

Yttrium (Y), the first of the rare-earth elements (REEs), was discovered in 1794 by the Finnish chemist Johan Gadolin in a mineral that was later named gadolinite for him. Gadolinite is a black, heavy mineral that was found seven years earlier by Arrhenius, a Swedish lieutenant, in a quarry near the village of Ytterby on an island near Stockholm. On this island pegmatite was being mined on feldspar and quartz for the production of ceramics and glass. Gadolin narrowly missed the discovery of another element when he ignored the then unknown beryllium, a main component of gadolinite, because he thought it was aluminum. Nevertheless, Ytterby went down in history as the place where more elements were discovered than anywhere else. Yttrium, ytterbium (Yb), terbium (Tb), and erbium (Er) are named for this village. Holmium (Ho) was named for Stockholm, whereas scandium (Sc) and thulium (Tm) were named for Scandinavia and for the mythical Thule, respectively. Gadolin had the element gadolinium (Gd) named for him. Although gadolinium is also contained in tiny traces in gadolinite, it was first discovered in a sample of samarskite from America.

Over and over again it turned out that an "element" was not pure but a mixture of several elements. Strictly speaking, Gadolin did not isolate yttrium but impure yttrium oxide called ytter-earth. This is where the term rare earths comes from. Separating rare-earth elements was so difficult that the story of their discovery took more than a century to unravel with some being found only by their spectral line and not isolated as an element for a long time. Moreover, every now and then elements were "discovered" that did not exist at all.

However, another mineral from another site entered the scene, in which cerium (Ce), the second rare-earth element, has been discovered. It was also found in Sweden, in the Bastnäs Mine near the village of Riddarhyttan about 150 km west of Stockholm. Iron and copper were mined here at that time in various hydro thermal deposits including skarns. The mineral today called cerite (($Ce,La,Ca)_9(Mg,Fe)(SiO_4)_6(SiO_3OH)(OH)_3$) had already been discovered in 1750 in skarns. It was first thought to contain tungsten. However, Carl Wilhelm Scheele who was the discoverer of tungsten could not confirm this so he called the mineral "false tungsten". However, he was unable to find anything conspicuous about the mineral. It was only in 1803 that the Swedes Wilhelm von Hisinger and Jöns Jakob Berzelius as well as the German Martin Heinrich Klaproth discovered the previously unknown cerium. A few decades later another rare-earth mineral called bastnäsite was found in this mine. Today it is much more important as ore. While gadolinite (Fig. 2.22) is a mineral of the heavy REEs, cerite is a mineral of the light REEs.

It took about 40 years for the Swedish chemist Carl Gustav Mosander to discover that cerite and gadolinite contained other unknown elements. He isolated lanthanum (La) from cerite and another element that he called "didymium." Didymium turned out to be a mixture of different rare-earth elements. He then separated the "yttrium" of gadolinite into yttrium, erbium, and terbium. Mosander was also the first to produce an elementary REE with metallic cerium.

There then followed a break of about 35 years when little happened other than the names of erbium and terbium being inadvertently swapped. Finally, further discoveries followed one after the other. In the meantime other REE minerals from other sites became known. In 1878 the Swiss Jean Charles Galissard de Marignac separated impure ytterbium from impure erbium and in 1879 Lars Frederik Nilson isolated scandium oxide from euxenite (a mineral first discovered in Norway) and gadolinite. In the same year the Swede Per Teodor Cleve found thulium and holmium in gadolinite. This was followed a few years later by gadolinium discovered by Marignac in samarskite and dysprosium (Dy) in impure holmium oxide discovered by the Frenchman

Fig. 2.22 Gadolinite–(Ce) from Tuftane Quarry, Frikstad (Norway) (*photo* © Rob Lavinsky/iRocks.com, CC-BY-SA, Wikimedia Commons)

Boisbaudran. In the same period "didymium" turned out to be a mixture of samarium (Sm), praseodymium (Pr), neodymium (Nd) and gadolinium. Samarium, discovered by Boisbaudran, was the first element named for a person, in this case Samarski, a relatively unknown Russian mining engineer. Praseodymium (Pr) and neodymium (Nd) were both discovered by von Welsbach. "Didynium" also contained europium (Eu) only discovered in 1901 by Damarçai. Lutetium (Lu), the last stable rare-earth element, was discovered in 1906 simultaneously and independently by the Frenchman Georges Urbain, the Austrian Carl Auer von Welsbach, and the American Georges Urbain in impure ytterbium. All that was now missing was radioactive promethium (Pm). It has a half-life of a few years and occurs only as a fission product of uranium, in nature in hardly measurable traces in pitchblende, and artificially in spent fuel rods of nuclear power plants. It was detected in 1945 by American nuclear scientists and closed the last gap in the periodic table. At that time rare-earth elements were considered exotic elements of little economic importance. It was not until 1947 that more effective separation of REEs was possible with ion exchangers followed a little later by solvent extraction. Since 1964 the demand for REEs grew rapidly when it was found that europium could be used for the red color in color television and ever-more applications were added with the passage of time.

Rare-earth elements (REEs) long led a shadowy existence at the edge of the periodic table (Box 2.4), but now they have suddenly become the talk of the town (Liedtke & Elsner 2009). The fact that they have been used for ever-more high-tech applications in recent decades is not surprising since they are after all 17 elements that had hardly been investigated in the past. Among them are lanthanum and the so-called lanthanides ranging from cerium to lutetium in the periodic table, and the very similar elements yttrium and scandium. Chemically, these elements behave so similarly that they always occur together in nature and their ratio is only slightly changed during fractionation. Consequently, they can only be extracted together. The similarity also makes it extremely time consuming and expensive to separate this group into individual elements. In the past this was done with conventional precipitation reactions requiring thousands of steps, which meant that only tiny quantities could be produced. Today, ion exchangers or solvent extraction are used.

Rare-earth elements are very reactive as metals. They are traded as oxides of varying degrees of purity or as mixed oxides. It has been found for many applications that using a mixture is significantly cheaper than using a pure rare-earth element. The consumption of respective elements by industry does not of course correspond to the relative proportions in ore minerals. This results in extreme price differences. In particular, europium is presently used in much larger quantities than its relative abundance. New technological developments can lead to a rapid increase in the consumption of a certain element and thus to enormous price fluctuations.

Cerium, the most common rare-earth element, is present in the Earth's crust at even higher concentrations than copper. In second place is yttrium, about half as common, closely followed by lanthanum and neodymium. The concentration of europium corresponds to only one sixtieth of the concentration of cerium and that of the rarest REEs thulium and lutetium not even one hundredth. Nevertheless, they are still significantly more abundant than antimony or cadmium and about a hundred times more abundant than gold. This does not take into account unstable promethium believed to be virtually non-existent, although it can arise during the fission of uranium. Even its most stable isotope has a half-life of only a few days.

In nature rare-earth elements always occur as trivalent cations. The only exceptions are europium and cerium, which can also appear as Eu^{2+} and Ce^{4+}. The ion radii of the REE^{3+} are very similar, with the radius continuously decreasing from La^{3+} to Lu^{3+}, known as lanthanide contraction. Although the chemical behavior is very similar for all of them, it changes slightly in the range from lanthanum to lutetium corresponding to the change in mass and radius. In the periodic table adjacent lanthanides are particularly similar, whereas the two ends of the series lanthanum and lutetium are somewhat different. Yttrium can be classified between dysprosium and holmium according to its ionic radius. The very light scandium is a bit out of the range. Along the range from lanthanum to lutetium the distribution coefficients of fractionation processes also change. In magmatic systems where no special minerals are to be considered lanthanum is very incompatible (see Box 3.4), while the smaller lutetium is only slightly incompatible. The distribution coefficients of the others lie on a simple curve in between. Accordingly, light REEs (LREEs) are easily enriched, while heavy REEs (HREEs) are hardly enriched. The boundary between both groups is drawn between gadolinium and terbium, although the transition is rather continuous. In hydrothermal systems light REEs are more mobile because the stability of their REE fluorocomplexes and REE chlorocomplexes is higher (Migdisov et al. 2009).

Depending on the size of their lattice sites many REE minerals (Fig. 2.23, Table 2.16) prefer either light REEs (such as monazite, bastnäsite, allanite, loparite, and parisite) or heavy REEs (such as xenotime, gadolinite, samarskite, and fergusonite). The same applies to minerals that contain REEs only in traces, which of course is noticeable during fractionation. For example, if garnet, which preferably contains heavy REEs, is present in the rock during partial melting, then heavy REEs remain in the solid rock. If

Table 2.16 Selected important REE minerals. Currently monazite and bastnäsite are the most important ore minerals, as well as xenotime for heavy REEs

Monazite (Fig. 2.24)	$(La,Ce,Nd,Sm,Th)PO_4$	Up to 65% REE_2O_3
Xenotime	$(Y,Yb)PO_4$	61% REE_2O_3
Bastnäsite	$(Ce,La,Y,Nd)CO_3F$	Up to 75% REE_2O_3
Parisite	$Ca(Ce,Nd,La)(CO_3)_2F_2$	
Synchisite	$Ca(Y,Ce,Nd)(CO_3)_2F$	
Gadolinite	$(Ce,Y)_2FeBe_2O_2(SiO_4)_2$	
Fergusonite	$(Y,Ce)NbO_4$	
Allanite (epidote group)	$Ca(Ce,La,Y)(Al_2Fe)(Si_3O_{11})O(OH)$	
Euxenite	$(Y,Ca,Ce,U,Th)(Nb,Ta,Ti)_2O_6$	
Samarskite	$YFe^{3+}U^{4+}(Nb,Ta)O_4$	
Loparite	$(Ce,Na,Ca)(Ti,Nb)O_3$	33% REE_2O_3
Steenstrupine	$Na_{14}Ce_6Mn_2Fe_2(Zr,Th)(Si_6O_{18})_2(PO_4)_7·3H_2O$	
Eudialyte	$Na_{15}Ca_6Fe_3Zr_3Si_{26}O_{72}(OH)_4Cl$	Up to 10% REE_2O_3

monazite (Fig. 2.24) crystallizes from magma, then light REEs are removed more quickly from the melt. If zircon crystallizes, then heavy REEs are removed. Zirconium, titanite, apatite, rutile, xenotime, monazite, and allanite are important minerals during fractionation. They are all relatively common accessory minerals (i.e., minerals that usually form only tiny crystals and account for less than 1% of the rock, but whose crystallization significantly changes the REE budget of the melt).

During fractionation the original pattern of lanthanide abundance can only be bent upward or downward, so to speak. This does not apply to cerium and europium that may

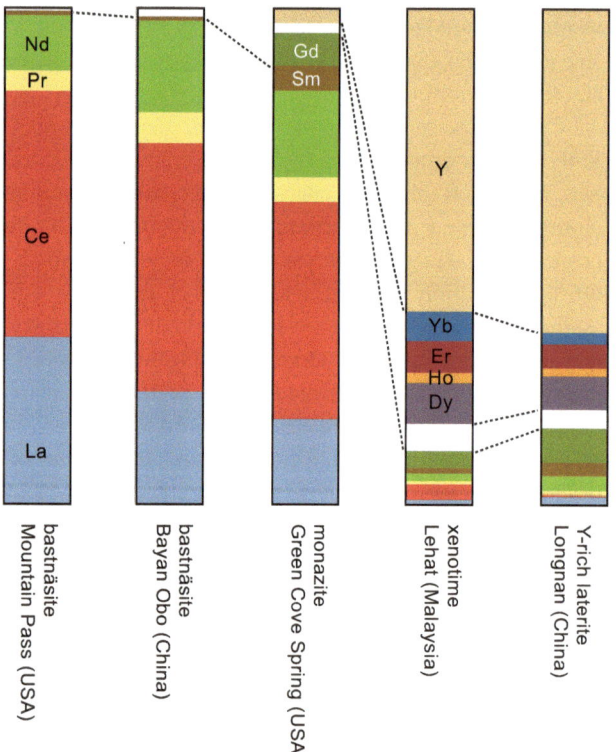

Fig. 2.23 Percentage shares of respective rare-earth elements (*white*, respective remainders of further REEs) in different ores (according to data from Long et al. 2010)

Fig. 2.24 Monazite from Roade Moss (Norway) (*photo* © F. Neukirchen/Mineralogical Collections of the TU Berlin)

occur in other valences. Since basaltic systems are somewhat reduced and Eu^{2+} can be well integrated into plagioclase, europium is depleted relative to other REEs in the melt (negative europium anomaly). Under oxidized conditions a cerium anomaly can form instead. For example, Ce^{4+} is absorbed much more by clay minerals than REE^{3+}.

For REE minerals the molar-dominant REE is usually given in parentheses such as monazite–(Ce), monazite–(La), or monazite–(Nd). The respective endmembers of the solid solution series are even considered independent minerals, whereas in reality other REEs are also present and possibly the concentration of the second most common REE is almost as high.

Many REE minerals also contain thorium and uranium as minor elements. This particularly applies to monazite, which often has a very high thorium content and which in India is even used as thorium ore for energy production by "breeding" uranium from it using nuclear technology. Radiation from REE minerals is problematic during the transport and processing of REEs and during the disposal of tailings. In Malaysia a plant processing monazite extracted from tin placers had to be closed in 1992 partly because of increased incidence of leukemia in the surrounding area. There are considerable environmental problems at REE mines in China due to lax legislation. Ironically, rare-earth elements are indispensable to green technologies such as wind turbines, electric cars, and energy-saving lamps.

Lanthanides are special because their electron configuration results from the fact that their 6s, 4f, and 5d orbitals (sorted by increasing energy) have very similar energy levels. In the elementary state they always have two electrons in the s orbitals of the sixth shell functioning as the conduction band. In contrast to elements with smaller atomic numbers where only the s, p, and d orbitals are filled shell by shell the f orbitals of the fourth shell now come into play. There are seven 4f orbitals that together can take up 14 electrons. These orbitals are filled starting with lanthanum (empty). The row is a bit confused because an electron is sometimes placed in the energetically slightly higher 5d orbital. One of the reasons for this is that an empty, half-filled, or completely filled 4f shell is more stable. Since 4f orbitals do not belong to the outer shells, the chemical properties of all these elements are virtually the same. REE^{3+} ions all have the very stable electron configuration $[Xe]4f^n$ with $n = 0–4$. Incomplete shielding of the 4f shell from the nuclear charge causes lanthanide contraction. In contrast to the conduction band the special and strictly localized arrangement of 4f orbitals and the amount of electrons they contain are responsible for the special optical and magnetic properties in which REEs differ far more than in their chemical behavior. Many REE ions show fluorescence (i.e., they glow when they are irradiated with light or UV). Although an electron is raised to a higher energy level by a photon, after some time it falls back and emits a photon with

exactly defined energy. Some applications of REEs are based on this effect.

Consumption of the respective REEs will change rapidly with further technical development. Around one fifth is currently used by the glass industry (doping, polishing). REEs are also used as catalysts, for magnets, for alloys (including batteries), and for other purposes (e.g., lamps, screens, ceramics) (Goonan 2011).

Historically, the first commercial application since 1884 used rare-earth oxides and thorium oxide to impregnate the incandescent mantle of gas lamps. Although these are rarely used nowadays, a second commercial application concerning the spark-releasing Auer metal or ferro rod used in lighters patented by Carl Auer von Welsbach in 1904 is still part of everyday life today. Auer metal is an alloy consisting of a mixture of REEs (about 70%) and iron (about 30%). The main component is cerium. In the past a natural REE mixture was used, whereas today the most expensive rare-earth elements are separated beforehand. An alloy consisting almost exclusively of an REE mixture is called mischmetal (after the German word for mixed). It is mainly used as a raw material for Auer metal, for steel production (as a reducing agent and alloy component that improves machinability and temperature resistance), and for special aluminum alloys. There are also superalloys such as the particularly temperature-resistant nickel–yttrium alloy used as an electrode in spark plugs.

Nickel metal hydride batteries (NiMH batteries) use a special alloy that absorbs hydrogen ions as a negative electrode. This is usually an alloy of REEs (lanthanum, cerium, or mischmetal) with nickel and cobalt. These batteries are often used in electric cars because they are way cheaper than the more powerful lithium–ion batteries. In contrast to the latter they can also be used as replacements for classic alkaline batteries.

Another important application was the creation of colored phosphors for computer screens and televisions since 1964. The first was Eu^{3+} emitting red light when excited. Color televisions could not be constructed before 1964 because previously only phosphors glowing in other colors were known; the red–green–blue combination was needed for additive color mixing. Eu^{3+} is utilized in the doping of other compounds such as yttrium oxide, yttrium oxide sulfide, or yttrium vanadate used in classical televisions or computer monitors that had a cathode ray tube. In addition, blue pixels can be created with Eu^{2+} and green pixels with Tb^{3+}. Combining the three phosphors (tri-phosphor) produces white light. This is exploited in energy-saving lamps and other gas discharge lamps because gas discharge in mercury vapor generates a lot of UV radiation that is converted into visible light by a corresponding coating on the inside of the glass. The backlighting of LCD flatscreens also works the same way as do the red, green, and blue pixels of plasma televisions. Green photoluminescent radar screens use gadolinium. Other REEs are used as phosphors making it possible to change the color of light-emitting diodes or the color temperature of lamps.

Yttrium aluminum garnet (YAG) is a synthetic gemstone with the composition $Y_3Al_5O_{12}$ and should be mentioned in this context. Doped with cerium it emits yellow light when excited by blue light, UV, X-ray, gamma radiation, or beta radiation. It is therefore used for color change in diodes, lamps, and various detectors (e.g., PET scanners). Neodymium-doped YAG (Nd:YAG) is even more important since it is the chief solid state laser. It is very powerful and used in industry for cutting and drilling and in medicine for surgery and many other purposes. Yttrium iron garnet (YIG) is also used in microwave technology.

Neodymium magnets have the composition $Nd_2Fe_{14}B$ in a tetragonal structure. Other REEs are usually added to improve certain properties (in particular, praseodymium, dysprosium, gadolinium, and terbium). These are by far the strongest permanent magnets known. They are used wherever strong magnets with a low weight or volume are required such as in electric motors (electric car motors, power steering, window drives, drills, computer hard disks, etc.), generators (especially, in wind turbines, which typically contain 1 t Nd per installed megawatt capacity, and in the power-generating brakes of hybrid cars), headphones and boxes, magnetic locks, magnetic bearings, and missile guidance systems. Alloyed REEs make magnets less sensitive to heat allowing them to be used directly without gears in wind turbines. The samarium cobalt magnet is also very strong and heat insensitive.

When cracking heavy crude oil zeolites are used as catalysts in which some of the aluminum has been replaced by lanthanum, small amounts of cerium, and traces of praseodymium and neodymium. Consumption is high because the zeolite has to be doped over and over again. Cerium(IV) oxide with small amounts of lanthanum and neodymium is used in automotive catalysts. In the future lanthanum perovskites could become important as catalysts; they are already used in the electrolysis of water and in fuel cells.

Doping with REEs or REE compounds can change the properties of glass so that it can absorb certain wavelengths (UV filters, infrared filters, car windows), amplify certain wavelengths (glass fiber, laser technology), change the refractive index (lenses), improve the strength (carrier disks of computer hard disks), and dye and decolorize. This involves using not only large quantities of lanthanum and cerium, but also smaller quantities of praseodymium, neodymium, yttrium, erbium, and various other REEs. They are not only used as dopants, but also in similar quantities for polishing glass and semiconductor wafers. In contrast to conventional polishing agents they act not only mechanically but also chemically on the surface dissolving the glass

and thus enabling perfect polishing. The main materials used are cerium oxide, lanthanum, and praseodymium.

REEs are used in ceramics to color glazes and to dope technical ceramics. Doping zirconium oxide with yttrium stabilizes the cubic crystal lattice (zirconia). Lanthanum-doped barium titanate ($BaTiO_3$) is a positive temperature coefficient (PTC) thermistor whose electrical resistance is temperature dependent. There are also ceramic capacitors containing REEs.

REEs such as $YBa_2Cu_3O_7$, $LaSrCuO_4$, and $LaBaCuO_4$ are major components in many important high-temperature superconductors. They are used to generate strong magnetic fields in particle accelerators, in (experimental) fusion reactors, and in medicine for magnetic resonance imaging (MRI). In addition, gadolinium compounds are injected into patients as contrast agents prior to examinations with MRI. There are many other applications such as in the composition of drugs and REE additives in animal feed intended to increase the milk production of cows and the egg production of hens. Large quantities of REEs are also used in armaments such as night vision devices, the coating of stealth bombers, missile guidance systems, radar systems, and lasers for guided weapons (Table 2.17).

Table 2.17 Overview of the most important applications of rare-earth elements sorted by atomic number (Z). Commercial applications of the heaviest REEs are limited by their rarity (see text for abbreviations)

Z	Element	Applications
21	Scandium (Sc)	Ultralight alloys, additive in mercury vapor lamps
39	Yttrium (Y)	Nd:YAG as laser crystal and base material for phosphors (television, monitors); YIG as microwave filter; Y-stabilized zirconia; high-temperature superconductors; white LEDs; energy-saving lamps
57	Lanthanum (La)	Catalysts, glass with high light refraction, polishing agent, NiMH batteries, doping of steel, Auer metal, PTC thermistors
58	Cerium (Ce)	Coloring of glass, polishing agents, catalysts, oxidizing agents, alloys, Auer metal, NiMH batteries
59	Praseodymium (Pr)	Magnets, NiMH batteries, alloys, ceramics, glass
60	Neodymium (Nd)	Magnets, Nd:YAG laser, alloys, ceramics, glass
61	Promethium (Pm)	No stable isotope
62	Samarium (Sm)	NiMH rechargeable batteries, Sm-Co magnet
63	Europium (Eu)	Red and blue phosphors (televisions, computer screens, lamps)
64	Gadolinium (Gd)	Magnets, phosphors (radar)
65	Terbium (Tb)	Green phosphors, magnets
66	Dysprosium (Dy)	Magnets

(continued)

Table 2.17 (continued)

Z	Element	Applications
67	Holmium (Ho)	Glass, doping for microwave lasers
68	Erbium (Er)	Doping for lasers, fiber optics
69	Thulium (Th)	Portable X-ray machines
79	Ytterbium (Yb)	Doping for lasers, fiber optics
71	Lutetium (Lu)	Doping

As far as the supply situation is concerned medium and heavy REEs are particularly critical such as europium, terbium, and dysprosium. Thulium and lutetium are so rare that they are hardly used commercially. Although heavy REEs are much rarer than light REEs anyway (except Y, which is classified as a heavy REE because of its chemical behavior), they are also enriched to a much lesser extent than light REEs during magmatic fractionation. In many deposits the total amount of heavy REEs account for less than 1% of REEs. Therefore, deposits with a comparatively high content of heavy REEs can be worthwhile even at lower ore grades and lower concentrations.

The most important deposits of REEs were formed directly or indirectly by alkaline magmatism such as carbonatites (Sect. 3.10) and agpaitic rocks (Sect. 3.11) and associated hydrothermal formations (veins, skarns, metasomatic replacements). The lateritic weathering products of carbonatites (Sect. 5.11) are also significant. The most important REE minerals of carbonatites are monazite, bastnäsite, parasite, and synchisite. Light REEs are very strongly enriched, whereas heavy REEs occur only in traces. With the exception of Lovozero (Russia) agpaitic rocks have not yet been mined for REEs, but they are expected to become important in the near future. Such rocks hold many different minerals containing REEs. Important are loparite (in Lovozero), steenstrupine (in rocks called hyperagpaitic lujavrites), and eudialyte (which contains a low percentage of REEs, but is a major mineral of most agpaitic rocks and can easily be dissolved in hydrochloric acid). Although light REEs dominate in agpaitic rocks, heavy REEs can also be enriched.

Lateritic REE deposits (Sect. 5.11.4) are another important type of deposit being easy to mine. Sometimes they have very high levels of heavy REEs. In exceptional cases heavy REEs even predominate. The metals are absorbed on the surface of clay minerals.

Finally, the minerals resistant to chemical wheathering, notably monazite and the less common xenotime, which is important for heavy REEs, can be enriched in placers (Sect. 5.9). Both minerals occur in many igneous (e.g., granite) and metamorphic (e.g., gneiss) rocks as accessory minerals. Monazite placers are occasionally exploited and both minerals are also obtained as by-products e.g. from tin placers.

Pegmatites (Sect. 3.8) sometimes have very high REE content. However, these deposits are usually so small that their economic significance is low. There are pegmatites in which heavy REEs predominate. REE minerals may include gadolinite, fergusonite, allanite, monazite, xenotime, euxenite, and samarskite.

Some skarns (Sect. 4.9) and IOCG deposits such as Olympic Dam (Sect. 4.7) should also be mentioned. Uraninite (pitchblende) contains some REEs that can sometimes be separated as a by-product in uranium production. Today scandium is mainly extracted from uranium ores. Zircon (containing heavy REEs) found in placers and apatite (containing medium REEs) from phosphorites (Sect. 5.8), phoscorites (Sect. 3.10.1), and Kiruna-type iron apatite ore (Sect. 3.6) also show potential since both minerals sometimes contain several percent REE_2O_3 and are extracted anyway. Although bauxite shows increased contents, at present they are not economically recoverable (see also Box 2.5).

Box 2.5 Rare-earth elements can be found even in the gravel pit next door!

Despite their name the REE metals are not as rare as some would believe. Cerium, lanthanum, and neodymium are much more abundant in the Earth's crust than lead and even the rarest stable REE thulium is 100 times more abundant than gold (Taylor 1964). Although everyone knows the form in which gold is found and many are familiar with the lead mineral galena, even geoscientists often only think of distant deposits of REEs in China and the United States.

Glaciers in the Ice Age transported so much gold that people search for gold in German gravel pits as a hobby (Kühne 1976, 1983; Lierl and Jans 1990). Rocks with significant contents of REEs have been found in perfectly normal gravel pits. On an excursion organized to celebrate the annual meeting of the Gesellschaft für Geschiebekunde in 1999 a rather unusual yet conspicuous rock was found in a gravel pit at Offlumer See, near Münster, Germany. It was white and had angular to elongated black–gray nests of a dark mineral. It was only under the microscope that the rock offered up its secrets. The dark nests turned out to be a mineral of the epidote group known to incorporate REEs. It was allanite (Fig. 2.25) $(CaREE(Al_2Fe^{2+}) (Si_3O_{11})O(OH))$. Analysis with a microprobe showed contents of cerium up to 18% and lanthanum around 11%. In addition, neodymium was present at around 4.5% and praseodymium at around 1.5% (Ries 2005). It was therefore allanite–(Ce). The white ground mass of the rock was mainly quartz and Ca–Mg silicates. Most likely it was a skarn (Sect. 4.9).

(a)   **(b)**

Fig. 2.25 This striking rock, an REE-rich skarn, was found in a gravel pit at Offlumer See, near Münster (Germany). **a** Hand specimen and **b** thin section under the microscope. The dark mineral is allanite–(Ce) (*photos* © G. Ries)

Of course, a single find is not a deposit since the rock did not form where it was found; it was transported there by glaciers during the Ice Age. Since there is clearly more where it came from what would a prospecting geologist do? He would simply have to retrace the glacier's path to find where it crossed the deposit and abducted the rock. In this way new deposits could be found in a quasi-backward way by looking at the legacies of Ice Age glaciers. Of course, such glaciers "sampled" a much larger area than a geologist could do on foot (they also had much more time). Indeed, new deposits have already been found this way. The same also applies to pebbles from rivers and so on. However, this method is less promising if such a rock was transported over such enormous distances (Meyer 1990).

2.6 Other Metals and Metalloids

2.6.1 Lithium (Li)

The alkali metal lithium is rarely used in its elementary form. It is mainly used in compounds produced directly from lithium minerals or the intermediate lithium carbonate. Lithium–ion batteries are the most important because they have very high energy density and no memory effect. They are therefore used, for example, in mobile phones, laptops, cameras, and electric vehicles where the positive electrode consists of $LiCoO_2$ or a similar compound and the negative electrode consists usually of graphite. In between there is an electrolyte solution with lithium salts and other ions. Lithium ions of the electrode dissolve in the electrolyte during charging, while positive ions accumulate on the graphite. The opposite happens during discharging. Since electric vehicles are likely to become more important in the future demand for lithium can be expected to rapidly increase.

Should energy generation by nuclear fusion ever work demand will increase again. This is because the heavier hydrogen isotopes deuterium (^2H) and tritium (^3H) are fused to helium (^4He)—not "normal" hydrogen (^1H). While deuterium can be obtained relatively easily from water (which has a certain amount of heavy water), this is not the case with tritium. Some tritium is formed in heavy-water reactors (a not very common type of nuclear power plant) by capturing a neutron. Another possibility is to bombard ^6Li (the second most common lithium isotope) with neutrons to produce helium and tritium.

Lithium carbonate is mainly obtained by precipitation from brines of salt lakes and salt playas (Sect. 5.7.2). Lithium minerals such as spodumene ($LiAlSi_2O_6$, known as a gemstone under the names kunzite and hiddenite), petalite ($LiAlSi_4O_{10}$), amblygonite (($Li,Na)Al(PO_4)(F,OH)$), lepidolite ($KLi_2AlSi_4O_{10}(F,OH)_2$), and zinnwaldite ($K(Li,Fe,Al)_3(F,OH)_2(AlSi_3O_{10})$) can be extracted from lithium pegmatites (Sect. 3.8). In most cases however they are only by-products, while tin and tantalum minerals are more important economically. Hydrothermal tin deposits (Sects. 4.5 and 4.6) can also contain lithium minerals such as zinnwaldite.

Lepidolite and zinnwaldite are the most important ores for rubidium, which they contain in traces. In addition, the cesium ore pollucite ($CsAlSi_2O_6 \cdot H_2O$) occurs in lithium pegmatites.

2.6.2 Boron (B)

The metalloid boron is rarely used in its elementary form. Perborates used as bleaching agents in detergents are produced in large quantities. Boron compounds are used in the manufacture of temperature-resistant and chemical-resistant glass. Boron is also used as a neutron absorber in nuclear power plants. Cubic boron nitride (BN) is the second hardest material after diamond and is often used as an abrasive.

The most important boron minerals are borax ($Na_2B_4O_5(OH)_4 \cdot 8H_2O$; Fig. 2.26), ulexite ($NaCaB_5O_6$ $(OH)_6 \cdot 5H_2O$), kernite ($Na_2B_4O_6(OH)_2 \cdot 3H_2O$), and colemanite ($CaB_3O_4(OH)_3 \cdot H_2O$). They are deposited in salt lakes (Sect. 5.7.2) (especially, in active volcanic areas). The skarn at Dalnegorsk, near Vladivostok (Russia) should be mentioned.

2.6.3 Beryllium (Be)

The alkaline earth metal beryllium is mainly used in special alloys employed in the aerospace industry that are light, hard, robust, and very heat resistant. Beryllium copper is the most important since it is also used in electrical switches, contacts, and other components (especially, when sparking has to be avoided). In some nuclear power plants beryllium is used as a moderator. Beryllium foil is used as a window offering high transmissivity to X-rays (Boland 2012a).

Beryllium ore minerals include beryl ($Al_2Be_3Si_6O_{10}$; Fig. 2.27; in gemstone quality known as aquamarine,

Fig. 2.27 Beryl (emerald) (*photo* © F. Neukirchen/Museum Reich der Kristalle Munich)

emerald, heliodor, and morganite) and bertrandite ($Be_4(OH)_2Si_2O_7$). The most important deposits are pegmatites (Sect. 3.8) in the vicinity of which hydrothermal and metasomatic deposits may also occur. In rare cases beryl was formed by hydrothermal processes within black shale as is the case with the famous emeralds from Colombia.

2.6.4 Germanium (Ge)

The metalloid germanium is a semiconductor with good properties, but its application is limited by the high price. It is used in high-frequency technology and as an X-ray detector. Germanium is mainly used in fiber optic cables in which germanium oxide is enriched in the core of the glass fiber providing a higher refractive index ensuring accurate

Fig. 2.26 Borax from Tibet (*photo* © F. Neukirchen/Mineralogical Collections of the TU Berlin)

guidance of light. Germanium oxide is also contained in catalysts for the production of polyethylene terephthalate (PET) plastic. Germanium and compounds of germanium with arsenic, antimony, and selenium (chalcogenide glasses) are processed into windows and lenses for infrared optics (night vision devices, thermal-imaging cameras).

Germanium is only present in very small amounts in the Earth's crust. Its geochemical behavior is similar to that of silicon such that there is hardly any significant enrichment. Sometimes it is contained in sphalerite, in smaller amounts in other non-ferrous metal ores, and in fly ash from coal combustion.

2.6.5 Indium (In)

Indium tin oxide (90% In_2O_3, 10% SnO_2) is a transparent semiconductor used in LCD screens (computer monitors), televisions, touchscreens, and thin-film solar cells. Indium-containing alloys are used in storage media such as CD-RW, in sprinkler thermostats, and in high-temperature thermometers. Indium phosphide is a semiconductor used for laser diodes.

Indium is a rare metal that almost never occurs significantly enriched. It is sometimes contained in sphalerite (ZnS) in low concentrations (up to 0.1%) and sometimes in even smaller concentrations in other ore minerals such as chalcopyrite ($CuFeS_2$), digenite (Cu_9S_5), stannite (Cu_2FeSnS_4), and cassiterite (SnO_2). It is mainly extracted from corresponding ore concentrates in specialized smelters and refiners.

2.6.6 Gallium (Ga)

Gallium has such a low melting point (30 °C) that it melts in the hand. Gallium alloys are used in thermometers instead of mercury. In some computers gallium-containing thermal pastes dissipate the heat from microchips. Gallium arsenide is a semiconductor that is clearly superior to silicon in many respects. Laser diodes, light-emitting diodes, and components for high-frequency technology are built with it. Solar cells made of gallium arsenide are used in space travel. Gallium nitride is used in blue, white, and green light-emitting diodes, while gallium phosphide is used in green and red light-emitting diodes. The semiconductor gallium antimonide is used in laser diodes and photodetectors.

At an average 15 ppm gallium is as abundant as lead in the Earth's crust. However, there are no fractionation processes that could lead to significant enrichment. Its ionic radius is similar to that of aluminum. Traces of gallium are incorporated in aluminum-containing minerals including many silicates. Traces are highest in bauxite (30–80 ppm). They can be separated using the Bayer process (Sect. 1.16)

in aluminum production, which accounts for almost all world production of gallium. Sphalerite also sometimes contains some gallium, which can be extracted as a by-product of zinc mining. Some coal and oil ashes have elevated gallium contents as is sometimes the case with spodumene and lepidolite from pegmatites. Recycling of gallium, zinc, and aluminum scrap accounts for a large part of production.

2.6.7 Selenium (Se) and Tellurium (Te)

The drums of photocopiers and laser printers are coated with selenium, As_2Se_3, or other semiconductors. This layer acts as an insulator in darkness and becomes electrically conductive when exposed to light. Toner only adheres to charged surfaces. Selenium semiconductors are also used in photodiodes and photo elements such as light meters and light barriers. It is used for coloring in glass and ceramics. Although selenium is an important trace element for humans and animals added to animal feed, in higher concentrations it is highly toxic. Selenium compounds are used in vulcanizers, catalysts, lubricants, selective oxidizers, and drugs.

The metalloid tellurium is mainly used as an additive in special steels. Storage media such as CD-RW contain alloys like silver indium antimony tellurium. Some semiconductors are used as photodiodes as is the case with CdTe, whereas bismuth telluride provides cooling in thermoelectric elements. Tellurium oxide glasses have very high light refraction. Some catalysts also contain tellurium.

Selenium and tellurium are almost exclusively obtained from the anode sludge of copper production and partly from the fly ash of non-ferrous metal smelting. Tellurium is about as rare as gold and very expensive, whereas selenium is not quite as rare. Both are contained in traces in many non-ferrous metal and gold deposits; selenium is also contained in some uranium deposits. They are mainly found as trace elements in sulfides. Selenides such as clausthalite (PbSe) and naumannite (Ag_2Se) as well as tellurides such as calaverite ($AuTe_2$), krennerite ($AuTe_2$), petzite (Ag_3AuTe_2), sylvanite ($AgAuTe_4$), and melonite ($NiTe_2$) can occur in tiny amounts. Native selenium and native tellurium are also known. Although contents of 50 g/t Se are regarded as high, there are ores with extreme contents of 1000 g/t Se like those in the VMS deposit in Boliden (Sweden).

2.6.8 Thallium (Tl)

Thallium is a highly toxic metal used in special electrical components such as infrared sensors and superconductors, as a dopant in glass, and in diagnostics in medicine. It is a by-product of some non-ferrous metal or gold deposits.

2.6.9 Mercury (Hg)

Mercury is the only metal liquid under standard conditions. It is very toxic (especially, mercury vapor) and has therefore been replaced in many applications. It has very high surface tension and rolls over the floor in the form of small beads since it is not readily absorbed by other substances.

Mercury thermometers and mercury switches are used less and less. Another important application of mercury is in energy-saving lamps where gas is discharged into mercury vapor. Mercury is used in chlorine production as a cathode for the electrolysis of NaCl. An **amalgam** is an alloy of mercury with gold or silver. If the mercury content is high, then the amalgam is liquid. Mercury dissolves gold very effectively. Mercury evaporates when heated, a property that can be used to facilitate gold extraction and gilding (fire gilding).

Large quantities of mercury were used for gold and silver extraction in the amalgam process, but today this is largely restricted to artisanal mining. Amalgams are still used in dentistry as a filling.

The most important mercury mineral is cinnabar (a.k.a. cinnabarite (HgS)). Native mercury occurs in the form of small droplets and some sulfosalts contain mercury. Deposits can occur in the coolest zones of hydrothermal systems (especially, as impregnations in sandstones; Box 4.10). Carlin-type deposits (Sect. 4.11) are also important.

2.6.10 Antimony (Sb)

The metalloid antimony is used as an alloying metal to harden lead (automobile batteries, ammunition, plain bearings, and formerly in lead pipes and lead letters in printing). Textiles impregnated with antimony oxide are flame retardant. Brake linings are made from antimony sulfide. Antimony is also used in ceramics, glass, as a dopant in semiconductors, and in fireworks. The most important ore mineral is stibnite (a.k.a. antimonite (Sb_2S_3); Fig. 2.28). The ancient Egyptians used antimony as make-up. Antimony is

also contained in many sulfosalts like tetrahedrite ($Cu_{12}Sb_4S_{13}$). Antimony ochre is a mixture of different antimony oxides. Native antimony (Sb) also occurs.

The minimum ore grade is 3%. Stibnite often occurs together with gold or mercury. Important types of deposits are hydrothermal veins (orogenic quartz veins; Sect. 4.2 and epithermal systems; Sect. 4.3), some SEDEX deposits (Sect. 4.17; especially, Woxi and Xikuanshan in China), and metasomatic replacements in limestones comprising a spectrum including skarns (Sect. 4.9), epithermal systems, Carlin-type gold (Sect. 4.11), and similar systems. In Zajača (Serbia) there are replacements with massive sulfide, stibnite ore and various Pb–Cu–Zn–As–Sb minerals.

Many sulfide deposits contain antimony and arsenic minerals. **Sulfosalt** is the term given to complex sulfide minerals containing $(AsS_3)^{3-}$, $(SbS_3)^{3-}$, or $(BiS_3)^{3-}$ such as tennantite, tetrahedrite, enargite, stannite, pyrargyrite, proustite, bournonite, boulangerite, and berthierite.

2.6.11 Arsenic (As)

The highly toxic metalloid arsenic is relatively frequently present in sulfide ores. As_2O_3 evaporates when the corresponding ores (mined for non-ferrous metals, silver or gold) are roasted. Arsenic is deposited as a white powder when the metallurgical smoke cools down. Significantly more arsenic is produced than is consumed and what is left is hazardous waste. The VMS deposit in Boliden, Sweden alone produced 566,000 t of arsenic, which is about five times more than all the metals mined there combined (copper, gold, silver).

Arsenic is important for the production of gallium arsenide (see gallium; Sect. 2.6.6). It is also used in pesticides (banned in Germany), wood preservatives, medicines, and as rat poison. It increases the hardness of lead and copper alloys (arsenic bronze was frequently used in the Bronze Age).

Arsenopyrite (FeAsS) is very common (Fig. 2.29). Further arsenic minerals include löllingite ($FeAs_2$), native arsenic (As), realgar (AsS), orpiment (As_2S_3), nickeline (NiAs), and rammelsbergite ($NiAs_2$), as well as sulfosalts such as enargite (Cu_3AsS_4) and tennantite ($Cu_{12}As_4S_{13}$). Arsenates such as scorodite ($FeAsO_4 \cdot 2H_2O$), erythrite ($Co_3(AsO_4)_2 \cdot 8H_2O$), and annabergite ($Ni_3(AsO_4)2 \cdot 8H_2O$) are formed in oxidation zones. They are colorful and used in exploration as pathfinder minerals.

2.6.12 Bismuth (Bi)

Although the metal bismuth does not have a stable isotope, its half-life is so long that radioactive decay can be neglected. Bismuth alloys are used as non-toxic substitutes for lead

Fig. 2.28 Stibnite (a.k.a. antimonite (Sb_2S_3)) from the lead–zinc–copper vein deposit Herja (Romania) (*photo* © F. Neukirchen)

Fig. 2.29 Arsenopyrite from Freiberg, Ore Mountains (Germany) (*photo* © F. Neukirchen/Mineralogical Collections of the TU Berlin)

in certain steel grades and as solder. Bismuth compounds are found in medicines and cosmetics (BiOCl as a pearlescent pigment).

Important bismuth minerals are native bismuth (Bi) and bismuthinite (Bi_2S_3). Bismuthite (($BiO)_2CO_3$) and bismite (Bi_2O_3) are formed in oxidation zones. Some sulfosalts also contain bismuth. It is a common minor component in pegmatites (Sect. 3.8), skarns (Sect. 4.9), polymetallic hydrothermal veins (Sect. 4.1), and in some porphyry tungsten-molybdenum (W–Mo) deposits and SEDEX. Although it is mainly extracted as a by-product of copper and lead mining, it is also extracted in the mining of zinc, tin, molybdenum, tungsten, cobalt, gold, and silver. There are also a few deposits where bismuth is one of the main products.

In the former East Germany the mining company SDAG Wismut mined deposits containing bismuth (*Wismut*) in an attempt to conceal uranium mining commissioned by the Soviet Union.

2.6.13 Uranium (U) and Thorium (Th)

Nuclear fission of the uranium isotope ^{235}U releases an enormous amount of energy used in nuclear power plants and nuclear weapons. One problem is the not inconsiderable risks associated with the release of radioactive substances into the biosphere. This is not only about the radiation of particles containing uranium since nuclear fission also produces

radioactive nuclides such as iodine-131, cesium-137, and cesium-134 that are taken up by organisms, incorporated into the body, and significantly increase the risk for cancer. Nuclear power plants have periodically experienced uncontrollable chain reactions as evidenced by Three Mile Island (in 1979), Chernobyl (in 1986), and Fukushima (in 2011). This is in addition to the long list of minor incidents during "normal" operation. Other major accidents have occurred in reprocessing plants (in particular, Kyschtym in 1957 and Tokaimura in 1999). The perhaps larger problem, however, is the nuclear waste produced. Such waste must be safely stored over geological time periods due to its long half-life, although it is questionable whether this is technically possible.

Fissionable ^{235}U represents only 0.7% of uranium, while 99.3% is non-fissionable ^{238}U. ^{235}U must first be enriched to 3–5% for energy production or 80% for nuclear weapons. The remaining "depleted" uranium is sometimes used in armor-piercing ammunition because of its high density, which is of course problematic because it still radiates after hostilities have ceased. Bombardment with fast neutron converts ^{238}U into fissionable plutonium.

Thorium also has the potential to be used for energy generation in nuclear power plants since ^{232}Th is converted to ^{235}U when bombarded with neutrons (thermal neutrons suffice). In the past the incandescent mantles of gas lamps used to turn a glass flame into a bright light source were impregnated with thorium oxide and rare-earth oxides.

Although uranium and thorium behave very similarly in magmatic systems and occur together, they are found in variable proportions. Granites have comparatively high uranium and thorium contents despite being found in accessory minerals such as zircon and monazite. In alkaline magmas the contents are often even higher. Frequent companions of thorium are the rare-earth elements.

Uranium and thorium behave differently in hydrothermal systems. Complexes with U^{6+} are relatively soluble in water, while U^{4+} is insoluble. Although thorium can be dissolved in magmatic hydrothermal systems as a fluorocomplex and transported together with uranium, it is normally insoluble.

The main uranium ore is uraninite (pitchblende (approx. UO_2); Fig. 2.30). It occurs as cubic crystals and more frequently as kidney-shaped colloform aggregates or crusts. Pitchblende is often more or less amorphous because the crystal lattice is destroyed by radiation. The older the pitchblende the higher the lead content formed by radioactive decay. Other primary uranium minerals are coffinite ($USiO_4$) and carnotite ($K(UO_2)_2(VO_4)_2 \cdot 3H_2O$). The latter is also an important vanadium ore.

Weathering produces a large number of secondary uranium minerals (so-called uranium micas) that are intensively colored. They were used in the nineteenth century to produce uranium colors. Since some secondary minerals consist of UO_2^{2+} together with suitable complexing

Fig. 2.30 Pitchblende from Niederschlema-Alberoda, near Schnee-berg, Ore Mountains (Germany) (*photo* © Geomartin, CC-BY-SA, Wikimedia)

agents they are comparatively water soluble. Common examples are torbernite ((Cu(UO$_2$)$_2$(PO$_4$)$_2$•10–12H$_2$O); Fig. 2.31), uranocircite (Ba(UO$_2$)$_2$(PO$_4$)$_2$•10–12H$_2$O), zeunerite (Cu(UO$_2$)$_2$(AsO$_4$)$_2$•10–12H$_2$O), and autunite (Ca(UO$_2$)$_2$(PO$_4$)10–12H$_2$O).

The most important thorium ores are monazite (REE,Th) (PO$_4$), thorite (Th,U)(SiO$_4$), and thorianite (Th,U)O$_2$.

The most important uranium deposits are unconformity-related deposits (Sect. 4.14), sandstone-hosted deposits (Sect. 4.14), and IOCG deposits (Sect. 4.7), followed by calcrete (Sect. 5.12), hydrothermal veins (Sect. 4.1), pegmatites

(Sect. 3.8), agpaitic rocks (Sect. 3.11), and conglomerates of the Witwatersrand (Box 5.18) and the Central African Copper Belt (Sect. 5.1.2). Thorium is enriched together with uranium in magmatic deposits such as in pegmatites and carbonatites (Sect. 3.10). It is also enriched in some alkaline rocks and sometimes in related magmatic–hydrothermal deposits. Monazite is found in placer deposits (Sect. 5.9).

2.6.14 Zirconium (Zr) and Hafnium (Hf)

Zirconium is the main component of the alloy Zircaloy (a.k.a. zirconium alloy) from which the shells of uranium fuel elements for nuclear power plants are manufactured. The alloy is corrosion resistant and barely intercepts thermal neutrons needed for the nuclear fission of uranium. Zirconium is also alloyed in some steel grades and burnt in some fireworks and signal lights.

Cubically stabilized zirconium oxide is better known as zirconia. Traces of yttrium oxide or calcium oxide ensure that the high temperature phase of ZrO$_2$ remains stable even at low temperatures. Zirconia is produced in large quantities as an inexpensive synthetic diamond substitute and as special "glass." Zirconium oxide is used in the production of refractory ceramics.

The production of hafnium is very costly, which restricts its use. It is used in control rods of nuclear power plants.

Zirconium and hafnium have such similar chemical properties that they always occur together. The most important ore minerals for both elements are zircon (ZrSiO$_4$) and baddeleyite (ZrO$_2$). Zircon is a typical accessory mineral in granites and occurs in pegmatites even as large crystals (Fig. 2.32). Baddeleyite can occur in igneous rocks that have a lower SiO$_2$ content such as carbonatite, kimberlite, and

Fig. 2.31 Torbernite from Lachaux (France) (*photo* © F. Neukirchen/Mineralogical Collections of the TU Berlin)

Fig. 2.32 Zircon from a syenite pegmatite in Seiland (Norway) (*photo* © F. Neukirchen/Mineralogical Collections of the TU Berlin)

syenite, and in LMIs and anorthosite. With a few exceptions (Kovdor; Sect. 3.10.1 and Phalaborwa; Box 3.15) the primary deposits have no economic significance. Both minerals are mainly extracted from placer deposits (Sect. 5.9) together with other heavy minerals. Agpaitic rocks (Sect. 3.11) containing minerals such as eudialyte have potential to extract Zr as a by-product.

Box 2.6 Semiconductors

Semiconductors are solid substances with certain electrical properties that differ from both electrical conductors and electrical insulators. Apart from the diffusion of ions, which can be neglected in a crystal lattice, electron movements are responsible for electrical conduction. The interactions of electrons with the atomic nucleus and with neighboring atoms are complicated. Quantum mechanics can only describe the position of electrons in the orbitals of a single atom using probabilities arrived at by wave equations. In a crystal lattice the outer orbitals of neighboring atoms overlap.

The band model allows a simplified approach. In contrast to a single atom in which (simply put) two electrons are in each orbital with a precisely defined energy level the interaction with neighboring atoms leads to the orbitals expanding into wider bands with permitted energy states. The outermost band occupied by electrons called the valence band is particularly wide due to its strong interaction with neighboring atoms. Although electrons in the valence band are also responsible for the chemical bond, in covalent bonds they cannot be clearly assigned to one atom. The next higher band is called the conduction band. The outermost electrons can be moved from one atom to the next only if the corresponding band is partially filled.

In metals the valence band and conduction band partially overlap; hence the conduction band is partially filled. In alkali metals where the valence band and conduction band are virtually identical the valence band is only partially filled. Insulators, on the other hand, have a wide forbidden zone between the valence band and the next permitted energy level representing a bandgap that cannot be overcome. Thus, all attainable energy levels are already occupied.

Semiconductors have a fully occupied valence band and a bandgap between the valence and conduction bands much like insulators. However, the bandgap is so narrow (between 0.1 and 4 eV) that it can be overcome by energy supply (heat, light). As soon as electrons are lifted into the conductive band these electrons are freely movable (n-conductor). The atom now has an electron hole that can be filled by electrons from the valence band of neighboring atoms leaving them in turn with an electron hole. Such movement of electron holes corresponds to the movement of a positive charge (p-conductor), even if it is triggered by the movement of negatively charged electrons in the opposite direction. Conductivity is temperature dependent and increases with increasing temperature (with metals it is the other way round).

A semiconductor can be additionally equipped with charge carriers by doping (i.e., targeted "contamination" of the substance with foreign atoms of different valency). Additional energy levels arise at the foreign atom that can help in overcoming the bandgap. A semiconductor with n-doping contains atoms of an element with an additional electron in the valence band (e.g., silicon with phosphorus doping) where the foreign atom serves as an electron donor. A semiconductor with p-doping contains atoms of an element that has one electron less in the valence band (e.g., silicon with boron doping) where the foreign atom serves as an electron acceptor. Undoped semiconductors are called intrinsic (i-semiconductors). Diodes (a sandwich of one p-layer and one n-layer is sufficient), transistors, and other parts can be built and assembled into complex integrated circuits within a semiconductor crystal by combining the n, p, and i regions. Solar cells (Fig. 2.33) can generate electricity from light, whereas light-emitting diodes and laser diodes emit light the other way round.

A number of elements are semiconductors (in particular, metalloid elements such as silicon and germanium with four valence electrons crystallized in a diamond structure). Compound semiconductors are often found with a very similar zincblende structure (Fig. 2.34) where on average four valence electrons are present. These compounds are made up of elements from the III and V main group of the periodic table (III–V semiconductors) such as GaAs, GaSb, GaP, GaN, InAs, and InSb; from the II and VI main group (II–VI semiconductors) such as ZnS, ZnSe, ZnTe, CdS, and CdTe; and IV–IV semiconductors such as SiC and SiGe. Semiconductors are also found in other structures (e.g., in the wurtzite structure, the hexagonal variant of ZnS) and in combinations such as GaSe, InSe (III–VI semiconductor), and Cu(In,Ga) $(S,Se)_2$ (I–III–VI semiconductor). Some organic compounds have semiconductor properties.

All such semiconductors differ in the size of their bandgap, in their temperature-dependent conductivity (undoped), and in many other properties making them suitable for various applications.

Fig. 2.33 Polycrystalline silicon that can be used in solar cells (*photo* © Achim Kübelbeck, CC-BY-SA, Wikimedia Commons)

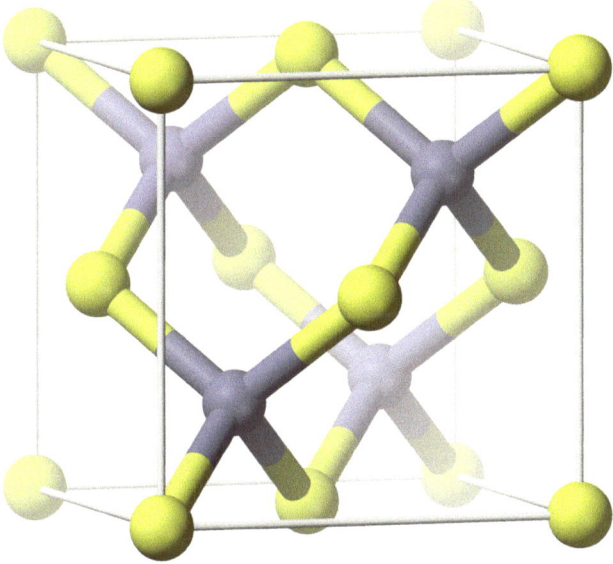

Fig. 2.34 Compound semiconductors are mostly present in the zincblende structure. The figure shows an elementary cell of the crystal lattice of sphalerite (zincblende)—also a semiconductor—in which zinc (*gray*) and sulfur (*yellow*) are arranged in a cubic surface-centered lattice. Although the diamond structure is identical, both lattice sites are occupied by atoms of the same element (*picture* Benjah-bmm27, public domain, Wikimedia Commons)

2.6.15 Silicon (Si)

Silicon is a very good semiconductor (Box 2.6) and is relatively cheap; hence almost all microchips and solar cells are made of silicon. An aluminum–silicon alloy with a eutectic composition is used for aluminum casting because it has low thermal contraction. Silicon is also added to cast iron and steel grades optimized for electrical purposes. The very hard silicon carbide (SiC) is used as an abrasive and in particularly solid ceramics. Silicones are polymers containing silicon, oxygen, and hydrogen.

Silicon is the second most common component of the Earth's crust after oxygen. The silicon ore quartz (SiO_2) occurs in large quantities. Quartz can be reduced using carbon to elemental silicon in an electric arc. Although such raw silicon is sufficient for alloys, for semiconductor technology it must be refined to high-purity silicon. The standard method is conversion to trichlorosilane by silicon powder reacting with HCl at high temperature. After distillation ultrapure silicon is deposited on heated silicon rods.

Literature

Boland, M.A. 2012a. Beryllium—Important for National Defense: U.S. Geological Survey Fact Sheet 2012–3056.

Boland, M.A. 2012b. Nickel—Makes Stainless Steel Strong: U.S. Geological Survey Fact Sheet 2012–3024.

Boland, M.A., and S.J. Kropschot. 2011. Cobalt—For Strength and Color: U.S. Geological Survey Fact Sheet 2011–3081.

Burt, R. 2010. Tantalum—a rare metal in abundance? *TIC Bulletin* 141: 2–7.

Claasen, D. 2007. Spekulationsgewinne: Nickelpreis bricht Rekorde. http://www.handelsblatt.com/finanzen/rohstoffe-devisen/rohstoffe/spekulationsgewinne-nickelpreis-bricht-rekorde/2793404.html. Accessed 3.27.13.

Dill, H.G. 2010. The "chessboard" classification scheme of mineral deposits: Mineralogy and geology from aluminum to zirconium. *Earth-Science Reviews* 100: 1–420.

Doebrich, J. 2009. Copper—A Metal for the Ages: U.S. Geological Survey Fact Sheet 2009–3031.

Elsner, H., F. Melcher, U. Schwarz-Schampera, and P. Buchholz. 2010. Elektronikmetalle—zukünftig steigender Bedarf bei unzureichender Versorgungslage? BGR Commodity Top News 33.

Frimmel, H.E. 2008. Earth's continental crustal gold endowment. *Earth and Planetary Science Letters* 267: 45–55.

Goonan, T.G. 2011. Rare Earth Elements—End Use and Recyclability. USGS Scientific Investigations Report 2011–5094.

Kropschot, S.J. 2010. Molybdenum—A Key Component of Metal Alloys: U.S. Geological Survey Fact Sheet 2009–3106.

Kropschot, S.J., and J.L. Doebrich. 2010. Chromium—Makes Stainless Steel Stainless: U.S. Geological Survey Fact Sheet 2010–3089.

Kropschot, S.J., and J.L. Doebrich. 2011. Lead—Soft and Easy to Cast: U.S. Geological Survey Fact Sheet 2011–3045.

Kropschot, S.J., and J.L. Doebrich. 2011. Zinc—The Key to Preventing Corrosion: U.S. Geological Survey Fact Sheet 2011–3016.

Kühne, W.G. 1976. Goldtransport durch Inlandeis; dem Andenken von Egon Erwin Kisch (1885–1948) gewidmet. *Der Aufschluss* 27: 165–169.

Kühne, W.G. 1983. Gold für uns aus der Kiesgrube. *Der Aufschluss* 34: 215–218.

Liedtke, M., and H. Elsner. 2009. Seltene Erden. BGR Commodity Top News 31.

Lierl, H.-J., and W. Jans. 1990. Geschiebegold aus Schleswig-Holstein. *Geschiebekunde Aktuell* 6: 47–57.

Long, K.R., B.S. Van Gosen, N.K. Foley, and D. Cordier. 2010. The Principal Rare Earth Elements Deposits of the United States—A Summary of Domestic Deposits and a Global Perspective. USGS Scientific Investigations Report 2010–5220.

Meyer, K.-D. 1990. Geschiebetransport im kanadischen und europäischen Inlandeis—ein Vergleich. *Eiszeitalter und Gegenwart* 40: 126–138.

Migdisov, A.A., A.E. Williams-Jones, and T. Wagner. 2009. An experimental study of the solubility and speciation of the rare earth elements (III) in fluoride- and chloride-bearing aqueous solutions at temperatures up to 300 °C 71: 3056–3096.

Naldrett, A.J., J. Kinnaird, A. Wilson, and G. Chunnett. 2008. Concentration of PGE in the earth's crust with special reference to the bushveld complex. *Earth Science Frontiers* 15: 264–297.

Ries, G. 2005. Ein Cer-Orthit-haltiger Quarzit als Geschiebe. *Geschiebekunde Aktuell* 21: 29–30.

Taylor, S.R. 1964. Abundance of chemical elements in the continental crust: a new table. *Geochimica et Cosmochimica Acta* 28: 1273–1285.

USGS. Minerals Yearbooks. http://minerals.usgs.gov/minerals/pubs/commodity/myb/.

Magmatic Deposits

Magma is fused rock (i.e., melt including dissolved gases and any crystals floating in it). Magma can cool and solidify at depth to form a large rock body such as a pluton, extrude at a volcano as lava, or rise finely fragmented as an ash cloud. Corresponding igneous (i.e., magmatic) rocks are called plutonic (or intrusive) and volcanic (or extrusive) rocks respectively (Figs. 3.1, 3.2 and 3.3).

Orthomagmatic deposits (a.k.a. magmatic igneous-related deposits) formed directly by magmatic processes from melt unlike the magmatic–hydrothermal deposits discussed in Chap. 4 (including porphyry copper deposits, greisen, and high-sulfidation epithermal veins whose ores were precipitated from a hydrothermal solution released by the magma). Magmatic processes discussed in this chapter are also relevant for hydrothermal systems because they affect the composition of hydrothermal solutions.

During partial melting of rock certain elements preferentially enter the melt while others remain behind. As soon as crystals form during cooling the composition of the remaining melt changes (Sect. 3.1.1) and continues to do so as crystallization progresses. Separating crystals from the melt (fractional crystallization) results in rock with a completely different composition. Sometimes liquid immiscibility occurs (Sect. 3.1.2) in which magma is separated into two differently composed melts. Such processes lead to effective fractionation that can enrich certain elements to such an extent that an ore deposit is formed (Fig. 3.4). Many hydrothermal deposits are also the result of initial enrichment by magmatic processes.

The terms acidic and basic refer to the SiO_2 content of magmatic rocks. These somewhat unfortunate terms have nothing to do with the pH value since they allude to the old-fashioned term silicic acid. Magmas with >66% SiO_2 are described as acidic, with 52–66% as intermediate, with 45–52% as basic, and with <45% as ultrabasic.

Magmatic deposits can roughly be divided into three groups. The first group includes deposits associated with primitive magmas that originate directly from the Earth's mantle such as basalt. Such basic and ultrabasic magmas can form deposits of chromium, nickel, platinum, iron, titanium, and vanadium. This group includes the huge layered mafic intrusions (LMIs; Sect. 3.3) such as Bushveld (South Africa), Great Dyke (Zimbabwe), and Sudbury (Canada) that are among the most important deposits of all. It also includes komatiites (Sect. 3.4), anorthosites (Sect. 3.5), and Kiruna-type deposits (Sect. 3.6).

The second group is related to granites (i.e., acidic melts). They can be formed by strong fractionation from a primitive melt originally derived from the mantle or by melt formation in the crust. Of economic interest are the late magmatic residual melts that may contain high contents of rare elements such as lithium, beryllium, rare-earth elements (REEs), niobium, tantalum, uranium, and thorium. The main focus of such interest is on rocks similar to granite called pegmatites that have particularly large crystals and sometimes exotic minerals. Granites have a high water content that is released during magma emplacement and crystallization. Granites are therefore responsible for many hydrothermal deposits (discussed in Chap. 4).

The third group comprises alkali-rich magmas (i.e., alkaline magmas and alkaline rocks) that occur at continental rifts and hotspots (Sect. 3.9). Such magmas can form under special conditions in the Earth's mantle. Fractionation allows them to develop into very special and diverse alkaline rocks some of which have very high contents of rare elements such as REEs, niobium, zirconium, and uranium. So-called agpaitic rocks are particularly highly enriched (Sect. 3.11). In addition to silicate magmas (such as nephelinite, phonolite, and nepheline syenite), carbonatites (i.e. carbonate magmas; Sect. 3.10) also play a role. They are responsible for the most important REE and niobium deposits. Phosphate, copper, iron, zirconium, and other substances are also extracted from carbonatites.

Before taking a closer look at such deposits it is important to get a feel for the necessary basics outlined in the

© Springer Nature Switzerland AG 2020
F. Neukirchen and G. Ries, *The World of Mineral Deposits*,
https://doi.org/10.1007/978-3-030-34346-0_3

Fig. 3.1 Magmatic dike at the Caldera de Taburiente (La Palma, Spain). Such dikes used to be fissures through which magma could rise (*photo* © F. Neukirchen)

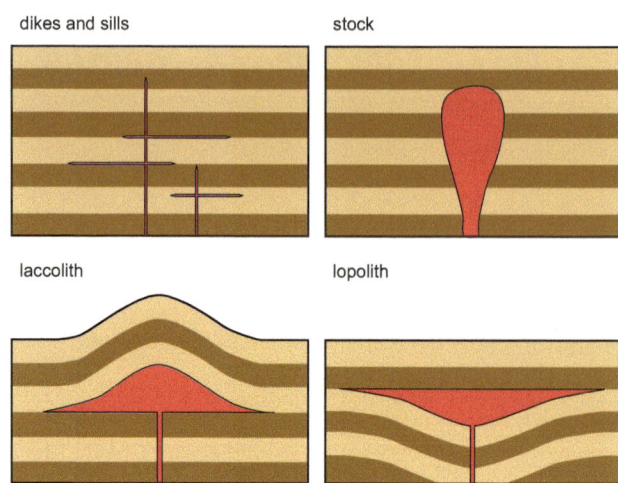

Fig. 3.3 Plutons can have different shapes depending on the way in which magma was emplaced. Stock, laccolith, and lopolith are idealized types. Shapes are generally more irregular and asymmetrical. Vertical dikes and horizontal sills are formed when smaller quantities of magma connected with plutons or volcanoes intrude

Fig. 3.2 Pillow lavas form during submarine volcanic eruptions. Here they are part of an ophiolite complex (Box 3.7) in the Lesser Caucasus (*photo* © F. Neukirchen/Blickwinkel)

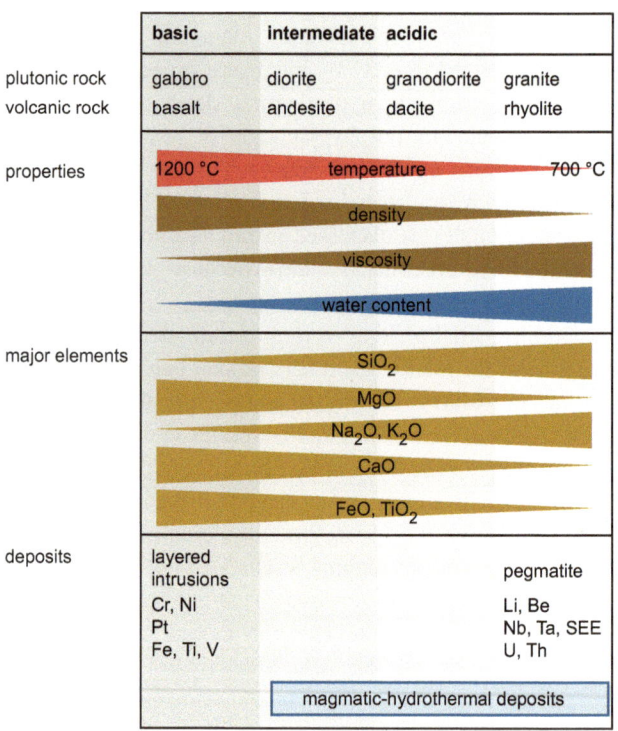

Fig. 3.4 When basic magmas (basalt) fractionate to form acidic magmas the composition of the melt changes and its physical properties change. Layered mafic intrusions and pegmatites are two important magmatic deposit types that can be found at the respective ends of this spectrum. Alkaline magmas are ignored in such a 1D scheme

following two sections to understand the respective fractionation processes (see Boxes 3.1 and 3.2).

Box 3.1 Silica saturation

SiO_2 saturation (silica saturation) is fundamental to understand igneous rocks and concerns the content of SiO_2 in relation to other elements such as Mg, Ca, Na, and K. Such elements are incorporated in silicate minerals such as olivine, pyroxene, mica, and feldspar (plagioclase and potassium feldspar). If magma contains more SiO_2 than is necessary for the formation of these minerals, then quartz also crystallizes and the rock is oversaturated with respect to silica (*upper triangle* in Fig. 3.5). If magma contains less SiO_2 or a lot of Na, K, and so on, then no quartz is formed and the rock is undersaturated (*lower triangle* in Fig. 3.5). SiO_2 deficiency is partly compensated by the formation of

minerals that are particularly silica poor like the feldspathoids (a.k.a. foids) such as nepheline, leucite, and sodalite. These minerals have a similar composition to feldspar but with less SiO_2. Feldspathoids therefore

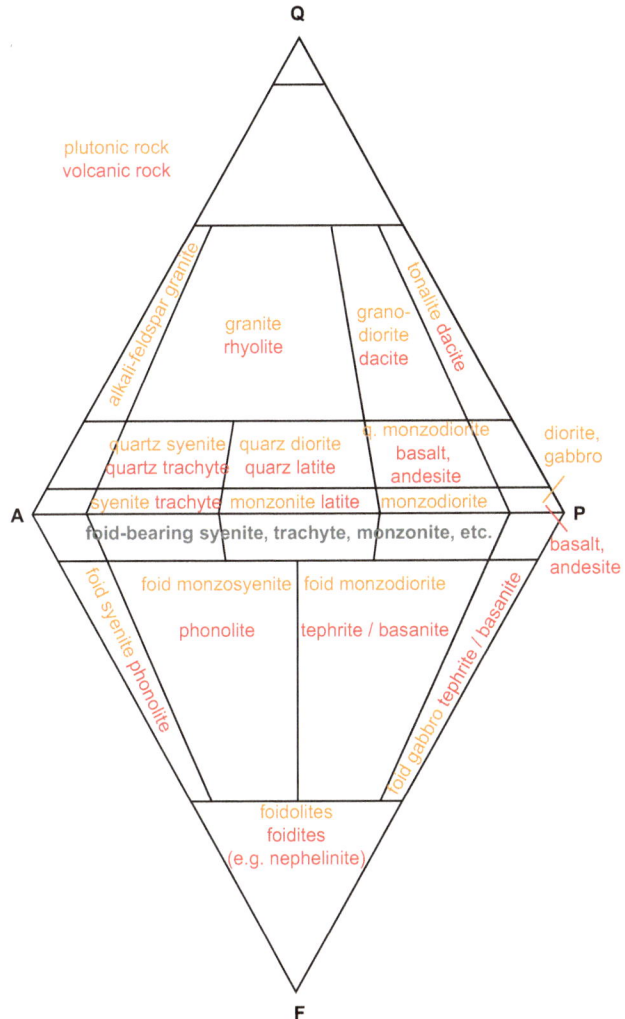

Fig. 3.5 QAPF diagram (a.k.a. Streckeisen diagram) used to classify igneous rocks. Classification is based on the proportions of the so-called felsic minerals quartz (Q), alkali feldspar (A), plagioclase (P), and feldspathoids (F) such as nepheline and leucite. The fields of volcanic rocks are wider (corresponding names occur several times in the figure)

(sedimented as tuff; Sect. 7.3). Plutonic (or intrusive) rocks are igneous rocks formed at depth that cooled and crystallized slowly; therefore they are coarse-grained. Such texture is not only due to slow crystal growth from the melt but also to subsequent recrystallization in the solid state. Volcanic rocks often contain isolated larger crystals called phenocrysts enclosed in a fine-grained or even glassy groundmass. For each volcanic rock there is a corresponding plutonic rock that has an identical composition but a different name. Parts of plutons, in particular, can be strongly enriched in certain crystals as a result of the crystals sinking or floating. Although corresponding rocks mainly made of accumulated crystals are called cumulates, their chemical composition does not match that of any melt.

A more detailed classification can be based on mineral content. This can be done by determining the proportions of the light-colored so-called felsic minerals quartz (Q), alkali feldspar (A), plagioclase (P), and feldspathoids (F) such as nepheline, sodalite, and leucite and determining their name from the so-called QAPF diagram (Fig. 3.5). So-called mafic minerals (mostly dark-colored minerals such as olivine, pyroxene, amphibole, and mica) are ignored. Only so-called ultramafic rocks where mafic minerals account for more than 90% of the rock use a different diagram (Fig. 3.6). The plagioclase corner of the QAPF diagram is further divided into diorite (extrusive: andesite), gabbro (extrusive: basalt), norite, troctolite, and anorthosite depending on the composition of plagioclase

never occur in rocks containing quartz. Olivine only occurs in silica-undersaturated rocks and is replaced in SiO_2-rich magmas by orthopyroxene (enstatite) whose composition lies between olivine and quartz. Feldspar-bearing rocks that contain neither feldspathoids nor quartz are silica saturated.

Box 3.2 Classification of igneous rocks

Igneous rocks can be coarsely classified according to whether magma cooled at depth in a so-called pluton (or intrusion), extruded at a volcano as lava, or with an explosive eruption as fine fragments called ash

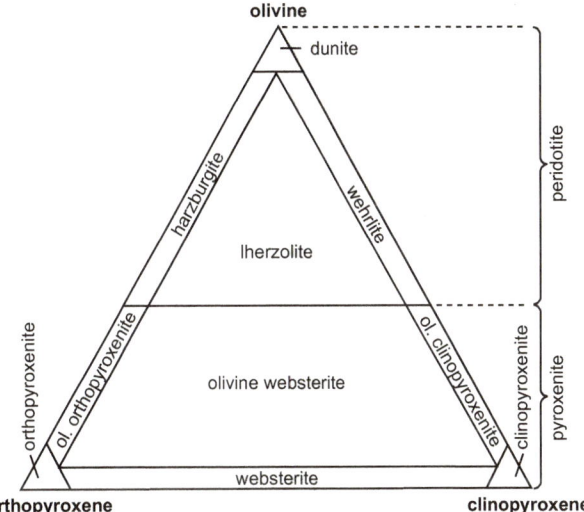

Fig. 3.6 Classification of ultramafic rocks (proportion of felsic minerals <10%). Peridotites and pyroxenites can both form as cumulates from basic magmas. The Earth's mantle consists of peridotite (lherzolite in particular)

(anorthite–albite) and the ratio of plagioclase, olivine, orthopyroxene, and clinopyroxene.

When it comes to igneous rocks silica saturation (Box 3.1) plays an important role. Rocks containing quartz (upper half of the QAPF diagram) are silica oversaturated, while rocks containing feldspathoids (so-called alkaline rocks in the lower half of the diagram) are silica undersaturated.

Basaltic magma is a melt formed in the Earth's mantle that solidifies at depth to form a gabbro or extrudes at a volcano as basalt lava. Gabbro consists mainly of plagioclase with some olivine and pyroxene (*right* corner of the diagram). Slight fractionation results in the composition of plagioclase changing and the corresponding rock diorite (extrusive: andesite) plots in the same field in the diagram. Further fractionation leads in the direction of granite. Granite contains approximately equal parts of quartz, alkali feldspar, and plagioclase. Rocks in surrounding fields are often also referred to as granite (in a broader sense). Also common is a plutonic rock called syenite consisting mainly of alkali feldspar.

It is often not possible to use the QAPF diagram for volcanic rocks as a result of the groundmass being fine-grained or glassy. Instead, the classification can be made on the basis of chemical composition in the total alkali silica (TAS) diagram (Fig. 3.7). SiO_2 content is plotted against the content of alkalis ($Na_2O + K_2O$) and can be used to divide igneous rocks into acidic (SiO_2-rich), intermediate, and basic

(SiO_2-poor) igneous rocks. Of course, boundary lines in the TAS diagram only approximate boundary lines in the QAPF diagram. The boundary between silica-oversaturated and silica-undersaturated rocks can only be drawn approximately in the TAS diagram.

Granites (and similar plutonic rocks) are often additionally classified according to their origin. I-type granites (igneous) can be produced either by fractional crystallization from mantle-derived basaltic magma or by partial melting of a gabbro that once crystallized from basaltic magma. They are typical of magmatism in subduction zones. S-type granites derive from sediments that have melted and form in the thick crust of mountain ranges. A-type granites (anorogenic) are formed in extensional settings where melts derived from a previously enriched mantle and melts formed in continental crust play a role.

Rocks can further be classified as peraluminous, metaluminous, and peralkaline (see Box 3.13).

3.1 Diversification of Magmas (Introduction)

3.1.1 Generation of Magmas and Fractional Crystallization

Instead of melting completely at a certain temperature (melting point) rocks melt continuously over a temperature interval that lies between solidus and liquidus and depends on respective composition and pressure. Solidus is the temperature at which the first melt is generated when a rock is heated (or when the last residual melt solidifies when magma cools). Liquidus is the temperature at which the last crystals disappear during heating and the rock is completely fused (or the first crystals form when magma cools down).

Temperatures prevailing in the Earth's mantle are normally below the solidus of peridotite (Fig. 3.8); hence the mantle is not molten. A melt is generated when the temperature of the mantle is unusually high or when the melting temperature is reduced by the presence of water. The former is the case under mid-ocean ridges and so-called hotspots where hot mantle material rises (decompression melting). The latter happens at subduction zones where magmas have an increased water content right from the start. However, the mantle region affected is only partially melted with some melt (maximum 10–25%) occurring between mineral grains. Melt has the composition of basalt corresponding to the eutectic composition (Box 3.3) of peridotite. It contains significantly more CaO, Al_2O_3, and SiO_2 than peridotite and less MgO. Peridotite is depleted of basalt mainly at the expense of diopside and the respective aluminum phase

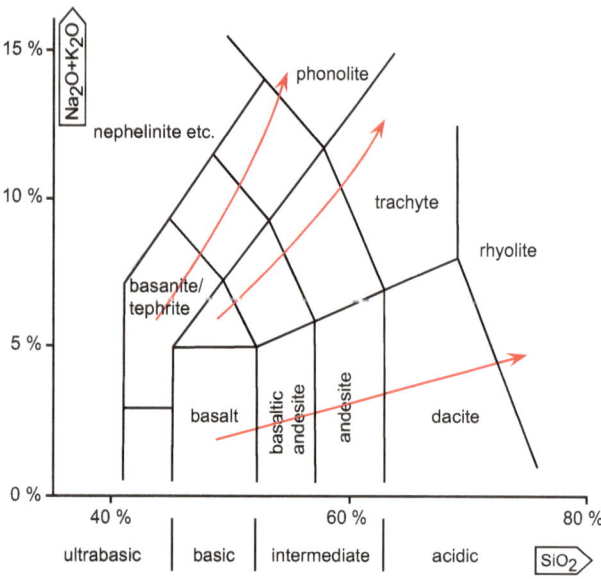

Fig. 3.7 Total alkali silica (TAS) diagram used to classify volcanic rocks (here only the most important rock names are given). Typical fractionation trends are shown schematically

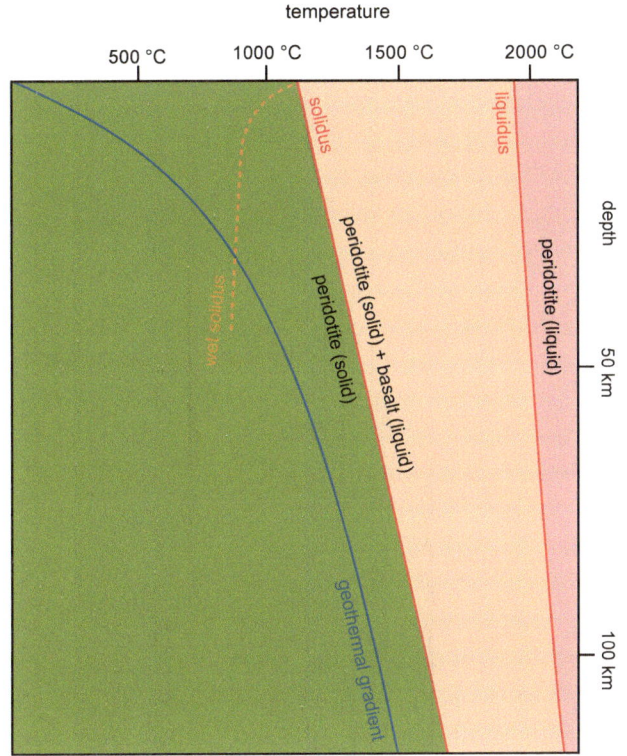

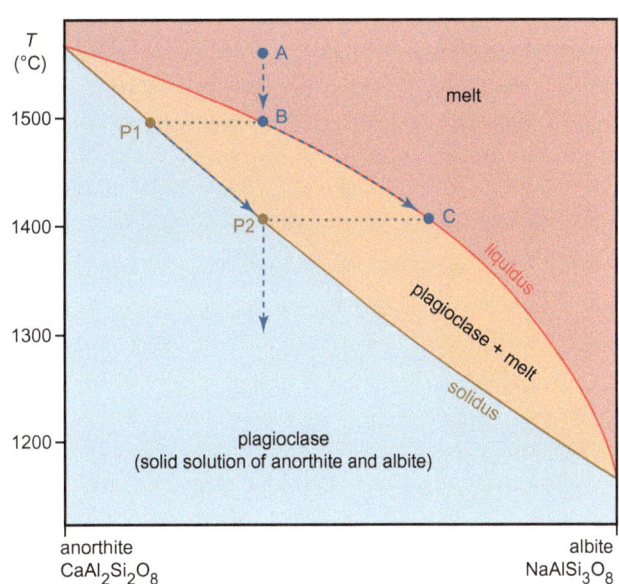

Fig. 3.9 Melting diagram of plagioclase (*x*-axis is the composition of plagioclase) at 0.1 MPa (air pressure at the Earth's surface). Between the stability fields of melt and solid plagioclase lies a field in which plagioclase and melt coexist. *Dots* and *arrows* represent crystallization of a melt of a certain composition (see text)

Fig. 3.8 In the Earth's interior the temperature increases (along the geothermal gradient) with depth (i.e., increasing pressure), but remains lower than the solidus at which partial melting would commence in the mantle (generating basaltic magma). A simple temperature increase is not possible in this case. Instead, hot material can rise from the depth without significant cooling (starting on the *blue curve* pointing vertically upward) and reach the solidus. The addition of water has a special effect in that it strongly depresses the solidus (*orange*) and can trigger melt formation

(plagioclase, spinel, or garnet). Such fractionation is not restricted to the main elements. A distribution coefficient can be specified for all elements according to which they either preferentially enter the melt or remain in the peridotite. Elements that tend to remain in the rock are called compatible such as chromium and nickel. Elements enriched in the melt are called incompatible (Box 3.4).

Box 3.3 Phase diagrams

Although rock does not have a specific melting point, it does melt over a long temperature interval during which the compositions of melt and rock change continuously. The temperatures required for this are well below the respective melting point of the minerals present in the rock. What exactly happens during melting and vice versa during crystallization can best be visualized using phase diagrams that are not easy to understand at first glance.

An important effect occurs with minerals whose composition is variable between two theoretical

endmembers in a so-called solid solution. Plagioclase, for example, is a continuous series of mixtures between albite (Na feldspar) and anorthite (Ca feldspar). In the melting diagram of plagioclase (Fig. 3.9) there are three fields. Between the field for solid plagioclase and the field for liquid melt there is a two-phase field in which plagioclase and melt coexist. The lines that bound this field are called the solidus and liquidus. If we have a melt with a temperature and composition of point A and let it cool down, then crystallization starts as soon as we reach the liquidus at point B. The first crystals formed at this temperature have the composition P1 corresponding to the solidus of our temperature. When the two-phase field is subjected to cooling the proportion of crystals increases constantly. At the same time the composition of crystals and melt changes with both following the respective line downward in the direction of albite. Finally, the solidus and thus the plagioclase correspond to our initial composition (P2) and the last melt droplets (with composition C) disappear. The same happens in natural magmas, early and hot crystallized plagioclase is anorthitic, whereas later plagioclase is albitic. It can also be seen that the melt loses calcium during crystallization and becomes richer in sodium. This is more extreme if the crystals formed are continuously separated from the melt and can no longer react with it. In this case the change in melt composition goes far beyond point C. Since liquidus and solidus meet at the endmembers both pure albite and pure anorthite actually have melting points.

The melting diagram of olivine is very similar. It is a series of minerals from Mg olivine to Fe olivine in which olivine is getting ever-richer in iron during fractionation. The melting of plagioclase or olivine naturally follows the same *arrows* but in the opposite direction (bottom-up).

The anorthite (Ca plagioclase)–diopside (clinopyroxene) system is a good example of two minerals between which there is no solid solution (Fig. 3.10). At low temperature we have solid rock with two phases (anorthite and diopside). Although this does not occur in nature, it can be used to get a rough approximation for basalt. The composition of the crystals corresponds to the edge of the diagram and the total composition of the rock lies somewhere in between. Again there is a liquidus and a solidus. The temperature at which the liquidus hits the side edges of the diagram corresponds to the melting point of the mineral (strictly speaking the solidus too jumps to the melting point at the edges of the diagram). The so-called eutectic point (a.k.a. the eutectic) is conspicuous in being the lowest point of the liquidus and at the same time the intersection of liquidus and solidus.

If a melt of temperature and composition A is allowed to cool down, then crystallization restarts as soon as the liquidus is reached (point B). Since the

two-phase anorthite + melt field has been entered, anorthite (whose composition is shown at the *left* edge) now crystallizes. Which mineral crystallizes first does not depend on the melting point of the mineral but on the composition of the melt! As the melt continues to cool so ever-more anorthite crystallizes, while the composition of the melt changes along the liquidus line. Finally, the eutectic at the eutectic temperature is reached and the remaining melt is of the eutectic composition. The more distant total composition (A) is from the eutectic composition the less melt is present at that moment. The remaining melt crystallizes completely to anorthite and diopside at the eutectic temperature such that rock is obtained according to the original total composition.

In this case fractional crystallization inevitably ends at the eutectic point. Conversely, when heating and melting, as soon as we reach the solidus a melt of the eutectic composition is produced first. Only at higher temperatures does it approach the initial composition of the rock. A special case is when a system already has the eutectic composition and the entire rock is melted at the eutectic temperature.

A rock consisting of more than two minerals also has a eutectic. Figure 3.11 schematically shows a system comprising three minerals that is close to that of natural rocks. Three eutectic systems are combined to form a triangle. The liquidus is now a curved surface with three peaks, valleys in between, and a ternary eutectic at the lowest point. When a melt cools down and reaches the *blue* point, A is the first mineral to crystallize and the melt develops downhill directly away from mineral A. On reaching the valley bottom at the so-called cotectic line additional crystallization of mineral B starts. The quantity ratio of simultaneously crystallizing minerals depends on the initial melt composition. The melt now develops downstream to the ternary eutectic where all three minerals crystallize simultaneously. Conversely, when a rock consisting of three minerals is melted a melt with the eutectic composition is always created first. If more minerals are added, then the system can only be represented as a projection. However, the principle remains the same.

There are three ternary eutectic systems that are of particular importance when it comes to natural magmas. Melt generation in the Earth's mantle brings about temperatures that are only slightly above the solidus of peridotite. Therefore, melts originating from the mantle almost always have the composition of basalt, which corresponds to the eutectic composition of peridotite. There are small differences in the details. The temperature and composition of the eutectic

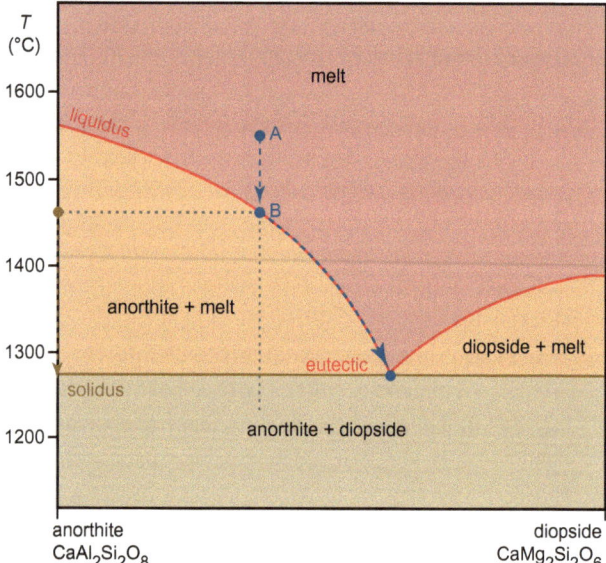

Fig. 3.10 Melting diagram of the anorthite–diopside system at 0.1 MPa. Side edges correspond to the composition of the crystals

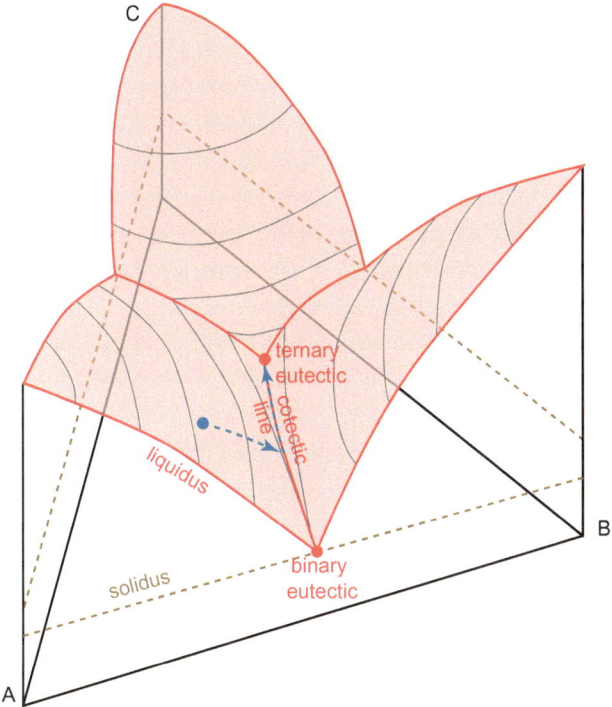

Fig. 3.11 Schematic of a ternary eutectic system with three phases A, B, and C

depend on the pressure, the CO_2 content, various trace elements, and other factors. Moreover, the degree of partial melting (i.e. the ratio liquid / (liquid + solid)) also has an effect on melt composition, hence the different types of basalt such as alkali olivine basalt and quartz-bearing tholeiitic basalt. Although they have a very similar chemical composition, they are just silica-undersaturated and silica-oversaturated, respectively (Box 3.1). In extreme cases alkaline magmas such as nephelinites can form.

A second eutectic concerns all rocks consisting mainly of quartz, alkali feldspar, and plagioclase—thus most of the rocks of the Earth's crust. In this case magma with the composition of granite is formed during partial melting. Again the exact temperature and composition depend on other factors such as water content and even low levels of fluorine, boron, and bromine. The granite eutectic is also the endpoint for fractional crystallization of silica-oversaturated basalt.

However, fractionation of silica-undersaturated, alkali-rich magmas leads to a different eutectic that typically corresponds to nepheline syenite or phonolite. The two systems are separated by a thermal barrier from which downhill fractionation leads in one or the other direction depending on the initial composition. Since basalts by and large lie on this thermal

barrier, basaltic magmas of similar composition can develop as a result of fractional crystallization following very different trends.

A thermal barrier can also occur in a binary system if there is a third mineral whose composition lies exactly between those of the two minerals. One such case is orthopyroxene (enstatite) that has a composition between olivine (forsterite) and quartz. Accordingly, a reaction is possible:

$$Mg_2SiO_4 + SiO_2 = 2\ MgSiO_3$$

1 forsterite(olivine) + 1 quartz = 2 enstatite(orthopyroxene)

At high pressure the forsterite–quartz phase diagram consists of two adjacent normal eutectic systems with enstatite as the thermal barrier in between. At low pressure the system is somewhat more complicated (Fig. 3.12). Keeping to the *left* half of the diagram for now only one eutectic (E) can be seen. However, there is another point that remains to be explained called the peritectic point (P). A melt that develops along the liquidus line by crystallization of forsterite inevitably hits this point. In accordance with the above reaction, melt reacts with forsterite at this point and converts it into enstatite at least partially. This reaction occurs during the crystallization of tholeiitic basalts.

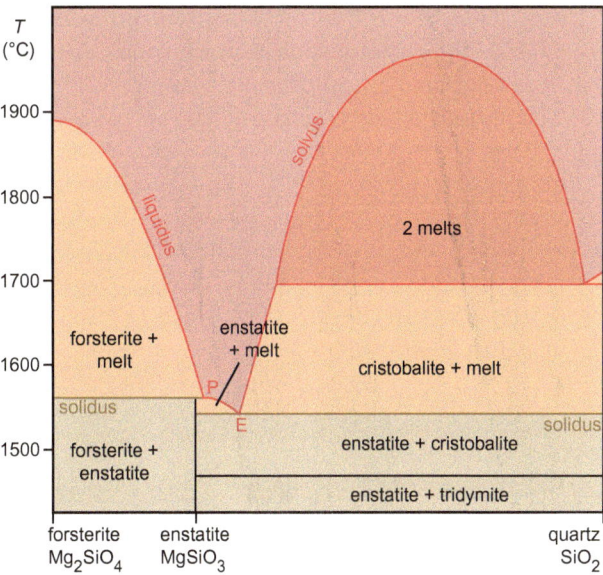

Fig. 3.12 Phase diagram of the forsterite (Mg olivine)–quartz system at low pressure (0.1 MPa). Enstatite (orthopyroxene) has a composition intermediate between those of the two minerals. In addition to a eutectic (E) there is a peritectic point (P) in this system where forsterite reacts with the melt to form enstatite. On the *right* side of the diagram there is a field with two immiscible melts. Tridymite and cristobalite are high-temperature phases of SiO_2

In the *right* half of the diagram the solvus stands out. Below the solvus lies a field with two immiscible melts. This case is explained in Sect. 3.1.1. Since quartz transforms into β-quartz, tridymite, and finally cristobalite (all SiO_2) at high temperature there are further phase boundaries in the diagram.

Box 3.4 Compatible and incompatible trace elements

Those elements that do not really fit into the crystal lattice of olivine, pyroxene, and so on (here the peridotite–basalt system is mainly referred to) are enriched in the melt both during melting and crystallization. Called incompatible these elements are mainly found in the Earth's crust. Compatible elements fit well into the crystal lattice of at least one of the minerals involved. During melting they remain largely in solid rock and during crystallization they are quickly removed from the melt. Compatible elements are mainly found in the mantle of the Earth and in primitive (i.e., unfractionated) magmas.

Each trace element can be given a partition coefficient defined as the ratio between the concentration in a mineral and that in the melt. If this coefficient is less than 1, then the element is incompatible and is enriched in the melt; if it is greater than 1, then it is compatible and is preferably incorporated into the mineral. No fractionation takes place with a distribution coefficient of 1, which is approximately the case with platinum-group elements (PGEs).

There are two properties that can make an element incompatible (Fig. 3.13). First, ions with a large radius namely large ion lithophile elements (LILEs) cannot be incorporated into the crystal lattice of olivine, pyroxene, and so on since these minerals are relatively densely packed—otherwise they would not be stable under the high pressure of the Earth's mantle. This group includes potassium, barium, and lithium. Some of these elements behave somewhat more compatibly in granite magma.

High-field-strength elements (HFSEs) make up the second group of incompatible elements; they have a small radius but a high charge. According to their charge such cations should be surrounded by a corresponding number of anions in the crystal lattice, but this is not possible in relatively densely packed minerals. Such elements include zirconium, titanium, niobium, uranium, and REEs.

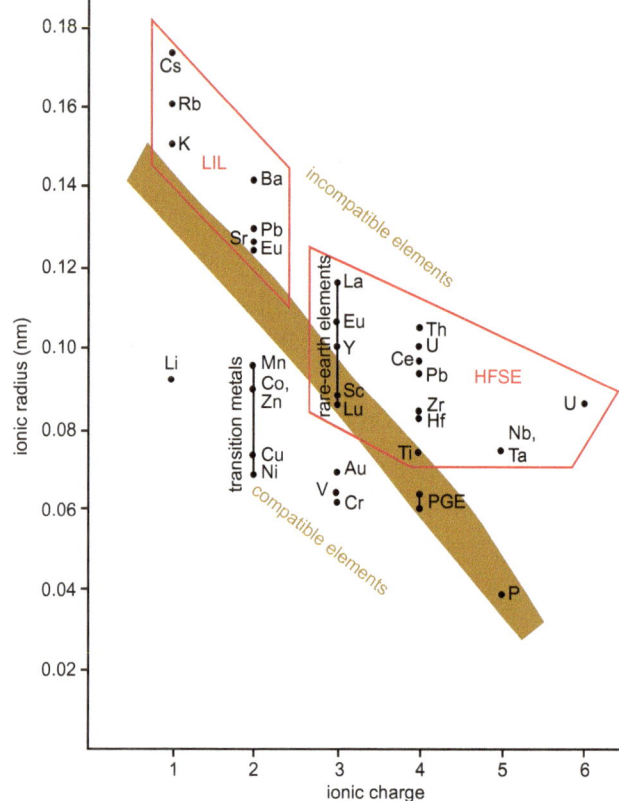

Fig. 3.13 Incompatible elements are enriched in the melt during partial melting and fractional crystallization, while compatible elements are depleted. Whether an element can be integrated into the crystal lattice of the minerals present (i.e., whether it is compatible) depends primarily on the ionic radius and the ionic charge. Incompatible elements include large ion lithophile elements (LILEs) and high-field-strength elements (HFSEs). The *brown stripe* marks the border between compatible and incompatible (from Okrusch and Matthes 2009)

Basaltic magma has a significantly lower density than the mantle and therefore rises. Crystallization commences during cooling during which fractionation occurs between the melt and the crystals formed. Again compatible elements preferably go into early crystallized crystals and are strongly depleted in the melt, while incompatible elements are further enriched in the melt. This is particularly effective when the crystals are separated from the melt either because they sink to the bottom of a magma chamber or because the melt is pressed out of the magma chamber and the crystals remain (as in a coffee filter). This process is called fractional crystallization. Since basaltic magma can rise relatively undisturbed in the case of oceanic crust there is minimal fractionation. However, if magma intrudes into continental crust that has a lower density, then it can literally get stuck and fractionate until its density is so low that it rises further. At subduction zones where oceanic crust is subducted under a continent the entire spectrum of more or less strongly fractionated melts occurs. In principle, magmatic deposits rich in compatible elements are associated with primitive basic magmas directly originating from the mantle, while deposits of incompatible elements may be formed in association with strongly fractionated acidic melts.

Water is also enriched in melts during fractionation. Moreover, water solubility in magma is higher in acidic melts. Strongly fractionated acidic melts therefore typically have a high water content that is separated at some point during crystallization. In the case of volcanoes at subduction zones the foaming of magma triggered in this process is the reason behind explosive volcanic eruptions. The water released can also cause hydrothermal deposits to form (Chap. 4). Water-rich residual melt of almost solidified granite can form pegmatites (Sect. 3.8) and other unusual rocks that likewise are important deposits.

The crystallization of magma begins with a single mineral phase (e.g., olivine) to which further mineral phases are added during further cooling (Box 3.3). The exact order depends on the exact composition of the magma and factors such as oxygen content (Box 3.5 and 3.6) and pressure. The mineral with the highest melting point is not necessarily the first to crystallize. Accordingly, there are different fractionation trends. The so-called calc-alkaline trend from basalt to rhyolite (or in the case of plutonic rocks from gabbro to granite) shown in Fig. 3.4 is typical of subduction zones. Such a trend leads to continuous enrichment of SiO_2 and alkalis.

Box 3.5 Redox state, redox potential, and oxygen fugacity

Many elements occur in several oxidation states and corresponding ions behave very differently. Therefore, the redox state of a system plays an important role in fractionation not only in magmatic but also in hydrothermal systems. In general, a system in contact with the atmosphere is rather oxidized, while the deep interior of the Earth is rather reduced. For example, magnetite can only crystallize in magma if some Fe^{2+} has been oxidized to Fe^{3+}. For magmatic and hydrothermal sulfide deposits the oxidation state of sulfur is also important.

While oxygen is contained in the atmosphere as gaseous O_2, free oxygen does not exist in the interior of the Earth. Almost all minerals such as oxides, silicates, and carbonates contain oxygen and so do magma, water, CO_2, and other fluids. Instead of oxygen concentration (or rather oxygen partial pressure) oxygen fugacity fO_2 must be used under the conditions prevailing there. Oxygen fugacity is a kind of thermodynamically corrected version of the concentration in which the free energy of the various phases is taken into account.

The oxygen fugacity of a system can change through metamorphic reactions or through magma rising into other rocks. As soon as a redox reaction takes place oxygen fugacity is buffered to a fixed value (depending on the temperature) until the reaction is complete. Fayalite–magnetite–quartz (FMQ) buffer is particularly important in magmatic and metamorphic systems (fayalite is a component in olivine):

$$Fe_2^{2+}SiO_4 + O_2 \leftrightarrow 2Fe^{2+}Fe_2^{3+}O_4 + 3SiO_2$$

The oxygen fugacity of magmatic systems is often given as difference relative to this reaction:

$$\Delta fO_2(FMQ)$$

Although oxygen fugacity determines which minerals may crystallize from a given magma, the crystallization of some minerals also changes the oxygen fugacity in the remaining melt (Markl et al. 2010).

For the redox state of hydrothermal solutions the redox potential Eh is normally used instead. The specification refers to electrochemical reduction or oxidation relative to the standard hydrogen electrode. In aqueous solutions ions can be reduced or oxidized by absorbing or releasing electrons thereby precipitating or dissolving various minerals. However, water can itself also serve as a source of oxygen with the release of H^+, which in turn changes the pH value.

Basalts of the mid-ocean ridges instead follow the tholeiitic trend in which iron and titanium are strongly enriched while SiO_2 and alkalis hardly change; hence basalts become rich in iron. It is only later that SiO_2 and alkalis are enriched (i.e., after magnetite starts to be crystallized). The tholeiitic trend assumes low oxygen fugacity (Box 3.5), which delays the oxidation of iron and thus the crystallization of magnetite.

Alkali basalts and other alkaline melts typical of hotspot volcanoes such as Hawaii and continental rift systems follow a third trend. They are continuously enriched with SiO_2 and alkalis despite being and remaining silica-undersaturated (Box 3.1). Fractionation leads to trachyte or phonolite (volcanic rocks) and syenite or nepheline syenite (plutonic rocks), respectively.

To reiterate, a melt formed in the mantle almost always has the composition of basalt. The exact composition whether it is alkali olivine basalt, olivine tholeiite, or quartz tholeiite depends on the degree of melting, the pressure, and the composition of the Earth's mantle.

Although the mantle may already be depleted by earlier melt generation (it then contains less diopside and less spinel), it may also be enriched by migrating melts or fluids in connection with subduction or mantle diapirs. Enriched mantle may also contain minerals such as amphibole, mica, and carbonates.

Apart from basalt there are exotic melts characterized by significantly higher alkali contents such as nephelinite or olivine melilitite that can also occur (especially, in continental rift systems). Several factors play a role in their formation such as extremely low degrees of melting at very great depths, high CO_2 content, and last but not least a highly enriched mantle. The most important deposits of REEs and niobium were formed in connection with these alkaline magmas (Sect. 3.9).

However, in the Archean when the Earth was much hotter the mantle could be melted much more strongly. This resulted in magmas that contain so little SiO_2 that they are called ultrabasic (especially, komatiite) (Sect. 3.4).

Melt generation in the Earth's crust basically follows the same logic as in the mantle except that we are dealing with a different eutectic system. Water also plays an important role here by lowering the solidus considerably. Most rocks of the crust consist of quartz, potassium feldspar, and plagioclase together with other minerals such as mica or pyroxene. The resulting eutectic melt has the composition of granite whereby a temperature of less than 700 °C is sufficient for water-saturated rocks to melt. Basaltic magma is significantly hotter such that large amounts of it can lead to melt formation in the crust. The result is so-called bimodal magmatism in which basic and acidic magmas occur often without intermediate compositions. Both melts can also of course mix with each other and the so-called assimilation of crustal material must be added to fractional crystallization as an important contribution to diversification. Granites can arise in different tectonic situations and can have different compositions (for more information see Sect. 3.7).

Fig. 3.14 Native iron in basalt from Bühl, near Kassel (Germany) (*photo* © F. Neukirchen/Mineralogical Collections of the TU Berlin)

happened in a magma chamber from whose floor shale fragments became loose, rose in the magma, and heated up very fast. At first they released H_2O and CO_2 and then they released CH_4, H_2, and CO as heat increased. Fluid inclusions in olivine have shown that CH_4 and H_2 dominated the fluid of this basalt (Solovova et al. 2002). This fluid was able to reduce some of the Fe^{2+} contained in the magma to native iron.

In the Bühl Quarry near Kassel iron lumps weighing several kilograms were found in basalt (Fig. 3.14). In this case a lignite seam was responsible for the reduction of iron.

3.1.2 Liquid Immiscibility

Under certain circumstances magma can be separated into two different melts that are not miscible with each other much like water and oil. Liquid segregation is known to occur relatively frequently. For example, the glassy solidified groundmass of tholeiitic basalts often consists of two differently colored components: droplets of dark-colored glass in a light-colored glass. The light-colored glass has the composition of granite and the dark-colored glass that of an iron-rich pyroxenite (Philpotts 1982).

A miscibility gap between two immiscible melts is shown in the schematic phase diagram in Fig. 3.15. When magma of a certain composition encounters the so-called solvus while cooling down it segregates. The mixing of two magmas that have compositions on both sides of the miscibility gap leads to the same effect. Segregation results in an

Box 3.6 Native iron in basalts
Even iron, which is known to rust easily, occurs in elementary form on Earth. The best known native iron is found in basalts on Disko Island, Greenland where meter-sized iron blocks weighing up to 22 t have been found. Millimeter-sized iron globules (a.k.a. solidified melt drops) are relatively common in the lavas with the larger chunks being accumulations of droplets that have sunk. Strictly speaking, it is a natural steel with a high carbon content that contains tiny inclusions of other highly reduced minerals such as fayalite (Fe_2SiO_4) and wüstite (FeO).

The composition of basalts is explained as originally tholeiitic magma that has assimilated 15–40% carbon-rich shale (Ulff-Møller 1990). It is believed this

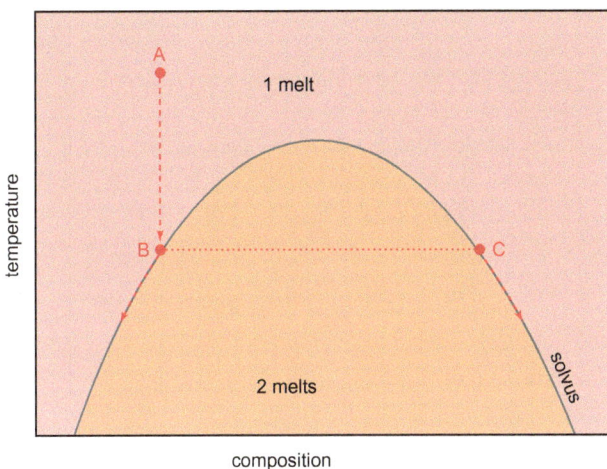

Fig. 3.15 Schematic representation of an immiscibility gap between two immiscible melts as a function of temperature. During cooling, magma of composition A reaches the so-called solvus (B). Here, melt droplets of composition C are separated. As the miscibility gap increases with decreasing temperature, a continuous fractionation between the two immiscible melts occurs during further cooling. Their respective composition changes along the solvus. Alternatively, segregation can be initiated by mixing two magmas located on both sides of the miscibility gap

emulsion made up of two differently composed melts. Emulsified melt droplets have a different density than the rest of the melt and can accumulate on the floor or under the roof of a magma chamber. However, the segregated melt often remained in the emulsion until the pluton had solidified; hence only inclusions in crystals indicate immiscibility.

The segregation of immiscible magmas plays an important role in the formation of deposits because it leads to the very effective fractionation of certain elements. Apart from some exotic examples of little interest here there are three important miscibility gaps in natural magmas:

- Silicate melt–iron oxide melt
- Silicate melt–sulfide melt
- Silicate melt–carbonatitic melt.

The first miscibility gap concerns basaltic magmas that follow tholeiitic fractionation in which iron accumulates strongly in the melt. The size of the gap and thus the composition of the two melts depend on the oxygen content and the content of phosphorus, titanium, and iron. Experiments have shown that a melt can segregate and solidify to magnetite and apatite in extreme cases, while the rest has the composition of diorite. Another example is segregation in the groundmass of tholeiitic basalts where iron-rich melt has a significant silicate content (as mentioned earlier). Yet another example is segregation of an iron-rich melt in the

Skaergaard intrusion (Sect. 3.3). However, this remained in emulsion and was therefore only detectable in inclusions in crystals (Veksler et al. 2007; Holness et al. 2011; Jakobsen et al. 2011). Such segregation may play a role in the formation of Kiruna-type iron deposits (Sect. 3.6). Segregation along these lines could lead to the formation of phoscorite (Sect. 3.10.1) in alkaline magmas.

The segregation between silicate and sulfide melt is of greater importance here and often happens when it comes to basic magmas. A sulfur content of 0.1% can be sufficient to bring the composition to the solvus of the miscibility gap during fractional crystallization. Sulfide melt has a composition approximating FeS and therefore solidifies predominantly to pyrrhotite. During segregation elements such as copper, nickel, cobalt, gold, and platinum (or chalcophile and siderophile elements; Sect. 1.20) effectively fractionate into the sulfide melt as long as of course they are present in the melt. This is the most important process in the formation of nickel and platinum deposits (Sect. 3.3.2) and provides initial enrichment for some magmatic–hydrothermal deposits (Sect. 4.4).

For elements like platinum that are only present in small amounts in the magma but fractionate strongly in segregated melts the so-called R factor plays a role in addition to the distribution coefficient. The R factor describes the quantitative ratio between silicate magma and sulfide magma in equilibrium. If a small amount of sulfide magma exchanges elements with a lot of silicate magma, then elements such as platinum are enriched in the sulfide melt to particularly high concentrations.

A significant deposit forms when sulfide melts segregate relatively early during fractional crystallization of the original magma. Nickel is a very compatible element and is therefore quickly removed from the melt during fractional crystallization. Although the concentration of PGEs changes minimally during fractional crystallization, continued fractionation also reduces the melt quantity and thus the PGE budget to be absorbed by the sulfide melt.

The solubility of sulfur in silicate magma depends not only on the temperature but also on the redox state (oxygen fugacity) and FeO content of the melt. Accordingly, other processes can trigger segregation in addition to cooling. A simple possibility is contamination with external sulfur. Many sediments have a significant sulfur content. If they are melted by and mixed with basic magma, then sulfur saturation can easily be exceeded. Under certain circumstances the mixing of two different magmas can also lead to a composition that is exactly right for segregation. Finally, segregation can also be triggered by crystallization of an FeO-rich phase such as magnetite.

The third miscibility gap occurs mainly in alkali-rich and silica-undersaturated (alkaline) magmas. If they have a very high CO_2 content, then a carbonate melt (carbonatite; Sect. 3.10) may be segregated.

3.2 Podiform Chromite Deposits in Ophiolites

The second most important type of chromium deposits are the so-called podiform chromite deposits found in ophiolites (i.e., pieces of oceanic lithosphere; Box 3.7). Typical are lenticular (a.k.a. podiform) rock bodies consisting mainly of the mineral chromite (Box 3.8) together with olivine or serpentine and occasionally other minerals. Chromite-rich rock is called chromitite. Such lenses are almost always surrounded by dunite (a rock consisting of more than 90% olivine). They are typically found in mantle rock just below the Moho (the boundary between the mantle and crust). There are also some within dunites that are interpreted as crystal cumulates of gabbros and are therefore located above the Moho.

Box 3.7 Ophiolites

An ophiolite is a piece of oceanic lithosphere (i.e., oceanic crust and the rigid lithospheric mantle that sticks underneath) that has strayed onto a continent. It was not until the 1960s when the formation of new oceanic crust on mid-ocean ridges was understood that the typical sequence of rocks known from the Alps became clear. However, it is now known that very few ophiolites really originate from a mid-ocean ridge, but this new insight is left aside for a moment.

The exact sequence of rocks that makes up an ophiolite has been redefined several times as a result of lack of completeness in many studies. The modern view holds that an ophiolite from top to bottom consists of:

(1) Deep-sea sediments
(2) Basalt in the form of pillow lavas
(3) Basalt in the form of a sheeted dike complex
(4) Gabbros with peridotitic cumulates at the base
(5) Mantle rock (peridotite that has possibly converted to serpentinite).

Such rocks reflect the processes occurring at mid-ocean ridges (Fig. 3.16). Mid-ocean ridges are high mountains in the deep sea with a deep rift system in the center. Basalt in the rift extrudes, solidifies, and continuously creates new oceanic crust moving away from the rift several centimeters per year like a conveyor belt. The Earth's mantle is pulled along leading to an upwelling of hot material in the asthenosphere below the rift. About 20% of upwelling hot mantle rock eventually melts. Melt with the composition of basalt rises leaving behind a depleted mantle (typically, harzburgite). The further the young crust moves away from the mid-ocean ridge the more it cools down and the thicker the rigid lithospheric mantle sticking underneath becomes.

Rising basaltic magma in the crust below mid-ocean ridges accumulates in magma chambers some of which solidify into plutons. Accordingly, the lower part of the oceanic crust consists of gabbro. Early formed crystals can accumulate on the bottom of the magma chamber during crystallization of these plutons. They are the same minerals that make up the Earth's mantle; hence the cumulates are peridotites as well.

The middle part of the oceanic crust is the so-called sheeted dike complex that consists almost exclusively of parallel basalt dikes. Extension causes fissures to be torn open over and over again facilitating magma to rise further through them. Finally, basalt appears at the bottom of the sea. On contact with seawater the melt is quenched at its surface to form a glassy shell, while liquid melt continues to flow. Lava takes on the form of a pillow filled with melt that becomes ever-bigger like a balloon before it rolls away from the vent and cools down further, while a new pillow is formed at the vent. Accordingly, the uppermost part of the oceanic crust consists mainly of pillow lavas. Over time deep-sea sediments are deposited on them.

Ophiolites were first thought to be typical results of the processes occurring at mid-ocean ridges when the theory of plate tectonics gained acceptance. However, it is now known that most ophiolites were instead created in connection with subduction zones. Such zones can also lead to strong extension that is usually noticeable behind the volcanic arc in so-called back-arcs where a miniature version of a mid-ocean ridge also forms new oceanic crust. The extension is even stronger during initiation of a new intraoceanic subduction zone before normal island arc volcanism sets in (Fig. 3.17). On one side of the new plate boundary the crust sinks into the mantle, while the plate boundary rolls back and the other plate is stretched so much that new crust is formed here. Later this new crust typically forms the fore-arc between the deep trench at the plate boundary and the volcanic arc. However, normal volcanism only begins when the subducted plate has reached a depth of about 100 km where dehydration reactions occur. Corresponding ophiolites are called supra-subduction zone (SSZ) ophiolites. Most large ophiolite complexes

Fig. 3.16 Mid-ocean ridges are constructive plate boundaries at which new oceanic crust forms

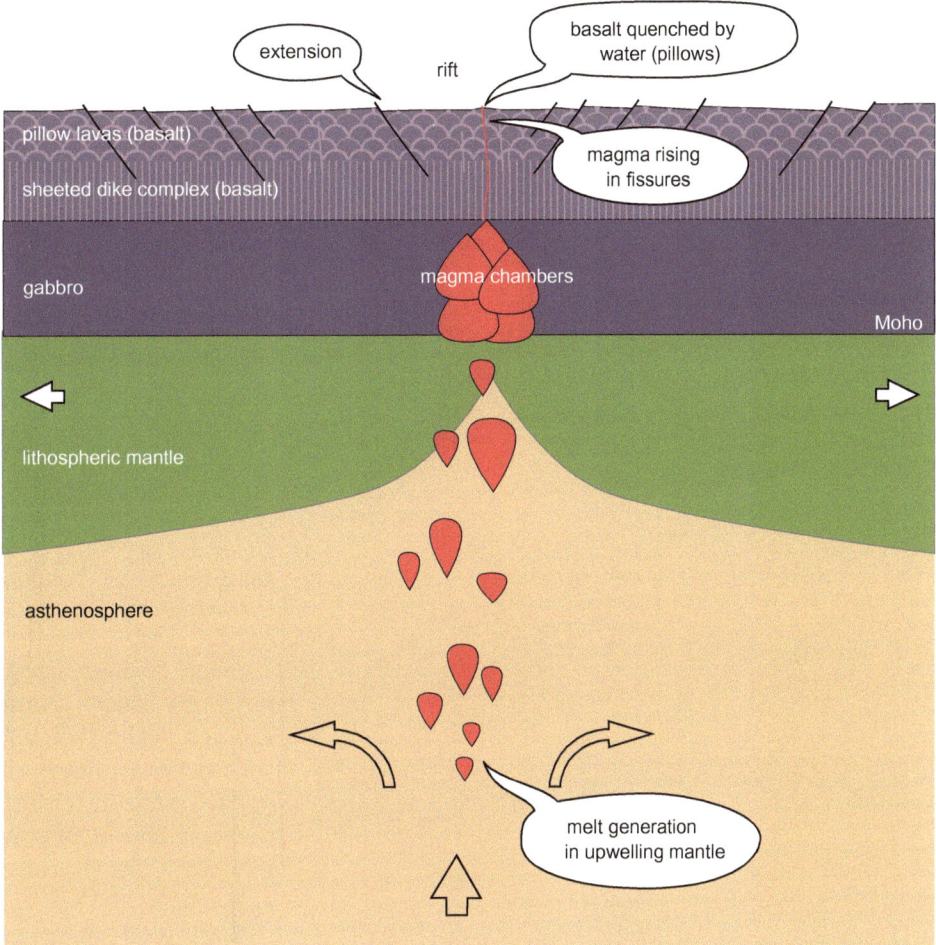

clearly belong to this type. This is probably because their connection with a subduction zone makes it much more likely that a contiguous piece of the oceanic lithosphere strays on to the edge of a continent. The simplest process is the collision of an island arc with a passive continental margin. Since continental crust cannot really be subducted due to its low density a piece of oceanic lithosphere is almost inevitably pushed onto the edge of the continent (obduction) before movement stops.

Alpine ophiolites found in the Alps and other mountain ranges differ in that they trace the seam (suture) between two colliding continents where oceanic lithosphere was partly subducted to great depths during the collision and transformed into high-pressure metamorphic rocks before they rose again along the plate boundary as a nappe.

A further type of ophiolites are found in active continental margins. Strongly deformed chips of oceanic lithosphere (i.e., ophiolites) can accumulate in the accretionary wedge of a subduction zone.

In addition to forming podiform chromite deposits in mantle rocks, ophiolites also contain Cyprus-type volcanogenic massive sulfide (VMS) deposits (Sect. 4.16.1) located in basalts (mined for copper). There may also be manganese deposits in deep-sea sediments (Sect. 5.5). The weathering of mantle rocks in the tropics can lead to the formation of nickel laterites (Sect. 5.11.2).

Two different textures are typical of the ore. So-called leopard ore (Fig. 3.18) consists of round aggregates of chromite crystals several millimeters in size found in a light-colored dunite groundmass (or in serpentinite). In contrast, cockade ore (a.k.a. ring ore) (Fig. 1.8) consists of light-colored dunite spheres surrounded by a dark-colored matrix of chromite and olivine. Sometimes there are also dunite layers with finely distributed chromite.

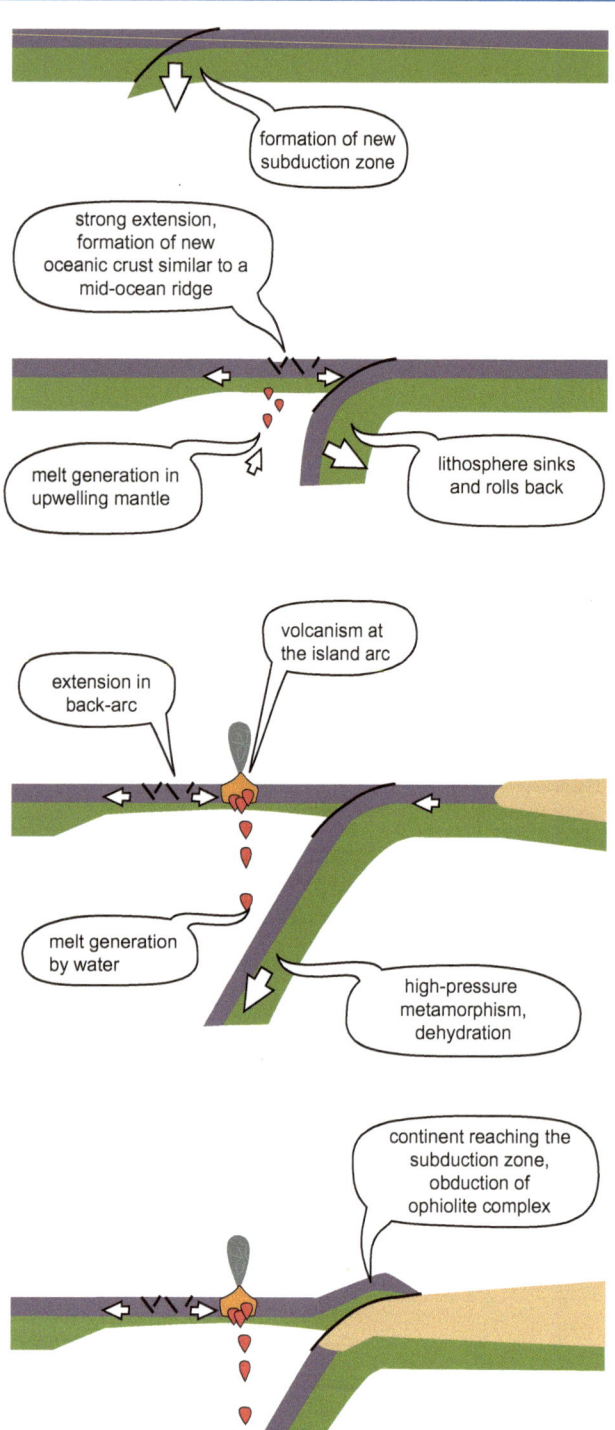

Fig. 3.18 Chromite (leopard ore) from Goleman, Turkey (*photo* © F. Neukirchen/Mineralogical Collections of the TU Berlin)

Some podiform chromite deposits also contain PGE metals. Generally, osmium, iridium, and ruthenium are enriched, but in rare cases rhodium, platinum, and palladium are enriched instead. Such metals occur as natural alloys, in sulfides, sulfosalt minerals, and other PGE minerals. They take the form of tiny inclusions in chromite and as tiny minerals in the interstices between mineral grains of the groundmass.

Most podiform chromite deposits are small and get depleted after a short period of mining. However, some ophiolites contain very many such inclusions that are unfortunately not easy to track down. There are many deposits in the countless ophiolites of the former Tethys, the ocean that disappeared during the Alpine orogeny (a.k.a. Alpide orogeny). The Alpide mountain belt stretched from the Alps to the Himalayas and beyond passing through the Balkans, Turkey, Cyprus, and Iran. Further occurrences can be found in the Philippines, New Caledonia, Japan, Cuba, Pakistan, Sudan, Canada, the United States, Norway, the Shetland Islands, Australia, Russia, and Kazakhstan.

Unusually large deposits can be found in the southern Urals in the Kepirsai ophiolite (Melcher et al. 1997; Distler et al. 2008). One celebrated Kazakhstan belonging to the Soviet Union and is called the "40 Years of the Kazakh Soviet Republic" deposit. Such deposits have made Kazakhstan the second largest chromium producer after South Africa accounting for about 15% of world production. These occurrences also show high contents of PGEs such as iridium, osmium, and ruthenium.

A number of contradictory models were used to explain the origin of these deposits. Most researchers today assume that they form when basalt or similar mantle melts rise through the upper mantle.

Fig. 3.17 Although many ophiolites are typical of the oceanic lithosphere, they did not usually originate from a mid-ocean ridge. Instead they originated during the formation of an intraocean subduction zone. This can initially be accompanied by rapid rolling back of the subduction zone leading to strong extension in the upper plate and the formation of new oceanic crust. Only later does usual island arc volcanism develop. Obduction into continental crust may happen as soon as a continent reaches the subduction zone

A simple explanation could be that olivine and chromite are already crystallizing in fissures through which basalt melt rises to mid-ocean ridges where they accumulate in tiny magma chambers, while most of the melt continues to rise. Later this part of the mantle becomes part of the lithospheric mantle due to the lithosphere cooling.

However, it is not that simple. In normal conditions a lot of olivine and only little spinel would crystallize in basaltic melt, and this would be a chromium spinel (mostly Al instead of Cr) rather than chromite. The simple model must be modified accordingly. It has been suggested that the mixing of different magmas is important (Ballhaus 1998), whereas many researchers think there is a reaction between the rising melt and the adjacent rock of the mantle (harzburgite). The pyroxene of the mantle is dissolved by the melt transforming the harzburgite along the ascent path into dunite (the residual olivine) (Prichard et al. 2008; Caran et al. 2010). An unusually high water content in the melt probably plays a role (Matveev and Ballhaus 2002; Büchl et al. 2004; Distler et al. 2008). Although this is atypical of normal basalts, it is not surprising in connection with SSZ ophiolites (Box 3.7). In addition, the typical composition of chromites might suggest that we are not dealing with normal basalt, but with boninite. This basalt-like rock is typical of young subduction zones in which hot young lithosphere is being subducted (Caran et al. 2010).

All such processes are believed to lead to the enrichment of PGEs. The crystallization of PGE minerals may then be triggered indirectly by chromite crystallization (Prichard et al. 2008). More about chromite and its relation to PGEs can be found in Sect. 3.3.2.

Box 3.8 Chromite

The mineral chromite ((Fe^{2+}, Mg)(Cr, Al)$_2$O$_4$) is the only mineral from which chromium can be extracted. It belongs to a group of oxide minerals called the spinel group that have a certain structure with solid solution between the theoretical endmembers. In the narrower sense spinel (MgAl$_2$O$_4$) is a gemstone found in metamorphic rocks. Chromium spinel (MgAl$_2$O$_4$ with some Cr and Fe) contained (at moderate pressure) in the peridotite of the Earth's mantle has a significantly lower chromium content than chromite. Magnetite ($Fe^{2+} Fe_2^{3+} O_4$) and ulvöspinel ($Ti Fe_2^{2+} O_4$) also belong to the spinel group. Chromite usually contains 45–55% Cr$_2$O$_3$. During crystallization of a melt the composition of the spinel (in the broader sense) formed depends on the composition of the melt and on factors such as pressure, temperature, and oxygen fugacity.

Chromite can be found in podiform chromite deposits and in LMIs. The Bushveld Complex (South Africa) (Sect. 3.3.3) is by far the largest occurrence.

The higher the chromium content the better the ore for the production of chromium-hardened stainless steel. Ore that has a low chromium content is used as a raw material for the chemical industry to produce various chromates. Chromite ore with high aluminum content is used to produce refractory materials (Sect. 7.10) such as fireclay bricks.

3.3 Layered Mafic Intrusions

Layered mafic intrusions (LMIs) are just very large plutons consisting of basic magma. However, a pluton does not solidify into a homogeneous gabbro in LMIs, but has developed into differently composed layers some of which are reminiscent of sediment layers. This layering affects both the small scale (e.g., constant repetition at decimeter intervals called rhythmic layering) and systematic variation at the large scale. In some cases individual layers can be tracked for more than 100 km.

Mafic rocks are igneous rocks such as gabbro and norite that have a relatively small content of the bright so-called felsic minerals quartz, feldspar, and feldspathoids. They have a correspondingly high proportion of other minerals that are predominantly dark in color called mafic minerals. In contrast, felsic rocks are igneous rocks like granite that have a high content of felsic minerals. Rocks with more than 90% mafic minerals are called ultramafic such as pyroxenite and peridotite. The terms mafic and ultramafic roughly correspond to basic and ultrabasic due to the low SiO$_2$ content of olivine. However, pyroxenite contains significantly more SiO$_2$ and would therefore be intermediate while being ultramafic. However, it is not the composition of any melt, but the cumulate of an (ultra)basic magma.

The layers of LMIs consist largely of the same minerals as an average gabbro but in strongly varying proportions. They are therefore given rock names such as pyroxenite, peridotite, and anorthosite.

The respective proportion of the minerals in LMIs changes (modal layering) as does their composition. Although early olivine and pyroxene are very Mg-rich, later on they become ever-more Fe-rich. In plagioclase the Ca content (anorthite) decreases, while the Na content (albite) increases. This invisible layering (cryptic layering) ultimately affects the chemical composition of the respective rock.

Table 3.1 Overview of the best known layered mafic intrusions

Name	Locality	Age	Area (km^2)
Bushveld	South Africa	Precambrian	66,000
Dufek	Antarctica	Jurassic	50,000
Duluth	Minnesota, USA	Precambrian	4700
Stillwater	Montana, USA	Precambrian	4400
Muskox	Northwest Territories, Canada	Precambrian	3500
Great Dyke	Zimbabwe	Precambrian	3300
Sudbury	Ontario, Canada	Precambrian	1100
Kiglapait	Labrador, Canada	Precambrian	560
Skaergaard	East Greenland	Eocene	100

Some LMIs are economically interesting. There are layers consisting almost exclusively of chromite (see Box 3.8) called **stratiform chromite deposits**. By far the most important deposits of chromite are found in the Bushveld Complex (South Africa) (Sect. 3.3.3), which accounts for almost half of global chromium production. There are also noteworthy occurrences in many other intrusions such as in Zimbabwe (Great Dyke; Sect. 3.3.4), Madagascar, the United States (Stillwater Complex), Finland (Kemi), India, Russia, and Brazil.

In some layers platinum is enriched together with other PGEs. In some cases these are chromite layers like the UG2 chromitite in the Bushveld Complex. In other cases they are pyroxenites or other (ultra)mafic rocks such as the Merensky Reef in the Bushveld Complex and the J-M Reef in the Stillwater Complex. These **stratiform PGE deposits** are also known as **reef-type** deposits. "Reef" has nothing to do with corals in this context, but means hard rock layers with a high content of precious metals. **Contact-type** deposits are platinum deposits that can occur at the contact of mafic intrusions with the host rock. This type includes the Platreef Mine in the northern part of the Bushveld Complex.

The Bushveld Complex is said to contain about 75% of the Earth's known platinum resources. The Great Dyke (Zimbabwe) and Norilsk (Russia) are mafic intrusions containing the rest (except for a few percent).

There may also be layers with worthwhile contents of nickel and copper sulfides. The Sudbury Complex (Canada) (Sect. 3.3.5) and Norilsk (Russia) are the most important nickel deposits of all, while Pechenga (Russia) and Jinchuan (China) are also important.

Finally, there may be layers made up predominantly of magnetite (titanomagnetite) or ilmenite that not only contain iron and titanium, but also quite large amounts of vanadium. The Bushveld Complex's main magnetite layer is 2 m thick and merits a mention here as being one of the most important vanadium deposits. Further examples are two intrusions at Emei Shan (China) (Zhang et al. 2009), Chineyskoye (Russia), and Stillwater and Duluth (USA).

LMIs occur in various shapes and sizes (Table 3.1). However, their horizontal extent is typically much greater than their vertical extent. The various shapes they take have been described as disk-shaped (or sill-shaped), saucer-shaped, funnel-shaped, canoe-shaped, etc.

By far the biggest LMI is the Bushveld Complex (South Africa) whose magma volume reaches an order of magnitude that is comparable with that known from flood basalt provinces such as the Deccan Traps (India).

Flood basalts are the result of exceptionally strong basic volcanism. In a very short time geologically an enormous area is buried lava flow by lava flow under a basalt layer hundreds of meters thick. In the case of the Deccan Traps (India) the area is the size of Spain. Lava erupted from fissures and individual lava streams often flowed several hundred kilometers. Other examples are the Karoo Basalt (South Africa), Paraná (Brazil), Etendeka (Namibia), the Siberian Trapps (Russia), and the Emei Shan Trapps (China).

Although most LMIs are much smaller, interestingly they often occur together with flood basalts of the same age. The formation of such vast quantities of basaltic magma in a geologically short period of time is usually explained by the ascent of a mantle diapir. As the mantle diapir rises its head widens and takes on the shape of a mushroom. If this head hits the crust, extensive melt formation takes place for a short time. Only later does normal hotspot magmatism take place. A mantle diapir can also trigger formation of a continental rift. On the other hand, stretching and the formation of a rift system lead to upwelling in the asthenosphere and therefore very similar magmatism. Thus, some intrusions have been interpreted as rift systems. All LMIs are connected in some way with hotspots, rift systems, or both. The only exception is the Sudbury Complex (Sect. 3.3.5) in Canada, which is an impact crater. It is noticeable that most of the larger LMIs had already formed by the end of the Precambrian.

It is difficult to reconstruct the exact composition of the original magma and how it evolved through fractionation. In principle the composition of the respective layers never corresponds to that of a liquid melt since here we are dealing with cumulates (Fig. 3.19). According to the simplest theory crystals growing in magma sink to the bottom of the magma chamber due to their higher density. A crystal slurry accumulates there with some melt in the gaps. The melt solidifies only after further cooling. In the meantime the slurry can compact whereby some of the remaining melt is squeezed out. The rock thus consists of minerals crystallized at different times from different melts. To top it all subsequent recrystallization often takes place (McBirney 2009) such that not even the texture can be interpreted with certainty. Since the mere sinking of crystals cannot explain the layering a number of further processes are presented below that might clarify matters. In any case most crystals do not rain down from a large volume but crystallize close to the ground, the roof, and the sides where it is cooler.

Many intrusions are not caused by a single magma pulse but by repeated injections of fresh magma that mixes with already fractionated melt. A somewhat simpler case is the small and economically insignificant Skaergaard intrusion in eastern Greenland (Fig. 3.20) considered by geologists to be a classic example. It was created in the Eocene by the unique influx of basic magma. The melt then solidified from all sides into the interior: the Layered Series from bottom to top, the Marginal Border Series from the sides, and the Upper Border Series from top to bottom. The sandwich horizon between the Layered Series and Upper Border Series is where the last remaining melt solidified.

The composition of the rocks in the Layered Series corresponds to the crystallization of different minerals during fractional crystallization (Thy et al. 2006). The Upper Border Series is also layered corresponding roughly to a

mirror-inverted sequence of the Layered Series. This already shows that the sinking of crystals proposed as an explanation for layering is too simple. Moreover, plagioclase has a lower

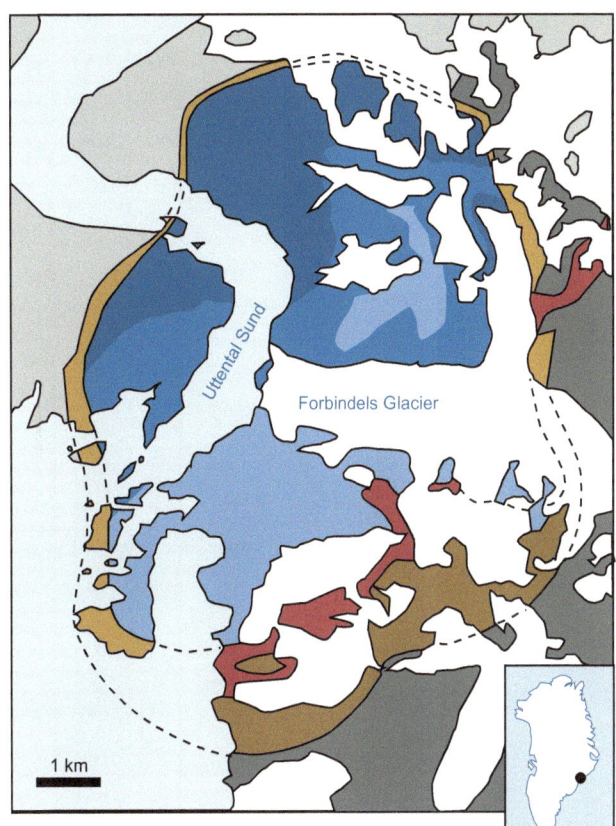

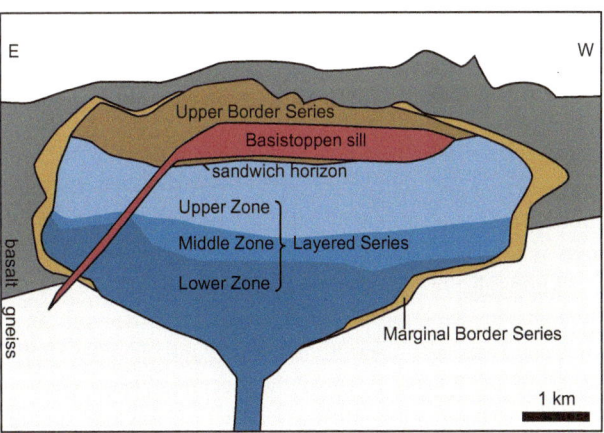

Fig. 3.20 Simplified geological map (**a**) of the Eocene Skaergaard intrusion (Greenland) and a schematic cross-section (**b**) in the same color scheme before it was tilted. The intrusion solidified from outside to inside. The marginal series solidified from the sides, the layered series from bottom to top, and simultaneously the upper border series from top to bottom. The last melt residues solidified as the sandwich horizon forming the boundary layer between the layered series and the upper border series. Only later did the Basistoppen sill intrude, which also has pronounced layering. In the upper part of the middle zone there is a PGE-rich horizon (Platinova Reef) considered potentially minable. *PGE*, Platinum-group element (**a** modified from McBirney 1989; **b** from Hoover 1978)

Fig. 3.19 Typical cumulate texture that develops from a slurry of crystals such as plagioclase (*gray*), pyroxene (*brown*), olivine (*yellow*), chromite (*black*), and melt (*red*). During further cooling the melt present in interspaces crystallizes. This happens through continued growth of existing crystals and through growth of new crystals that are often so large that they enclose many older crystals (poikilitic texture). Some older crystals can be dissolved and replaced by new ones

Fig. 3.21 Change in mineral composition and mineral content in the Skaergaard intrusion. The layered series solidified from bottom to top and the upper border series from top to bottom. The line between them corresponds to the sandwich horizon that solidified last from the most strongly fractionated melt (from Wager and Brown 1968; Naslund 1983)

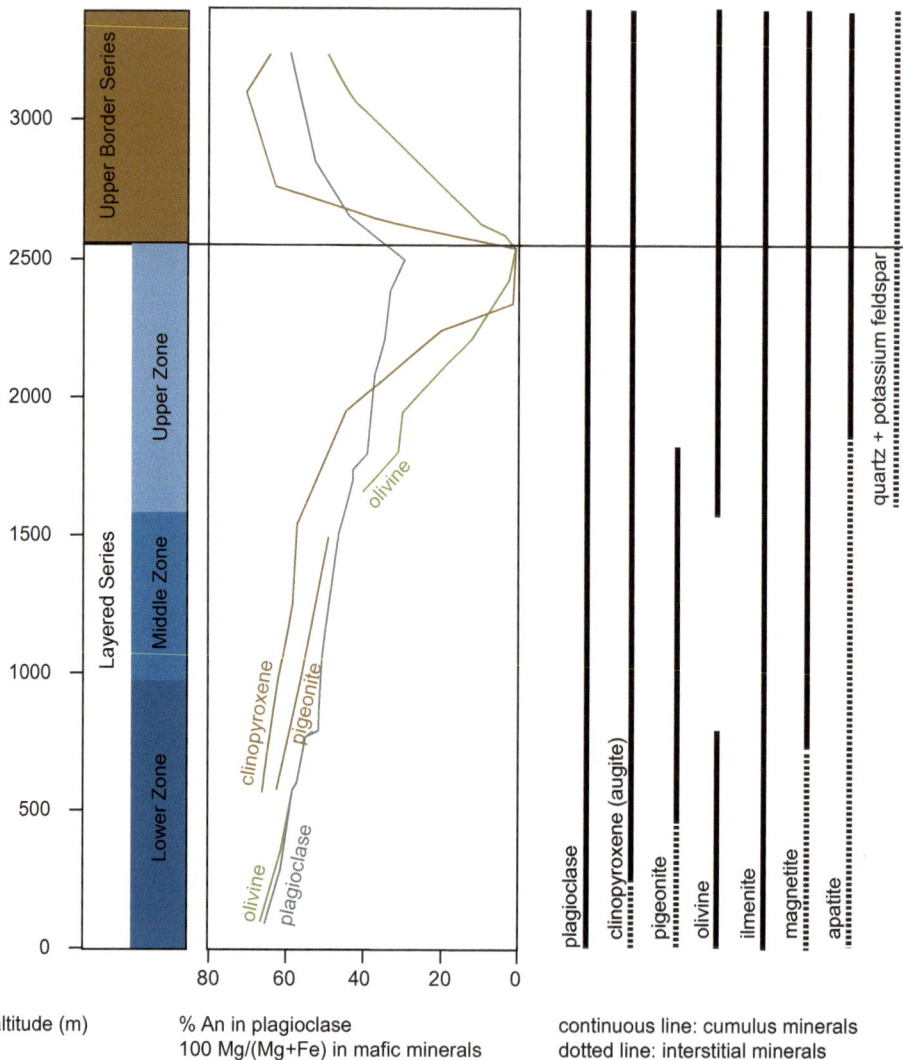

altitude (m)

% An in plagioclase
100 Mg/(Mg+Fe) in mafic minerals

continuous line: cumulus minerals
dotted line: interstitial minerals

density than the melt and should float up, yet it is also contained in the cumulates of the Layered Series. The Upper Border Series also contains the heavy minerals olivine and pyroxene.

Although fractionation at Skaergaard is exemplary (Fig. 3.21), here too the details are controversial. In the deepest exposed layers olivine and plagioclase crystallized (base of the Lower Zone of the Layered Series) and later pyroxenes (first augite, then pigeonite) were added. Initially, fractional crystallization hardly changed the SiO_2 content of the melt, but very strong enrichment of iron occurred. This is the typical tholeiitic trend. Finally, the crystallization of magnetite began. However, it is unclear at which moment this happened. Magnetite appears together with ilmenite and hematite in the upper part of the lower zone, but it is questionable whether they are cumulus phases or phases crystallized later in the interstitial spaces (Jang et al. 2001). The crystallization of magnetite is mainly controlled by oxygen content (more precisely, by oxygen fugacity), which

is not an independent variable because it is buffered by the other phases.

However, it was only now that the SiO_2 content of the melt slowly increased, the crystallization of olivine stopped (base of the Middle Zone), and some olivine even dissolved again. In the Upper Zone olivine reappears but this time with a very iron-rich composition while pigeonite disappears. Finally, interstitial apatite, alkali feldspar, and quartz crystallized.

The late crystallization of olivine and quartz breaks the rule explained in Box 3.1. This rule states that since the composition of orthopyroxenes lies exactly between the two minerals both never occur together. However, in the case of magma with very high iron content this rule does not apply because orthopyroxene is no longer stable. In such a case olivine also crystallizes in silica-oversaturated rocks, but with a very iron-rich composition (fayalite).

Researchers relatively recently realized that the melt had been segregated into two different magmas. The greater part

Fig. 3.22 Magmatic layering in gabbro on the Lyngen Peninsula (Norway). This is modal layering with varying proportions of the minerals pyroxene (*black*) and plagioclase (*white*) (*photo* © F. Neukirchen)

consisted of a melt rich in iron and low in silicates, while the other part was low in iron and rich in silicates. Initially, it was thought that this happened very late and only affected the last melt residues. However, melt inclusions in plagioclase show that this obviously happened toward the end of the Lower Series (Jakobsen et al. 2011). The droplets of separated melt initially remained small and in emulsion. It was only after further fractionation that they accumulated to form larger droplets.

3.3.1 Magmatic Layering and Its Causes

Many gabbros and similar basic intrusive rocks show intensive layering reminiscent of the layers of sediments (Fig. 3.22). In the case of LMIs such as Skaergaard, Bushveld, Great Dyke, and Stillwater this is so pronounced that layers of very different rocks have formed. Similarly strong layering is sometimes observed in alkaline rocks spectacular examples of which are Ilimaussaq (Sect. 3.11.1) and Lovozero (Sect. 3.11.2). A prerequisite for intensive layering is that magmas have a low viscosity (i.e., are thin fluid).

There are different types of magmatic layering in which variations relate to a different scale. There can be both gradual and sharp transitions between the layers.

Such intrusions can be roughly divided largely based on where a mineral first appears or disappears within the stratigraphy (a.k.a. phase layering). This is mainly about cumulus phases (i.e., larger crystals that accumulated as slurry at the bottom of the magma chamber) because the melt remaining in the interstices solidified later. The sequence of

the cumulus phases corresponds to fractional crystallization from an increasingly fractionated melt. This alone creates layers of different rocks in which the lower layers are usually ultramafic cumulates consisting mainly of olivine and pyroxene, followed by mafic rocks with additional plagioclase, and finally magnetite, although olivine eventually disappears.

Such rocks can even be layered at a smaller scale. Some of these layers are internally homogeneous both in their mineralogy and their texture (uniform layering). In other layers the proportions of respective minerals change (modal layering). This can be a slow change from bottom to top, a sudden change, or a fast gradual change on a small scale. Particularly striking is a sequence that is constantly repeated called rhythmic layering. This is usually a gradual change from predominantly mafic to predominantly felsic minerals at a distance of a few centimeters (microrhythmic) or decimeters or meters (macrorhythmic). In profile this is a rock with alternate dark and light stripes. Sometimes rhythmic layering alternates with relatively homogeneous layers (intermittent layering). There may also be rhythmic repetition of sharply defined layers consisting of a single mineral.

Apart from these visible variations there is also invisible variation (cryptic layering) in the chemical composition of the minerals and accordingly the rock. The composition of the minerals changes with increasing fractionation.

How does such layering come about? The diverse forms of layering already described suggests that a single process is not enough to explain the problem (Fig. 3.23). The simplest explanation is that gravity causes crystals growing in magma

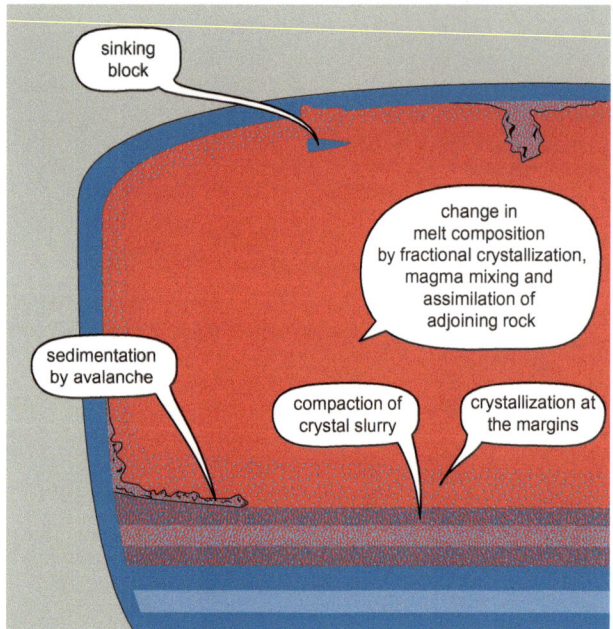

Fig. 3.23 Manifold layering of mafic intrusions generated by combining several processes

to sink to the bottom of the magma chamber. This should result in sorting according to their density. Rhythmic repetition can be explained by periodic turnover as a result of convection. However, this process may be at most of minor importance for a number of reasons. For example, plagioclase occurs in cumulates at the bottom of the magma chamber despite having a lower density than the melt and should therefore float upward instead. Another problem is that according to Stokes' law the size of particles should have a greater effect on the speed at which they sink than their density. However, magmatic layering in which sorting accords with grain size can only be found in a few exceptional cases. Significant sinking of crystals only occurs in very hot and preferably SiO_2-poor (ultrabasic) magmas that have a low viscosity and behave approximately as a Newtonian liquid. However, crystals are more or less kept in suspension in most cases.

Instead, let us assume that plutons solidify from the outside to the inside and there is a corresponding temperature gradient. The nucleation of crystals and their growth therefore take place mainly at the margins of the magma chamber. As a rule heat loss is greatest either on the floor or on the ceiling. The crystal slush that accumulates at the bottom of the magma chamber, on the sidewalls, and near the ceiling consists mainly of crystals that have grown more or less in situ and have sunk or risen only slightly.

Layering is now achieved through a series of processes that can be divided into three types:

- Change in the respective crystallizing minerals depending on temperature, oxygen content, composition of the melt, and so on;
- Sedimentation of crystal slush by mass flows that can slide down the sidewalls and spread over the ground; and
- Processes occurring during the compaction of crystal slush.

The first type naturally involves long-term change in mineralogy that occurs during fractional crystallization. Although the exact sequence varies from intrusion to intrusion (e.g., plagioclase can crystallize very early or late), it typically follows tholeiitic fractionation and results in the sequence of different rocks already described.

This sequence can be confused by renewed injection of primitive melt that mixes with already fractionated melt of the intrusion. In most cases this must have happened repeatedly as demonstrated by the multiple jumps in cryptic layering. Mixing results in the sudden crystallization of completely different minerals, even monomineralic layers may form. The almost pure chromite layers, for example, can be explained by the mixing of magmas leading to compositions finding themselves temporarily in the stability field of chromite (Irvine 1977; Murck and Campbell 1986).

Incidentally, new magma can penetrate between older layers and solidify as a sill partly reacting with the neighboring rock (Féménias et al. 2005b).

A similar short-term change in crystallization deviating from the normal cotectic line (see Box 3.3) may also be caused by other processes. Another form of magma mixing occurs when convective circulation begins in a magma chamber that was previously more static and therefore heterogeneous. The assimilation of adjoining rock also changes magma composition. The segregation and release of fluids (water, gases) can also play a role as can the supply of fluids from heated host rock. Finally, changes in factors such as temperature or oxygen content can also make a contribution.

The second type of processes relate to mass flows. They concern avalanches and landslides of crystal slush that spread from the steep sides of the magma chamber over the bottom. Flow dynamics in the avalanche causes the minerals to be sorted according to their density (Irvine 1980; Irvine et al. 1998). Even a less dynamic slide causes sorting. Crystal slush behaves very idiosyncratically under mechanical stress and any deformation is almost exclusively concentrated on melt in the interstices. This is why shear zones formed from melt later solidify into lighter rock than the crystal-rich parts that move in one piece (McBirney and Nicolas 1997). In Skaergaard mass flows are clearly visible at the edge of the Layered Series. Here there are structures well known from clastic sediments such as cross-stratification and erosion channels (later backfilled) cut into older layers by currents. These structures are particularly clear in the lower part of the Layered Series because the height of the sidewall of the still liquid magma chamber became not only ever-smaller over time, but also reduced possible flow dynamics. Cross-stratification turns to horizontally stratified deposits of mass flows towards the interior of the intrusion. Its structure still shows signs of deformation such as crystals being aligned and preferred movement at molten shear zones leaving light stripes in the rock.

Irvine et al. (1998) also blame mass flows for rhythmic stratification inside the intrusion, while others consider mass flows only a marginal phenomenon and point to the lack of traces of deformation in the central parts. Sidewall dynamics clearly has little impact on the distant center with a much larger intrusion like Bushveld. However, it is also conceivable that a lump of crystal slush from the roof region sinks like a diapir to the ground and spreads there.

Processes of the third type are believed to be largely responsible for rhythmic layering further away from margins (Boudreau and McBirney 1997). Constant burial of crystal slush causes it to be compressed (compaction). The size of gaps decreases and the remaining melt is partially squeezed out and rises upward. How much melt remains in the gaps determines the type of rock. Melt can also locally displace crystal slush and form streaks or layers. Moreover, rising melt reacts with crystals with which it is not in chemical equilibrium resulting in certain crystals possibly being dissolved thus changing the modal proportions of the cumulate.

However, it is not only chemical equilibrium that plays a role in this case but also the surface energy of the crystals. If solid rock is exposed to high temperature for a long period of time, then grain enlargement occurs in which large crystals grow at the expense of smaller crystals of the same mineral because the surface-to-volume ratio is energetically more favorable. This process known as Ostwald ripening occurs in metamorphic rocks and in plutonic rocks that have already solidified. The surface energy of a crystal also depends on its neighbors in that the more neighboring crystals resemble it the lower it is. This has little effect in solid rock because a crystal cannot choose its neighbors. However, in crystal slush this leads to crystals that are only present in small quantities being dissolved preferentially by flowing melt, while the most frequent phase has an energetic advantage. This effect reinforces modal variations that already exist and can go so far as to change a gradual rhythmic stratification into a rhythmic stratification with sharply limited monomineralic layers. For example, in one part of the Stillwater Complex (Montana, USA) layers consisting almost exclusively of plagioclase or pyroxene repeat at intervals of a few centimeters.

It is also possible for compaction to lead to mechanical sorting according to density. Geologists are only aware of this once they have separated crystals with a heavy liquid. When the differences in density are only small and there is little melt, floating and sinking crystals block each other and accumulate in coarse layers. This seems to happen in smaller sills too. If it also occurs in LMIs, then the corresponding texture has to have been modified by other processes that took place during compaction.

A further process has at times been considered by researchers: the formation of magma layers of varying density that are only circulated internally by convection. Such layered magma chambers are more likely to be the rule than the exception, but this is probably not (at least in the case of low-viscosity basic melts) directly "frozen" in the layering. However, direct stratification in liquid magma can cause other combinations of minerals to crystallize than would be the case in a homogeneous magma and raises many possibilities for magma mixing.

Last but not least is the process in which fluids segregated from the magma can flow through already solid rock and alter it (Zingg 1996; McBirney 2009) when certain minerals are replaced by others in a process called metasomatism. In some cases water has even led to remelting. There are also pegmatite-like coarse-grained layers and streaks that have developed from water-rich residual melts. Basic magmas have a very low water content compared with acidic magmas; hence these effects play only a minor role and then mainly in the last layers to solidify.

Spectacular stratifications have also formed in the extremely alkaline intrusions of Ilimaussaq and Lovozero both of which comprise agpaitic nepheline syenites (Sect. 3.11). In both complexes there are rocks whose proportions of cumulus minerals are subject to cyclic repetition whereby the composition of minerals hardly changes (somewhat surprisingly).

The lujavrite at Ilimaussaq has microrhythmic layering (dark–light). It has been suggested that in a magma layer not subject to dynamic circulation at the floor of an intrusion the crystallization of sodalite, alkali feldspar, and arfvedsonite (dark layer) could cause the remaining melt of the immediate surroundings to be enriched with certain elements such that nepheline and eudialyte might crystallize instead (light layer) for a short time. When further cooling takes place a dark layer follows once more because the crystallization front has migrated into melt of average composition (Bailey et al. 2006).

In contrast, Ilimaussaq's kakortokite has macrorhythmic layering in which the ratios of alkali feldspar, nepheline, eudialyte, and arfvedsonite change. One relatively plausible model proposes that during crystallization the vapor pressure in magma increased episodically until it exceeded the rock pressure and the vapor could escape. Since vapor pressure affects the liquidus temperature such episodic fluctuations led to an episodic change in the minerals crystallized (Pfaff et al. 2008). However, this must have been accompanied by repeated injections of magma as well.

As far as the cyclic layers of Lovozero are concerned Féménias et al. (2005a) argue for a completely different model on the basis of the rock texture in which each layer is an individual sill.

In contrast to basic or alkaline magmas the physical properties of acidic magmas cause granites to solidify into relatively homogeneous rocks. Only in rare cases is less developed layering observed such as a granodiorite in Morocco where rhythmic layering is attributed to the flow motion that took place during continued bloating of the pluton by intruding magma (Pons et al. 2006).

3.3.2 Chromium, Nickel, and Platinum in Basic Magmas

The peridotite of the Earth's mantle has a significantly higher content of chromium and nickel than rocks of the Earth's crust. Chromium is found mainly in pyroxene (chromium diopside) and at moderate pressure in chromium spinel, while nickel is found mainly in olivine. Both elements behave compatibly (i.e., largely remain in the mantle during partial melting). Their content is correspondingly lower in basalt, but still significantly higher than in average crust. In the fractionated crystallization of pyroxene and olivine both elements are effectively removed from the melt; hence the concentration is already low in slightly fractionated melts. Since olivine and pyroxene are not of course suitable as ores there is a need for chromite and nickel sulfides instead.

In the mantle PGEs are located in sulfides that make up only a tiny fraction of the rock. Normally, they largely remain in the mantle during melting. With a high degree of melting or repeated melting the sulfides are lost and the melt becomes PGE bearing (Naldrett et al. 2008). The distribution coefficient is about 1 so there is only minimal fractionation between rock and melt in which platinum and palladium are slightly incompatible, osmium and iridium are slightly compatible, and ruthenium and rhodium are in between. The same applies to the fractional crystallization of magma whose contents change minimally but in different directions. Although PGEs therefore behave completely differently from the highly compatible elements chromium and nickel, in LMIs they are nevertheless conspicuously often enriched in chromite layers or together with nickel sulfides (see also Box 3.9 for platinum placer deposits in the Urals).

Box 3.9 Platinum belt in the Urals

Platinum was extracted almost exclusively from placer deposits in the northern Urals (Russia) before the discovery of the Merensky Reef. Platinum from the northern Urals originates from a series of smaller ultramafic intrusions (Alaska–Urals type) lined up in a 500-km-long belt a little east of the main overthrust of the mountain range. In contrast to typical LMIs these probably originated at a subduction zone. PGE contents in the intrusions themselves are usually too low for mining. There are isolated secondary enrichments by hydrothermal fluids.

Although chromite layers are often only a few millimeters or centimeters thick, sometimes they are a few decimeters or a few meters thick. Platinum-rich UG2 chromitite in the Bushveld Complex, for example, is 70 cm thick. An extreme case is the 20-m-thick chromitite in the Kemi intrusion (Finland). Normally, chromite only crystallizes simultaneously (cotectic) with other phases such as olivine and in much smaller amounts than these. Therefore, it should only occur as subordinate in the cumulate. Several processes can explain why chromite and only chromite crystallized in the short term. We know from experiments that under certain circumstances an increase in pressure or an increase in oxygen fugacity can lead to this. However, the most elegant model preferred by many researchers is the mixing of different magmas. In fact, in chromite layers there is often an abrupt change in the composition of cumulates indicating the injection of new magmas. There are two possibilities.

The first possibility is the mixing of slightly fractionated melt of the intrusion with newly injected primitive melt (Irvine 1977; Murck and Campbell 1986). Box 3.3 explains fractional crystallization as following a cotectic line that can be imagined as a valley (with a low melting temperature) between the stability fields of different minerals. In the cotectic crystallization of olivine and chromite this valley describes a curve (Fig. 3.24). If a fractionated melt mixes with a more primitive melt, then the composition of the mixture is within the stability field of chromite. Therefore, chromite and only chromite crystallizes until the composition meets the cotectic line again.

The second possibility can be arrived at from the same phase diagram. The mixing of a primitive magma with an SiO_2-rich magma can also lead to a composition within the chromite stability field. The effectiveness is limited since SiO_2-rich melts generally contain hardly any chromium and the mixture can crystallize correspondingly less chromite.

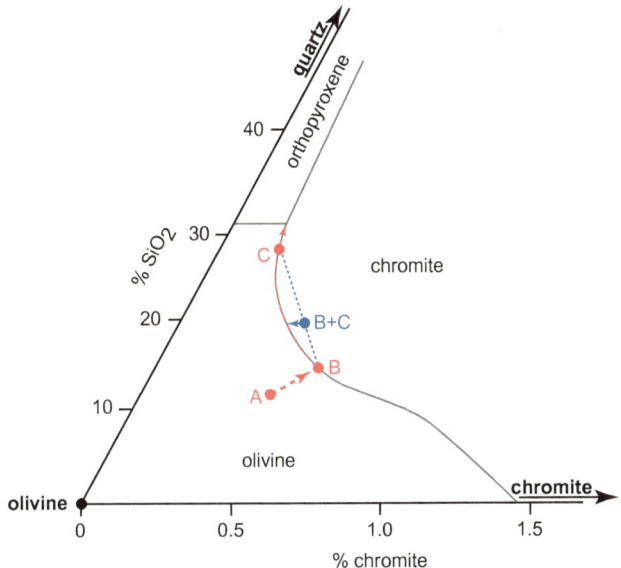

Fig. 3.24 Detail of the phase diagram of the olivine–chromite–quartz system that explains the formation of chromite layers (see also Box 3.3). Magma with composition A is in the stability field of olivine, which therefore crystallizes first. The composition of the melt evolves downhill from olivine to the cotectic line (magma B). Now olivine and a smaller amount of chromite crystallize at the same time. The melt develops downstream along the cotectic line until it meets the stability field of orthopyroxene. When a slightly fractionated melt of composition C is mixed with the more primitive melt B the resulting composition falls in the chromite stability field (magma B + C whose exact position on the connecting line depends on the mixing ratio). Now chromite and only chromite crystallizes and the melt composition changes downhill until it meets the cotectic line again. Moreover, the contamination of magma B with SiO_2 (or acidic magma) would push the composition into the chromite field (from Irvine 1977)

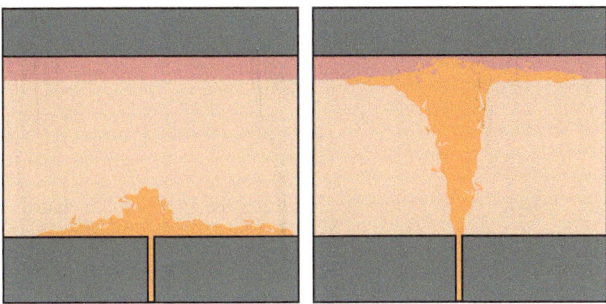

Fig. 3.25 The density of tholeiitic magmas temporarily increases during fractionation. As soon as the crystallization of magnetite begins, it decreases. When new magma of higher density than the existing magma is injected it spreads on the ground where mixing takes place (*left*). If the density is lower, then it rises (*right*) and mixes with magma under the roof possibly comprising acidic crustal melt

Nevertheless, the process is plausible (Kinnaird et al. 2002; Spandler et al. 2005). The heat of basic magmas can lead to significant melting of continental crust allowing a layer of granitic magma to accumulate under the roof of the intrusion. If primitive magma is injected, then it may have a lower density than mafic magma already fractionated (the density of tholeiitic magma increases temporarily during fractionation) and therefore rise to the roof where mixing with granitic magma leads to the crystallization of chromite (Fig. 3.25). The crystals would then have to sink to the ground through mafic magma possibly in the form of heavy lumps of crystal slush.

In the case of the Bushveld Complex (Sect. 3.3.3) it was pointed out that the Critical Zone contains significantly more chromium than would be expected from an estimate of magma quantity. According to an alternate model the crystallization of magma already began during its ascent on the way to the intrusion. Chromite is said to have accumulated in smaller magma chambers. The cumulate was cleared out again by later magma injections and brought into the intrusion as chromite sludge (Eales 2002; Mondal and Mathez 2006). Even an injection of chromite sludge into older cumulates in the form of sills has been proposed (Voordouw et al. 2009). Lowering of the central part of an intrusion that has not yet solidified completely brought about by the high density could cause the crystal slush to slide inward. This not only allows thicker layers to be formed, but can also lead to dynamic segregation of different suspensions that solidify into different layers (Maier et al. 2013). However, according to Naldrett et al. (2012) the variation in chromite composition cannot be explained by the movement of crystal slush. According to models proposed by these authors magma mixing on its own cannot explain the unusual Bushveld magmas. Instead, they suggest that the magma already had a significantly higher chromium content than assumed, although it remains unclear how it could have come about.

Nickel quickly disappears from the melt during the fractional crystallization of olivine. However, if early segregation of a sulfide magma occurs (liquid immiscibility; Sect. 3.1.2), then nickel will effectively fractionate into it along with copper and other elements. An increased sulfur content even has an effect on fractional crystallization since nickel behaves less compatibly and is therefore removed from the melt less quickly (Li et al. 2003). Segregated sulfide melt has a significantly higher density than silicate melt so the droplets sink to the bottom of the magma chamber where segregated sulfide melt penetrates the crystal slush or even accumulates in puddles and solidifies into a massive sulfide body. Frequently, larger numbers of crystals are deposited at the same time such that sulfides make up only a small part of a cumulate.

During the crystallization of sulfide melt, fractionation also occurs of course between mineral and melt (Li et al. 1996; Mungall et al. 2005). First, an iron–nickel–copper sulfide that is only stable at high temperature crystallizes. Elements such as copper, platinum, and gold are strongly enriched in the melt. Platinum minerals and native gold may crystallize from the last melt residues. Fractionation only has

Table 3.2 Sulfide minerals typically formed from sulfide melts

Pyrrhotite	FeS
Chalcopyrite	$CuFeS_2$
Pentlandite	$(Ni, Fe)_9S_8$

an effect if crystals and melt are separated. Sudbury (Sect. 3.3.5) and Norilsk (Box 3.10) are two examples where this has led to zoning within sulfide bodies. The original iron–nickel–copper sulfide is transformed into pyrrhotite, pentlandite, and chalcopyrite during further cooling possibly together with other sulfides (Table 3.2).

Box 3.10 Norilsk

The Siberian mining city of Norilsk is the northernmost city in the world and one of the most polluted places on the planet. The history of the city began in the days of Stalin when forced labor from gulags built the mines and metallurgical plants that eventually developed into the world's largest non-ferrous metal industry complex. The plants release large quantities of sulfur dioxide and heavy metals into the atmosphere thereby producing smog and acid rain—trees no longer grow in the surrounding area. Norilsk is the world's worst offender when it comes to SO_2 emission. The city is said to be responsible for 1% of global emissions. Soils in the area have extremely high concentrations of heavy metals and in places they contain so much palladium that it may even be worthwhile to mine them.

The Norilsk-Talnakh mine holds one of the world's largest deposits of nickel, palladium, and platinum and is also an important copper deposit. Cobalt, gold, silver, and other metals are also extracted. The Pd/Pt ratio is unusually high and the deposit contains approximately four times more palladium than platinum.

The ores are not found in a typical LMI but in relatively small sills located below the flood basalts of the Siberian Trapps that erupted at the Permian–Triassic boundary. The sills are considered to be part of the flood basalt vent system (Naldrett 1992; Arndt et al. 2003; Lightfoot and Keays 2005). In these sills the basaltic magma was contaminated with evaporites and other sedimentary rocks thus increasing the sulfur content of the magma. Droplets of segregated sulfide melt (liquid immiscibility) were able to absorb nickel, copper, and PGEs from the large melt volume. Over and over again magma was replenished from below, while the magma chamber was constantly tapped and its magma erupted on the surface as flood basalt. Flood lavas are correspondingly depleted in these elements.

Sulfide melt accumulated on the floor of magma chambers. During the crystallization of molten sulfide further fractionation took place resulting in the zonation of ore bodies.

In Jinchuan in China significant crystallization and segregation of sulfide melt happened even before injection into the magma chamber. When viscous crystal slush finally intruded the currents ensured that the sulfide melt was enriched in certain zones (De Waal et al. 2004).

PGEs are also enriched in sulfide magma during segregation of immiscible sulfide melt (Naldrett et al. 2008). This is so effective that later sulfides in later rock layers often contain very few PGEs because they have already been completely removed. In the case of Norilsk and Sudbury significant fractionation even occurred within the sulfide melt with PGEs being enriched in certain zones.

PGEs are often enriched in a certain rock horizon within normal cumulates. A good example is the Merensky Reef (Fig. 3.26) in the Bushveld Complex. It is a pyroxenite layer typically 1 m thick containing finely distributed sulfides (Schoenberg et al. 1999; Seabrook et al. 2005; Naldrett et al. 2008). These are predominantly Fe–Cu–Ni sulfides such as chalcopyrite, pyrrhotite, and pentlandite in addition to which

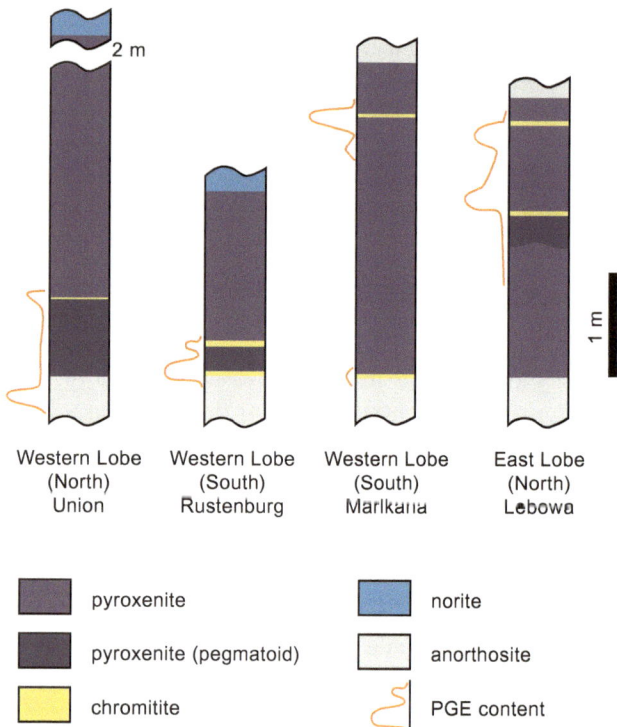

Western Lobe (North) Union   Western Lobe (South) Rustenburg   Western Lobe (South) Marikana   East Lobe (North) Lebowa

- pyroxenite
- pyroxenite (pegmatoid)
- chromitite
- norite
- anorthosite
- PGE content

Fig. 3.26 The Merensky Reef in the Bushveld Complex comprises a pyroxenite with a high PGE content. The profiles from four mines showcase this layer as not being completely homogeneous (from Naldrett et al. 2008)

PGE sulfides such as braggite and laurite and native PGE alloys also occur. Only PGEs are mined as the ore grade for nickel and copper is not economical. The immiscible sulfide melt thus segregated during the crystallization of pyroxene sucked the PGEs out of a large magma volume and sank together with pyroxene to the bottom of the magma chamber. In contrast to the Merensky Reef and the very similar J-M Reef of the Stillwater Complex, PGEs in the Great Dyke (Sect. 3.3.4) in Zimbabwe are not enriched within a specific rock layer that is underlain and superimposed by other rock types, but in thin horizons (main/lower sulfide zone) within a thick pyroxenite that is otherwise quite homogeneous.

It is generally assumed that the segregation of immiscible sulfide melt was triggered by the injection of fresh magma that mixed with slightly fractionated melt either directly because the new composition was in the miscibility gap or indirectly because the composition of the mixture can lead to crystallization of an iron-rich mineral (e.g., chromite), which in turn reduces the solubility of sulfur in the melt.

If the injection of fresh magma can trigger the formation of chromite layers and the segregation of a sulfide melt, then it is not surprising that there are also PGE-rich chromite layers such as the UG2 chromitite in the Bushveld Complex. PGE minerals such as PGE sulfides and native alloys occur there both as tiny inclusions in chromite crystals and finely distributed in the groundmass. However, Fe–Cu–Ni sulfides would be expected at the same time, which is not the case. Two different theories have been proposed to explain this. One of them assumed that a sulfide melt was also segregated here and corresponding sulfides were deposited together with chromite. However, these sulfides dissolved during cooling because chromite absorbs the iron contained in them. The other theory is based on the realization that PGEs are not homogeneously dissolved in magma but accumulate into clusters consisting of about 10–100 atoms. Not only can these be absorbed by a sulfide melt they can also be enclosed by growing crystals such as chromite (Ballhaus and Sylvester 2000). This may have resulted in a PGE-rich chromite layer without the aid of a sulfide melt.

Horizons with elevated PGE contents are relatively common but not always of economic interest. The thickness of the affected layer is often not constant; the highest PGE concentrations are usually found where the thickness of the layer is lowest. Of great importance for the PGE budget is the volume of magma from which the sulfide melt could absorb PGEs. Of course, the timing of segregation also plays a role. There are mafic intrusions in which this happened very late when magnetite had already crystallized. However, these enrichments are (at least for the time being) not economic. In other cases the sulfide melt segregated while the first magma was emplaced in the intrusion possibly triggered by contamination of the magma with the surrounding rock. These are the so-called contact-type PGE deposits. An example is the Platreef at the base of the northern lobe of the Bushveld Complex. The surrounding rocks are sediments such as shales, banded iron formations, and dolomite. The magma injected into sills reacted with the sediments or fused trapped blocks. This increased the sulfur content of the melt leading to segregation of immiscible sulfide melt. The cumulate deposited on the floor was a pyroxenite containing sulfide drops with a diameter of up to 3 cm and a high PGE content. Their isotopic composition clearly indicates the contribution of sulfur from sediments. However, there is also evidence that segregation already began before intrusion. PGEs may already have been enriched in segregated drops during magma ascent in smaller magma chambers only to be later brought into the sills in emulsion (Holwell et al. 2011). The nickel content in the Platreef is teetering on the brink of profitability.

PGEs can also be hydrothermally mobilized (especially, in hot saltwater). In some intrusions there are dunite tubes or streaks formed by solid rock reacting with late-magmatic fluids and sometimes showing increased PGE concentrations. However, these secondary processes only play a subordinate role.

3.3.3 Bushveld Complex

The Bushveld Complex (South Africa) (Fig. 3.27) accounts for almost half of global chromium production, a good two thirds of global platinum production, and almost a quarter of vanadium production. The shares of known worldwide resources of these metals are similarly high. It is therefore clear that this is a unique and exceptional deposit. With an area of 66,000 km^2 (only slightly smaller than Bavaria) and a thickness of 6–8 km it is by far the largest LMI. Although the shape resembles a flat disk (a gigantic sill), its thickness corresponds to one fifth of the continental crust!

Mafic magmas derived from the Earth's mantle intruded about 2.06 billion years ago along the boundary between sediments of the Transvaal Supergroup and the overlying Rooiberg Group consisting of (at that time still young) volcanic rocks. Some connection with a mantle diapir is usually assumed. Alternately, a position in the back-arc of a subduction zone was proposed (Clarke et al. 2009) related to the collision between the Kaapvaal Craton and the Zimbabwe Craton, which began at about the same time.

At the Earth's surface the Bushveld Complex (more precisely, the Rustenburg Layered Series) forms several lobes that are arranged in a ring and plunge toward the center. The Western Lobe and Eastern Lobe are the two most important and are almost identical. In fact, geophysical data suggest that they are or were connected at depth (Cawthorn and Webb 2001). The intrusion has a significantly higher density than average continental crust. It is therefore

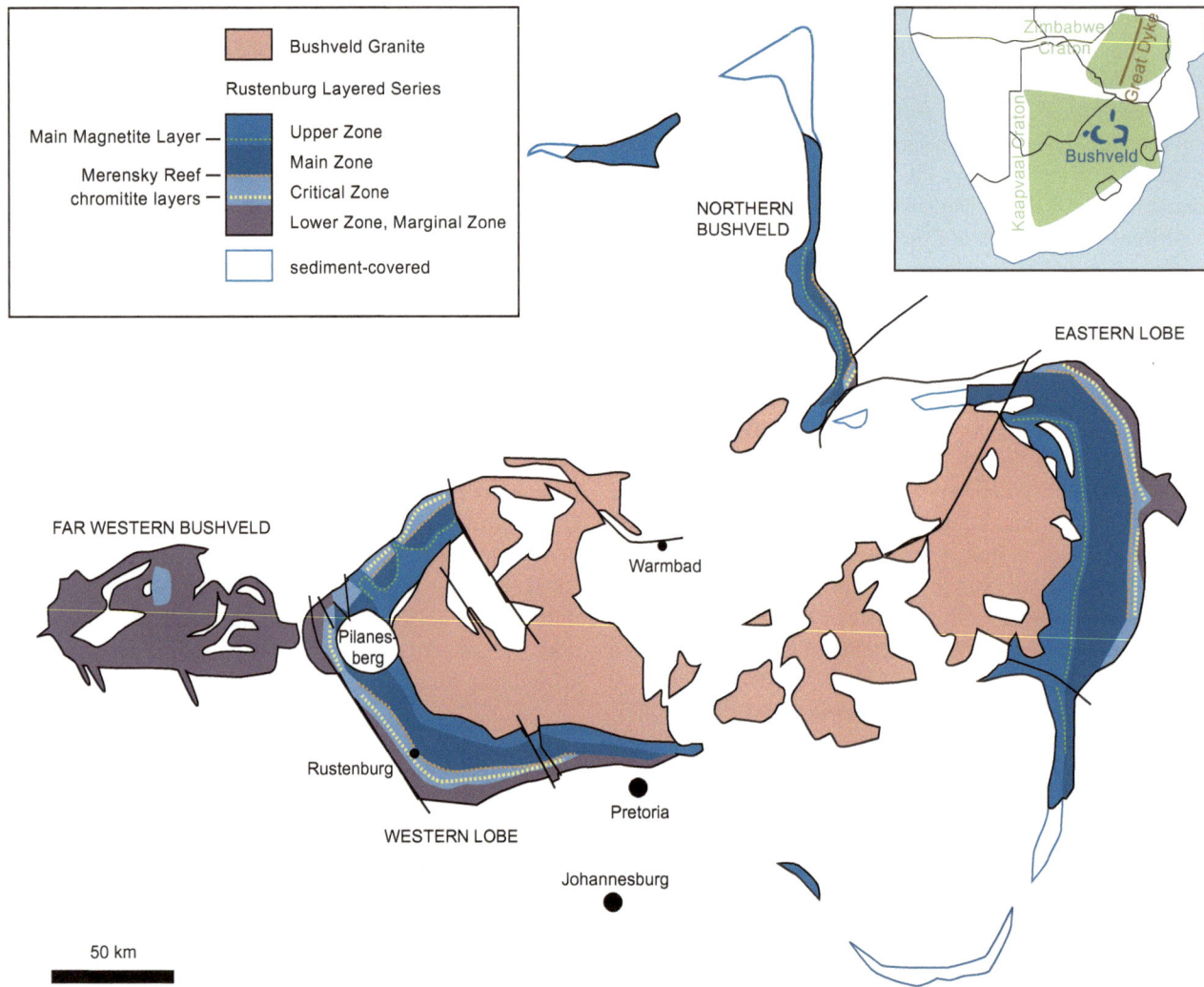

Fig. 3.27 The Bushveld Complex is a disk-shaped mafic intrusion 6–8 km thick whose central part is arched downward. This results in the ring-shaped arrangement of individual lobes on the surface the most important of which are the Western Lobe and Eastern Lobe. Platinum, chromium, and vanadium mines are located along the *dotted lines* roughly representing the position of the most important layers. The Bushveld granite located in the center above the mafic intrusion is slightly younger than the mafic intrusion. Pilanesberg is a younger alkaline intrusion (modified from Naldrett et al. 2008)

assumed that the entire crust subsided under the weight of the massive mafic intrusion (Fig. 3.28) resulting in today's bowl shape. Subsequently, the Bushveld granite located in the center above the mafic intrusion intruded. Its magma originates from the base of the continental crust. It is worth mentioning that it contains tin deposits (tin granite; Sect. 3.7.1), but at this point we are more interested in the mafic intrusion (Fig. 3.29).

The base of the Marginal Zone consists of quenched norite (a rock similar to gabbro, but with orthopyroxene instead of clinopyroxene) that is more or less heavily contaminated by the surrounding rock. The remainder of the intrusion is a sequence of rocks that at first glance look like a sequence of cumulates formed from an increasingly fractionated melt. The Lower Zone is made up of ultramafic cumulates of orthopyroxene and olivine. In the Critical Zone chromite is added, in the Upper Critical Zone plagioclase, while olivine is only present in individual layers. The following mafic cumulates of the Main Zone consist of plagioclase and pyroxene (gabbronorite). In the Upper Zone these become significantly enriched in iron with the introduction of magnetite as a cumulus phase.

However, the details are much more complicated. Cryptic layering shows above all that fresh magma was injected repeatedly and mixed with the already fractionated melt of the intrusion. These injections must have come from at least two different types of magma, but the exact compositions are controversial (Eales 2000; Cawthorn 2007). Additionally, the magmas are more or less contaminated as a result of assimilating continental crust. The giant sill was thus still

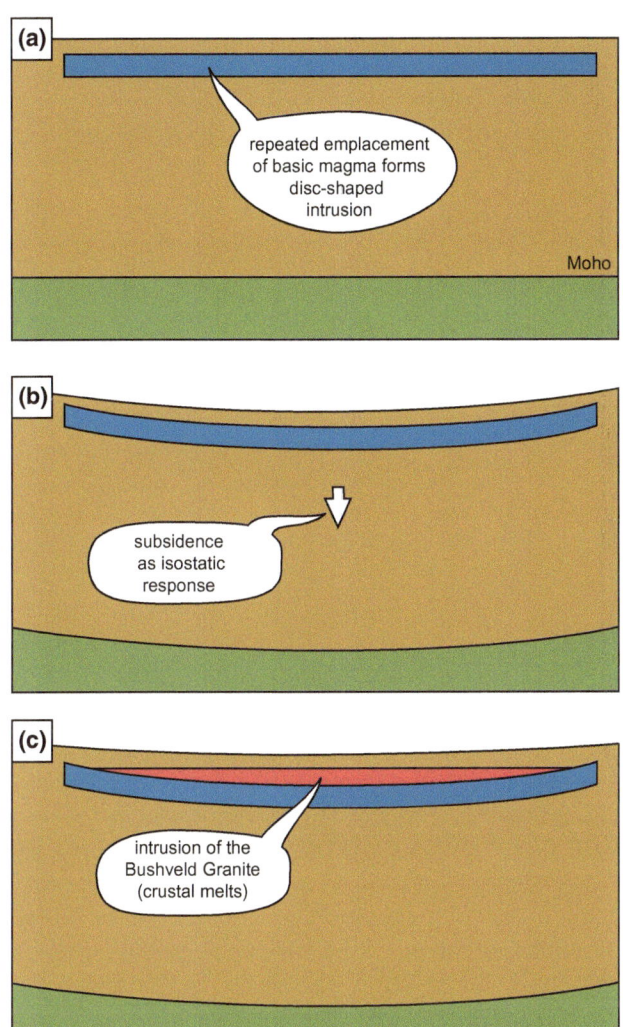

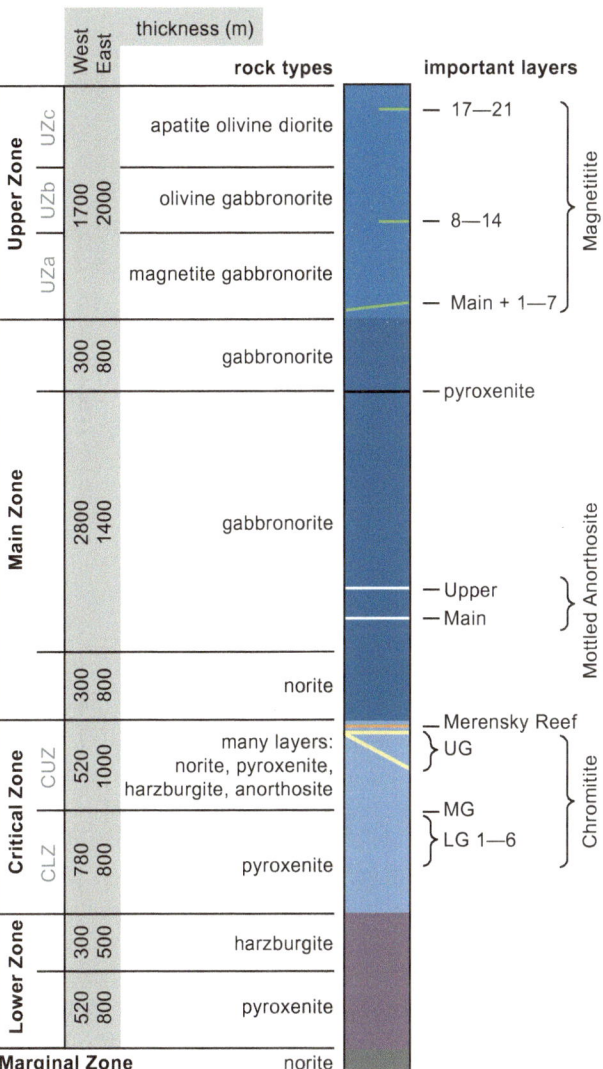

Fig. 3.28 Development of the Bushveld Complex. **a** The mafic intrusion was disk-shaped, about 7 km thick on average, and about 400 km wide. The magmas that intruded in several batches and made the intrusion ever-thicker were derived from the mantle. **b** The intrusion had a significantly higher density than average continental crust; hence the crust subsided. **c** Subsequently, the Bushveld granite, a melt derived from the base of the crust, intruded (from Cawthorn and Webb 2001)

Fig. 3.29 Profile of the Bushveld intrusion. Its total thickness is 7200 m in the west and 8100 m in the east. The marginal zone at the base consists of quenched norite. The lower part of the intrusion (lower zone and CLZ) consists largely of ultramafic cumulates. The very heterogeneous and strongly stratified upper part of the critical zone (CUZ) is overlain by the homogeneous mafic cumulates of the main zone. In the upper zone these cumulates are increasingly iron-rich and more strongly fractionated felsic layers occur. Chromium (mainly from mining the UG2, LG, and MG chromitite layers) and platinum (Merensky Reef and UG2) are enriched in the critical zone. The magnetite layers of the upper group are primarily mined as vanadium ore. Important marker horizons are also shown (from Cawthorn 2007)

being filled during crystallization and not only became ever-thicker, but also increased its surface area (Kruger 2005). This is the reason the lower zones are missing in the Northern Lobe of the complex.

The Lower Zone and Critical Zone formed when magma was more or less continuously injected and the sill inflated (Kruger 2005). Fractional crystallization was partly compensated by constant magma mixing. Most likely it was a basaltic andesite (i.e., intermediate SiO_2 content) with a very high magnesium content (Cawthorn 2007; Godel et al. 2011), which is a rather unusual melt. Going into greater detail the magma pulses produced cyclic sequences that are particularly noticeable in the Critical Zone where they begin with a chromitite layer, ideally followed by pyroxenite,

norite, and anorthosite (Fig. 3.30) (see also Van der Merwe and Cawthorn 2005). These layers can usually be traced for more than 100 km, but their composition and thickness do change. This depends mainly on the distance to the dikes through which new magma was injected. There are also erosion structures caused by currents. The chromitite layers are subdivided into Lower Group, Middle Group, and Upper Group (LG, MG, UG) and numbered according to their position within the Critical Zone.

Fig. 3.30 Chromitite layers
(*dark*) in anorthosite (*light*) on the
Mononono River (South Africa).
They belong to the UG1
chromitites of the critical zone of
the Bushveld Complex (*photo* ©
Kevin Walsh, CC-BY,
Wikimedia Commons)

Completely different magma from the Middle Zone then intruded (possibly via a different dike system) with a composition that was basically basaltic. It follows from the composition of the Middle Zone that there was no significant mixing with the remaining melt of the Critical Zone. This is astonishing because it follows from the composition of the crystals in the Critical Zone that only about a quarter was crystallized, so there must have been a remaining magma layer several kilometers thick. Presumably, the new Middle Zone magma had a higher density than that of the Critical Zone, it spread on the bottom of the intrusion, and then pressed the remaining melt of the Critical Zone out of the intrusion (Cawthorn 2007; Naldrett et al. 2012). So far it can only be speculated where this melt went. Did it disappear in dikes, sills, or smaller intrusions in the surrounding rocks or in the northern lobe of the complex into which the first magma intruded at this time, or did it erupt from long gone volcanoes?

Subsequently, magma of the Main Zone crystallized quietly until finally a significant amount of magma (presumably iron-rich basalt) intruded and mixed with the remaining melt forming the pyroxenite at the base of the Upper Main Zone. Continued fractional crystallization formed the Upper Main Zone and after the onset of magnetite crystallization the Upper Zone.

The heat of the intrusion was high enough to partly melt the roof. Therefore, a layer of acidic magmas occasionally existed above the mafic magmas solidifying to the fine-grained granite (or granophyre) that can be found in places above the Upper Zone.

As explained in Sect. 3.3.2 the chromitite layers and the enrichment of PGEs in the Merensky Reef are due to newly injected primitive magma mixing with the slightly fractionated melt of the intrusion. The Merensky Reef (Fig. 3.26) is a pyroxenite located at the base of the last cycle of the Critical Zone. This cycle differs from all previous ones in that this time magma of the Main Zone was injected for the first time. Surprisingly, the isotopic composition of the crystals in the cumulates of this cycle corresponds to either the Critical Zone or the Main Zone. Therefore, it seems there was hardly a mixture of magmas; rather the crystals grew in two different magma layers and accumulated together on the bottom of the magma chamber (Seabrook et al. 2005). Perhaps this model is still too simple. This could be addressed by the suggestion that an alternating influx of both magma types occurred (Mitchell and Scoon 2007).

A completely different model for the Bushveld Complex resulted from a study of pyroxenite located at the base of the Upper Main Group. As already pointed out it is usually explained by the injection of primitive magma into the fractionated magma of the Main Zone. However, according to Maier and Barnes (2010) the composition is more in line with the Critical Zone. They therefore believe that the Upper Main Zone is in reality the uppermost part of the Critical Zone, while magma of the Main Zone was emplaced only later as a sill into not yet completely solidified cumulates. The central area of the intrusion had been pressed downward and the cumulates of the Critical Zone slid to the center following gravity. Melt residues between the crystals were partially emplaced into other layers.

The Northern Lobe of the Bushveld Complex differs from the stratigraphy described above. Here the Lower Zone and Critical Zone are missing. They are replaced by the PGE-rich Platreef located below the Main Zone (Sect. 3.3.2). This may be magma from the Critical Zone that has entered from the side in the form of a sill. However, there are also arguments for the hypothesis that it was originally a small independent intrusion that was only later connected to the main intrusion during the injection of Main Zone magma (Holwell et al. 2011).

3.3.4 Great Dyke

The Great Dyke (Zimbabwe) at 550 km long and 4–11 km wide is the second largest platinum deposit in the world and an important chromium deposit. Strictly speaking this LMI consists of two large magma chambers (and a third small one) that in turn are subdivided into individual subchambers (Fig. 3.31). In profile the chambers are reminiscent of a funnel with the layers curved downward in the center. The intrusion consists largely of ultramafic cumulates. In the lower part olivine predominates with cyclic layering of chromitite, dunite, and harzburgite. In the upper part pyroxene predominates and there is cyclic layering in the sequence chromitite, dunite, harzburgite, and pyroxenite. Mafic cumulates (gabbro and norite) then follow above the ultramafic cumulates. The cycles can be traced back to the repeated injection of magma. Platinum, palladium, and other PGEs are mainly enriched in the Main Sulfide Zone. This is a horizon within the uppermost pyroxene layer with finely distributed sulfides between the pyroxene crystals. The intrusion was emplaced 2.58 billion years ago (Oberthür et al. 2002). It runs across the entire Zimbabwe Craton, which was a small continent at the time. It is conceivable that a mantle diapir not only triggered strong magmatism, but also led to extension and the formation of a rift system.

3.3.5 Sudbury Igneous Complex

There is only one deposit in the world known to have been formed by meteor impact. This is the Sudbury Igneous Complex (Ontario, Canada) dated as being 1.85 billion years old. It is one of the two most important nickel deposits (the other being Norilsk; Box 3.10) and produces copper, cobalt, platinum, palladium, gold, and silver. The nickel does not of course come from the meteorite. The diameter of the meteorite crater is estimated at 200–250 km. This would make it the second largest known meteorite crater on Earth (Figs. 3.32 and 3.33), about 10 times larger than the Nördlinger Ries (Germany). Since several kilometers of rock have been eroded since then only the central area of the crater

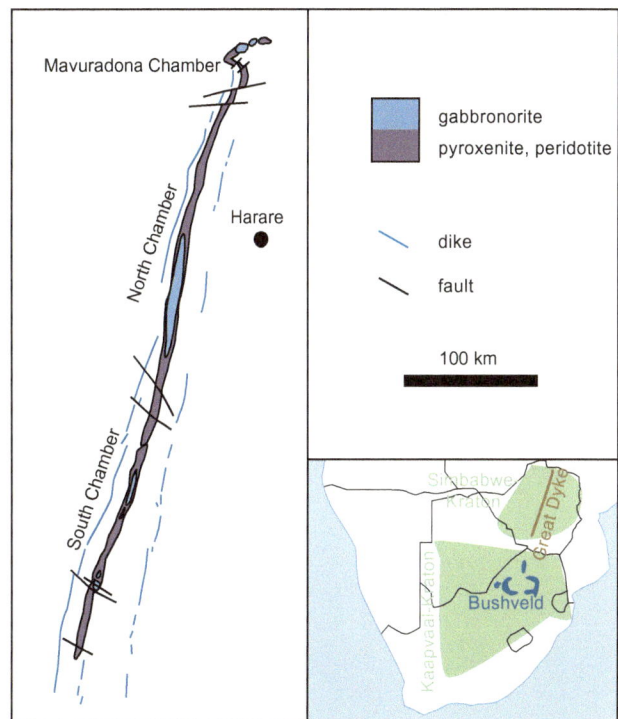

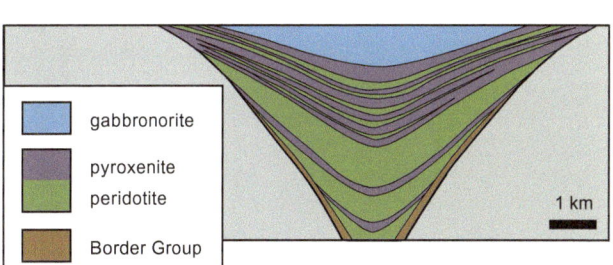

Fig. 3.31 The Great Dyke (Zimbabwe) is an intrusion that at first glance appears to be a giant magmatic dike. It is composed of three magma chambers that in turn are subdivided into several subchambers. In profile their shape is reminiscent of a funnel (from Naldrett and Wilson 1990; Wilson et al. 2000)

bowl remains (Roussel et al. 2003). Where the edge of the crater was once located a deeper zone with footwall rocks is exposed. In this area the rocks are partly broken into impact breccias and violent movement in faults has even led to melting by frictional heat (pseudotachylites). During a later orogeny the structure was deformed to an ellipse shape. Thus, a shear zone with thrust faults runs through the complex.

The impact of the 10-km-wide meteorite was so violent (Box 3.11, see also Box 5.4) that it more or less penetrated the entire continental crust. A huge magma lake on average 2.5 km deep formed in the crater. This developed into a kind of layered mafic intrusion. However, "intrusion" is not quite correct in this case so we better call it an impact melt sheet—it differs from other LMIs in some important points. Similar

Fig. 3.32 The sudbury igneous complex was formed by meteorite impact. Nickel deposits are mainly located at the base of the norite and in the offset dikes. Lake Wanapitei is yet another impact crater but much younger (modified from Ames et al. 2008)

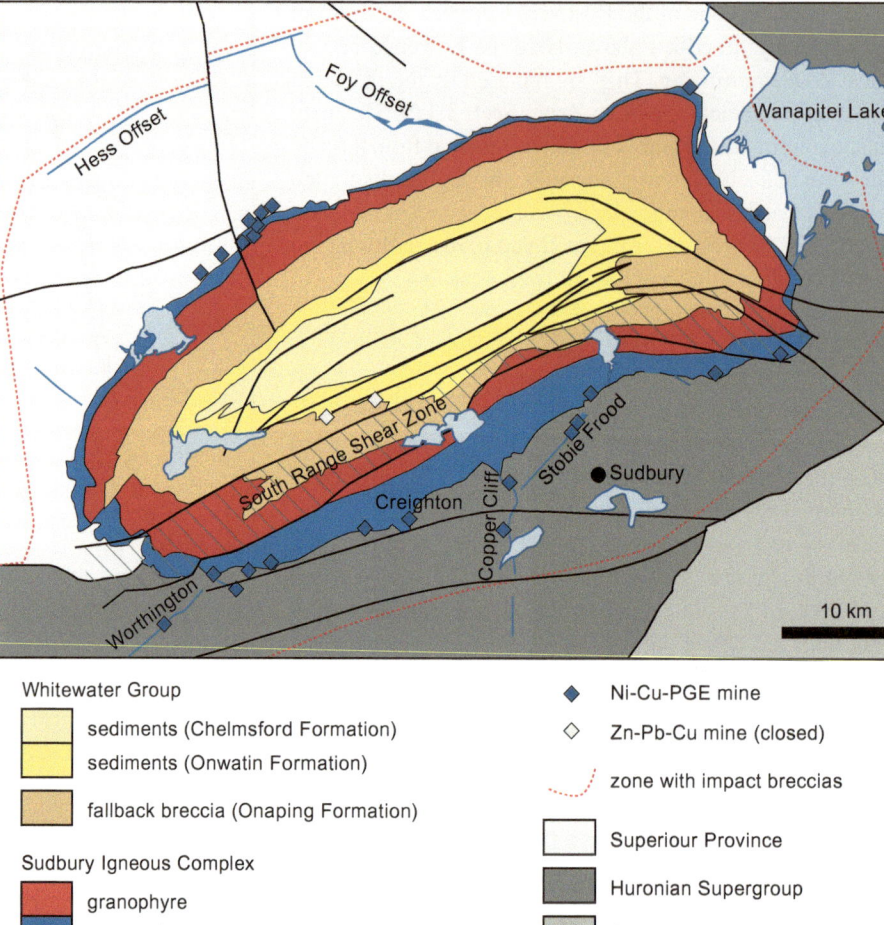

Whitewater Group

- 🟨 sediments (Chelmsford Formation)
- 🟨 sediments (Onwatin Formation)
- 🟫 fallback breccia (Onaping Formation)

Sudbury Igneous Complex

- 🟥 granophyre
- 🟦 gabbro / norite

- ◆ Ni-Cu-PGE mine
- ◇ Zn-Pb-Cu mine (closed)
- ⌇ zone with impact breccias

- ⬜ Superiour Province
- ⬛ Huronian Supergroup
- ⬜ Grenville Province

Fig. 3.33 Satellite image of the sudbury igneous complex (*oval structure*). This is the remnant of the world's second largest known meteorite crater. Lake Wanapitei (*upper right*) was created by another meteorite impact (*photo* © NASA)

melt sheets are known from other impact craters, but they are clearly smaller and not differentiated to different layers.

Magma in all other LMIs comes from the Earth's mantle, whereas here it is melted crust. To the north of the complex there is Archean gneiss of the Superior Province and to the south there are basalt, gabbro, granite, and Paleozoic sediments. The average melt composition of Sudbury is approximately intermediate between granite and gabbro making it much more acidic than other LMIs, but more basic than normal crustal melts. The unusual composition is due to extremely high temperatures far above the liquidus being reached suddenly. The rock melted completely, which normally never happens. In this way even compatible elements such as nickel that would otherwise remain in the solid rock when gabbro was melted ended up in the melt. While earlier researchers blamed a mantle melt or melted older deposits for the nickel ores, there is no geochemical evidence for this even though a small contribution of mantle melts cannot be ruled out.

In the center of the basin above the igneous rocks the crater fill is exposed (Whitewater Formation). In the hours following the impact the ash cloud originally comprising melt droplets solidified into glass rained down and formed a tuff-like rock (Onaping Formation) called suevite (a.k.a. fallback breccia). This layer is 1.5 km thick. The lower part recrystallized by heat from the covered magma lake to solid rock and some material even melted and extruded in the form of mini granite plutons into the upper parts of the suevite. The crater filled with water and magmatic heat from below caused repeated steam explosions. In the upper part of the Onaping Formation large hydrothermal zinc–lead–copper deposits similar to VMS (Sect. 4.16) were formed. At the bottom of the lake various sediments were deposited including limestone, silt, and greywacke (Fig. 3.34).

The covered superheated melt sheet probably took millennia to cool down from the original 1700 °C to the liquidus before crystallization began. In the beginning it was so hot that the melt sheet even grew by melting the wall rock. As a result the melt became increasingly heterogeneous because only small convection cells were formed; hence the entire melt sheet was not subject to mixing (Darling et al. 2010). Since mafic and felsic rocks melted at different rates an emulsion may have been formed from different melts resulting in stratification of the intrusion. The upper half has the composition of granite (granophyre). The mafic lower half consists of several layers mostly called gabbro and norite in the literature. Strictly speaking they fall into the quartz monzodiorite field in the QAPF diagram (Therriault et al. 2002).

The unusually high temperature of the melt had a further effect that does not occur with other LMIs. So much sulfur was dissolved in the original melt that most likely immiscible sulfide melt had already segregated before crystallization had begun (Lightfoot et al. 2001; Keays and Lightfoot

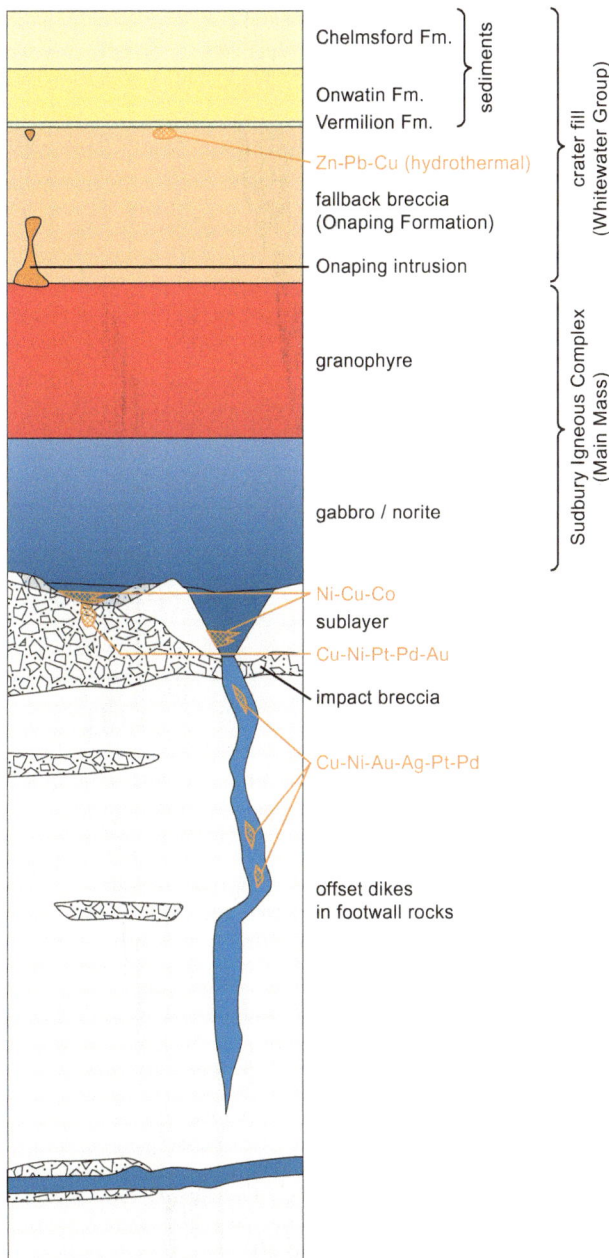

Fig. 3.34 Profile of the sudbury igneous complex. Magmatic sulfide deposits are located in troughs at the base of igneous rocks (sublayers), in the underlying country rock (footwall), and in dikes cutting through country rock (offset) each with a different metal content. Breccia zones are numerous in country rock. Above the igneous rocks the impact crater is filled with suevite and sediments (from Ames et al. 2008)

2004). Sinking droplets of sulfide melt ensured that elements such as nickel, copper, and PGEs were almost completely removed from the upper layers and were effectively enriched at the base. There the sulfide melt accumulated mainly in depressions and troughs (the so-called sublayer). Sulfide was still being segregated when magma had cooled down enough for crystallization to commence.

The sublayer (i.e., troughs at the base of mafic rocks) contains most of the ores. It consists largely of very heterogeneous norite with high sulfide content together with many wall rock fragments. The ore forms zones with massive sulfides. It is also disseminated in norite. Sulfide melt also partly penetrated the breccias below and formed dikes and (in former pores) disseminated ores. The amount of sulfide that has accumulated in a depression is related to the thickness of the igneous rocks above it, which varies between 300 and 5000 m. The ore in the sublayer (contact ore) and the ore in the breccias (footwall ore) have different compositions due to fractional crystallization of sulfide melt (Li et al. 1996; Prevec et al. 2000; Mungall et al. 2005), on the one hand, and to the continued accumulation of further sulfide droplets with a changing composition, on the other hand. Finally, later hydrothermal remobilization also occurred to a certain extent (Molnár et al. 2001).

About a quarter of the ores are located in magmatic dikes outside the actual igneous complex (offset ore). These dikes (Lightfoot and Farrow 2002; Scott and Benn 2002) run either radially or concentrically around the complex and mainly follow the breccia zones. Horizontally, they are not continuous and it is suspected that magma penetrated from above from the melt sheet (later removed by erosion in these areas). It is believed that the first melt was injected a few seconds after impact when the edges of the original crater slid into the interior. Accordingly, the quartz diorites of these dikes can be considered the average melt composition of the complex. In the center of these dikes (especially, where they are particularly thick) there is a zone that contains a large number of wall rock fragments and sulfides. It is probably a melt that was later injected into the not yet completely solidified dikes at a time when sulfides were already accumulating at the base of the melt sheet. In dikes the ores occur in the form of massive sulfides, vesicles, veinlets, and disseminated ores.

Box 3.11 Impact crater
The impact of a large meteorite releases unimaginable forces that act in the blink of an eye (Fig. 3.35). First, shock waves spread and break the surrounding rock. Fracture surfaces are typically arranged in the form of so-called shatter cones. Some minerals are converted into high-pressure minerals such as stishovite, coesite, and diamond. Broken rock is ejected to the sides, the center bulges downward, and extremely high temperatures prevail as a result of which country rock is melted abruptly. In the case of the Sudbury Igneous Complex this superheated melt sheet was particularly thick and even partially evaporated. An ash cloud of small melt droplets rises above the crater. Instantly a deep, hemispherical crater forms the unstable rims of

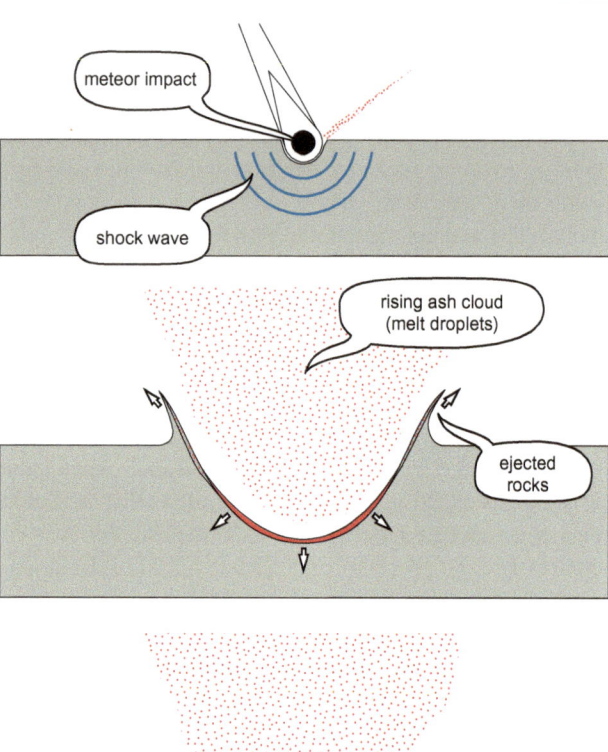

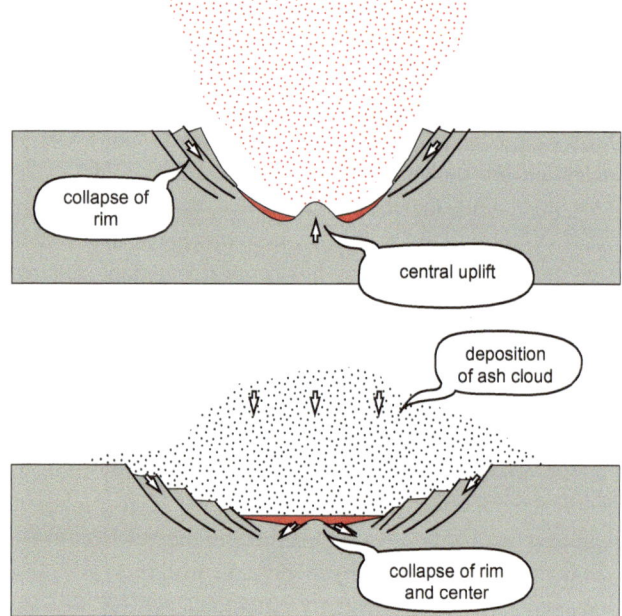

Fig. 3.35 It takes just a few seconds after impact of a meteorite for a crater to realize its final shape modified by the collapse of the rims. The rock is broken by shock waves and subsequent movement and is partly melted (*red*) With large craters the center rebounds to form a central uplift. This is known to collapse again with the result that very large craters have multiple rings. Finally the ash cloud rains down and forms a tuff-like fallback breccia (suevite)

which immediately slide into the interior. With large craters the center rebounds at the same time and forms a central mountain (that may collapse again with the largest craters). A few seconds after the impact things begin to settle down. However, the ash continues to rain down forming a tuff-like rock called suevite. The name is derived from Swabia and refers to the Nördlinger Ries (Germany).

Nothing remains of the meteorite itself with such a violent impact. One trace of the meteorite is the so-called iridium anomaly that often occurs in sediments of the same age. These contain unusually high concentrations of iridium and other PGEs. In the case of the Cretaceous–Paleogene boundary (formerly known as the Cretaceous–Tertiary boundary, a.k.a. the K-T boundary) this anomaly even occurs worldwide. It was caused by the Chicxulub impact in Mexico, which is blamed for the extinction of the dinosaurs. However, these iridium anomalies are not a mineral deposit. Sudbury (Canada) is the only known mineral deposit formed during an impact, the ores formed in the melt sheet of fused crustal rocks. Moreover, the Chicxulub impact is indirectly related to an oil deposit (Sect. 6.3).

The largest known meteorite crater on Earth is the heavily eroded Vredefort Crater (South Africa) that is 2 billion years old and has a diameter of 250–300 km. The next largest are Sudbury and Chicxulub. The Nördlinger Ries at 14 million years old with a diameter of 24 km is of medium size.

3.4 Komatiite Hosted Deposits

In the Precambrian (Box 3.12) the Earth was even hotter than it is today and the mantle could be melted to a higher degree (40–70%) resulting in ultrabasic magma (<45% SiO_2) with a high MgO content. Named for a river in South Africa this volcanic rock is called komatiite. It consists predominantly of olivine and orthopyroxene. Such magma is very fluid and extremely hot at a temperature of more than 1600 °C. When cooled quickly its typical spinifex texture with elongated olivine crystals is formed reminiscent of *Spinifex* grass growing in the Australian steppe.

Box 3.12 The world in the Archean

"The present is the key to the past." (Charles Lyell)

The fact that in the past the same processes took place as in the present was one of the most important dogmas of geology. However, the further back we go in the history of the Earth the less this dogma applies. In the Archean, the oldest period from which rocks are preserved (3.8–2.5 billion years ago), the Earth differed greatly from the present. The Earth was even hotter than today and the processes of plate tectonics were faster and somewhat different. At hotspots and mid-ocean ridges temperatures and degree of partial melting were so high that ultrabasic magmas like komatiites could develop. Subducted oceanic crust could be melted directly so that plagioclase-rich plutonic rocks such as tonalite, trondhjemite, and granodiorite formed—not the magmas typical of today's subduction zones. Unlike today there may also have been types of vertical tectonics such as crustal diapirs and sagduction.

The first continents, still preserved today as so-called cratons, originated and grew in the Archean. Their embryos had been formed by strong magmatism over mantle diapirs and subduction zones, and the rapid movement of tectonic plates caused deep-sea mountain chains and island arcs to collide with each other. The results are the so-called greenstone belts consisting of (low-grade metamorphic) volcanic and plutonic rocks and sediments such as the Abitibi Greenstone Belt (Box 4.27). Gneiss and (high-grade metamorphic) granulite are found when a deeper part of the crust has been exposed by erosion.

The first life-forms such as cyanobacteria appeared in the Archean. Through photosynthesis these unicellular organisms released oxygen slowly changing the originally hostile, reduced atmosphere and creating the conditions for higher life-forms to evolve. This oxygen also played an important role in the formation of banded iron formations (Sect. 5.2).

Komatiitic pillow lavas and large lava flows are very common in the Archean greenstone belts. In the Proterozoic there were fewer komatiites and only one known occurrence formed in more recent times.

Greenstone belts are zones in Archean cratons (Box 3.12) that consist mainly of igneous rocks and sediments that were later subjected to low-grade metamorphism. A characteristic rock is greenschist that is formed by low-grade metamorphism of basalt and contains green minerals such as epidote, actinolite, and chlorite. Greenstone belts can contain a variety of deposits such as nickel in komatiite and gabbro; they also contain VMS deposits (Sect. 4.16), banded iron formations (BIFs; Sect. 5.2), orogenic gold veins (Sect. 4.2), and gold placers (Sect. 5.9).

In some komatiites there are nickel deposits that also contain some copper and in places PGEs. Although the deposits are typically several orders of magnitude smaller than those in LMIs, they are also significantly higher in grade. Very many such deposits are located in Western Australia (60% of the world's komatiite-bound nickel reserves) (especially, in the Eastern Goldfields of the Yilgarn Craton) including the unusually rich deposits of Kambalda and Mount Keith (Hoatson et al. 2006; Barnes 2007). In Canada there are such deposits in greenstone belts around the Slave Craton, in the Thompson Nickel Belt (Manitoba), the Cape Smith Belt (northern Quebec), and the Abitibi Greenstone Belt (Ontario and Quebec; Box 4.27). There are others in Zimbabwe. It is astonishing that all these deposits were formed in a relatively short period of time in the late Archean and early Proterozoic. The vast majority of the world's komatiites, on the other hand, contain no deposits. One reason could be a change in mantle dynamics; another could be a change in the composition of sedimentary rocks because assimilation with sediments seems to play an important role.

Komatiite lava flows were fed by fissure eruptions (Fig. 3.36). The highly fluid lava spread mainly in the form of sheets less than 20 m thick that flowed dozens to hundreds of kilometers cooling relatively quickly and solidifying into their typical spinifex texture. In the center of these lava flows channels formed in which lava flowed faster and more turbulently. The extreme heat of the lava coupled with turbulence was even capable of partly melting underlying rocks making the channels ever-deeper. In some cases these channels cut through sediments thus increasing the sulfur content in the magma and triggering the segregation of

immiscible sulfide melt. Olivine crystallizing in the melt also accumulated on the bottom of the channels.

The highest ore grades (Kambalda type) were produced when the sulfide melt sank to the bottom of the channel and accumulated in depressions. These puddles, which can be up to 2500 m long and 300 m wide, solidified into massive sulfide consisting mainly of pyrrhotite and pentlandite (as well as pyrite, chalcopyrite, magnetite, and chromite). In Kambalda there are several such massive sulfide bodies containing about 3.3% nickel. Former channels above these bodies are filled with cumulates consisting almost exclusively of olivine (dunite). Sulfides are sometimes dispersed in this cumulate.

The Mount Keith type consists instead of disseminated sulfides within olivine cumulates. Although the ore grade is thus significantly lower (0.1–1.5% nickel), the ore volume is much larger. These deposits are particularly economical due to their comparatively low iron sulfide content. In addition to pentlandite other nickel sulfides such as millerite, heazlewoodite, godlevskite, and polydymite also occur.

In contrast to these types the Thompson deposits (Canada) are not lava flows but subvolcanic sills that were emplaced into sedimentary rocks.

Fig. 3.36 Komatiite lava was highly fluid and extremely hot. Lava flows spread largely in a sheet-like manner, but deep channels were also formed as a result of faster turbulent flow and assimilation of footwall rock by thermal erosion. Some of these channels cut through sediments thus increasing the sulfur content of the melt. Segregated sulfide melt sank to the bottom of the channel and accumulated into puddles (Kambalda type). In other cases olivine crystals and sulfide melt droplets were deposited together (Mount Keith type) (from Dowling and Hill 1998; Hoatson et al. 2006)

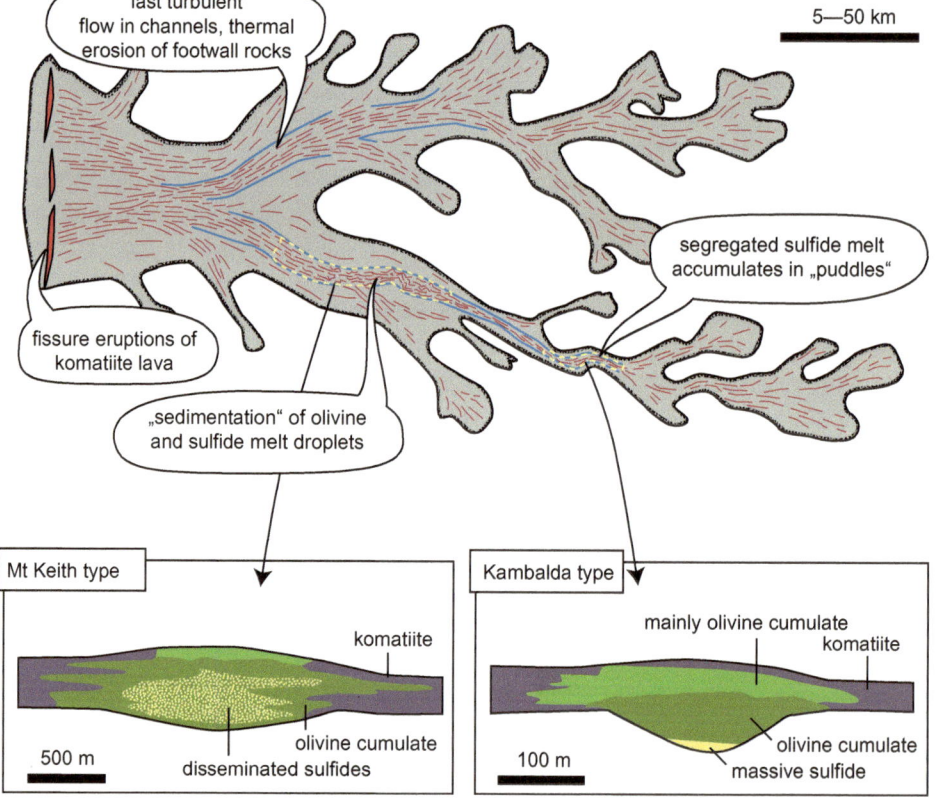

All these deposits were later more or less modified by tectonic movement, penetrating hydrothermal solutions, and metamorphism. This eventually led to secondary nickel enrichment in that pyrrhotite was replaced by pyrite, marcasite, and magnetite and at the same time the sulfides reequilibrated with olivine. Olivine in the cumulates often converted to serpentine by hydration turning dunite into serpentinite. Hydrothermal solutions led to strong mobilization and reprecipitation in the Thompson deposits.

3.5 Anorthosite-Hosted Deposits

Major primary titanium deposits are located in Proterozoic anorthosite complexes. Anorthosites are plutonic rocks consisting of more than 90% plagioclase (corresponding volcanic rocks are not known). Even though the rock can take on a very light color, it has nothing to do with granites. The mineralogical and chemical composition correspond most closely to gabbro or norite. The bright parts of the Moon consist of anorthosite. On Earth anorthosite occurs in the form of cumulate layers in ophiolites and in LMIs (Sect. 3.3). There are also large anorthosite intrusions that were formed exclusively in the Precambrian. Since the melting temperature of this composition is far too high even for the Archean, it has been assumed that a plagioclase-rich crystal slush had been emplaced. The melt between the crystals was probably fractionated tholeiitic basalt and thus rich in iron and titanium.

Archean and Proterozoic anorthosites differ greatly from each other. Archean anorthosites are smaller intrusions or sills found in oceanic crust or at island arcs and have an extremely Ca-rich plagioclase often in the form of snowflake-like clusters. On the other hand, the so-called massive anorthosites from the Proterozoic period form gigantic plutons within the continental crust and almost always occur together with anorogenic granites (or charnokites). Their plagioclase is relatively Na rich, explained by high-pressure crystallization, and forms several centimeter large (up to 1 m) crystals. Let us restrict our interest here to these massive Proterozoic anorthosites. In a reconstruction of Proterozoic continents all massive anorthosites lie in two broad belts. One (in the northern hemisphere) stretches from eastern Canada via Greenland and Scandinavia to the Ukraine; the other (in the southern hemisphere) stretches from India via Madagascar and Antarctica to East Africa.

The formation of massive anorthosites (Fig. 3.37) can be imagined as follows. Basic mantle melt (basaltic or similar) accumulated in large quantities under the base of the Earth's crust (underplating) because of the higher density it had relative to crustal rocks. The heat led to melt formation in the crust and thus to the formation of granites typically associated with anorthosites. The basic melt fractionated

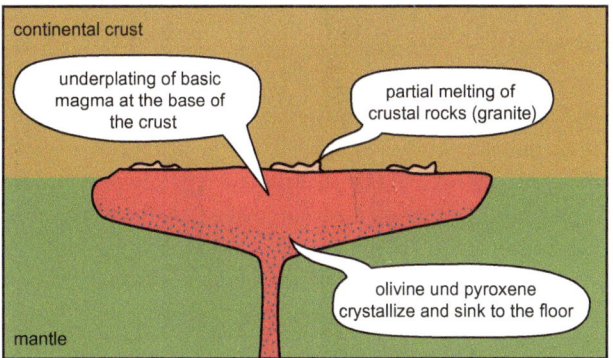

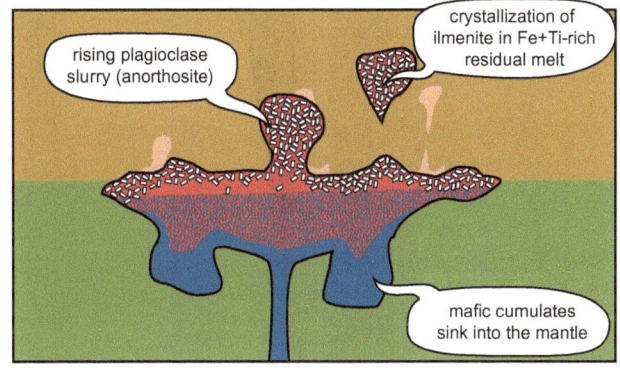

Fig. 3.37 Massive anorthosites were created exclusively during the Proterozoic. They are intrusions of plagioclase-rich crystal slush derived from basic magma as a floating cumulate. After emplacement in the crust, ilmenite rich in iron and titanium crystallized from the residual melt. Significant titanium deposits could have been created in the process (from Ashwal 1993)

following the tholeiitic trend enriching iron and titanium in the melt. Mafic minerals descended, accumulated on the floor of the intrusion, and finally sank into the mantle. Lighter plagioclase rose and accumulated as crystal slush under the roof of the melt sheet. Plagioclase mud had a lower density than the crust and could rise diapir-like along weak zones into the crust. During further crystallization of the iron-rich melt still present between the plagioclase crystals ilmenite ($FeTiO_3$) was formed in addition to plagioclase and pyroxene. Such ore minerals are normally disseminated between plagioclase crystals. Residual melt within the anorthosite was sometimes squeezed into a dike or sill. By sorting crystals as they sink cumulate layers with a high proportion of ilmenite can result. The largest titanium deposit in the world is Lac Tio (Quebec, Canada) in the Allard Lake anorthosite complex that contains an almost pure ilmenite body. In second place is Tellnes (southern Norway) in the Rogaland Anorthosite Province (Charlier et al. 2006). Here a tilted sill with ilmenite-rich norite is mined containing 14% of known global ilmenite reserves.

Anorthosite complexes also contain mafic and ultramafic rocks some of which contain sulfide deposits similar to LMIs (Sect. 3.3). The mine at Voisey's Bay (Canada) is an important source of nickel, copper, cobalt, and PGEs.

3.6 Kiruna-Type Magnetite Apatite Iron Ore

Around Kiruna (in the extreme north of Sweden) there are several large deposits of particularly high-quality iron ore consisting almost exclusively of magnetite and some fluorapatite (Fig. 3.38). The city was founded around 1900 after completion of the railway line when industrial mining began. Since then it has been one of the most important ore suppliers for the steel industry of Western Europe. This is especially true of the Kiirunavaara mountain that is home to the deepest underground iron mine in the world. The ore contains on average about 60% iron.

Other Kiruna-type deposits in Scandinavia are Malmberget, near Gällivare (northern Sweden) and Grängesberg (southern Sweden). Worldwide examples are Avnik (Turkey), Bafq (Iran), Yangtze River (lower reaches) (China), Coast Range (northern Chile), and Missouri (USA).

However, how these massive magnetite–apatite ore bodies were formed is controversial. There are two theories. According to one the blame can be laid at the door of an unusual phosphorus-rich iron oxide magma (Frietsch and Perdahl 1995; Harlov et al. 2002; Naslund et al. 2002; Hou et al. 2011). According to the other the deposits were formed by hydrothermal solutions (Sillitoe and Burrows 2002; Jami et al. 2007, 2009) and should therefore be discussed in the following chapter.

According to the first theory during the crystallization of a diorite pluton the remaining melt reached a composition where liquid immiscibility (Sect. 3.1.2) led to the segregation of a phosphorus-rich iron oxide melt. Iron is generally enriched in the melt during the fractional crystallization of tholeiitic basalts since the crystallization of magnetite begins relatively late. Unmixing of an iron oxide melt occurs when the concentration of elements such as phosphorus is increased at the same time (Hou et al. 2011; Charlier and Grove 2012). Phosphorus also reduces the melting point of iron oxide melt. This melt was finally squeezed upward, partly flowed out as a lava flow at a volcano, partly got stuck at shallow depth, and solidified.

However, the alleged iron oxide lava flow at El Laco (Chile) contains tuff inclusions some with a diameter of several meters. This rather suggests that a hydrothermal solution reacted with the tuff and partly replaced it with magnetite (Sillitoe and Burrows 2002). This process could also apply to other Kiruna-type deposits, although it is still conceivable that segregation of an iron oxide melt was an intermediate step in the generation of an iron-rich hydrothermal fluid (Jami et al. 2009).

The similarity between Kiruna-type iron deposits and iron oxide copper gold (IOCG) deposits is striking (Sect. 4.7). IOCG contain less iron and instead have increased contents of copper sulfides, gold, uranium, and REEs. Some researchers believe that both deposits are iron-rich or copper-rich endmembers of a continuum. IOCG deposits have been formed by hydrothermal solutions that are known in some cases to also originate from diorite magmas. In northern Chile and northern Sweden both types of deposits occur together, so it may well be that there is a genetic connection.

Fig. 3.38 Kiruna-type ore containing magnetite and apatite from Malmberget, near Gällivare (Sweden) (*photo* © F. Neukirchen/Mineralogical Collections of the TU Berlin)

3.7 Granite (Introduction)

Granites and similar plutonic rocks (a.k.a. granites in the broader sense) can be formed by fractional crystallization from a basaltic magma or normally by partial melting in the Earth's crust. Water plays an important role in melt generation as it significantly lowers the melting point (or solidus). In water-saturated crustal rocks less than 700 °C is sufficient to start melting. Small concentrations of phosphorus, fluorine, and boron lower the melting point even further. Water can be released in metamorphic reactions by the collapse of water-bearing minerals such as mica or amphibole.

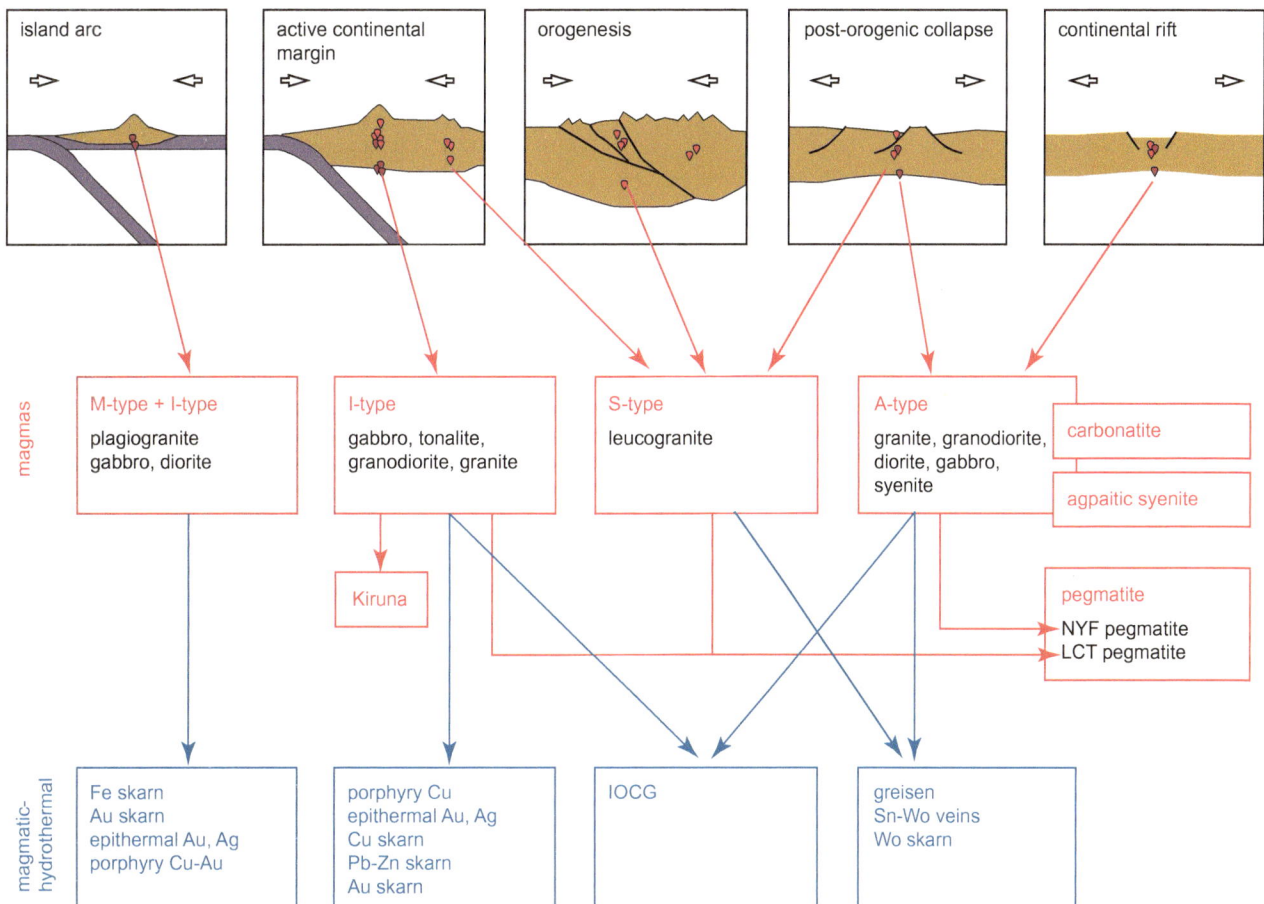

Fig. 3.39 Relationship between the plate tectonic setting and granite types (or granite-like plutonic rocks) formed there. Such a relationship can lead to the formation of different magmatic (*red*) and hydrothermal deposits (*blue*). See text for explanation and Chap. 4 for hydrothermal deposits

How crustal rocks melt can be observed in so-called migmatites. They are high-grade gneisses in which streaks of light-colored melt have formed, while dark-colored minerals have remained solid and in place. Thus, migmatites form the transition between metamorphic and igneous rocks.

Granite magma is very viscous because SiO_2^{4-} tetrahedrons combine to form polymers. It often transports crystals that have never been melted. It also has a comparatively high content of dissolved water. In the case of migmatites the degree of melting was so low that granite magma was formed but remained in the parent rock. Granite magma only rises when the amount of water is sufficient and the temperature is high enough to achieve a higher degree of melting.

Granite magmas can arise in different plate tectonic settings and can also be composed differently depending on the type of parent rock. Geologists distinguish between S-type, I-type, M-type, and A-type granites (Fig. 3.39) mainly based on geochemical criteria but these are attributed to different origin. Hybrids between these types are the rule rather than the exception. This classification is nevertheless helpful and we will use it for the following magmatic deposits as well as for some hydrothermal deposits.

I-type granites ("I" for igneous) are typical of subduction zones. They can be traced back to a basaltic melt derived from the mantle either directly by fractional crystallization or indirectly by basaltic magma solidifying into gabbro that subsequently partly melted. The material that erupts at volcanoes in subduction zones is only a fraction of the magma formed there; most of it remains stuck at depth in the form of large plutons. The result is a so-called batholith, a broad belt consisting only of plutonic rocks such as gabbro, tonalite, granodiorite, and granite that have intruded each other. Compared with other granite types I-type granites are more strongly oxidized (they contain magnetite) and the ratio of aluminum to alkalis plus calcium is in the medium range (metaluminous or slightly peraluminous; see Box 3.13).

Box 3.13 Peraluminous, metaluminous, and peralkaline rocks

When it comes to the mineral content of an igneous rock the molar ratio of aluminum to alkalis and calcium also plays a role that is virtually analogous to silica saturation (Box 3.1). These elements are contained in alkali feldspar and plagioclase. Peraluminous rocks (including many S-type granites) contain more Al_2O_3 than $Na_2O + K_2O + CaO$; hence they also contain aluminum-rich minerals such as muscovite (common mica), cordierite, andalusite, or garnet. Peralkaline rocks contain less Al_2O_3 than $Na_2O + K_2O$, hence they contain alkali-rich minerals such as aegirine (alkali pyroxene) or arfvedsonite (alkali amphibole). Peralkaline rocks are further subdivided into miaskitic and agpaitic rocks the latter with important deposits (Sect. 3.11). Most granites have an Al_2O_3 content between $Na_2O + K_2O$ and $Na_2O + K_2O + CaO$ that is called metaluminous. No special minerals need to be formed.

M-type granites ("M" for mantle where they originate) are plagiogranites that can be found on young island arcs. They are formed by fractional crystallization from mantle-derived basaltic melts.

S-type granites ("S" for sediment in which they are created) are generated by melting sediments in the strongly thickened crust of a mountain range. This can be an orogenic belt formed by continental collision or at an active continental margin such as a subduction zone like the Andes (especially, at a certain distance from the volcanic arc). Due to the thickened crust, sediments (more precisely, metamorphic rocks formed from sediments) reach a greater depth where the temperature is higher. In addition, heat formation is intensified by the decay of radioactive isotopes. In some cases the heat of basaltic magma that has accumulated under the crust (underplating) also plays a role. S-type granites are typically light-colored (leucogranite). They are somewhat reduced (with ilmenite) and have a high aluminum content (strongly peraluminous); hence they contain Al rich minerals such as muscovite or cordierite.

A-type granites ("A" for anorogenic) are not generated in a compressional setting like the others but by extension. This can be the case in a continental rift or during so-called post-orogenic collapse. A large, high mountain range often literally flows apart after orogeny. Extension causes hot material to rise rapidly and cool little possibly triggering melt generation. Melts with higher alkali content such as nephelinites, olivine melilitites, and a multitude of other exotic magmas can form in the mantle—not just basalts. This is due not only to a very low degree of melting at very great depths, but also to the fact that the mantle has often been enriched beforehand in connection with subduction or mantle diapirs. Such mantle-derived melts can now ascend and trigger melt formation in the crust due to their heat whereby a large number of mixtures is possible. In many cases the lower crust was probably enriched shortly before with alkalis and other elements by magmatic water released by ascending alkaline mantle-derived melts (Martin 2006). A-type granites in the broader sense exist in many very different flavors including both silica-oversaturated rocks such as granite and granodiorite and silica-undersaturated rocks such as syenite and nepheline syenite. They often have very high alkali contents (metaluminous to strongly peralkaline) and above-average contents of Cl, F, and HFSEs.

Fractional crystallization occurs only to a certain extent during the crystallization of granite magma as a result of being close to the eutectic composition. After slight cooling almost all minerals therefore crystallize simultaneously. In addition, the melt is highly viscous preventing crystals from sinking. Although granites consist largely of quartz, potassium feldspar, and plagioclase, they have a low content of mafic minerals such as mica (biotite, muscovite), pyroxene, or amphibole. They also contain small quantities of minerals such as ilmenite, magnetite, and pyrite. For the purposes of this chapter the so-called accessory minerals that make up less than 1% of the rock also play a role. They are rarely larger than 1 mm and often so tiny that they can only be seen under a microscope. These minerals include zircon ($ZrSiO_4$), titanite ($CaTiSiO_5$), apatite ($Ca_5(PO_4)_3(OH)$), allanite (rare-earth silicate of the epidote group), and monazite and xenotime (rare-earth phosphates). They contain highly incompatible elements that cannot be incorporated into normal granite minerals such as zirconium, REEs, uranium, and thorium. In common granites the concentration of interesting elements is of course much too low to be mined, but accessory minerals may accumulate in a secondary placer deposit (Sect. 5.9) after a granite has been weathered.

During crystallization granites release a large amount of water (more precisely, water-rich fluids) from which hydrothermal deposits (Chap. 4) can develop. The metal content of granite melt naturally plays a role in the composition of this fluid as does the timing of water saturation during crystallization.

The last residual melt of almost solidified granite contains a lot of water together with a very high content of incompatible elements. Pegmatites (Sect. 3.8) or tin granites may be formed from residual melt.

3.7.1 Tin Granite

So-called tin granites are formed by particularly strongly fractionated granite melts that have a high tin content. They often occur as small irregularly shaped plutons (a.k.a. stocks) that have been formed in connection with larger intrusions of S-type or A-type granites from late residual melts. Tin granite is sometimes simply the last solidified part of a large intrusion. Such rocks are enriched in tin, fluorine, and other incompatible elements and contain cassiterite and topaz (sometimes even tantalum ores). Their composition is similar to that of some pegmatites from which they differ mainly by their smaller grain size. The mining of tin granites is rarely worthwhile due to the usually low ore grade. Nevertheless, in Egypt the mining of tantalum from a tin granite is planned (Küster 2009). Moreover, tin granites often lead to the formation of important hydrothermal tin deposits such as tin veins (Sect. 4.5) and greisen (Sect. 4.6). Cassiterite from tin granites can also play a role in tin placers (Sect. 5.9).

3.8 Pegmatite

The last melt residues of an almost solidified granite pluton are highly enriched in water and incompatible elements. They can solidify into a comparatively small rock body with unusually large crystals called a pegmatite. Although crystals several centimeters or decimeters in size are the rule, even crystals several meters in size are not uncommon. A single potassium feldspar in a pegmatite in Colorado (USA) is said to have been 50 m long, 14 m wide, and 36 m high.

Most pegmatites have more or less the same composition as granite and consist mainly of quartz, feldspar, and mica (Fig. 3.40). However, incompatible elements are often so highly enriched that exotic minerals are formed (Table 3.3). Pegmatites are important deposits of niobium and tantalum (coltan; Box 3.14), REEs, beryllium, lithium, and other metals and provide large quantities of gemstones (Table 3.4)

Fig. 3.40 Mica and potassium feldspar from a beryl-bearing pegmatite in Tanzania (*photo* © F. Neukirchen)

Table 3.3 Important minerals that can occur in pegmatites together with feldspar, quartz and mica some of which can be mined as ore

Beryl	$Be_3Al_2Si_6O_{18}$
Topaz	$Al_2SiO_4F_2$
Zircon	$ZrSiO_4$
Rutile	TiO_2
Ilmenite	$FeTiO_3$
Cassiterite	SnO_2
Uraninite	UO_2
Thorianite	ThO_2
Niobium–tantalum minerals	
Columbite–tantalite (a.k.a. coltan)	$(Fe, Mn)(Nb, Ta)_2O_6$
Wodginite	$(Mn, Sn, Ti)Ta_2O_6$
Microlite	$Ca_2Ta_2O_6(O, OH, F)$
Lithium minerals	
Spodumene (Li pyroxene)	$LiAlSi_2O_6$
Petalite (phyllosilicate)	$LiAlSi_4O_{10}$
Lepidolite (Li mica)	$K(Li, Al)_3(Si, Al)_4O_{10}(F, OH)_2$
Zinnwaldite (Li mica)	$KLiFeAl(AlSi_3)O10(OH, F)_2$
Elbaite (Li turmaline)	$NaLi_{1,5}Al_{1,5}Al_6Si_6O_{18}(BO_3)_3(OH)_4$
Amblygonite (Li phosphate)	$(Li, Na)AlPO_4(F, OH)$
Rare-earth element minerals	
Gadolinite	$(Ce, Y)_2FeBe_2O_2(SiO_4)_2$
Fergusonite	$(Y, Ce)NbO_4$
Allanite (epidote group)	$Ca(Ce, La, Y)(Al_2Fe)Si_3O_{12}(OH)$
Monazite	$(La, Ce, Nd, Sm)PO_4$
Xenotime	$(Y, Yb)PO_4$
Euxenite	$(Y, Ca, Ce, U, Th)(Nb, Ta, Ti)_2O_6$
Samarskite	$YFe^{3+}U^{4+}(Nb, Ta)O_4$

and industrial minerals such as feldspar, quartz, and mica (Sect. 7.4). Thus, pegmatites can be classified according to economic criteria such as gem-bearing, REE, and lithium pegmatites.

Table 3.4 Gemstones from pegmatites

Mineral	Varieties
Beryl	Aquamarine, heliodor, morganite, goshenite
Chrysoberyl	
Spodumene	Hiddenite, kunzite
Tourmaline (especially, elbaite)	Indigolite, rubellite, verdelite
Topaz	
Zircon	
Brazilianite	
Quartz	Smoky quartz, morion, rose quartz
Potassium feldspar	Amazonite

The so-called abyssal pegmatites are special in that they did not develop in connection with granites but with high-grade metamorphism. They are found in migmatites (i.e., partly melted gneiss) and formed on the spot by low degree of partial melting. They are usually granite pegmatites of simple composition. However, certain elements can also be enriched if the corresponding rocks are melted. A good example of this type is found in the Rössing uranium mine (Namibia).

Nepheline syenites and other plutonic rocks can also form pegmatites that then have a correspondingly different composition.

Fluorine, boron, and phosphorus are elements enriched in the late residual melt of a granite intrusion that play a special role in the formation of pegmatites because they serve as fluxes (e.g., Rickers et al. 2006). They form complexes with alkalis and other elements and thus indirectly suppress the polymerization of silica tetrahedrons in the melt. They thus lower the melting point (or shift the eutectic), which can be well below 500 °C, and reduce the viscosity. At the same time they increase the solubility of normally less soluble elements. Minerals containing one of these three elements are common in pegmatites such as topaz, tourmaline, and various phosphates.

Water-rich residual melt can be squeezed out of the crystal slurry of a granite pluton that has almost solidified. Most of the time it accumulates in streaks within the granite forming smaller pegmatites within the granite. Sometimes the residual melt is squeezed to the edge of the pluton and solidifies there to form a marginal pegmatite. On other occasions the melt penetrates through fissures into the surrounding rock and solidifies there as a dike or in the form of a small pluton called a pegmatite stock. Moreover, there may be dike swarms with systematically arranged dikes around a granite intrusion. Since the associated granite of many pegmatite dikes is not exposed at the surface it is assumed that it is hidden at depth.

The formation of giant crystals is not due to slow growth since there is evidence they grow very quickly. This is because small magma volumes cool down faster than a large pluton. It is estimated that pegmatites take between a week and a few months to crystallize.

London (2005, 2008, 2009) argues that the melt is substantially undercooled and that it cooled around 150–250 °C below its melting point before crystallization started. This is possible because water and any fluorine, boron, and phosphorus present suppress the nucleation of silicate minerals. At the same time diffusion (with the exception of alkalis) in cool silica-rich melt is very slow due to the high viscosity. The more the melt is undercooled the slower the diffusion. Finally, crystallization begins at the edges of a magma body where it is even colder. In contrast to granite in which crystals grow distributed, in the melt of a pegmatite crystals grow from the sidewalls into the interior.

Diffusion in the melt is so slow that (according to the model of London) incompatible elements and water accumulate in a thin film immediately ahead of the crystallization front growing inward. This thin boundary layer has a different composition and thus different physical properties than the rest of the melt. It contains a lot of water making it less viscous and accelerating diffusion. In certain areas the concentration of rare elements can become so high that exotic minerals are formed. Those that are also contained in the granite as tiny accessory minerals suddenly reach considerable size and other minerals only occur in pegmatites. If these are minerals that contain boron, phosphorus, or fluorine, then the content of fluxes in the melt decreases in turn accelerating the crystallization of other minerals.

The center of the pegmatite crystallizes last. This is where the enriched layers from both sides meet. Exotic minerals are therefore most frequently found in the center where the water content of the melt often becomes so high that water is released as gas bubbles—at least if the pressure is not too high (i.e., at shallow depth). This leaves so-called miarolitic cavities that are sometimes filled with gems. In other cases the transition from silicate melt to a hot aqueous fluid is continuous. It is interesting to note that the crystallization of undercooled melt is metastable with the result that minerals crystallize in a different order than in granite. Although potassium feldspar, plagioclase, quartz, and mica form simultaneously at the edge, later only potassium feldspar and quartz are formed and finally only quartz in the center. Undercooling can also cause intergrowth of quartz and potassium feldspar. Large feldspar crystals with quartz inclusions are common and resemble cuneiform writing, hence the name graphic granite.

This model of pegmatite generation is quite controversial. An investigation of solidified melt inclusions trapped in crystals indicates that the viscosity of the pegmatite melt is by no means so high (Thomas et al. 2012). These authors found melt inclusions in various pegmatites from all over the world that not only contained very high contents of water and the abovementioned fluxes, but also other ingredients such as alkali carbonates that also reduce the viscosity. In addition, these authors often found two different inclusions consisting of a granite-like melt with moderate water content and an alkali-rich melt with a very high content of water and fluxes. It is possible that liquid immiscibility resulted in segregation into two different melts (Sect. 3.1.1) with the former suspended as droplets in the latter. The viscosity of the suspension was similar to that of chocolate syrup. However, the former initially crystallizes into quartz and feldspar thus increasing the proportion of exotic melt in the center. Thomas et al. (2012) also emphasize that sudden changes in consistency may occur as the pegmatite cools. For example, the last residual melt in the center of the pegmatite may develop into a gel that ultimately crystallizes into quartz. According to this model further melt injection and fractional crystallization can also lead to further diversity.

Which elements are enriched in a pegmatite mainly depends on the composition of the parent rock, where the magma was generated, and thus indirectly the tectonic setting (Martin and De Vito 2005; Černý and Ercit 2005). The two pegmatite families that are distinguished based on different enrichment trends are lithium–cesium–tantalum (LCT) and niobium–yttrium–fluorine (NYF) pegmatites (Fig. 3.41). The first family includes lithium, tantalum, and many gem-bearing pegmatites; the second family includes REE pegmatites and some gem-bearing pegmatites. There are also LCT–NYF hybrids.

LCT pegmatites (lithium–cesium–tantalum) form in compressional settings, at subduction zones, or in the thickened crust of high mountains ranges formed by the collision of two continents (in connection with S-type and I-type granites). Li, Rb, Cs, Be, Sn, Ga, Ta, Nb (Ta > Nb), B, P, and F are most strongly enriched.

NYF pegmatites (niobium–yttrium–fluorine) form in extensional settings, in a continental rift, or during post-orogenic collapse (in connection with A-type granites). Nb, Ta (Nb > Ta), Ti, Y, REE, Zr, U, Th, and F are strongly enriched here.

Whether any rare element can be strongly enriched at all hinges above all on the pressure and thus the depth (Černý 1992; Černý and Ercit 2005). This becomes clear when comparing the type of pegmatite with the pressure and temperature conditions determined for the respective host rock (Fig. 3.42).

Muscovite pegmatites are formed at great depth (high-pressure amphibolite facies) and contain high-quality mica but no ores. They are generated either by direct melting of crustal rocks or by minor fractional crystallization of the granitic melt. The content of rare elements such as muscovite

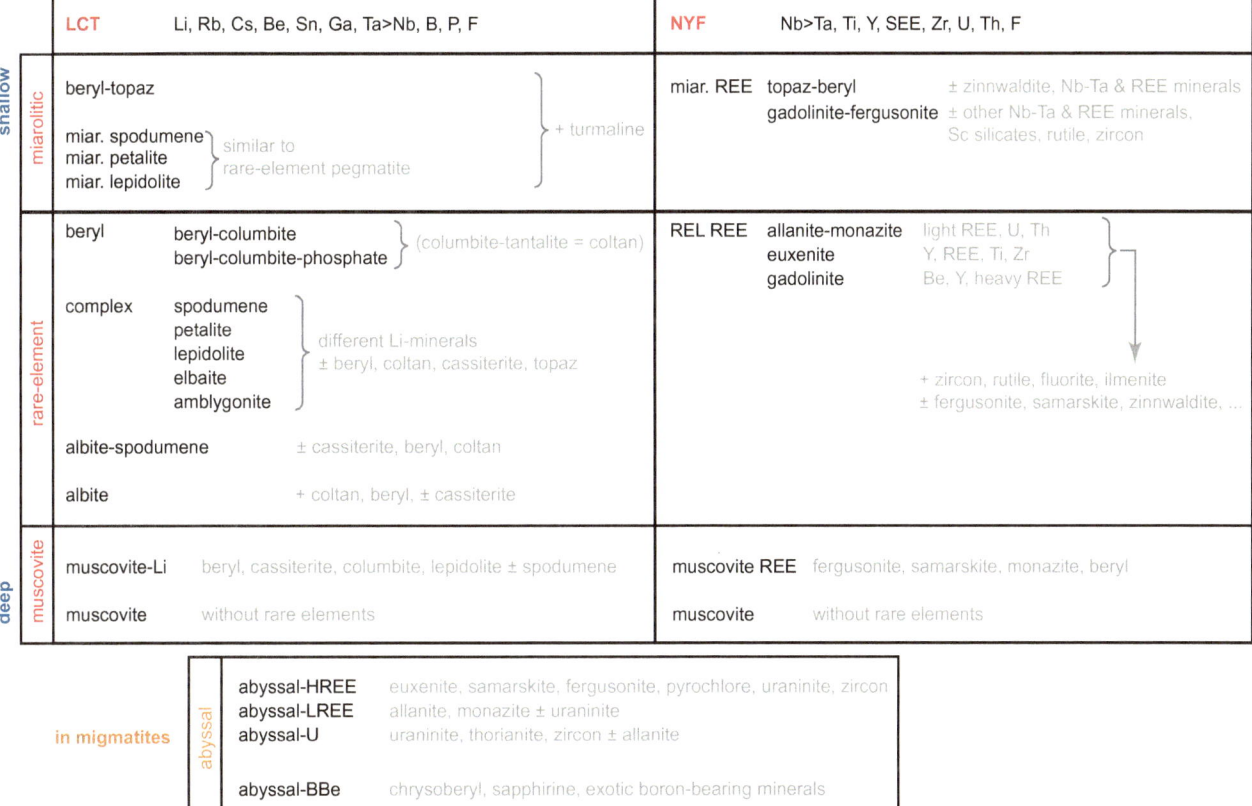

Fig. 3.41 Classification of pegmatites. The lithium–cesium–tantalum (LCT) and niobium–yttrium–fluorine (NYF) families follow different enrichment trends as a result of different parent granites and the tectonic setting. Depth also has an effect on mineral content. Rare elements are particularly enriched in shallow miarolitic pegmatites and in rare element pegmatites formed at medium depth. Within these pegmatite classes different types and subtypes can be distinguished. There are gradual transitions between the different types. Abyssal pegmatites are different because they are not formed as the residual melt of a granite but in situ by a low degree of melting (from Černý 1992; Černý and Ercit 2005)

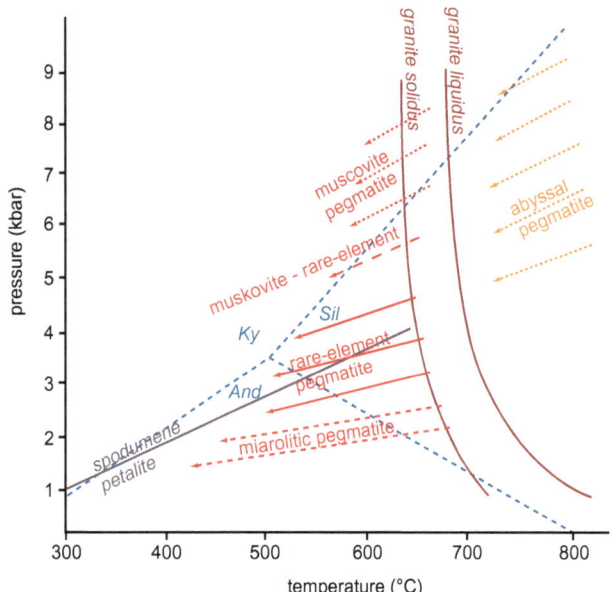

Fig. 3.42 Pressure and temperature of the surrounding rock for different pegmatite classes. *Arrows* show the fractionation trends of pegmatites observed in different regions compared with the respective host rock. Ore-free muscovite pegmatites form at greater depths, muscovite and rare-element pegmatites form at shallower depths, and rare-element pegmatites of different compositions where the enrichment of rare elements is strongest form at even shallower depths. Miarolitic pegmatites form at low pressure due to the segregation of water vapor. Abyssal pegmatites are formed by direct melting at temperatures above the granite liquidus (wet melting). The stability fields of spodumene and petalite have an effect on lithium pegmatites. The stability fields of kyanite, andalusite, and sillimanite serve only for orientation purposes (from Černý and Ercit 2005)

Fig. 3.43 Gemstone quality spodumene (kunzite is pink and hiddenite is green) from Minas Gerais (Brazil). More frequent are inconspicuous sparry crystals (light-colored, opaque) (*photo* © F. Neukirchen/Mineralogical Collections of the TU Berlin)

lithium pegmatite or muscovite rare-earth element pegmatite increases with decreasing depth.

Enrichment is strongest at medium depth and involves **rare-element pegmatites** (not to be confused with rare-earth element pegmatites). They are in turn subdivided into types and subtypes with gradual transitions. For example, in the LCT family increasing fractionation trends from beryl columbite pegmatite via beryl columbite phosphate pegmatite to complex lithium pegmatites. The latter in turn are divided into different subtypes depending on the predominant lithium mineral. Of these spodumene (Li pyroxene) is the most common (Fig. 3.43). This is an important lithium ore and the colored varieties kunzite and hiddenite are expensive gemstones. At lower pressure petalite is formed instead. The presence of Li minerals such as elbaite (Li tourmaline, an important gemstone), lepidolite (Li mica), and amblygonite (Li phosphate) is controlled by the presence of boron, fluorine, and phosphorus.

Rare-element rare-earth pegmatite is a member of the NYF family that mainly contains various REE minerals such as allanite, monazite, euxenite, gadolinite, and fergusonite. Since some of these minerals contain niobium and tantalum

at the same time no special niobium–tantalum minerals are formed.

At low pressure water vapor may be released forming miarolitic cavities in the form of gas bubbles within the pegmatite that are often lined with large crystals. A high boron content supports the formation of miarolitic cavities because the solubility of water in the melt drops abruptly during the crystallization of tourmaline. Water separation can lead not only to loss of certain elements from the system, but above all to faster solidification of the remaining melt. **Miarolitic pegmatites** otherwise largely resemble the respective rare-element pegmatites of both families.

Abyssal pegmatites are not included in this classification because they are formed at high temperatures (above the melting point of granite).

It is strange that different pegmatite dikes are often arranged concentrically around a granite intrusion. Although proximal pegmatites lack exotic minerals, a little further away beryl pegmatite is found and yet further still distal lithium pegmatite. Fractional crystallization leading from one dike to the next can be ruled out since the highly viscous undercooled melt can hardly be moved. Presumably, they represent melt fractions from different parts of the granite reflecting a zoned pluton (Černý 1992).

Box 3.14 Coltan

High tech depends on a resource called coltan for such items as cellphones, computers, and airplanes. The term is an abbreviation for the solid solution of the minerals columbite $(Fe, Mn)Nb_2O_6$ (Fig. 3.44) and tantalite $(Fe, Mn)Ta_2O_6$. Columbite–tantalite is found primarily in pegmatites of the LCT family and their weathered products and secondarily in placer deposits formed from them. Originally, coltan designated

Fig. 3.44 Columbite with the composition $Fe_{0.85}Mn_{0.15}Nb_{1.5}Ta_{0.5}O_6$ from Hühnerkobel, Bayerischer Wald (Germany) (*photo* © Monika Günther/Archive of the Mineralogical Collections of the TU Berlin)

corresponding ore concentrates from Central Africa, but now the term is used more generally.

Coltan is the most important tantalum ore and the second most important niobium ore after pyrochlore. The two metals have very similar properties and therefore always occur together in nature. They are not very reactive, hardly susceptible to corrosion, resist acids, and due to their extremely high melting point (tantalum: 2996 °C) they can also be used at very high temperatures. Both are used for the production of special alloys and particularly hard and resistant steel grades. Tantalum screws and prostheses are used in medical technology. Tantalum's extremely high electrical capacity (i.e., the ability to store electrical charge) is of even greater importance. Tantalum electrical capacitors can be built such that they have a high capacity but nevertheless are very small thus making the production of many electrical devices possible. Well over half the tantalum produced is processed to make electrical capacitors.

In addition to modern mining of pegmatite (especially, in Australia, Brazil, Canada, China, and several African countries) some tantalum production is via artisanal mining in Central Africa. Artisanal mining's relative share rose sharply in 2009 when the Wodgina Mine in Australia that had long supplied more than half the world's production was closed due to high production costs and slump in demand caused by the economic crisis. As even more mines were abandoned world production fell to less than half the former level and demand was partly met from stocks. Wodgina later reopened, closed, reopened again, and finally closed for good.

Artisanal tantalum mining in Central Africa (especially, in the east of the Democratic Republic (DR) of the Congo, Rwanda, Burundi, and Uganda) involves coltan being mined together with cassiterite mainly from placer deposits and heavily weathered pegmatites. According to estimates of the German Geological Survey (BGR) around 2 million people work in artisanal mining (coltan, tin, and diamonds) in the DR Congo alone. Owing to the extreme poverty of the population and correspondingly low wages ore mined by hand is even cheaper than mechanized production in other countries. Although factors such as safety at work, child labor, forced labor, and ecological over-exploitation pose major problems, informal small-scale mining offers an important source of income and is often the only prospect for work.

Artisanal mining also takes place in nature reserves where the largest remaining gorilla populations live. Deforestation of the rainforest and river pollution pose serious threats to their habitat and thus survival.

The darkest chapter in the saga of artisanal mining involved the role played by so-called blood coltan in the Congolese civil war. The war cost the lives of more than 5 million people and came about as a result of the genocide in Rwanda. Ultimately, the conflicts can be traced back to the policies of the colonial powers that had privileged certain sections of the population. Resources such as coltan, gold, and diamonds were the most important source of money used by parties to the civil war to finance it. Many battles were fought simply to control mines the proceeds of which were used to buy weapons. The participants benefited from the high demand for tantalum due to the simultaneous boom in microelectronics in industrialized countries.

Today attempts are being made to prevent as far as possible the trade in blood coltan. Since the adoption of the Dodd Frank Act in 2010 the United States has demanded companies prove the clean origin of coltan. The German BGR has been developing a certification scheme for a few years now with the intention of enabling a controlled retail chain. In addition, attempts are being made to use the geochemical fingerprint of ores to prove their origin.

3.9 Alkaline Rocks (Introduction)

The partial fusion of enriched mantle by a very low degree of partial melting generates alkali-rich silica-undersaturated magmas (Fig. 3.45) such as alkali basalt, nephelinite, or

Fig. 3.45 In alkaline complexes very different alkaline magmas were emplaced in several pulses. Pyroxenite (*black*) from the Tamazeght Complex (Morocco) is crossed by countless veins of syenite (*white*) (*photo © F. Neukirchen*)

melilitite. Depending on the nature of mantle enrichment, the depth of melting, and amount of CO_2 present at the time other alkaline magmas can form. Strongly alkaline magmas are particularly common in rift systems and on hotspots within Archean cratons where the lithosphere is particularly thick thus affecting the depth of melt generation. The Na/K ratio can vary greatly depending on the type of mantle enrichment. Although sodium mostly predominates, there are also melts rich in potassium in which completely different minerals are formed.

Alkaline magmas can in turn develop into a multitude of different rocks. Fractional crystallization typically leads to a eutectic with the composition of phonolite (or nepheline syenite) or (starting from alkali basalt) to trachyte (or syenite). However, the possibilities are more diverse than in basaltic systems. Exotic minerals can crystallize early in alkaline magmas resulting in fractionation leading in a different direction. In extreme cases the enrichment of alkalis and incompatible elements is so strong or the content of SiO_2 and Al_2O_3 so low that a range of exotic minerals is formed. Moreover, water released by the magma plays an important role because large quantities of alkalis (together with chlorine, fluorine, and other elements) can disappear from the magma. This water can in turn react with older igneous rocks and transform them into completely different rocks (fenitization). Finally, the separation of immiscible carbonatite melt can also occur. Alkaline complexes as is the case with mafic intrusions are made up of crystal cumulates whose composition may not represent a melt composition.

Accordingly, there is a whole catalog of different alkaline rocks. What they have in common is that they contain minerals low in silica such as olivine, feldspathoids

(nepheline, sodalite, leucite) or melilite. Minerals are typically rich in alkalis such as aegirine (alkali pyroxene) or arfvedsonite (alkali amphibole). Other typical minerals are melanite (a garnet) and perovskite.

Alkaline magmas occur (in addition to basalts) at hotspots and continental rift systems. Depending on the age and degree of erosion this can be an active or an extinct volcano, the intrusions underneath, or even isolated dikes. There are also alkaline complexes widespread worldwide that normally consist of a number of different alkaline rocks (Woolley 1987, 2001; Kogarko et al. 1995). For example, many active volcanoes with alkaline magmas exist in the East African Rift System.

Only a fraction of these rocks are economically interesting such as carbonatites (Sect. 3.10) and agpaitic nepheline syenites (Sect. 3.11). The water released by alkaline magmas can also form exotic hydrothermal or metasomatic deposits. In addition, unusual pegmatites can develop. Thor Lake (Northwest Territories, Canada) is worth mentioning because its pegmatites and metasomatic zones that formed in connection with syenite and peralkaline granite have high concentrations of REEs, thorium, niobium, and beryllium.

3.10 Carbonatite

Carbonatites are exotic igneous carbonate rocks. Although they are not rare with more than 500 known occurrences, they have only relatively small volumes (Table 3.5). There are small intrusions (stocks), sills and dikes, and sometimes tuffs or lava flows. They mostly occur together with alkaline rocks. Almost all carbonatites are found in continental rift systems. However, there are some exceptions in orogens (continental collision) and two on ocean islands (hotspots).

They are of particular interest because they have a high content of REEs and niobium. In addition, phosphates such as apatite or fluorapatite (especially, for fertilizer production) and the swellable clay mineral vermiculite (Sect. 7.5) are extracted. An unusual carbonatite is the Phalaborwa copper deposit (Box 3.15).

Table 3.5 Carbonatite types and their main minerals

Carbonatite	Major minerals
Calcite carbonatite:sovite (coarse-grained), alvikite (fine-grained)	Calcite ($CaCO_3$)
Dolomite carbonatite	Dolomite ($CaMg(CO_3)_2$)
Ferrocarbonatite	Ankerite ($Ca(Fe, Mg)(CO_3)_2$) or siderite ($FeCO_3$)
Natrocarbonatite	Nyerereite ($Na_2Ca(CO_3)_2$) and gregoryite ((Na_2, K_2, $Ca)CO_3$)

Box 3.15 Phalaborwa

Phalaborwa (a.k.a. Palabora) in South Africa is a carbonatite deposit that is particularly important economically (Fig. 3.46). It is part of a tubular ring complex consisting of several intrusions. Each magma pulse was injected into the center of older ones creating a structure similar to onion rings. Although there are syenites, the largest part of the complex consists of different pyroxenites (i.e., ultramafic cumulates). Some of these rocks are strongly metasomatically transformed (fenitized) including zones almost exclusively of mica or vermiculite that are also mined. The central part contains two different carbonatite intrusions surrounded by phoscorite (Sect. 3.10.1). The carbonatites consist mainly of calcite together with magnetite, dolomite, apatite, and other minerals. Subsequently, a large amount of copper sulfide precipitated from magmatic–hydrothermal solutions in a certain zone within the carbonatite is extracted from South Africa's largest open-pit mine. Mineralization has certain similarities to IOCG deposits (Sect. 4.7; Groves and Vielreicher 2001). The phoscorite is mainly mined for apatite the by-products of which are magnetite, baddeleyite (zirconium ore), gold, silver, and PGEs.

Carbonatite is defined as igneous rock consisting of more than 50% carbonate minerals, but often this percentage is significantly higher. One or the other of calcite (calcite carbonatite: coarse-grained, sovite; fine-grained, alvikite) or dolomite (dolomite carbonatite) typically dominates. Sometimes ankerite or even siderite (ferrocarbonatite; Gittins and Harmer 1997) dominates. An exotic special case is the natrocarbonatite from Oldoinyo Lengai (Fig. 3.47) containing the sodium carbonates nyerereite and gregoryite. Apart from carbonate minerals carbonatites also contain silicate minerals such as olivine, clinopyroxene, mica, and amphibole in variable proportions; oxides such as magnetite, hematite, ilmenite, rutile, perovskite, and pyrochlore; phosphates such as apatite and monazite; and sulfides such as pyrite.

Carbonatite magmas are highly fluid and even small amounts of melt are mobile in the siderock due to their good wettability. The melting temperature of dolomite carbonatite and calcite carbonatite is of a similar order of magnitude to that of silicate magmas, while that of natrocarbonatite is significantly lower.

The existence of igneous carbonate rocks had long been hypothesized by a few researchers. However, most researchers considered these rocks to be former limestone or dolomite that had been transformed into marble by magmatic heat. That did not change until the 1960s. A milestone discovery was Oldoinyo Lengai (Tanzania), the only active carbonatite volcano on Earth (Guest 1956; Dawson 1962; Bell and Keller 1995), and its unusual composition. It is the only known occurrence of natrocarbonatite in the world. However, carbonatite tuffs—proof of an igneous origin— have also been found in occurrences of normal carbonatites (consisting predominantly of calcite or dolomite). A good example is Kaiserstuhl (Keller 1981), a volcano in the Rhine rift valley, near Freiburg (Germany) that was active in the Neogene. Moreover, initial experiments had shown that igneous carbonate rocks could actually exist. The explanation that they could be molten sedimentary rocks seemed to be obvious at first. However, isotope data proved that the

Fig. 3.46 Aerial view of the Phalaborwa open-cast mine (*photo* © Getty Images/iStockphoto)

Fig. 3.47 Oldoinyo Lengai (Tanzania) is the only active carbonatite volcano on Earth and the only occurrence of natrocarbonatite. Fresh lava is black in color, but changes within a few days to a white loose mass. This picture from 2003 shows the crater with hornitos about 10 m high (*photo* © F. Neukirchen)

melt has nothing to do with the Earth's crust but is derived from the mantle. More precisely, it is about melt formation in ascending material in the asthenosphere (e.g., in a mantle diapir) and to a lesser extent about interaction with the enriched lithospheric mantle (Bell 1989; Bell and Tilton 2001; Lee et al. 2006b).

In the mantle carbon occurs mainly as fluid (carbon dioxide or methane), as carbonate (magnesite or carbonatite melt), and sometimes as diamond—depending among other things on redox conditions (Rohrbach and Schmidt 2011). Most carbon has been in the mantle since the birth of the Earth. Only a tiny part is organic matter or carbonates originating from the Earth's surface that have been transported to depths at a subduction zone. In average mantle rock the carbon content is low, but in connection with subduction zones or mantle diapirs enrichment by CO_2 or carbonatite melt is possible—both before melt formation or simultaneously.

Experiments show that there are three fundamentally different ways to generate carbonatite magma from this enriched mantle (Fig. 3.48). The first possibility is direct formation by partial melting in which there is a very low degree of melting just above the solidus of the enriched carbonate-containing mantle (Harmer and Gittins 1997, 1998). This melt normally corresponds to a dolomite carbonatite with 5–7% alkalis whereby the exact composition depends on the mantle phases present. For example, an enriched mantle can contain mica (phlogopite) or amphibole (pargasite). In a reasonably normal mantle this melt does not get far because it reacts with mantle rock (metasomatism) as a result of the lower pressure. Orthopyroxene of the mantle and carbonatite melt react to clinopyroxene, olivine, and carbon dioxide. The result is a highly enriched mantle (wehrlite). By contrast, carbonatite can ascend into the crust or even to the Earth's surface if the mantle is already heavily enriched. Slight metasomatism of already enriched mantle may cause the dolomite carbonatite melt to develop into a calcite carbonatite melt.

The other two possibilities start with partial melting of the mantle to generate a CO_2-rich silicate melt. Although the degree of melting is a little higher than in the first case, it is still low and the magma formed is rich in alkalis and is silica-undersaturated (e.g., nephelinite or melilitite). The magma rises, cools down, and begins to crystallize whereby

Fig. 3.48 Assuming an enriched, carbon-bearing mantle carbonatites can be formed in different ways: **a** directly in the mantle by a very low degree of melting; **b** by liquid immiscibility and exsolution in CO_2-rich silicate melt (e.g., nephelinite); and **c** as late residual melt in the fractional crystallization of a CO_2-rich silicate melt. In all three cases the carbonatites can evolve further through fractional crystallization and the release of water

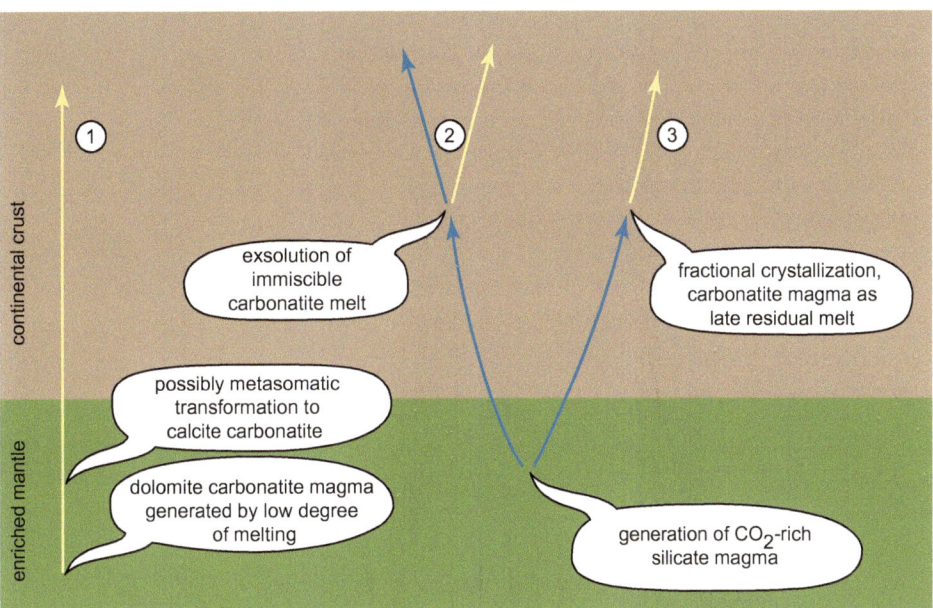

CO_2 and alkalis are continuously enriched in the melt during fractional crystallization.

This can continue up to liquid immiscibility when carbonatite melt is segregated (Kjarsgaard and Hamilton 1989; Kjarsgaard and Peterson 1991; Halama et al. 2005; Brooker and Kjarsgaard 2011). This can happen if the magma is oversaturated with CO_2 (which is unlikely at high pressure but can happen in a magma chamber in the crust) and contains at least 5% alkalis. Figure 3.49 shows a field with two immiscible melts under pressure prevailing in the middle crust. It shows that segregated carbonatite cannot be a pure calcite carbonatite since it has a high content of alkalis and silica. However, carbonatite magma could subsequently develop into a calcite-dominated composition by fractional crystallization of silicate minerals and by the release of water that has a high alkali content. The figure also shows that natrocarbonatite can be segregated in certain compositions. Liquid immiscibility leads to strong fractionation of trace elements. In carbonatite melt Sr, Ba, and the lightest REEs are strongly enriched, while Zr, Hf, Nb, Ta, Th, U, and most other REEs remain in the silicate melt (Veksler et al. 1998). When the fluorine content is high, Nb and possibly other incompatible elements prefer carbonatite (Jones et al. 1995).

If liquid immiscibility does not occur, then fractional crystallization may result in the sudden crystallization of a carbonate mineral such as calcite, dolomite, ankerite, or siderite in late residual melts due to the high CO_2 content. In this case carbonatite is formed as a late result of the fractional crystallization of silicate magma. In this process there is a high probability that incompatible elements such as Nb, Zr, and REEs are enriched to economically relevant concentrations.

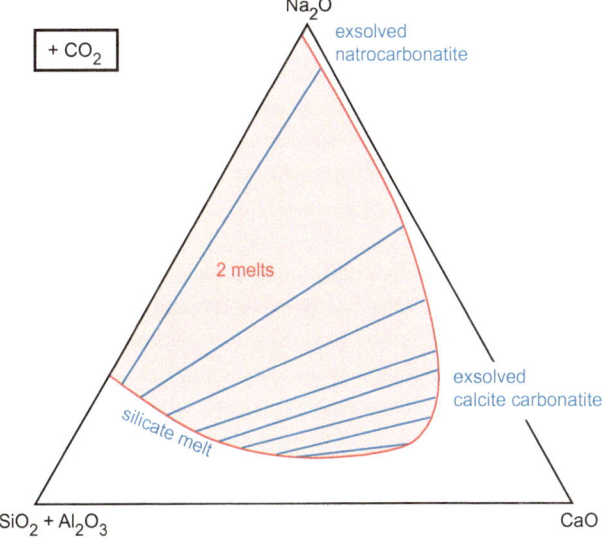

Fig. 3.49 A field with two immiscible melts (silicate magma and carbonatite) in the experimental system SiO_2–Na_2O–A_2O_3–CaO–CO_2 at 0.5 GPa, 1250 °C, and CO_2 saturation. *Lines* connect segregated melts that are in equilibrium. The position and size of the field depends on such factors as pressure, temperature, and fluorine content (from Brooker and Kjarsgaard 2011)

It is not always possible to say with certainty which of the three processes has taken place in each individual case since carbonatites formed in different ways could occur in a single alkaline complex. Moreover, carbonatite melts can develop further through fractional crystallization leading to, say, dolomite carbonatite converting to calcite carbonatite (Gittins et al. 2005).

The very high water content of carbonatite magmas is another important factor (Keppler 2003). Such water is released during ascent (as pressure decreases) and during crystallization. Soluble components such as alkalis, chlorine, and fluorine are then partially lost. Aggressive water penetrates the surrounding rock and reacts with it. This type of metasomatism is called fenitization and can produce rocks consisting almost exclusively of alkali feldspar or nepheline. Carbonates cool in the presence of their own fluids and may recrystallize and reequilibrate (Marks et al. 2009).

The multitude of possibilities show that there is a whole world of different carbonatites as reflected in their very different compositions. Many carbonatites occur together with other igneous rocks. Most often these belong to the nephelinite clan (extrusive: nephelinite, phonolite; intrusive: melteigite, ijolite, urtite, nepheline syenite); in some cases they belong instead to the melilitite clan (extrusive: melilitite; intrusive: okaite, turjaite, afrikanandite, perovskite pyroxenite, and others); or in still further cases to other alkaline rocks (Mitchell 2005). However, there is not necessarily a genetic connection. Thus, different magmas can come from different depths or from different areas of an enriched and therefore heterogeneous mantle (Bell and Tilton 2001). Perhaps melt generation took place at different times depleting the mantle of certain components one after the other and then different magmas used the same path for the ascent. However, there are also carbonatites (especially, dolomite carbonatites) that are not associated with silicate magmas.

Carbonatites (Fig. 3.50) have high albeit very variable contents of REEs, Nb, and other incompatible elements. The most important niobium ore is pyrochlore ($Ca_2Nb_2O_7$; Fig. 3.51), a mineral that may also contain some thorium and REEs. It occurs frequently though not always in economic

Fig. 3.51 Pyrochlore in carbonatite from the Vishnevogorsky Complex in the Ilmen Mountains (southern Urals, Russia) (*photo* © F. Neukirchen/Mineralogical Collections of the TU Berlin)

quantities. By far the most important niobium deposit is the open-cast mine at Araxá (Minas Gerais, Brazil). The complex consists of intrusions of dolomite carbonatite, some calcite carbonatite, and ultramafic strongly fenitized rocks such as micaceous rocks. The ore zone in the carbonatites consists of crystal cumulates of magnetite, mica (phlogopite), pyrochlore, ilmenite, and apatite in which the average Nb_2O_5 is 1.5%. Reserves are estimated at 460 Mt sufficient to cover current world needs for about 500 years. Phosphate is also produced. Carbonatites are also mined from pyrochlore ore in Catalão (Brazil) and Saint-Honoré (Quebec, Canada). In Kaiserstuhl (Germany) pyrochlore was briefly mined from the calcite carbonatite intrusion in the center of the extinct and eroded volcano.

REEs in some calcite carbonatites are enriched to several percent and there are exceptional cases even containing 10% REE_2O_3. This is particularly true of light REEs such as lanthanum, cerium, and neodymium, while heavy REEs are virtually absent. In carbonatites with a low REE content, REE are mainly found in apatite. Monazite (REE phosphate) is also known to sometimes crystallize. At very high contents the last remaining melt of carbonatite is extremely enriched in REEs when interstitial REE fluorocarbonates such as bastnäsite, parisite, and synchisite are formed. Some researchers no longer regard these as late magmatic formations but as magmatic–hydrothermal formations crystallized by the water released by carbonatite magma (Ngwenya 1994). REE minerals (especially, monazite) often contain significant amounts of thorium (and slightly less uranium). The fact that the ores are therefore radioactive must be taken into account during mining and processing.

One of the most important REE deposits is Mountain Pass (California, USA), a carbonatite consisting of 10% bastnäsite together with calcite, dolomite, baryte, quartz, and

Fig. 3.50 Carbonatite veins (sovite) in the Tamazeght Complex (Morocco) (*photo* © F. Neukirchen)

apatite. Siliceous igneous rocks are also a special feature here since they are extremely rich in potassium (Castor 2008).

The so-called REE carbonatites of Maoniuping (Sichuan, China) are hydrothermal veins containing fluorite, baryte, and REE fluorocarbonates (Xu et al. 2004, 2010; Yang and Woolley 2006). The fluid probably originated from nearby calcite carbonatite containing hardly any REE minerals such that magmatism was followed by effective hydrothermal fractionation. The world's most important REE deposit at Bayan Obo (China) was created by similar fluids (Box 3.16).

Box 3.16 Bayan Obo

The world's largest known deposit of REEs is Bayan Obo (Inner Mongolia, China). More than 100 Mt of REE_2O_3 are held in this deposit together with 600 Mt of iron oxide and significant amounts of niobium. The most important ore minerals are monazite (REE phosphate), bastnäsite (REE fluorocarbonate), and magnetite. They occur together with more than 150 different minerals. This is very likely a hydrothermal or metasomatic deposit (see Sect. 4.10) where ore-forming fluids were probably released from carbonatite magma.

The deposit (Fig. 3.52) consists of two large (main and eastern ore body) zones and several small (western ore body) zones in which REEs are highly enriched (1–12% REE_2O_3). They are located in a so-called dolomite marble interpreted by some researchers as a dolomite carbonatite sill rather than a former sediment transformed by metamorphism (Yang et al. 2011). The whole deposit is located within similarly old sediments (sandstone, slate) that had been deposited in a continental rift. This is located at the northern edge of the

North China Craton and was possibly formed when the hypothetical supercontinent Columbia broke up in the Mesoproterozoic.

Several carbonatite dikes are found in the wider area ranging from early dolomite carbonatite via calcite–dolomite carbonatite to late calcite carbonatite now interpreted as a fractionation trend. Since REEs were increasingly enriched the calcite carbonatites have extremely high contents of them (10% REE_2O_3) —about as high as the best ores of Bayan Obo. Hydrothermal fluid from this strongly fractionated calcite carbonatite was presumably released metasomatically thus altering the marble (or carbonatite sill) (Yang et al. 2011).

Metasomatism involving differently composed fluids clearly occurred several times (Smith et al. 2000). The first stage began with the formation of monazite and magnetite (and some bastnäsite) as disseminated minerals throughout the "marble". The subsequent main stage was limited to the ore zones. Initially banded mineralizations with monazite, bastnäsite, apatite, magnetite, and hematite formed along the fluid pathways. Over time dolomite was more or less completely replaced by aegirine, aegirine–augite, fluorite, amphibole, and phlogopite. In the final stage hydrothermal veins comprising aegirine (with apatite, calcite, fluorite, bastnäsite) were formed and fluorite–baryte veins comprising parisite (Ca-REE fluorocarbonate), huanghoite (Ba-REE fluorocarbonate), and other minerals were formed. Niobium minerals such as Nb–rutile, columbite, aechymite, and pyrochlore are found in veins and irregularly shaped pockets.

Hydrothermal transport probably took place mainly as REE fluorocomplexes. The hot, salt-rich aqueous solution reacted with the marble consuming F and releasing CO_2 at the same time thus diluting the fluid. Both effects reduced the solubility of REEs and increased precipitation. While most REE minerals are mainly enriched in light REEs, the late veins have elevated middle REEs.

3.10.1 Phoscorite

Carbonatite in some alkaline complexes occurs directly together with stocks and dikes of another exotic rock called phoscorite. It consists predominantly of magnetite, apatite, and olivine (or diopside) whose respective proportions can vary strongly even at the small scale. They are crystal cumulates in which crystals rained down in an iron–phosphorus melt that was probably segregated by liquid

Fig. 3.52 Satellite image of the rare-earth mine at Bayan Obo (China) showing two open-cast mines, dumps, and tailings (*photo* © NASA)

immiscibility (Sect. 3.1.2) from a late residual magma (Wall and Zaitsev 2004; Lee et al. 2006a). Carbonatites and phoscorites in such alkaline complexes are derived from the same Fe–P–CO_2-rich magma. Although the complexes often also contain plutonic rocks containing melilite, they consist mainly of ultramafic cumulates such as pyroxenite and peridotite. Important deposits are Phalaborwa (South Africa; the name of the rock derives from Foscor, the local mining company), Kovdor (Kola Peninsula, Russia), Maymeicha-Kotui (Siberia, Russia), and Jacupiranga (Brazil).

Some phoscorites are mainly mined for apatite, a phosphate mineral primarily used for fertilizer production that in some cases also contains significant amounts of REEs. Magnetite (often rich in titanium), mica, and other minerals are usually just by-products. Moreover, many phoscorites also contain PGEs and other precious metals that may be obtained as by-products. Kovdor (Fig. 3.53) on the Kola Peninsula (Russia) is important here in that its phoscorites are mined in various mines for magnetite and apatite where baddeleyite (ZrO_2), cerium-rich perovskite, and nepheline are by-products. There are also zones in ultramafic rocks that consist almost exclusively of mica. Kovdor holds the world's largest known deposit of industrially exploitable phlogopite and vermiculite. On the Kola Peninsula there are a number of other alkaline ultramafic complexes such as Afrikanda (Russia) and Sokli (Finland) that also contain carbonatites and phoscorites all of which intruded during the Devonian into several rift systems.

3.11 Agpaitic Rocks

Nepheline syenite and similar rocks frequently occur in alkaline complexes. They are silica-undersaturated and have a high content of alkali metals (especially, sodium). Other incompatible elements are also enriched. Such elements are usually located in relatively simple accessory minerals such as zircon and respective titanium minerals such as titanite, ilmenite, and perovskite.

In some peralkaline nepheline syenites the content of sodium and rare incompatible elements such as zirconium and REEs is so high that complex Na–(Zr, Ti, REE) silicates like eudialyte (Fig. 3.54), rinkite, and aenigmatite are formed instead (Table 3.6). Such exotic rocks are called agpaitic (Sørensen 1997) and normal nepheline syenites are called miaskitic. There is also an intermediate form of rocks containing both types of minerals.

The three known alkaline complexes consisting predominantly of agpaitic rocks are Ilimaussaq (Greenland), Khibina, and Lovozero (the latter two on the Kola Peninsula, Russia). Other alkaline complexes with smaller amounts of agpaitic rocks include Mont Saint-Hilaire (Quebec, Canada), Pilanesberg (South Africa), Saima (Liaoning, China), Poços de Caldas (Minas Gerais, Brazil). Then there are some predominantly miaskitic complexes where only the latest pegmatites and hydrothermal veins are agpaitic.

Peralkaline A-type granites also occur. Although these are silica-oversaturated, they are undersaturated with aluminum. In rare cases they also contain complex minerals

Fig. 3.53 Kovdor (Russia) is a ring complex of several alkaline intrusions. In addition to ultramafic cumulates such as dunite, peridotite, and pyroxenite there are also various plutonic rocks containing melilite and feldspathoids. There are also several carbonatite and phoscorite stocks and dikes. Phoscorites are important phosphate and iron deposits. Mica-rich ultramafic rocks are mined for phlogopite and vermiculite (from Zaitsev and Bell 1995)

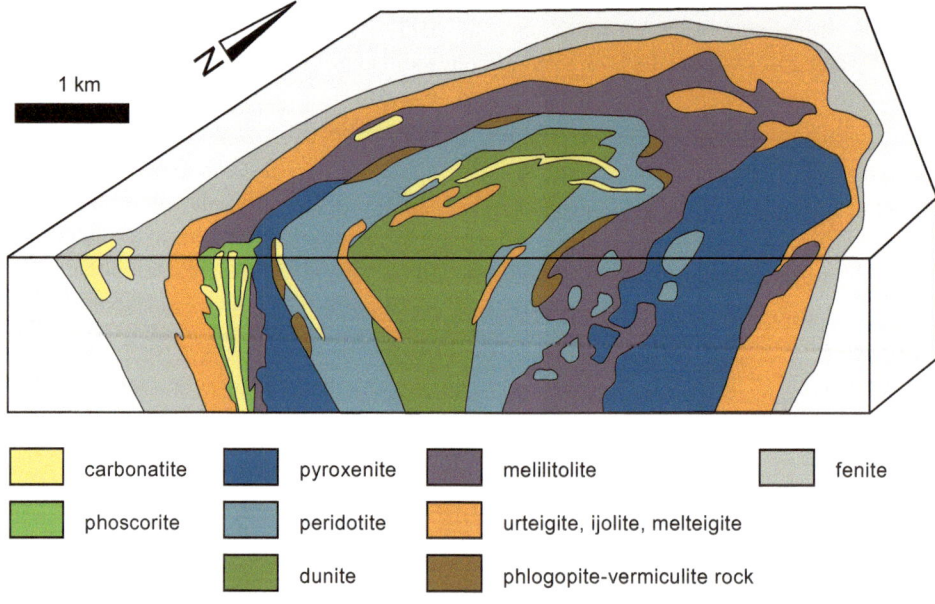

carbonatite	pyroxenite	melilitolite	fenite
phoscorite	peridotite	urteigite, ijolite, melteigite	
	dunite	phlogopite-vermiculite rock	

1 km

Fig. 3.54 Eudialyte (*red*) together with potassium feldspar and nepheline (*white*) and arfvedsonite (*black*) from a late streak in the kakortokite of the Ilimaussaq Complex (Greenland) (*photo* © F. Neukirchen)

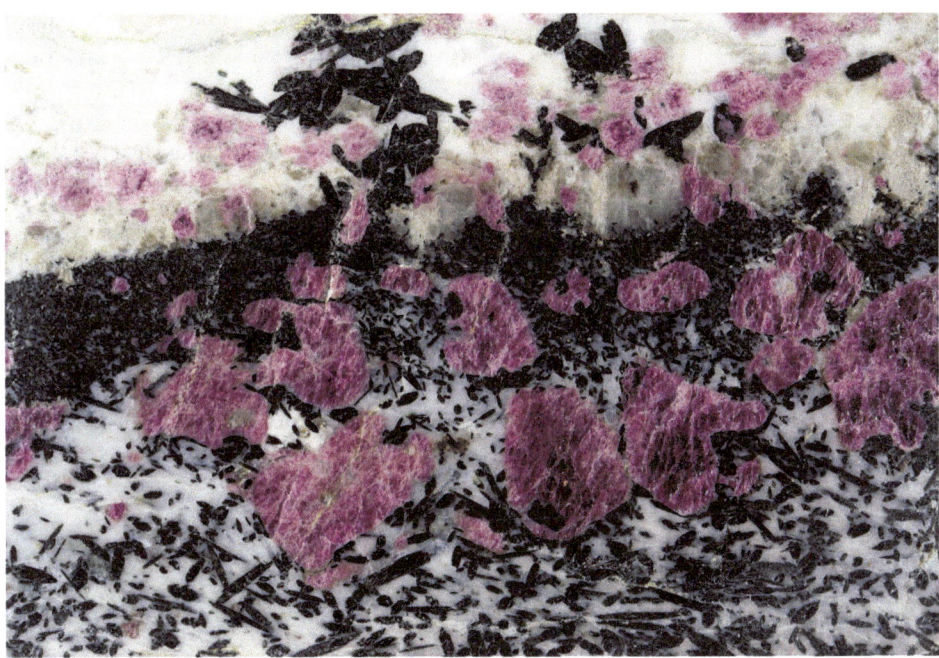

Table 3.6 Selection of minerals with complex compositions found in agpaitic rocks (simplified formulas, variable compositions, and incompatible elements)

Mineral	Formula
Eudialyte (Fig. 3.54)	$Na_{15}Ca_6Fe_3Zr_3Si_{26}O_{72}(OH)_4Cl$
Catapleiite	$Na_2ZrSi_3O_9 \cdot 2H_2O$
Aenigmatite	$Na_2Fe_5TiSi_6O_{20}$
Låvenite	$Na_2MnFe(Zr, Ti)Si_2O_7(O, OH, F)_2$
Rinkite	$Ti(Na, Ca)_3(Ca, Ce)_4(Si2O7)(O, F)_4$
Wöhlerite	$Na_2Ca_4ZrNb(Si_2O_7)O_3F$
Rosenbuschite	$(Ca, Na)_3(Zr, Ti)(Si_2O_7)OF$
Astrophyllite	$K_3Fe_7Ti_2Si_8O_{26}(OH)_5$
Ussingite	$NaAlSi_3O_8 \cdot NaOH$
Lovozerite	$Na_2Ca(Zr, Ti)Si_6(O, OH)_{18}$
Murmanite	$Na_2(Ti, Nb)_2Si_2O_9 \cdot nH_2O$
Steenstrupine	$Na_{14}Ce_6Mn_2Fe_2(Zr, Th)(Si_6O_{18})_2(PO_4)_7 \cdot 3H_2O$
Naujakasite	$Na_6(Fe, Mn)Al_4Si_8O_{26} \cdot H_2O$
Lomonosovite	$Na_5Ti_2(Si_2O_7)(PO_4)O_2$
Loparite	$(Ce, Na, Ca)(Ti, Nb)O_3$

typical of agpaitic rocks. A good example is the Strange Lake intrusion (Quebec, Canada) (Salvi and Williams-Jones 1995).

Although agpaitic rocks have never played a major role economically, this is expected to soon change. The reason is the high content of REEs in which—despite light REEs predominating—heavy REEs are also significantly enriched in contrast to deposits in carbonatites. In Lovozero the mineral loparite, which resembles perovskite, was temporarily mined. Since it contains 30% REE_2O_3 as well as titanium, niobium, and tantalum, extraction may be resumed. Ilimaussaq is considered the world's second largest REE deposit and the largest for heavy REEs. Mining of the most highly enriched rocks is planned (Kvanefjeld area) in which U and Zn are to be extracted in addition to REEs and Y. Some agpaitic rocks are already being mined for uranium (Saima and Poços de Caldas). There are also some large deposits of phosphate (apatite) (especially, in Khibina).

In general, agpaitic rocks have high or extremely high contents of Na, Zr, Cl, F, Li, Be, Rb, Ga, REEs, Nb, Ta, Hf, Zn, Sn, U, and Th. In contrast, Mg, Cr, Ni, and other compatible elements are virtually absent. The mineral content of agpaitic rocks varies widely with hundreds of different minerals occurring in some complexes. Some of them are magmatic phases, others only occur in late-stage igneous rocks such as pegmatites, and yet others are late magmatic–hydrothermal formations found in veins and as a result of metasomatic reactions in cooling rocks. Different agpaitic nepheline syenites and associated cumulates are often given their own rock name depending on their mineralogy, texture, and where the unique corresponding rock occurs in only one location in the world such as naujaite, kakortokite, lujavrite, khibinite, and foyaite.

Eudialyte is the most common of the complex minerals. It is cherry red in color and occurs as a major mineral in many agpaitic rocks. In contrast to the formula given in Table 3.6 its composition is very variable (Pfaff et al. 2010; Schilling et al. 2011) since it also contains REEs, niobium, uranium, thorium, and other incompatible elements in varying amounts. The mineral easily dissolves in hydrochloric acid and hence is easily extracted. The rocks in which it is found mainly contain nepheline, sodalite (due to the high chlorine content), alkali feldspar, aegirine (alkali pyroxene), and arfvedsonite (alkali amphibole).

Nepheline and eudialyte are absent in the most highly enriched hyperagpaitic lujavrites. Instead analcime (a hydrous feldspathoid) occurs with exotic minerals such as steenstrupine, lomonosovite, murmanite, ussingite, naujakasite, and the water-soluble villiaumite (NaF). Such minerals were probably not formed magmatically but during the cooling of the rock when it reacted with late-magmatic fluids previously released during solidification of the last residual melt (Markl and Baumgartner 2002).

A special feature of agpaitic melts is the extremely large temperature interval from the beginning of crystallization at the liquidus (approx. 900 °C) to the solidification of the last residual melt at the solidus (approx. 450 °C). The high content of Cl^- and F^- plays a role here in lowering the solidus and at the same time increasing the solubility of rare elements in the magma.

Agpaitic melts were formed as a result of extreme fractional crystallization from alkaline mantle-derived melts by means of one of two possible processes (Marks et al. 2011). Ilimaussaq is an example of the first. The initial magma was presumably silica-undersaturated alkali basalt melt from which plagioclase was fractionated together with pyroxene and olivine. Sodium, iron, and incompatible elements were enriched in the melt, while calcium and magnesium were removed. At some point the melt corresponded to a syenite rich in iron and sodium, which developed further through the crystallization of nepheline, alkali feldspar, aegirine, or arfvedsonite. Magma reached an extremely agpaitic composition when a small percentage of the original melt quantity was left.

Khibina and Lovozero represent the second process. There the initial melt was nephelinite that may have been peralkaline from the beginning. Since no plagioclase was fractionated in this case the calcium content remained high.

The point at which melt is so strongly enriched that complex agpaitic minerals are formed depends among other things on the content of Na, K, Ca, H_2O, and Cl (Marks et al. 2011). The rarity of agpaitic rocks suggests that special conditions must prevail for such extreme enrichment to be enabled at all. It will soon be apparent that the development of oxygen fugacity plays an important role. Also important is that there is no significant contamination with crustal rocks. If less extreme compositions have already ascended on the same path and sealed the wall rock, so to speak, then agpaitic melts would be able to retain their extreme composition as they ascend.

Moreover, there must be no early release of water because Na^+, Cl^-, F^-, and various rare elements would be partially lost. The initial melt must therefore be low in water. Surprisingly, agpaitic rocks often do not contain any $H_2O–CO_2$ fluid. However, they do contain methane and other hydrocarbons. Methane in Khibina occurs both as free gas in rock pores and in inclusions within the minerals and gas even escapes at the surface (Nivin et al. 2005; Ryabchikov and Kogarko 2006). Most researchers assume that it formed in Khibina during cooling of the rock by a reaction corresponding to Fischer–Tropsch synthesis. Accordingly, the hydration of certain minerals released H_2 that reacted with the still present magmatic CO_2:

$$CO_2 + 4H_2 = CH_4 + 2H_2O$$

Certain mineral surfaces in the rock could function as catalysts. Since CH_4 and H_2O are immiscible, gas rises while water reacts with other minerals and releases further H_2. Maybe magmatic methane was already there in small amounts.

Fluid inclusions in early magmatic minerals in Ilimaussaq already consisted of methane instead of $H_2O–CO_2$ (Krumrei et al. 2007). This was only possible because the magma was extremely reduced according to the reaction:

$$CO_2 + H_2O = CH_4 + O_2$$

In fact, Ilimaussaq is home to some of the most reduced rocks known. According to Markl et al. (2010) the nature of enrichment in the mantle already plays a role in the redox state of the magma whereby metasomatism by sodium-rich fluids finally leads to more strongly reduced magmas than happens with enrichment by potassium-rich fluids. During fractional crystallization of this melt the high Na content and low SiO_2 content eventually lead to the crystallization of aegirine or arfvedsonite despite these containing Fe^{3+} that is hardly present in reduced magma. The remaining melt is reduced even further. An oxygen donor is necessary to provide the necessary Fe^{3+}; hence methane is formed after the above reaction. The result is that no water can be released from the magma and therefore there is no loss of sodium, chlorine, fluorine, and so on. This process may have been an important prerequisite for the formation of extremely enriched agpaitic melts. Experiments confirm that such enrichment can take place in a reduced and anhydrous melt (Giehl et al. 2013). This is also due to the crystallization of uncommon minerals under these conditions, which does not occur with normal magmas.

Oxygen fugacity in Ilimaussaq increased slightly during fractional crystallization within the intrusion. Finally, water continued to be released from the last residual melt leading to the formation of hydrothermal veins whose composition was extreme and providing additional enrichment in some rocks.

3.11.1 Ilimaussaq

The Ilimaussaq Complex (Sørensen 2001; Markl et al. 2001) in southern Greenland is 1.16 billion years old, and it is the classic example of agpaitic rocks, and the type locality of countless minerals. It is the most extremely fractionated of the many other alkaline intrusions of the Gardar rift system. Four magma pulses (Sørensen 2001) were injected successively to a depth of 3–4 km (Fig. 3.55). Such magmas are attributed to fractional crystallization at depth starting with an alkali basalt magma derived from the enriched mantle whereby the composition from one intrusion to the next corresponded to ever-stronger fractionation. With the exception of the second pulse (alkali granite) there was no significant contamination by crustal rocks (Marks et al. 2004). Individual plutons consist of differently fractionated zones.

The first pulse was an augite syenite (Marks and Markl 2001) with a relatively normal (miaskitic) composition. What is particularly unusual is the extremely low oxygen fugacity shown by its more strongly fractionated parts. This rock is found at the edges and in the roof of the complex. The second pulse was an alkali granite whereby the assimilation of crustal rocks presumably led to silica-oversaturation.

The third and fourth pulse involved extremely strongly fractionated agpaitic magmas. They developed a pronounced magmatic stratification of differently composed crystal cumulates (see also Sect. 3.3). First, the roof cumulates crystallized solidifying from top to bottom in the order pulaskite, foyaite, sodalite foyaite, and naujaite. Pulaskite consists predominantly of alkali feldspar and the others have varying contents of feldspathoids, aegirine, arfvedsonite, eudialyte, and so on. Naujaite is regarded as a flotation cumulate and has the largest volume (Fig. 3.56). It consists to a large extent of sodalite that had accumulated under the roof of the magma chamber due to its low density. Inclusions in the core of sodalites prove that their crystallization had already begun with the rise of magma at a depth of more

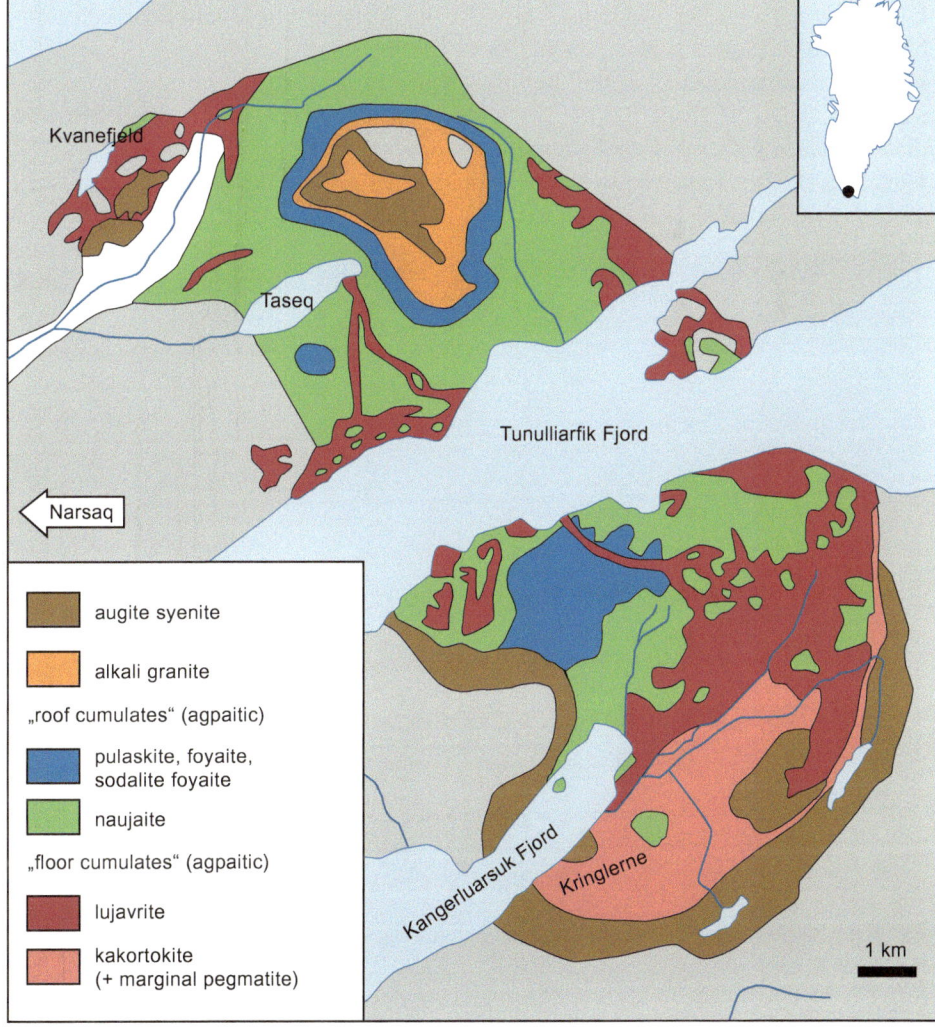

Fig. 3.55 The Ilimaussaq Complex was created by four magma pulses of which agpaitic rocks of roof cumulates and bottom cumulates make up the largest part. The highest levels of rare elements are found in lujavrite in the north of the intrusion (Kvanefjeld and Taseq) where mining is planned. Kakortokite in the south of the intrusion (Kringlerne) is also a potential deposit. The town of Narsaq is just off the map (from Markl et al. 2001)

than 10 km (Krumrei et al. 2007). They also contain methane inclusions instead of H_2O–CO_2 fluid. Large slabs of naujaite were also found in deeper rocks.

Once the roof cumulates had solidified the floor cumulates followed solidifying from bottom to top. The deepest rock that has been exposed at the surface is the 300-m-thick so-called kakortokite the lower two thirds of which show spectacular magmatic layering (Figs. 3.57 and 3.58) where layers on average 8 m thick alternate 28 times as black, red, and white. The rock consists of alkali feldspar and nepheline (white), eudialyte (red), and arfvedsonite (black) where the layers correspond to changing proportions of these minerals. One model put forward to explain this suggests a process working in a similar way to a pressure cooker (Pfaff et al. 2008). Accordingly, the water content in the melt increased during crystallization until the vapor pressure became greater than the confining pressure and water was able to escape from the system. Since vapor pressure affects the liquidus temperature, fluctuating vapor pressure could well lead to cyclic crystallization. However, such a model would require cyclic influx of new magma.

Above the kakortokite is internally layered aegirine lujavrite green in color and finally arfvedsonite lujavrite black in color (Sørensen et al. 2006). These are the most strongly fractionated rocks of the Ilimaussaq Complex and have a thickness of about 500 m. The composition of the mineral phases indicates that it is made up of residual melt remaining after the formation of the kakortokite (Pfaff et al.

Fig. 3.56 Naujaite regarded as a flotation cumulate of sodalite (*green*) is a constituent of the roof cumulates of Ilimaussaq. Sodalite is in part enclosed by arfvedsonite crystals (*black*) approx. 8 cm large (i.e., has a poikilitic texture). Red mineral: eudialyte (*photo* © F. Neukirchen)

2008). The rock is fine-grained, relatively dark in color, and predominantly contains aegirine or arfvedsonite, alkali feldspar, nepheline, and eudialyte (or other complex minerals). The crystals are often aligned.

During the crystallization of lujavrite a late magmatic–hydrothermal fluid was finally released (Graser et al. 2008; Markl and Baumgartner 2002) whose pH increased during further cooling due to reaction with the rock. This fluid is responsible for the formation of hydrothermal veins with Na–Be silicates and other exotic minerals. The reaction of the fluid with the rock resulted in additional metasomatic enrichment of water-soluble elements. In extreme cases so-called hyperagpaitic rocks were formed.

The most highly enriched rocks are located in the north of the complex (e.g., in the Kvanefjeld area) where uranium has at times been extracted from a small mine. There are plans to set up an open-pit mine to produce REEs, yttrium, uranium, and zinc. It is said to be the second largest REE deposit in the world (and maybe the largest for heavy REEs). In the surroundings there are further strongly enriched zones that are planned to be mined. Kakortokite in the south of the complex also represents a potential deposit despite being not as highly enriched.

3.11.2   Khibina and Lovozero

The two largest agpaitic alkaline complexes are Khibina (a. k.a. Chibiny) and Lovozero located on the Kola Peninsula (Russia) in close proximity to each other (Fig. 3.59). They originated in a short period of time in the Devonian (Arzamastsev et al. 2007) in a subvolcanic area below large caldera volcanoes. Different magmas intruded in several phases and formed ring-shaped complexes with different plutonic rocks the youngest of which intruded in the center. Agpaitic magmas of these complexes probably originated from extreme fractionation of a nephelinite magma.

Khibina consists of agpaitic nepheline syenites, feldspar-free rocks of the nephelinite clan (melteigite, iolite, urtite), and carbonatite. The main phase of the complex began with two intrusions of agpaitic nepheline syenite that differ in their texture. These rocks are called khibinite. They are relatively light in color, coarse-grained, and contain titanite—which formed because of the high Ca content of the magma—in addition to some eudialyte and other typical agpaitic minerals.

Subsequently, nephelinitic magmas intruded and formed layers of different cumulate rocks such as urtite, ijolite, and melteigite today exposed in a sickle-shaped zone. Agpaitic minerals such as eudialyte and aenigmatite occur as accessories. These rocks also contain apatite-rich cumulates (apatite–nepheline, apatite–titanite) that belong to the most important phosphate deposits in the world and are mined in

Fig. 3.57 Ilimaussaq's kakortokite is one of the most spectacular examples of magmatic layering. A sequence comprising black, red, and white layers is repeated regularly in which the minerals eudialyte, arfvedsonite, nepheline, and alkali feldspar are contained in varying proportions (*photo* © F. Neukirchen)

Fig. 3.58 Kakortokite (*red layer*) consisting of the minerals eudialyte (red), arfvedsonite (black), and nepheline and alkali feldspar (white) (*photo* © F. Neukirchen)

open-cast mines. A further intrusion consists of an exotic K-rich agpaitic nepheline syenite (rischorrite) with low-Na potassium feldspar, leucite, and the potassium zirconium silicate wadeite.

Another agpaitic nepheline syenite similar to khibinite but known as foyaite later intruded into the center. Finally, carbonatites and a large number of dikes with alkaline rocks followed. Although the enrichment of incompatible elements in the average rocks of Khibina is just enough to make them agpaitic, there are a multitude of small pegmatites and hydrothermal veins that have an extreme composition and diverse mineralogy.

Although Lovozero is smaller, it consists of more strongly fractionated rocks. The complex is also composed of several intrusions. One of the oldest rocks is poikilitic syenite (an agpaitic rock similar to naujaite in Ilimaussaq).

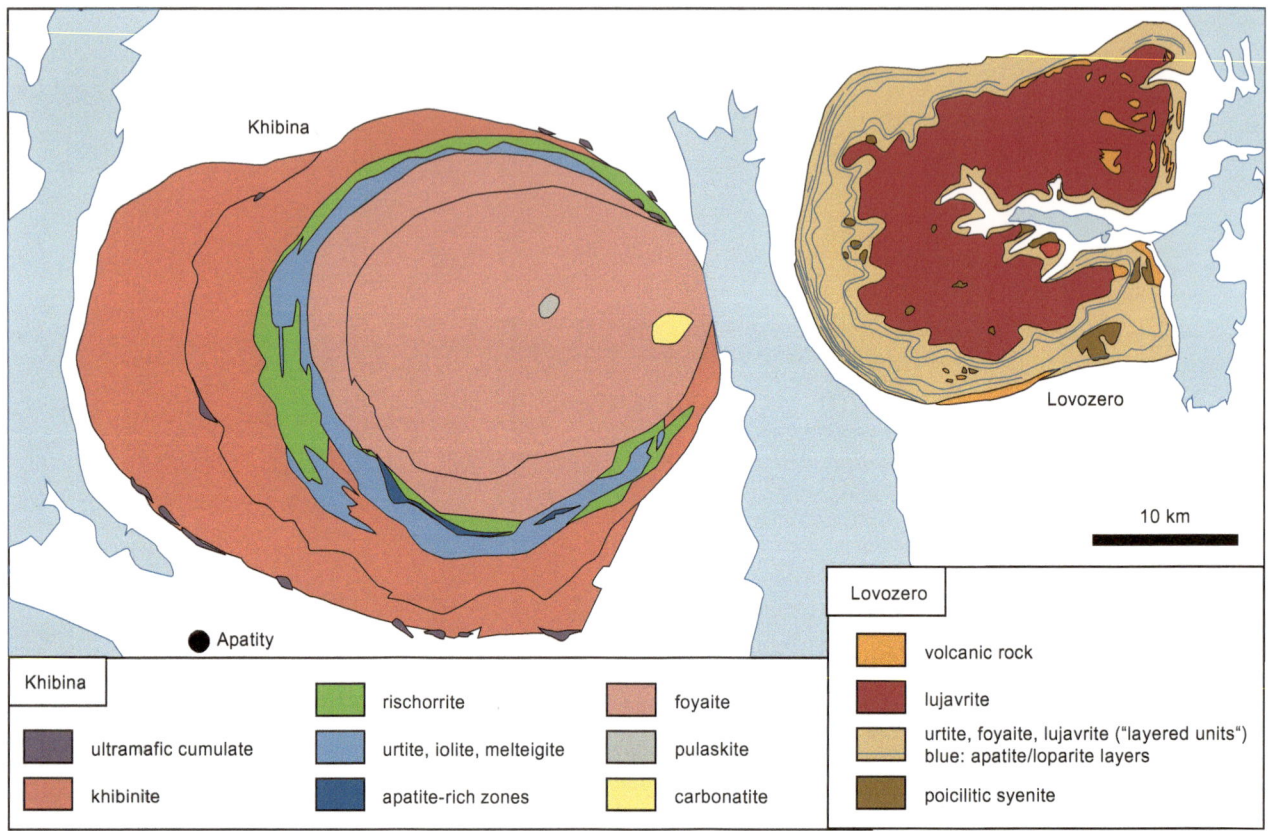

Fig. 3.59 Geological map of the agpaitic alkaline complexes at Khibina and Lovozero (Russia) (from Arzamastsev 1994)

The agpaitic intrusion that formed afterward makes up the largest volume and shows pronounced magmatic stratification (layered units) comparable with stratification in LMIs (Sect. 3.3). Cyclic layers (Fig. 3.60) are each several meters or dozens of meters thick. Ideally, they begin with a rock containing predominantly nepheline (urtite). Toward the top the content of potassium feldspar gradually increases (foyaite) and even further up aegirine becomes a main mineral (lujavrite). Minerals such as loparite, eudialyte, aenigmatite, astrophyllite, and murmanite are found as accessories in the entire layer. Sometimes there is a horizon at the base of the urtite layer with a high content of apatite (>10%) and loparite (>1%). These were mined at times in small quarries where phosphate, tantalum, niobium and REEs were also extracted. In contrast to LMIs there is no cryptic layering concerning the composition of the minerals. Féménias et al. (2005a) argue that each layer is an independently intruded sill.

Lujavrites of a different composition later intruded forming a flat body in the center above the layered intrusion. They are the most strongly fractionated rocks of the intrusion. While most of these lujavrites contain eudialyte, there are murmanite lujavrites that are more enriched.

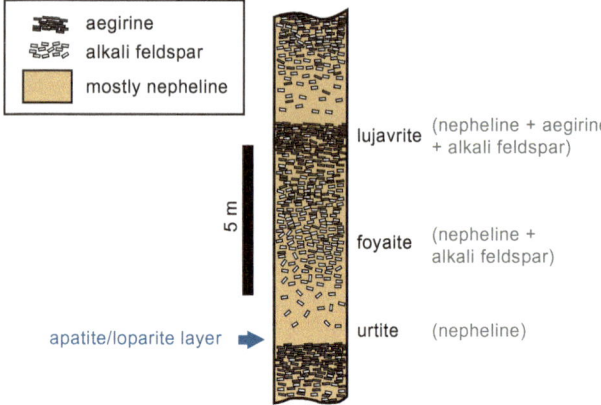

Fig. 3.60 Schematic profile of the cyclically stratified part of Lovozero. Repeated many times (though not always completely) this stratification reaches down several kilometers. On average, the rock corresponds to an agpaitic nepheline syenite. It also contains other minerals such as apatite, loparite, eudialyte, and aenigmatite. There are economically interesting amounts of apatite and loparite (REE–Nb ore) at the base of the cycles (from Arzamastsev 1994)

The Kola Peninsula is home to a number of other alkaline complexes and countless magmatic dikes (Fig. 3.61). They were formed in the Devonian in a system of continental rifts

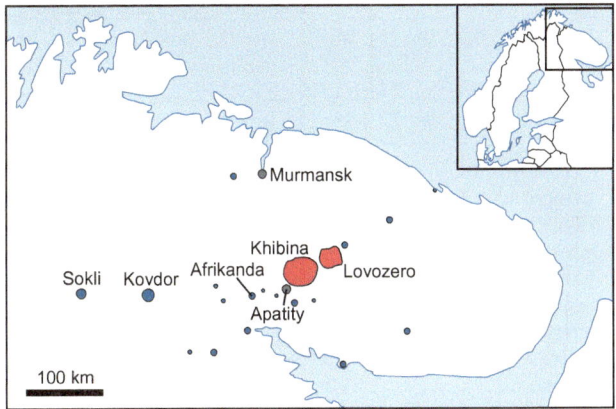

Fig. 3.61 On the Kola Peninsula there is a series of alkaline complexes that formed in a rift system in the Devonian. Khibina and Lovozero are the largest consisting largely of agpaitic nepheline syenites (*red*). The others (*blue*) mainly consist of (ultra-)mafic cumulates and partly contain carbonatites and phoscorites

(Downes et al. 2005). Most complexes consist predominantly of ultramafic cumulates (nephelinite clan or melilitite clan) and in some cases together with carbonatite and phoscorite (Sect. 3.10.1). A good example being Kovdor, which is particularly economically significant.

3.12 Ivigtut

The Ivigtut (a.k.a. Ivittuut) cryolite deposit in southern Greenland has long been exhausted, but it is of great historical importance. It is the only place where the rare mineral cryolite (Na_3AlF_6) occurs in large amounts (Fig. 3.62).

Fig. 3.62 Cryolite (*white*) with siderite (*brownish*) and sphalerite (*black*) from Ivigtut (Greenland) (*photo* © F. Neukirchen/Mineralogical Collections of the TU Berlin)

Cryolite is indispensable to aluminum production. When a mixture of aluminum oxide and cryolite is melted, aluminum can be obtained by electrolysis in the melt. The role cryolite plays is in lowering the melting point. Al_2O_3 melts at 2072 °C and the eutectic temperature of the mixture is half that. Today cryolite is synthetically produced, but for a long time cryolite from Greenland was the only choice.

The deposit has been mined since 1856—first, for the production of NaOH; then, for the direct chemical production of aluminum that was initially more expensive than gold. With the invention of the process described above in 1884 the deposit developed into an important economic sector of the Danish economy.

The deposit is a late-magmatic or hydrothermal–metasomatic formation in the roof region of a small fluorine-rich A-type granite (Pauly and Bailey 1999; Goodenough et al. 2000; Köhler et al. 2008) that like Ilimaussaq formed in the Proterozoic Gardar rift system. The main bulk of the deposit is a banded ore of cryolite + siderite ($FeCO_3$). However, there is also a zone of pure cryolite, a zone of cryolite + fluorite, a zone of fluorite + topaz, and a zone of siderite + quartz. The granite is partially converted to greisen (Sect. 4.6).

According to Pauly and Bailey (1999) the magmatic fluid released by granite was an aggressive fluorine–water mixture. Once the upper part of the granite had solidified it reacted with fluid flowing in from the depths to form greisen. The fluid absorbed sodium and aluminum and developed into a fluorine-rich melt brought about by the melting point of fluorine-rich systems being very low. In this melt further cooling (500–600 °C) led to liquid immiscibility and segregation into a fluorine-rich and an SiO_2-rich melt. The former formed cryolite siderite ore, the latter formed siderite quartz rock. The fluorine-rich, water-rich melt developed into a hydrothermal fluid that led to explosions and the formation of fluorite–cryolite breccias and the fluorite–topaz zone. Finally, hydrothermal veins were formed. The pure cryolite body is said to have been formed by glacier-like flows.

Köhler et al (2008) state that liquid inclusions in all parts of the deposit are mixtures of a NaCl–H_2O fluid and a CO_2 fluid in which the water was originally meteoric groundwater and the CO_2 originated from the mantle or the magma. However, it is not clear when these fluids mixed. It is possible that the water made segregation in two melts possible. Köhler et al (2008) also assume that there was a continuous transition from the alkaline and fluorine-rich melt to a fluorine-rich hydrothermal fluid.

All igneous rocks of the Gardar rift generally have a high fluorine content with the highest contents around Ivigtut (Köhler et al. 2009). This has to be down to corresponding enrichment of the mantle.

Literature

Ames, D.E., A. Davidson, and N. Wodicka. 2008. Geology of the giant Sudbury polymetallic mining camp, Ontario, Canada. *Economic Geology* 103: 1057–1077.

Arndt, N.T., G.K. Czamanske, R.J. Walker, C. Chauvel, and V.A. Fedorenko. 2003. Geochemistry and origin of the intrusive hosts of the Noril'sk-Talnakh Cu-Ni-PGE sulfide deposits. *Economic Geology* 98: 495–515.

Arzamastsev, A.A. 1994. *Unique Paleozoic Intrusions of the Kola Peninsula*. Apatity: Geological Institute of the Kola Science Centre.

Arzamastsev, A.A., L.V. Arzamastseva, A.V. Travin, B.V. Belyatsky, A.M. Shamatrin, A.V. Antonov, A.N. Larionov, N.V. Rodionov, and S.A. Sergeev. 2007. Duration of formation of magmatic system of polyphase paleozoic alkaline complexes of the central Kola: U-Pb, Rb-Sr, Ar-Ar Data. *Doklady Earth Sciences* 413A: 432–436.

Ashwal, L.D. 1993. *Anorthosites*. Berlin: Springer.

Bailey, J.C., H. Sørensen, T. Andersen, L.N. Kogarko, and J. Rose-Hansen. 2006. On the origin of microrhythmic layering in arfvedsonite lujavrite from the Ilímaussaq alkaline complex, South Greenland. *Lithos* 91: 301–318.

Ballhaus, C. 1998. Origin of podiform chromite deposits by magma mingling. *Earth and Planetary Science Letters* 156: 185–193.

Ballhaus, C., and P. Sylvester. 2000. Noble metal enrichment processes in the Merensky Reef, Bushveld complex. *Journal of Petrology* 41: 545–561.

Barnes, S.J. 2007. Cotectic precipitation of olivine and sulfide liquid from komatiite magma and the origin of komatiite-hosted disseminated nickel sulfide mineralization at Mount Keith and Yakabindie, Western Australia. Economic Geology 299–304.

Bell, K. (ed.). 1989. *Carbonatites: Genesis and evolution*. London: Chapman & Hall.

Bell, K., and J. Keller. 1995. *Carbonatite volcanism: Oldoinyo Lengai and the petrogenesis of natrocarbonatites*. Heidelberg: Springer.

Bell, K., and G.R. Tilton. 2001. Nd, Pb and Sr isotopic compositions of East African carbonatites: Evidence for mantle mixing and plume inhomogeneity. *Journal of Petrology* 42: 1927–1945.

Boudreau, A.E., and A.R. McBirney. 1997. The Skaergaard layered series. Part III. Non-dynamic layering. *Journal of Petrology* 38: 1003–1020.

Brooker, R.A., and B.A. Kjarsgaard. 2011. Silicate-carbonate liquid immiscibility and phase relations in the System SiO2-Na2O-Al2O3-CaO-CO2 at 0.1–2.5 GPa with applications to carbonatite genesis. *Journal of Petrology* 52: 1281–1305.

Büchl, A., G. Brügmann, and V.G. Batanova. 2004. Formation of podiform chromitite deposits: Implications from PGE abundances and Os isotopic compositions of chromites from the Troodos complex, Cyprus. *Chemical Geology* 208: 217–232.

Caran, Ş., H. Çoban, M.E.J. Flower, C.J. Ottley, and K.Y. lmaz. 2010. Podiform chromitites and mantle peridotites of the Antalya ophiolite, Isparta Angle (SW Turkey): Implications for partial melting and melt-rock interaction in oceanic and subduction-related settings. Lithos 114: 307–326.

Castor, S.B. 2008. The Mountain Pass rare-earth carbonatite and associated ultrapotassic rocks, California. *Canadian Mineralogist* 46: 779–806.

Cawthorn, R.G. 2007. Cr and Sr: Keys to parental magmas and processes in the Bushveld Complex, South Africa. *Lithos* 95: 198–381.

Cawthorn, R.G., and S.J. Webb. 2001. Connectivity between the western and eastern limbs of the Bushveld Complex. *Tectonophysics* 330: 195–209.

Černý, P. 1992. Geochemical and petrogenetic features of mineralization in rare-element granitic pegmatites in the light of current research. *Applied Geochemistry* 7: 393–416.

Černý, P., and T.S. Ercit. 2005. The classification of granitic pegmatites revisited. *The Canadian Mineralogist* 43: 2005–2026.

Charlier, B., and T.L. Grove. 2012. Experiments on liquid immiscibility along tholeiitic liquid lines of descent. *Contributions to Mineralogy and Petrology* 164: 27–44.

Charlier, B., J.-C. Duchesne, and J. Vander Auwera. 2006. Magma chamber processes in the Tellnes ilmenite deposit (Rogaland Anorthosite Province, SW Norway) and the formation of Fe-Ti ores in massif-type anorthosites. *Chemical Geology* 234: 264–290.

Clarke, B., R. Uken, and J. Reinhardt. 2009. Structural and compositional constraints on the emplacement of the Bushveld Complex, South Africa. *Lithos* 111: 21–36.

Darling, J.R., C.J. Hawkesworth, P.C. Lightfoot, C.D. Storey, and E. Tremblay. 2010. Isotopic heterogeneity in the Sudbury impact melt sheet. *Earth and Planetary Science Letters* 289: 347–356.

Dawson, J.B. 1962. Sodium carbonate lavas from Oldoinyo Lengai, Tanganyika. *Nature* 195: 1075–1076.

De Waal, S.A., Z. Xu, C. Li, and H. Mouri. 2004. Emplacement of viscous mushes in the Jinchuan ultramafic intrusion, Western China. *Canadian Mineralogist* 42: 371–392.

Distler, V.V., V.V. Kryachko, and M.A. Yudovskaya. 2008. Ore petrology of chromite-PGE mineralization in the Kempirsai ophiolite complex. *Mineralogy and Petrology* 92: 31–58.

Dowling, S.E., and R.E.T. Hill. 1998. Komatiite-hosted nickel sulphide deposits, Australia. *Special Jubilee Issue of Australian Geological Survey Organisation Journal* 17: 121–127.

Downes, H., E. Balaganskaya, A. Beard, R. Liferovich, and D. Demaiffe. 2005. Petrogenetic processes in the ultramafic, alkaline and carbonatitic magmatism in the Kola Alkaline Province: A review. *Lithos* 85: 48–75.

Eales, H.V. 2000. Caveats in defining the magmas parental to the mafic rocks of the Bushveld Complex, and the manner of their emplacement: review and commentary. *Mineralogical Magazine* 66: 815832.

Eales, H.V. 2002. Implications of the chromium budget of the Western Limb of the Bushveld Complex. *South African Journal of Geology* 103: 141–150.

Féménias, O., N. Coussaert, S. Brassinnes, and D. Demaiffre. 2005a. Emplacement processes and cooling history of layered cyclic unit II-7 from the Lovozero alkaline massif (Kola Peninsula, Russia). *Lithos* 83: 371393.

Féménias, O., D. Ohnstetter, N. Coussaert, J. Berger, and D. Demaiffre. 2005b. Origin of micro-layering in a deep magma chamber: Evidence from two ultramafic-mafic layered xenoliths from Puy Beaunit. *Lithos* 83: 347–370.

Frietsch, R., and J.-A. Perdahl. 1995. Rare earth elements in apatite and magnetite in Kiruna-type iron ores and some other iron ore types. *Ore Geology Reviews* 9: 489–510.

Giehl, C., M. Marks, and M. Nowak. 2013. Phase relations and liquid lines of descent of an iron-rich peralkaline phonolitic melt: an experimental study. *Contributions to Mineralogy and Petrology* 165: 283–304.

Gittins, J., and R.E. Harmer. 1997. What is ferrocarbonatite? A revised classification. *Journal of African Earth Sciences* 25: 159–168.

Gittins, J., R.E. Harmer, and D.S. Barker. 2005. The bimodal composition of carbonatites: Reality or misconception? *Lithos* 85: 129–139.

Godel, B., S.-J. Barnes, and W.D. Maier. 2011. Parental magma composition inferred from trace element in cumulus and intercumulus silicate minerals: An example from the lower and lower critical zones of the Bushveld Complex, South-Africa. *Lithos* 125: 537–552.

Goodenough, K.M., B.G.J. Upton, and R.M. Ellam. 2000. Geochemical evolution of the Ivigtut granite, South Greenland: a fluorine-rich "A-type" intrusion. *Lithos* 51: 205–221.

Graser, G., J. Potter, J. Köhler, and G. Markl. 2008. Isotope, major, minor and trace element geochemistry of late-magmatic fluids in the peralkaline Ilímaussaq intrusion, South Greenland. *Lithos* 106: 207–221.

Groves, D.I., and N.M. Vielreicher. 2001. The Phalaborwa (Palabora) carbonatite-hosted magnetite-copper sulfide deposit, South Africa: An end-member of the iron-oxide copper-gold-rare earth element deposit group? *Mineralium Deposita* 36: 189–194.

Guest, N.J. 1956. The volcanic activity of Oldoinyo L'Engai, 1954. *Rec. Geological Survey Tanganyika* 4: 56–59.

Halama, R., T. Vennemann, W. Siebel, and G. Markl. 2005. The Grønnedal-Ika carbonatite-syenite compex, South Greenland: Carbonatite formation by liquid immiscibility. *Journal of Petrology* 46: 191–217.

Harlov, D.E., U.B. Andersson, H.-J. Förster, J.O. Nyström, P. Dulski, and C. Broman. 2002. Apatite-monazite relations in the Kiirunavaara magnetite-apatite ore, northern Sweden. *Chemical Geology* 191: 47–72.

Harmer, R.E., and J. Gittins. 1997. The origin of dolomitic carbonatites: Field and experimental constraints. *Journal of African Earth Sciences* 25: 5–18.

Harmer, R.E., and J. Gittins. 1998. The case for primary, mantle-derived carbonatite magma. *Journal of Petrology* 39: 1895–1903.

Hoatson, D.M., S. Jaireth, and A.L. Jaques. 2006. Nickel sulfide deposits in Australia: Characteristics, resources, and potential. *Ore Geology Reviews* 29: 177–241.

Holness, M.B., G. Stripp, M.C.S. Humphreys, I.V. Veksler, T.F.D. Nielsen, and C. Tegner. 2011. Silicate liquid immiscibility within the crystal mush: Late-stage magmatic microstructures in the Skaergaard intrusion, East Greenland. *Journal of Petrology* 52: 175–222.

Holwell, D.A., I. McDonald, and I.B. Butler. 2011. Precious metal enrichment in the Platreef, Bushveld Complex, South Africa: Evidence from homogenized magmatic sulfide melt inclusions. *Contributions to Mineralogy and Petrology* 161: 1011–1026.

Hoover, J.D. 1978. Petrologic features of the Skaergaard Marginal Border Group. *Carnegie Institution Washington Yearbook* 77: 732–739.

Hou, T., Z. Zhang, and T. Kusky. 2011. Gushan magnetite-apatite deposit in the Ningwu basin, lower Yangtze River Valley, SE China: Hydrothermal or Kiruna-type? *Ore Geology Reviews* 43: 333–346.

Irvine, T.N. 1977. Origin of chromitite layers in the Muskox intrusion. *Geology* 5: 273–277.

Irvine, T.N. 1980. Magmatic density currents and cumulus processes. *American Journal of Science* 280A: 1–58.

Irvine, T.N., J.C.Ø. Andersen, and C.K. Brooks. 1998. Included blocks (and blocks within blocks) in the Skaergaard intrusion: Geologic relations and the origin of rhythmic modally graded layers. *Geological Society of America Bulletin* 110: 1398–1447.

Jakobsen, J.K., I.V. Veksler, C. Tegner, and C.K. Brooks. 2011. Crystallization of the Skaergaard intrusion from an emulsion of immiscible iron- and silica-rich liquids: Evidence from melt inclusions in plagioclase. *Journal of Petrology* 52: 345–373.

Jami, M., A.C. Dunlop, and D.R. Cohen. 2007. Fluid inclusion and stable isotope study of the Esfordi apatite-magnetite deposit, central Iran. *Economic Geology* 102: 1111–1128.

Jami, M., A.C. Dunlop, and D.R. Cohen. 2009. Fluid inclusion and stable isotope study of the Esfordi apatite-magnetite deposit, central Iran—A reply. *Economic Geology* 104: 140–143.

Jang, Y.D., H.R. Naslund, and A.R. McBirney. 2001. The differentiation trend of the Skaergaard intrusion and the timing of magnetite crystallization: Iron enrichment revisited. *Earth and Planetary Science Letters* 189: 189–196.

Jones, J.H., D. Walker, D.A. Pickett, M.T. Murrel, and P. Beattie. 1995. Experimental investigations of the partitioning of Nb, Mo, Ba, Ce, Pb, Ra, Th, Pa, and U between immiscible carbonate and silicate liquids. *Geochimica et Cosmochimica Acta* 59: 1307–1320.

Keays, R.R., and P.C. Lightfoot. 2004. Formation of Ni-Cu-Platinum group element sulfide mineralization in the sudbury impact melt sheet. *Mineralogy and Petrology* 82: 217–258.

Keller, J. 1981. Carbonatitic volcanism in the Kaiserstuhl alkaline complex: Evidence for highly fluid carbonatitic melts at the earth's surface. *Journal of Volcanology and Geothermal Research* 9: 423–431.

Keppler, H. 2003. Water solubility in carbonatite melts. *American Mineralogist* 88: 1822–1824.

Kinnaird, J.A., F.J. Kruger, P.A.M. Nex, and R.G. Cawthorn. 2002. Chromitite formation—A key to understanding processes of platinum enrichment: Institution of mining and metallurgy transactions, section B. *Applied Earth Science* 111: B23–B35.

Kjarsgaard, B.A., D.L. Hamilton. 1989. The genesis of carbonatites by immiscibility. In *Carbonatites: Genesis and evolution*, ed. K. Bell. London: Chapman & Hall.

Kjarsgaard, B.A., and T.D. Peterson. 1991. Nephelinite-carbonatite liquid immiscibility at Shombole volcano, East Africa: Petrographic and experimental evidence. *Mineralogy and Petrology* 43: 293–314.

Kogarko, L.N., V.A. Kononova, M.P. Orlova, and A.R. Woolley. 1995. Alkaline rocks and carbonatites of the world. Part 2: Former USSR. London: Chapman and Hall.

Köhler, J., J. Konnerup-Madsen, and G. Markl. 2008. Fluid geochemistry in the Ivigtut cryolite deposit, South Greenland. *Lithos* 103: 369–392.

Köhler, J., J. Schönenberger, B. Upton, and G. Markl. 2009. Subduction-related mantle metasomatism and fluid exsolution from alkalic melts. *Lithos* 113: 731–747.

Kruger, F.J. 2005. Filling the Bushveld Complex magma chamber: lateral expansion, roof and floor interaction, Magmatic uniformities, and the formation of giant chromite, PGE and Ti-V magnetite deposits. *Mineralium Deposita* 40: 451–472.

Krumrei, T., E. Pernicka, M. Kaliwoda, and G. Markl. 2007. Volatiles in a peralkaline system: Abiogenic hydrocarbons and F-Cl-Br systematics in the naujaite of the Ilimaussaq intrusion, South Greenland. *Lithos* 95: 298–314.

Küster, D. 2009. Granitoid-hosted Ta mineralization in the Arabian-Nubian shield: Ore deposit types, tectono-metallogenetic setting and petrogenetic framework. *Ore Geology Reviews* 35: 68–86.

Lee, M.J., J.I. Lee, D. Garcia, J. Moutte, C.T. Williams, F. Wall, and Y. Kim. 2006a. Pyrochlore chemistry from the Sokli phoscorite-carbonatite complex, Finland: Implications for the genesis of phoscorite and carbonatite association. *Geochemical Journal* 40: 1–13.

Lee, M.J., J.I. Lee, S.D. Hur, Y. Kim, J. Moutte, and E. Balaganskaya. 2006b. Sr-Nd-Pb isotopic compositions of the Kovdor phoscorite-carbonatite complex, Kola Peninsula, NW Russia. *Lithos* 91: 250–261.

Li, C., S.-J. Barnes, E. Makovicky, J. Rose-Hansen, and M. Makovicky. 1996. Partitioning of nickel, copper, iridium, rhenium, platinum, and palladium between monosulfide solid solution and sulfide liquid: Effects of composition and temperature. *Geochimica et Cosmochimica Acta* 60: 1231–1238.

Li, C., E.M. Ripley, and E.A. Mathez. 2003. The effect of S on the partitioning of Ni between olivine and silicate melt in MORB. *Chemical Geology* 201: 295–306.

Lightfoot, P.C., and C.E.G. Farrow. 2002. Geology, geochemistry, and mineralogy of the Worthington Offset Dike: A genetic model for offset dike mineralization in the Sudbury Igneous Complex. *Economic Geology* 97: 1419–1446.

Lightfoot, P.C., and R.R. Keays. 2005. Siderophile and chalcophile metal variations in flood basalts from the Siberian Trap, Noril'sk Region: Implications for the origin of the Ni-Cu-PGE ores. *Economic Geology* 100: 439–462.

Lightfoot, P.C., R.R. Keays, and W. Doherty. 2001. Chemical evolution and origin of nickel sulfide mineralization in the Sudbury Igneous Complex, Ontario, Canada. *Economic Geology* 96: 1855–1875.

London, D. 2005. Granitic pegmatites: an assessment of current concepts and directions for the future. *Lithos* 80: 281–303.

London, D. 2008. Pegmatites. Special Publication 10, Mineralogical Association of Canada.

London, D. 2009. The origin of primary textures in granitic pegmatites. *Canadian Mineralogist* 47: 697–724.

Maier, W.D., and S.-J. Barnes. 2010. The petrogenesis of platinum-group element reefs in the Upper Main Zone of the Northern Lobe of the Bushveld Complex on the farm Moordrift, South Africa. *Economic Geology* 105: 841–854.

Maier, W.D., S.-J. Barnes, and D.I. Groves. 2013. The Bushveld Complex, South Africa: Formation of platinum-palladium, chrome- and vanadium-rich layers via hydrodynamic sorting of a mobilized cumulate slurry in a large, relatively slowly cooling, subsiding magma chamber. *Mineralium Deposita* 48: 1–56.

Markl, G., and L. Baumgartner. 2002. PH changes in peralkaline late-magmatic fluids. *Contributions to Mineralogy and Petrology* 144: 331–346.

Markl, G., M. Marks, G. Schwinn, and H. Sommer. 2001. Phase equilibrium constraints on intensive crystallization parameters of the Ilimaussaq Complex, South Greenland. *Journal of Petrology* 42: 2231–2258.

Markl, G., M.A.W. Marks, and B.R. Frost. 2010. On the controls of oxygen fugacity in the generation and crystallization of peralkaline melts. *Journal of Petrology* 51: 1831–1847.

Marks, M., and G. Markl. 2001. Fractionation and assimilation processes in the alkaline augite syenite unit of the Ilimaussaq intrusion, South Greenland, as deduced from phase equilibria. *Journal of Petrology* 42: 1947–1969.

Marks, M.A.W., T. Vennemann, W. Siebel, and G. Markl. 2004. Nd-, O-, and H-isotopic evidence for complex, closed-system fluid evolution of the peralkaline Ilimaussaq intrusion, South Greenland. *Geochimica et Cosmochimica Acta* 68: 3379–3395.

Marks, M.A.W., F. Neukirchen, T. Vennemann, and G. Markl. 2009. Textural, chemical, and isotopic effects of late-magmatic carbonatitic fluids in the carbonatite-syenite Tamazeght complex, High Atlas Mountains, Morocco. *Mineralogy and Petrology* 97: 23–42.

Marks, M.A.W., K. Hettmann, J. Schilling, B.R. Frost, and G. Markl. 2011. The mineralogical diversity of alkaline igneous rocks: Critical factors for the transition from miaskitic to agpaitic phase assemblages. *Journal of Petrology* 52: 439–455.

Martin, R.F. 2006. A-type granites of crustal origin ultimately result from open-system fenitization-type reactions in an extensional environment. *Lithos* 91: 125–136.

Martin, R.F., and C. De Vito. 2005. The patterns of enrichment in felsic pegmatites ultimately depend on tectonic setting. *The Canadian Mineralogist* 43: 2027–2048.

Matveev, S., and C. Ballhaus. 2002. Role of water in the origin of podiform chromitite deposits. *Earth and Planetary Science Letters* 203: 235–243.

McBirney, A.R. 1989. The Skaergaard layered series: I. Structure and average compositions. *Journal of Petrology* 30: 363–379.

McBirney, A.R. 2009. Factors governing the textural development of Skaergaard gabbros: A review. *Lithos* 111: 1–5.

McBirney, A.R., and A. Nicolas. 1997. The Skaergaard Layered Series. Part II. Magmatic flow and dynamic layering. *Journal of Petrology* 38: 569–580.

Melcher, F., W. Grum, G. Simon, T.V. Thalhammer, and E.F. Stumpfl. 1997. Petrogenesis of the ophiolitic giant chromite deposits of Kempirsai, Kazakhstan: A study of solid and fluid inclusions in chromite. *Journal of Petrology* 38: 1419–1458.

Mitchell, A.A., and R.N. Scoon. 2007. The Merensky Reef at Winnaarshoek, Eastern Bushveld Complex: A primary magmatic hypothesis based on a wide reef facies. *Economic Geology* 102: 971–1009.

Mitchell, R.H. 2005. Carbonatites and carbonatites and carbonatites. *The Canadian Mineralogist* 43: 2049–2068.

Molnár, F., D.H. Watkinson, and P.C. Jones. 2001. Multiple hydrothermal processes in footwall units of the North Range, Sudbury Igneous Complex, Canada, and implications for the genesis of vein-type Cu-Ni-PGE deposits. *Economic Geology* 96: 1645–1670.

Mondal, S.K., and E.A. Mathez. 2006. Origin of the UG2 chromitite layer, Bushveld complex. *Journal of Petrology* 48: 495–510.

Mungall, J.W., D.R.A. Andrews, L.J. Cabri, P.J. Sylvester, and M. Tubrett. 2005. Partitioning of Cu, Ni, Au, and platinum-group elements between monosulfide solid solution and sulfide melt under controlled oxygen and sulfur fugacities. *Geochimica et Cosmochimica Acta* 69: 4349–4360.

Murck, B.W., and I.H. Campbell. 1986. The effects of temperature, oxygen fugacity and melt composition on the behaviour of chromium in basic and ultrabasic melts. *Geochimica et Cosmochimica Acta* 50: 1871–1887.

Naldrett, A.J. 1992. A model for the Ni-Cu-PGE ores of the Noril'sk region and its application to other areas of flood basalt. *Economic Geology* 87: 1945–1962.

Naldrett, A.J., and A.H. Wilson. 1990. Horizontal and vertical variations in noble-metal distribution in the Great Dyke of Zimbabwe: A model for the origin of the PGE mineralization by fractional segregation of sulfide. *Chemical Geology* 88: 279–300.

Naldrett, A.J., J. Kinnaird, A. Wilson, and G. Chunnett. 2008. Concentration of PGE in the Earth's crust with special reference to the Bushveld complex. *Earth Science Frontiers* 15: 264–297.

Naldrett, A.J., A. Wilson, J. Kinnaird, M. Yudovskaya, and G. Chunnet. 2012. The origin of chromitites and related PGE mineralization in the Bushveld Complex: New mineralogical and petrological constraints. *Mineralium Deposita* 47: 209–232.

Naslund, H.R. 1983. Petroloy of the upper border series of the Skaergaard intrusion. *Journal of Petrology* 25: 185–212.

Naslund, H.R., F. Henriques, J.O. Nystrom, W. Vivallo, and F.M. Dobbs. 2002. Magmatic iron ores and associated mineralization: Examples from the Chilean high Andes and coastal Cordillera. In *Hydrothermal iron oxide-copper-gold and related deposits: A global perspective*, ed. T.M. Porter. Australia: Australian Mineral Foundation, Adelaide.

Ngwenya, B.T. 1994. Hydrothermal rare earth mineralisation in carbonatites of the Tundulu complex, Malawi: Processes at the fluid/rock interface. *Geochimica et Cosmochimica Acta* 58: 2061–2072.

Nivin, V.A., P.J. Treloar, N.G. Konopleva, and S.V. Ikorsky. 2005. A review of the occurrence, form and origin of C-bearing species in the Khibiny Alkaline Igneous Complex, Kola Peninsula, NW Russia. *Lithos* 85: 93–112.

Oberthür, T., D.W. Davis, T.G. Blenkinsop, and A. Höhndorf. 2002. Precise U-Pb mineral ages, Rb-Sr and Sm-Nd systematics for the Great Dyke, Zimbabwe—Constraints on late Archean events in the Zimbabwe craton and Limpopo belt. *Precambrian Research* 113: 293–305.

Okrusch, M., and S. Matthes. 2009. *Mineralogie: Eine Einführung in die spezielle Mineralogie, Petrologie und Lagerstättenkunde*, 8th ed. Heidelberg: Springer.

Pauly, H., and J.C. Bailey. 1999. Genesis and evolution of the Ivigtut cryolite deposit, South Greenland. Meddelelser om Grønland, Geoscience 28. Kopenhagen.

Pfaff, K., T. Krumrei, M. Marks, T. Wenzel, T. Rudolf, and G. Markl. 2008. Chemical and physical evolution of the 'lower layered sequence' from the nepheline syenitic Ilímaussaq intrusion, South Greenland: Implications for the origin of magmatic layering in peralkaline felsic liquids. *Lithos* 106: 280–296.

Pfaff, K., T. Wenzel, J. Schilling, M. Marks, and G. Markl. 2010. A fast and easy-to-use approach to cation site assignment for eudialyte-group minerals. *Neues Jahrbuch für Mineralogie* 187: 69–81.

Philpotts, A.R. 1982. Compositions of immiscible liquids in volcanic rocks. *Contributions to Mineralogy and Petrology* 80: 201–218.

Pons, J., P. Barbey, H. Nachit, and J.-P. Burg. 2006. Development of igneous layering during growth of pluton: The Tarçouate Laccolith (Morocco). *Tectonophysics* 413: 271–286.

Prevec, S.A., P.C. Lightfoot, and R.R. Keays. 2000. Evoluton of the sublayer of the Sudbury Igneous Complex: Geochemical, Sm-Nd isotopic and petrologic evidence. *Lithos* 51: 271–292.

Prichard, H.M., C.R. Neary, P.C. Fisher, and M.J. O'Hara. 2008. PGE-rich podiform chromitites in the Al 'Ays Ophiolite Complex, Saudi Arabia: An example of critical mantle melting to extract and concentrate PGE. *Economic Geology* 103: 1507–1529.

Rickers, K., R. Thomas, and W. Heinrich. 2006. The behavior of trace elements during the chemical evolution of the H2O-, B-, and F-rich granite-pegmatite-hydrothermal system at Ehrenfriedersdorf, Germany: A SXRF study of melt and fluid inclusions. *Mineralium Deposita* 41: 229–245.

Rohrbach, A., and M.W. Schmidt. 2011. Redox freezing and melting in the Earth's deep mantle resulting from carbon-iron redox coupling. *Nature* 472: 209–214.

Roussel, D.H., J.S. Fedorowich, and B.O. Dressler. 2003. Sudbury Breccia (Canada): A product of the 1850 Ma Sudbury event and host to footwall Cu-Ni-PGE deposits. *Earth-Science Reviews* 60: 147–174.

Ryabchikov, I.D., and L.N. Kogarko. 2006. Magnetite compositions and oxygen fugacities of the Khibina magmatic system. *Lithos* 92: 35–45.

Salvi, S., and A.E. Williams-Jones. 1995. Zirconosilicate phase relations in the Strange Lake (Lac Brisson) pluton, Quebec-Labrador, Canada. *American Mineralogist* 80: 1031–1040.

Schilling, J., F.-Y. Wu, C. McCammon, T. Wenzel, M.A.W. Marks, K. Pfaff, D.E. Jacob, and G. Markl. 2011. The compositional variability of eudialyte-group minerals. *Mineralogical Magazine* 75: 87–115.

Schoenberg, R., F.J. Kruger, T.F. Nägler, T. Meisel, and J.D. Kramers. 1999. PGE enrichment in chromitite layers and the Merensky Reef of the western Bushveld Complex; a Re-Os and Rb-Sr isotope study. *Earth and Planetary Science Letters* 172: 49–64.

Scott, R.G., and K. Benn. 2002. Emplacement of sulfide deposits in the copper cliff offset dike during collapse of the sudbury crater rim: Evidence from magnetic fabric studies. *Economic Geology* 97: 1447–1458.

Seabrook, C.L., R.G. Cawthorn, and F.J. Kruger. 2005. The Merensky Reef, Bushveld Complex: Mixing of minerals not mixing of magmas. *Economic Geology* 100: 1191–1206.

Sillitoe, R.H., and D.R. Burrows. 2002. New field evidence bearing on the origin of the El Laco magnetite deposit, Northern Chile. *Economic Geology* 97: 1101–1109.

Smith, M.P., P. Henderson, and L.S. Campbell. 2000. Fractionation of the REE during hydrothermal processes: Constraints from the Bayan Obo Fe-REE-Nb deposit, Inner Mongolia, China. *Geochimica et Cosmochimica Acta* 64: 3141–3160.

Solovova, I.P., I.D. Ryabchikov, A.V. Girnis, A. Pedersen, and T. Hansteen. 2002. Reduced magmatic fluids in basalt from the island of Disko, central West Greenland. *Chemical Geology* 183: 365–371.

Sørensen, H. 1997. The agpaitic rocks—An overview. *Mineralogical Magazine* 61: 485–498.

Sørensen, H. 2001. Brief introduction to the geology of the Ilimaussaq alkaline complex, South Greenland. *Geology of Greenland Survey Bulletin* 190: 7–24.

Sørensen, H., H. Bohse, and J.C. Bailey. 2006. The origin and mode of emplacement of lujavrites in the Ilímaussaq alkaline complex, South Greenland. *Lithos* 91: 286–300.

Spandler, C., J. Mavrogenes, and R. Arculus. 2005. Origin of chromitites in layered intrusions: Evidence from chromite-hosted melt inclusions from the Stillwater Complex. *Geology* 33: 893–896.

Therriault, A.M., A.D. Fowler, and R.A.F. Grieve. 2002. The Sudbury Igneous Complex: A differentiated impact melt sheet. *Economic Geology* 97: 1521–1540.

Thomas, R., P. Davidson, and H. Beurlen. 2012. The competing models for the origin and internal evolution of granitic pegmatites in the light of melt and fluid inclusion research. *Mineralogy and Petrology* 106: 55–73.

Thy, P., C.E. Lesher, T.F.D. Nielsen, and C.K. Brooks. 2006. Experimental constraints on the Skaergaard liquid line of descent. *Lithos* 92: 154–180.

Ulff-Møller, F. 1990. Formation of native iron in sediment-contaminated magma: I. A case study of the Hanekammen Complex on Disko Island, West Greenland. *Geochimica et Cosmochimica Acta* 54: 57–70.

Van der Merwe, J., and R.G. Cawthorn. 2005. Structures at the base of the Upper Group 2 chromitite layer, Bushveld Complex, South Africa, on Karee Mine (Lonmin Platinum). *Lithos* 83: 214–228.

Veksler, I.V., C. Petibon, G.A. Jenner, A.M. Dorfman, and D.B. Dingwell. 1998. Trace element partitioning in immiscible silicate-carbonate liquid systems: An initial experimental study using a centrifuge autoclave. *Journal of Petrology* 39: 2095–2104.

Veksler, I.V., A.M. Dorfman, A.A. Borisov, R. Writh, and D.B. Dingwell. 2007. Liquid immiscibility and the evolution of basaltic magma. *Journal of Petrology* 48: 2187–2210.

Voordouw, R., J. Gutzmer, and N.J. Beukes. 2009. Intrusive origin of Upper Group (UG1, UG2) stratiform chromitite seams in the Dwars River area, Bushveld Complex, South Africa. *Mineralogy and Petrology* 97: 75–94.

Wager, L.R., and G.M. Brown. 1968. *Layered igneous rocks*. San Francisco: Freeman.

Wall, F., and A.N. Zaitsev (ed.). 2004. Phoscorites and carbonatites from mantle to mine. Mineralogical society series, vol. 10.

Wilson, A.H., C.Z. Murahwi, and B. Coghill. 2000. Stratigraphy, geochemistry and platinum group element mineralisation of the central zone of the Selukwe subchamber of the Great Dyke, Zimbabwe. *Journal of African Earth Sciences* 30: 833–853.

Woolley, A.R. 1987. Alkaline rocks and carbonatites of the World. Part 1: North and South America. British Museum, London and University of Texas Press.

Woolley, A.R. 2001. Alkaline rocks and carbonatites of the World. Part 3: Africa. London: The Geological Society.

Xu, C., H. Zhang, Z. Huang, C. Liu, L. Qi, W. Li, and T. Guan. 2004. Genesis of the carbonatite-syenite complex and REE deposit at Maoniuping, Sichuan Province, China: Evidence from Pb isotope geochemistry. *Geochemical Journal* 38: 67–76.

Xu, C., L. Wang, W. Song, and M. Wu. 2010. Carbonatites in China: A review for genesis and mineralization. *Geoscience Frontiers* 1: 105–114.

Yang, K.-F., H.-R. Fan, M. Santosh, F.-F. Hu, and K.-Y. Wang. 2011. Mesoproterozoic carbonatitic magmatism in the Bayan Obo deposit, Inner Mongolia, North China: Constraints for the mechanism of super accumulation of rare earth elements. *Ore Geology Reviews* 40: 122–131.

Yang, Z., and A. Woolley. 2006. Carbonatites in China: A review. *Journal of African Earth Sciences* 27: 559–575.

Zaitsev, A., and K. Bell. 1995. Sr and Nd isotope data of apatite, calcite and dolomite as indicators of source, and the relationships of phoscorites and carbonatites from the Kovdor massif, Kola peninsula, Russia. *Contributions to Mineralogy and Petrology* 121: 324–335.

Zhang, Z., J. Mao, A.D. Saunders, Y. Ai, Y. Li, and L. Zhao. 2009. Petrogenetic modeling of three mafic-ultramafic layered intrusions in the Emeishan large igneous province, SW China, based on isotopic and bulk chemical constraints. *Lithos* 113: 369–392.

Zingg, A.J. 1996. Recrystallization and the origin of layering in the Bushveld Complex. *Lithos* 37: 15–37.

Further Reading

Best, M.G., and E.H. Christiansen. 2001. *Igneous petrology*. Malden, Massachussetts: Blackwell Science.

Guilbert, J.M., and C.F. Park. 1986. *The geology of ore deposits*. New York: WH Freeman.

Laznicka, P. 2010. *Giant metallic deposits: Future sources of industrial metals*, 2nd ed. Heidelberg: Springer.

Markl, G. 2008. *Minerale und Gesteine: Mineralogie—Petrologie—Geochemie*, 2nd ed. Heidelberg: Spektrum Akademischer Verlag.

Misra, K.C. 2000. *Understanding mineral deposits*. Dordrecht, Niederlande: Kluwer Academic Publishers.

Naldrett, A.J. 2004. *Magmatic sulfide deposits*. Heidelberg: Springer.

Neukirchen, F. 2012. *Edelsteine: Brillante Zeugen für die Erforschung der Erde*. Heidelberg: Springer Spektrum.

Pohl, W.L. 2011. *Economic geology*. Chichester: Wiley-Blackwell.

Robb, L. 2005. *Introduction to ore-forming processes*. Malden, Massachussetts: Blackwell Science.

Rothe, P. 2010. *Schätze der Erde*. Darmstadt: Primus Verlag.

Winter, J.D. 2001. *Igneous and metamorphic petrology*. New Jersey: Prentice Hall.

Hydrothermal Deposits

Hot water is a very effective medium for the transport and enrichment of certain elements that are subsequently precipitated from the water. This can happen in very different places such as at a hot spring on the seabed, along a fault, in fine cracks above a granite pluton, in a cave, or in the pores of the rock. The space necessary for precipitation can even be created by simultaneously dissolving the surrounding rock in a process called replacement. Another possibility is metasomatism in which rock can transform when supplied with certain elements. For example, magmatic–hydrothermal solutions can react so violently with limestone that it is transformed into a completely different rock called a skarn. Although deposits presented in this chapter are correspondingly diverse, they all have one thing in common in that hot water plays the main role. More precisely, hydrothermal solutions or fluids play the main rule since of course it is not pure water. Such fluids (Box 4.1) also contain varying amounts of dissolved salts (especially, NaCl and CaCl), carbon dioxide, hydrogen carbonate, hydrogen sulfide, sulfur dioxide, and various other substances the presence of some of which has a strong effect on the solubility of metals.

Anyone swimming in a hot spring is in direct contact with a hydrothermal fluid albeit in a relaxed way. Anyone who has visited a volcanic region gets a good initial impression of what is going on. Steam escapes at fumaroles sometimes together with sulfur-bearing gases from which elementary sulfur may sublimate to bizarre tree-shaped crystal aggregates. Near active volcanoes there are often hot springs, boiling mud pools (Fig. 4.1), and sometimes geysers. In some cases the water at a spring is even hotter than boiling point made possible by the strong pressure drop during ascent and at the spring itself. However, metals are only deposited in tiny amounts in such geothermal fields. Metals together with microorganisms contribute to the wide range of colors that can be observed in the springs and their surroundings. Slightly deeper down gold may be precipitated (Sect. 4.3). A frequent formation at hot springs is silica sinter consisting of crusts of amorphous SiO_2 that contain some water (opal). This is because the solubility of SiO_2 decreases greatly when water cools down such that the water is oversaturated with SiO_2. Opal only forms at or just below the Earth's surface; at depth SiO_2 crystallizes as quartz instead. Perfectly formed crystals can grow as evidenced by rock crystals that have formed in alpine clefts. In many hydrothermal deposits quartz is the most abundant mineral, whereas in others calcite, fluorite, or baryte are the most abundant. In former times miners could do nothing with these "useless" minerals that they called gangue. However, fluorite and baryte are today coveted resources themselves (Sect. 7.14).

Box 4.1 What is a fluid?

This chapter is about water or, more precisely, aqueous solutions that also contain dissolved solids such as salts and dissolved gases such as CO_2. However, at ever-increasing depths into the Earth's crust the less it makes sense to talk about water. If pressure and temperature exceed the so-called critical point, then it is no longer possible to even distinguish between liquid water and gaseous vapor and instead of a boiling point there is only a slow change in density. We are dealing with a supercritical fluid that can be mixed with CO_2 (Lowenstern 2001) and other gases in any ratio.

Fluid (Roedder 1984; Kyser 2007; Audetat et al. 2008; Pirajno 2009) is an umbrella term used in geology for all liquids, gases, and supercritical fluids. This can be almost pure water or steam, a brine saturated with salt, an acid or a base, crude oil, a gas such as CO_2, SO_2, H_2S, and CH_4—and any possible mixtures thereof. In extreme cases even magma can be counted as fluid. Sometimes there is a smooth transition from a water-rich residual melt to a hydrothermal fluid. In most cases, however, this term refers to a fluid rich in water.

© Springer Nature Switzerland AG 2020
F. Neukirchen and G. Ries, *The World of Mineral Deposits*,
https://doi.org/10.1007/978-3-030-34346-0_4

Fig. 4.1 Sol de la Mañana (Bolivia) geothermal field at sunrise (*photo* © F. Neukirchen)

Box 4.2 Catathermal, mesothermal, epithermal, pneumatolytic

In the past, when cooling was still considered the most important factor in the formation of hydrothermal deposits, such deposits were classified according to their formation temperature such as epithermal (below 200 °C), telethermal (below 100 °C), mesothermal (200–300 °C), and catathermal or hypothermal (>300 °C). However, now that other factors are deemed more important this classification not only makes little sense but also many examples classified according to this scheme have been found to fall out of the corresponding temperature range. Although the terms are still used, their meaning has changed. Epithermal now means formations at shallow depths with water temperature between 100 and 320 °C, mesothermal at medium depths (a few kilometers), and hypothermal at great depths (inevitably at high temperatures). Some geologists prefer to speak of epizonal, mesozonal, and hypozonal to emphasize independence from temperature. The terms are not precisely defined in today's use and serve only as a rough orientation.

The word pneumatolytic used to be employed to describe hydrothermal mineralizations hotter than 400 °C such as deposits formed from water released from magma in the immediate vicinity of a pluton. This term was mainly used to describe greisen (Sect. 4.6) and some skarns (Sect. 4.9). Pneumatolytic deposits were even distinguished from hydrothermal deposits in German literature as an independent category evoking a transition from the late magmatic stage (pegmatites) via the pneumatolytic stage to the hydrothermal stage. The distinction was based on these high-temperature fluids having a low density comparable with vapor. It was therefore thought that a change in pressure (during ascent) would have a greater effect on the precipitation of ores than cooling. Now that it is known that pressure has little to do with the formation of these deposits the term pneumatolytic is hardly ever used. Corresponding deposits are now counted as hydrothermal deposits as has always been the case in English-speaking countries.

The terms vapor or gas are misleading when discussing supercritical fluids. There is no boiling point at which there is a transition between gaseous and liquid state at a defined temperature, but a continuous change in density over a large pressure or temperature interval. Accordingly, there is no sudden change in the behavior of such fluids as suggested by the distinction between pneumatolytic and hydrothermal. A supercritical fluid can be separated into a liquid and a vapor phase during ascent. This happens, for example, at the formation of porphyry copper deposits (Sect. 4.4) and ensures separation and enrichment of the metals, but does not lead to the precipitation of ores. The situation is different in epithermal systems near the Earth's surface (Sect. 4.3) where rising water can boil due to the reduced pressure and actually trigger the precipitation of ores. However, such systems have never been described as pneumatolytic.

Gangue and the ores of hydrothermal deposits were formed by two steps: 1) certain elements were dissolved in a fluid followed by 2) oversaturation (supersaturation) and precipitation. Precipitation reactions are equilibrium reactions that in principle can take place in both directions depending on whether the water is undersaturated or oversaturated with the corresponding substances. Oversaturation is not always caused by water cooling: in fact cooling only plays a role when magma has risen to shallow areas of the Earth's crust or when water rises very quickly from great depths (Box 4.2). In all other cases there is no corresponding temperature gradient and fast cooling of hot water is not possible. There are however other ways to bring a solution to oversaturation.

A change in the pH value of water is very effective in many cases. This also applies to the precipitation of quartz. The solubility of SiO_2 in an alkaline fluid is considerably higher than in water with a neutral pH at the same temperature. A change in the pH value has a similar effect on the solubility of metals partly because the stability of complexes (Box 4.3) depends on the pH value.

The pH value (i.e., the concentration of H^+ ions) in groundwater is buffered by the rock. An acid is neutralized when it reacts with limestone or feldspar. In contrast, the oxidation of sulfide minerals releases sulfuric acid. Hot, salty, almost neutral deep groundwater can develop into an acid as it rises and cools because the balance of the dissociation of HCl shifts. The mixing of two different fluids leads to a sudden change in pH. Reaction with a rock that is not in equilibrium with the fluid also changes the pH.

Redox reactions can also trigger precipitation. Many metals have several oxidation states that differ greatly in their solubility. Fe^{2+} is quite soluble while Fe^{3+} is not. Sulfur also has several oxidation states and it can be either reduced (sulfides: S^{2-}), elemental, or oxidized (S^{4+}, S^{6+}) (e.g., in sulfate: SO_4^{2-}; or SO_2, HSO_4^-, etc.). Since these ions together with metals can form soluble complexes the oxidation or reduction of sulfur also affects the solubility of metals. Accordingly, a change in the oxygen content or rather the redox potential (Eh) can lead to precipitation. The redox potential changes, for example, through reaction of the fluid with minerals in the surrounding rock, with graphite, organic carbon or methane; through ascent to more strongly oxidized areas; through mixing with another fluid; or through the activity of bacteria.

The redox potential (Eh) is a measure of how oxidized or reduced the ions and complexes dissolved in water are. Its specification refers to electrochemical reduction or oxidation relative to the standard hydrogen electrode. Water itself can also serve as a source of oxygen with the release of H^+ that in turn changes the pH value. The redox potential is of course related to the oxygen content that is expressed as oxygen fugacity (fO_2) (Box 3.5).

The precipitation of many hydrothermal deposits is due to different fluids mixing: especially mixing a hot and salty solution (brine) rising from the depth with meteoric water (groundwater originating from percolated rain) or with the connate water of sediments (partly former seawater). Precipitation reactions are similar to experiments in a test tube well known to readers who have undergone a practical course in inorganic chemistry. Which ore mineral is formed during precipitation also depends on temperature, pH, and Eh. These in turn are controlled by the properties of the fluids involved and the mixing ratio. Some minerals such as pyrite have a very large stability field and therefore occur in deposits formed under very different conditions. Other minerals have a small stability field and can only be formed under very specific conditions.

An example of a precipitation reaction is the formation of sphalerite through:

$$Zn^{2+} + H_2S = ZnS + 2H^+$$

This equilibrium reaction shifts to the right when water is oversaturated with sphalerite (i.e., the concentration of Zn^{2+} and H_2S is higher than water can dissolve under given conditions such as temperature, pH, Eh, and salt content). This reaction releases H^+ thus reducing the pH value. The H^+ can then dissolve limestone, for example, and thus create space in which further ores can be precipitated. On the other hand, if a merely saturated but already acidic fluid enters limestone and dissolves it, then the acid is neutralized. This triggers the precipitation of sphalerite because the balance of the precipitation reaction shifts to the right. In both cases ore replaces the limestone. Neutralization of the acid can of course also be achieved by alteration (see Box 4.14) of another rock. However, the reaction is not only dependent on the pH but also on the Eh because this controls the balance

Table 4.1 Vertical zoning in a hypothetical hydrothermal vein

Depth	Metal	Ore minerals
Near surface	Hg	Cinnabar
	Sb	Stibnite
	Au–Ag	Gold, electrum, acanthite
	Ag–Mn	Acanthite, rhodochrosite
	Pb	Galena
	Zn	Sphalerite
	Cu–As–Sb	Chalcopyrite, tennantite–tetrahedrite
	Cu	Chalcopyrite
Deep	Mo, W, Sn	Molybdenite, scheelite, cassiterite

This empirically determined scheme can only serve as a rough orientation. Normally, there are only a few zones in any vein and the order may vary (modified from Robb 2005; Emmons 1936)

between sulfide (or H_2S) and sulfate. If this balance changes, then so does the concentration of H_2S as well. A reduction of sulfate thus also shifts the balance of our precipitation reaction to the right side. On the other hand, the precipitation of sphalerite affects other equilibrium reactions with H_2S or H^+ and can therefore lead to the precipitation or dissolution of other minerals.

Since the dependence of solubility on factors such as temperature, Eh, and pH value is somewhat different for each metal the metals are well sorted in hydrothermal systems. Within a deposit the composition of ores often varies with depth. Table 4.1 shows a hypothetical sequence empirically determined from a combination of many veins. The scheme is a rough simplification because it is not a linear system and water does not necessarily contain all the metals mentioned. However, it gives a good overview. As a rule there are only a few zones (or only one) in any vein. The order can also be different. Sometimes several zones are shifted into each other (telescoping).

Once a deposit is near the Earth's surface an oxidation zone is formed in which primary ores are oxidized and replaced by secondary ore minerals (Box 4.16) in addition to primary zoning. This results in secondary zoning with the depletion of soluble metals in the near-surface gossan and their enrichment in the so-called cementation zone at the groundwater level.

Some deposits contain several generations of ores that were formed at different times from different fluids and differ in their metal content. Metals can of course be remobilized in later episodes by water dissolving minerals and precipitating other minerals elsewhere. By examining fluid inclusions (Box 4.4) details about these fluids can be learned.

Many hydrothermal systems are related to magmatism in which it is not only water released from the magma that plays a role but also meteoric water or seawater. Magmatic heat leads to convection of the groundwater in cracks and pores, soluble substances being leached out of the rock, and the heated water becoming ever-richer in salt and metals.

However, there are hydrothermal systems far away from magmatically active regions (Fig. 4.2). For example, water can be released by metamorphic reactions. In fact, many of the reactions a rock undergoes on its way into the depths are dehydration reactions where minerals containing water are replaced by minerals containing less water. Connate water— often still ancient seawater—long trapped in the pores of sedimentary rock is involved in other cases. It exchanges certain substances with the rock until an equilibrium is reached that is dependent on pressure, temperature, pH, and Eh. If the sediment pile also contains layers of salt, then this water can be salty (brine). Water can also penetrate into the basement and react with the rock there over millions of years until finally rock and water reach chemical equilibrium.

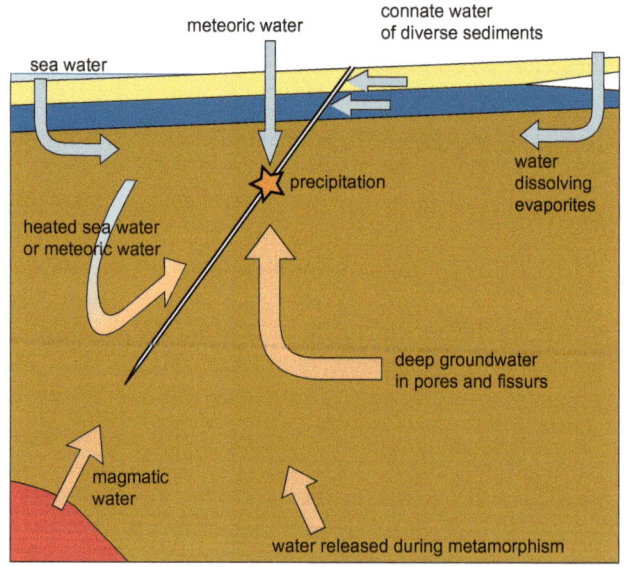

Fig. 4.2 Precipitation in a hydrothermal system often occurs when hot water rising from the depth mixes with cool water near the surface. The fluids involved can be of different origin and composition; mixtures are also possible. Water percolating through the pores and fissures reacts with the rock and changes in composition

Water in the pores and fissures of rock must be pumped up into higher areas so that the dissolved freight can be precipitated again somewhere else (Fig. 4.3). In the sediment basin in the foreland of a mountain range gravity-driven fluid flow directed away from the mountains driven by the hydraulic potential of the mountains can occur. Another possibility is the burial of rocks either by deposition of further sediments or by thrusting of tectonic nappes in an orogeny. The sediment in question is now at a greater depth —and burial is associated with compaction and reduction in the pore volume. Pore water must inevitably move upward. If a tectonic nappe is thrusted over the foreland basin, then water in the sediments is driven away from the tectonic front.

At first glance it is surprising that a similar effect can be triggered by *rising* rocks. Erosion exposes rocks in a mountain range that were previously at greater depths. The pore volume of the rocks does not change, but the pressure of the water contained in them does. As pressure decreases the water expands and must therefore rise. Something very similar happens in a rift system. Although the surface of a continental rift is noticeable by a sinking tectonic block, this does not play a role here. It is more a matter of what happens to the middle and lower crust. The entire crust is thinned out by extension meaning that the middle and lower parts of the crust are moved to shallower depths. The pore volume of rock in the middle crust is quite small; hence correspondingly little water is available to expand. However, an enormous volume of rock is affected. Accordingly, significant quantities of hydrothermal water rise at the faults of a rift system (Staude et al. 2009).

Another effect plays a role in active faults. In the aftermath of earthquakes significantly more water is often observed pouring from sources in the immediate vicinity in the short term. The reason for this is that overpressured water at depth flows into the newly formed cracks, which in turn increases the formation of cracks. Such seismic pumping can lead to the episodic rise of hydrothermal fluids.

The fluid pathways available to hydrothermal water are very important for the formation of deposits. Economic deposits are only formed when the largest possible quantities of fluid precipitate their freight in a relatively small volume of rock. Local variations in the water conductivity of faults and fissures and the geometry of faults and clefts are important to focusing the hydrothermal system. There are therefore tectonic situations that are advantageous for the formation of larger deposits. Examples are the smaller branches of faults between laterally offset segments of a large fault.

It would actually be appropriate to follow a chapter on magmatic deposits by continuing with hydrothermal mineralizations directly related to magmatism in which porphyry

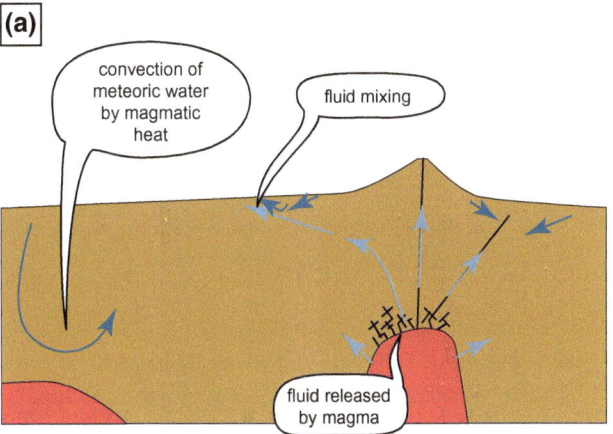

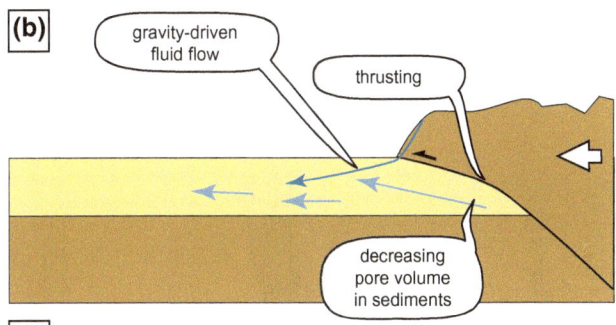

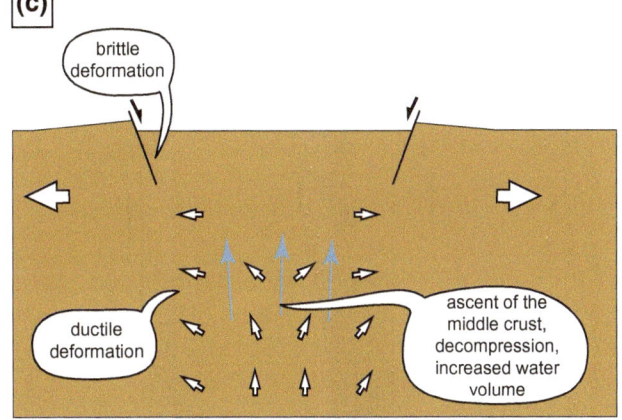

Fig. 4.3 Hydrothermal water can be pumped upward in different ways. **a** Water can be released from magma; such magmatic water rises automatically due to the heat. Pore water can also rise due to magmatic heat. **b** When sediments are buried (either by deposition of further sediments or by overthrusting during orogeny) the pore volume decreases and pore water rises or is driven away from the front of a mountain range. In addition, water flow can pass through sediments in foreland basins caused by the topography of the mountains and even continue so after overthrusting. **c** Strong extension results in the formation of a continental rift where water rises from the middle crust that has thinned because pore water expands there due to pressure relief. (*Not shown*) Water can also rise at earthquake faults (seismic pumping)

copper, skarn, and greisen would probably come first. Instead, we start with hydrothermal veins since they can provide the necessary basics to understanding the other types. Magma will play an important role in the following sections. As the chapter progresses it will move ever-further from magmatism to finally discuss quiet sediment basins and in so doing provide a smooth transition to the processes of sedimentation and diagenesis discussed in the following chapter.

Box 4.3 Solution as complexes

Typical hydrothermally formed ore minerals such as sphalerite (ZnS) can only be dissolved in pure water in tiny traces as Zn^{2+} and S^{2-} even if the water is hot. Solubility increases by several orders of magnitude if, for example, common salt (NaCl) is dissolved in water (Na^+ and Cl^-). The chloride combines with the metal ion to form a complex such as $(ZnCl)^+$ by transferring an electron pair to the zinc ion. The resulting bond resembles a covalent bond with the difference that both electrons originate from one of the partners. Lewis base is the name Chemists call the electron pair donor and Lewis acid the name they call the electron pair acceptor.

The chloride attached to the metal ion is called a ligand. There are also complexes such as $(ZnCl_2)$, $(ZnCl_3)^-$, and $(ZnCl_4)^{2-}$ in which several ligands or whole molecules are linked to the metal ion. F^-, O^{2-}, OH^-, H^-, S_2^-, SO_4^{2-}, NO_3^-, HS^-, CN^-, HCO_3^-, CO_3^{2-}, PO_4^{3-}, CH_3COO^-, $C_2O_4^{2-}$, NH_3, and CO may also be used as ligands if they are present in addition to Cl^-. Chloride is particularly important in hydrothermal solutions because it is often present in high concentrations and can form complexes with many metals.

Metal ions generally prefer certain ligands according to their ion potential (charge/radius). So-called hard Lewis acids such as W^{6+}, Mo^{x+}, U^{6+}, and V^{4+}, alkalis, and earth alkaline metals prefer hard Lewis bases such as O^{2-}, OH^-, HCO_3^-, CO_3^{2-}, and SO_4^{2-}. Soft Lewis acids such as Au^+, Ag^+, Cu^+, and Hg^{2+} prefer soft Lewis bases such as HS^-. Bivalent transition metals such as Zn^{2+}, Pb^{2+}, and Fe^{2+} and as the ligand Cl^- are placed in between and can be combined with hard, medium, and soft partners.

As a rule the solubility of metal complexes is significantly higher than that of pure ions. The solubility often increases exponentially with increasing temperature. Some metal ions such as Fe^{3+} and U^{4+} (both hard Lewis bases) are virtually insoluble even if the appropriate ligands are present. The metal content of metal-rich water is in an order of magnitude that is best expressed as parts per million (1 ppm = 0.0001%).

Hydrothermal water usually contains a whole catalog of different metal complexes in different concentrations. Both the stability and solubility of respective complexes depend on temperature, pH, Eh, and salinity. A change in these factors can lead to instability of complexes and thus to oversaturation and precipitation of certain minerals. Sometimes a large number of reactions are triggered that take place simultaneously or one after the other. Under certain circumstances contact of the fluid with a previously formed mineral may be sufficient to trigger small-scale precipitation reactions resulting in typical sequences of overgrown minerals (Staude et al. 2012b).

Box 4.4 Fluid inclusions

Minerals often contain tiny fluid inclusions that indicate the fluids involved in mineralization (Roedder 1984). They often look like a spirit level because at room temperature they contain a gas bubble. If the sample is heated to the so-called homogenization temperature, then the gas bubble disappears. The homogenization temperature gives the density of the fluid and (depending on the pressure estimate) the approximate temperature at the time of inclusion. By freezing the liquid and then thawing the ice the melting temperature of the ice can be determined, which in turn gives an indication of the salt content. Both measurements are performed on a special sample chamber mounted on the stage of a microscope called a heating and cooling stage. It is now even possible to measure the exact composition of individual liquid inclusions. For this purpose laser ablation inductively coupled plasma mass spectrometry (LA-ICP-MS) is used in which a laser beam burns into the sample. The material "evaporated" to plasma is analyzed in a mass spectrometer.

4.1 Hydrothermal Veins

An open fissure offers not only an ascent pathway for hydrothermal fluids, but also the space in which to precipitate minerals. Such minerals are often simply quartz or calcite, but sometimes they are baryte or fluorite (Sect. 7.14). Some veins also contain ore minerals in addition to these so-called gangue minerals.

Veins come in various sizes. There are hair-thin veinlets, centimeter-wide veins, and finally veins that are dozens of centimeters or even several meters wide. Many veins can be traced hundreds of meters or even a few kilometers. Big quartz veins are also called lodes and small veins are also called veinlets. A single vein rarely occurs. Vein systems are typical with many veins staggered beside and behind each other. A famous example is Mother Lode, a 190-km-long system of gold and silver bearing quartz veins at the edge of the Sierra Nevada in California, where individual veins are up to 15 m wide and a few kilometers long.

Some veins consist almost entirely of one mineral. Others are composed of more or less regular layers or bands (Fig. 4.4) that are arranged mirror-symmetrically or irregularly. Others are breccias (broken rock), often called ring ore, or cockade ore, that were subsequently cemented with hydrothermally formed minerals (Fig. 4.5).

Most hydrothermal veins were formed in the uppermost kilometers of the Earth's crust where the rock is so cool that it reacts to stress (directional pressure) by brittle deformation (opposed to ductile deformation in hot rocks, where no open fissures can form). In continental crust brittle deformation changes to ductile deformation at a depth of roughly 10–15 km.

Clefts are a variant of open fissures that form by decompression during erosion and related uplift of a mountain range. Although beautiful rock crystals can sometimes be found on these alpine clefts, there are no ores worth mentioning.

In contrast, hydrothermal veins primarily arise in connection with faults where two crustal blocks move past each other. Normally, there are no cavities because such movement takes place on a surface. However, this is not the case if the fault is not a smooth plane but has bends or curvatures. The reason for this are inhomogeneities in the rock such as layers that react differently to deformation or existing structures that have developed in a previous deformation acting in a different direction. In such a case small open fissures may occur (Fig. 4.6). Multiple curvatures create a fissure whose width increases and decreases (pinch and swell) or that thins out at the end and possibly opens again a little further.

This is particularly the case with small faults that have only minimal displacement. These are often small faults in connection with larger shear zones (Fig. 4.6c). Such faults occur when there are smaller branches or when movement jumps from one segment to another one via a hinge or stepover. Two segments of a major fault are connected by networks of small fractures at which either strong extension or strong compression occurs depending on the geometry.

Water is capable of entering even small, isolated cracks from the surrounding rock. Far-reaching cracks lead to a concentrated flow of water in turn promoting crack formation. Movement at a fault can be a slow creep or a sudden jerky shift that makes itself felt as an earthquake. An earthquake can lead to complicated interplay between fracturing, increased permeability, release of overpressured

Fig. 4.4 Banded ore with vein-parallel bands of galena (*gray*), sphalerite (*brown*), chalcopyrite (*golden*), and quartz (*white*) from Bad Grund, Harz (Germany) (*photo* © F. Neukirchen/Mineralogical Collections of the TU Berlin)

Fig. 4.5 Breccia from a vein in the Harz Mountains (Germany). The broken rock was cemented with galena (*metallic gray*), quartz, and calcite (*white*) (*photo* © F. Neukirchen/Mineralogical Collections of the TU Berlin)

fluids originally trapped in the rocks, fluid flow, and mineral precipitation (Oliver and Bons 2001; Micklethwaite et al. 2010).

While looking at a bigger vein, it is a mistake to imagine a meter-wide gap being torn open and then subsequently filled. The precipitation of hydrothermal minerals takes place simultaneously with the opening of the fissure. It is filled relatively quickly so water flow is not necessarily long lasting through wide fissures but rather through repeated

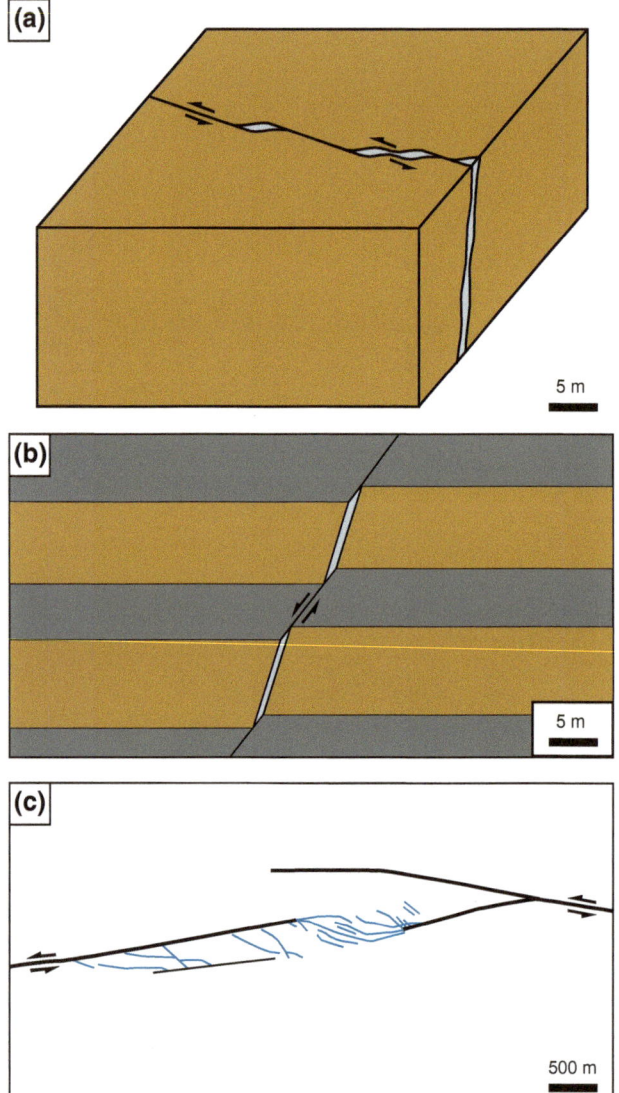

Fig. 4.6 Hydrothermal veins are fissures that have opened up only to be immediately filled by minerals precipitated from percolating water. Opening and filling can be repeated several times. **a** Open gaps can occur on a fault (here a strike-slip fault) with a small displacement. Gaps open here is because the fault is not a smooth surface but one with curvatures. The width of the fissure increases and decreases. **b** This example shows a vein in a normal fault within rock layers deforming differently and at a different angle. **c** Veins are often small fractures with a small displacement located in the hinge between two segments of a larger fault (**c** from Micklethwaite et al. 2010)

opening and sealing. Cavities that are not filled in this way are the exception. Which minerals are precipitated in a fissure depends on the composition of the hydrothermal water as well as on the temperature, pH, Eh, and changes in these factors. The decisive trigger is often the mixing of hydrothermal water rising from depth with water of the neighboring rock. This often happens near the boundary between basement and sediment cover because contrasting

fluids inevitably meet here. The crystal morphology of veins often depends on movement of the fault (Bons 2001b).

The mixing ratios of the fluids involved and factors such as temperature, pH, and Eh vary within the vein and over time. Thus, veins are composed of minerals of different generations and each generation is zoned with depth. Older minerals can of course be dissolved again and metals can be precipitated elsewhere under different conditions in the form of other minerals.

Clearly, there are a large number of different veins with very different metal contents. Despite all the variety however, there are typical combinations of metals or minerals. Paragenesis is the term given to minerals occurring together. Typical parageneses can be used to divide vein deposits. There are a number of important combinations such as lead–zinc (or lead–silver–zinc that includes most medieval silver mines), iron–manganese, cobalt–nickel–arsenic–silver–bismuth–uranium, tin–silver–bismuth–copper, and lead–zinc–copper.

Hydrothermal water can have very different origins. This chapter gives some examples of hydrothermal water directly related to magmatism such as epithermal veins (Sect. 4.3), cordilleran polymetallic veins (Box 4.9), polymetallic veins in Cornwall (Box 4.19), and tin veins in Bolivia (Sect. 4.5). Orogenic gold veins (Sect. 4.2), on the other hand, were usually formed by metamorphic water.

However, many veins were created by deep groundwater that was originally rainwater (meteoric water) or seawater (connate water trapped in sediments) that leached metals and other substances out of the rock. Which metals were leached naturally depends on the rock from which the fluid originated. The solubility of respective minerals and saturation in water also play roles here. For example, certain elements that are only present in small concentrations in the rock can be leached out preferentially.

It has already been demonstrated that deep groundwater is literally pumped upward when the crust is stretched, while at the same time active tectonics can lead to the opening of fissures. Accordingly, there are many hydrothermal veins along continental rifts such as those on both sides of the Rhine Rift Valley (Oberrheingraben), in the Black Forest (Germany) (Sect. 4.1.1), and in the Vosges (France). Water pouring out of a spring is sometimes thermal or mineral water (Box 4.5). In fact, many hydrothermal veins in the vicinity of a granite pluton were not formed by magmatic fluids, but much later when the crust was streched. Granites often have many clefts making them much better aquifers than neighboring gneisses or other metamorphic rocks. Nevertheless, the veins still contain a combination of metals typical of granite because they were leached out of the granite. Of course, in a mining district there may be old veins formed by magmatic or metamorphic fluids side by side with younger veins formed by deep groundwater.

Historically, the importance of hydrothermal vein deposits has been much greater than it is today. Although they often have a very high ore grade, the volume is far too small to warrant opening a large open-pit mine and their geometry (narrow and long) makes it difficult to use large machines underground. The high costs of mining make veins less interesting despite the high ore grade.

This is an appropriate point to take a closer look at veins in the Black Forest and the Ore Mountains in Germany. They are good examples of important vein types consisting in these instances of lead–silver–zinc veins and polymetallic veins, respectively. The Upper Harz was another important mining district with lead–silver–zinc veins; similar occurrences are widespread throughout the world.

Iron and iron–manganese veins mostly contain hematite with more or less large amounts of oxidic manganese ores such as pyrolusite or psilomelan. However, there are also veins that contain the iron carbonate siderite instead such as those in Siegerland (Germany). During the course of this chapter further hydrothermal mineralizations in veins will be encountered.

4.1.1 Veins in the Black Forest

The Black Forest (*Schwarzwald*) in Germany has approximately 1000 hydrothermal veins. It is a good example to use to discuss hydrothermal veins because it is well researched and relatively clear despite its diversity (Walenta 1992; Meyer et al. 2000; Schwinn et al. 2006; Staude et al. 2007, 2009, 2011; Pfaff et al. 2009; Danisik et al. 2010). Although the Romans began the mining of silver ores, the heyday of silver mining followed in the Middle Ages and ended in the fourteenth century. After a long pause mining began again in the eighteenth century when zinc, cobalt, and copper were also in demand in addition to silver. In the twentieth century mining of the once worthless gangue minerals fluorite and baryte started. In the Clara Mine near Oberwolfach they are still extracted.

The Black Forest and the Vosges are rift shoulders of the Rhine Rift Valley. Rifting commenced around 50 million years ago (Paleogene) and accelerated in the Neogene. While the rift valley was sinking in the center, the shoulders were uplifted as tectonic blocks. The Mesozoic sediment cover of the Black Forest was eroded. Thin remains can be found in the northern Black Forest and in the south the basement is exposed (Fig. 4.7). It consists largely of gneiss and granite that can be traced back to the Variscan orogeny (in the Devonian and Carboniferous). One of the sutures (a. k.a. seams) of this orogenic belt runs across the southern Black Forest: The so-called Badenweiler–Lenzkirch zone consists of sediments and island arc volcanic rocks that were subducted in a northerly direction. Another suture runs largely hidden under sediments through the northern Black Forest near Baden-Baden.

A few quartz veins with tin, copper, tungsten, and tourmaline had already formed with the emplacement of granite magma in the late phase of the Variscan orogeny. Later, as a result of continued stretching and erosion of the Variscan Mountains further quartz veins (orogenic quartz veins; Sect. 4.2) were formed throughout the Black Forest often containing antimony minerals and occasionally silver, gold, bismuth, or uranium (e.g., the uranium deposit in Menzenschwand). These veins were formed by the cooling of hot water that was about 240–400 °C previously released during metamorphic reactions (Baatartsogt et al. 2007; Staude et al. 2011).

By the Triassic the Variscan Mountains were long gone, at times rivers deposited sand, later the sea invaded, and a sequence of evaporites (salt, gypsum) and above all thick limestones and shales covered the region. Further sediments were deposited in the Jurassic sea.

Almost all the Black Forest's veins were formed long after the Variscan orogeny in phases when the region was stretched. Extension is known to lead to water rising from the middle crust. At the same time cracks opened through which water could rise as a result of such extension. This happened along old (Variscan) faults (especially, strike-slip faults that were reactivated with simultaneous extension) and later also along new faults parallel to the Rhine Rift Valley. This pulse-like hydrothermal activity peaked in two phases (Pfaff et al. 2009; Staude et al. 2009). The first took place in the Jurassic and was related to the opening of the North Atlantic; the second phase took place in the Paleogene and Neogene (both formally known as the Tertiary) and led to the formation of the Rhine Rift Valley. The responsible fluids remained surprisingly similar throughout the Black Forest despite the long period of time (Schwinn and Markl 2005; Schwinn et al. 2006). There are still some springs where mineral or thermal water upwells (Box 4.5).

Box 4.5 Mineral and thermal water in the Black Forest

The basement of the Black Forest is traversed to a depth of more than 10 km (until the transition from brittle to ductile deformation) by a network of fissures filled with water. Deep groundwater rises in many places and pours from a spring as mineral or thermal water (Stober and Bucher 1999a, b). In contrast to shallow groundwater that contains hardly any dissolved substances, mineral water is rich in CO_2, Ca^{2+}, Na^+, Mg^{2+}, and HCO_3^- (i.e., typical sparkling water). Warm thermal water upwells from greater depths and has an even higher content of dissolved substances—mainly chlorides (salt) (especially, Na^+ and Cl^-).

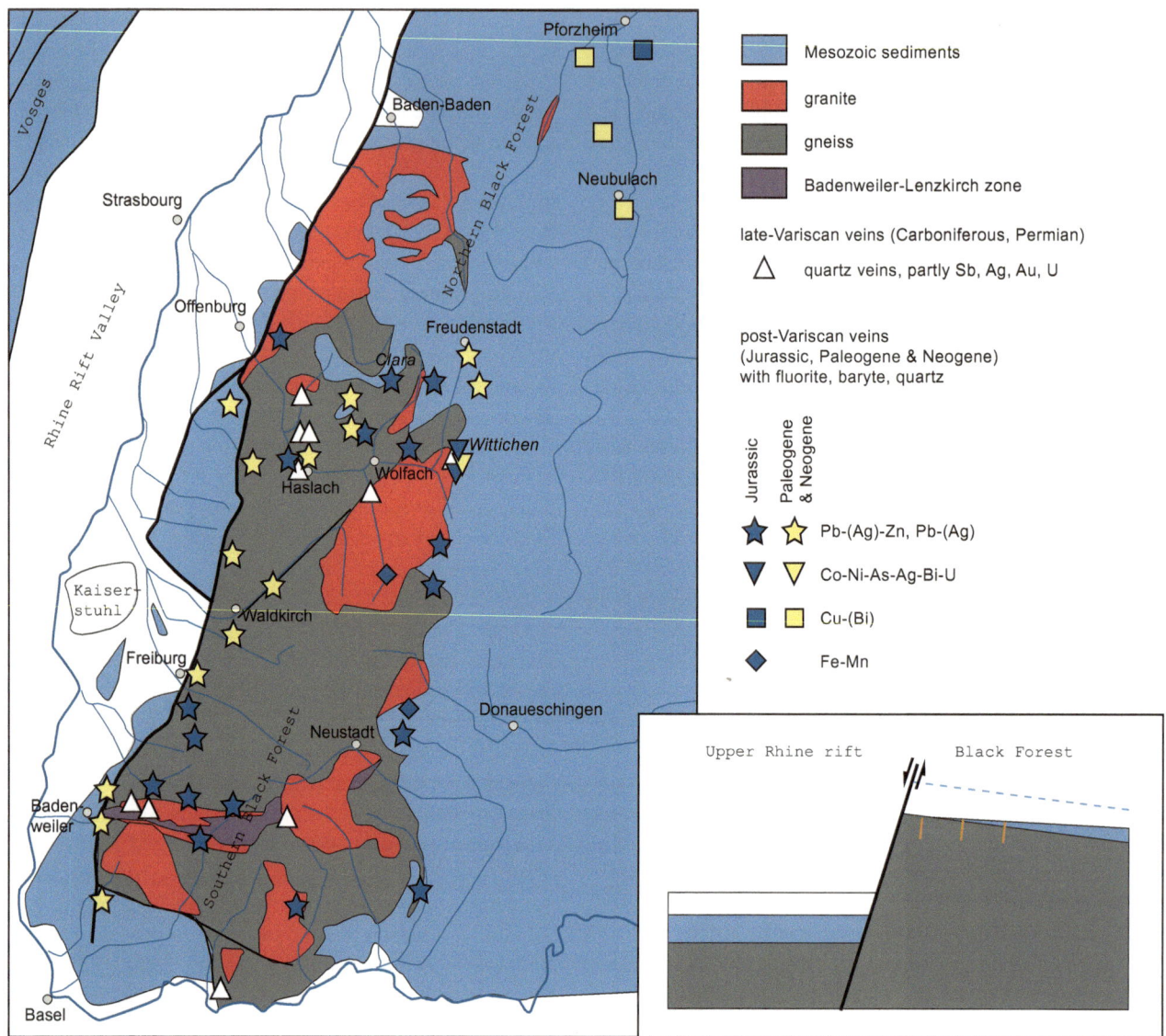

Fig. 4.7 Simplified geological map of the Black Forest with selected hydrothermal veins most of which opened as a result of extension in the Jurassic or in connection with the Rhine Rift Valley in the Paleogene and Neogene when they were filled with fluorite, baryte, quartz, and ore minerals (based on Schwinn et al. 2006; Pfaff et al. 2009; Staude et al. 2009, 2011)

Mixing of the two types of water do not occur in the Black Forest so the hydraulic systems appear to be separated. Although thermal water can be found mainly along the faults of the Rhine Rift Valley, they are also found in granites. This is not due to magmatic heat they have not had in a long time, but rather to the density and orientation of the fissure network that allows water to rise from greater depths compared with the fissure network of gneisses that typically contains mineral water springs.

CO_2 in mineral water probably comes from greater depths (e.g., the Earth's mantle). Dissolved in water it forms a weak acid that attacks the surrounding rock.

Plagioclase in granite and gneiss is highly susceptible to chemical weathering. It is partly replaced by clay minerals, while Ca^{2+}, Na^+, HCO_3^-, and SiO_2 are dissolved in water. As depth increases so do the contributions of Na^+, SO_4^{2-}, and Cl^- from less vulnerable minerals such as potassium feldspar and mica since the fissure network is less permeable and water spends correspondingly more time in it. The greater the amount of CO_2, the higher the content of dissolved substances in the water. At the same time water is consumed during chemical weathering since the weathering products formed such as clay minerals incorporate water or $(OH)^-$. This indirectly increases the concentration of dissolved ions in the water. Minerals such as zeolites or calcite are precipitated from the water at some point, which not only changes the composition of the water but also seals the fissure walls and protects the rock from further alteration.

Thermal water rising from a depth of several kilometers owes its composition to the weathering of plagioclase and mica that releases Na^+ and Cl^- at this depth—in mafic rocks such as gabbro or amphibolite Ca^{2+} and Cl^- would be released instead (Bucher and Stober 2010). The Cl/Br ratio of water in the Black Forest reveals that some of the salt was already present before water reacted with the rock; hence it must be fossil seawater.

Fig. 4.8 Two generations of galena (PbS) from the Friedrich–Christian Mine, Wildschapbach, Black Forest (Germany) (*photo* © F. Neukirchen, Markl Collection/Tübingen)

Upwelling water had a high content of dissolved substances (especially, salt). Presumably, it was originally meteoric water or seawater that spent a long time in the basement at a depth of 7–10 km and evolved due to reactions with the rock (Schwinn et al. 2006; Baatartsogt et al. 2007) until it was in equilibrium with granite or gneiss (Schwinn and Markl 2005). There is a narrow strip in the southern Black Forest where it was also in equilibrium with subducted rocks of the Badenweiler–Lenzkirch zone (Staude et al. 2011). Certain elements were preferably leached out of the rock and in altered rock minerals such as feldspar and mica were replaced by clay minerals.

At shallow depths (especially, near the contact between basement rocks and sediment cover) rising deep groundwater mixed with cool meteoric water or with the connate water of the sediments (Schwinn et al. 2006; Staude et al. 2010, 2011, 2012a), which also contained dissolved substances. Mixing of different solutions triggered the precipitation of baryte, fluorite, quartz, and ore minerals. As soon as a fissure opened it was filled again relatively quickly only to open again later in a process that was often repeated several times.

The minerals precipitated differ from vein to vein and from one ore generation to the next. This is due to the composition of the fluids involved (and the respective rocks with which they were in equilibrium) and the respective mixing ratio. The surrounding rock is also important as ingressing water reacts with it and its content changes as a result of increasing alterations in the rock (Pfaff et al. 2012). The so-called gangue minerals baryte, fluorite, and quartz usually make up the largest part of the filling. Nevertheless, it is normally the case that one of the three minerals predominates. Regional differences can be identified when it comes to ore minerals. In the southern Black Forest the combination of silver-bearing galena (PbS; Fig. 4.8) and sphalerite (ZnS) is typical (Pb–Ag–Zn veins). Similar veins

in the central Black Forest mainly contain galena without sphalerite (Pb–Ag veins). There are Fe–Mn veins at Eisenach and Triberg. The Co–Ni–As–Ag–Bi–U veins in Wittichen (Box 4.6) are unusual for the Black Forest. Copper veins cutting the sandstones in the northern Black Forest mainly contain chalcopyrite and tennantite, as well as bismuth ores.

Box 4.6 Cobalt–nickel–arsenic–silver–bismuth–uranium veins in Wittichen

In the mining district of Wittichen (Staude et al. 2012b) no sulfides occurred before the last ore generation. Instead, there are different cobalt–nickel–arsenic–silver–bismuth–uranium minerals (especially, native bismuth, native silver, and pitchblende). Thus, these polymetallic veins differ completely from the relatively simple veins of the rest of the Black Forest.

It is nothing short of remarkable that four large faults, all continuously active since the late Variscan period, run like the beams of a cross toward Wittichen (Elztal, Kinzigtal, Schramberger Fault, and Swabian Lineament). In the area of Wittichen they branch into a multiplicity of smaller faults. The different behavior of granite (in Wittichen) and gneiss on deformation also plays a role. This tectonic situation allowed frequent rupturing and intense fluid flow in fissures. Accordingly, the veins were filled with many generations of minerals formed at different times at least eight of which can be distinguished by dating. In contrast to the rest of the Black Forest, where each vein can be assigned to one of three main phases (late Variscan, Jurassic, or Paleogene–Neogene), the veins of Wittichen were active in all three phases.

Hydrothermal minerals were precipitated mainly by mixing ascending water (hot, salty) with connate water of the sediments (cool, salty) as in the rest of the Black Forest. Some of the metals (especially, copper) originate from connate water of sandstones that overlay the granite. In recent generations there has been a change to lead and zinc-rich water attributed to increasing reaction with Triassic limestones from which corresponding diagenetic minerals are known. Bismuth, on the other hand, was leached out of granite either from the surrounding rock or at greater depth from the source region of ascending water.

The occurrence of native bismuth and silver is typical for Wittichen. They are in turn overgrown with pitchblende or cobalt and nickel arsenides. Small-scale changes in the pH value during mixing seem to have played an important role whereby the precipitation of certain minerals changed the pH value and led to the precipitation of further minerals. However, the pH value of the water must have been unusually high because bismuth and silver can only be precipitated together as native metals from alkaline water. Such a reaction reduces the pH value (e.g., for bismuth):

$$4Bi^{3+} + 6H_2O \rightarrow 4Bi + 12H^+ + 3O_2$$

The lower pH in turn destabilizes uranyl complexes present in water such that native bismuth is immediately followed by the precipitation of pitchblende.

Wittichen was the largest silver and cobalt producer in southwest Germany in the eighteenth century. In the twentieth century the mining of cobalt and uranium was considered.

Differences between the northern and southern Black Forest are primarily due to a different level being uncovered brought about by varying degrees of uplift and erosion. A deep area of basement is exposed in the south, the middle Black Forest is only just below the former sediment cover, and in the north there is still some sandstone overlaying the basement with veins within tectonic horsts. The influence of sedimentary connate water increases to the north as the depth shallows (Staude et al. 2010, 2011; Ströbele et al. 2012).

The ingredients of hydrothermal minerals come from ascending deep groundwater and the connate water of sediments. Both have leached out certain substances from the corresponding rocks (Staude et al. 2010, 2011). For example, the fluorine contained in fluorite was mainly a component of mica or amphibole that are common in granite and gneiss. Calcium and barium were components of feldspar. The weathering of sulfides finely distributed in the rock released metal ions and S^{2-} that was partly oxidized to SO_4^{2-}.

However, sulfur can also come from the connate water of sediments in which layers of gypsum, for example, are intercalated (Staude et al. 2011). Baryte is most abundant in the most recent veins (Staude et al. 2009), which is explained by the increasing influence of connate water either due to sediments buried in the adjacent rift valley or in the northern Black Forest because mineralization took place within the sediments themselves. Some of the metals can also come from the connate water of sediments. For example, when it comes to lead it can be shown that it was present in both deep groundwater from basement rocks and connate water from sediments whereby the different mixing ratio in the respective depth levels also had an effect here (Ströbele et al. 2012). Copper in the veins in Wittichen probably comes from sandstone (Staude et al. 2012b).

Occasionally, there are cavities in the veins where beautiful crystals can be found. They generally formed at a much later time than most of the other parts of vein minerals through local remobilization. For example, it is assumed that the beautiful baryte crystals of the Clara Mine known as chisel spar did not grow until the Pleistocene (the Ice Age) (Staude et al. 2011).

4.1.2   Polymetallic Veins in the Ore Mountains

The Ore Mountains known in German as *Erzgebirge* (Fig. 4.9) run along the German–Czech border. Historically, this is one of the most important mining regions (Sebastian 2013). In the Freiberg district alone there are well over 1000 hydrothermal veins (Fig. 4.10) characterized by having a very diverse composition. Silver was discovered in 1168 and led in 1185 to foundation of the town of Freiberg that promptly became the richest town in Saxony. More than 5000 t of silver were produced between the end of the twelfth century and the end of the nineteenth century along with copper and tin. Between 1950 and 1969 lead, zinc, silver, and uranium were extracted as were germanium, cadmium, bismuth, gold, thallium, and indium as by-products. There are still ore reserves estimated at just less than 5 Mt. The Freiberg Mining Academy discovered the elements indium and germanium in 1863 and 1886, respectively, in local ores (sphalerite and argyrodite) (Seifert and Sandmann 2006).

In Freiberg silver occurs not only in silver-bearing galena and silver-rich tetrahedrite (freibergite), but also in the form of silver minerals such as proustite (Ag_3AsS_3), pyrargyrite (Ag_3SbS_3), miargyrite ($AgSbS_2$), and as native silver.

Further deposits with very similar silver-rich polymetallic veins have been discovered and exploited such as Schneeberg (from 1471), Annaberg (from 1491), Jáchymov (a.k.a. St. Joachimsthal, from 1516), Marienberg (1520), and

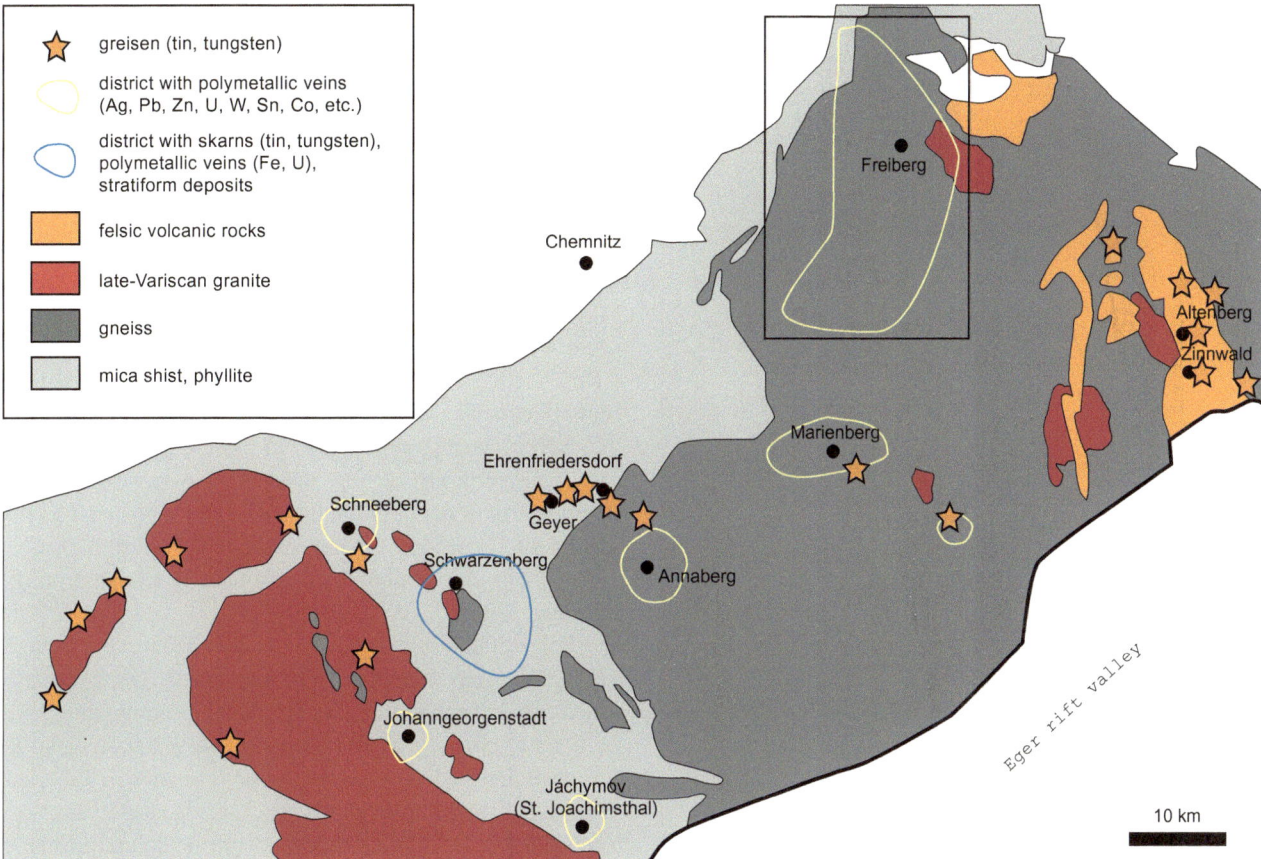

Fig. 4.9 Simplified geological map of the Ore Mountains (Erzgebirge) showing important mining districts (from Seifert and Sandmann 2006)

Johanngeorgenstadt (1654). The Joachimstaler, a coin minted in Jáchymov, gave its name to the thaler and the dollar.

Cobalt mining became increasingly important in Schneeberg and Annaberg after the decline of silver mining in the sixteenth century. Cobalt was used for the deep blue pigment cobalt blue. The name of the element derives from the German word for goblins (*Kobold*) that were said to live in deep parts of the veins. The ores (especially, skutterudite) look similar to silver minerals. Since no metal could be extracted from them using methods of the time the miners believed that the silver had been eaten by the goblins. Later on nickel was also mined in the Erzgebirge. The by-product bismuth was used as an antiseptic and in letterpress printing in the alloy from which lead letters were cast. The Schwarzenberg district is particularly well known for the mining of iron ore.

After the Second World War the German–Soviet company SDAG Wismut began mining uranium throughout the Ore Mountains. The company also mined uranium in Thuringia and the Elbsandsteingebirge. The GDR (a.k.a. East Germany) was then the third largest uranium producer in the world and the 231,000 t of uranium mined there made the Soviet Union a nuclear power. Mining took place not

only in former silver mines that had been reopened using modern mining methods but also in newly discovered deposits. The Schneeberg district was particularly productive: especially, near Schlema one of the largest uranium vein deposits in the world was mined. Johanngeorgenstadt was also of importance in that uranium minerals had already been mined there for paint production in the nineteenth century. The pitchblende specimen in which the element uranium was discovered by Klaproth in 1789 also originated from there. Marie Curie used pitchblende from the Czech town of Jáchymov (a.k.a. St. Joachimsthal) for her research. Back in the day miners were exposed to radiation since no one knew about radioactivity. Radiating dust particles and the radioactive gas radon released during decay are now known to be particularly harmful. Exposure brought on bouts of chronic coughing and diseases like lung cancer referred to as Schneeberg disease were common.

Apart from polymetallic veins there are other important deposits. Tin may have been panned from rivers as early as the Bronze Age, but the intensive mining of tin placers (Sect. 5.9) in the Middle Ages blurred the traces. Later on the mining of primary deposits began in small open-cast mines and underground. These are mainly greisen (Sect. 4.6)

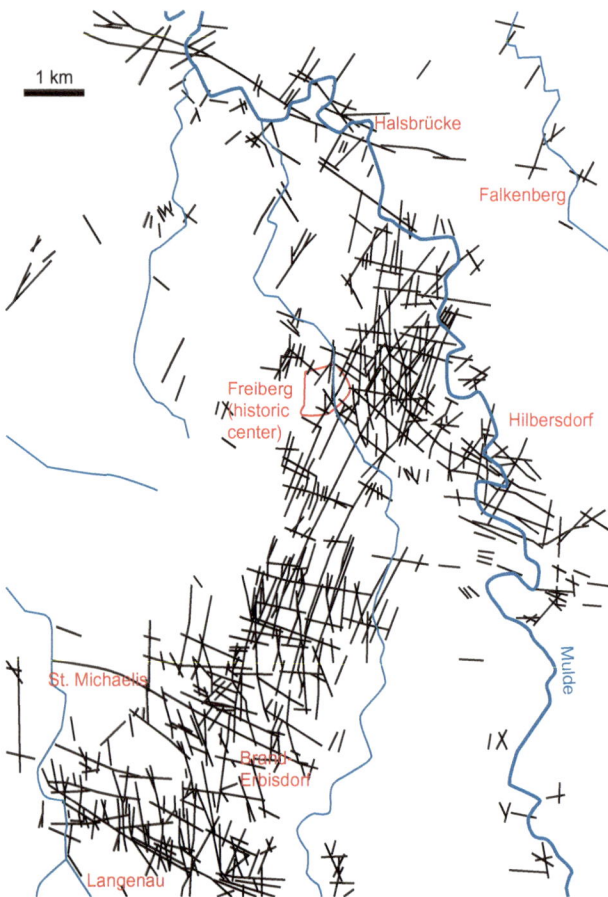

Fig. 4.10 In the Freiberg district there are more than 1000 hydrothermal veins of different sizes forming an entire vein network. They contain silver, lead, zinc, uranium, and a number of other metals (from Seifert and Sandmann 2006)

rocks. During this time some tiny granite intrusions (A-type) were emplaced that were very rich in fluorine and tin (tin granites). Greisen and tin veins can be found in their roof regions (Breiter 2012). In the eastern Ore Mountains they formed in the tuffs of a caldera volcano that had just erupted. There are also S-type granites with greisen in the Ore Mountains.

The numerous lamprophyre dikes of the Ore Mountains probably played an important role in the further development of veins. Such dikes are potassium-rich alkaline and ultramafic igneous rocks containing a lot of mica. This magma was obviously created by metasomatically highly enriched lithospheric mantle melting. The veins were created in several generations in the late phase of the Variscan orogeny and during subsequent extension. They are particularly frequent in mining areas with polymetallic veins and a good portion of metals precipitated hydrothermally in later times are assumed to originate from lamprophyres (Seifert 2008).

Polymetallic veins were created by a sequence of different mineralizations forming at different times. On the basis of their mineral content (paragenesis) they can be divided into different vein formations. Sequences of mineralizations can be distinguished in some of these formations (Table 4.2). Several of these formations can occur in a single vein. Veins generally show strong changes in ore grade and in mineral and metal content laterally and vertically. Silver content decreases with depth, while uranium and cobalt content increases.

Although the tin veins were still formed by hydrothermal systems in connection with granites, there is probably no direct connection with magmatism in subsequent mineralizations. Several extensional phases led to a number of formations such as *kiesig-blendige Formation* ("pyritic lead formation"), *edle Braunspatformation* ("noble carbonate formation"), and *Uran-Quarz-Karbonat-Formation* ("uranium-quartz-carbonate formation") occurring one after the other all of which are still referred to as late Variscan.

Further ores were precipitated during even later (post-Variscan) extensional phases including Bi–Co–Ni–As formation. Older formations were remobilized over and over again and precipitated elsewhere. This was especially the case in the Mesozoic during the opening of the Atlantic Ocean. The most recent formations originated in the Cenozoic in connection with the Eger Rift that is also responsible for uplift of the Ore Mountains as a tectonic block.

Mining in Erzgebirge stopped after German reunification. However, high commodity prices in recent years have led to increasing interest. Mining of fluorite and baryte has recently been resumed. Moreover, lithium, tin, indium, silver, and other metals are also expected to be extracted soon (Box 1.8).

and some tin veins. The most important tin mines were Altenberg (one of the largest tin deposits on Earth), Zinnwald, and several deposits near Ehrenfriedersdorf and Geyer. In the western Ore Mountains and in the Vogtland there are also tungsten–molybdenum greisen deposits. Also worth mentioning are skarns (Sect. 4.9) consisting of different minerals such as tin, tungsten, and iron (especially, at Geyer and in the Schwarzenberg area) and stratiform deposits with magnetite or copper sulfides.

Deposit formation began toward the end of the Variscan orogeny. At this time the supercontinents Laurussia and Gondwana collided as did some smaller terranes (e.g., small continents or island arcs) in between. Older sediments and granites were transformed into mica slate, phyllite, and various gneisses in the area of today's Ore Mountains. Several S-type granites were emplaced as a result of the collisions.

After the collisions the crust was strongly stretched (post-orogenic collapse) with fast denudation of deeper

Table 4.2 Vein formations in the Freiberg area and their composition (in chronological order)

Late-Variscan mineralization

Cassiterite formation (Sn–W)

Gangue	Quartz
Ore	Cassiterite, wolframite, scheelite, molybdenite, hematite
Note	Only subordinated in Freiberg; more frequent in districts with greisen

Kiesig-blendige Formation (kb) (i.e., pyritic lead formation)

Gangue	Quartz
Ore	(1) Pyritic sequence: arsenopyrite, pyrite (marcasite, pyrrhotite, gold)
	(2) Zinc–tin–copper sequence: sphalerite, chalcopyrite, tetrahedrite (chalcocite, bornite, stannite)
	(3) Lead sequence: galena (sphalerite, arsenopyrite, pyrite, marcasite)
Note	Ag-rich, In-bearing, Cd-bearing (approx. 400 veins in Freiberg; additional veins in other districts)

Edle Braunspatformation (eb) (i.e., noble carbonate formation)

Gangue	Carbonates (siderite, rhodochrosite, ankerite, dolomite, calcite, magnesite)
Ore	(1) Sulfide carbonate sequence: arsenopyrite, pyrite, sphalerite, chalcopyrite, galena, freibergite
	(2) Ag–Sb sequence: freibergite, jamesonite, berthierite, boulangerite, stibnite, bournonite, freieslebenite, miargyrite, pyrargyrite, stephanite, polybasite, dyskrasite, argentite, proustite, native silver
Note	Very rich in silver (approx. 400 veins in Freiberg; additional veins in other districts)

Uran-Quarz-Karbonat-Formation (uqk) (i.e., uranium–quartz carbonate formation)

Gangue	Quartz, carbonates (fluorite)
Ore	Pitchblende, hematite
Note	Only subordinated in Freiberg; important in Schneeberg, Johanngeorgenstadt, etc.

Post-Variscan mineralization

Iron baryte formation (eba)

Gangue	Quartz, chalcedony (also agate, amethyst), baryte, (fluorite)
Ore	Hematite or siderite (or barren)
Note	Approx. 100 veins in Freiberg

Fluorite baryte formation (fba = bafl)

Gangue	(1) Quartz (baryte, fluorite)
	(2) Fluorite, baryte (quartz)
Ore	Galena, sphalerite (honey-colored or colloform), tetrahedrite, chalcopyrite, pyrite, marcasite
Note	Approx. 200 veins in Freiberg

Bi–Co–Ni–Ag formation

Gangue	(1) Quartz (baryte, fluorite)
	(2) Carbonates: siderite, ankerite, dolomite, calcite (quartz)
Ore	(1) Arsenidic sequence: native bismuth, native silver, skutterudite, rammelsbergite, safflorite, native arsenic, gersdorffite, pitchblende
	(2) Silver sulfide sequence: galena, chalcopyrite, sphalerite, pyrite, freibergite, jamesonite, miargyrite, proustite, stephanite, polybasite, argyrodite, argentite, native silver, bravoite, millerite, breithauptite
Note	This germanium-bearing formation mainly occurred at crossings of veins in Freiberg; important in Schneeberg, Marienberg, Annaberg, Johanngeorgenstadt, and Jáchymov

4.2 Orogenic Gold Veins

Apart from secondary placer deposits (Sect. 5.9) gold occurs primarily in quartz veins that were either formed in connection with volcanism (epithermal gold veins of the following section) or during continental collision and mountain building (orogeny). Orogenic gold veins (Groves et al. 1998, 2003; Goldfarb et al. 2001; Bierlein et al. 2006) are referred to as mesothermal gold veins in early literature. Larger quartz veins are also called lodes.

Quartz veins (Fig. 4.11) are a few centimeters to a few meters wide and a few kilometers long. Entire systems of veins staggered side by side and one behind the other (Fig. 4.12) are common. The depth at which they were formed is very variable. They are known to form in the upper and middle crust, from just below the Earth's surface, to 15 or even 20 km deep. The temperature range of 130–700 °C is correspondingly wide. Most form at a depth of 5–10 km and a temperature of 250–350 °C at about the limit between

brittle and ductile deformation or just above it. Accordingly they are located in metamorphic rocks. The most common siderocks are those of greenschist facies, but there are also veins in subgreenschist, amphibolite, and even granulite facies rocks.

The veins consist almost exclusively of quartz, up to 15% carbonates, and a few percent sulfides (mainly pyrite, pyrrhotite, or arsenopyrite). Native gold (Fig. 4.13) occurs in the form of tiny grains between quartz grains or as inclusions in sulfides ("invisible gold"). Rarely, gold sheets or tree-shaped structures occur. On average, the gold contains about 10% silver but sometimes significantly more (electrum). Particularly gold-rich veins are sometimes called bonanzas. In addition, individual cases of albite, mica, chlorite, scheelite, tourmaline, stibnite, cinnabar, various gold–silver tellurides, bismuth, cobalt, uranium, copper, lead, or zinc minerals can also occur in orogenic quartz veins depending on the depth. There are also orogenic quartz veins without gold such as the relatively common antimony quartz veins with stibnite.

Fig. 4.11 The Homestake Mine (South Dakota, USA) was one of the largest gold mines in the United States. Gold-containing quartz veins can be seen in the wall of the open-cast mine (*photo* © Rachel Harris, CC-BY, Wikimedia Commons)

Fig. 4.12 Open-pit mining is worthwhile when many orogenic gold veins occur side by side. Super Pit gold mine, near Kalgoorlie, Eastern Goldfields of Yilgarn Craton (Western Australia, Australia) (*photo* © Getty Images/iStockphoto)

Fig. 4.13 Native gold in quartz (Nevada, USA) (*photo* © Jeffrey Daly/Fotolia)

Unfortunately, the vast majority of quartz veins do not contain any interesting minerals.

The quartz veins are mainly located in the hinges (Fig. 4.6c) of laterally offset segments of large faults and shear zones that have experienced either brittle or ductile deformation depending on the temperature. Often these faults are originally the seams on which terranes were "glued" to an active continental margin (subduction zone). However, there are also examples in the fold axis of strongly deformed turbidites (rocks formed by submarine turbidity currents with so-called turbidite-hosted gold).

Water released by rocks in metamorphic reactions is primarily responsible for the formation of orogenic quartz veins. Continental collisions lead to the overthrusting of rock nappes and significant thickening of the crust. Underlying rocks are located at increasing depths and thus at a higher pressure and temperature; hence they transform into metamorphic rocks. Many of these reactions relate to dehydration. Minerals containing water (or OH^-) in the rock are replaced by drier minerals. The water released rises through the crust.

The release of metamorphic water begins at shallow depths and continues in rocks that are descending. Many of these reactions depend more on temperature than pressure. Since it always takes some time for a rock that has reached its maximum depth to be heated to the corresponding temperature some reactions only take place when the main phase of orogeny has already passed. Accordingly, most orogenic quartz veins were formed at a very late stage of mountain formation (late orogenic). Further quartz veins can form after mountain formation during extension of the mountain range (post-orogenic).

The water released is hot (about as hot as the rock), has a near-neutral pH, and usually has a low salt content. Its CO_2 content is very variable. If pressure and temperature are high enough as happens at a depth of a few kilometers, then H_2O and CO_2 can mix at will. There are also of course metamorphic reactions in which CO_2 is released. The water comes from a large volume of rock from which it leaches out gold and other substances. The gold goes into solution as Au (HS) and $Au(HS)_2^-$ complexes. The gold content in the water does not have to be particularly high, but it is important that there is a focused water flow and effective precipitation. This could be due to the fact that a CO_2-rich fluid eventually separates gaseous CO_2 from the water during ascent. Another possible reason is the reaction of the fluid with the surrounding rock; veins are often surrounded by an alteration halo (see Box 4.14) comprising sulfides, carbonates, sericite, chlorite, albite, or other minerals depending on the depth. A third possible reason (outlined in the model in Box 4.7) is that water rises extremely quickly before it gets stuck at a certain depth and cools down there. The latest model argues that earthquakes are responsible for the precipitation of gold (Weatherley and Henley 2013). Seismic pumping of water along active faults has already been mentioned. Weatherley and Henley (2013) assume that pressure fluctuations in the fluid are so great that it evaporates in a flash at low pressure during rupturing and opening of fissures in an earthquake. The dissolved substances precipitate abruptly and fill the gap with quartz, some gold, and other minerals. The process is repeated as long as new cracks open and water flows in.

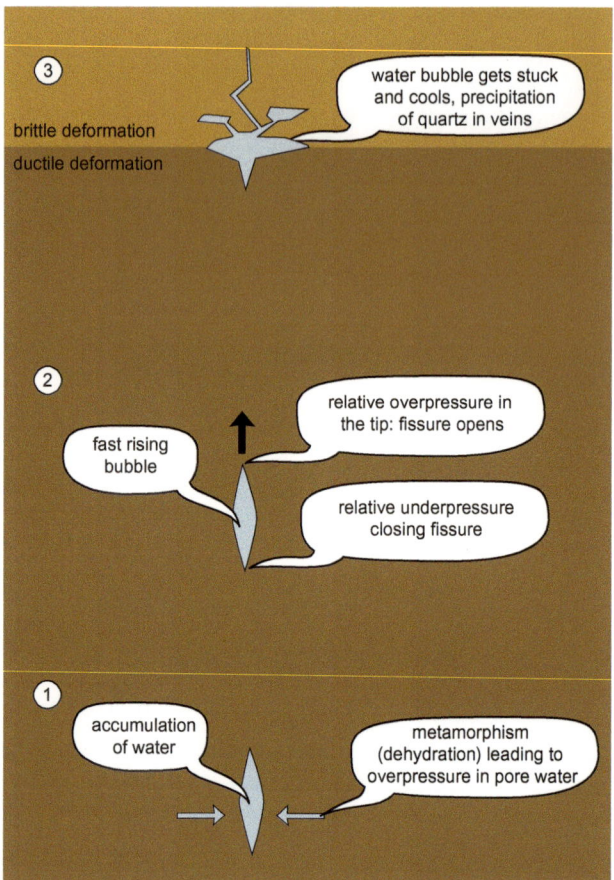

Fig. 4.14 Water bubbles in ductile deformed rock can rise at high speed as mobile fractures. They remain stuck in the brittle deformation part of the transition zone, which leads to cooling and the precipitation of quartz (based on Bons 2001a)

Box 4.7 Water-filled mobile fractures

Water not only rises in open fractures, it can also create the fracture itself. This happens at shallow depths when water under overpressure breaks the rock into a breccia. Something similar might happen in the middle and lower crust. The water and the open fissures rise together as a kind of bubble (Bons 2001a; Oliver and Bons 2001). This process is likely to play an important role in the formation of orogenic gold veins.

The transition zone from brittle to ductile deformation of crustal rocks is at a depth of about 10–15 km. Ductile deformation does not result in open fractures and the volume of water-filled pores is very small. Water in the pores is therefore under high pressure roughly corresponding to the pressure of the rock (lithostatic pressure).

If deformation is severe, then this pore water can accumulate to form a kind of bubble. Once the bubble has reached a critical size it begins to rise (Fig. 4.14). The reason is the vertical pressure gradient in the water relative to the pressure gradient in the rock; there is relative overpressure at the top of the bubble and relative underpressure at the bottom. The fissure opens up ever wider at the top, while it closes again at the bottom.

Despite the high rock pressure such ascending bubbles allow water formed by metamorphic reactions in the rock to drain through these bubbles. Water pressure in the pores drops abruptly and then slowly builds up again through metamorphic reactions.

The bubble follows the path of least resistance during its ascent often corresponding to the path taken by earlier bubbles. The bubble can remain temporarily attached to irregularities in the rock until it combines with later bubbles to form a larger one that can overcome the resistance.

At 1 m/s such bubbles rise so quickly that the water does not cool down; hence only a few dissolved substances are precipitated on the way. Such bubbles could get stuck at the transition to brittle deformation because of cracks in the rock (which are not necessarily interconnected) where the water pressure is lower. Another reason could be that the bubble meets a rock in which it is more likely to spread horizontally. In both cases the tip of the bubble loses relative overpressure and the water now slowly flows away through small cracks giving it time to cool down and precipitate a large amount of quartz. The rise of such water bubbles can be repeated any number of times such that quartz veins can slowly become thicker.

According to this model large quartz veins are where water that quickly ascended from below got stuck—not the ascent pathway of hydrothermal water. The model explains why quartz veins are preferentially formed at the transition from brittle to ductile deformation or just above it. It also requires significantly less water than would be the case if the water were to rise slowly.

Gold veins in the Alps such as those of the High Tauern and around Monte Rosa in the Alps are among the youngest. Orogenic veins in Europe were also formed during the Variscan and Caledonian orogenies.

The collision of small terranes such as mini continents, island arcs, and basalt plateaus with a subduction zone is another very effective means of exposing relatively large quantities of sediments to metamorphism. Such terranes welded to the edge of a continent are called accretionary orogens. Such welding was brought about by the overthrusting of small nappes or often by strike-slip faults with strong compression (transpression). Metamorphic water preferentially rises at the seams (sutures) and can deposit quartz veins there. Subduction then continues at the new continental margin. In the Mesozoic a whole series of terranes were welded to the western edge of North America. This led to the development of a number of rich vein systems including Mother Lode (in the foothills of the Sierra Nevada in California). At the same time something similar happened in eastern Siberia and East Asia. Moreover, large gold deposits had formed in Central Asia (especially, in Tian Shan and Kazakhstan) and Mongolia during the accretion of terranes as early as the Paleozoic.

However, the vast majority of gold-bearing quartz veins (Table 4.3) originated in the Precambrian. The most

Table 4.3 Selected regions with orogenic gold veins and amount extracted in millions of ounces	Province	Extracted gold (millions of ounces)
	Archean	
	Barberton Greenstone Belt (Kapvaal Craton, South Africa)	>10
	Eastern Goldfields Superterrane (Yilgarn Craton, Australia)	90
	West Yilgarn Superterrane (Yilgarn Craton, Australia)	18
	Southern Superior Province (especially, Abitibi Greenstone Belt) (Canada)	180
	Slave Province (Canada)	16
	Greenstone belt in the eastern block of the Dharwar Craton (India)	27.6
	Greenstone belt in the Zimbabwe Craton	17
	Rio das Velhas Greenstone Belt (Minas Gerais, Brazil; Box 4.8)	30
	Proterozoic	
	Birimian Greenstone Belt (Ghana, Mali, Guinea, Burkina Faso)	50
	Dakota segment with the Homestake Mine (USA) (Fig. 4.11)	>40
	Nubian Shield (Egypt, Sudan, Ethiopia)	100?
	Paleozoic	
	Lachlan Fault Belt (Australia)	34
	Southern Tian Shan with the Muruntau gold deposit (Uzbekistan, Kyrgyzstan, Tajikistan)	50
	Ural Mountains (Russia)	>28
	Mesozoic	
	Sierra Nevada foothills with the Mother Lode system (USA)	35

Much of the gold from the Nubian Shield had already been extracted in early history (modified from Goldfarb et al. 2001)

important from the Archean period are found in the cratons of Canada, Brazil, Guyana, Africa (south, east, and west), Australia, and India in granite and greenstone belts (greenstone-hosted gold) that were themselves created by subduction zones and repeated collisions between and with terranes (see Box 3.12). The veins cut through all the rocks found in these cratons such as greenschists (i.e., basalts modified by metamorphism), granite, banded iron formations (Sect. 5.2), conglomerates, and turbidites. In the Phanerozoic further gold veins were formed by collisions between and with the first continents. This often happened due to overthrusting and movement on strike-slip faults with concurrent compression in shales or slates lying on the edges of cratons (slate belt–hosted gold).

An additional heat source was likely added to normal metamorphism in many cases. Several processes have been proposed all of which lead to local upwelling of molten asthenosphere in the short term that may even lead to partial melting. The oceanic plate subducted before the collision of two continents initially still adheres to one of the two continents. While it broke off, there was rapid uplift and melt generation due to molten asthenosphere flowing in. Peeling of the heavy lithospheric mantle (a.k.a. delamination) has a similar effect. This can happen under a mountain range with a very thick crust because the lower crust is then very hot and soft. Other possibilities include subduction of a mid-ocean ridge where a window opens in the subducted plate or rolling back of a subducted plate could lead to extension in the volcanic arc. Yet another possibility is a mantle diapir may have risen regularly under a subduction zone in the Archean.

In some cases quartz veins are formed to a certain extent by magmatic fluids—not just metamorphic fluids. Quartz veins are even almost exclusively caused by magmatic fluids in the vicinity of porphyry gold deposits (Sect. 4.4.2). Unfortunately, the distinction between orogenic and intrusion-related quartz veins is nigh on impossible since both types can occur together and the processes mentioned above not only lead to increased metamorphism, but possibly also to melt generation.

Box 4.8 Minas Gerais

Gold, diamonds, aquamarine, topaz, and tourmaline are found in the Brazilian province of Minas Gerais ("General Mines") where treasures abound. In the middle of the sixteenth century the first Portuguese explorers invaded the region and set up a small settlement. However, there were only sporadic finds. Little changed until the beginning of the eighteenth century when gold was found triggering a gold rush. Magnificent baroque cities sprang up like mushrooms and one of them Ouro Preto ("Black Gold") developed into America's richest city (Fig. 4.15). For a century the gold of Ouro Preto played a similar role to the silver of Potosí: it literally formed the world currency. In the nineteenth century production declined sharply.

The deposits comprise orogenic quartz veins and placer deposits formed from them. Compared with other orogenic gold deposits found later in Australia, California, and Siberia this is only a medium-sized deposit. The name black gold derives from the frequent black coloring of the surface caused by iron hydroxides. This is a greenstone belt where there are banded iron formations.

The quartz veins were formed in the late Proterozoic during the Brasiliano orogeny when the São Francisco craton collided with the Congo craton; more precisely, the veins only formed after mountain building during strong extension of the crust (Chauvet et al. 2001). The mountain belt literally flowed apart in a process called post-orogenic collapse. The process involves some of the nappes that had previously thrusted sliding back a bit. Within the nappes fissures were formed by extension and filled with quartz. The nappes consist mainly of metamorphic former sediments.

During the same orogeny numerous pegmatites also developed many of which contain precious stones. It is the world's largest collection of gemstone pegmatites from which several million carats (carat = 200 mg) of beryl, chrysoberyl, topaz, and kunzite have been extracted. Two types of kimberlites responsible for the diamonds did not appear until the Cretaceous (Neukirchen 2012).

4.3 Epithermal Gold and Gold–Silver Deposits

This section deals with deposits that arise relatively near the surface (at a depth of 50–1000 m) in connection with volcanism (Fig. 4.16). The main focus is on the acidic and water-rich magmas of subduction zones. Such hydrothermal systems are noticeable at the Earth's surface as hot springs, geysers, fumaroles, or acidic crater lakes and sometimes elemental sulfur is deposited in a volcanic crater. Contrary to the original meaning of "epithermal" (Box 4.2) the temperature of the water responsible can be very different. Although the range is typically between 160 and 270 °C, there are also cases when it is more than 300 °C. Epithermal deposits are small in volume but very high in grade

Fig. 4.15 The gold town of Ouro Preto (Minas Gerais, Brazil) (*photo* © Rosino, CC-BY-SA, flickr.com)

(especially, for gold and silver). They also contain copper, lead, zinc, arsenic, antimony, selenium, tellurium (there are even deposits containing tellurides), mercury, and other metals.

The two different types that are generally distinguished are high-sulfidation and low-sulfidation epithermal systems (Table 4.4, Fig. 4.17). This mainly refers to whether the sulfur in the fluid is predominantly oxidized (SO_2, SO_4^{2-}, HSO^{4-}) or predominantly reduced (H_2S, HS^-). Both types differ not only in the minerals precipitated, but also in the way the surrounding rock is altered (Cooke and Simmons 2000; Hedenquist et al. 2000; Sillitoe and Hedenquist 2003; Heinrich 2005). There are also sulfidation systems that are intermediate between the two endmember types.

High-sulfidation epithermal deposits exist mainly along subduction zones. Important examples of such deposits are El Indio (Chile), Pueblo Viejo (Dominican Republic), Pascua-Lama (Chile, Argentina), Yanacocha (Peru), Goldfield (Nevada, USA), Lepanto (Philippines), Chelopech (Bulgaria), and Bor (Serbia).

The fluids responsible are somewhat oxidized and comprise magmatic fluid released by andesitic or granitic magmas. Such systems often develop in connection with

porphyry copper deposits. Relatively deeply a brine (salt-rich fluid) is separated from the steam-like fluid where gold, silver, copper, arsenic, and other metals as well as CO_2, SO_2, and so on are enriched in the steam. This is discussed in greater detail in the following section in connection with porphyry copper (Sect. 4.4). As the ascent of supercritical steam continues it contracts to liquid hot water that is normally relatively low in salt but rich in SO_2. Some of the SO_2 combines with water to form sulfuric acid; some hydrochloric acid is also formed. The cooling fluid thus becomes increasingly acidic and aggressive on its way to the Earth's surface. The acid attacks feldspars and other minerals in the rocks and converts them into clay minerals and other minerals. This process is called argillic alteration (see Box 4.14). Illite (a clay mineral) is formed first and then kaolinite, pyrophyllite, and alunite are formed as argillic alteration continues. Argillic alteration neutralizes the acid. However, the further the alteration progresses, the less acid is neutralized and the more acidic the water becomes in the upper parts of the system achieving typically a pH between 1 and 3. At very low pH even aluminum can be dissolved and carried away; in extreme cases all that remains is a porous rock rich in quartz called vuggy silica. This aggressive fluid

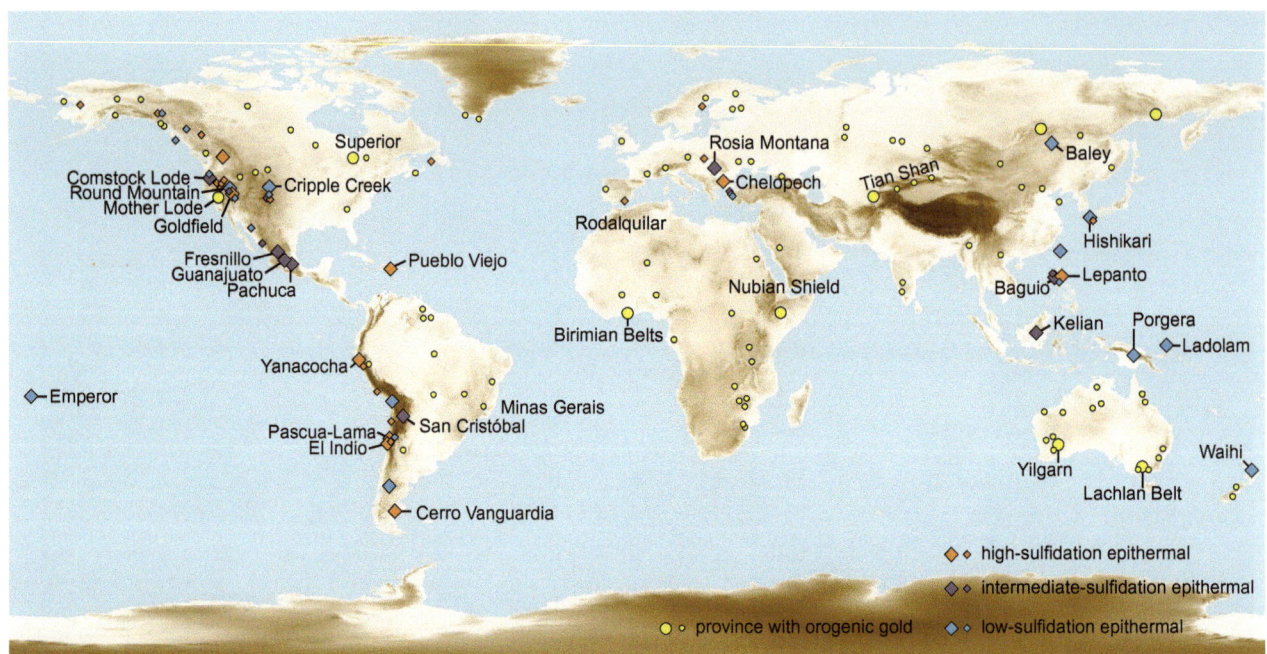

Fig. 4.16 Regions with orogenic gold (from Goldfarb et al. 2001) and important epithermal gold–silver deposits. Orogenic gold is found mainly in the greenstone belts of Archean cratons (the oldest cores of continents) and in Proterozoic collision zones. The most important younger occurrences are on the western edge of North America and in East and Central Asia. In Europe there are smaller occurrences related to Alpine, Variscan, and Caledonian orogeny. Epithermal deposits are mostly very young and are found along active subduction zones

Table 4.4 Differences between high-sulfidation and low-sulfidation epithermal systems

	High sulfidation	Low sulfidation
Origin of the fluid	Magmatic fluid	Predominantly heated groundwater
Fluid properties	Oxidized, very acidic	Reduced, near-neutral pH
Alteration	Kaolinite, pyrophyllite, alunite (advanced argillic alteration), distal zone illite	Sericite (or illite), orthoclase
Site of mineralization	Disseminated in the rock, rare veins	Veins, sometimes disseminated
Gangue	Quartz, baryte, alunite	Quartz, chalcedony, orthoclase, calcite, baryte
Ore minerals	Pyrite, gold, electrum, enargite, tennantite–tetrahedrite, covellite, chalcopyrite, tellurides, selenides	Pyrite, gold, electrum, arsenopyrite, sphalerite, galena, tennantite–tetrahedrite, pyrargyrite, acanthite, cinnabar, tellurides, selenides, arsenopyrite

corresponds to the vapor of high-temperature fumaroles in some volcanic craters.

Gold is dissolved in this fluid mainly as $AuCl_2$ and $AuHS$. Once argillic alteration is well advanced, the low pH allows transport even in relatively cool water. Although there is some precipitation as a result of simple cooling during the ascent, gold precipitates increasingly near the Earth's surface as tiny grains together with pyrite, quartz, alunite, baryte. The trigger is either mixing of the water with shallow groundwater or boiling as a result of the pressure drop shortly before discharge at a source. This sometimes happens in small veins but more frequently disseminated in rock pores.

In addition to gold other metals are transported and precipitated (especially, silver, copper, and arsenic). The silver content of gold increases with depth from native gold to the gold–silver alloy electrum. In deeper parts the copper content increases and the gold content decreases. Copper arsenic minerals such as enargite and tennantite–tetrahedrite as well as covellite and chalcopyrite are typical here. The exact

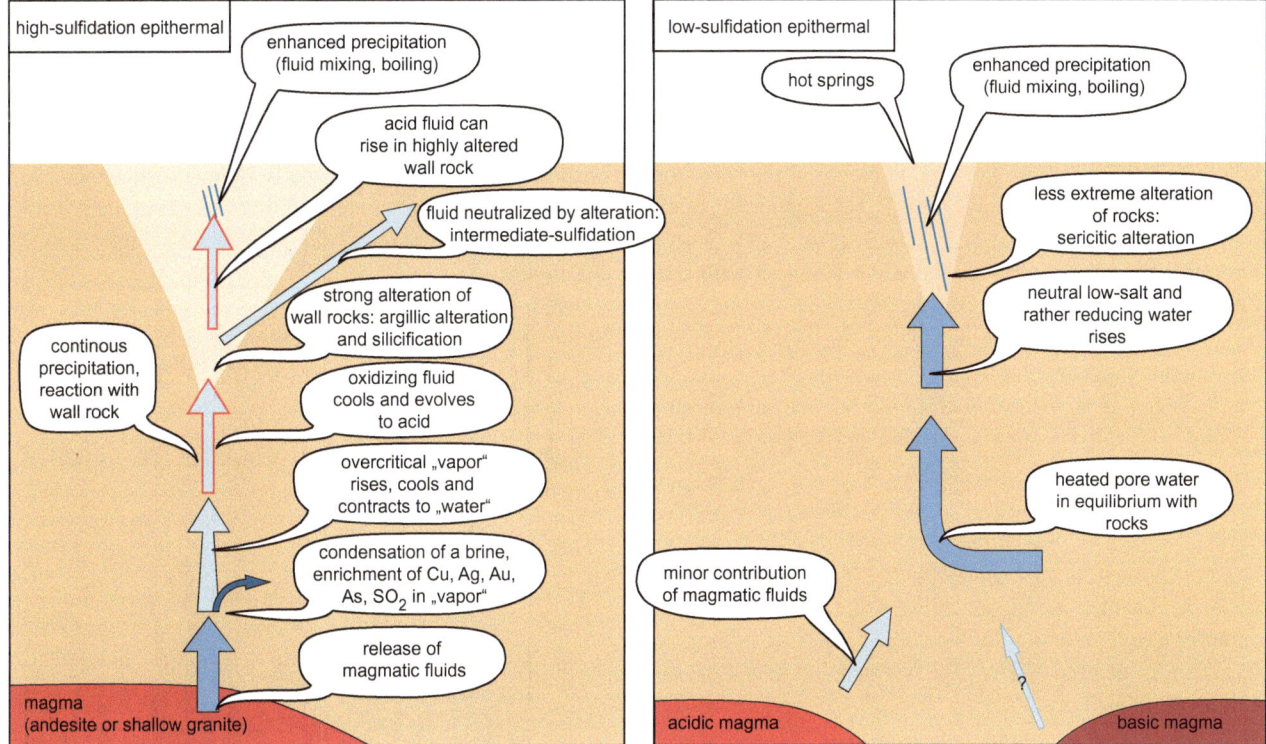

Fig. 4.17 Schematic representation of the origin and further evolution of hydrothermal fluids in high-sulfidation and low-sulfidation epithermal systems. The precipitation of gold and silver occurs in both cases mainly near the Earth's surface

composition also depends on the metal content of the magma and on what the water has already lost on the way. In such systems the composition of incoming water is constantly changing as are the minerals that are precipitated. At even greater depths copper (gold–silver) veins (a.k.a. cordilleran polymetallic veins) are found instead. They mainly contain enargite and tennantite–tetrahedrite (Box 4.9).

Box 4.9 Cordilleran polymetallic veins

A porphyry copper deposit is often surrounded by numerous veins containing a variety of metals such as Cu, Ag, Au, As, Sb, Hg, Te, Zn, Pb, and Mn. In addition to sulfides these veins mainly contain so-called sulfosalt minerals (compounds with AsS_3^{3-} or SbS_3^{3-}) such as enargite or tennantite–tetrahedrite. These are found in the deeper part of high-sulfidation to intermediate-sulfidation epithermal systems.

Increasing distance from the porphyry stock often results in zoning as a result of fluids changing on their way through cooling, precipitation of minerals, and alteration of the surrounding rock. A typical sequence would be a proximal copper-rich zone surrounded by Pb–Zn further surrounded by Au–Ag–As (Sillitoe 2010).

However, veins are also zoned internally and reflect large-scale zoning on a small scale demonstrating how the hydrothermal system in question changed in the course of its life. Investigations of fluid inclusions from a zoned vein of Morococha (Peru) have shown that zones of the vein formed in chronological order at ever lower pressure. The telescoping of the zones into a single vein was therefore caused by simultaneously eroding several kilometers of overlying rock (Catchpole et al. 2011).

If high-sulfidation fluid penetrates rocks that have not yet been altered, then the acid is partially neutralized and **intermediate-sulfidation** veins are formed whose mineralogy stands between the two main types. They include large silver deposits such as Fresnillo (Zacatecas, Mexico) and San Cristóbal (Bolivia) and some large gold–silver deposits such as Guanajuato, Pachuca–Real del Monte, Tayoltita (all Mexico), Tonopah, Comstock Lode (Nevada, USA), Creede (Colorado, USA), Baguio (Philippines), and Rosia Montana (Romania).

Low-sulfidation epithermal systems are the second main type. The hydrothermal water of such systems is predominantly groundwater heated by magma (usually meteoric water, i.e., originally rainwater filling rock pores that was in equilibrium with the rock) (Morishita and Nakano 2008). In addition, there are small amounts of magmatic steam containing CO_2, SO_2, and $NaCl$ that have mixed with the groundwater. The water has approximately a neutral pH and usually a low salt content; the sulfur is predominantly reduced. The water from hot springs found in the surroundings of many active volcanoes mostly belongs to this type.

Gold is mainly dissolved as Au(HS) (Fig. 4.18). Gold or electrum is precipitated when the water is mixed with shallow groundwater or when H_2S is degassing during ascent (especially, when water boils due to the lower pressure). This time it happens more often within veins or so-called ore shoots (Páez et al. 2011). Such veins mainly contain chalcedony and quartz; sometimes they also contain calcite, orthoclase (potassium feldspar), illite (clay mineral), or baryte. Moreover, the most common sulfide this time is pyrite. Copper is only present in small quantities as tennantite–tetrahedrite and chalcopyrite. Instead, arsenopyrite (FeAsS), galena (PbS), sphalerite (ZnS), and sometimes cinnabar (HgS) occur. Silver minerals such as pyrargyrite (Ag_3SbS_3), acanthite (Ag_2S), naumannite (Ag_2Se) and native silver also occur.

Low-sulfidation epithermal deposits are widespread. Some examples of particularly large deposits are Ladolam (Papua New Guinea), Baley (Russia), Cripple Creek (Colorado, USA), Hishikari (Japan), Emperor (Fiji), Cerro Vanguardia, Esquel (Argentina), and El Peñón (Chile). This type also includes many gold-rich individual quartz veins containing more than 30 t of gold called bonanzas most of which are located in Nevada.

High-sulfidation and low-sulfidation veins rarely occur together (an exception is Bodie in California; Fig. 4.19). While high-sulfidation systems are typical of subduction zone volcanoes, low-sulfidation veins tend to form by extension. This can be in the back-arc of a subduction zone, in a stretched volcanic arc, and sometimes in other tectonic situations (Sillitoe and Hedenquist 2003). They often accompany bimodal volcanism with both basic and acidic magmas. It may be that basic magmas contributed some of the sulfur and possibly all the metals. There are also low-sulfidation veins associated with alkaline volcanism that were formed somewhat unusually by magmatic fluids. Most epithermal gold and silver deposits are geologically young and found in active subduction zones because they are easily lost by erosion.

Before discussing porphyry copper deposits it is worth taking a look at mercury deposits (Box 4.10) and native copper deposits at Lake Superior (Box 4.11).

Box 4.10 Mercury and cinnabar of Almadén (Spain)

Mercury is the only metal that is liquid under standard conditions. Elementary mercury actually occurs in nature in the form of small droplets, native mercury (Hg) is the only liquid mineral (Fig. 4.20). However, the most important mercury mineral is mercury sulfide cinnabar (HgS). Native mercury is formed in the oxidation zone from cinnabar.

Mercury is the most volatile metal and is often lost in hydrothermal systems such as hot springs that have a high mercury content or volcanic areas where mercury-containing vapor is released at fumaroles. Significant crystallization of cinnabar occurs in epithermal systems that are relatively close to the Earth's surface. Sometimes mercury occurs together with antimony or arsenic minerals such as stibnite, realgar, and orpiment. Mercury is typically found impregnated in the pores of sandstone or other porous rocks, in the sinter of hot springs, or in Carlin-type replacement deposits (Sect. 4.11).

The largest known mercury deposit in the world is Almadén (Spain) home to approximately one-third of

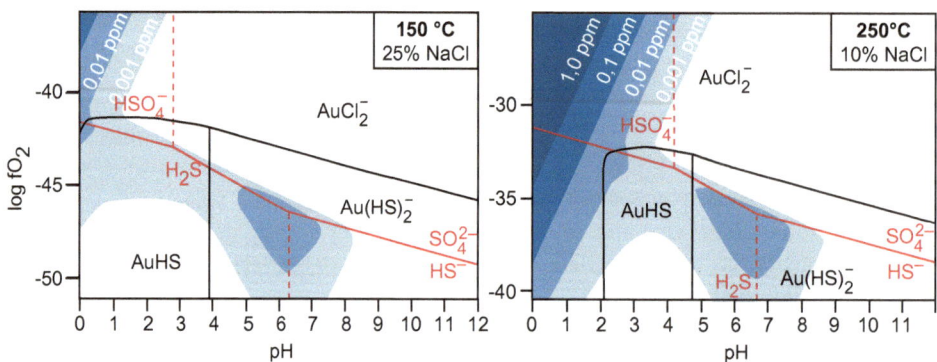

Fig. 4.18 The solubility of gold strongly depends on the temperature, pH value, and oxygen content (shown here as fO_2) of the water. The graph shows (*blue*) the solubility at two selected temperatures together with the predominant complexes. The redox reaction between sulfate and sulfide (*red*) buffers the oxygen content in water as it cools (from Cooke et al. 2000)

Fig. 4.19 After the discovery of epithermal gold ores Bodie grew into one of California's largest towns within a few years. The gold rush lasted a few decades. Today it is a ghost town and tourist attraction (*photo* © Tom Hilton, CC-BY, flickr.com)

all mercury ever mined. The region consists of sediments such as sandstones and shale (Ordovician to Devonian) that were later folded and transformed into metamorphic rocks during the Variscan orogeny. Volcanism (alkali basalts, volcanic breccias) occurred during sedimentation and hydrothermal systems developed. Whether the mercury originates from the magmas or from older sediments has not been clarified (Hall et al. 1997; Saupé and Arnold 1992).

In several small deposits around Almadén cinnabar occurs in veins within volcanic rocks. However, most (and the largest) deposits are located in a specific layer of quartzite (metamorphic sandstone) whose pores are impregnated with cinnabar. It is possible that this had already begun during deposition in not-yet-consolidated sand when water escaping from hot springs mixed with seawater. Cinnabar was partially remobilized and precipitated again in veins together with baryte and dolomite during the Variscan orogeny. In addition to cinnabar there are also droplets of native mercury. Some of the mercury even dripped out of gallery walls.

Cinnabar has to be roasted for mercury to be extracted. At around 350 °C cinnabar oxidizes to SO_2 and to mercury vapor that had to be condensed in ceramic tubes. Mining probably began in early antiquity. Cinnabar was used as a color pigment called vermilion red. Even Pliny the Elder wrote that the cinnabar of Almadén was best suited to coloring ceremonial robes. In the sixteenth century the Fuggers (German bourgeois family of bankers) leased the mines and benefited from the high mercury demand needed for silver mining in South America before the Huancavelica deposit was discovered in Peru. Due to the risk for mercury poisoning hardly anyone volunteered to work in the mines, so prisoners were used as forced labor. In the nineteenth century new technologies made safe mining possible. As a consequence of the low price of mercury the mines were closed down in 2000.

There are two other large mercury deposits in Europe. In Idrija (Slovenia) cinnabar is located in a sedimentary exhalative (SEDEX) deposit (Sect. 4.17) mainly in limestone. In Tuscany (Italy) (especially, at Monte Amiata) hot springs connected with young

Fig. 4.20 Droplets of native mercury on cinnabar (*photo* © Parent Géry, public domain, Wikimedia Commons)

volcanism have deposited cinnabar in sandstone, limestone, and volcanic rocks. Silver and mercury are also found in acidic volcanic rocks in Moschellandsberg and Stahlberg, Rhineland-Palatinate (Germany) where cinnabar occurs together with minerals like moschellandsbergite (Ag_2Hg_3).

Box 4.11 Native copper from Lake Superior

On the Keweenaw Peninsula of Lake Superior (Michigan, USA) there are very special copper deposits in basalts. The top of respective superimposed basalt flows is broken into breccias and rich in gas bubbles. Hydrothermal solutions within these breccias precipitated native copper (Fig. 4.21), chlorite, epidote, zeolites, apophyllite, prehnite, pumpellyite, quartz, and calcite. Copper was found in the form of filigree trees and massive lumps some of which weighed hundreds of tonnes. The water may have been meteoric water that had dissolved salts and leached copper in the sediments of the rift system (Braun 2006). Other spectacular specimens of native copper can be found nearby in shale (particularly, in the White Pine Mine).

Fig. 4.21 Native copper from the Caledonia Mine (Michigan, USA) (*photo* © F. Neukirchen)

0–0.04% molybdenum, 0–1.5 g/t gold and some silver, rhenium, palladium, tellurium, selenium, bismuth, zinc, lead, arsenic, antimony—there are huge volumes on the order of dozens of cubic kilometers (Sillitoe 2010). Mining is worthwhile despite the low ore grade because huge quantities can be mined relatively inexpensively in open-pit mining. Chuquicamata (Chile) and Bingham (USA) are the largest and deepest holes dug by humans. Such holes have a diameter of several kilometers and a depth of more than 1 km (Figs. 4.23 and 4.24). The most copper-rich deposit is El Teniente (Chile) with 94 Mt of copper in 12 Gt ore (Fig. 4.25). One of the most productive is currently La Escondida (Chile). In some porphyry copper deposits gold or molybdenum is so important that porphyry copper is classified into Cu–Mo, Mo–Cu, Cu–Au, and Cu–Au–Mo according to the most important metals economically. Although all other metals are by-products, some of them are still of economic importance. Another important by-product produced in huge quantities during smelting is sulfuric acid.

4.4 Porphyry Copper Deposits

About three-quarters of all copper produced worldwide, one-half of the molybdenum, one-fifth of all gold, and almost all rhenium and selenium come from porphyry copper deposits (for the term see Box 4.12). These are gigantic deposits (Fig. 4.22). Although they are low grade—typical ore grades are 0.5–1.5% copper (often less than 1%),

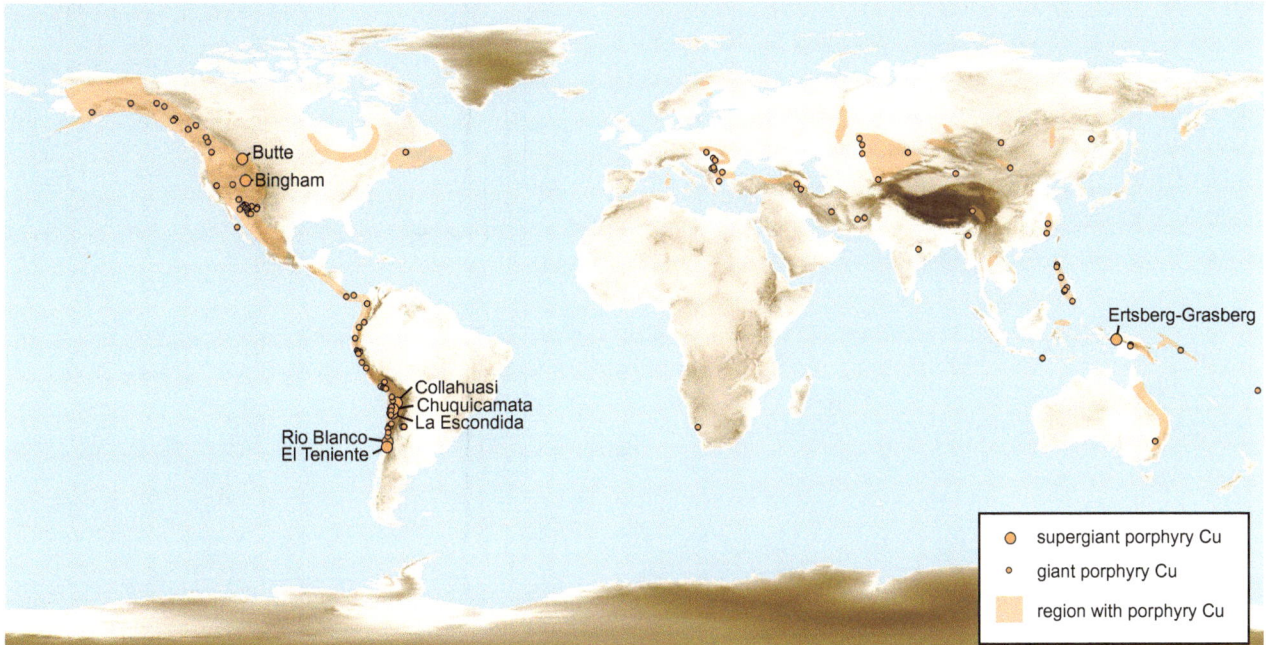

Fig. 4.22 Map of the largest porphyry copper deposits. Most are located along the Ring of Fire around the Pacific. There are more than 200 giants (more than 2 Mt copper or more than 100 t gold) and 12 supergiants (more than 24 Mt copper or more than 1200 t gold). Chile has the largest concentration of porphyry copper deposits making it the largest copper producer with about one-third of world production. Further smaller porphyry copper deposits are found in the *orange* regions. Occasionally, there are occurrences outside these regions (data from World Minerals Project, Geological Survey of Canada)

Box 4.12 Porphyry and porphyry copper

Unfortunately, the term porphyry copper is quite misleading because it has almost nothing to do with porphyry. Porphyry is the textural name given to volcanic rocks that contain large crystals (phenocrysts) in a fine groundmass. The crystals were usually already growing when the magma cooled down in a magma chamber. During a volcanic eruption the residual melt quickly solidified and formed the fine-grained groundmass. Typical phenocrysts are quartz (especially, in rhyolite; sometimes called quartz porphyry), feldspar (e.g., in phonolite), or feldspathoids (such as nepheline and leucite in rocks such as nephelinite and tephrite). Many rocks known as quartz porphyry were not lavas but pyroclastic flows of pumice and ash (ignimbrite) that were still hot and welded together to form solid rock. The glass of pumice and ash crystallized over time into fine-grained minerals.

Porphyry copper refers to copper ore bodies formed from hydrothermal mineralization in the uppermost zone of a pluton or a finger-shaped stock (granite, granodiorite, monzonite, etc.) and its surrounding rock —not to volcanic rocks. In appearance the sulfide-rich and strongly altered rock resembles porphyry.

Sometimes there are also larger phenocrysts between the coarse-grained minerals of a plutonic rock. In some granites the crystals of potassium feldspar, in particular, are much larger than quartz and plagioclase. The resulting texture is called porphyritic. Since potassium feldspars are not crystallized before other minerals their size is a result of slow nucleation and rapid growth. Also some porphyry copper stocks have a porphyritic texture in which large potassium feldspar or large hornblende crystals are embedded.

Box 4.13 Stockwork, massive and disseminated ores

Stockwork (a.k.a. stringer zones) consists of a finely branched network of veins that looks in profile like the branches of a tree. They are cracks in which hydrothermal water could rise. Stockwork is typical of porphyry copper deposits or the lower part of volcanogenic massive sulfide (VMS) deposits.

Hairline cracks and pores of usually strongly altered country rock are often filled with sulfides (i.e., finely distributed ore minerals). The rock is then an ore

Fig. 4.23 Although porphyry copper deposits have a low ore grade, they have a huge volume. The Chuquicamata open-pit mine (Chile) is 4.3 km long, 3 km wide, and about 1000 m deep. The system consists of several granite stocks. A large strike-slip fault runs through the open-pit mine (*photo* © James Byrum, CC-BY-ND, flickr.com)

with a low metal content known as disseminated ore. This is the main ore of porphyry copper deposits. There may also be disseminated ores in VMS, SEDEX, and other deposits.

Large ore bodies consisting almost exclusively of sulfides are referred to as massive ore or massive sulfide. They are typical of VMS, SEDEX, and some magmatic deposits.

Most porphyry copper deposits are found along the Ring of Fire around the Pacific. They are magmatic–hydrothermal systems related to the magmatism of subduction zones. However, this is not about volcanoes. Rather it concerns small finger-shaped intrusions called granite stocks (see Fig. 4.28) with a typical diameter of 1–2 km that rise up from a much larger granite pluton (I-type granite or granodiorite,

diorite, monzonite) that are stuck at a depth of 1–3 km. They are surrounded by a large zone with extremely strongly altered country rocks. Ore minerals whose primary formations are mainly pyrite, chalcopyrite, and bornite are located at the top of the granite stock, in adjacent altered country rock, in small veins cutting through the rocks (stockwork), and disseminated in former pores (Box 4.13). The ore grade decreases continuously toward the outside and the bottom. Sometimes even the center is hardly mineralized because the temperature there was too high such that the ore zone takes on a bell shape. Further related deposits are often formed in the vicinity such as high-sulfidation epithermal veins (Sect. 4.3), cordilleran polymetallic veins (Box 4.9), and sometimes skarns (Sect. 4.9), chimneys, and mantos (Sect. 4.8).

Granite actually contains much less copper than basalt. The reason the largest copper deposits are nevertheless formed in connection with fractionated magmas is that in these cases the copper of a very large magma volume

Fig. 4.24 Aerial photograph of the Bingham Canyon Mine (Utah, USA) (*photo* © PhotoQ, CC-BY, Wikimedia Commons)

was effectively mobilized by hydrothermal fluids and precipitated in a rather small rock volume.

The story begins with an oceanic plate sliding down into the mantle in the subduction zone. Basalts of the oceanic crust are transformed with increasing depth into high-pressure rocks such as blueschist and finally eclogites. The lithospheric mantle sticking underneath often had been partially hydrated to serpentinite before reaching the subduction zone and is converted to anhydrous peridotite. These metamorphic reactions release large amounts of water that rise into the mantle wedge above the subducting slab. This water contains dissolved substances originating from basalt and smaller amounts of sediments from the subducted plate. Mantle rocks are metasomatically enriched by these fluids. Subducted deposits such as Cyprus-type VMS (Sect. 4.16) and manganese nodules (Sect. 5.6) also contribute to the dissolved cargo.

The wedge-shaped mantle above the subducted plate is hot enough to partially melt in the presence of water. The result is basalt magma. This differs from other basalts in its high water content and higher content of certain elements supplied by the water. This magma rises into the crust. Since continental crust has a similar density some of the magma gets stuck under the crust and cools to a gabbro

(underplating), but at the same time partial melting of the lower crust can generate granite magma that in turn can mix with basalt magma. Basalt magma evolves by fractional crystallization (Sect. 3.1.1) to an acidic composition (gabbro to diorite to granodiorite to granite) and rises further. During fractional crystallization there may be repeated mixing with less fractionated melts and small amounts of crustal melts. The water content of the residual melt continues to increase at the same time. Large quantities of these magmas get stuck at a depth of a few kilometers and cool down there to become plutons. The result is an enormous elongated body of countless plutons that have intruded each other called a batholith. Some of the magma reaches the Earth's surface and builds up to form the explosive stratovolcanoes of the subduction zone.

Although readers should now have a good understanding of normal magmatism of subduction zones, it is also important for them to understand that magma must have a high sulfur content and should be relatively oxidized (i.e., sulfur is predominantly contained as sulfate) for the formation of porphyry copper deposits. This prevents metals and sulfur from being removed from the magma at an early stage of fractionation by crystallization of sulfides or by segregation of a sulfide melt (Chap. 3). A relatively new finding is that

Fig. 4.25 The world's largest porphyry copper deposits in terms of their copper content (**a**) and their gold content (**b**). They differ greatly in their ore grade and total amount of ore (from Cooke et al. 2005)

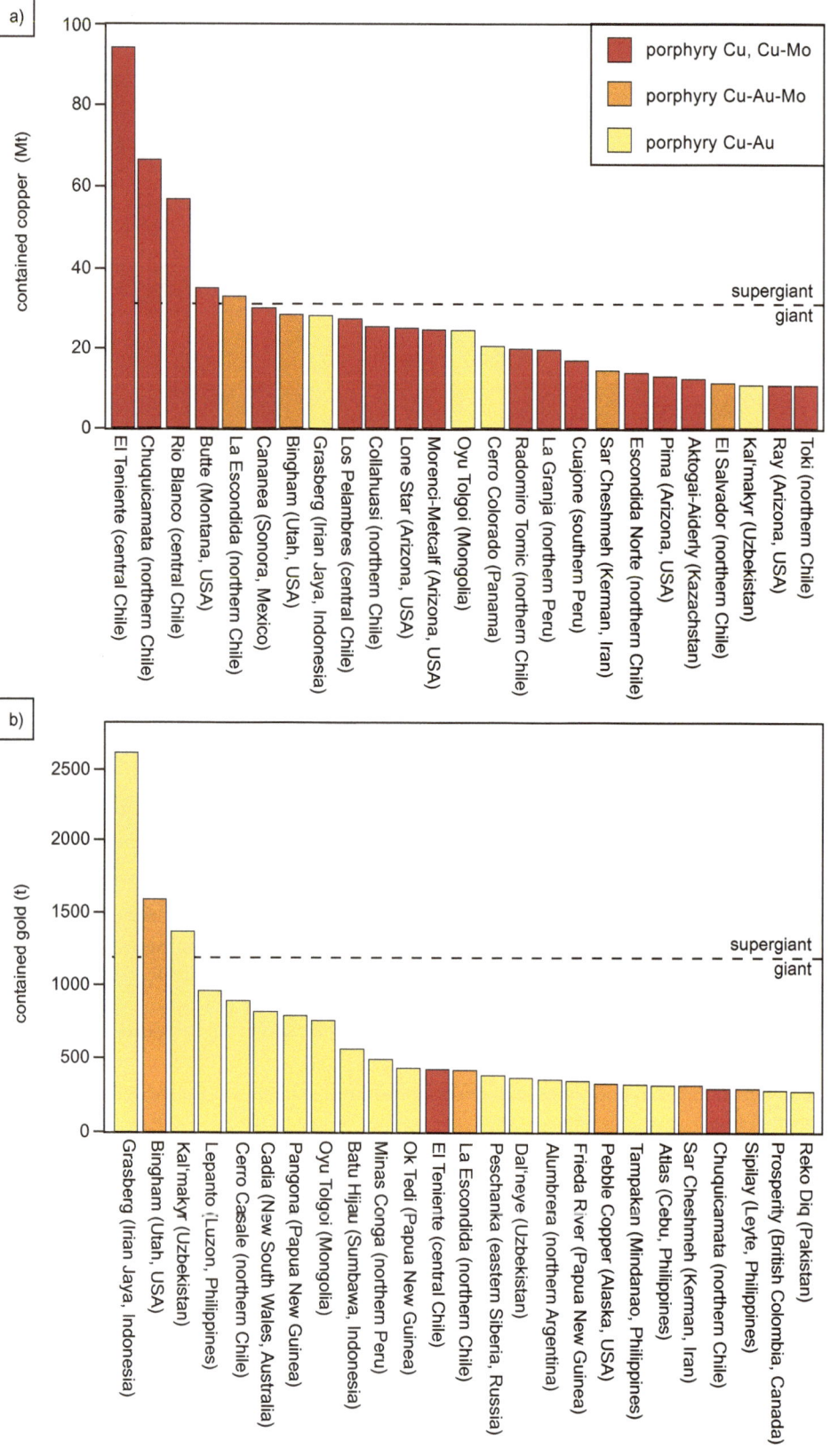

segregation of a sulfide melt can nevertheless occur as an intermediate step (Halter et al. 2002, 2005; Nadeau et al. 2010), but only with advanced fractionation in a pluton at shallow depth. Copper and gold fractionate very effectively in a sulfide melt. Even if further (less fractionated) magma is injected into the pluton the molten sulfide can absorb all of its metals and sulfur. Metals are thus preconcentrated and later such sulfides dissolve completely in hydrothermal water.

The thickness and type of crust also play a role. Basaltic melts at a young island arc can rise almost unhindered up to the Earth's surface and only minimal fractionation occurs, which is not sufficient for the formation of porphyry copper deposits. Fractionation is much stronger at active continental margins where oceanic crust slides under continental crust (especially, if there is very thick crust due to long-lasting magmatism and overthrusting as is the case with the Andes). In island arcs that have been active for a long time a kind of continental crust has already formed. Under certain circumstances porphyry copper deposits can develop here, typically they have a high gold content.

It was once believed that porphyry copper deposits formed under a volcano. However it is now clear that they don't form by normal subduction zone volcanism because a volcano would lose much of its sulfur as gaseous SO_2 and for a deposit to form magma would have to remain stuck at depth even though a lot of water is released. This would normally lead to foaming and thus to an explosive eruption. There needs to be a phase of compressive tectonics that prevents active volcanism. The switch to compressive tectonics is often assumed to be the decisive trigger. In fact, porphyry copper deposits occur together with large quantities of volcanic rocks that have already erupted in earlier times. But during or shortly after the emplacement of porphyry copper stocks, there was almost no volcanism, at the utmost small lava domes and maars.

Compressive tectonics has another effect in that it leads to overthrusting and the crust becoming thicker. This leads to fast uplift and in turn to stronger erosion (Fig. 4.26).

All this has an effect on a pluton that has partially solidified whether it is diorite, granodiorite, granite, or monzonite. Rapid erosion of overlying rock reduces the pressure as does flank collapse of a large volcanic cone. The pressure drop results in the release of large amounts of magmatic fluids from the residual melt. Such fluids consist mainly of water, but there are also SO_2, H_2S, CO_2, HF, as well as dissolved salt and metals (mainly as chloride complexes). Any sulfide melts or magmatic sulfides present are dissolved by the water.

The water rises together with some magma and leads to the sudden hydraulic breakage of adjacent rock in the roof of the pluton. This effect is strongest in dome-shaped apexes of a pluton (Guillou-Frotter and Burov 2003). It is here that a finger-shaped body made of magma and large quantities of water acting as a kind of pressure relief valve begins to rise. At a depth of 1–4 km below the Earth's surface it gets stuck and the magma solidifies into granite stock while water continues to flow from the deeper pluton.

In the upper part of porphyry stock the pressure is so low that water is partitioned into a brine (liquid and salty) and a

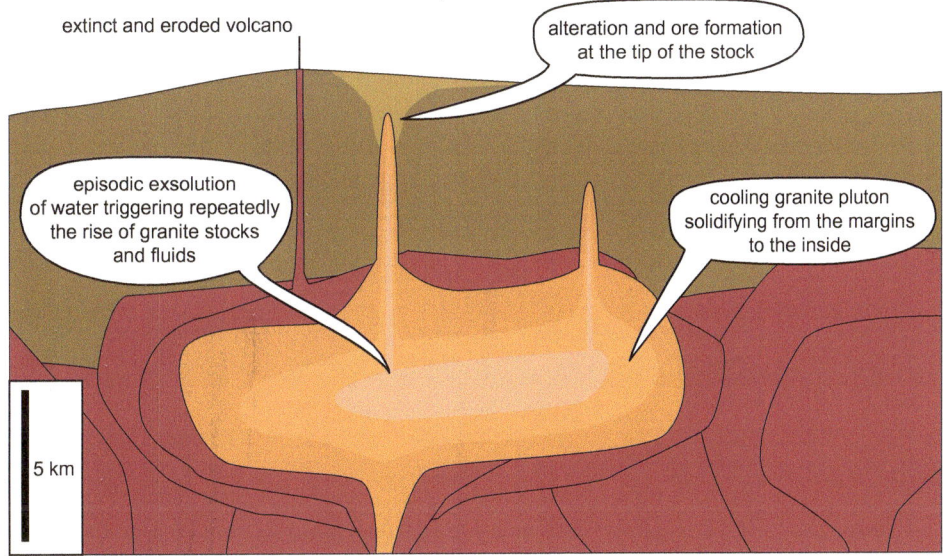

Fig. 4.26 Rapid erosion reduces the pressure in a deep granite pluton. This can trigger episodic release of large amounts of magmatic fluids that rise together with some magma into the highest areas of the pluton (e.g., into dome-shaped apexes) where the side rock is broken hydraulically and magma and fluids continue to rise as granite stock. With compressive tectonics the magma gets stuck and there is no volcanic eruption. Instead, the fluids at the top of the granite stock lead to alteration and ore formation. This process is repeated over wide intervals of time and new stock rises within the older one thus tapping unsolidified parts of the pluton. Older plutonic rocks are shown in *red* (from Sillitoe 2010)

vapor phase with low salt content. Although ores have not yet precipitated, the metals are fractionated. Iron, zinc, lead, manganese, and molybdenum are enriched in the brine, while copper, gold, silver, arsenic, antimony, tellurium, and bismuth are enriched in the vapor phase together with SO_2, H_2S, SO_2, and HF. Although the concentration of metals in steam is lower than in hot brine due to the low density, there is a much larger amount of steam. Thus, pressure and depth affect the Cu/Au ratio at this point (Murakami et al. 2010). Due to the low density the steam rises further and the heavier brine remains behind. The fluids tear open countless cracks through which they penetrate the surrounding rock by breaking the rocks hydraulically. Sometimes hydraulic crushing above the intrusion is so intense that a breccia is formed comprising rock fragments, rock flour, and hydrothermally precipitated cement. Such a breccia is sometimes the preferred place for ore mineralization.

Box 4.14 Hydrothermal alteration

Alteration is the process in which surrounding rock is changed by hydrothermal fluids (Fig. 4.27). Although this can be discreet and give the rock the appearance of being a little weathered, it can also completely transform the rock so that it is no longer recognizable. Alteration involves minerals being transformed by reaction with H^+ or $(OH)^-$ from the water and an exchange of cations (metasomatism). The reactions that take place depend on the composition of the rock and fluid and on pressure and temperature. The strength of alteration also depends on the amount of water flowing per rock volume. Alteration zones can

---

be indications of deeper ores and are therefore important for prospecting.

Alteration in its various forms changes not only the rock but also the fluid. The fluid cools down, changes the pH value (the weathering of feldspar neutralizes acids), the redox state, and so on. This in turn affects the solubility of metals and can even trigger the precipitation of ore minerals. Overall it is a dynamic system that changes constantly over time with older alterations being modified to new ones. A halo often forms in which there is typical zoning of differently altered rocks whose exact geometry is somewhat different in each deposit (Fig. 4.28).

Potassic alteration: The center of the alteration zone of porphyry copper deposits at the tip of the porphyry stock and the surrounding host rock. It is transformed by large amounts of magmatic fluids at very high temperatures. The rock absorbs potassium from the water. The original granite now consists almost entirely of potassium feldspar and biotite (with some sericite, chlorite, and quartz). Pyrite, chalcopyrite, and bornite are the main ore minerals found in veins and former pores.

Phyllic alteration or **sericitic alteration**: Sericite is fine-grained muscovite (common mica) that is formed by the reaction of potassium feldspar with H^+. The corresponding alteration zone also contains chlorite (from the transformation of plagioclase, pyroxene, hornblende) as well as quartz and pyrite. This type of alteration occurs at a wide temperature range in felsic rocks and is typical of porphyry copper, VMS

Fig. 4.27 Alteration zone with silicification at Cerro Palla Palla (Peru). The *white* zone is almost pure SiO_2 (*photo* © Gerhard Wörner)

Fig. 4.28 Schematic section through a porphyry copper deposit. Sulfides (disseminated and in veins) are located in the *yellow* shaded zone at the upper end of a granite stock. The surrounding rocks are strongly altered forming a halo made up of different alteration zones. Other deposits may form in the vicinity: in particular, high-sulfidation epithermal gold veins and (if limestone is present) skarns (from Sillitoe 2011)

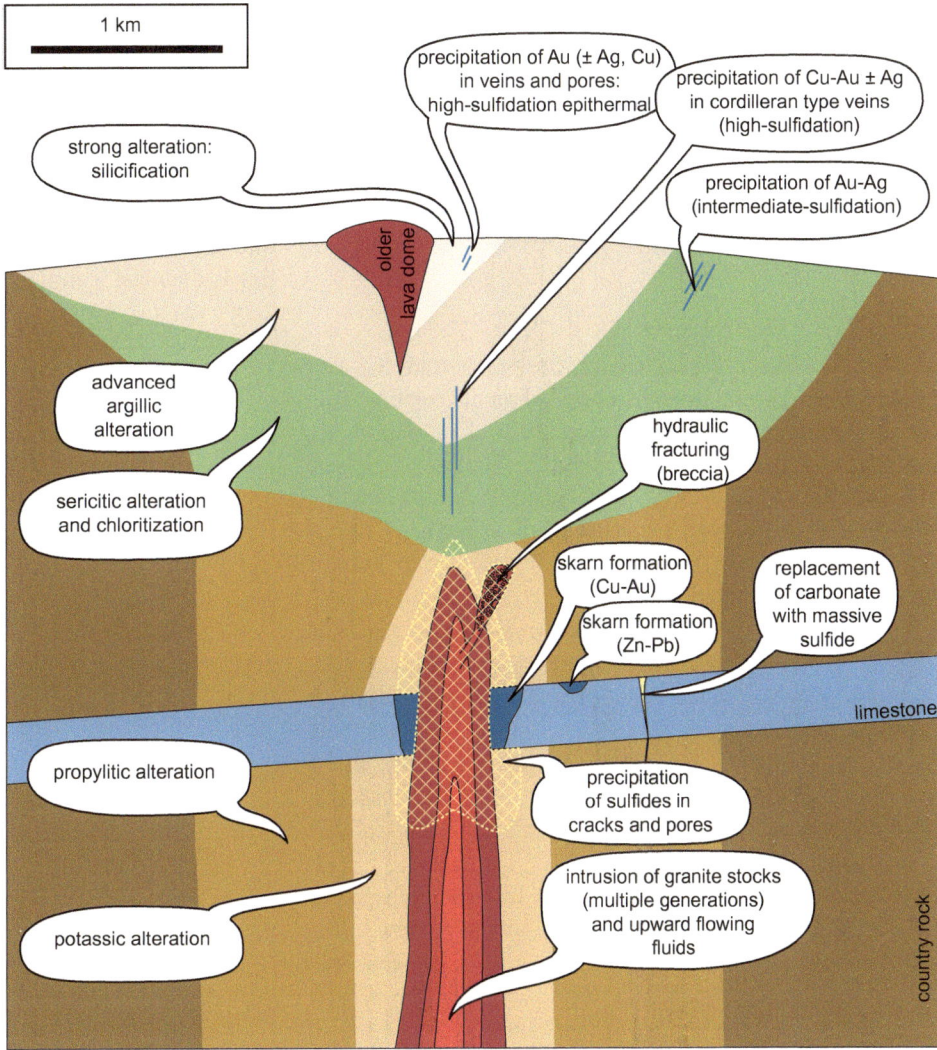

deposits, and mesothermal veins. The chlorite content can gradually increase and alteration can gradually change to chloritization. Enargite, tennantite–tetrahedrite, bornite, chalcocite, and sphalerite can also occur in addition to pyrite and chalcopyrite.

Chloritization: Formation of chlorite, sericite, and pyrite. This alteration zone is strongly colored green as a result of chlorite.

Propylitic alteration: Chemical alteration of rock at a lower temperature mostly caused by heated groundwater and only minor amounts of hydrothermal water. The alteration is less pronounced and resembles metamorphism in greenschist facies. Chlorite and epidote (together with such minerals as clinozoisite, zoisite, calcite, albite, and pyrite) are formed. This type of alteration is typical of the wider environment of porphyry copper deposits. This alteration can also occur along veins. The sulfide content is low (pyrite, sphalerite, galena).

Argillic alteration: At shallow depths strongly acidic fluids convert feldspar into clay minerals such as smectite (montmorillonite), illite, or kaolinite. Alkali metals are leached out and transported away. Initially, plagioclase is mainly affected. Advanced argillic alteration results in rock consisting only of kaolinite, pyrophyllite (a mineral similar to talc), and alunite (a sulfate) with some quartz and pyrite. There may be topaz or tourmaline. This alteration has been observed in many hydrothermal deposits such as high-sulfidation epithermal gold veins (Sect. 4.3) and porphyry copper systems. Pyrite (with gold), enargite, chalcocite, and covellite can occur locally in small quantities.

Silicification: Formation of new quartz, chalcedony, or opal and the removal of all other substances. Silicification occurs mainly in high-sulfidation epithermal systems often together with argillic alteration. It is the counterpart to the alteration of feldspar in which SiO_2 is dissolved. Often rocks with many cavities are formed consisting almost exclusively of quartz (vuggy silica).

As the system cools, the granite stock and the surrounding rocks are strongly altered through reaction with fluids resulting in alteration zones (Box 4.14, Fig. 4.28). Sulfides such as pyrite, chalcopyrite, bornite, chalcocite (Fig. 4.29), digenite, and molybdenite are simultaneously precipitated in pores and cracks. Strictly speaking, a sulfide with a composition between chalcopyrite and bornite is formed at high temperature and then segregates into both minerals when cooled down. Some gold can be mixed in bornite and to a lesser degree in chalcopyrite; it also occurs as tiny grains of native gold. After potassic alteration the granite stock consists almost exclusively of potassium feldspar and biotite. When the temperature decreases in time and space chlorite, sericite, and other minerals replace the original rocks.

Simultaneously with alteration innumerable veins are being formed one after the other at different temperatures cutting through each other with different compositions. The earliest high-temperature veins still contain little or no sulfides, while sulfides predominate in later veins formed over a wide temperature range. Which sulfides are deposited changes over time and in different areas of the deposit, creating zones enriched with different metals. Remobilization of older sulfides can also occur leading to local enrichment or depletion of metals (e.g., remobilization has an effect on gold content) (Kesler et al. 2002).

Fig. 4.29 Chalcocite ore from a porphyry copper deposit in Chile (*photo* © F. Neukirchen/Mineralogical Collections of the TU Berlin)

The extent to which mineralization also occurs in the surrounding rock also depends on how permeable the rock is to water. As a rule these rocks are predominantly volcanic and include porous tuffs. Since it is desirable for metals to be concentrated in a small volume the surrounding rock should be as impermeable to water as possible. However, a small volume of highly porous rocks can lead to metal concentration. Sometimes this happens in breccias that have only just hydraulically formed—a spectacular example being El Teniente (Box 4.15).

Box 4.15 El Teniente

El Teniente (Vry et al. 2010) is 70 km southeast of Santiago de Chile, it contains 94 Mt of copper, and is by far the largest copper deposit on Earth (Fig. 4.30). It is a cluster of several porphyry copper stocks that have intruded in the immediate vicinity. However, the more than 12 Gt of ore have an average grade of only 0.65% copper and 0.019% molybdenum. Higher ore grades are found in breccias located at the tips of porphyry stocks. In contrast to most other porphyry copper deposits mining takes place underground.

As can be seen in other porphyry copper deposits, country rock can be hydraulically fractured into a breccia whose fragments are cemented by hydrothermal minerals. Sometimes a magmatic breccia is formed cemented by a granite-like matrix. The special thing about El Teniente is that these breccias do not branch off from the porphyry stock like a small vermiform appendix. They have a larger diameter than the actual stock and are attached to its tip like the bristles of a brush. They are so important in this deposit that El Teniente is classified by some as a breccia deposit instead of a porphyry copper deposit.

In addition to magmatic breccias there are also hydrothermal breccias cemented by very different minerals such as potassium feldspar, biotite, anhydrite, or tourmaline. The youngest and by far largest breccia is the Braden Pipe breccia consisting of rock fragments and rock flour. Apart from small tourmaline breccias at the edges (with Cu–As–Sb minerals) it contains no ores. Right next to it is the largest stock called the El Teniente dacite porphyry. It is an elongated intrusion 1.5 km long and 200 m wide. Although the other ore-bearing porphyry stocks and breccias are significantly smaller, they are so numerous that alteration zones and veins overlap. Only a fraction is exposed at the Earth's surface.

As with normal porphyry copper deposits there is a dense network of hydrothermal veins that intersect the stock, breccias, and altered host rocks. As usual, there are several generations of different veins cutting

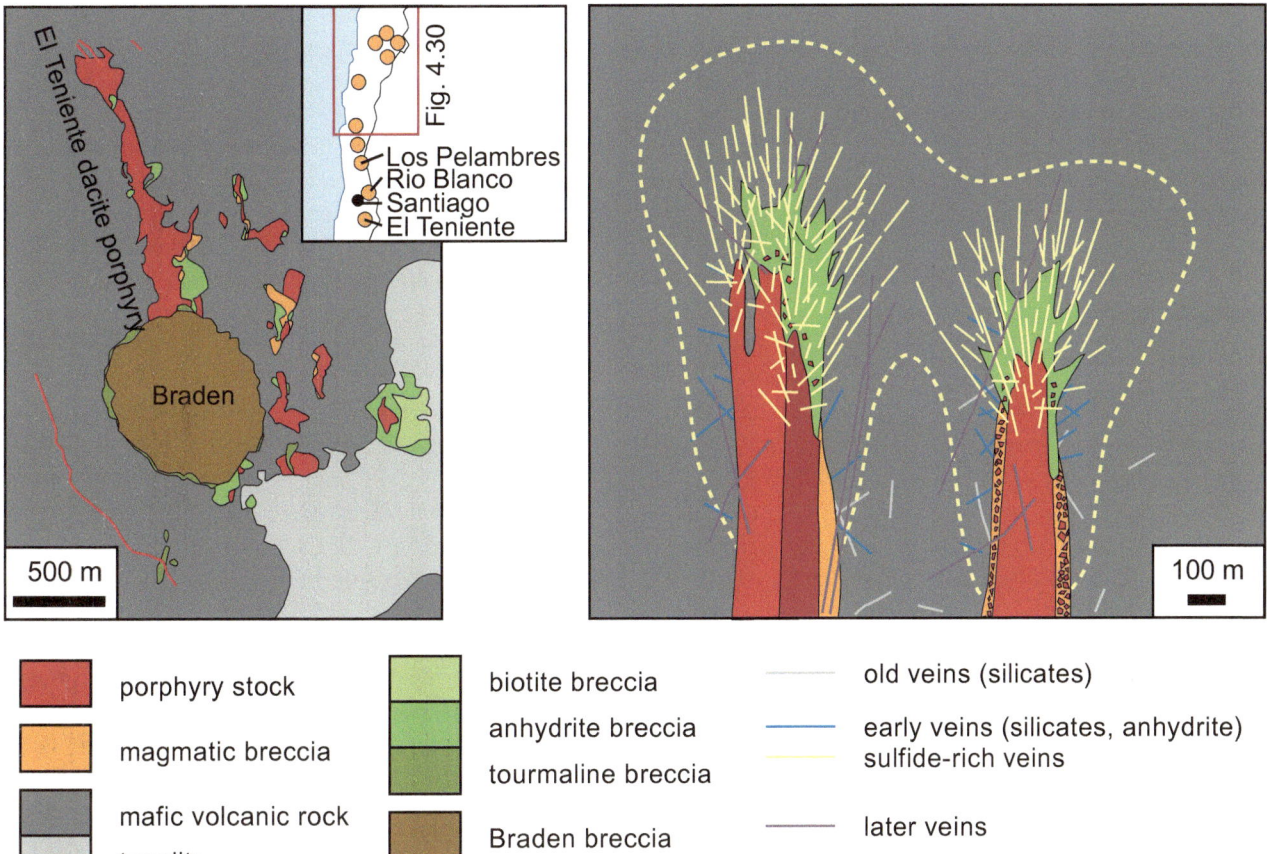

through each other. The most important phase began with gangue-dominated veins (quartz, biotite, potassium feldspar, anhydrite) followed by differently composed sulfide-rich veins.

El Teniente, Río Blanco–Los Bronces, and Los Pelambres–El Pachón are the three largest porphyry copper deposits in central Chile and the youngest in the Chilean copper belt. They were created when flat slab subduction was triggered by subduction of the Juan Fernández Ridge. East of El Teniente there are still active volcanoes, while north of Santiago the volcanic arc has a large gap. The side rocks of El Teniente consist of subvolcanic basalts and andesites from an extinct and deeply eroded volcano. Their physical properties clearly favored breaking into breccias. They also acted as a kind of chemical trap because the Fe^{2+} obtained in the rocks could reduce the sulfate dissolved in the water.

The presence of limestone or dolomite in the surroundings of deposits can have a very positive effect. In this case skarn (Sect. 4.9) is formed with ores of much higher grade than the actual porphyry copper deposit. A little farther away where acidic hydrothermal water flows into the limestone from a crack in a focused manner, limestone is completely replaced by massive Zn–Pb–Ag–Cu sulfides (chimney or manto; Sect. 4.8).

Gabbro and other surrounding rocks that contain a lot of Fe^{2+} also favor a high ore grade. This is because during alteration of the rock sulfate dissolved in water is reduced to sulfide by oxidation of the iron.

Above the porphyry copper deposit, rising supercritical vapor contracts and develops into an aggressive, acidic, and oxidized fluid on its way to the Earth's surface. This leads to extreme types of alteration such as argillic alteration and silicification (Box 4.14). Rocks are largely converted into clay minerals and a vuggy rock consisting almost exclusively of SiO_2 is formed in places where aluminum itself is dissolved. Such places are called lithocaps. They are typically made up of soft, colorful hills. Sometimes they surround a somewhat older lava dome. Inside lithocaps gold may precipitate in pores and veins as is the case with high-sulfidation epithermal deposits (Sect. 4.3). At depth the veins become richer in silver and copper and merge into cordilleran-type polymetallic veins. Outside lithocaps there may be epithermal systems near the surface that are mainly fed by heated groundwater and only small amounts of magmatic fluids.

A porphyry copper deposit does not usually consist of a single granite stock, but of several that have risen one after the other in the center of older stocks whose alteration and mineralization transformed older alterations and mineralizations. The time intervals are large since the activity can extend over hundreds of thousands or even several million years. A large pluton not only cools down very slowly, but fresh magma can also be injected into a pluton that has not completely solidified. A pluton solidifies from the margins into the interior; therefore the mixture of water and melt rising as a stock is increasingly derived from the interior of the pluton. Often the strongest alteration and ore formation took place in episodes temporally in the middle of the formation of rising granite stocks. Since every intrusion results in new veins also being formed in older stocks the oldest stocks tend to contain the most metals. In many cases erosion has lowered the Earth's surface in the meantime such that the stocks remain stuck at a greater depth relative to older ones and the zoning shifts.

Often there is a cluster of several porphyry copper stocks, in other cases a single huge stock is emplaced, and sometimes instead of a round or oval stock a swarm of granite dikes has been intruded. This geometry is determined by the respective tectonic situation. Stocks almost always rise along large strike-slip faults (especially, where they cross other faults and local extension can occur despite prevailing compression). In the Atacama Desert in northern Chile (Ossandón et al. 2001; Padilla Garza et al. 2001; Richards et al. 2001; Richards 2003; Sillitoe 2003), for example, the largest stocks are located along the Domeyko Fault. There are often strike-slip faults parallel to subduction zones since the plates usually move obliquely toward each other and the margin of the upper plate is pulled sideways (Fig. 4.31).

Before a porphyry copper deposit can be reached by humans the overlying 1–3 km of rock must be removed. Accordingly, deposits are not found at an active volcanic arc but in a belt parallel to it at a former position of the volcanic arc. This can shift for several reasons. On the one hand, the angle at which the oceanic plate is subducted can change. On the other hand, the plate boundary itself can shift. When terranes (mini continents, island arcs, basalt plateaus) collide with the subduction zone and are welded to the continent the continent grows and both plate boundary and volcanic arc

migrate seaward. The same happens somewhat more slowly when sediments are scraped off the subducting plate and accumulate in an accretionary wedge at the plate boundary.

However, there are also subduction zones in which the subducted plate grinds itself into the continent like abrasive paper. This is happening in the central Andes where a strip more than 200 km wide has already fallen victim to subduction erosion and the volcanic arc has accordingly shifted ever-further into the continent; hence the several copper belts in northern Chile of different ages are located where former volcanic arcs used to be (Fig. 4.31). The first had already formed in the Mesozoic in the area of today's Coast Range. These are rather small except for Andacollo probably due to the fact that extension prevailed—not compression. They occur together with significant copper mantos (Sect. 4.8), skarns (Sect. 4.9), iron oxide copper gold (IOCG) deposits (Sect. 4.7), and Kiruna-type iron deposits (Sect. 3.6); hence this strip is also referred to as the Chilean Iron Belt (Oyarzun et al. 2003). Two belts with some large porphyry copper deposits were formed in the Paleogene the largest being in the younger belt that was active from the Eocene to the Oligocene.

As soon as a porphyry copper deposit reaches the surface by erosion another process begins in which an oxidation zone is formed (Box 4.16). This results in strong enrichment of copper and in many cases only this secondary enrichment made a deposit economically interesting.

There is increasing evidence that the change from normal to flat slab subduction (Fig. 4.32) is particularly favorable for the formation of porphyry copper deposits (Oyarzun et al. 2001, 2002; Rabbia et al. 2002; Richards 2002; Cooke et al. 2005; Hollings et al. 2005; Rosenbaum et al. 2005). This is especially true of giant porphyry copper deposits. Flat subduction occurs, for example, when a chain of seamounts is subducted. Due to the thicker crust the density of the subducted slab is lower and it folds upward. Now it slides directly under the lithosphere of the upper plate causing the end of magmatism and leading to compression in the upper plate. This is exactly what is happening today in northern Peru and in Chile just north of Santiago where the subduction of seamount ranges has led to flat subduction and a gap in the volcanic arc. The decisive moment is probably reached when magmatism has not yet stopped, but the increasing compression is already hampering the ascent of magma to the Earth's surface and the degassing of SO_2. At the same time the crust is thickened by overthrusting causing rapid ascent and increased erosion. For magmatic intrusions this means pressure relief that can lead to the exsolution of magmatic fluid, sudden hydraulic fracturing of the surrounding rock, and thus to the ascent of porphyry stocks. A further effect may also play a role. While the subducted plate folds upward, in exceptional cases melt formation may occur in the subducting plate. These so-called adakites differ

Fig. 4.31 Map showing porphyry copper deposits and associated deposits in the Atacama Desert in northern Chile. The subduction zone grinds itself into the continent along the plate boundary via a process called subduction erosion; hence the volcanic arc has shifted several times. There are therefore several copper belts of different ages that run parallel to the active volcanic arc. The deposits are often located at large strike-slip faults (especially, where they intersect with other shear zones). The Domeyko Fault corresponds to the Mesozoic back-arc that was later pushed together again and finally became active as a strike-slip fault. In the south of the map the active volcanic arc ends because here the subduction of a chain of seamounts led to flat slab subduction (based on Richards 2003; Sillitoe 2003)

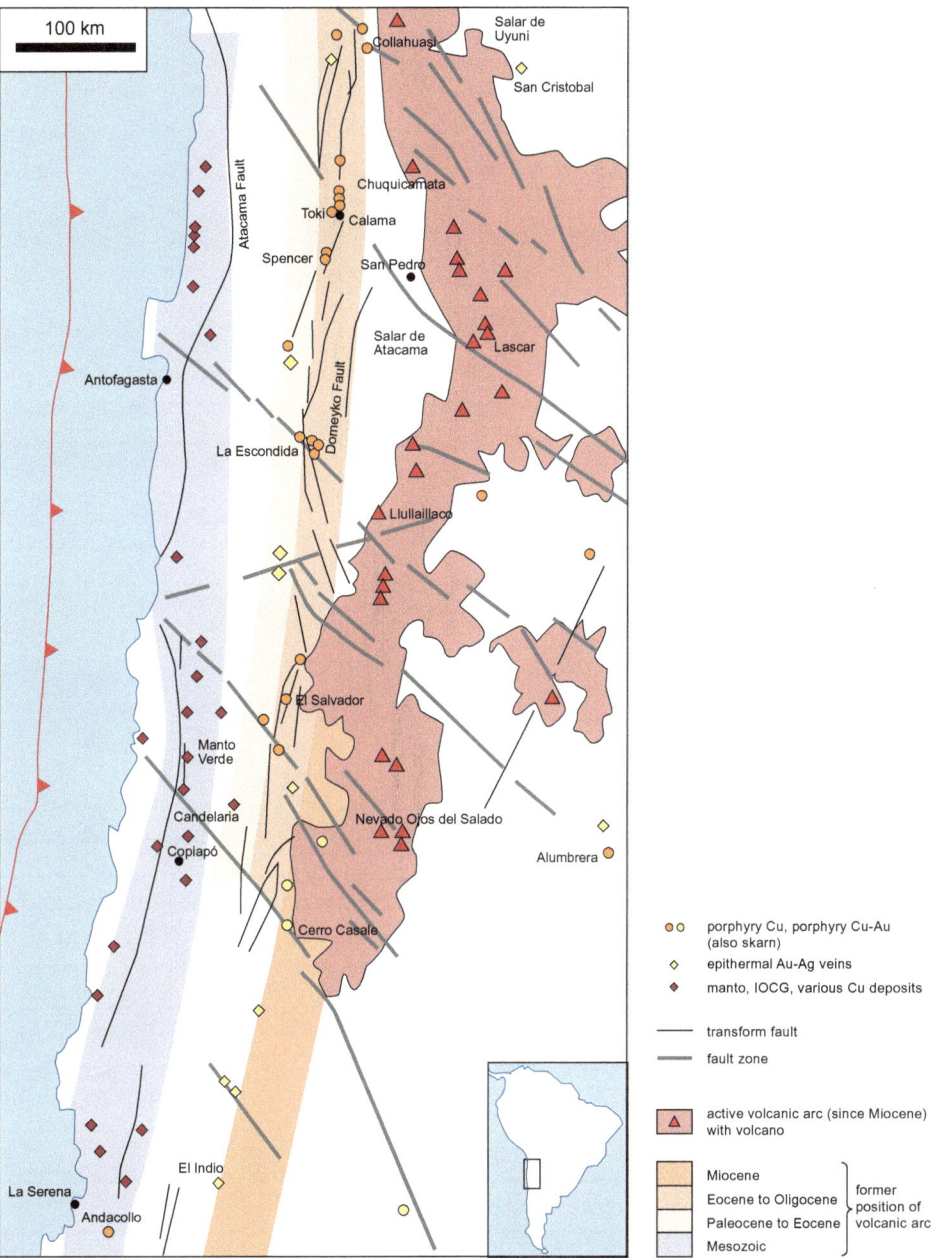

from normal subduction zone magmas only in their trace elements and isotope compositions. Nevertheless, this signature is present in some cases. It has also been suggested that the subduction of seamounts intensifies subduction erosion and therefore fragments of the Earth's crust are transported into the mantle wedge where they alter melt formation.

Box 4.16 Gossan (oxidation zone) in copper deposits

When a sulfide deposit reaches the Earth's surface by erosion an oxidation zone called gossan is formed (Fig. 4.33) where sulfides are oxidized and water-soluble components are leached out and precipitated again at a deeper level. At the Earth's surface the so-called iron cap (*eiserner Hut*) is formed, while below it a secondary enrichment of certain metals such as copper and silver occurs. This is known as supergene enrichment (i.e., near-surface enrichment).

This process is particularly important for copper deposits. For example, porphyry copper deposits have a low ore grade and many are profitable only thanks to secondary enrichment (Sillitoe 2005). In addition, oxidation zone ores are easier to smelt than sulfides, which is why they were so important in the early history of humankind.

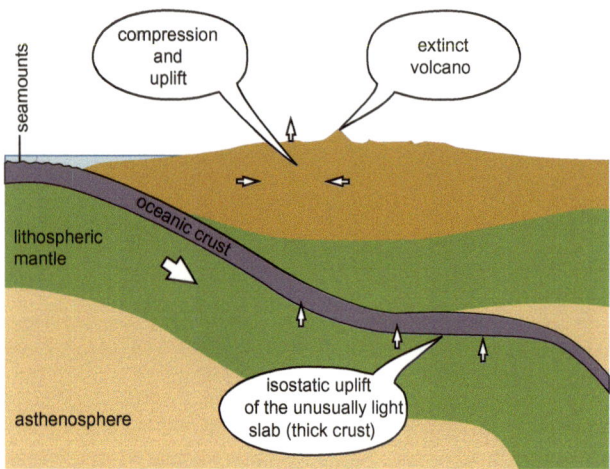

Fig. 4.32 Many porphyry copper deposits were formed during the change from normal to flat slab subduction. This can, for example, be triggered by the subduction of a seamount range since the corresponding lithosphere is of lower density. The subducted slab slides flat under the upper plate. Magmatism stops and compression increases accompanied by fast uplift and increased erosion

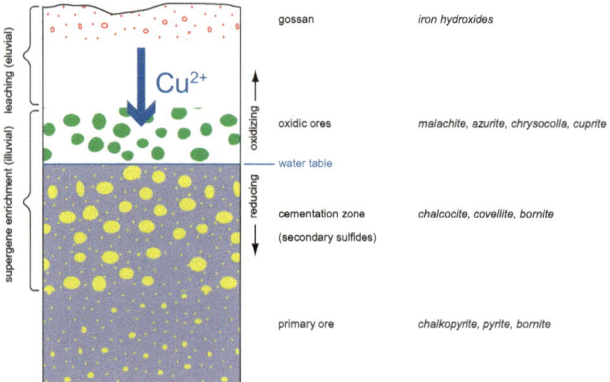

Fig. 4.33 An oxidation zone (gossan) is formed in exposed sulfide deposits. This can lead to strong secondary enrichment (especially, in the case of copper deposits). During the oxidation of primary sulfides, copper dissolves in increasingly acidic water and iron remains in the form of hydroxides in the iron cap. By neutralizing the acid so-called oxidic ores such as the copper carbonates malachite and azurite can be precipitated from downward-seeping water. In the cementation zone below the groundwater level primary sulfides are displaced by secondary sulfides that have a significantly higher copper content (from Robb 2005)

Primary sulfides such as pyrite and chalcopyrite are oxidized by the infiltration of water. Cu^{2+}, SO_4^{2-}, and H^+ are dissolved, while the iron oxidized to Fe^{3+} remains in the form of hydroxides—usually as limonite, a rust-colored mixture of different iron hydroxides such as goethite. Exemplary is the oxidation of chalcopyrite by the reaction:

Fig. 4.34 Iron crusts in an alteration zone at Cerro Palla Palla (Peru) (*photo* © Melanie Brandmeier)

$$4CuFeS_2 + 17O_2 + 10H_2O \rightarrow 4Fe(OH)_3$$
$$+ 4Cu^{2+} + 8SO_4^{2-} + 8H^+$$

Through this reaction (and even more through the oxidation of pyrite) the pH of the water drops rapidly, which in turn increases the solubility of copper.

Insoluble substances remain near the surface. They mainly comprise iron hydroxides from which the shape of the original sulfides can often be ascertained. They are located in strongly altered rock consisting mainly of clay minerals or quartz (Box 4.14). Sulfates such as alunite and jarosite are also typical. Under certain circumstances gold may be enriched (see also Sect. 5.11.3). This zone is called the iron cap (Fig. 4.34). Many sulfide deposits have been found thanks to its striking rust color.

Water seeps down carrying its dissolved metals until it reaches the so-called cementation zone below the groundwater level where conditions are rather reduced and the solution reacts with primary sulfides to form secondary sulfides. These are minerals such as chalcocite (Cu_2S) and covellite (CuS) that have a significantly higher copper content than primary sulfides. On the one hand, the Fe^{2+} of the primary ores dissolves and is replaced by Cu^{2+} and, on the other hand, the ratio of metal to sulfur also increases. Even the amount of sulfide minerals increases.

If the pH of the downward seeping water changes, then copper minerals can also precipitate above the groundwater level because such a change affects the stability and solubility of complexes (Box 4.2). Acid can, for example, be neutralized by reaction with feldspar or limestone. Depending on which complexes become unstable and which minerals are stable under given conditions such as pH, Eh, and temperature a large number of minerals can be formed. The copper carbonates malachite and azurite are very common (especially, if there is limestone in the area). The water-containing copper silicate chrysocolla also frequently occurs. Less common are the copper oxides cuprite and tenorite and phosphates like libethenite. In the case of a low oxygen content native copper can form at the transition between the cementation zone and the oxidation zone. In arid climates like that of the Atacama Desert minerals that dissolve readily can also form including copper sulfates such as chalcanthite, brochantite, and antlerite and even chlorides such as atacamite ($Cu_2Cl(OH)_3$).

Silver behaves similarly in the oxidation zone where native silver, in particular, can be formed. The Atacama Desert is again worthy of mention because unusual minerals like embolite (Ag(Cl,Br)) occur in relatively large quantities.

Although zinc is also leached out during the oxidation of sulfides, it is more frequently lost instead of being enriched elsewhere. Typical formations are the zinc carbonate smithsonite or the zinc–copper carbonate aurichalcite. Although lead is hardly mobilized, lead sulfate (anglesite), lead carbonate (cerussite), and lead phosphate (pyromorphite) are formed.

The extent of secondary enrichment of copper and silver depends among other things on the climate and the rate of erosion. When the groundwater level drops due to slow erosion such enrichment migrates ever-further into the deposit resulting in the leached zone being constantly removed. Particularly spectacular is secondary enrichment in the porphyry copper deposits in the Atacama Desert where the enriched zone is several hundred meters thick. In the current hyperarid climate of the Atacama Desert there is not enough water for further enrichment. This was, however, different in the past (Hartley and Rice 2005). Special features here are water-soluble minerals such as atacamite (Cu_2Cl ($OH)_3$) that have only been preserved thanks to the hyperarid climate. Atacamite is formed by reaction with very salty groundwater (Reich et al. 2008).

There are a few porphyry copper deposits that did not form at a subduction zone but in an extremely thick crust after continental collision (Hou et al. 2011). The most important examples are in Tibet, eastern China, and Iran. In southern Tibet the extremely thickened mountain range diverges laterally marking the beginning of so-called post-orogenic collapse that led to the formation of rifts intersecting the structures of the orogen. Such conditions led to the

emplacement of granite from which the porphyry copper deposits of Gangdise Shan were created. Interestingly, the surrounding rocks are also granites but were formed before the collision with India in an active volcanic arc by subduction of the Tethys. In eastern Tibet there is the Yulong copper belt created in connection with large strike-slip faults and evasive movements of crustal blocks squeezed away sideways by the continued collision with India. In both cases regional upwelling of the asthenosphere seems to have led to melt generation in the lithospheric mantle and lower crust. The necessary water for porphyry copper deposits was probably released by a metamorphic reaction in which hornblende was replaced by pyroxene and garnet. The lower crust is mainly gabbro that had previously accumulated there during subduction of the Tethys (underplating).

Although porphyry copper deposits have been formed throughout the Earth's history, the vast majority originate from the Cenozoic. A particularly old example is the small porphyry Cu–Au deposits in the Abitibi greenstone belt in Canada (see Box 4.27). The reason is they develop near the Earth's surface during rapid uplift and therefore fall victim to erosion relatively quickly. Thus, older ones are more likely to have disappeared.

4.4.1 Porphyry Molybdenum Deposits (Climax Type)

Climax (Colorado, USA) is the most important molybdenum deposit in the world. It resembles a porphyry copper deposit but differs in containing hardly any copper. There are many more porphyry molybdenum deposits in the wider vicinity of Climax (Klemm et al. 2008).

Although many porphyry copper deposits contain a similar amount of molybdenum, it is distributed over a much larger volume of rock. As discussed in the preceding Section, brine eventually separates from the magmatic fluid leading to the enrichment of copper, gold, and other metals in vapor that rises due to its low density. However, molybdenum is enriched in brine, which is of course heavier. If this cools down, then the mineral molybdenite crystallizes. In porphyry copper deposits this occurs in veins that intersect the same rock in which there are copper-rich veins too, although there are naturally zones that are more or less molybdenum rich.

However, in Climax-type deposits copper-rich vapor has disappeared from the system leaving the brine behind and the latter crystallizes quartz and molybdenite as it cools. When a breccia is formed over the porphyry stock during an early surge of magmatic fluids, mineralization is concentrated in a small volume of rock that can result in a high ore grade. Flat, plate-shaped ore bodies with stockwork veins are typical (Box 4.13) within a breccia.

Nevertheless, in Colorado the magma itself seems to be unusually rich in molybdenum. Perhaps ascending basaltic magmas met unusually molybdenum-rich rocks in the lower crust.

4.4.2 Porphyry Gold (Intrusion-Related Gold)

There are also porphyry gold deposits (a.k.a. intrusion-related gold) (Lang and Baker 2001). In addition to gold and bismuth they also contain W, As, Mo, Te, Sb, and Sn but hardly any copper. The most important are the Tintina Gold Belt in Alaska (USA) and Yukon (Canada). Further North American examples are Fort Knox, Donlin Creek, Pogo, Dublin Gulch, True North, and Brewery Creek. Other major examples are Vasilkovskoe (Kazakhstan), Kori Kollo, near Oruro (Bolivia), and Kidston (Queensland, Australia).

Compared with porphyry copper deposits, porphyry gold deposits are significantly smaller and contain only a few sulfides (arsenopyrite, pyrrhotite, pyrite). Instead of a stockwork there is often a crowd of parallel veins. Although the alteration halo is relatively small, pronounced contact metamorphism can be seen in neighboring sediments. In the environment there are often skarns (Sect. 4.9) with W $\pm$ Cu $\pm$ Au or Cu–Bi–Au $\pm$ W. Farther away there are quartz veins (with Au–As–Sb $\pm$ Hg) that are very similar to orogenic quartz veins (Sect. 4.2).

Of course, porphyry gold deposits are formed by magmas and fluids that have a different composition from that of porphyry copper systems. They form at active continental margins (subduction zones) at a greater distance from the volcanic arc in a similar position to the tin belts of the following section. Many are found in an accretionary orogen (i.e., collision of terranes with a subduction zone. Typically they are related to subduction under old continental crust (craton).

The fluids responsible for such deposits are very rich in CO_2 and separation into CO_2-rich steam and a water-rich fluid probably plays an important role in precipitation. According to one model, melt generation occurred relatively deeply in the Earth's mantle triggered by CO_2-rich fluids. Magma rose into the crust, fractionated, and also led to partial melting of the crust.

4.5 Tin–Tungsten Deposits

In the central Andes (especially, in Bolivia) there are a number of large tin deposits (Fig. 4.37). They also contain tungsten, bismuth, and silver. In particular, the Cerro Rico of Potosí (Box 4.17) is famous for its exceptional silver wealth. The tin belt follows the Eastern Cordillera parallel to the subduction zone (Fig. 4.38). The up-to-200-km-wide Altiplano separates it from the active volcanic arc and the

porphyry copper deposits of the copper belt. The deposits show clear similarities to porphyry copper and their associated deposits. Moreover, there are porphyry tin deposits related to tin granites (Sect. 3.7.1) and associated breccias, veins, and greisen (discussed in the following section). Instead of the I-type granites of the copper belt these are S-type granites (i.e., melts derived from sediments).

Box 4.17 The Cerro Rico of Potosí

The Cerro Rico of Potosí (Bolivia) is an unusual deposit in every respect. More silver has been mined there than anywhere else (Figs. 4.35 and 4.36). In the 16th and seventeenth centuries it formed the economic center of the Spanish colonial empire and only a few decades after its foundation Potosí was almost as big as London then the largest city in Europe. Emperor Charles V awarded the city a coat of arms with the inscription: "I am the rich Potosí, treasury of the world, king of the mountains, envied by kings." Even the paving stones lining the route of the Corpus Christi procession were replaced by silver ingots in 1658. Silver was so plentiful that it is said that back in the day even horseshoes were made of it. While the colonial masters indulged in luxury in the city, the indigenous people were forced to work in the mines.

It is said that at times about 1 t of silver came out of the mountain every day. This accounted for 70% of world production. The total quantity produced is estimated at 30,000–60,000 t. Using primitive methods thousands of miners (*mineros*) organized in collectives are still digging shafts and galleries in search of silver in the almost exhausted upper part of the mountain. In the lower part a state enterprise is mining tin using modern methods. The baroque city of Potosí and the mines are one of the most important tourist attractions in Bolivia.

The Cerro Rico is a 13.8-million-year-old dacite lava dome that rose on the edge of the somewhat older Kari-Kari Caldera. Beneath it was a large magma chamber in which zoning consisting of different strongly fractionated melts developed during crystallization. Additional primitive melts were likely injected a number of times into the magma chamber and partly mixed the zones again. The dacite lava dome emerged relatively early from a less fractionated zone. Hydrothermal activity began shortly after the ascent of the lava dome and lasted at least 200,000 years (Rice et al. 2005). The volcanic vent served as an ascent route for fluids released from more strongly fractionated magmas (Dietrich et al. 2000). The result is a kind of hybrid between porphyry tin, polymetallic veins, and high-sulfidation epithermal veins (Sect. 4.3).

Fig. 4.35 The Cerro Rico of Potosí is an unusually rich silver and tin deposit (*photo* © Neils Photography, CC-BY, flickr.com)

Fig. 4.36 Profile of the Cerro Rico of Potosí. The strongly altered dacite lava dome is cut by countless hydrothermal veins with vertical zoning (from Bartos 2000)

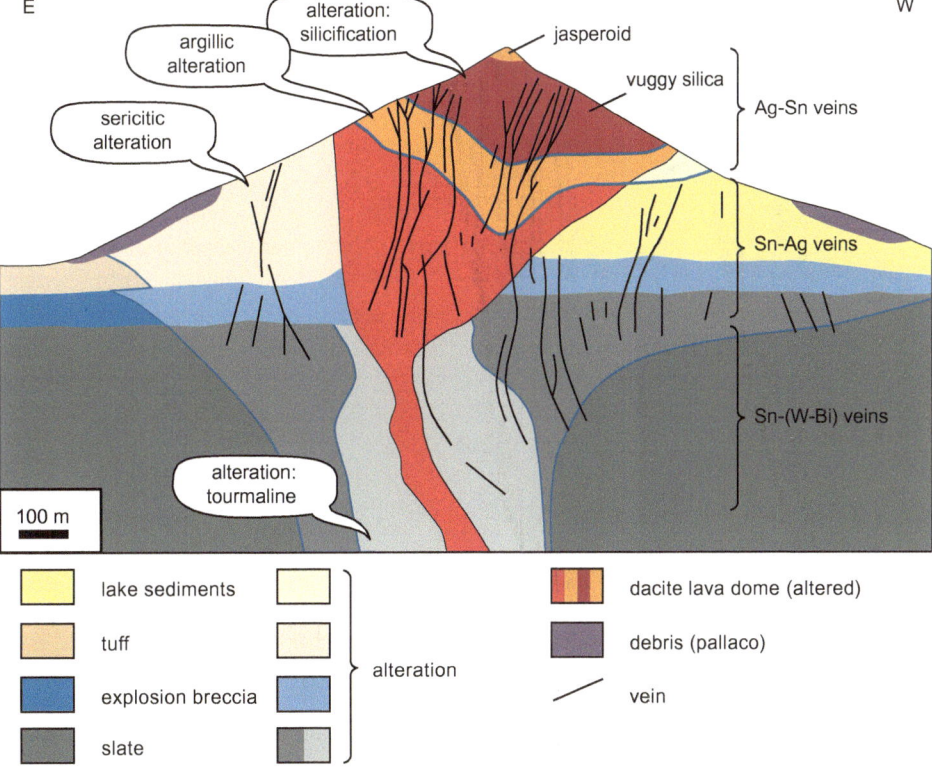

Table 4.5 Selected ore minerals of Cerro Rico

Early stage	+ Quartz, pyrite
Cassiterite	SnO_2
Arsenopyrite	FeAsS
Wolframite	$FeWO_4$
Bismuthite	Bi_2S_3
Intermediate stage	+ Quartz, chalcopyrite, sphalerite
Stannite	Cu_2FeSnS_4
Tetrahedrite	$Cu_{12}Sb_4S_{13}$
Late stage	
Jamesonite	$Pb_4FeSb_6S_{14}$
Boulangerite	$Pb_5Sb_4S_{11}$
Acanthite	Ag_2S
Pyrargyrite	Ag_3SbS_3
Myargyrite	$AgSbS_2$
Stephanite	Ag_5SbS_4
Matildite	$AgBiS_2$
Andorite	$PbAgSb_3S_6$
Oxidation zone	
Native silver	Ag
Chlorargyrite	AgCl

The lava dome was strongly altered by high-sulfidation fluids. Its tip consists of little more than quartz or chalcedony (jasperoid and vuggy silica). Since the fluids had a high boron content, tourmaline was formed in the deep part of the alteration halo. Strong sericitic alteration occurred in the environment.

Ore minerals were precipitated from the hydrothermal fluid in a series of larger veins, countless smaller veinlets, and disseminated in rock pores (Table 4.5). Precipitation began with the crystallization of quartz, pyrite, cassiterite, and arsenopyrite and in deeper sections with wolframite and bismuthite. This was followed by quartz veins with sulfides such as stannite, sphalerite, chalcopyrite, and tetrahedrite. Later mineralizations are increasingly rich in antimony and silver with numerous silver sulfides and sulfosalts (acanthite, pyrargyrite, stephanite, and many others), jamesonite, and boulangerite.

Subsequently, an oxidation zone was formed that in this case led to the formation of large quantities of native silver and chlorargyrite (silver chloride) together with alunite—not to the usual redistribution (Box 4.16). Even the slope debris (a.k.a. pallacos) at the foot of the mountain contains so much tin and silver that mining is worthwhile (Bartos 2000).

Legend has it that the Inca king Huayna Cápac discovered silver in the "beautifully shaped mountain, dyed in different shades of red." However, when the Incas began to dig they heard a thundering voice from the mountain telling them the wealth was not for them, but for those who came "from over there." The Incas took flight and the Spaniards were not long in coming. Contrary to the legend silver mining began around the year 1000 albeit on a small scale. In colonial times silver was mainly extracted by the amalgam method and the mercury required for this process was mined in Huancavelica in Peru. In Potosí about 32,000 t of mercury was released into the atmosphere and rivers. The urban population was exposed to high concentrations and many died as a result of poisoning (Hagan et al. 2011).

In the Central Andes (Fig. 4.38) subduction is strongly compressive, which leads to overthrusting. The strongest movement takes place at the eastern edge of the mountain range where nappes are thrust over the Amazon Basin and the Andes expand to form a mountain range nearly 700 km

Fig. 4.37 Tin and tungsten deposits (porphyry, vein, greisen, skarn, and placer) throughout the world. The largest reserves are located in Southeast Asia, China, the Central Andes, and Brazil (data from World Minerals Project, Geological Survey of Canada)

Fig. 4.38 The tin belt of the Central Andes is located in the Eastern Cordillera. The Altiplano separates it from the active volcanic arc and the copper belt. The oldest deposits (Triassic–Jurassic boundary) are located near La Paz. To the north and south they become increasingly younger. Among the most important deposits are San Rafael (Peru), Oruro, Llallagua, Cerro Rico of Potosí, Tasna, and Chorolque (Bolivia) (from Dietrich et al. 2000; Mlynarczyk and Williams-Jones 2005)

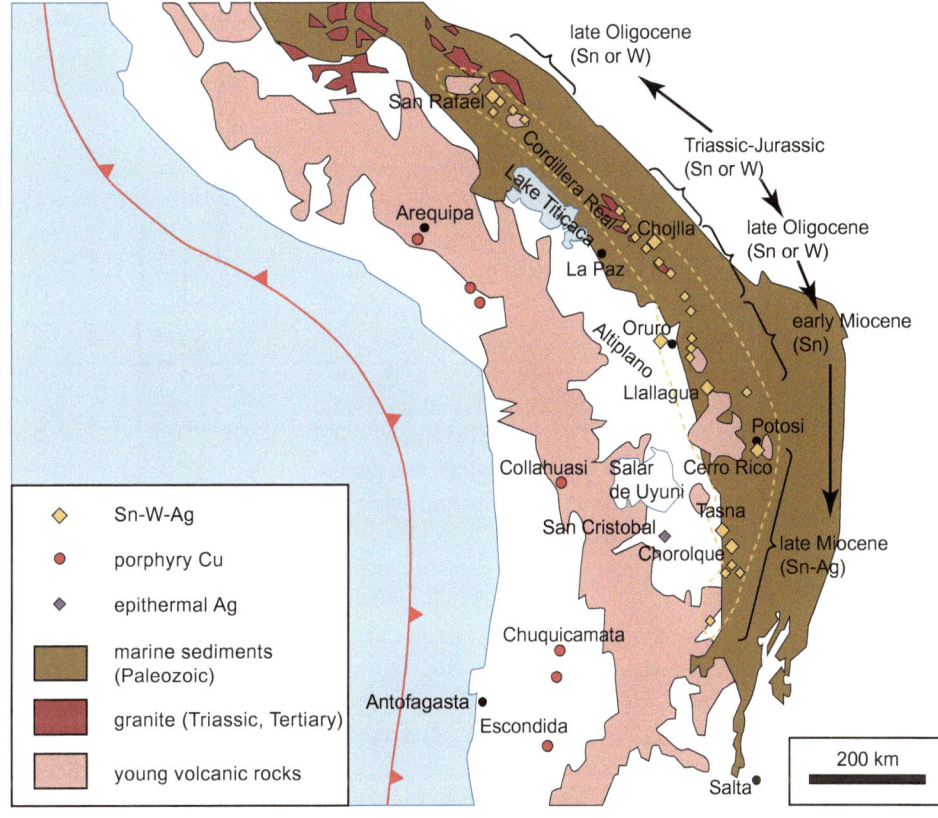

wide. At more than 70 km the continental crust of these mountains is about twice as thick as the neighboring Brazilian shield. The rocks of the Eastern Cordillera are predominantly marine sediments such as shales and corresponding metamorphic rocks such as schist and gneiss that reach down to the lower crust.

Several times these rocks melted to form granite magmas. It is believed this happened when compression was strongest (Mlynarczyk and Williams-Jones 2005) because of flat slab subduction. This not only led to increased thrusting, but temporarily also to relocation of the volcanic arc into the interior of the continent. However, the basaltic magmas got stuck below the continental crust triggering partial melting of crustal rocks.

The first time this happened was in the Mesozoic when granite plutons were formed. Today these plutons form the giant mountains of the Cordillera Real north of La Paz and already contain a few small tin veins. However, most of the deposits originated in the Cenozoic during the course of which the Altiplano was raised. Interestingly, no new deposits were added in the area occupied by older veins but only north (in Peru) and south (from Illimani to Argentina). Activity shifted increasingly to the north and south. In the case of older deposits a deeper level with tin granites is exposed as a result of erosion. These granites and especially superimposed porphyry tin stocks, and breccias are cut by veins. Younger deposits from Cerro Rico de Potosí to northern Argentina are volcanoes that have eroded little. Volcanic vents enabled the rise of hydrothermal fluids from the larger magma chambers below.

Since shales were the starting material for the granites the latter have a very high aluminum content (peraluminous). They are also greatly reduced thanks to the organic components of the shales. This results in an enrichment of tin during fractionation but only if tin is present as Sn^{2+}—not as Sn^{4+}. Moreover, ilmenite crystallizes instead of other minerals such as titanite or magnetite that would incorporate Sn^{4+} and remove it from the melt. In this way tin granites with a very high tin content were produced (Mlynarczyk and Williams-Jones 2005). Since the water content of this granite is relatively low, magmatic fluid is only released when the granite has already largely solidified.

Water rises to the highest parts of the system in much the same way as happens in porphyry copper systems. Precipitation occurs in the cupolas of granite stocks or in volcanic vents and breccias. Tin is mainly soluble as Sn^{2+} in $SnCl_2$, $SnCl^+$, and SnF^+ complexes and precipitation occurs by cooling or by increasing the pH or the oxygen content (especially, by mixing with meteoric water).

The tin and tungsten minerals are disseminated in former pores and in quartz-rich veins within the granite stock and its immediate vicinity. Zoning with a tin-rich zone in the center made up of quartz, cassiterite (Fig. 4.39), and tourmaline is typical. This is surrounded by a tungsten-rich zone made up

Fig. 4.39 Cassiterite from a tin vein east of La Paz (Bolivia) (*photo* © F. Neukirchen)

of quartz, wolframite, and pyrite and at a greater distance a zone made up of various sulfides and sulfosalts of copper, zinc, lead, silver, and antimony. Since the tin granites of the central Andes have a very high boron content, tourmaline forms in the alteration zone.

Porphyry tin deposits from the Llallagua Mine have supplied the largest quantity of tin in the world to date. The Cerro Rico of Potosí (Box 4.17) was long the most important silver mine in the world. San Rafael (Box 4.18) is a large vein system with the highest known ore grade of any tin deposit. Oruro is another important tin district; it also has a porphyry gold deposit (Sect. 4.4.2).

Box 4.18 San Rafael

The richest known tin veins in the world are in San Rafael (Peru) (Mlynarczyk et al. 2003). The average ore grade of 4.7% is uniquely high and has a tin ore content of about 1 Mt making San Rafael one of the largest known tin deposits as well.

It is a vein system in the hinge between two segments of a strike-slip fault. Local extension that took place here also made it possible for a granite stock to rise into the surrounding metasediments. Magmatic water released by the granite in several pulses mixed in the vein system with cool, oxidizing meteoric water. This resulted in strong vertical zoning of tin and copper. Cassiterite crystallized in deep areas of the vein within the granite, while copper ore (chalcopyrite) was deposited on contact with metasediments and within metasediments.

The cassiterite occurs partly as wood tin (Box 4.21) indicating extreme oversaturation in mixed fluids. In deeper areas fine-grained massive cassiterite occurs and at shallower depths needle-shaped cassiterite (a.k.a. needle tin, typical of low temperatures) occurs together with chalcopyrite.

Tin-rich granites can also form by continental collision when shales are transported to great depth—not just in subduction zones as is the case in the Central Andes. Melting to S-type granite magma occurs relatively late when tectonic movements have already come to a standstill and the rocks transported downward slowly heat up (late orogenic). If the crust is subsequently stretched A-type granites (anorogenic) can form that are enriched in elements such as F, Li, and Sn due to their low degree of partial melting; they also form tin deposits. Europe's large tin deposits in the Ore Mountains (Sects. 4.1.2 and 4.6), in Cornwall (Box 4.19), and in Portugal were formed during the Variscan orogeny.

Box 4.19 Cornwall (UK)

Cornwall and Dartmoor in neighboring Devon together form one of the most important and oldest mining districts in Europe (Fig. 4.40). Since the early Bronze Age metals have been mined here. Initially, mainly tin and copper were mined in the form of secondary tin placers and in shallow trenches within the oxidation zone of hydrothermal veins. Rumor has it that the Phoenicians imported tin from Cornwall. At the time of the Roman Empire the region was the most important tin producer in Europe. In the nineteenth century the Industrial Revolution led to a renewed upswing when the demand for metal was very high and steam engines made it possible to mine at ever-greater depths. This led to the production of iron, lead, silver, arsenic, manganese, zinc, and tungsten in addition to tin and copper.

The tin and copper deposits were formed during the Variscan orogeny. In the process some S-type granites rose to form the so-called Cornubian Batholith, which is mostly hidden under sediments. Six large plutons are exposed by erosion one of which is located off-shore near the coast.

Hydrothermal veins are located in the upper part of the dome-shaped granites and in the sediments around them. The composition has strong zoning in which tin-rich veins cut through the granites and their immediate surroundings. They are surrounded by a zone of copper-rich veins followed by a lead–zinc zone. Zoning within the veins is vertical with tin in the deep areas, then predominantly copper above, and finally lead and zinc if this zone has not eroded away. In individual cases zoning is more complicated and in places metals were remobilized by later fluids. An oxidation zone also formed in addition to the original zoning.

Aggressive hydrothermal fluids have partly transformed the roof region of the granite into greisen (Sect. 4.6). Some of the rock has also been altered to kaolinite (Sect. 7.5) mined for the production of ceramics. Tin-rich skarns were also formed as a result of fluids coming into contact with limestone (Sect. 4.9). Within the granites there are also small pegmatite streaks a minority of which also contain ore minerals. To top it all off there are also older deposits in the sediments such as massive sulfides, veins, and manganese-containing chert layers. The Lizard ophiolite complex contains further copper occurrences.

The Southeast Asian tin belt has long been the most important tin region and dates back to the collision of several terranes. It is even more important than the tin belt of the Central Andes. It stretches from the Indonesian tin islands of

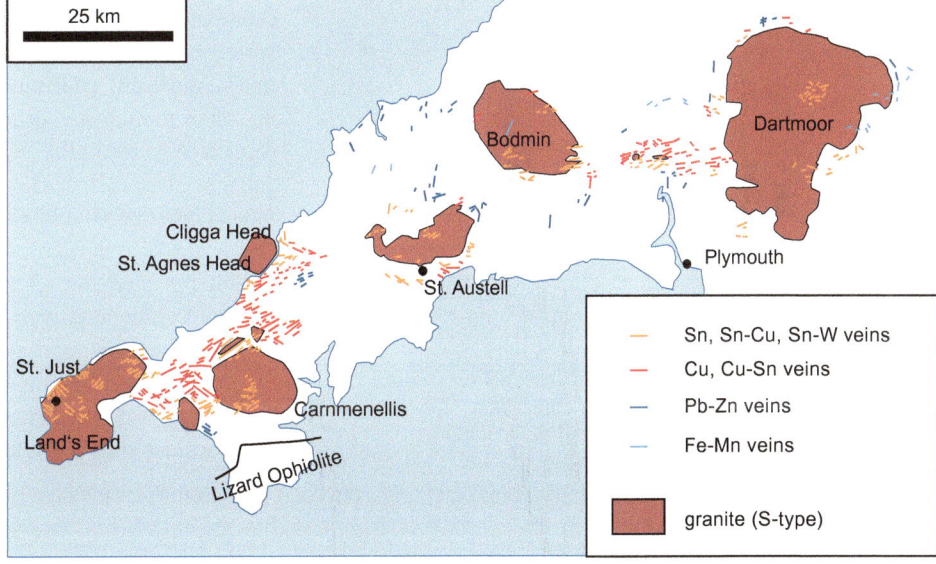

Fig. 4.40 Polymetallic veins in Cornwall and Devon (UK) were formed during the Variscan orogeny by granite fluids. Zoning with tin, copper, and lead–zinc veins occurs with increasing distance from the respective granite pluton (from Dunham et al. 1978)

Belitung (a.k.a. Billiton) and Bangka (northeast of Sumatra) via Malaysia to Thailand. Secondary placer deposits (Sect. 5.9) are mainly mined there. The cassiterite originally comes from tin granites and associated greisen and veins. More than half the historical production of this region is related to the granites of the Main Range in Malaysia (Schwartz et al. 1995).

4.6 Greisen

The fluid released by certain granites can be so acidic and aggressive that it attacks feldspar and even reacts with already solidified parts of a granite pluton or with older granites. Greisen is formed (Fig. 4.41): new rock consisting mainly of quartz together with mica, topaz, and tourmaline. Often cassiterite and wolframite are present as well as fluorite, apatite, scheelite, molybdenite, and hematite. Greisen is an old German mining term. Greisen can be important tin deposits and have significant contents of tungsten and molybdenum. Tin veins with quartz and cassiterite are often found in the surroundings.

An important reaction is the conversion of potassium feldspar to muscovite and quartz:

$$3KAlSi_3O_8 + 2H^+ = KAl_3Si_3O_{10}(OH)_2 + 6SiO_2 + 2K^+$$

Fig. 4.41 Greisen with cassiterite and fluorite from Krupka (a.k.a. Graupen), Ore Mountains (Czech Republic) (*photo* © F. Neukirchen/Mineralogical Collections of the TU Berlin)

Since the fluid often also contains a high amount of lithium, muscovite is converted into a lithium mica called zinnwaldite. Topaz can be produced, for example, by the reaction of plagioclase with HF (hydrofluoric acid). Tourmaline is formed when the fluid contains a lot of boron. The reactions increase the pH of the fluid (neutralization) and make complexes such as $SnCl_2$, $SnCl^+$, and SnF^+ unstable; hence cassiterite is precipitated at the same time.

Related granites are strongly fractionated peraluminous granites (S-type or A-type) (they have already been discussed in the previous section; Bettencourt et al. 2005). Tin, fluorine, boron, lithium, and other incompatible elements can be enriched during crystallization to high concentrations in the residual melt. The residual melt can also solidify into topaz granite or tin granite.

Greisen are formed in the roof region of plutons. They can be large flat bodies in contact with the surrounding rock, smaller lenticular bodies, or vein-shaped bodies. In other cases the entire roof region of the granite is interspersed with more or less altered zones. Cassiterite quartz veins are also often surrounded by a narrow greisen zone. The cassiterite of greisen deposits often occurs as a network of fine-grained metasomatic veinlets called *Zwitter* by German miners. In English they are sometimes called vein stuff or tin stuff.

The Zwitterstock tin deposit in Altenberg (Germany) in the Ore Mountains (see also Sect. 4.1.2) is one of the largest on Earth (Fig. 4.42). During post-orogenic extension after the Variscan orogeny a small A-type granite (outer granite) with a high content of fluorine, tin, and lithium was emplaced here into the tuffs of a caldera volcano. The upper 250 m of this granite are completely transformed into fine-grained greisen with a dense network of *Zwitter* veins. It is interesting to note that another granite (inner granite) was injected into the first one. The inner granite has a pegmatite called Stockscheider in its roof region in contact with the greisen. Some of the feldspars of the pegmatite several centimeters in size were replaced with topaz. This beautiful variety of greisen is called pyknite (specimens can be seen in many museum collections). Since the fifteenth century intensive mining has been carried out in the Altenberg tin deposit. As a result the cavities collapsed in 1620 and a large collapse shaft called Altenberger Pinge emerged at the surface. Until 1990 mining continued without interruption. There are still significant amounts of ore left. In the Ore Mountains there are further greisen deposits in the roof region of different A-type and S-type granites (Breiter 2012). Copper, tungsten, and germanium could be extracted as by-products in addition to tin. Currently, the greisen in Gottesberg, Vogtland, Saxony (Germany) is being explored and considered as a world-class deposit.

(a)

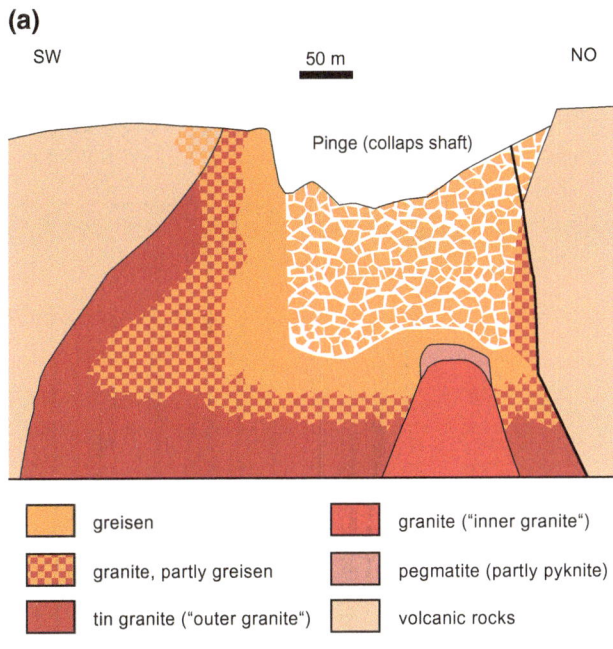

(b)

Fig. 4.42 Altenberg, Ore Mountains (Germany). **a** A-type granite with high tin content that has been injected into older tuffs. The roof region of the granite was transformed into greisen by aggressive magmatic fluids forming fine-grained bands (*Zwitter*) of cassiterite. Later a second granite was injected including a pegmatite in which topaz replaced the feldspar (pyknite). **b** The collapse shaft (Altenberger Pinge) formed in 1620 due to intensive mining and the interior filled with debris from the collapse (**a** from Okrusch and Matthes 2009; **b** © Norbert Kaiser, CC-BY-SA, Wikimedia Commons)

4.7   Iron Oxide Copper Gold Deposits

One of the largest known deposits was coincidentally discovered at Olympic Dam (South Australia) in 1975 (Box 4.20). It had an unusual composition and did not fit into the then established schemata. So a new type of deposit had to be defined called iron oxide copper gold (IOCG)

deposits. Since then further IOCG deposits have been found and some known occurrences have been reinterpreted as such. Other important examples are the Cloncurry district in Queensland (Australia) (including Ernest Henry), the Carajás district in Brazil (with Salobo, Cristallino, Sossego, and Alemão), the Coast Range of northern Chile and southern Peru (with Candelaria–Punta del Cobre, Manto Verde, and others), and Aitik, near Gällivare (Sweden).

Box 4.20 Olympic Dam

Hidden beneath younger sediments more than 300 m thick a huge copper–gold–uranium deposit was discovered at Olympic Dam (South Australia) in 1975. This happened rather accidentally because actually geologists were searching for deposits in the overlying sediments, but nothing was found in them.

The deposit (Haynes et al. 1995; Reynolds 2000; Drummond et al. 2006) consists of breccias cemented with hematite and other ore minerals. Although the deposit is one of the largest iron deposits in the world, the ore grade at 26% iron is not sufficient for economic iron production. The contents of copper, uranium, gold, and silver are of interest economically. As already pointed out Olympic Dam is one of the largest known deposits of each of these metals. Sulfides are disseminated in the cement of breccias. In addition to chalcopyrite there are sulfides with a high copper and low sulfur content such as bornite and chalcocite. Uranium occurs as finely dispersed pitchblende (uraninite or amorphous uranium oxide). Rare-earth elements (REEs) are also enriched but have not yet been extracted.

The breccias originated about 1.5 billion years ago when an A-type granite rose up to just below the Earth's surface. As crystallization continued the magma released several batches of hot brines that rose and fractured the already solidified roof region of the granite. Breccias deemed economically interesting are grouped around a tubular, ore-free hematite–quartz breccia that may be a diatreme (a volcanic vent in which steam explosions occured due to contact with groundwater).

Precipitation occurred mainly when magmatic water mixed with cooler, salty, and strongly oxidized water. Where this came from and how large the respective contribution of the fluids to metals and sulfur was is disputed. According to one theory it was water from a salt lake that may have leached metals from volcanic rocks overlying the granite during infiltration (Haynes et al. 1995). It could also be deep groundwater of originally meteoric origin. Perhaps it was magmatic fluid that was released by more mafic magmas (Reynolds 2000).

Fig. 4.43 IOCG deposits compared with related deposits. IOCG deposits can be caused by magmatic, meteoric, or metamorphic water or mixtures. There may be a smooth transition to Kiruna-type iron deposits. Other deposits shown have certain similarities and partly occur together with IOCG deposits. See text for further explanation. *IOCG*, Iron oxide copper gold

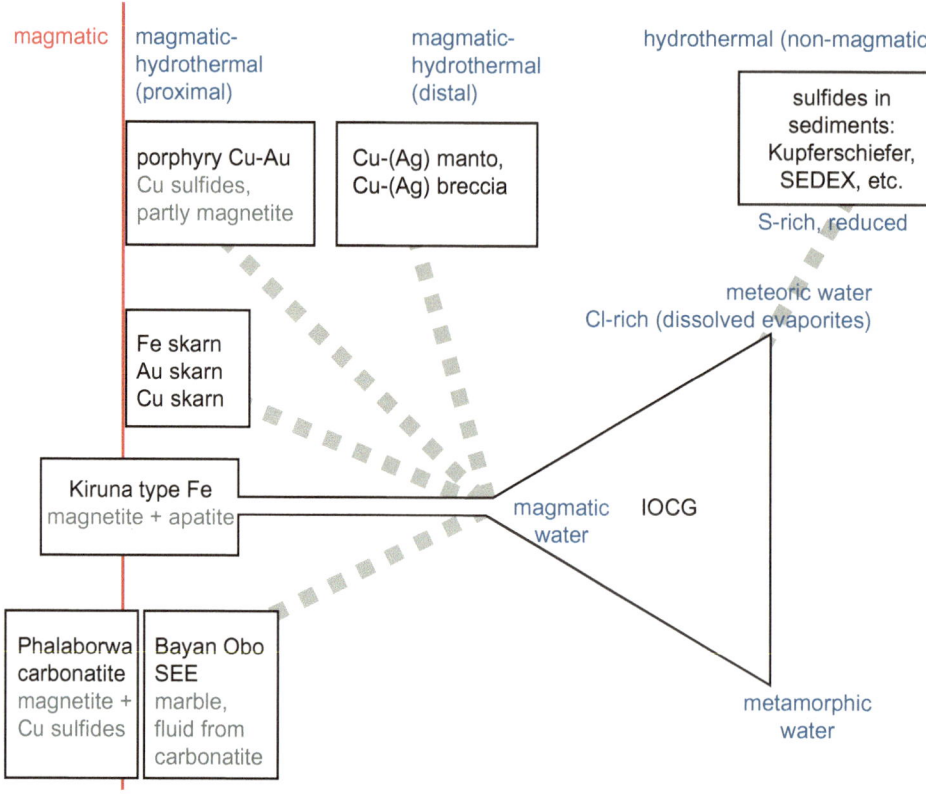

So far those zones with the highest ore grade have been mined underground. There are plans to remove the sediment cover and continue mining in a huge open-pit mine. To extract the high uranium content requires complex processing of ores. Although flotation can separate most non-sulfide minerals, the copper concentrate produced still contains some uranium. Flotation produces 10 million tonnes of radioactive sludge known as tailings per year that are stored on site. Uranium is obtained from the tailings and the copper concentrate by leaching using acids.

There are other IOCG deposits of various sizes in the vicinity of Olympic Dam including Prominent Hill.

Some IOCG deposits at Olympic Dam consist of breccias cemented with ore (breccia-type deposits) where the rock was fractured either by tectonic movement, hydraulically by water, or explosively by magma. In other cases there are veins (forming a stockwork), disseminated ores in former pores, or replacements (e.g., in the form of mantos; Sect. 4.8). They are often located along or at least near major faults.

They contain a lot of magnetite or hematite and remarkably little hydrothermal quartz. The sulfide content is comparatively low but rich in copper (chalcocite, bornite,

chalcopyrite, and a small amount of pyrite) and gold. The sulfides occur in veins, as matrix in breccias, or sometimes as massive sulfides. In addition U, Ag, Co, Ni, As, Mo, light REEs, F, and P may also be enriched.

Liquid inclusions reveal that salty, CO_2-containing, low-sulfur water played a role here. The temperatures determined span a very wide spectrum, so temperature does not seem to play a major role. In many cases there are different fluids that obviously mixed or flowed through the system at different times.

Moreover, individual IOCG deposits differ greatly from each other (Porter 2000; Williams et al. 2005; Pollard 2006; Groves et al. 2010). Consensus has not yet been reached on how the term IOCG should be defined, partly because there are also more or less gradual transitions to other types of deposits (Fig. 4.43) some of which also occur together with IOCG deposits. The way the term is used is analogous to a basket full of different types of fruit.

Although there is often a connection to magmatism, it is at a greater distance than is the case with porphyry copper deposits. Nevertheless, it seems that it was directly and exclusively formed by hot magmatic brine in only a few cases. Often fluids of different origins were involved, although it is not entirely clear how large the respective contributions of salt, metals, and sulfur were. In other cases magmatism does not seem to have played a role at all. For example, surface water or metamorphic water could have dissolved salt from evaporites and leached metals from the Earth's crust. Magmatic heat or metamorphism are not necessarily needed because in Canada there is an example where tectonic movements within sediments were obviously sufficient. This probably caused impermeable layers to leak allowing pressurized pore water from deeper layers to rise along faults and dissolved metals to precipitate (Hunt et al. 2007).

Even within a single mining district there may be IOCG deposits that formed at different times by different fluids. In the Cloncurry district (see also Fig. 4.80) the first deposit (Osborne) was formed during metamorphism and orogeny, while the others (including Ernest Henry, Eloise, and Mount Elliott) followed later when a series of A-type granites were injected (Duncan et al. 2011).

The mixing of different fluids is often assumed to be the most important reason for precipitation, but evolution of the fluid by reaction with the host rock can also play a role. There is generally strong alteration of the surrounding rock, namely sericitic alteration and metasomatic enrichment of iron, sodium, and potassium.

Most large IOCG deposits were formed in the Precambrian in connection with A-type granites (i.e., mantle diapirs rising under the continents or as a result of extension). It could have been important that the lithospheric mantle had previously been metasomatically enriched (Groves et al. 2010).

In contrast, IOCG deposits in the Coast Range of northern Chile and southern Peru were formed in the late Jurassic and early Cretaceous at a subduction zone (Sillitoe 2003). This was still young at that time. Instead of the Andes there were large back-arc basins filled with seawater and sediments. In southern Peru, even within the volcanic arc, extension was so strong that there were only submarine volcanoes (partly with Kuroko-type VMS deposits; Sect. 4.16.3). Continental crust had not yet thickened; accordingly average magmas were less fractionated. At depth a batholith was formed by many stuck plutons. In connection with diorite intrusions and diorite veins several IOCG and Kiruna-type iron deposits were formed (Sect. 3.6), while smaller porphyry copper deposits were simultaneously formed by plutons that were slightly more strongly fractionated. The same fluids also produced copper (gold) skarns (Sect. 4.9) and copper mantos (Sect. 4.8). In the Middle Cretaceous the tectonics changed to compression (triggered by a mantle diapir in the Pacific and the opening of the South Atlantic). According to Oyarzun et al. (2002) this was a particularly important moment for Kiruna-type deposits because large amounts of diorite magma were squeezed upward along faults for a short time.

Kiruna-type deposits occur together with IOCG deposits in northern Chile and Sweden. Moreover, they have a very similar composition (magnetite, apatite, and sometimes copper sulfides in very small amounts) despite the proximity. Some researchers regard them as a low-copper, iron-rich endmember of the IOCG spectrum and speculate whether there is a genetic connection according to which IOCG deposits would be formed at a greater distance from the diorite intrusion.

Gold-rich porphyry copper deposits also have similarities to some IOCG deposits. Early veins of porphyry copper systems also contain magnetite that is later partially replaced (Sillitoe 2010). The biggest differences are therefore the much higher sulfur content and the lack of enrichment of uranium and REEs.

Skarns often occur together with IOCG deposits. Although they may have originated from the same magmatic–hydrothermal fluid, they form at the direct contact of magma with limestone. In northern Chile there are copper (gold) skarns and in the Urals there are iron and gold skarns related to IOCG.

The carbonatite Phalaborwa (Box 3.15) differs from other carbonatites in its high content of magnetite and copper sulfides. Its mineralogy and enriched elements such as REEs, phosphorus, and uranium are strongly reminiscent of IOCG deposits (Groves and Vielreicher 2001). The same applies to the Bayan Obo REE deposit (Box 3.16). Since both carbonatites and A-type granites occur in continental rift systems there may be a remote relationship, but this does not necessarily mean that carbonatites play a role in the formation of IOCG deposits.

Finally, there are similarities and possibly genetic connections between IOCG deposits formed by non-magmatic fluids and sediment-bound deposits such as SEDEX (Sect. 4.17) or copper-bearing shale (e.g., Kupferschiefer; Sect. 5.1.1). In contrast to these deposit types, IOCG deposits are generally formed by fluids that are rich in salt, low in sulfur, and less reduced.

4.8 Chimneys and Mantos

Sometimes hydrothermal fluids can move relatively undisturbed along faults, breccias, or older magmatic dikes until they encounter a porous or reactive rock layer such as limestone or certain volcanic rocks in which precipitation occurs. Depending on the rock this happens in the pores or is accompanied by dissolution of the rock. Limestone can even be completely displaced by massive sulfides. Water flow determines whether ore bodies take on the shape of a chimney cutting through a layer or of a mantle lying on the layer boundary; hence they are called chimneys or mantos. The name of these special strata-bound ore bodies refers to the shape regardless of the metals they contain or the type of hydrothermal system. Such deposits are mostly associated with other deposits, particularly those in the vicinity of

Table 4.6 Selection of typical skarn minerals

Garnet	
Grossular	$Ca_3Al_2Si_3O_{12}$
Andradite	$Ca_3Fe^{3+}{}_2Al_2Si_3O_{12}$
Spessartine	$Mn_3Al_2Si_3O_{12}$
Pyroxene	
Diopside	$CaMgSi_2O_6$
Hedenbergite	$CaFeSi_2O_6$
Amphibole	
Tremolite	$Ca_2Mg_5Si_8O_{22}(OH)_2$
Ferro-actinolite	$Ca_2Fe_5Si_8O_{22}(OH)_2$
Hornblende	$(Na,K)_{0-1}Ca_2(Fe,Mg)_4Al_2Si_7O_{22}(OH)_2$
Wollastonite	$CaSiO_3$
Epidote	$Ca_2FeAl_2Si_3O_{12}(OH)$
Vesuvianite	$Ca_{10}(Mg,Fe,Mn)_2Al_4Si_9O_{34}(OH)_4$
Ilvaite	$CaFe^{2+}{}_2Fe^{3+}Si_2O_8(OH)$

porphyry copper (Sect. 4.4) and skarns. Due to their high ore grade they can be the most important part of a system economically.

Cu mantos and Pb–Ag–Zn mantos are the most common. For example, copper mantos with predominantly copper sulfides occur frequently in the Coast Range of northern Chile. Lead-zinc mantos often contain silver, they are also called Leadville-type mineralization (LTM), they contain galena and sphalerite. Important examples can be found in Mexico and near Leadville (Colorado, USA). There are tin mantos with cassiterite in Bolivia. Manto Verde in northern Chile is a manto-shaped IOCG deposit (Sect. 4.7) with magnetite and copper sulfides.

4.9 Skarn

Limestone can react strongly with hydrothermal water. This is particularly true when a granite intrusion is emplaced into limestone and heats the limestone up while the magmatic fluids released are simultaneously flowing through it. There is an exchange of substances between water and rock called metasomatism. Metasomatism thus differs from metamorphism. The latter is a transformation with almost no change in chemical composition.

Metasomatism occurs when very different rocks meet. For example, when peridotite contacts granite a so-called blackwall of dark minerals characteristically forms. Although diffusion is sufficient for this, mass transport through mobile water is much more effective. When reactive hydrothermal fluids meet certain rocks effective mass transfer can occur. Metasomatic deposits include skarns, greisen (Sect. 4.6), Carlin-type deposits (Sect. 4.11), and other replacement deposits.

When granite intrudes limestone the result is a completely new rock consisting mainly of various calcium silicate minerals. Such rock is very hard and called skarn for an old Swedish mining term. Typical skarn minerals (Table 4.6) are garnets of different composition such as grossular, andradite, and spessartine. Other typical skarn minerals include pyroxene (diopside, hedenbergite), amphibole (tremolite, actinolite), wollastonite, vesuvianite, and epidote. Some are intensely green, others red, orange, or brown. The rock itself can be correspondingly colorful and may also contain ore minerals. There are hundreds of skarn deposits worldwide (Meinert et al. 2005) some of which have very high ore grades. Fe, Cu, Au, Zn–Pb, Mo, Sn, and W skarns are distinguished according to metals of interest economically. REEs, U, and other elements are also enriched. Wollastonite (Sect. 7.7) is also mined as an industrial mineral.

Skarns can be very different in appearance and composition (Ciobanu and Cook 2004; Meinert et al. 2005; Meinert 2009). They are also very heterogeneous internally with zones of different composition, different textures, and grain sizes. Inhomogeneities in the rock and the geometry of fissures have a strong effect on water flow and thus on the reaction. There are sharp or diffuse reaction seams along fissures and fine-grained or coarse-grained areas with striped or dotted rocks or needle-like minerals arranged like bundles of sheaves.

The exact composition and zoning depends on many factors. Most important is fluid composition and thus the type of magma. However, carbonate rock can be very different. Limestone never consists exclusively of calcite since dolomite, quartz, clay minerals, and other minerals may also be present in varying quantities affecting the result. Any organic carbon present has a reducing effect and thus controls, for example, whether minerals are more likely to be

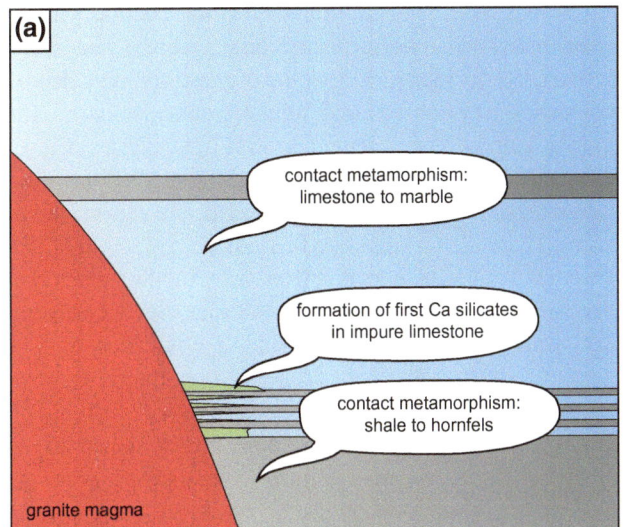

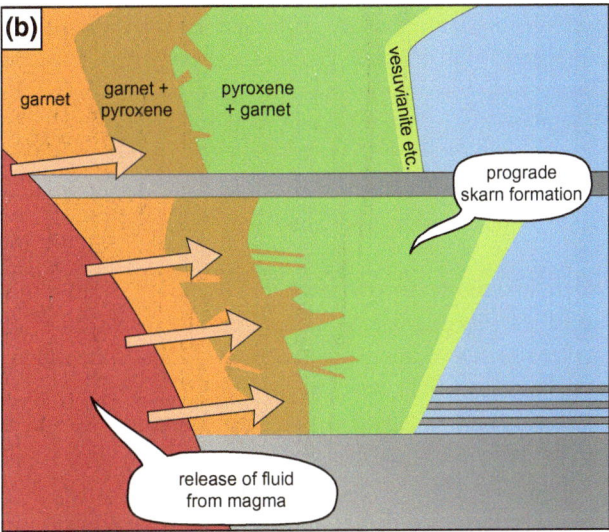

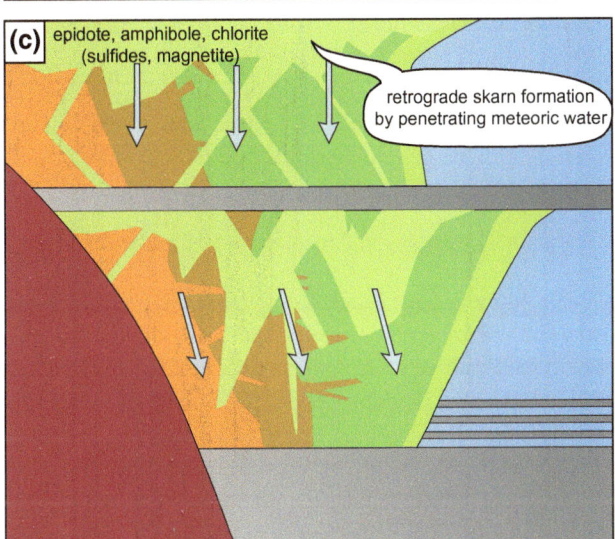

◀ **Fig. 4.44** Skarn formation generally takes place in three steps. **a** Sediments are transformed into contact metamorphic rocks when magma is emplaced. **b** Water released by magma provides SiO_2, iron, and so on. Water reacts with carbonates that are replaced by various calcium silicates (prograde skarn formation). The result is zoning with different minerals that changes again with further cooling of the system. **c** Meteoric water penetrates the system and the silicates previously formed at high temperatures are partially replaced by other silicates (retrograde skarn formation). Precipitation of sulfides and magnetite occurs

Fig. 4.45 Initial skarn formation at the margin of the Adamello Intrusion in the Southern Alps (Italy). Magmatic fluids penetrating the limestone along cracks reacted with it forming grossular garnet (red) and diopside (green) (*photo* © F. Neukirchen)

formed with Fe^{3+} (andradite) or with Fe^{2+}. On the other hand, if the rock is dolomite rather than limestone, then Mg skarn with a completely different mineralogy is formed (e.g., with talc and serpentine—often with a lot of magnetite). Finally, the depth at which the reaction takes place is also important. In deeply formed skarns the temperature is already high (even without magma) and sediments have already transformed into metamorphic rocks such as marble and slate. Skarns formed here are smaller and tend to be vertical in orientation.

Skarn formation generally takes place in several stages (Figs. 4.44 and 4.45). Initially (i.e., before magmatic water is released) the heat can cause contact metamorphism during which no substances are exchanged (isochemical). Relatively pure limestone is turned into marble, interlayered shales are turned into hornfels bands, and impure limestone produces the first Ca silicates. This releases CO_2 and (from the clay minerals) water.

With the influx of hot, often salty, and acidic water released from the magma there follows the main phase of metasomatism called prograde skarn formation. Substances transported in water such as SiO_2, iron, and copper react with the limestone or marble and the carbonate minerals are completely replaced by calcium-rich silicates. Typical zoning is predominantly garnet on contact with the pluton, farther away it is increasingly pyroxene, and still yet farther at the transition to unaltered marble it is vesuvianite or wollastonite. There are also zonings in the composition of garnet and pyroxene.

Water released by magma as it cools becomes ever cooler and changes its composition. Early minerals are partially replaced by later minerals such as magnetite, actinolite, and first sulfides.

The more the system cools down, the greater the amount of meteoric water penetrating the skarn and transforming it yet again at least along fissures. Retrograde skarn formation involves former silicates being partially replaced by other silicates such as epidote, amphibole, plagioclase, and chlorite whereby older zoning is merely overstamped. Precipitation of ore minerals such as sulfides and magnetite is intensified in this step. These too are zoned. Typical zoning is pyrite, chalcopyrite, and magnetite near the pluton, higher content of chalcopyrite and bornite in the center, and bornite, galena, and sphalerite at the margin.

The transport of CO_2, Ca^{2+}, and water in the opposite direction can also occur resulting in the edge of the granite being then transformed into a so-called endoskarn. However, deposits are rarely formed. Skarns formed by the transformation of carbonate rocks are correspondingly referred to as exoskarn. Farther away from the pluton mantos sometimes appear in marble (Sect. 4.8).

Which metals are enriched in the skarn depends first and foremost on the metal content of magma at the time fluid is released. Important factors are the kind of magmatic fractionation (gabbro, diorite, granite), the type of granite (I-type, S-type, A-type), and the oxygen content. The very common **copper skarns** are formed in connection with the I-type granites of subduction zones (especially, in the vicinity of porphyry copper deposits; Sect. 4.4). Ore minerals such as chalcopyrite, digenite, bornite, enargite, and tennantite–tetrahedrite are found. At a greater distance there can be related **lead–zinc skarns** with galena and sphalerite. Bingham (Utah, USA) is home to one of the largest porphyry copper deposits in the world and one of the largest copper skarns. Probably the largest copper–zinc skarn system is the Antamina Mine (Peru) where molybdenum and silver are also produced as by-products.

Tungsten skarns are the most important tungsten deposits. In addition to the main ore mineral scheelite ($CaWO_4$) there are also various sulfides such as pyrrhotite, molybdenite, chalcopyrite, sphalerite, and arsenopyrite. The most important W skarns are located in the west of North America, in Australia, and in East Asia. Tin likewise can be enriched in skarns.

Iron skarns tend to be formed in connection with less fractionated magmas such as diorite, gabbro, and syenite (especially, on island arcs). Some of these are huge deposits containing several hundred million tonnes of iron. They consist largely of magnetite and iron-rich silicates. Cu, Co, Ni, and Au may also be present in smaller quantities. Endoskarns can make up a good part of the deposit in such cases. There are several large Fe skarns in the southern Urals such as at Magnitogorsk (Russia). Another giant skarn deposit is Sarbai-Sokolov (Kazakhstan). In former times iron skarn was mined on the island of Elba (Italy).

Gold skarns are also of economic importance. However, the term does not refer to a particular type of skarn but to a large number of skarns with different compositions and geneses. They are most frequently associated with only slightly fractionated but reduced magmas such as diorite and granodiorite. Others arose in connection with gold-rich porphyry systems. Finally, gold skarns may also occur in connection with orogenic gold veins (Sect. 4.2). Canada and China are particularly rich in gold skarns. Navachab (Namibia) has to date been the most important.

4.10 Metasomatic Siderite Deposits

About half the Erzberg Mountain in Styria (Austria) is made up of the iron carbonate siderite together with other carbonates such as dolomite, ankerite, magnesite, and calcite. It also contains some manganese, which is very desirable in steel production. About 1 to 2 million tonnes of ore are mined annually in Central Europe's largest open-pit mine. The neighboring town is consequently called Eisenerz ("iron ore"). However, siderite has a lower iron content than hematite or magnetite.

Ores within normal carbonate rocks form ore bodies up to 70 m thick that have a length and width of 1–3 km. They were most likely formed by metasomatic replacement in which limestone and dolomite reacted as a result of iron-rich water penetrating:

$$CaCO_3 + Fe^{3+} \rightarrow FeCO_3 + Ca^{2+}$$

The permeability of rocks controlled where metasomatism occurred. However, it is not clear when this happened or where the fluids came from (Laube et al. 1995; Schulz et al. 1997; Pohl and Belocky 1999). This may have happened shortly after deposition of the limestones (late diagenetic), perhaps later (epigenetic), but most probably before formation of the Variscan Mountains. It could have been hydrothermal fluids from the sediment basin (e.g., formed during the diagenesis of other sediments) and these fluids probably rose through the volcanic rocks below the carbonates. However, magmatic fluids have also been proposed. During the

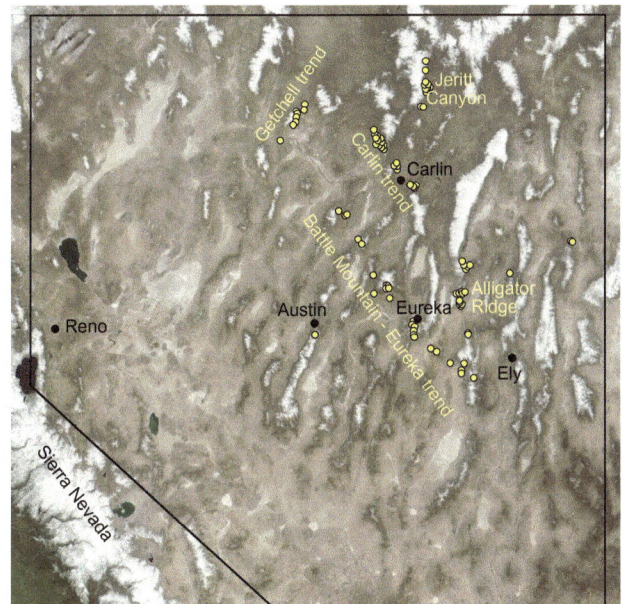

Fig. 4.46 Almost all Carlin-type gold deposits are located in Nevada (USA) on faults of the Basin and Range Province that consists of parallel rifts and horsts. The satellite image shows snow-covered horsts clearly standing out from rifts (map from Cline et al. 2005; satellite image © NASA)

Variscan and Alpine orogenies there was partial recrystallization, which makes interpretation difficult. Ore deposits were also folded. Erzberg is considered to be the largest known siderite deposit in the world. There are other metasomatic siderite and magnesite deposits worldwide such as those in northern Spain, Tunisia, and Algeria.

4.11 Carlin-Type Gold

Carlin-type gold deposits (Cline et al. 2005) are replacements found in limestone. The ore consists mainly of quartz, jasper (microcrystalline quartz), and pyrite in which hydrothermal fluid has completely dissolved the limestone and removed Ca^{2+} and CO_2. Fine-grained, arsenic-bearing pyrite of the main stage of ore formation contains gold both as tiny invisible inclusions of native gold and in the form of Au^+ embedded in the crystal lattice. These are large deposits containing hundreds or even thousands of tonnes of gold. Almost all (Fig. 4.46) are located in the northern half of Nevada (USA). Together they form the second largest known gold occurrence after the Witwatersrand (Box 5.18). They were created in an amazingly short period of time in the Eocene. Their geological history clearly had to have been unique for such gold wealth to emerge. This type of deposit is traditionally regarded as a formation in the back-arc of a subduction zone, but this is probably a gross simplification.

Such deposits are generally tied to normal faults that allowed hydrothermal water to rise. The limestone was mainly displaced where overlying impermeable layers prevented the water from rising further. These traps (located at a depth of a few kilometers) date back to early orogeny in the Devonian and Carboniferous affecting sediments lying on a passive continental margin of an Archean craton. They were thrust as individual nappes over each other and a nappe of impermeable clastic sediments slid over the limestones.

Further mountain building (Laramide orogeny) began in the late Cretaceous when the Farallon Plate, subducted off the west coast at that time, folded upward and was therefore pushed as a flat slab under the continent for hundreds of kilometers. This brought an end to the volcanic arc in today's Sierra Nevada and brought about compression resulting in the uplift of the Rocky Mountains and the Colorado Plateau.

The deposits were formed in the Eocene when the Farallon Plate peeled off from the base of the continental crust and sank into the asthenosphere (Fig. 4.47). Hot asthenosphere flowed in and triggered strong bimodal volcanism in which both mantle and crustal melts erupted. At the same time there was strong extension. Part of the originally larger Colorado Plateau was pulled apart. Today this covers more than twice the area (post-orogenic collapse) forming a region with countless parallel rift valleys and uplifted horst blocks called the Basin and Range Province. In the process much older normal faults were once again set in motion. Peeling took place from the northwest corner of the region toward the southeast. Volcanism and extension become ever younger in this direction. This also applies to Carlin-type deposits that were formed 42 to 36 million years before present.

The normal faults of rift systems served as ascent pathways for hydrothermal fluid. Unfortunately, it is not known exactly where the water and gold it carried came from. The isotopes sometimes indicate meteoric, magmatic, or metamorphic water. Gold and sulfur could come from magma or from sediments or corresponding metamorphic rocks. In any case the water was not very hot (about 180–240 °C) during precipitation. It contained CO_2 and a lot of H_2S but little salt. The iron contained in the pyrite comes largely from the limestone itself, which contains iron-bearing dolomite.

Carbonic acid was formed when the water containing CO_2 cooled down. Carbonic acid was able to dissolve the calcite of the limestone and in so doing increased the water flow. Jasper or quartz were precipitated in the cavities. In some cases large cavities even collapsed and fragments were then cemented into a breccia. Silicates contained in the limestone (sand, silt, clay) were strongly altered (advanced argillic alteration). Pyrite was formed by reacting with iron-containing carbonates releasing CO_2 and consuming

Fig. 4.47 Carlin-type deposits in Nevada were formed in the Eocene. When the previous flat slab of the subducted Farallon Plate peeled off it triggered short-term magmatism. Parts of the Colorado Plateau were affected by severe extension and the rift valleys of the Basin and Range Province were formed. Hydrothermal fluids rose along normal faults. In traps under impermeable layers limestone was completely displaced by SiO_2 and gold-bearing pyrite

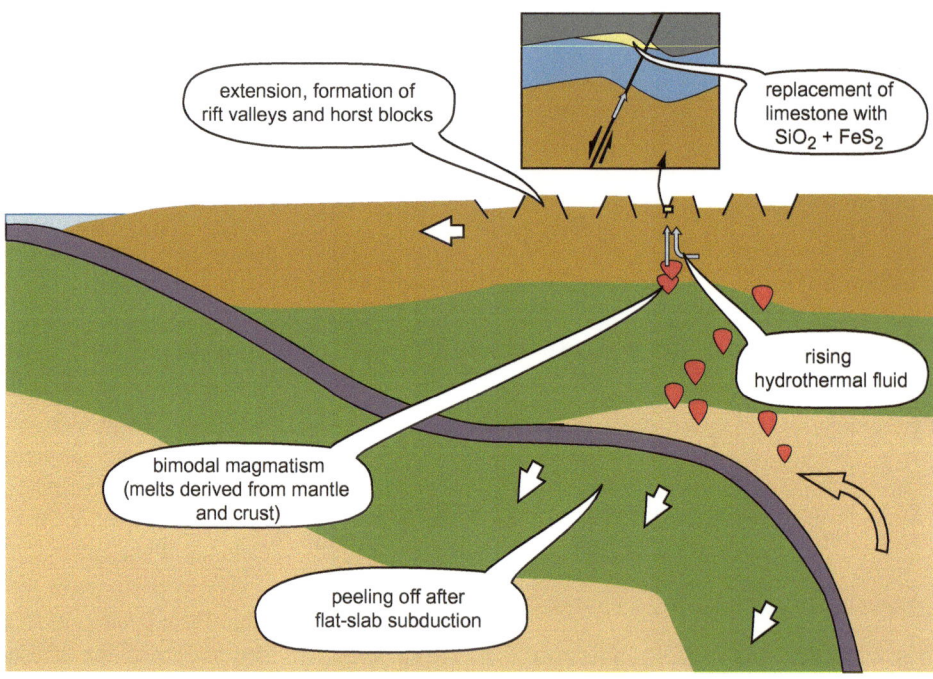

H_2S. This in turn led to destabilization of gold complexes and therefore to effective precipitation of the gold. Further elements such as antimony, mercury, thallium, tellurium, and copper were enriched. Later minerals such as orpiment, realgar, and stibnite formed. Baryte and fluorite can also occur.

A new model (Muntean et al. 2011) has been proposed to explain the origin of the fluid. This argues that the change to flat slab subduction brought about metasomatic enrichment of the lithospheric mantle with water, CO_2, and elements such as Au, As, Tl, Te, Cu, and Sb. As the Farallon Plate peeled off it was partially melted to correspondingly enriched basalts that rose into the lower crust. The lower crust in turn was partially melted generating water-rich, gold-bearing magmas that could rise further. The formation of copper sulfide during fractional crystallization removed most of the copper from the melt before water saturation occurred. Finally, magmatic water was released and during further ascent a brine separated. The gold was enriched in steam while iron was removed from it. On yet further ascent the steam mixed with hot pore water (meteoric water that had already leached out substances from the rock). Additional sulfur and gold may have been mobilized from slates. However, according to other authors (Large et al. 2011) the gold comes from black shales whose diagenetically formed sulfides were recrystallized during metamorphism and were leached out by hydrothermal water.

Further deposits classified as Carlin-type deposits are Mesel (Sulawesi, Indonesia), Lucky Hill (Sarawak, Malaysia), several deposits in China (especially, in Guizhou; Zhuang et al. 1999), the Suzdal Trend (Kazakhstan;

Kovalev et al. 2012), and Allchar (Macedonia). They differ greatly in the mineralized host rock (carbonates, black shale, turbidites, volcanic rocks), the precipitation reaction (e.g., reduction by organic substances), the location (strata bound or at faults), the mineralogy (e.g., arsenopyrite instead of pyrite; different contents of Hg, Te, etc.), and even in the geodynamic setting (back-arc, island arc, passive continental margin). Hg–As–Sb deposits similar to Carlin-type deposits are mined for mercury in Ukraine.

4.12 Mississippi Valley Type

In carbonate rocks such as dolomite and limestone there are significant zinc–lead deposits that often also contain silver. They are referred to as Mississippi Valley type (MVT) deposits named for the large deposits in the United States (Leach et al. 2010; Figs. 4.48 and 4.49). There are also important deposits in Europe such as those in Upper Silesia (Poland), in the Alps—e.g., Bleiberg ("lead mountain") in Austria—and in Ireland. The deposits are epigenetic (i.e., younger than carbonate rocks). They are the result of fluid flow that has on occasion traveled several hundred kilometers in sediment basins. In many cases this happened in a foreland basin of a mountain range. However, there are also examples that were not created by compression but by extension in a rift system. There is generally no connection with magmatism.

Regional names such as Irish type (IRT), Alpine type (APT), Bleiberg type, and Silesian type are used in addition to Mississippi Valley type (MVT). This is particularly true of

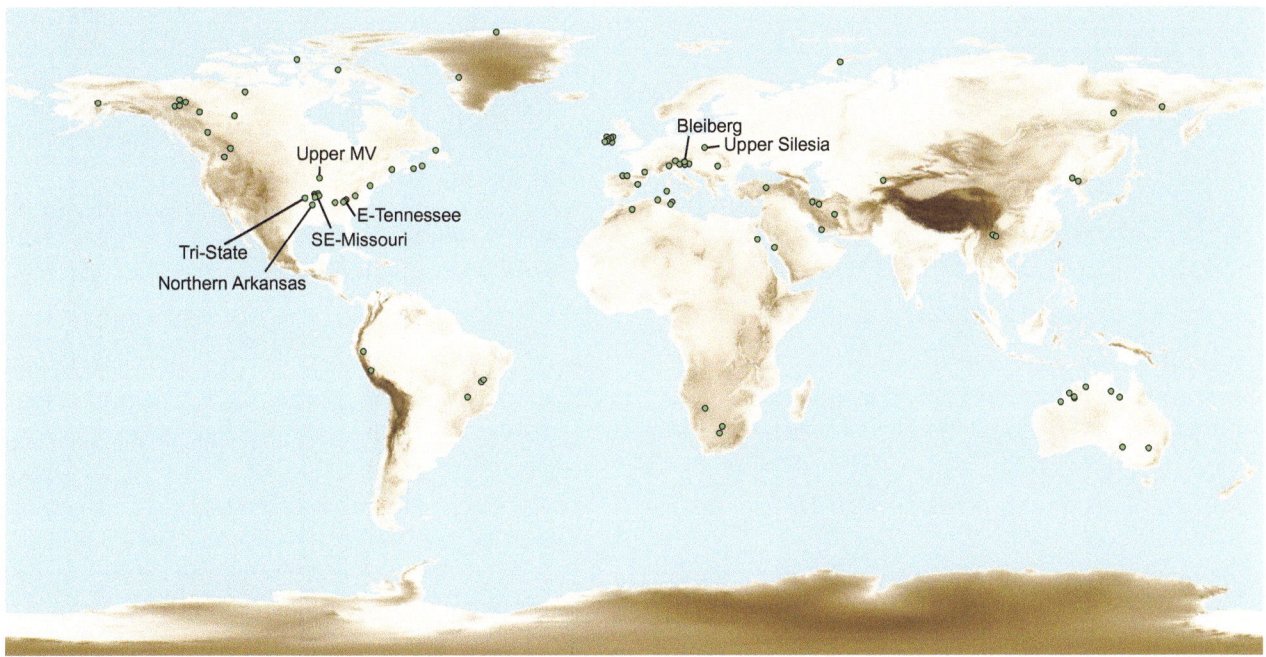

Fig. 4.48 Map of Mississippi Valley–type (MVT) deposits. The most important are located in the wide basin of the Mississippi Valley (USA) (data from Leach et al. 2001, 2010; World Minerals Project, Geological Survey of Canada)

Fig. 4.49 Ore from the Mississippi Valley type deposit at Nanisivik (Baffin Island, Canada) containing sphalerite (dark gray), pyrite (yellowish), and dolomite (white) (*photo* © Mike Beauregard, CC-BY, flickr.com)

the older literature. However, it makes more sense to group them under a single term as MVT.

MVT deposits contain sphalerite and galena almost exclusively. Minor iron sulfides (pyrite and marcasite), carbonates (dolomite, siderite, ankerite, calcite), and baryte are also present. The average ore grade is 6% Zn, 1.9% Pb, and 32.5 g/t Ag. There are also examples where lead outweighs zinc. Some copper minerals are also present in small quantities. Indium, germanium, gallium, and cadmium are also extracted in some cases. The sulfides are coarse-grained to fine-grained and can be massive or disseminated. Colloform

sphalerite (Box 4.21) is typical. Ores are predominantly replacements of carbonate rock. Sometimes existing cavities such as fissures or even cave systems are also filled. Although individual deposits are comparatively small, often a large number occur in a small area. The Upper Mississippi Valley mining district, for example, comprises around 400 individual deposits of 0.2–3 Mt ore each.

Box 4.21 Colloform aggregates

A number of ore minerals frequently occur in the form of fine-grained colloform (or botryoidal) aggregates whose shape is best described as kidney like or cauliflower like. Comprising lumps or coatings that are often internally banded they are usually characterized by a very shiny or even reflective surface. They typically cover areas of a few square centimeters or decimeters. In exceptional cases they have even be traced over hundreds of meters.

Kidney ores are mainly found in iron-rich hydrothermal veins. Such ores consist of colloform hematite (Fe_2O_3) (Fig. 4.50) (a.k.a. kidney iron ore, fibrous red iron ore, or wood hematite) and colloform goethite (FeOOH) (a.k.a. fibrous brown iron ore). Colloform manganese oxides such as pyrolusite and psilomelan are also called black manganese or fibrous manganese oxide.

Sphalerite often occurs as colloform banded aggregates (Fig. 4.51) together with wurtzite (also

Fig. 4.50 Hematite (kidney ore) from Bad Lauterberg, Harz (Germany) (*photo* © F. Neukirchen/Mineralogical Collections of the TU Berlin)

Fig. 4.51 Collophorm sphalerite and wurtzite from Altenberg, near Aachen (Germany) (*photo* © F. Neukirchen/Mineralogical Collections of the TU Berlin)

ZnS) and galena. They are named colloform or botryoidal sphalerite, botryoidal blende, fibrous sphalerite, or gel sphalerite. Colloform sphalerite is particularly common in MVT deposits.

Wood tin is a brown colloform aggregate of cassiterite that is common in tin veins of Bolivia.

Uranium oxide (uraninite or amorphous uranium oxide) typically forms pitch-black coatings called pitchblende in uranium-bearing veins. Pyrite (gel pyrite) and many other minerals also occur as colloform aggregates.

It used to be believed that small crystals grew in a colloidal suspension, were transported as such, and this suspension developed into a viscous gel that eventually dried out.

Today we know that they are formed by extreme oversaturation of a solution (Roedder 1968). This suddenly leads to mass nucleation of mineral grains that form so quickly that diffusion in the solution cannot keep pace and growth of the grains slows down. Such an imbalance results in a dynamic, self-organizing sorting process that leads to fine banding of the ore. Extreme oversaturation can be triggered by contrasting fluids mixing (Barrie et al. 2009) and by the activity of sulfate-reducing bacteria (Pfaff et al. 2011).

Rocks of different permeability play an important role in focusing water flow and the place of precipitation. Particularly important are the facies boundaries of a former carbonate platform between the reef slope, reef, and lagoon that are built up of rocks of different permeability. In some cases the ores are in solution cavities and collapse structures. Some of these are former karst systems; in other cases they were probably formed shortly before or during ore formation by hydrothermal fluids. Veins in the hinges between faults are typical (particularly, normal faults or strike-slip faults). Deposits can also form under a superimposed shale layer. MVT deposits in Tunisia are located above salt domes in structures that function much like an oil trap. All deposits of a mining district are generally located within the same rock stratum (strata bound), while individual deposits usually cut through the corresponding layer. However, there are also stratiform ore bodies.

Hydrothermal water is typically a brine (Na–Ca–Cl with 10–30% salt content) in which metals are transported as chloride complexes. The Cl/Br ratio of liquid inclusions often indicates that it is a brine coming directly from a partially evaporated sea basin—not from evaporites dissolved on the way. The temperature of the fluid can be very different (between 75 and 200 °C). Many researchers point out that the composition of the fluid is similar to that of brines known from oil fields. Such brines are typical of sediment basins. Liquid inclusions showed that different fluids arrived one after the other in the Tri-State and Northern Arkansas districts. Initially, a brine with low metal content precipitated dolomite, then a brine with high metal content precipitated the ores, and then another brine with low metal content precipitated calcite. Such fluids clearly originated from different areas of the sediment basin (Stoffell et al. 2008).

The trigger for precipitation is often assumed to be mixing with another fluid of contrasting composition such as the connate water of sediments (Wilkinson et al. 2005; Stoffell et al. 2008; Pfaff et al. 2010). This leads to sudden changes in pH, temperature, and oxygen content. The sulfur presumably comes predominantly from the sulfate of connate water, which is reduced during or before mixing. Perhaps in some cases precipitation was triggered instead by the dissolution of carbonates and the associated change in pH.

In some deposits there are indications that sulfate-reducing bacteria have played a more or less important role (Pfaff et al. 2011). This particularly applies to Irish deposits that were probably formed by mixing with connate water at shallow depths (Wilkinson et al. 2005). In other cases sulfate was probably reduced to a large extent by reaction with organic matter. It is even conceivable that kerogen contained in limestone (cf. Sect. 6.2) was heated by hydrothermal water and gaseous CH_4 was formed in turn reducing sulfate to H_2S (Anderson 2008). As happens with gas fields a mixture of CH_4 and H_2S could accumulate in a trap. If hydrothermal, metal-rich water hits the gas, then it becomes oversaturated and precipitates. If the gas is CH_4 rich, then sulfate from the water is reduced in a process that is relatively slow and leads to slow growth of coarse-grained sphalerite. If the gas is already H_2S-rich, then the water is immediately highly oversaturated and fine-grained sphalerite is formed. This model could explain stratiform MVT deposits in structures resembling gas traps.

Unfortunately, it is not easy to date ores and the results are not always clear (Leach et al. 2001, 2002, 2010; Kesler and Carrigan 2002; Muchez and Heijlen 2003; Kesler et al. 2004; Bradley et al. 2004). According to Leach et al. (2001) almost all MVT deposits were formed in two periods corresponding to important phases of mountain building. Most of them originated between the Devonian and the Permian when the supercontinent Pangaea was formed by the Variscan (Europe) and Acadian (North America) orogenies. The second period consisted of the Cretaceous to the Neogene when deposits were formed in connection with the Alpine (e.g., Upper Silesia in the foothills of the Carpathians) or the Laramide orogeny of the Canadian Rocky Mountains. It is astonishing how few MVTs were formed in the Precambrian. This may well be due to the lower sulfate content of seawater or to the fact that carbonate rocks of that time were made up of stromatolites and were less permeable to water.

Obviously, it does not matter whether mountains were created by continental collision, by strike-slip faults with concurrent compression, by subduction similar to the Andes, or by an island arc colliding with a passive continental margin. However, it does matter that corresponding continental margins spent enough time in a hot climate such that reefs could build up a thick carbonate platform before mountain building commenced.

MVT deposits were generally formed in the late phase of orogeny on the outer edge of the foreland basin in which material eroded from the mountain range was deposited. Water in the sediment basin was presumably set in motion by hydraulic potential built up by the topography of the mountains and the rainfall there. This caused water in the foreland basin to flow from the mountains to the foreland.

Foreland basins can be flooded by seawater. Seawater evaporating in a hot climate is known to generate a brine capable of seeping into sediments and leaching out metals that precipitated again at the margin of the basin in carbonate rocks. In the case of deposits whose fluids were particularly hot seawater probably even flowed through the basement, while a crust thinned by extension provided a higher geothermal gradient at the same time.

Some of the most important MVT deposits such as Tri-State, Viburnum Trend, Old Lead Belt, Southeast Missouri, Central Missouri, and Northern Arkansas originated in the foreland of the Ouachita Mountains (Bradley and Leach 2003). Here an island arc collided with a passive continental margin in the Carboniferous (Fig. 4.52). Something similar happens today in Indonesia at the eastern end of the Sunda Arc where the island arc between Timor and New Guinea is colliding with the Australian Shelf. Continental crust is being pulled down into the subduction zone. Bending the crust before diving down can lead to a forebulge where the sediments of the shelf even rise out of the water forming the Ari Islands in the case of Indonesia. In the process cave systems form in limestones in which deposits can develop at a later stage. The pull of a subducted slab causes normal faults where further ores can be formed later. Platform carbonates migrate much like an assembly line across the forebulge into the trench and further into the subduction zone. In the trench increasingly clastic sediments are deposited (flysch) that can later serve as a trap for later fluids. As soon as the former island arc is uplifted due to the thickened crust to form a mountain range a foreland basin develops and replaces the former trench where debris from erosion of the mountain range is deposited. When the sea finally disappears, evaporites can form and brine can seep into the sediments. Finally, the fluids of the sediment basin are set in motion by the hydraulic potential of the mountain range. In the case of the Ouachita Mountains MVT deposits were formed not only on the edge of the foreland basin (Northern Arkansas), but also outside the basin hundreds of kilometers from the mountains (Central Missouri).

The history of other deposits such as those on the edge of the Appalachian Mountains cannot be reconstructed so well. This is because several phases of mountain building followed each other and the dating of ores is not clear. There

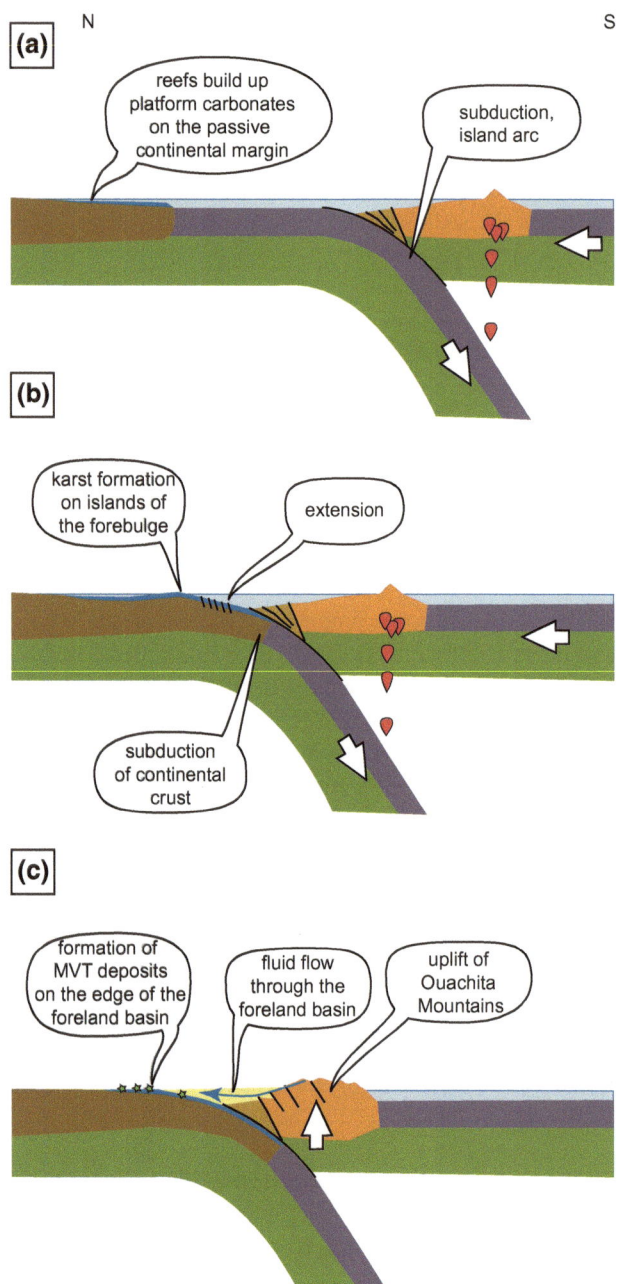

Fig. 4.52 Some important Mississippi Valley–type deposits of the United States were formed in the foreland of the Ouachita Mountains where an island arc collided with a passive continental margin. **a** Reefs built up thick platform carbonates on the continental margin before the collision. **b** Structures were formed in which ores later precipitated as the continental margin entered the subduction zone. At the forebulge (a ridge caused by flexible bending of the crust) islands rose where caves were formed. Moreover, continental crust was stretched and normal faults were formed. The carbonate platform moved much like an assembly line through the forebulge and the trench into the subduction zone. **c** Continued collision resulted in thrusting, crustal thickening, and uplift. Erosion debris was deposited in the foreland basin where the last seawater evaporated to form a brine seeping into the sediments. Hydraulic potential caused the flow of groundwater toward the foreland. At the edge of the basin Pb–Zn ores precipitated within platform carbonates (based on Bradley and Leach 2003)

are probably some MVT deposits that formed within the mountain range after the overthrusting of nappes—not at the edge of the foreland basin. An example is Picos de Europa (Spain).

A few MVT deposits undisputedly formed during extension in a rift system such as a number of large deposits in Australia and the small Wiesloch deposit in the Rhine Rift Valley (Germany) (Pfaff et al. 2010). It is also possible that some of the major European MVT deposits formed during extension (Muchez et al. 2005). There are good arguments for extension to be responsible for the emergence of deposits in the Eastern Alps toward the end of the Triassic. In many cases the results of dating are still too contradictory to determine definitively whether they arose during extension or compression. It is noteworthy that other Pb–Zn deposits such as SEDEX deposits and Pb–Zn veins are also formed by very similar fluids during extension. A type of deposit with similarities to MVT deposits is presented in Box 4.22.

Box 4.22 Tsumeb (Namibia), Kipushi (Congo), Kabwe (Zambia)

The now-depleted copper deposit of Tsumeb (Namibia) is known to every mineral collector because of its wealth of different secondary minerals. Such minerals are exhibited in many museums. More than 200 minerals have been described and 30 are not known elsewhere. In addition to copper Tsumeb also contained lead, zinc, silver, arsenic, germanium, gallium, and cadmium. It is a hydrothermal deposit in platform carbonates (in particular, dolomite). The ores fill cavities or replace carbonates. They are found in tubular solution breccia and other karst structures, in manto-like massive sulfides, and in veins. The primary ores are massive sulfides with pyrite, chalcopyrite, bornite, germanite ($Cu_{13}Fe_2Ge_2S_{16}$), sphalerite, gallite ($CuGaS_2$), enargite, tennantite, briartite ($Cu_2(Fe,Zn)GeS_4$), renierite (($(Cu,Zn)_{11}Fe_4(Ge,As)_2S_{16}$), and galena. Gangue minerals are calcite and quartz. Kombat and Khusib Springs are similar deposits (both in Namibia). They were probably formed in a late phase of the Damarra orogeny (Neoproterozoic). Fluids of the main phase at approx. 400 °C were clearly hotter than in MVT deposits (Chetty and Frimmel 2000) even though replacement reactions show certain similarities to these. Salinity of the hydrothermal fluid is explained by evaporites dissolving.

At a similar time and under similar conditions comparable hydrothermal Zn–Pb–Cu deposits such as Kabwe (Zambia) and Kipushi (DR Congo) were formed that also contain Cd, Co, Ge, Ag, Re, As, Mo, Ga, and V (Kampunzu et al. 2009). Although they are located in the Central African copper belt (Sect. 5.1.2), they are younger than stratiform copper deposits and are found somewhat higher in stratigraphy in platform carbonates. Mineralization is mainly bound to tectonic breccias and karst systems. They originated during and after orogeny of the Lufilian Arc in the Neoproterozoic and at the beginning of the Paleozoic.

4.13 Sandstone-Hosted Copper and Lead–Zinc Deposits

Sandstones and similar sedimentary rocks (red beds) are permeable to water (aquifers) and the pores provide space in which minerals can precipitate. Sandstones oxidized or reduced to varying degrees (e.g., with hematite as cement between the sand grains) can determine where ores precipitate. A mixture of fluids can also play a role. There are corresponding copper deposits and lead–zinc deposits. However, such deposits are usually not economic today but are more interesting from a historical point of view. Although some had already formed during diagenesis (similar to copper-bearing shales like Europe's Kupferschiefer; Sect. 5.1.1), many others did so only through later fluids (e.g., during extension within rift–horst structures). Some metals originate from the sediment itself in which they are contained as detritic heavy minerals (derived from rock weathering and transported by rivers). Ultimately, hydrothermal leaching and precipitation lead to enrichment in certain zones.

4.14 Unconformity-Related and Sandstone-Hosted Uranium Deposits

Redox reactions play a central role in the formation of hydrothermal uranium deposits since U^{4+} is insoluble. However, U^{6+} is soluble even at low temperatures and independent of pH. This happens in the form of charged or uncharged complexes (whose stability depends on the pH) such as UO_2^{2+} with F^-, Cl^-, OH^-, CO_3^{2-}, PO_4^{3-}, or SO_4^{2-}. The reduction of uranyl complexes thus leads to the precipitation of pitchblende. Under certain circumstances a change in the pH value can also lead to precipitation if a suitable complexing agent is missing. Dissolved uranium can also be absorbed by some minerals and organic substances.

The formation of hydrothermal uranium deposits always proceeds in a similar way. Meteoric water (relatively oxidized) leaches out uranium from granite, gneiss, or tuff (possibly also from detritic minerals in sediments), flows on, and precipitates the uranium elsewhere due to a change in Eh or pH.

Box 4.23 Oklo: a natural nuclear reactor

Today only 0.7% of uranium consists of the fissionable isotope ^{235}U. However, the ratios of isotopes change over time because they have different half-lives. Two billion years ago it was still 3% ^{235}U, which corresponds to the amount in fuel rods of nuclear power plants. Since there is no significant natural fractionation of uranium isotopes the isotope ratio is the same all over the world at any given time. Exceptions are small zones in Oklo (Gabon) in a roll front–type sandstone-hosted deposit 2 billion years old with zones of unusually high grade.

In 1972 French workers discovered that enrichment of the Oklo ore produced unusually little ^{235}U. The only possible explanation was that nuclear fission naturally occurred in the deposit. Even fission products such as curium, americium, and traces of plutonium have been found (i.e., elements that were previously only known as artificial products of nuclear power plants). The zone within the deposit clearly had to have once been a natural nuclear reactor. A total of 16 reactors have now been discovered in Oklo and in neighboring Bangombé.

Thanks to a more detailed investigation using xenon isotopes much more is known (Meshik et al. 2000; Meshik 2005). Water clearly had to be available because it has the important property of slowing fast neutrons down to thermal neutrons. Although fast neutrons are released during fission, only thermal neutrons can cause the fission of further uranium nuclei. Accordingly, a chain reaction could commence in high-grade-ore zones. The heat released let the water evaporate and escape switching the reactor off again. When it had cooled enough new water could penetrate and restart the chain reaction. It is estimated that the reactor switched on for 30 min each time and off for at least 2.5 h thereafter. Such pulses seem to have taken place for a few hundred thousand years.

The last step often takes place in sandstones. Since sandstones are aquifers they are easily traversed by water. Sandstone-hosted uranium deposits are widespread worldwide. Although they are relatively small, they are usually found in large numbers in a single region. The grade is low with an average of 0.05–0.4% U_3O_8. The most important examples are found in the United States. Although there are especially many on the Colorado Plateau (Utah, Colorado, Arizona, New Mexico), they are also found in Texas and Wyoming. There are further examples in Niger and in Germany at Königstein in Saxony. A special case is the Oklo natural reactor (Box 4.23).

Three types of sandstone-hosted uranium deposits can be distinguished: tabular, roll front, and tectono-lithological uranium deposits.

The **tabular** type is common on the Colorado Plateau. These are small tabular mineralizations that are arranged more or less parallel to the flow direction of groundwater (i.e., almost parallel to the layers). On the Colorado Plateau coffinite ($USiO_4$) is the most abundant ore mineral replacing organic matter including former tree trunks. Since vanadium was precipitated at the same time uranium–vanadium minerals such as carnotite and tujamunite also occur. Often several such mineralizations are to be found in the sandstone above and next to each other. It is assumed that these horizons were formed by water mixing at the interface between a stagnant brine and oxidized meteoric water flowing through the sandstone (Northrop and Goldhaber 1990). The mixing triggered a whole series of reactions in which organic substances present in the sandstone and H_2S (released by sulfate-reducing bacteria) played a role as reducing agents. Moreover, changes in pH triggered precipitation reactions.

Roll front deposits formed on oxidation fronts within sandstone (Fig. 4.53). Such reaction fronts are clearly visible in sandstones as a sharp boundary separating red, hematite-containing sandstone (oxidized) from greenish (reduced) sandstone, which typically contains some carbonate, organic matter, and pyrite or marcasite between the sand grains. In a sandstone aquifer oxygen-rich meteoric water leads to oxidation, but this effectively consumes oxygen at the redox front. The reaction front therefore only moves slowly through the sandstone. It often takes on the shape of a cylinder (more precisely C-shaped in cross section) since flow in the center of an aquifer is faster. U^{6+} dissolved in water is reduced at the front and precipitated as pitchblende. Other metals dissolved in the water such as V, As, Se, Mo, Cu, and Co are also precipitated as a result of which zonation with different mineralization is formed within the roll. The fluid mixture may also be important since rolls are often located near faults where it is assumed reducing water flowed into the aquifer (similar to the next type). In some cases incoming methane seems to have led to reduction.

Roll front deposits resemble **tectono-lithological** deposits in this regard. In tectono-lithological deposits water penetrated a sandstone aquifer along permeable faults where tongue-shaped ore zones were formed by fluid mixing.

Unconformity-related deposits are the most important uranium deposits (Fig. 4.54). Unconformity is the name given to a surface where sediments lie at a different angle above much older tilted and partially eroded rocks. The underlying principle of these deposits is similar to that of sandstone-hosted deposits. However, unconformity-related deposits are much larger and have a higher ore grade measurable in the percentage range (in some cases even more than 20% U_3O_8). The most important unconformity-related deposits are located in the southeastern part of the Athabasca Basin (Saskatchewan, Canada) and in the Alligator River District (Northern Territory, Australia). They are found in Proterozoic rocks just above or just below the unconformity between "young" (in comparison, but actually very old) sandstones (Mesoproterozoic in Canada, a little older in Australia) and the underlying basement of folded Paleoproterozoic metasediments (such as schists) and Archean granites and gneisses. Their consistently great age is due to the very incompatible uranium once also contained in the Earth's mantle ultimately ending up in the crust in those times.

Uranium was mainly leached out of granite or gneiss by hydrothermal fluids (Hecht and Cuney 2000). The precipitation of pitchblende was probably caused by redox reaction with graphite that is widespread in metasediments (Jefferson et al. 2007). Mixing with the connate water of the sediments may also have played a role (Derome et al. 2005). Another possible reaction is the simultaneous oxidation of Fe^{2+} and precipitation of hematite. Ores are located along fluid pathways (especially, at faults and breccias within graphite-bearing metasediments close to the unconformity). They are surrounded by an alteration halo. The highest grades are found in the sandstone (particularly, in Cigar Lake and McArthur River in the Athabasca Basin). At Cigar Lake alteration has resulted in a watertight clay cap over the ore zone. Silicification occurred toward the top of the alteration zone by precipitation of additional quartz.

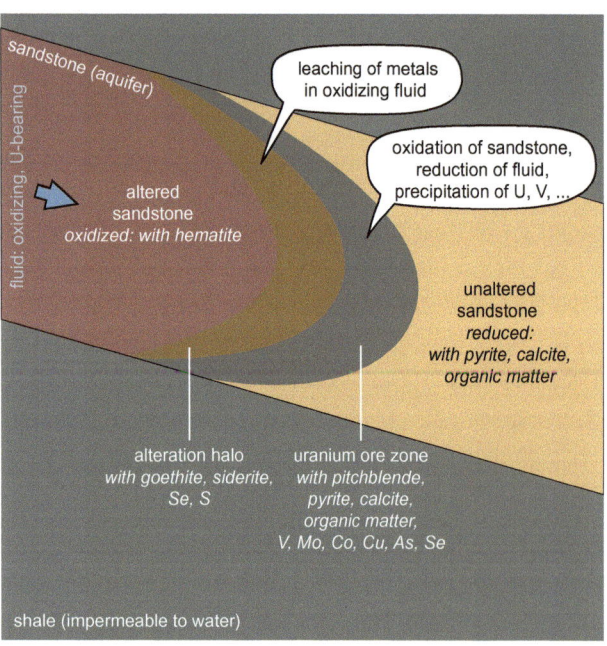

Fig. 4.53 Uranium, vanadium, and other metals are precipitated from water at a redox front in a sandstone aquifer. The front is often roll-shaped (roll front deposit)

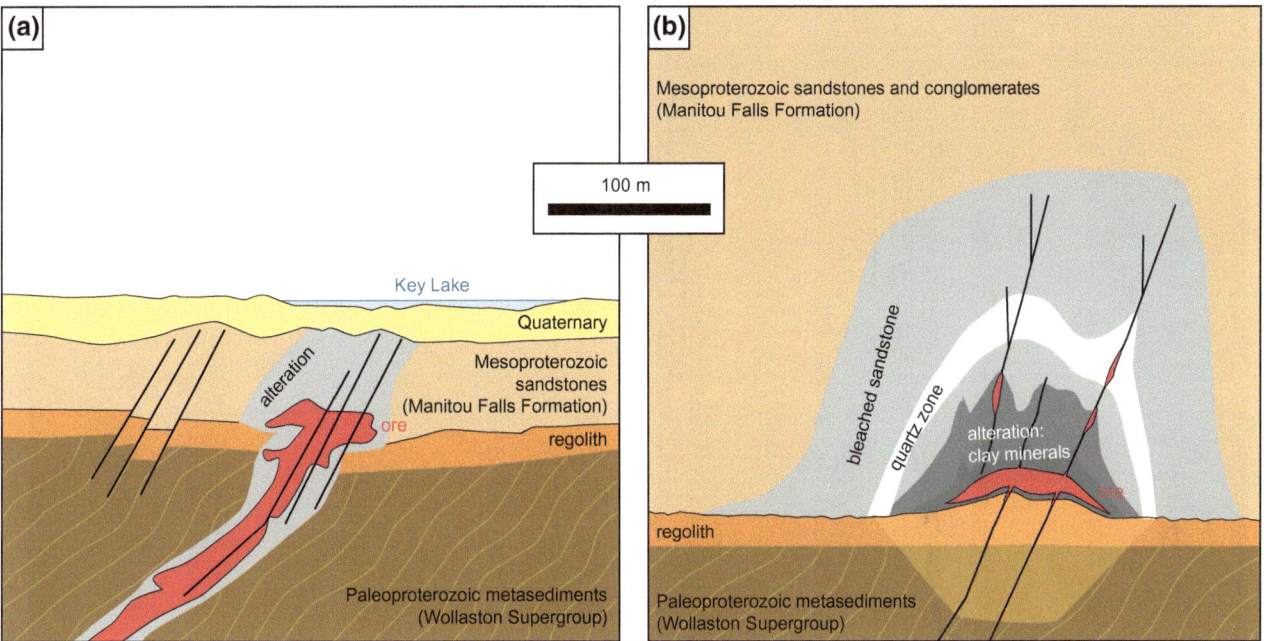

Fig. 4.54 Unconformity-related deposits are located in Proterozoic rocks just above or just below the unconformity between slightly younger sediments and even older metasediments. Two examples from the Athabasca Basin (Canada) are **a** Key Lake and **b** Cigar Lake (from Jefferson et al. 2007)

4.15 Hydrothermal Systems on the Seafloor (Introduction)

When hydrothermal fluid pours from a hot spring on the seafloor it mixes with cold seawater and the dissolved cargo precipitates abruptly. Deposits formed in this way are referred to as exhalative. The temperature of the fluid can be significantly higher than in epithermal systems and hot springs on land because the boiling point is increased by the water pressure. Accordingly, larger quantities of metals can be dissolved and iron sulfides, copper sulfides, lead sulfides, and zinc sulfides are mainly deposited. Manganese deposits form at a slightly greater distance from hot springs, they could be termed exhalative as well (Sect. 5.5).

Although many hot springs are directly related to volcanism (e.g., at mid-ocean ridges or island arcs), there are also such springs far removed from volcanoes in sediment basins. Those relating to volcanism are called volcanogenic massive sulfide (VMS) deposits (Sect. 4.16) and those relating to sediment basins are called sedimentary exhalative (SEDEX) deposits (Sect. 4.17) where there is fluent transition between endmembers. Hydrothermal water in both cases is originally seawater that has leached soluble substances from the seabed; hence volcanism was mainly the source of heat for VMS deposits. Metal content also depends on whether sediments, basalts, or acidic igneous rocks have been leached out. In the case of acidic igneous rocks related

VMS deposits often have an economically relevant content of gold enclosed in sulfides (especially, pyrite) as tiny grains.

What exactly happens at the source depends on the density of the fluid, its temperature, and the amount of dissolved substances. If the fluid is heavier than seawater, then it accumulates in depressions—an example is the Atlantis II Deep in the Red Sea (see below). If it is lighter, then black smokers are created (discussed in the following section). Moreover, ores precipitate even below the seabed in fissures where water rises and in pores of the surrounding rock that in extreme cases is completely replaced.

4.15.1 Black Smokers

The deep-sea submarine *Alvin* made a spectacular discovery in 1977. At a mid-ocean ridge running north of the Galápagos Islands in the Pacific Ocean strange hot springs were found at a depth of 2500 m. Looking like smoking factory chimneys they are settled by a community of living creatures that get along completely without light including heat-loving microbes, tubeworms, mussels, starfish, and crabs (Corliss et al. 1979). This mid-ocean ridge was a recent counterpart to VMS deposits. Such black smokers have since been found in the deep sea in all oceans (Figs. 4.55, 4.56 and 4.57). Most are located at mid-ocean ridges (see also Box 3.7) (especially, at faults in central rift valleys where there is extension and strong magmatism). Their hydrothermal systems play an important role in the cooling of newly formed

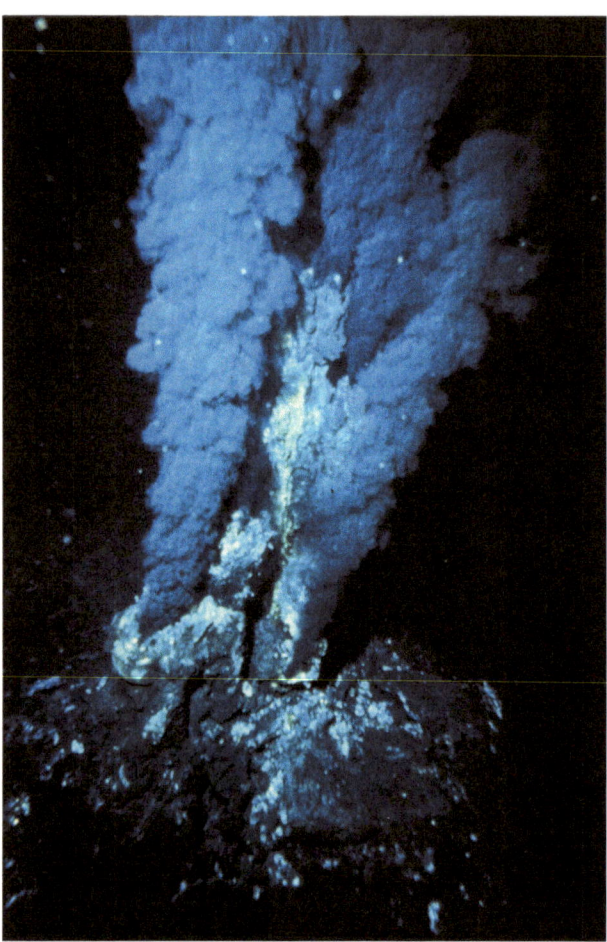

Fig. 4.55 Black smokers are hot springs in the deep sea where large quantities of sulfides and other substances precipitate from the water and rise as plumes of "smoke" above the chimneys (*photo* © NOAA, public domain, Wikimedia Commons)

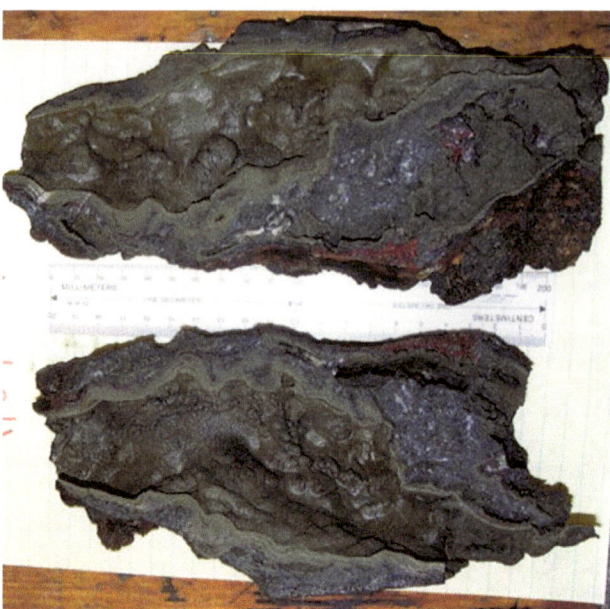

Fig. 4.56 Piece of a chimney (22 cm long) sawn off from a black smoker in the Mariana Trench. The dark interior is predominantly chalcopyrite and the gray exterior consists of pyrite, sphalerite, and baryte (*photo* © NOAA)

oceanic crust as they effectively dissipate heat. However, there are also black smokers in submarine rifts with continental crust, others in the back-arc basins of island arcs, and yet others at active seamounts (typically, in calderas) at hot spots, island arcs, and off the central axis of mid-ocean ridges (Hannington et al. 2005).

The water escaping black smokers is extremely hot. Although a temperature between 300 and 350 °C is typical, it can reach 400 °C. This is only possible because the enormous water pressure of the deep sea prevents evaporation. There are no black smokers at a water depth less than 1500 m. Instead hydrothermal activity is then taken up by epithermal (Sect. 4.3) and porphyry (Sect. 4.4) systems much like those that occur on land (Large et al. 2001).

On contact with cold water (2 °C) of the deep sea the dissolved freight precipitates abruptly. Ore chimneys grow several centimeters per day and can grow a few meters high (rarely even dozens of meters) before they collapse. The sharp drop in temperature causes small-scale zoning. The hot

interior of the chimneys consists mainly of zoned Fe–Cu–Zn sulfides (see Table 4.7), while on the cool outside and at the tops of high chimneys mainly anhydrite and baryte are deposited (the sulfate comes from the seawater). The chimneys contain finely distributed sulfides and amorphous SiO_2. Anhydrite is often dissolved again later and replaced by newly precipitated minerals.

More than 90% of the freight precipitated forms the black "smoke" of fine ore grains that rises in the seawater above the chimneys. A fraction sinks in the surroundings to the bottom where they can accumulate to form a sediment of iron oxide and quartz (exhalative chert) or iron oxide and carbonate. Most particles are transported great distances as suspended particles. Pyrite nanoparticles (Yücel et al. 2011) and metals incorporated in organic molecules (Sander and Koschinsky 2011) not only play a role in the formation of distant sedimentary deposits, they are also important trace elements in the ocean food chain.

White smokers are similar sources that are somewhat cooler (250–300 °C). The precipitation of anhydrite or baryte leads to their having a white smoke plume. Although there are often white smokers in the vicinity of black smokers, they can also occur at shallower water depths. They form large baryte deposits—not metal deposits. Moreover, in some hot hydrothermal fields there are isolated springs where extremely hot water escapes containing hardly any dissolved substances. This is interpreted as a vapor phase that has separated from the remaining hydrothermal water (Von Damm et al. 1997, 2003).

Fig. 4.57 Precipitation from black smokers—extremely hot chimney-like springs at the bottom of the deep sea (from Okrusch and Matthes 2009)

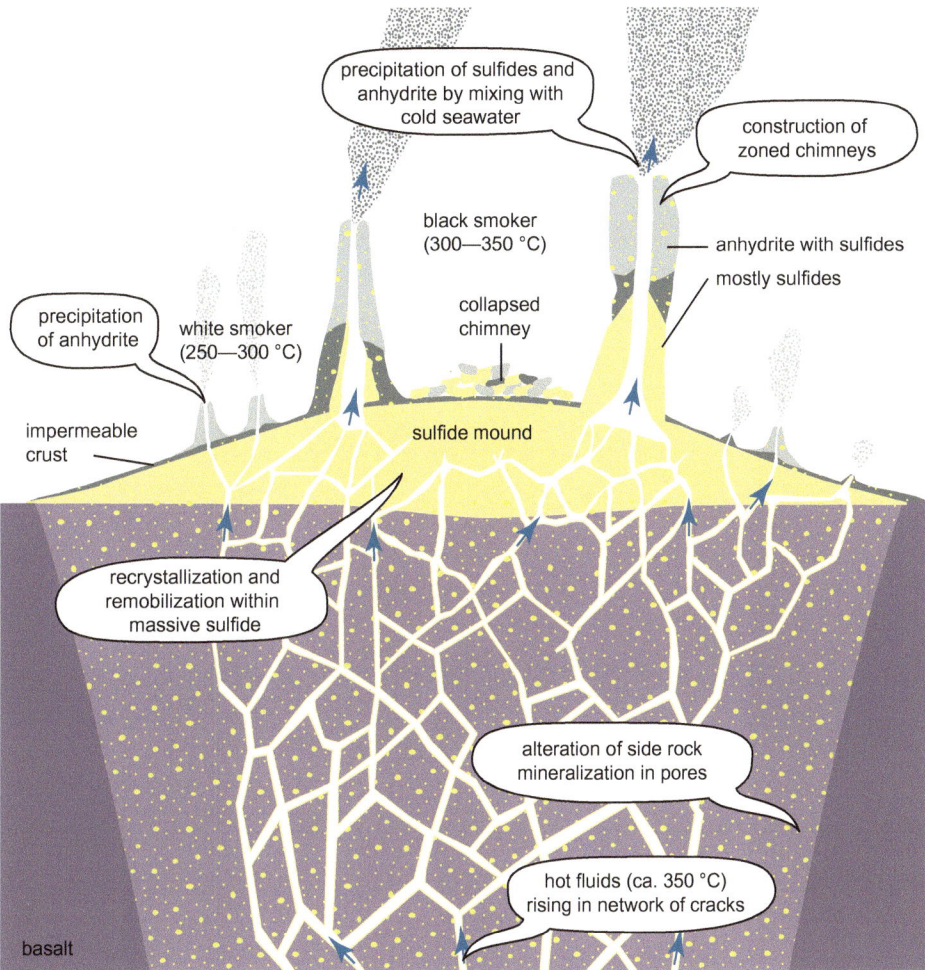

Table 4.7 Minerals of black smokers

Hot zone

Chalcopyrite	$CuFeS_2$
Isocubanite	$CuFe_2S_3$
Pyrrhotite	FeS
Pyrite	FeS_2
Bornite	Cu_5FeS_4
Anhydrite	$CaSO_4$

Cooler zone

Sphalerite	ZnS
Wurtzite	ZnS
Marcasite	FeS_2
Pyrite	FeS_2
Anhydrite	$CaSO_4$
Baryte	$BaSO_4$
Amorphous SiO_2	

Sulfide precipitation begins with the ascent of hydrothermal water in the fissure network under black smokers leading to a stockwork (a.k.a. stringer zone; Box 4.13) developing in the basalt of the ocean floor surrounded by an altered zone in which siderock strongly changes. Ores can also be disseminated as deposits in pores of the rock altered.

Black smoker chimneys stand on a hill of massive sulfide ore (with low contents of sulfates and silicates). Over time this hill builds up from collapsed chimneys, sulfide grains sunk from the plume ("smoke"), and sulfides precipitated directly on site. The internal temperature gradient causes recrystallization and constant remobilization within this ore body (zone refining) that also undergoes temperature-dependent zoning (Fig. 4.58, Box 4.24). Zinc and lead (particularly, lead) migrate outward into the cooler area, while the hot interior becomes increasingly rich in copper. At very high temperatures a core is formed consisting almost exclusively of pyrite. This zoning changes with temperature fluctuations

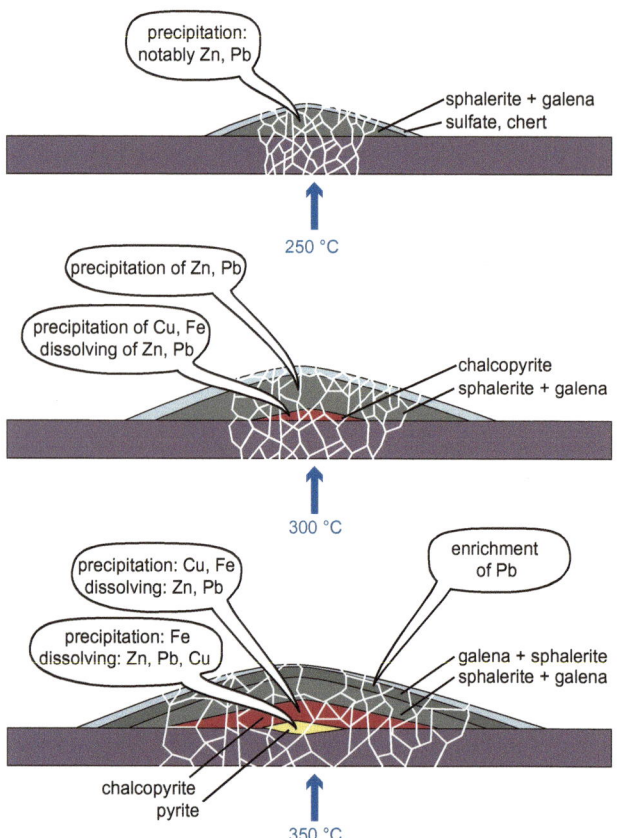

precipitation:
notably Zn, Pb

sphalerite + galena
sulfate, chert

250 °C

precipitation of Zn, Pb

precipitation of Cu, Fe
dissolving of Zn, Pb

chalcopyrite
sphalerite + galena

300 °C

precipitation: Cu, Fe
dissolving: Zn, Pb

enrichment
of Pb

precipitation: Fe
dissolving: Zn, Pb, Cu

galena + sphalerite
sphalerite + galena

chalcopyrite
pyrite

350 °C

Fig. 4.58 Zoning in an ore mound brought about by remobilization and precipitation (i.e., zone refining) (from Large 1992)

or changes in the composition of the hydrothermal fluid (Petersen et al. 2000; Houghton et al. 2004).

Box 4.24 "Chalcopyrite disease"

"That's not a texture, it's a disease." This saying led to the term "chalcopyrite disease" describing a texture often observed in sphalerite (Barton and Bethke 1987). The texture comprises fine-grained or microscopically small inclusions of chalcopyrite irregularly distributed in sphalerite. Such inclusions are interpreted as a replacement reaction in which hydrothermal water added copper and reacted with ferrous sphalerite resulting in zinc being removed. This reaction is typical of VMS deposits in which temperature-dependent zoning is formed by remobilization of metals.

Although such a hydrothermal field can deposit about 250 t of massive sulfides per year, these sulfide mounts are usually only a few meters high and contain only a few thousand tonnes of ore. One of the largest is the TAG hydrothermal

field in the Central Atlantic (Petersen et al. 2000). Its sulfide mount is 50 m high and has a diameter of 200 m. With 4.5 million tonnes of sulfide ore it is similar in size to typical VMS deposits.

Hydrothermal systems are often only active for a few decades. They then pause for a few millennia before the next pulse. As new oceanic crust including sulfide mounts moves away from mid-ocean ridges they eventually become inactive and covered with sediments over time. Often there is a magma chamber below active black smokers at a shallow depth (approx. 2–3 km). The injection of new magma into the magma chamber plays a particularly important role (Wilcock et al. 2009) since cracks form in the surrounding rock focusing the circulation of water. The sidewalls of a caldera are also well suited as an ascent path for fluids.

By contrast, magmatic water plays hardly any role in black smokers. Instead the hydrothermal water is former seawater that penetrates fissures of the seabed, heats up, and reacts with rock (alteration). New minerals such as zeolites, chlorites, and epidotes are formed in basalt. They incorporate $(OH)^-$ thus lowering the pH value of the water. Other substances such as metals are leached out of the rock. Oxygen dissolved in the water is lost on the way due to the oxidation of Fe^{2+} and the sulfate of seawater is reduced to sulfide at the same time. The result is hot water that is acidic and low in oxygen, has a high content of dissolved substances, and rises once again above the hot magma chamber.

Black smokers in the back-arc basins of island arcs mainly differ from those at mid-ocean ridges in that sediments and acidic volcanic rocks are leached out in addition to basalt of the ocean floor. The ores precipitated are more heterogeneous, have less iron but more zinc, lead, silver, gold, arsenic, and other metals. They comprise a correspondingly diverse list of minerals (Table 4.8).

Table 4.8 Average metal contents (wt% unless marked otherwise) in massive sulfides at mid-ocean ridges and back-arc basins of island arcs (from Herzig and Hannington 1995)

	Mid-ocean ridges	Back-arc basins
Fe	23.6	13.3
Zn	11.7	15.1
Cu	4.3	5.1
Pb	0.2	1.2
As	0.03	0.1
Sb	0.01	0.01
Ba	1.7	13.0
Ag (ppm)	143	195
Au (ppm)	1.2	2.9

4.15.2 Marine Brine Pools and Atlantis II Deep

The density of saltwater mainly depends on temperature and salt content. When water pouring from a hot spring on the seabed has a greater density than seawater it does not rise further as is the case with black smokers. Instead, it sinks and collects in depressions in which metal-rich sludge is deposited. This can happen near black smokers in the deep sea, at less hot springs, and at hydrothermal systems on the continental margin in relatively shallow water. Recent examples have been found in the Gulf of California, the Gulf of Mexico, the Red Sea, and the Black Sea. The best known is the Atlantis II Deep in the Red Sea.

The Red Sea (Fig. 4.59) was created by the breakup of a continent. The northern part of the sea is still an advanced continental rift system in which continental crust is being stretched and thinned. However, the southern part is more advanced and new oceanic crust is forming on a mid-ocean ridge along the central axis. In between there is a transition zone with small, quasi-embryonic mid-oceanic ridges within strongly thinned continental crust. Atlantis II and several other depressions are located immediately south of the transition zone in an area strongly disturbed by strike-slip

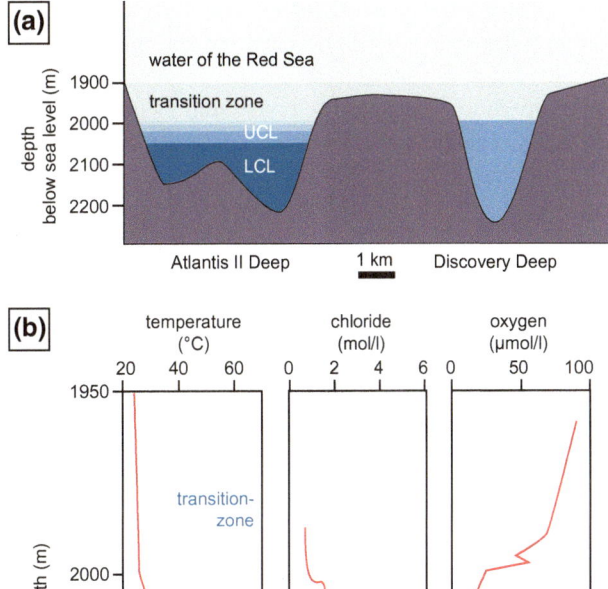

Fig. 4.60 **a** Schematic section (heightened tenfold) through the Atlantis II Deep and the adjacent Discovery Deep both filled with warm brine. In the Atlantis II Deep water layers of different temperature and composition have formed such as the lower convective layer (LCL) and several upper convective layers (UCLs). **b** Jumps in temperature, salinity, and oxygen concentration across water layers from measurements taken in 1997 (from Schmidt et al. 2003)

Fig. 4.59 Location of the Atlantis II Deep in the Red Sea

faults. Some are hydrothermally active and filled to the brim with warm brine (saltwater).

The Atlantis II Deep (Fig. 4.60) is about 6 km wide, 12 km long, and the bottom is between 200 and 300 m deeper than the surrounding seabed (water depth of 1900 m). It is filled with several layers of brine that do not mix. Sharp jumps in temperature and salt content occur between the layers. The thickness of the layers, their temperature and composition, and the number of layers has changed over three decades of repeated measurements. For example, the temperature of the lowest layer called the lower convective layer (LCL) has risen from 54 to 67 °C. This layer is anaerobic and its salinity at 27% is as high as that of the Dead Sea—the average salinity of water in the Red Sea is 4%. Nevertheless, such brine pools are not devoid of life (Box 4.25). Methane (Schmidt et al. 2003) and dissolved metals also accumulate in the depression. Hot springs have a temperature estimated at 195–310 °C, a salinity of 37%, and a flow rate between 700 and 1000 kg/s (Anschutz and Blanc

1996). The water was originally seawater that on its way through the crust not only leached out substances from the basalts, but also dissolved salt from evaporites. Water likely boils up before reaching the spring and is further concentrated by separation of a vapor phase (Winckler et al. 2000).

Over the last 15,000 years between 10 and 30 m of metal-rich finely laminated sediments have accumulated on the bottom of the brine pool (Anschutz and Blanc 1995). The sediments are mainly made up of chert (exhalative and biogenic) and carbonates (such as ankerite, dolomite, and siderite), shale, and anhydrite layers. The sludge on top of this has not yet solidified. Metals are precipitated not only at the sources themselves (sulfides such as sphalerite, chalcopyrite, and pyrite), but also at the boundary between seawater and brine. The sharp contrast in the oxygen content is particularly noticeable here; Fe^{2+} and Mn^{2+} dissolved in the brine are oxidized to Fe^{3+}, Mn^{3+}, and Mn^{4+}. Such metals precipitate immediately as oxides or hydroxides (goethite, hematite, manganite, groutite, todorokite). The particles percolate downward as a result of which some of the manganese is reduced and dissolved again. Which sulfides, oxides, and hydroxides are predominantly precipitated changes with the composition and temperature of both the hydrothermal water and the brine accumulated in the depression.

Even below the seabed ores are precipitated from water rising in fissures and pores. Since pores in the sediment are also saturated with brine the metals are remobilized within the sediment (Anschutz et al. 2000). Additional pyrite is formed diagenetically or by sulfate-reducing bacteria.

Brine pools similar to that of the Atlantis II Deep are necessary for SEDEX deposits to form. However, they are located in sedimentary basins far removed from active volcanism, but in the case of the Atlantis II Deep the basaltic volcanism of incipient ocean spreading is not far away. There are also VMS deposits that have formed similarly to deposits in the Atlantis II Deep (Solomon et al. 2004) including several examples in the Iberian pyrite belt (Sect. 4.16.4).

Box 4.25 Life in the brine pool

Depressions on the seabed where anoxic and hypersaline brines have accumulated are not devoid of life. For example, in a pool in the Gulf of Mexico there are 100 fold more microorganisms than in the seawater above. More than 10 strains are present and they metabolize in different ways. Some reduce sulfate, others degrade fatty acids to acetate, from which others produce methane (Joye et al. 2009).

4.16   Volcanogenic Massive Sulfide Deposits

Sulfides accumulated exhalatively by black smokers on the seabed of the deep sea are of little use in the first instance (even though one such deposit is soon to be mined in the deep sea; Sect. 1.13). It is only when they end up on a continent that they can be easily mined. Fortunately, this has sometimes happened in the course of Earth's history. Such deposits are known as volcanogenic massive sulfide (VMS) deposits. Even though the ores are located in volcanic rocks (Piercey 2011; Shanks and Thurston 2012) there is not always necessarily a direct connection to active volcanism. Some geologists therefore prefer the term volcanic-hosted massive sulfides (VHMS). The transition to massive sulfide deposits in sediments (i.e., SEDEX deposits; Sect. 4.17) is fluid.

Some VMS deposits are gigantic having more than 100 Mt of ore. However, smaller deposits with a few megatonnes are much more frequent. They commonly occur within certain regions. Several of the largest are located in the Iberian pyrite belt in the south of Portugal and Spain (Sect. 4.16.4). Further gigantic deposits are found in the Abitibi greenstone belt (see Box 4.27) in Ontario and Quebec (Canada) such as Noranda and Kidd Creek and at the northern end of the Appalachian Mountains near Bathurst (New Brunswick, Canada). There are also important occurrences in Japan, Australia, the Urals, Turkey, Cyprus, Oman, Scandinavia, the Balkans, and so on (Figs. 4.61, 4.62 and 4.63). The metals of interest include copper, lead, zinc, gold, and silver.

The chimneys of black smokers or their fragments are only rarely preserved in VMS deposits. The former ore mound below black smokers is usually the most productive part of a deposit. It must have been covered relatively quickly by pillow lavas, tuffs, or sediments and thus protected from oxidation. Fissures (stockwork or stringer zone) through which hydrothermal water has risen can still be found under the more or less lenticular massive sulfide body. Such fissures are filled with ores. The surrounding rock is strongly altered and partly contains disseminated ores that fill former pores. Sometimes plutons can be seen in deeper areas. They are interpreted as shallow magma chambers whose heat has driven the circulation of hydrothermal fluids.

The replacement of host rocks or older sulfides below the seabed has also played an important role in many VMS deposits (Doyle and Allen 2003). Tuffs made up of volcanic ash and pumice are common on island arcs. Mineralization is particularly effective in these because they contain many pores and rock glass is easily dissolved. Accordingly, massive ore bodies can form in tuffs consisting almost exclusively of sulfides. Limestone can be completely dissolved and replaced by sulfides that are then surrounded by a halo of skarn and dolomite. Moreover, clastic sediments (espe-

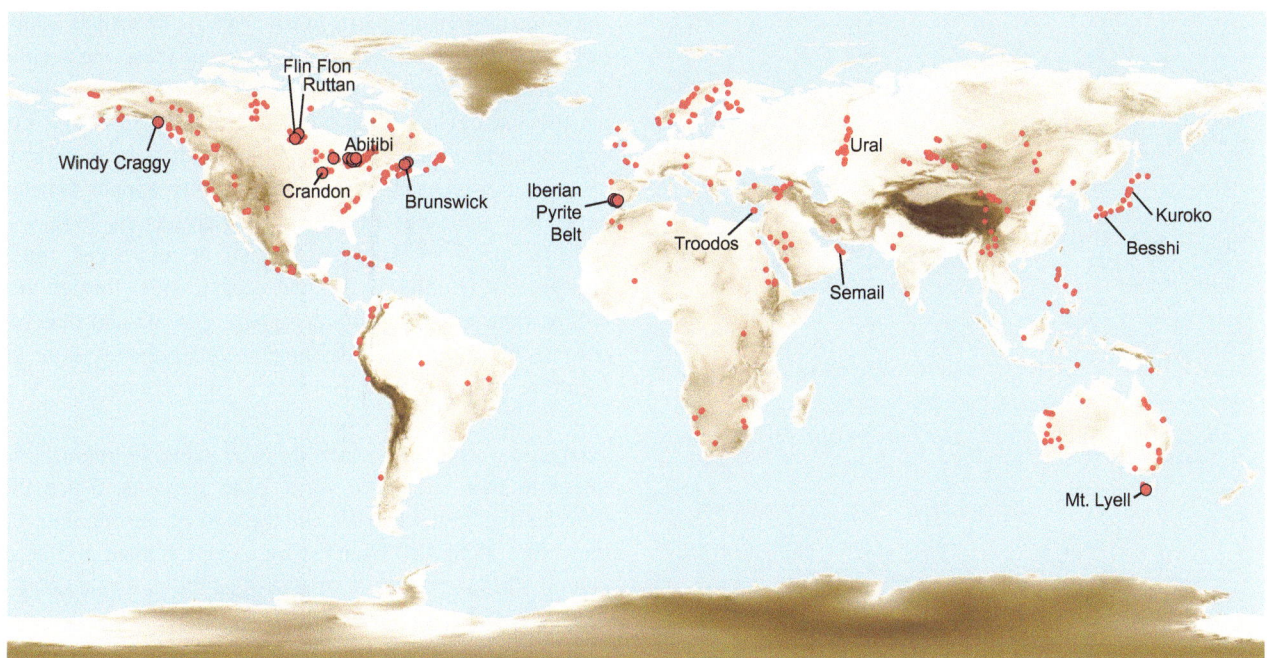

Fig. 4.61 Worldwide distribution of VMS deposits. Very large deposits containing more than 50 million tonnes of ore are highlighted (from Franklin et al. 2005)

Fig. 4.62 Historic copper smelter in Røros (Norway) where copper ores from a number of VMS deposits were mined between the seventeenth and twentieth centuries (*photo* © F. Neukirchen/Blickwinkel)

cially, if they have not yet solidified) and even lavas and plutonic rocks after strong alteration can be replaced by massive sulfide ores that still contain fragments of the altered rock. This is especially the case where water is focused by fracture zones or volcanic vents within particularly porous layers or below water-impermeable layers. Mineralization occurs on reactive contact (replacement of unstable minerals or rock glass) and in fissures some of which have only just been torn open by the water. There need not necessarily have been black smokers since replacement below the seabed (at a depth of a few to a maximum of 200 m) was the only or at least the predominant process for some VMS deposits and the resulting ore body can even be significantly larger than sulfide mounds on the seabed.

Fig. 4.63 Massive galena crystal from the VMS deposit at Izok Lake (Nunavut, Canada) (*photo* Mike Beauregard, CC-BY, flickr.com)

For example, lava flows covered sulfide mounds that formed on the seabed in Noranda (Canada). Hot water was able to partly break through the heat-insulating layer and deposit more sulfides on the new seabed; hence there are ore bodies on different horizons. However, water continued to flow through the ore bodies buried and their minerals were replaced by new sulfides that were richer in copper as a result of the higher temperature. A similar situation occurs by rapid burial by sediments (particularly, by turbidity currents) when turbidites are deposited.

However, it is not always easy to decide whether a massive ore body was formed on the seafloor or beneath it

since the overlying rocks were usually deposited shortly before or shortly after the ores. Some of these replacement deposits were formed at the bottom of a shallow sea—not in the deep sea. They have metal contents (e.g., a high gold content), alteration zones, and textures that are more reminiscent of epithermal veins and porphyry copper deposits; hence they represent a hybrid type (Large et al. 2001).

Most black smokers are located at mid-ocean ridges. However, it is unlikely that the oceanic crust formed there will be transported on to a continent. The old and thus cool oceanic lithosphere is heavier than the asthenosphere and therefore normally disappears in a subduction zone. More exotic black smokers occurring in the back-arc basins of subduction zones or other atypical tectonic regimes are therefore more likely to form VMS deposits. Since they differ in their metal content and the type of surrounding rock they were often also changed by metamorphism and deformation VMS deposits are subclassified into a whole range of deposit types.

Different concepts are used to classify VMS deposits (Table 4.9). They can be classified according to the content of economically interesting metal (Cu–Zn, Cu–Zn–Pb, Cu–Zn–Au, etc.); by comparison with type localities (Cyprus type, Besshi type, Kuroko type, and any number of other types); according to the tectonic setting (mid-ocean ridges with different spreading rates, back-arcs at different stages in oceanic or continental crust, etc.); or according to the host rock (mafic or felsic volcanic rocks with or without sediments). Classification according to the host rock has the advantage that it is not based on interpretation and is therefore more robust.

Cyprus-type VMS deposits (Sect. 4.16.1) correspond very closely to the black smokers at mid-ocean ridges. They are located in the pillow lavas of an ophiolite complex (i.e., a piece of oceanic lithosphere pushed onto a continent in a process called obduction). They mainly contain Fe–Cu–Zn sulfides and gold is sometimes extracted as a by-product.

Table 4.9 Different types of volcanogenic massive sulfide (VMS) deposits and their connection with the host rock and tectonic setting in which they were formed	Classification by host rock (Franklin et al. 2005)	Classification according to type examples	Metal content	Tectonic setting
	Mafic	Cyprus type	Cu–Zn	Mid-ocean ridges; intraoceanic back-arcs
	Mafic–siliciclastic (pelitic–mafic)	Besshi type	Zn–Cu	Intraoceanic back-arcs; mid-ocean ridges close to continents
	Bimodal mafic	Noranda type	Zn–Cu	Extension in island arcs; oceanic basalt plateaus
	Bimodal felsic	Kuroko type	Zn–Pb–Cu	Rift (continental crust); continental back-arcs
	Felsic–siliciclastic (bimodal–siliciclastic)	Bathurst type; Iberian pyrite belt	Zn–Pb–Cu	Continental back-arcs; oblique collisions

Despite all similarities to normal mid-ocean ridges such deposits usually formed in connection with subduction zones (e.g., in back-arcs; Box 4.26) or by stretching during the formation of a new subduction zone (e.g., a supra-subduction zone; see also Box 3.7).

The back-arc basin behind an island arc can be regarded as a miniature version of a mid-oceanic ridge. New oceanic crust is formed here by extension of the oceanic plate and the extrusion of basalt magma. However, this is very quickly covered by sediments supplied by the neighboring islands. Such sediments are deep-sea shales or the deposits of turbid currents that slide into the deep-sea called turbidites. Sometimes they comprise sandstones and isolated tuffs. Corresponding VMS deposits can be located in basalts and in sediments and they are usually referred to as Besshi type (Sect. 4.16.2).

If extension takes place directly at an island arc (e.g., because a real back-arc basin has not yet formed), then acidic volcanic rocks from the island arc occur in addition to basalts resulting in Noranda-type VMS deposits being formed. In the Urals two further types can be distinguished (Prokin et al. 1998; Herrington et al. 2005) consisting of the Urals type in bimodal volcanic rocks and the Baimak type in subvolcanic felsic intrusions.

Subduction under a continent can also cause stretching in the hinterland. This initially creates a rift system that over time can merge into a real back-arc basin in which new oceanic crust is formed. An elementary example is the Okinawa Trough (Fig. 4.68) and an advanced example is the Sea of Japan that came into being after Japan was separated from Eurasia by extension in the back-arc of a subduction zone. In contrast to the back-arc basins of island arcs in which basic volcanic rocks (especially, basalt) predominate, in this case acidic volcanic rocks such as andesite and rhyolite prevail. Basalts such as alkali basalt and mid-ocean ridge basalt (MORB) occur only subordinately. The VMS deposits in the Sea of Japan are more heterogeneous and contain significantly more zinc, lead, gold, and silver than those already mentioned. Since acidic magmas often have a high water content the hydrothermal fluid may also contain a magmatic component. Named for the most important example of a VMS deposit from a continental back-arc they are called Kuroko type (Sect. 4.16.3).

VMS deposits in the Iberian pyrite belt (Sect. 4.16.4) are a special case since they arose during oblique continental collision.

Many VMS deposits were subducted to a certain degree into the Earth's interior during orogeny before they rose again to the Earth's surface due to erosion. Secondary rocks were correspondingly transformed into metamorphic rocks like schists. Since some sulfides (galena, chalcopyrite, sphalerite, but not pyrite) can be plastically deformed at relatively low temperatures they are often cut through by shear zones in which different zones are deformed to different degrees. Therefore, the original geometry is often no longer preserved. In many cases deformation has resulted in remobilization of copper, gold, and silver that have been strongly enriched in the shear zone (Matignac et al. 2003; Gu et al. 2007; Castroviejo et al. 2011). Metals enriched in such a way are called ore shoots and make mining more profitable.

In the course of the Earth's history VMS deposits were formed in short episodes—not continuously. They correspond to phases when supercontinents such as Gondwana and Pangaea were forming (Huston et al. 2010). Plate tectonic situations in such phases allowing the formation and subsequent preservation of deposits (especially, local extension in connection with subduction zones) seem to occur more frequently. Changes in the composition of seawater also appear to have an impact. In the Archean it was poor in oxygen and sulfur but saltier and rich in iron. VMS deposits of this period contain almost no sulfates.

Box 4.26 Extension in back-arc basins

For extension to occur in the hinterland of a subduction zone is at first glance astonishing. Should not the upper plate be compressed? The reason for extension lies in an effect analogous to suction, the pull of the sinking slab. This is not passively pushed under the overriding plate but sinks due to its high density because the old and thus cool oceanic lithosphere has a greater density than the asthenosphere. Rocks that are subducted are then transformed into high-pressure rocks such as blueschist and eclogite. Such rocks are even heavier and pull on the slab much like a child tugging on a tablecloth. This pull is believed to be the most important engine for plate tectonics. Since there is no table edge in plate tectonics the subduction zone slowly rolls back in the direction of the sinking plate and the edge of the upper plate is pulled and stretched accordingly. This is especially noticeable directly behind a volcanic arc in the back-arc.

4.16.1 Cyprus-Type Volcanogenic Massive Sulfide Deposits in the Troodos Ophiolite

The Troodos Ophiolite (Fig. 4.64) is a piece of oceanic lithosphere from the Cretaceous that is 92 million years old. It slid on the edge of a continent (obduction) shortly after its formation. It consists of about 90 small and medium-sized VMS deposits first mined in the Bronze Age for copper. The largest are Mavrovouni (15 million tonnes) and Skouriotissa (6 million tonnes).

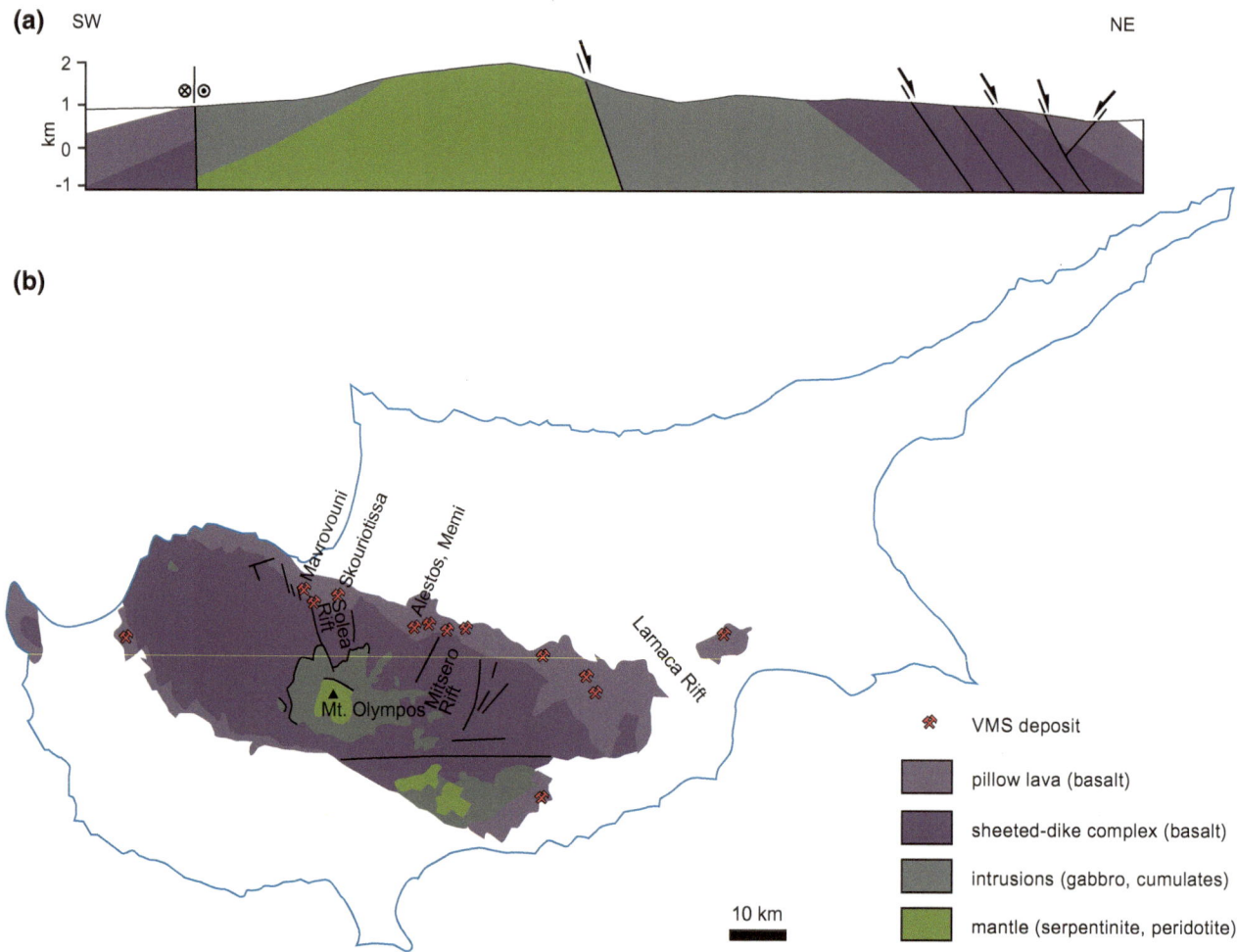

Fig. 4.64 Troodos Ophiolite in Cyprus. **a** Simplified profile through the ophiolite complex. Apart from minor faults the complex has hardly been deformed. **b** Map of the ophiolite complex. VMS deposits are located in pillow lavas (especially, on the northern edge of the Troodos Mountains) (from Eddy et al. 1998; Dilek and Furnes 2009; Pearce and Robinson 2010)

An ophiolite (see also Box 3.7) is a complete section through the oceanic crust and the underlying lithospheric mantle that was incorporated into the continental crust. The Troodos Ophiolite is regarded as a classic example of a relatively slow–spreading mid-ocean ridge; hence VMS deposits there are the exact counterpart to black smokers. The mantle rocks in Cyprus are largely transformed into serpentinite by hydration which, when accompanied by a strong increase in volume, led to a diapir-like rise of material (Schuiling 2011). Accordingly, Mt. Olympos, the highest mountain of the Troodos Mountains, consists of the deepest rocks. Parts of the mantle still consist of peridotite in which there are lenticular accumulations of chromite (podiform chromite deposits; Sect. 3.2). At the top are deep-sea sediments (chert, shales) that have been deposited above the pillow lava. At their base there are lenticular accumulations of iron and manganese oxides or hydroxides that have been used as color pigments such as umber since ancient times.

These are exhalative deposits from the smoke plume of black smokers.

The VMS deposits are located in pillow lavas (especially, on the north side of the Troodos Mountains). They were former black smokers located on the seafloor at normal faults of the rift system. There were also other black smokers off the central axis. The deposits at Alestos and Memi are located in the calderas of a seamount (Eddy et al. 1998).

Although the chimneys of the black smokers no longer exist, the former sulfide mounds are preserved as massive sulfide bodies (mainly, pyrite and chalcopyrite but a small amount of sphalerite and galena) within pillow lavas because they were covered by even younger pillow lavas. Below there is still the ore-filled fissure zone (a.k.a., stockwork) in which hydrothermal water has risen surrounded by a strongly altered zone in which the original basalt minerals are largely replaced by chlorite, prehnite, epidote, quartz, hematite, albite, sericite, pyrite, chalcopyrite, and so on.

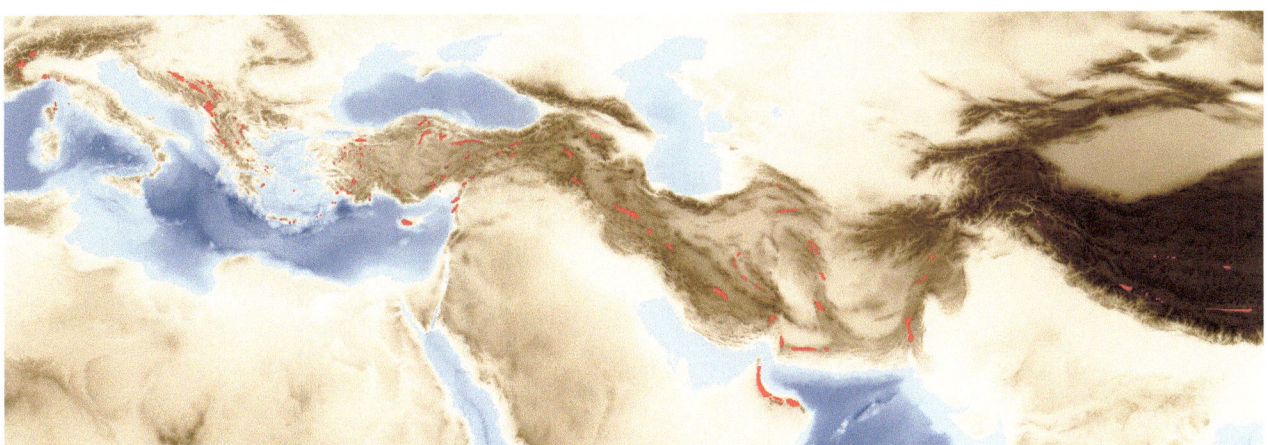

Fig. 4.65 Ophiolites are frequently found in the seams (sutures) between two continents after continental collision and are therefore usually lined up. The ophiolites shown here go back to several sea basins of the Tethys Ocean. Most have been strongly deformed and transformed into metamorphic rocks during continental collision. However, a few such as the Troodos Ophiolite on Cyprus and the Semail Ophiolite in Oman have been obducted in one piece. Some ophiolites contain Cyprus-type VMS deposits

Such deposits are typical of ophiolite complexes. VMS deposits in basalts of the ocean floor are generally referred to as Cyprus type. According to a new model it is now believed that even the Troodos Ophiolite was not created on a mid-ocean ridge but during the formation of a new sub-duction zone (Dilek and Furnes 2009; Pearce and Robinson 2010). Ophiolites from such a zone are called supra-subduction zone ophiolites. The composition of some basalts (calc-alkaline instead of MORB) and the process of obduction that took place shortly afterward are easier to explain with this model. The same is true of a number of other ophiolites that were obducted almost simultaneously all of which are located in a belt that runs from Cyprus through the eastern Taurus and the Zagros Mountains to the famous Semail Ophiolite in Oman (Fig. 4.65).

In the belt that continues on from Cyprus to the eastern Taurus in Turkey (Bölücek et al. 2004) there are further ophiolites with VMS deposits including Ergani Maden. The Semail Ophiolite in Oman is another classical occurrence. In the Balkans there are other ophiolites with VMS deposits such as Mirdita (Albania) and Pindos and Othrys (Greece) in which there are not only basaltic pillow lavas but atypically also andesitic pillows. The VMS deposits there have increased contents of Au, Ag, As, Se, Sb, Mo, and Hg (Economou-Eliopoulos et al. 2008). Ophiolites in the Balkans were formed either during the initiation of a subduction zone or in a back-arc. Such deposits can be regarded as a hybrid between the Cyprus type and the Besshi or Noranda type.

Box 4.27 Abitibi greenstone belt

In the Archean (3.8–2.5 billion years ago) the first continents were formed when countless island arcs and seamount ranges—built of material from the underlying mantle at hotspots—collided with each other. This resulted in cratons that today make up the cores of continents. Archean cratons consist mainly of granite and gneiss. However, they also contain areas made up of (former) volcanic rocks, sediments, and granites that are only slightly metamorphosed. As a result of low-grade metamorphism (greenschist facies) such rocks often have a green color.

The Abitibi greenstone belt (Fig. 4.66) in Ontario and Quebec is the largest greenstone belt in the Superior Province (the largest Archean craton in North America) and one of the most productive mining areas on Earth. The rocks were formed about 2.7 billion years ago by a mantle diapir forming a large basalt plateau and by subduction when several island arcs collided with the plateau.

Although igneous rocks predominate with basalt being the most common (pillow lava and dikes), there are also acidic volcanic rocks (andesite, rhyolite). Clastic sediments such as shales, turbidites, grey-wacke, and partly gold-bearing conglomerates were deposited between the volcanoes. Banded iron formations (Sect. 5.2) also occur.

Additional to volcanic rocks and sediments there are some large granite plutons and intrusions of

Fig. 4.66 Abitibi greenstone belt and its main deposits. The green area consists mainly of basalt (metamorphosed to greenschist) with some andesite, rhyolite, and komatiite as well as sporadic sediments including banded iron formations. VMS deposits are bound to acidic volcanic rocks. There are also magmatic Ni–Cu–Pt deposits and gold veins. The small map shows (*gray*) the Archean cratons of North America and (*green*) the location of the Abitibi greenstone belt in the Superior Province (modified from Thurston et al. 2008)

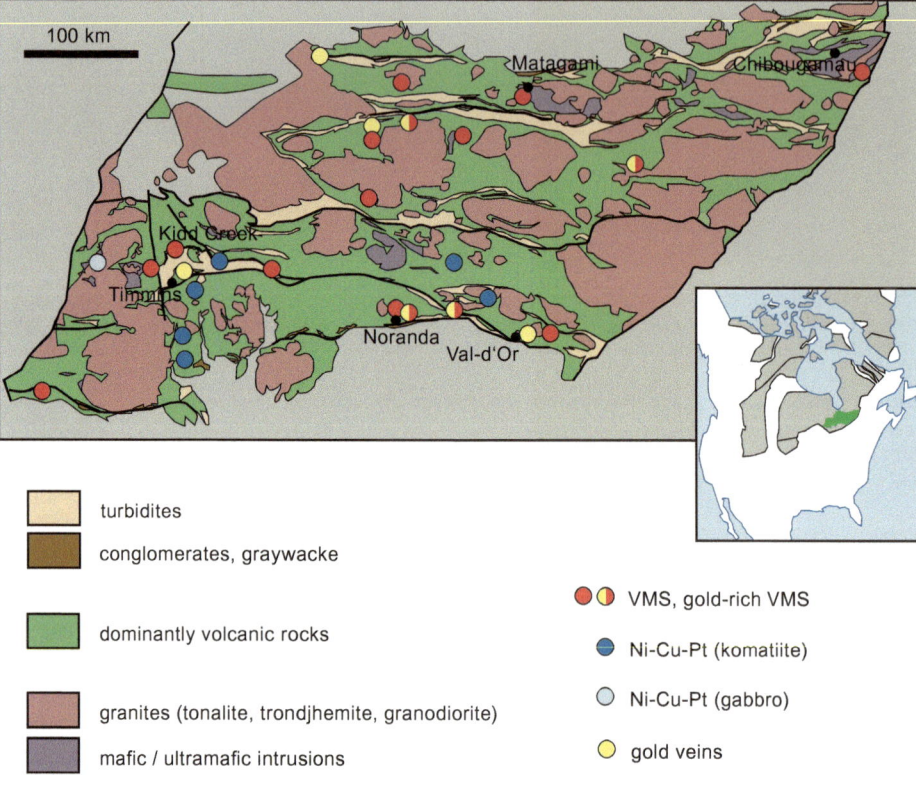

turbidites

conglomerates, graywacke

dominantly volcanic rocks

granites (tonalite, trondjhemite, granodiorite)

mafic / ultramafic intrusions

VMS, gold-rich VMS

Ni-Cu-Pt (komatiite)

Ni-Cu-Pt (gabbro)

gold veins

anorthosite. Strictly speaking, the granites are tonalite, trondhjemite, and granodiorite. This combination is typical of the Archean. Since the Earth at that time was even hotter slabs sinking in subduction zones may well have partially melted. The whole thing was then exposed to low-grade metamorphism (greenschist facies and partly amphibolite facies).

Large VMS deposits are generally bound to acidic volcanoes (Gaboury and Pearson 2008). Some ore bodies are unusually large and of high grade with Timmins (including Kidd Creek) and Noranda being the most significant mining districts. Although they mainly contain copper, zinc, lead, and silver, several VMS deposits also have a high gold content.

Komatiite (Sect. 3.4) is also found in the greenstone belt. In ultramafic lavas such as komatiite magmatic nickel deposits are found that also contain copper and platinum. In addition, there are countless orogenic gold veins (Sect. 4.2). Further gold veins and small gold-rich porphyry copper deposits were formed in connection with plutons (Sect. 4.4).

4.16.2 Besshi (Japan)

Two so-called paired metamorphic belts each consisting of a belt of high-pressure rock and a belt of high-temperature rock run lengthwise through Japan (Fig. 4.67). This is a section of a deep level of old subduction zones where a subducted oceanic plate transformed into eclogite (high-pressure rocks) and so on and crustal rocks partly melted (migmatite) as a result of heat from the volcanic arc (high-temperature rocks). The older subduction zone was active from the Permian to the Jurassic and the younger zone was active in the Cretaceous. Japan was still located on the continental margin of Asia during that period of time.

In the younger high-pressure belt termed the Sanbagawa Belt there are more than 100 VMS deposits called Besshi type for the largest deposit on the island of Shikoku. High-pressure rocks were once again transformed (retrograde) during their ascent; hence greenschists and amphibolites dominate instead of eclogites. In addition, there are large quantities of former shales and sandstones converted into various types of schists. Since deposits deformed together with these rocks the original geometry has not been preserved. For example, there is no stockwork zone. Stratiform or lenticular sulfide bodies occur in the metabasalts or

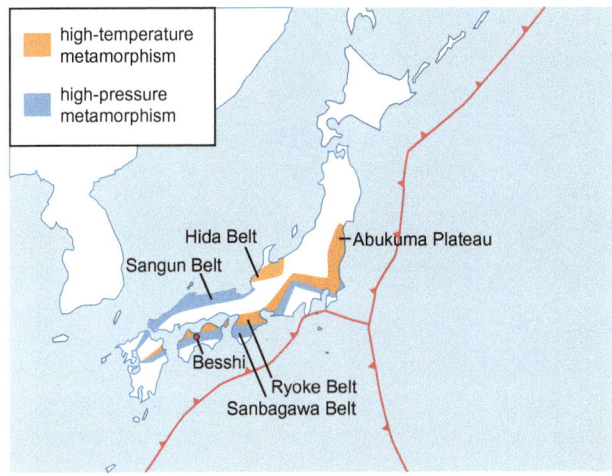

Fig. 4.67 The two so-called paired metamorphic belts in Japan. Each consists of a belt with high-pressure metamorphism and a belt with high-temperature metamorphism (Abukuma type). They are sections through a deep level of old subduction zones representing the sinking slab and the volcanic arc. The subduction zones were active from the Permian to the Jurassic (Hida and Sangun Belts) and during the Cretaceous (Ryoke and Sanbagawa Belts). There are more than 100 VMS deposits in the Sanbagawa Belt. They are named Besshi type for the largest deposit

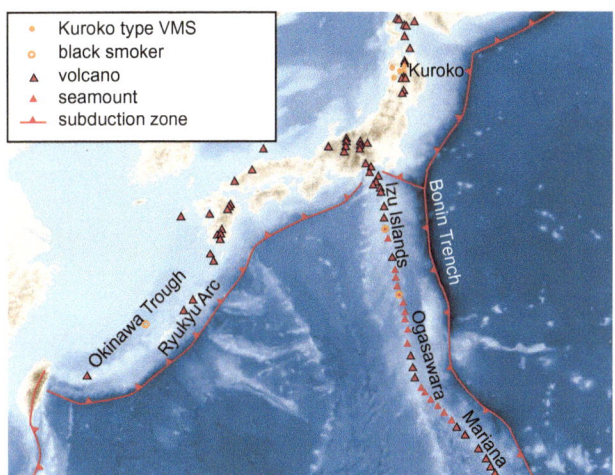

Fig. 4.68 Location of Kuroko-type deposits on the Japanese island of Honshu. Black smokers found in the Okinawa Trough (back-arc of the Ryukyu Arc where continental crust is stretched) and on seamounts of the Izu Ogasawara Arc are considered recent analogs

in their vicinity in metasediments. Often there are also chert layers containing iron and manganese and sometimes small bodies of exotic rocks that were formed during the metamorphism of altered country rocks or exhalative layers (e.g., coticule, a schist containing quartz and spessartine).

Their metal content is similar to that of Cyprus-type deposits in that the ores consist mainly of pyrite and chalcopyrite with small amounts of sphalerite and bornite; hence copper and some zinc have been mined. It is possible that Besshi-type deposits were formed in the late Jurassic in the back-arc of an island arc where new oceanic crust was formed similar to what happens at mid-ocean ridges, while sediments were supplied from the island arc at the same time. According to another model they were created on a real mid-ocean ridge that was subsequently covered with sediments (Nozaki et al. 2006, 2010). The corresponding oceanic plate called the Izanagi Plate was then subducted completely under the edge of Asia in the Cretaceous. Some of it resurfaced in the Sanbagawa Belt.

4.16.3 Kuroko (Japan)

Kuroko are stagehands in traditional Japanese theater. Despite being dressed in black they are not completely invisible as they move the scenery. Kuroko is also an old Japanese miner designation for black ore. In 1919 Ohashi developed the concept of submarine exhalative formation of VMS deposits. He was the first to explain how such deposits

formed at classical localities in the northeast of the Japanese island of Honshu (Ohmoto 1996; Shikazono 2003) (Fig. 4.68).

Kuroko-type deposits were formed 14–15 million years ago on the seafloor in a back-arc with thinned continental crust at a time of strong bimodal volcanism (i.e., simultaneously acidic and basic magmas). The former shallow sea had sunk to a depth of about 3500 m due to thinning only shortly before. Most ore bodies are located in and above lava domes or lava flows within acidic volcanic rocks (dacite or rhyolite) or in depressions next to them. The lava domes were probably located in calderas of submarine volcanoes. The basin was then filled with tuff and shales and finally uplifted and folded. At less than 1 million tonnes most ore bodies are small. Nevertheless, the Matsumine ore body of the Hanaoka Mine had about 60 million tonnes and many small ore bodies occur in clusters that were mined together between 1883 and 1994. Copper, lead, zinc, iron, gold, silver, baryte, and gypsum were produced.

Typical Kuroko-type deposit ore is a sequence of black, yellow, and siliceous ores together with large quantities of baryte, anhydrite, and gypsum (Fig. 4.69). The siliceous ores (*Keiko*) consist of quartz, pyrite, and chalcopyrite, which is typical of the stockwork zone. The yellow ores (*Oko*) lie directly above the lava and consist mainly of pyrite and chalcopyrite but also contain sphalerite, bornite, quartz, and baryte. The black ores (*Kuroko*) are located above (and also farther away) and form at lower temperatures. They consist of galena and sphalerite together with pyrite, chalcopyrite, tennantite–tetrahedrite, bornite, electrum, and baryte. Higher still is a thin layer of chert containing hematite and other ore minerals. Within the ore body there are also accumulations of sulfates such as gypsum ore (gypsum, anhydrite with

Fig. 4.69 Schematic of a Kuroko-type VMS deposit with siliceous ore in the stockwork zone and a massive sulfide body of yellow ore (pyrite, chalcopyrite) and black ore (galena, sphalerite) (based on Ohmoto 1996)

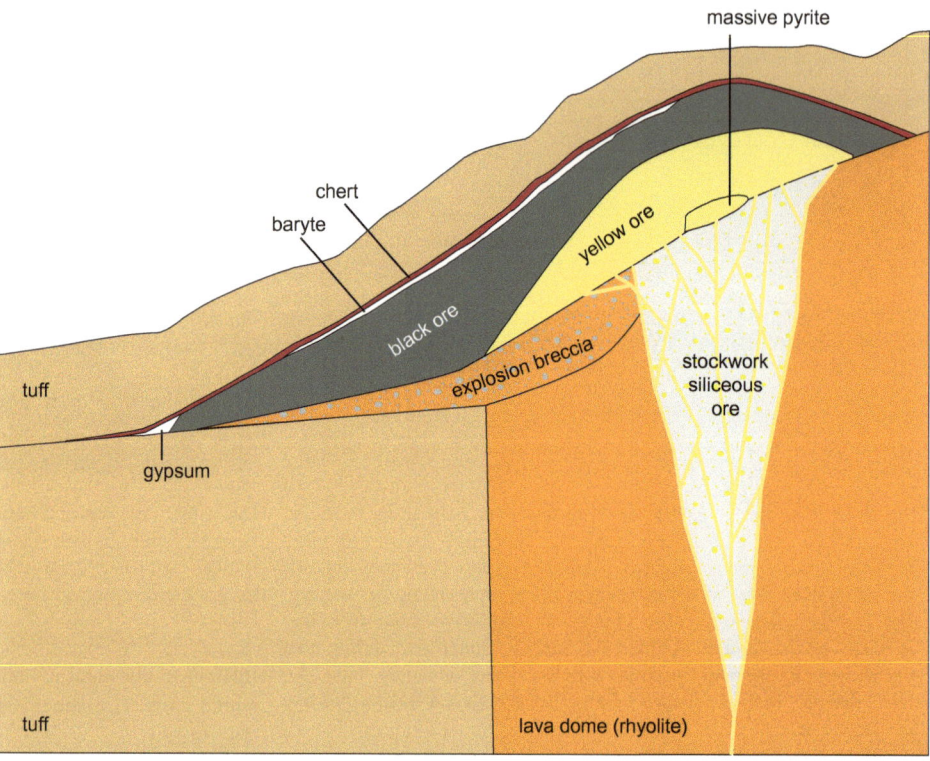

sulfides) and baryte ore (predominantly baryte, sometimes with calcite, dolomite, and siderite). Sometimes there are massive pyrite bodies within tuff or below the yellow ore.

The black ores were formed by hydrothermal water mixing with cold seawater. The yellow ores were formed later replacing black ores in hotter areas of the hydrothermal system. They are richer in copper as a result of the higher temperature. As with other VMS deposits the hydrothermal fluid was mainly seawater that had seeped through cracks into the depths, been heated there, and leached metals out of the rock. However, the continental crust below Kuroko-type deposits has a high proportion of felsic rocks that differ in their metal content from basaltic oceanic crust. In addition, magmatic water directly contributes to Kuroko-type VMS deposits.

A recent analog of Kuroko-type VMS deposits is a hydrothermal field with black smokers located in the Okinawa Trough between the Japanese island of Kyushu and Taiwan. The Okinawa Trough is the back-arc basin of the Ryukyu Arc (Lüders et al. 2001) where continental crust of the Eurasian Shelf is stretched. Such stretching and bimodal volcanism correspond well with classical Kuroko localities. However, some metals (Pb, Zn, Sb, As, Ag) are much more enriched than is the case with classical Kuroko ores.

There are also black smokers at several active seamounts of the Izu-Ogasawara island arc (Urabe and Kusakabe 1990; Glasby et al. 2008). They are located at lava domes (partly within calderas) and have very similar metal contents to

classical Kuroko deposits. However, in contrast to such deposits these black smokers are located on the active island arc within a section where the arc itself has extended—not in a back-arc with continental crust. No real back-arc basin has yet developed here—as is the case farther south with the Mariana Arc.

4.16.4 Iberian Pyrite Belt

The Iberian pyrite belt is home to 8 of the world's 20 largest VMS deposits. They are Río Tinto, Aznalcóllar–Los Frailes, Sotiel–Migollas, Tharsis, La Zarza, Masa Valverde, Aljustrel, and Neves Corvo each of which holds more than 100 Mt of ore. In addition there are 44 medium-sized and hundreds of small deposits located in this narrow belt in the south of Portugal and Spain. It runs through the South Portuguese Zone (SPZ), which is part of the long-eroded Variscan Mountains (Fig. 4.70).

Mining began as early as the Copper Age and Early Bronze Age. In Roman times metal production was already becoming industrial with enriched and easier-to-process ores of the oxidation zone being mined. At that time about half the worldwide production of copper, lead, and silver came from the pyrite belt as did gold. In Tharsis, Río Tinto, Aljustrel, and other districts there are still several million tonnes of slag on dumps from Roman times. After a few centuries of low activity mining resumed in the 19th and

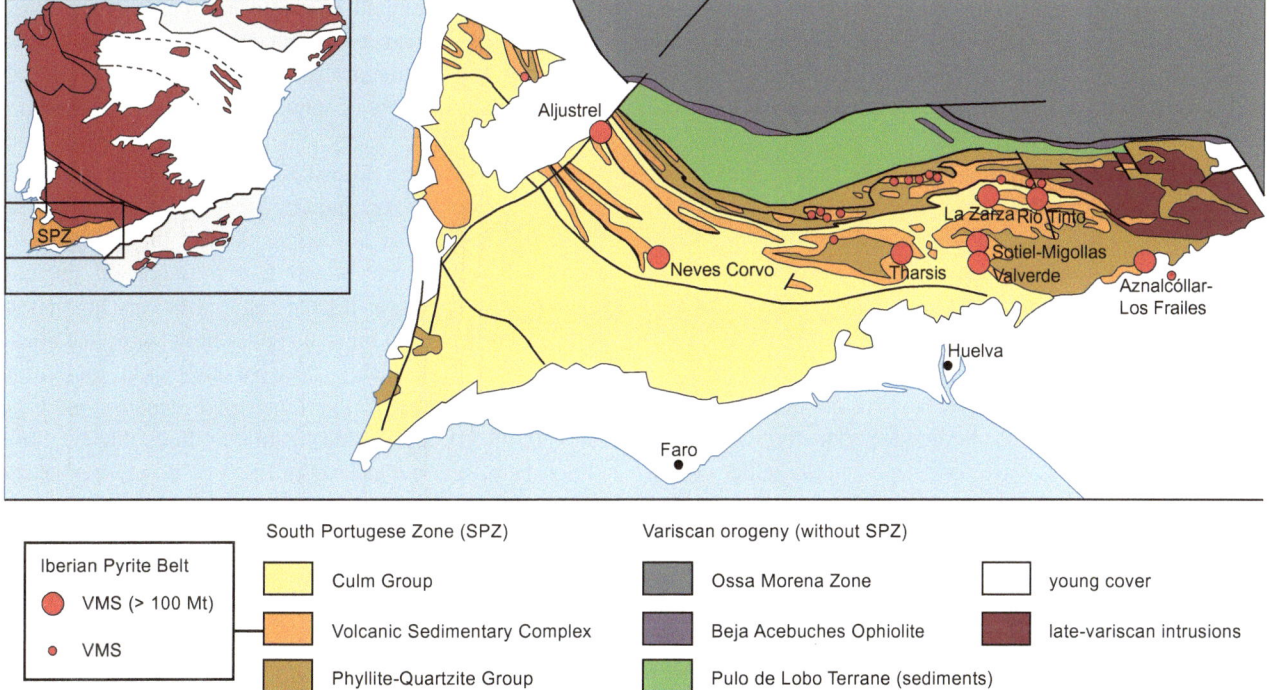

Fig. 4.70 The Iberian pyrite belt hosts a number of large VMS deposits. Located in the South Portugese Zone (SPZ) it is an exotic terrane that collided with the Ossa Morena Zone during the Variscan orogeny. The inset shows the location of the SPZ together with other units of the Variscan Mountains on the Iberian Peninsula (from Tornos 2006)

twentieth centuries. Neves Corvo and Masa Valverde were only discovered a few decades ago.

The Iberian pyrite belt is not only unusually rich it was also formed in an unusual way (Tornos et al. 2005; Tornos 2006). It was created during an oblique collision between a mini continent (the SPZ) and a subduction zone at the edge of a continent (Ossa Morena Zone) during the formation of the Variscan Mountains (Fig. 4.71). Although the basement of the mini continent is unknown, sediments (shale and sandstone) 2000 m thick were deposited on its shelf before collision. Being slightly metamorphic they are called the Phyllite–Quartzite Group of the SPZ. During the oblique collision, which happened toward the end of the Devonian, the mini continent was segmented and deformed by countless strike-slip faults. By local extension (transtension) pull-apart basins were torn open into which predominantly acidic crustal-derived melts (rhyolite and dacite lava domes and their pyroclastic flows and breccias) and some basalt poured as a result of bimodal volcanism. Shale (and some chert, jasper) was deposited at the same time forming the Volcanic Sedimentary Complex of the SPZ. The juxtaposition of magmatism and active strike-slip faults provided perfect conditions for strong hydrothermal activity. It was probably seawater that leached out metals from the

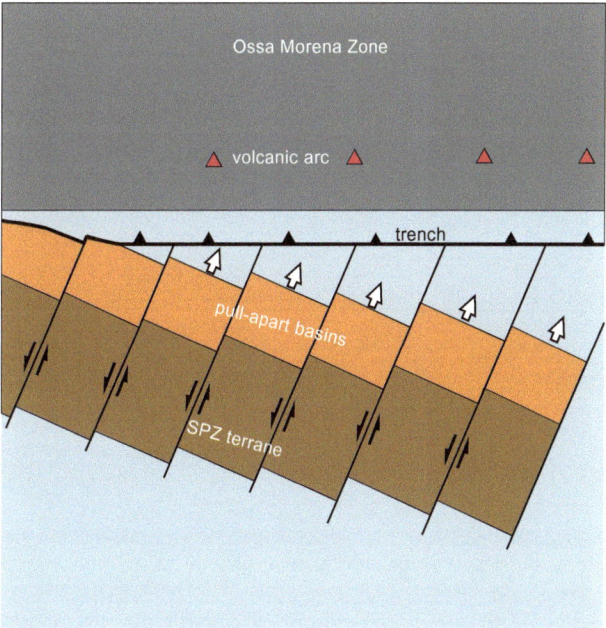

Fig. 4.71 Schematic representation of the oblique collision of the SPZ terrane with the Ossa Morena Zone. The terrane was deformed by strike-slip faults resulting in local stretching and opening of pull-apart basins (*orange*). Volcanism, sedimentation, and the formation of VMS deposits occurred in these areas

sediments of the Phyllite–Quartzite Group with the possible contribution of magmatic fluids (the significance of which is disputed). A little later the basins collided with the continental margin. The deposition of submarine landslides that resulted in the formation of turbidite and flysch (the Culm Group of the SPZ) occurred before the sediment stack was glued to the edge of the continent in the form of thin nappes pushed one on top of the other. Since this all happened in an oblique direction the original geometry is not known.

The ores located in the Volcanic Sedimentary Complex comprise lenticular or stratiform massive sulfides as well as stockwork zones. Gold is enriched in pronounced oxidation zones. La Zarza has the largest massive ore body at 170 Mt. Clusters of up to six ore bodies are more typical. In Río Tinto (Fig. 4.72) there are four zones that together contain 500 Mt of massive sulfides and 2000 Mt of low-grade stockwork ore—possibly the largest accumulation of sulfides in the entire crust of the Earth. The name derives from the Río Tinto, one of the rivers most heavily polluted with heavy metals on the planet. Its water is strongly acidic (pH 2.3) and red in color due to the weathering and oxidation of sulfides. The deposit also gave its name to the multinational Rio Tinto mining group.

Although the amount of ore in the Iberian pyrite belt is enormous, it consists predominantly of pyrite and almost invariably small amounts of chalcopyrite, sphalerite, and galena. Small amounts of tennantite–tetrahedrite, arsenopyrite, pyrrhotite, and many other minerals are also found. Although the average deposit has only 0.85% Cu, 1.13% Zn, 0.53% Pb, 38.5 g/t Ag, and 0.8 g/t Au, there are zones that are more enriched. A few deposits have higher ore grades such as Neves Corvo. Neves Corvo stands out in many respects. Not only does it have zones with very high copper and zinc contents, but tin is also present in large quantities in the form of high-quality massive cassiterite (and minor stannite). It is one of the largest tin deposits on Earth. Although it is possible that magmatic fluids made a contribution here, it is perhaps more likely that seawater leached out tin-bearing granites in the basement of the SPZ.

Ore bodies were strongly deformed during continental collision and overthrusting; hence most are located directly at faults. Often several lenses or parts of a large lens were stacked and combined to form duplex structures; in other cases large ore bodies were divided into smaller ones. Today's zoning of ore bodies is likely partly based on the different deformability of sulfides (Castroviejo et al. 2011).

Fig. 4.72 Río Tinto open-pit mine and its heavily altered rocks (*photo* © F. C. G./Fotolia)

Moreover, metals were internally remobilized by metamorphic fluids resulting in an accumulation of copper, zinc, lead, gold, and silver in ore shoots in the most deformed zones (Matignac et al. 2003).

In general, the deposits can be divided into two groups formed in different ways. In the north (La Zarza, Aljustrel, parts of Río Tinto, and some smaller deposits) they occur mainly within acidic volcanic rocks (especially, in tuffs and breccias near lava domes) (Tornos 2006; Rosa et al. 2010). They were probably formed underground when hot fluid mixed with cold seawater in pore-rich volcanic rocks made of soluble glass. They are strata bound because they formed either under a waterproof layer (shale or massive volcanic rocks) or in a layer rich in pores.

However, in the south (Aznalcóllar–Los Frailes, Sotiel–Migollas, Masa Valverde, Tharsis, Neves Corvo, and some parts of Río Tinto) they are almost exclusively found in shales with little or almost no volcanic rocks. Here the ore bodies are strata bound and tend to be larger and less zoned than in the north. They also contain carbonates such as siderite but hardly any sulfates. Sometimes typical sedimentary structures such as small landslides can be observed. They are interpreted as deposits of hot springs in a brine pool on the seafloor similar to the Atlantis II Deep (Tornos 2006; Solomon 2008) and as quasi-hybrids between VMS and SEDEX. In the case of Neves Corvo the brine pools were located near rhyolite domes that had been active a little earlier (Rosa et al. 2008).

Although the hydrothermal fluids were salt rich, they contained little sulfide and transported metals mainly as chloride complexes. The sulfur of the ores probably came partly from seawater whose sulfate was reduced by bacteria or by redox reactions with the surrounding rock.

Manganese-rich chert and jasper layers in the Volcanic Sedimentary Complex are also remarkable in having hundreds of deposits of manganese oxides, silicates, and carbonates (Sect. 5.5) making Spain the world's largest producer of Mn in the late nineteenth century.

About the same time as the VMS deposits of the Iberian pyrite belt were forming other deposits were also being formed farther north in the Ossa Morena Zone including skarns, magmatic nickel ores, gold veins, lead–zinc veins, impregnations with mercury, VMS deposits, and an IOCG deposit. They are related to a volcanic arc, to local extension at strike-slip faults, and to a mafic intrusive complex presumably located at depth (Tornos 2006).

4.17   Sedimentary Exhalative Deposits

On the seafloor there may also be hydrothermal systems in sediment basins far removed from volcanism where massive sulfide ores are deposited. They are referred to as sedimentary exhalative (SEDEX) deposits or alternately as stratiform lead–zinc deposits or sediment-hosted massive sulfides (SHMS). Being "far removed from volcanism" is relative. Actually there is a smooth transition to sediment-rich VMS deposits.

SEDEX deposits are usually larger than average VMS deposits and have a higher ore grade. The temperature of the fluid is lower (between 100 and 250 °C); hence lead and zinc contents are higher and the copper content is lower. An alteration zone is less pronounced and only rarely is there a kind of stockwork zone.

The hydrothermal system was located either in the sediments of a flooded continental rift or on a young passive continental margin that was further stretched after the opening of a new ocean and sank under the weight of the sediments deposited. In both cases hydrothermal water rose along active faults. At the same time sediments were deposited (especially, clay minerals and organic matter) from which shale or black shale was formed (see also Sect. 5.1.1). There are also SEDEX deposits with carbonate rocks (e.g., in Ireland where they occur together with MVT deposits) (Sect. 4.12).

Box 4.28 Rammelsberg (Germany)

Rammelsberg is a mountain in the Harz Range (Large and Walcher 1999; Liessmann 2010) just south of Goslar, Lower Saxony (Germany) where mining has been carried out almost continuously for more than a thousand years (Figs. 4.73 and 4.74). Sporadic mining of copper ore probably began as early as the Bronze Age. In the early Middle Ages there is clear evidence of ore from Rammelsberg being smelted in the fourth century long before the first written mention in the year 968. Silver was also in demand in the Middle Ages as demonstrated by the many famous Romanesque bronze works of art made of copper from Rammelsberg. By today's standards Rammelsberg would be a lead–zinc deposit with copper, silver, and gold as by-products.

Initially, the Rammelsberg mines belonged to the German Emperor and were one of the most important sources of income for the empire. Later it came into the hands of the Dukes of Braunschweig who in turn granted all rights to the city of Goslar. At that time there were still a large number of small trenches and galleries dug by hand with hammer and pick within the oxidation zone, which was enriched with copper and silver. In the following centuries primary sulfide ores were mined access to which was achieved by excavating ever-deeper shafts and galleries expedited by fire setting. In the mines vitriol (iron sulfate, zinc sulfate, copper sulfate) formed by oxidation. Such

Fig. 4.73 The Rammelsberg Mountain near Goslar, Lower Saxony (Germany) with the winding tower and buildings of the ore-processing plant (*photo* © F. Neukirchen)

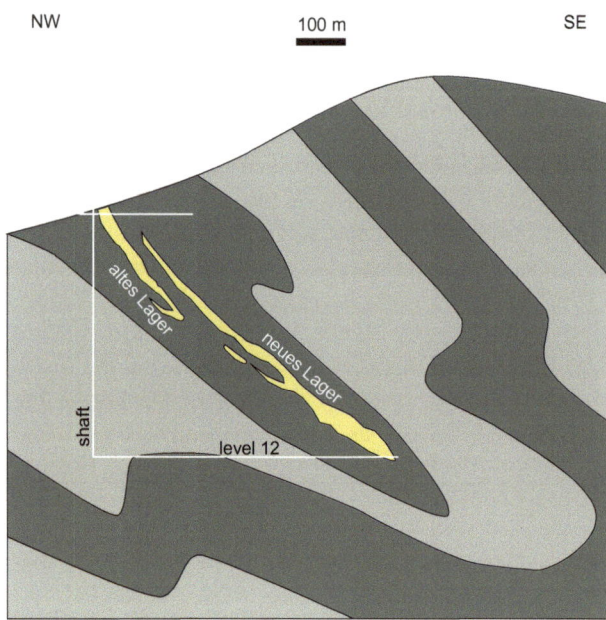

Fig. 4.74 Schematic section through the Rammelsberg Mountain. Contrary to this representation the two massive sulfide bodies called the altes Lager ("Old Camp") and the neues Lager ("New Camp") are laterally offset. In the Devonian they were formed exhalatively at hot springs on the seafloor as black shale was being deposited. In the Carboniferous during the Variscan orogeny they were folded; hence most of the ores are located in the overturned limb of a fold (from Liessmann 2010)

the Thirty Years' War brought about a break in production. Since the eighteenth century there has been a resurgence. In 1859 a second ore body was discovered. This was the neues Lager, which was about twice the size of the altes Lager. This marked the start of machines being increasingly used. During the Nazi period a new processing plant, a new shaft, and the use of forced labor enabled a further increase in production. In addition, zinc, gold, and baryte have also been extracted. After the war and up to around 1960 there were record years when up to 320,000 t of ore per year were extracted. The deposit was eventually exhausted and mining had to be stopped in 1988.

The nearly 30 Mt of ore at Rammelsberg had an unusually high grade: 14% zinc, 6% lead, 2% copper, 140 g/t silver, 1 g/t gold, and 20% baryte. Ore occurred as massive sulfides or as banded ore with alternate layers of ore minerals, shale, and carbonate. The remains of a kind of stockwork zone were also preserved called *Kniest* by the miners.

The sulfides were deposited exhalatively in the Devonian sea at sources on the seabed. They were located on the continental shelf of the Rhenohercynian Terrane (that had only recently been formed by the opening of a new ocean) at an active normal fault at the edge of the former Goslar Trough, which was simultaneously filled with clay minerals (black shale). A little later (i.e., in the Variscan orogeny) the sediments were bent into a large fold. The ore bodies are mainly located along the fold axis and in the

sulfates were exported to Flanders and used in tanneries and dye works. In its heyday in the sixteenth century up to 20,000 t of ore were mined per year until overturned limb; hence they are more or less lying on their heads. The deepest areas of the neues Lager are superimposed by folding that is twice as thick.

At about the same time as the Rammelsberg deposit a second SEDEX deposit was created on the same continental shelf at Meggen in the Rhenish Massif, Westphalia (Germany). Although it is twice as large, it has a low ore grade. It consists mainly of pyrite. Apart from a little zinc and lead there are hardly any other metals. Pyrite was temporarily mined for the production of sulfuric acid, together with baryte, and as by-products some zinc and lead.

SEDEX deposits contain more than half the known reserves of lead and zinc and significant amounts of silver. Among the most important examples are Rammelsberg in the Harz Mountains (Germany) (Box 4.28); McArthur River (HYC deposit), Mount Isa, and Broken Hill (Australia); Sullivan and Howard's Pass (Canada); Red Dog (Alaska, USA); Gamsberg (South Africa); and Rajpura-Dariba (India) (Fig. 4.75). The stratiform ore bodies are several meters to dozens of meters thick and hundreds or even thousands of meters wide. Typical are layered, finely banded ores with sulfides (especially, pyrite, sphalerite, and galena), carbonates (e.g., siderite), and clay minerals (Figs. 4.76, 4.77 and 4.78). The spectrum ranges from massive sulfides containing hardly any carbonates or clay minerals to sedimentary rocks with low sulfide content. Often large amounts of baryte also occur.

Their origin is believed to be much the same as happened in the Atlantis II Deep (Sect. 4.15.2) where salt-rich hydrothermal solutions rose along faults, emerged at a source on the seafloor, and collected in depressions because they have a greater density than seawater (Fig. 4.79). The mixing of different waters caused the precipitation of ore minerals. Since clay minerals were deposited at the same time metal-rich sludge collected at the bottom of depressions. In this case ore minerals formed simultaneously with sediments (syngenetic).

Hydrothermal water can also penetrate sludge that has not yet solidified and sediment that is only slightly compacted (a few meters below the seafloor) and precipitate sulfides or replace older sulfides (diagenetically). The reaction with pyrite previously precipitated by sulfate-reducing bacteria can play a role here that is very similar to diagenetically formed mineralizations in copper-bearing shales (Sect. 5.1). At greater depths the shale is already more or less impermeable to water where sulfides can be precipitated epigenetically (i.e., when the ores are younger than the sediment; see also Box 4.29) in porous sediment layers preferably under impermeable shale layers. It is not easy to distinguish whether an ore body was formed on the seafloor or below it (e.g., Chen et al. 2003; Ireland et al. 2004) and in many deposits both may have happened. Sometimes landslides in the sediment give an indication of formation on the seabed.

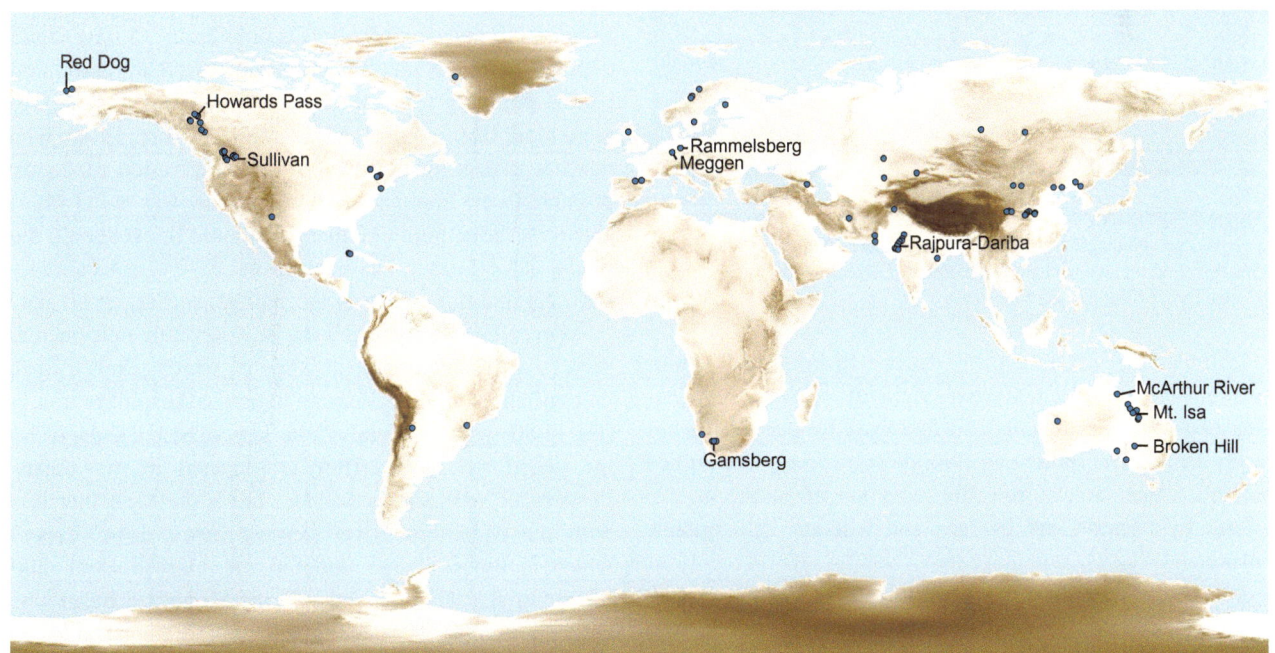

Fig. 4.75 World map of important sedimentary exhalative (SEDEX) deposits (data from World Minerals Project, Geological Survey of Canada)

Fig. 4.76 Banded sedimentary exhalative (SEDEX) ore with chalcopyrite, pyrite, galena, and sphalerite from Rammelsberg, Harz Mountains (Germany) (*photo © F. Neukirchen/Mineralogical Collections of the TU Berlin*)

Fig. 4.77 Massive sulfide ore (SEDEX) with chalcopyrite, pyrite, galena, and sphalerite from Rammelsberg, Harz Mountains (Germany). *SEDEX*, Sedimentary exhalative (*photo © F. Neukirchen/Mineralogical Collections of the TU Berlin*)

Similar to other deposit types, hydrothermal fluids responsible for SEDEX were originally seawater that has leached deeper sediments and dissolved evaporites. Metals were mainly transported as chloride complexes. The sulfide of ores came largely from the reduction of sulfate of seawater by bacteria and by heat and reaction with organic matter.

Since water was often predominantly in contact with shales and other clastic sediments it was somewhat reduced ($H_2S > SO_4^{2-}$) and acidic. Such water could transport barium that later precipitated as baryte and possibly gold. In

Fig. 4.78 Banded ore (SEDEX) with pyrite, chalcopyrite, sphalerite, galena, from Meggen, Westphalia (Germany). *SEDEX*, Sedimentary exhalative (*photo © F. Neukirchen/Mineralogical Collections of the TU Berlin*)

Sullivan (Canada) water was so reduced that tin could also be transported and precipitated. However, in the case of the large deposits in northern Australia water was slightly oxidized ($H_2S < SO_4^{2-}$) because it was in contact with carbonates, evaporites, and hematite-containing sandstone at depth (Cooke et al. 2000). These deposits only contain zinc, lead, silver, and siderite but no baryte or gold.

Almost all SEDEX deposits were formed between the Proterozoic and the Middle Carboniferous. A few smaller deposits from the Jurassic are exceptions. The reason could be that during this period anoxic or low-oxygen water layers were more likely to form in the oceans (Turner 1992), which is what makes the formation and preservation of deposits possible in the first place. Although there was not always an anaerobic brine pool like that of Atlantis II, eventually there would have been a sharp jump in the oxygen content of seawater and pore water in the sediments (Sáez et al. 2011).

Many SEDEX deposits were later severely deformed and subjected to metamorphism some of which resulted in further enrichment. A good example is the Australian zinc belt (Fig. 4.80) where several of the largest SEDEX deposits in the world are found. They are located in two adjacent Proterozoic sedimentary basins. One is the McArthur Basin with the McArthur River deposit (a.k.a. HYC deposit), which is by and large undeformed (Garven et al. 2001; Ireland et al. 2004). The other is the Mount Isa Basin where the metamorphic grade increases the more the basin extends to the southeast. Not only was sedimentary rock transformed into metamorphic rock (releasing large amounts of water),

Fig. 4.79 SEDEX deposits are formed at the same time as sedimentation (especially, of shale). Since hydrothermal fluids can ascend at an active fault sulfides are precipitated at hot springs on the seafloor (exhalative) where metal-rich sludge accumulates and below the seafloor by impregnation and replacement in porous or unconsolidated sediments. *SEDEX*, Sedimentary exhalative

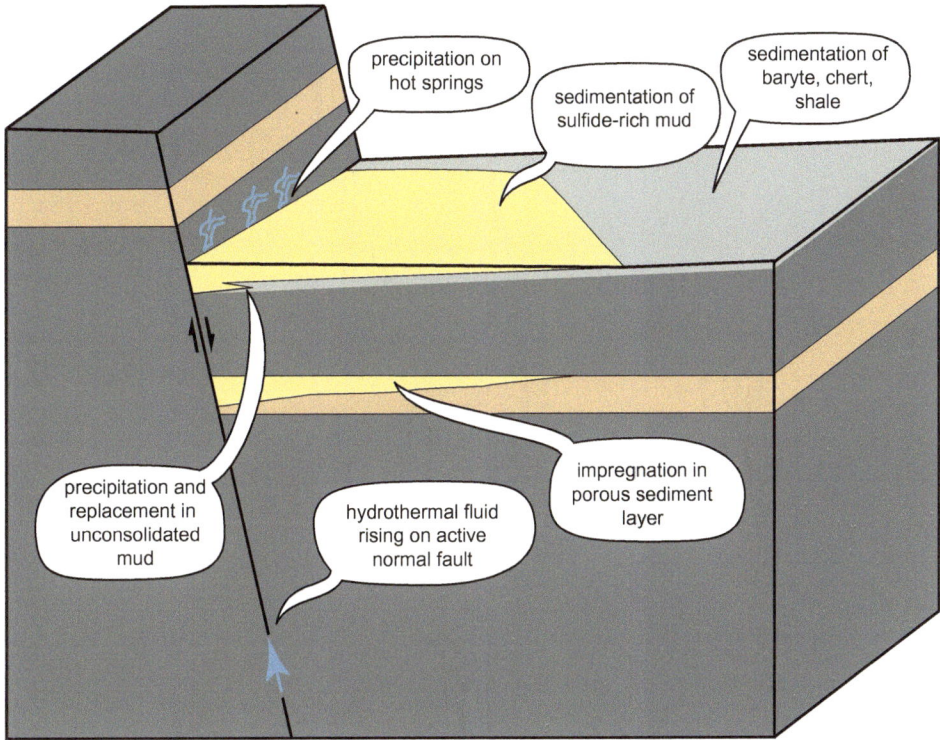

but sulfides were also remobilized. This happened both on a small scale where the original layering can still be ascertained to a certain extent (Feltrin et al. 2009) and on a large scale that led to new mineralization in shear zones and veins (Gessner et al. 2006). An extreme example is Mount Isa where there is also a high copper content. This contrasts with the less metamorphic deposits of the zinc belt. Copper ores were only produced as a result of replacement reactions with hydrothermal fluids during metamorphism. This may even apply to lead and zinc sulfides that are found there (Perkins 1997)—mainly in shear zones. Remobilization was so strong that Mount Isa can hardly be called a SEDEX deposit anymore.

At Broken Hill (New South Wales, Australia) metamorphism was even stronger. Together with the host rock this deposit has been strongly deformed at least five times and on occasion temperatures approached 800 °C at a depth of about 20 km (granulite facies). Although sulfides could have melted in the process, it is perhaps more likely that the temperature was sufficient to produce just a few melt droplets (Spry et al. 2008).

Box 4.29 Salton Sea (California, USA)
The Salton Sea is today a salt lake in the dry south of California that formed as a result of a dam breaking. Scientists use it as a kind of accidental experiment on hydrothermal deposits. It used to be a dry salt pan (Sect. 5.7.2) in a basin located directly on the San

Andreas Fault almost 100 m below sea level. In 1905 a flood broke the dam of an irrigation channel that supplied water from the Colorado River to the Imperial Valley diverting the entire river into the basin. It took two years to repair the dam and force the river back into its natural bed. In the interim the Salton Flat salt pan had become a big lake and was given the name the Salton Sea.

As a result of evaporation and salts dissolving on the bottom of the lake the water since then has become ever more salty. Water seeps deep down through older lake sediments and along faults and leaches soluble substances from the sediments. The region is magmatically active and water is accordingly heated. At a depth of 1000–3000 m hot brine at about 350 °C accumulates (Williams and McKibben 1989). The brine has a high density made up of 20–25% dissolved solids. Although these are mainly Na–Ca–K–Cl sulfides, there are also metals such as Fe, Mn, Pb, Zn, Cu, Li, and Ag. The water is used for geothermal power generation. Hydrothermal water also rises by itself at faults and emerges at hot springs. Drillings show that ores have already formed at depths of 750–1500 m by this salty thermal water mixing with cooler surface water (McKibben and Elders 1985). These are mainly Fe–Cu–Zn–Pb sulfides in fine veins and pores with pyrite predominant at the top. This process shows how strata-bound deposits can develop within older

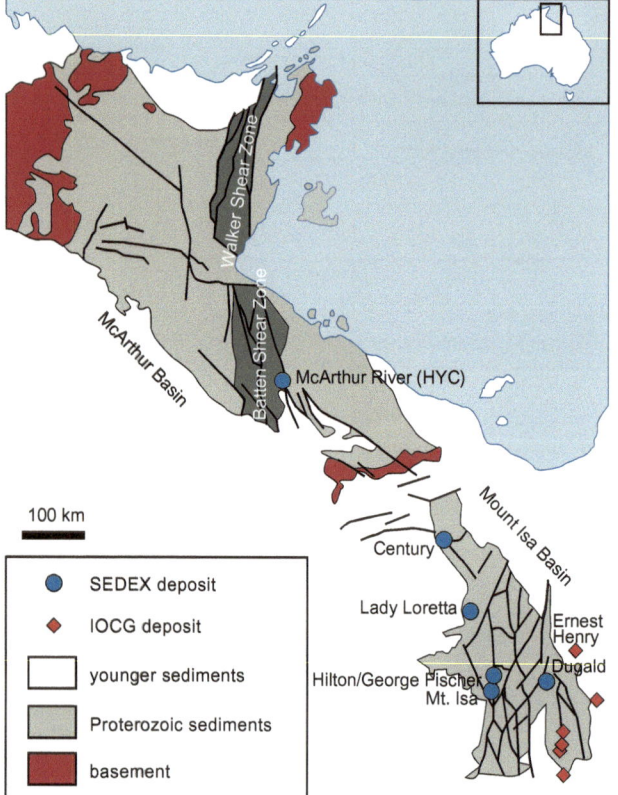

Fig. 4.80 The Australian zinc belt contains several Proterozoic SEDEX deposits. The McArthur River deposit is colloquially termed the "here's your chance" (HYC) deposit. Although it is the largest in the world with 227 Mt of ore, it has a relatively low grade. Four others have more than 100 Mt of ore and thus can be listed among the world's most massive deposits. They are located at faults in two adjacent Proterozoic sedimentary basins: the McArthur Basin and the Mount Isa Basin. The former is almost undisturbed, while the latter has been transformed into metamorphic rocks. The degree of metamorphism increases to the southeast (subgreenschist to amphibolite facies). There are also several IOCG deposits in the east of the Mount Isa Basin (Sect. 4.7) (from Cooke et al. 2000)

minerals such as iron sulfide, copper sulfide, sphalerite, galena, tennantite–tetrahedrite, and silver minerals (even native silver). The iron-rich silica sinter contains 20% copper and up to 6% silver (Skinner et al. 1967). Experiments have been carried out at times to extract zinc from brine using ion exchangers. Lithium is to be extracted in the future.

4.18 Lahn–Dill-Type Iron Deposits

Iron deposits considered exhalative are referred to as Lahn–Dill type after their occurrence in Hesse (Germany). The deposits comprise stratiform so-called bloodstone consisting mainly of hematite. The deposits are located within heavily altered basalts and tuffs. Comparable recent deposits can be observed on some volcanic islands at hot springs. After hydrothermal fluid mixes with seawater at such springs Fe^{2+} quickly oxidizes to Fe^{3+} in a process that is accelerated by certain bacteria. However, this can only happen if hardly any H_2S is present.

Although this has a certain resemblance to banded iron formations (Sect. 5.2), precipitation in today's oxidized oceans occurs directly at the source. Similar occurrences can also be found in sediments. For example, in carbonate facies siderite is formed instead. An example of an exhalative stratiform siderite deposit is Vares (Bosnia-Herzegovina). Lahn–Dill-type ores are widespread all over the world and were economically significant until the middle of the twentieth century. Further examples can be found in Germany (Siegerland, Harz Mountains, and Thuringian Forest), Ireland, and Hungary. In India, North and South America, South Africa, and Australia there are metamorphosed occurrences.

sediments. It also plays a role in SEDEX deposits. Although the salt lake is on land, it is an important analog to understanding hydrothermal systems on the seafloor.

At a depth of less than 760 m and a temperature less than 250 °C diagenetic mineralization takes place in which fine-grained iron sulfides form as cement in sandstone or are disseminated or banded in shales. The latter are comparable with copper-bearing shales like the Kupferschiefer (Sect. 5.1.1).

There are also deposits in the pipes of geothermal power plants since pressure relief during ascent leads to sudden evaporation in much the same way as happens with epithermal veins (Sect. 4.3). This produces 2–3 t of siliceous sinter monthly containing ore

Literature

Anderson, G.M. 2008. The mixing hypothesis and the origin of Mississippi Valley-Type ore deposits. *Economic Geology* 103: 1683–1960.

Anschutz, P., and G. Blanc. 1995. Chemical mass balances in metalliferous deposits from the Atlantis II Deep, Red Sea. *Geochimica et Cosmochimica Acta* 59: 4205–4218.

Anschutz, P., and G. Blanc. 1996. Heat and salt fluxes in the Atlantis II Deep (Red Sea). *Earth and Planetary Science Letters* 142: 147–159.

Anschutz, P.G., C.Monnin Blanc, and J. Boulège. 2000. Geochemical dynamics of the Atlantis II Deep (Red Sea): II. Composition of metalliferous sediment pore waters. *Geochimica et Cosmochimica Acta* 64: 3995–4006.

Audetat, A., T. Pettke, C.A. Heinrich, and R.J. Bodnar. 2008. The composition of magmatic-hydrothermal fluids in barren and mineralized intrusions. *Economic Geology* 103: 877–908.

Baatartsogt, B., G. Schwinn, T. Wagner, H. Taubald, T. Beitter, and G. Markl. 2007. Contrasting paleofluid systems in the continental basement: A fluid inclusion and stable isotope study of hydrothermal vein mineralization, Schwarzwald district, Germany. *Geofluids* 7: 123–147.

Barrie, C.D., A.J. Boyce, A.P. Boyle, P.J. Williams, K. Blake, J. J. Wilkinson, M. Lowther, P. McDermott, and D.J. Prior. 2009. On the growth of colloform textures: a case study of sphalerite from the Galmoy ore body, Ireland. *Journal of the Geological Society London* 166: 563–582.

Barton, P.B., and P.M. Bethke. 1987. Chalcopyrite disease in sphalerite: Pathology and epidemiology. *American Mineralogist* 72: 451–467.

Bartos, P.J. 2000. The Pallacos of Cerro Rico de Potosí, Bolivia: A new deposit type. *Economic Geology* 95: 645–654.

Bettencourt, J.S., W.B. Leite Jr., C.L. Goraieb, I. Sparrenberger, R.M. S. Bello, and B.L. Payolla. 2005. Sn-poymetallic greisen-type deposits associated with late-stage rapakivi granites, Brazil: fluid inclusion and stable isotope characteristics. *Lithos* 80: 363–386.

Bierlein, F.P., D.I. Groves, R.J. Goldfarb, and B. Dubé. 2006. Lithospheric controls on the formation of provinces hosting giant orogenic gold deposits. *Mineralium Deposita* 40: 874–886.

Bölücek, C., M. Akgül, and I. Türkmen. 2004. Volcanism, sedimentation and massive sulfide mineralization in a Late Cretaceous arc-related basin, Eastern Taurides, Turkey. *Journal of Asian Earth Sciences* 24: 349–360.

Bons, P.D. 2001a. The formation of large quartz veins by rapid ascent of fluids in mobile hydrofractures. *Tectonophysics* 336: 1–17.

Bons, P.D. 2001b. Development of crystal morphology during unitaxial growth in a progressively widening vein: I. The numerical model. *Journal of Structural Geology* 23: 865–872.

Bradley, D.C., and D.L. Leach. 2003. Tectonic controls of Mississippi Valley-type lead-zinc mineralization in orogenic forelands. *Mineralium Deposita* 38: 652–667.

Bradley, D.C., D.L. Leach, D. Symons, P. Emsbo, W. Premo, G. Breit, and D.F. Sangster. 2004. Reply to Discussion on „Tectonic controls of Mississippi Valley-type lead-zinc mineralization in orogenic forelands" by S. E. Kesler, J. T. Christensen, R. D. Hagni, W. Heijlen, J. R. Kyle, K. C. Misra, P. Muchez and R. Van der Voo, Mineralium Deposita. *Mineralium Deposita* 39: 515–519.

Braun, A.C. 2006. Genesis of native copper lodes in the Keweenaw district, Northern Michigan: A hybrid evolved meteoric and metamorphogenic model. *Economic Geology* 101: 1437–1444.

Breiter, K. 2012. Nearly contemporaneous evolution of the A- and S-type fractionated granites in the Krušné hory/Erzgebirge Mts., Central Europe. *Lithos* 151: 105–121.

Bucher, K., and I. Stober. 2010. Fluids in the upper continental crust. *Geofluids* 10: 241–253.

Castroviejo, R., C. Quesada, and M. Soler. 2011. Post-depositional tectonic modification of VMS deposits in Iberia and its economic significance. *Mineralium Deposita* 46: 615–673.

Catchpole, H., K. Kouzmanov, L. Fontboté, M. Guillong, and C.A. Heinrich. 2011. Fluid evolution in zoned cordilleran polymetallic veins – insights from microthermometry and LA-ICP-MS of fluid inclusions. *Chemical Geology* 281: 293–304.

Chauvet, A., P. Piantone, L. Barbanson, and P. Nehlig. 2001. Gold deposit formation during collapse tectonics: structural, mineralogical, geochronological, and fluid inclusion constraints in the Ouro Preto gold mines, Quadrilátero Ferrífero, Brazil. *Economic Geology* 96: 25–48.

Chen, J., M.R. Walter, G.A. Logan, M.C. Hinman, and R.E. Summons. 2003. The Paleoproterozoic McArthur River (HYC) Pb/Zn/Ag deposit of northern Australia: Organic geochemistry and ore genesis. *Earth and Planetary Science Letters* 210: 467–479.

Chetty, D., and H.E. Frimmel. 2000. The role of evaporites in the genesis of base metal sulphide mineralisation in the Northern Platform of the Pan-African Damara Belt, Namibia: geochemical and fluid inclusion evidence from carbonate wall rock alteration. *Mineralium Deposita* 35: 364–376.

Ciobanu, C.L., and N.J. Cook. 2004. Skarn textures and a case study: the Ocna de Fier-Dognecea orefield, Banat, Romania. *Ore Geology Reviews* 24: 315–370.

Cline, J.S., A.H. Hofstra, J.L. Muntean, R.M. Tosdal, and K.A. Hickey. 2005. Carlin-type gold deposits in Nevada: Critical geologic characteristics and viable models. Economic Geology100th Anniversary Volume, 451–484.

Cooke, D.R., and S.F. Simmons. 2000. Characteristics and genesis of epithermal gold deposits. *Reviews in Economic Geology* 13: 221–244.

Cooke, D.R., S.W. Bull, R.R. Large, and P.J. McGoldrick. 2000. The importance of oxidized brines for the formation of Australian proterozoic stratiform sediment-hosted Pb-Zn (Sedex) deposits. *Economic Geology* 95: 1–17.

Cooke, D.R., P. Hollings, and J.L. Walshe. 2005. Giant porphyry deposits: characteristics, distribution and tectonic controls. *Economic Geology* 100: 801–818.

Corliss, J.B., J. Dymond, L.I. Gordon, J.M. Edmond, R.P. von Herzen, R.D. Ballard, K. Green, D. Williams, A. Bainbridge, K. Crane, and T.H. van Andel. 1979. Submarine thermal springs on the Galápagos Rift. *Science* 203: 1073–1083.

Danisik, M., K. Pfaff, N.J. Evans, C. Manoloukos, S. Staude, B. J. McDonald, and G. Markl. 2010. Tectonothermal history of the Schwarzwald ore district (Germany): An apatite triple dating approach. *Chemical Geology* 278: 58–69.

Derome, D., M. Cathelineau, M. Cuney, C. Fabre, T. Lhomme, and D. A. Banks. 2005. Mixing of sodic and calcic brines and uranium deposition at McArthur River, Saskatchewan, Canada: A Raman and laser-induced breakdown spectroscopic study of fluid inclusions. *Economic Geology* 100: 1529–1545.

Dietrich, A., B. Lehmann, and A. Wallianos. 2000. Bulk rock and melt inclusion geochemistry of Bolivian tin porphyry systems. Economic Geology, 313–326.

Dilek, Y., and H. Furnes. 2009. Structure and geochemistry of Tethyan ophiolites and their petrogenesis in subduction rollback systems. *Lithos* 113: 1–20.

Doyle, M.G., and R.L. Allen. 2003. Subsea-floor replacement in volcanic-hosted massive sulfide deposits. *Ore Geology Reviews* 23: 183–222.

Drummond, B., P. Lyons, B. Goleby, and L. Jones. 2006. Constraining models of the tectonic setting of the giant Olympic Dam iron oxide-copper-gold deposit, South Australia, using deep seismic reflection data. *Tectonophysics* 420: 91–103.

Duncan, R.J., H.J. Stein, K.A. Evans, M.W. Hitman, E.P. Nelson, and D.J. Kirwin. 2011. A new geochronological framework for mineralization and alteration in the Selwyn-Mount Dore Corridor, Eastern Fold Belt, Mount Isa Inlier, Australia: Genetic implications for iron oxide copper-gold deposits. *Economic Geology* 106: 169–192.

Dunham, K., K.E. Beer, R.A. Ellis, M.J. Gallagher, M.J.C. Nutt, and B.C. Webb. 1978. United Kingdom. In *Mineral Deposits of Europe*, ed. S. H.U. Bowie, A. Kvalheim and H.W. Haslam Volume: Northwest Europe. London: Institution of Mining and Metallurgy and Mineralogical Society.

Economou-Eliopoulos, M., D.G. Eliopoulos, and S. Chryssoulis. 2008. A comparison of high-Au massive sulfide ores hosted in ophiolite complexes of the Balkan Peninsula with modern analogues: Genetic significance. *Ore Geology Reviews* 33: 81–100.

Eddy, C.A., Y. Dilek, S. Hurst, and E.M. Moores. 1998. Seamount formation and associated caldera complex and hydrothermal mineralization in ancient oceanic crust, Troodos ophiolite (Cyprus). *Tectonophysics* 292: 189–210.

Emmons, W.H. 1936. *Hypogene zoning in metalliferous lodes*. Report 1 of the 16th International Geological Congress, 417–432.

Feltrin, L., J.G. McLellan, and N.H.S. Oliver. 2009. Modelling the giant, Zn–Pb–Ag Century deposit, Queensland, Australia. *Computers & Geosciences* 35: 108–133.

Franklin, J.M., H.L. Gibson, I.R. Jonasson and A.G. Galley. 2005. Volcanogenic massive sulfide deposits. Economic Geology 100th Anniversary Volume, 523–560.

Gaboury, D., and V. Pearson. 2008. Rhyolite geochemical signatures and association with volcanogenic massive sulfide deposits: Examples from the Abitibi Belt, Canada. *Economic Geology* 103: 1531–1562.

Garven, G., S.W. Bull, and R.R. Large. 2001. Hydrothermal fluid flow models of stratiform ore genesis in the McArthur Basin, Northern Territory, Australia. *Geofluids* 1: 289–311.

Gessner, K., P.A. Jones, A.R. Wilde, and M. Kühn. 2006. Significance of strain localization and fracturing in relation to hydrothermal mineralization at Mount Isa, Australia. *Journal of Geochemical Exploration* 89: 129–132.

Glasby, G.P., K. Iizasa, M. Hannington, H. Kubota, and K. Notsu. 2008. Mineralogy and composition of Kuroko deposits from northeastern Honshu and their possible modern analogues from the Izu-Ogasawara (Bonin) Arc south of Japan: Implications for mode of formation. *Ore Geology Reviews* 34: 547–560.

Goldfarb, R.J., D.I. Groves, and S. Gardoll. 2001. Orogenic gold and geologic time: A global synthesis. *Ore Geology Reviews* 18: 1–75.

Groves, D.I., and N.M. Vielreicher. 2001. The Phalabowra (Palabora) carbonatite-hosted magnetite-copper sulfide deposit, South Africa: An end-member of the iron-oxide-copper-gold-rare earth element deposit group? *Mineralium Deposita* 36: 189–194.

Groves, D.I., R.J. Goldfarb, M. Gebre-Mariam, S.G. Hagemann, and F. Robert. 1998. Orogenic gold deposits: A proposed classification in the context of their crustal distribution and relationship to other gold deposit types. *Ore Geology Reviews* 13: 7–27.

Groves, D.I., R.J. Goldfarb, F. Robert, and C.J.R. Hart. 2003. Gold deposits in metamorphic belts: overview of current understanding, outstanding problems, future research, and exploration significance. *Economic Geology* 98: 1–29.

Groves, D.I., F.P. Bierlein, L.D. Meinert, and M.W. Hitzman. 2010. Iron oxide copper-gold (IOCG) deposits through earth history: implications for origin, lithospheric setting, and distinction from other epigenetic iron oxide deposits. *Economic Geology* 105: 641–654.

Gu, L.X., Y. Zheng, X. Tang, K. Zaw, F. Della-Pasque, C. Wu, Z. Tian, J. Lu, P. Ni, X. Li, F. Yang, and X. Wang. 2007. Copper, gold and silver enrichment in ore mylonites within massive sulphide orebodies at Hongtoushan VHMS deposit, N.E. China. *Ore Geology Reviews* 39: 1–29.

Guillou-Frotter, L., and E. Burov. 2003. The development and fracturing of plutonic apexes: Implications for porphyry ore deposits. *Earth and Planetary Science Letters* 214: 341–356.

Hall, C.M., P.L. Higueras, S.E. Kesler, R. Lunar, H. Dong, and A.N. Halliday. 1997. Dating of alteration episodes related to mercury mineralization in the Almaden district, Spain. *Earth and Planetary Science Letters* 148: 281–298.

Halter, W.E., T. Pettke, and C.A. Heinrich. 2002. The origin of Cu/Au ratios in porphyry-type ore deposits. *Science* 296: 1844–1846.

Halter, W.E., C.A. Heinrich, and T. Pettke. 2005. Magma evolution and the formation of porphyry Cu–Au ore fluids: Evidence from silicate and sulfide melt inclusions. *Mineralium Deposita* 39: 845–863.

Hagan, N., N. Robins, H. Hsu-Kim, S. Halabi, M. Morris, G. Woodall, T. Zhang, A. Bacon, D.B. Richter, and J. Vandenberg. 2011. Estimating historical atmospheric mercury concentrations from silver mining and their legacies in present-day surface soil in Potosí, Bolivia. *Atmospheric Environment* 45: 7619–7626.

Hannington, M.D., C.E.J. de Ronde, and S. Peterson. 2005. Sea-floor tectonics and submarine hydrothermal systems. Economic Geology 100th Anniversary Volume, 111–141.

Hartley, A.J., and C.M. Rice. 2005. Controls on supergene enrichment of porphyry copper deposits in the Central Andes: A review and discussion. *Mineralium Deposita* 40: 515–525.

Haynes, D.W., K.C. Cross, R.T. Bills, and M.H. Reed. 1995. Olympic Dam ore genesis: A fluid-mixing model. *Economic Geology* 90: 281–307.

Hecht, L., and M. Cuney. 2000. Hydrothermal alteration of monazite in the Precambrian crystalline basement of the Athabasca Basin (Saskatchewan, Canada): implications for the formation of unconformity-related uranium deposits. *Mineralium Deposita* 35: 791–795.

Hedenquist, J.W., A.R. Arribas, and E. Gonzalez-Urien. 2000. Exporation for epithermal gold deposits. *Reviews in Economic Geology* 13: 245–277.

Heinrich, C.A. 2005. The physical and chemical evolution of low-salinity magmatic fluids at the porphyry to epithermal transition: A thermodynamic study. *Mineralium Deposita* 39: 864–889.

Herrington, R., V. Maslennikov, V. Zaykov, I. Seravkin, A. Kosarev, B. Buschmann, J.-J. Orgeval, N. Holland, S. Tesalina, P. Nimis, and R. Armstrong. 2005. Classification of VMS deposits: Lessons from the South Uralides. *Ore Geology Reviews* 27: 203–237.

Herzig, P.M., and M.D. Hannington. 1995. Polymetallic massive sulfides at the modern seafloor. A review. *Ore Geology Reviews* 10: 95–115.

Hollings, P., D. Cooke, and A. Clark. 2005. Regional geochemistry of Tertiary igneous rocks in central Chile: Implications for the geodynamic environment of giant porphyry copper and epithermal gold mineralization. *Economic Geology* 100: 887–904.

Hou, Z., H. Zhang, X. Pan, and Z. Yang. 2011. Porphyry Cu (-Mo-Au) deposits related to melting of thickened mafic lower crust: Examples from the eastern Tethyan metallogenic domain. *Ore Geology Reviews* 39: 21–45.

Houghton, J.L., W.C. Shanks, and W.E. Seyfried Jr. 2004. Massive sulfide deposition and trace element remobilization in the Middle Valley sediment-hosted hydrothermal system, northern Juan de Fuca Ridge. *Geochimica et Cosmochimica Acta* 68: 2863–2873.

Hunt, J.A., T. Baker, and D.J. Thorkelson. 2007. A review of iron oxide copper-gold deposits, with focus on the Wernecke Breccias, Yukon, Canada, as an example of a non-magmatic end member and implications for IOCG genesis and classification. *Exploration and Mining Geology* 16: 209–232.

Huston, D.L., S. Pehrsson, B.M. Eglington, and K. Zaw. 2010. The geology and metallogeny of volcanic-hosted massive sulfide deposits: Variations through geologic time and with tectonic setting. *Economic Geology* 105: 571–591.

Ireland, T., S.W. Bull, and R. Large. 2004. Mass flow sedimentology within the HYC Zn Pb Ag deposit, Northern Territory, Australia: evidence for syn-sedimentary ore genesis. *Mineralium Deposita* 39: 143–158.

Jefferson, C.W., D.J. Thomas, S.S. Gandhi, P. Ramaekers, G. Delaney, D. Brisbin, C. Cutts, D. Quirt, P. Portella, and R.A. Olson. 2007. Unconformity-associated uranium deposits of the Athabasca Basin, Saskatchewan and Alberta. In *Mineral Deposits of Canada: A Synthesis of Major Deposit-Types, District Metallogeny, the Evolution of Geological Provinces, and Exploration methods*, ed. D. Goodfellow. Geological Association of Canada, Special Publication 5.

Joye, S.B., V.A. Samarkin, B.N. Orcutt, I.R. MacDonald, K.-U. Hinrichs, M. Elvert, A.P. Teske, K.G. Lloyd, M.A. Lever, J. P. Montoya, and C.D. Meile. 2009. Metabolic variability in seafloor brines revealed by carbon and sulphur dynamics. *Nature Geoscience* 2: 349–354.

Kampunzu, A.B., J.L.H. Cailteux, A.F. Kamona, M.M. Intiomale, and F. Melcher. 2009. Sediment-hosted Zn–Pb–Cu deposits in the Central African Copperbelt. *Ore Geology Reviews* 35: 263–297.

Kesler, S.E., and C.W. Carrigan. 2002. Discussion on "Mississippi Valley-type lead-zinc deposits through geological time: implications from recent age-dating research" by D. L. Leach, D. Bradley, M. T. Lewchuk, D. T. A. Symons, G. de Marsily, and J. Brannon (2001) Mineralium Deposita 36:711–740. *Mineralium Deposita* 37: 800–802.

Kesler, S.E., S.L. Chryssoulis, and G. Simon. 2002. Gold in porphyry copper deposits: Its abundance and fate. *Ore Geology Reviews* 21: 103–124.

Kesler, S.E., J.T. Chesley, J.N. Christensen, R.D. Hagni, W. Heijlen, J. R. Kyle, P. Muchez, K.C. Misra, and R. van der Voo. 2004. Discussion of "Tectonic controls of Mississippi Valley-type lead-zinc mineralization in orogenic forelands" by D.C. Bradley and D.L. Leach. *Mineralium Deposita* 39: 512–514.

Klemm, L.M., T. Pettke, and C.A. Heinrich. 2008. Fluid and Source magma evolution of the Questa porphyry Mo deposit, New Mexico, USA. *Mineralium Deposita* 43: 533–552.

Kovalev, K.R., Y.A. Kalinin, V.I. Polynov, E.L. Kydyrbekov, A.S. Borisenko, E.A. Naumov, M.I. Netesov, A.G. Klimenko, and M.K. Kolesnikova. 2012. The Suzdal gold-sulfide deposit in the black shale of Eastern Kazakhstan. *Geology of Ore Deposits* 54: 254–275.

Kyser, T.K. 2007. Fluids, basin analysis, and mineral deposits. *Geofluids* 7: 238–257.

Lang, J.R., and T. Baker. 2001. Intrusion-related gold systems: The present level of understanding. *Mineralium Deposita* 36: 477–489.

Large, D., and E. Walcher. 1999. The Rammelsberg massive sulphide Cu–Zn–Pb–Ba–Deposit, Germany: An example of sediment-hosted, massive sulphide mineralisation. *Mineralium Deposita* 34: 522–538.

Large, R.R. 1992. Australian volcanic-hosted massive sulfide deposits: Features, styles and genetic models. *Economic Geology* 87: 471–510.

Large, R.R., J. McPhie, J.B. Gemmell, W. Herrmann, and G. J. Davidson. 2001. The spectrum of ore deposit types, volcanic environments, alteration halos, and related exploration vectors in submarine volcanic successions: Some examples from Australia. *Economic Geology* 96: 913–938.

Large, R.R., S.W. Bull, and V.V. Maslennikov. 2011. A carbonaceous sedimentary source-rock model for Carlin-type and orogenic gold deposits. *Economic Geology* 106: 331–358.

Laube, N., H.E. Frimmel, and S. Hoernes. 1995. Oxygen and Carbon isotopic study on the genesis of the Steirischer Erzberg siderite deposit (Austria). *Mineralium Deposita* 30: 285–293.

Leach, D.L., D. Bradley, M.T. Lewchuk, D.T.A. Symons, G. de Marsily, and J. Brannon. 2001. Mississippi Valley-type lead-zinc deposits through geological time: Implications from recent age-dating research. *Mineralium Deposita* 36: 711–740.

Leach, D.L., D. Bradley, M. Lewchuk, D.T.A. Symons, W. Premo, J. Brannon, and G. De Marsily. 2002. Reply to Discussion on "Mississippi Valley-type lead-zinc deposits through geological time: implications from recent age-dating research" by S. E. Kesler and C. W. Carrigan (2001) Mineralium Deposita 36:711–740. *Mineralium Deposita* 37: 803–805.

Leach, D.L., R.D. Taylor, D.L. Fey, S.F. 2010. Diehl, and R.W. Saltus. 2010. A deposit model for Mississippi-Valley-Type lead-zinc-ores. In *USGS, Mineral Deposit Models for Resource Assessment: U.S.* Geological Survey Scientific Investigations Report 2010–5070.

Liessmann, W. 2010. *Historischer Bergbau im Harz*, 3rd ed. Heidelberg: Springer.

Lowenstern, J.B. 2001. Carbon dioxide in magmas and implications for hydrothermal systems. *Mineralium Deposita* 36: 490–502.

Lüders, V., B. Precejus, and P. Halbach. 2001. Fluid inclusion and sulfur isotope studies in probable modern analogue Kuroko-type ores from the JADE hydrothermal field (Central Okinawa Trough, Japan). *Chemical Geology* 173: 45–58.

Matignac, C., B. Diagana, M. Cathelineau, M.-C. Boiron, D. Banks, S. Fourcade, and J. Vallance. 2003. Remobilisation of base metals and gold by Variscan metamorphic fluids in the south Iberian pyrite belt: Evidence from the Tharsis VMS deposit. *Chemical Geology* 194: 143–165.

McKibben, M.A., and W.A. Elders. 1985. Fe–Zn–Cu-Pb mineralization in the Salton Sea geothermal system, Imperial Valley, California. *Economic Geology* 80: 539–559.

Meinert, L. D., 2009. Skarn Web Page. http://www.science.smith.edu/geosciences/skarn/.

Meinert, L.D., G.M. Dipple, and S. Nicolescu. 2005. World skarn deposits. Economic Geology 100th Anniversary Volume, 299–336.

Meshik, A.P. 2005. The workings of an ancient nuclear reactor. Scientific American. Online: http://www.scientificamerican.com/article.cfm?id=ancient-nuclear-reactor.

Meshik, A.P., H.J. Lippolt, and Y.M. Dymkov. 2000. Xenon geochronology of Schwarzwald pitchblendes. *Mineralium Deposita* 35: 190–205.

Meyer, M., O. Brockamp, N. Clauer, A. Renk, and M. Zuther. 2000. Further evidence for a Jurassic mineralizing event in central Europe. K–Ar dating of geothermal alteration and fluid inclusion systematics in wall rocks of the Käfersteige fluorite vein deposit in the northern Black Forest, Germany. *Mineralium Deposita* 35: 754–761.

Micklethwaite, S., H.A. Sheldon, and T. Baker. 2010. Active fault and shear zone processes and their implications for mineral deposit formation and discovery. *Journal of Structural Geology* 32: 151–165.

Mlynarczyk, M.S.J., and A.E. Williams-Jones. 2005. The role of collisional tectonics in the metallogeny of the Central Andean tin belt. *Earth and Planetary Science Letters* 240: 656–667.

Mlynarczyk, M.S.J., R.L. Sherlock, and A.E. Williams-Jones. 2003. San Rafael, Peru: Geology and structure of the worlds richest tin lode. *Mineralium Deposita* 38: 555–567.

Morishita, Y., and T. Nakano. 2008. Role of basement in epithermal deposits: The Kushikino and Hashiari gold deposits, southwestern Japan. *Ore Geology Reviews* 34: 597–609.

Muchez, P., and W. Heijlen. 2003. Origin and migration of fluids during the evolution of sedimentary basins and the origin of Zn–Pb deposits in Western and Central Europe. *Journal of Geochemical Exploration* 78: 553–557.

Muchez, P., W. Heijlen, D. Banks, D. Blundell, M. Boni, and F. Grandia. 2005. Extensional tectonics and the timing and formation of basin-hosted deposits in Europe. *Ore Geology Reviews* 27: 241–267.

Muntean, J.L., J.S. Cline, A.C. Simon, and A.A. Longo. 2011. Magmatic-hydrothermal origin of Nevada's Carlin-type gold deposits. *Nature Geoscience* 4: 122–127.

Murakami, H., J.H. Seo, and C.A. Heinrich. 2010. The relation between Cu/Au ration and formation depth of porphyry-style Cu-Au ± Mo deposits. *Mineralium Deposita* 45: 11–21.

Nadeau, O., A.E. Williams-Jones, and J. Stix. 2010. Sulphide magma as a source of metals in arc-related magmatic hydrothermal ore fluids. *Nature Geoscience* 3: 501–505.

Northrop, I., and M.B. Goldhaber. 1990. Genesis of the tabular-type vanadium-uranium deposits of the Henry Basin, Utah. *Economic Geology* 85: 215–269.

Nozaki, T., K. Nakamura, S. Awaji, and Y. Kato. 2006. Whole-rock geochemistry of basic schists from the Besshi Area, Central

Shikoku: Implications for the tectonic setting of the Besshi sulfide deposit. *Resource Geology* 56: 432.

Nozaki, T., Y. Kato, and K. Suzuki. 2010. Re-Os geochronology of the Iimori Besshi-type massive sulfide deposit in the Sanbagawa metamorphic belt, Japan. *Geochimica et Cosmochimica Acta* 74: 4322–4331.

Ohmoto, H. 1996. Formation of volcanogenic massive sulfide deposits: The Kuroko perspective. *Ore Geology Reviews* 10: 135–177.

Okrusch, M., and S. Matthes. 2009. *Mineralogie: Eine Einführung in die spezielle Mineralogie, Petrologie und Lagerstättenkunde*, 8th ed. Heidelberg: Springer.

Oliver, N.H.S., and P.D. Bons. 2001. Mechanisms of fluid flow and fluid-rock interaction in fossil metamorphic hydrothermal systems inferred from vein-wallrock patterns, geometry and microstructure. *Geofluids* 1: 137–162.

Ossandón, G., R. Fréraut, L.B. Gustafson, D.D. Lindsay, and M. Zentilli. 2001. Geology of the Chuquicamata mine: A progress report. *Economic Geology* 96: 249–270.

Oyarzun, R., A. Márquez, J. Lillo, I. López, and S. Rivera. 2001. Giant versus small porphyry copper deposits of Cenozoic age in northern Chile: Adakitic versus normal calc-alkaline magmatism. *Mineralium Deposita* 26: 794–798.

Oyarzun, R., A. Márquez, J. Lillo, I. López, and S. Rivera. 2002. Reply to Discussion on "Giant versus small porphyry copper deposits of Cenozoic age in northern Chile: adakitic versus normal calc-alkaline magmatism" by Oyarzun R, Marquez A, Lillo J, Lopez I, Rivera S (Mineralium Deposita 36:794–798, 2001). *Mineralium Deposita* 37: 795–799.

Oyarzun, R., J. Oyarzun, J.J. Ménard, and J. Lillo. 2003. The cretaceous iron belt of northern Chile: role of oceanic plates, a superplume event, and a major shear zone. *Mineralium Deposita* 38: 640–646.

Padilla Garza, R.A., S.R. Titley, and R. Pimentel. 2001. Geology of the Escondida porphyry copper deposit, Antofagasta Region, Chile. *Economic Geology* 96: 307–324.

Páez, G.N., R. Ruiz, D.M. Guido, S.M. Jovic, and I.B. Schalamuk. 2011. Structurally controlled fluid flow: High-grade silver ore-shoots at Martha epithermal mine, Deseado Massif, Argentina. *Journal of Structural Geology* 33: 985–999.

Pearce, J.A., and P.T. Robinson. 2010. The Troodos ophiolitic complex probably formed in a subduction initiation, slab edge setting. *Gondwana Research* 18: 60–81.

Perkins, W.G. 1997. Mount Isa lead-zinc orebodies: Replacement lodes in a zoned syndeformational copper-lead-zinc system? *Ore Geology Reviews* 12: 61–110.

Petersen, S., P.M. Herzig, and M.D. Hannington. 2000. Third dimension of a presently forming VMS deposit: TAG hydrothermal mound, Mid-Atlantic Ridge, 26°N. Mineralium Deposita 233–259.

Pfaff, K., R.L. Romer, and G. Markl. 2009. Mineralization history of the Schwarzwald ore district: U-Pb ages of ferberite, agate, and pitchblende. *European Journal of Mineralogy* 21: 817–836.

Pfaff, K., L.H. Hildebrandt, D.L. Leach, D.E. Jacob, and G. Markl. 2010. Formation of the Wiesloch Mississippi Valley-type Zn–Pb–Ag deposit in the extensional setting of the Upper Rhinegraben, SW Germany. *Mineralium Deposita* 45: 647–666.

Pfaff, K., A. Koenig, T. Wenzel, I. Ridley, L.H. Hildebrandt, D.L. Leach, and G. Markl. 2011. Trace and minor element variations and sulfur isotopes in crystalline and colloform ZnS: Incorporation mechanisms and implications for their genesis. *Chemical Geology* 286: 118–134.

Pfaff, K., S. Staude, and G. Markl. 2012. On the origin of sellaite (MgF2)-rich deposits in Mg-poor environments. *American Mineralogist* 97: 1987–1997.

Piercey, S.J. 2011. The setting, style, and role of magmatism in the formation of volcanogenic massive sulfide deposits. *Mineralium Deposita* 46: 449–471.

Pohl, W., and R. Belocky. 1999. Metamorphism and metallogeny in the Eastern Alps. *Mineralium Deposita* 34: 614–629.

Pollard, P.J. 2006. An intrusion-related origin for Cu-Au mineralization in iron oxide-copper-gold (IOCG) provinces. *Mineralium Deposita* 41: 179–187.

Porter, T.M. (ed.). 2000. *Hydrothermal Iron Oxide Copper-Gold and Related Deposits: A Global Perspective*. Adelaide, Australia: Australian Mineral Foundation.

Prokin, V.A., F.P. Buslaev, and A.P. Nasedkin. 1998. Types of massive sulphide deposits in the Urals. *Mineralium Deposita* 34: 121–126.

Rabbia, O.M., L.B. Hernández, R.W. King, and L. López-Escobar. 2002. Discussion on "Giant versus small porphyry copper deposits of Cenozoic age in northern Chile: adakitic versus normal calc-alkaline magmatism" by Oyarzun et al. (Mineralium Deposita 36:794–798, 2001). Mineralium Deposita 37, 791–794.

Reich, M., C. Palacios, M.A. Parada, U. Fehn, E.M. Cameron, M.I. Leybourne, and A. Zúñiga. 2008. Atacamite formation by deep saline waters in copper deposits from the Atacama Desert, Chile: Evidence from fluid inclusions, groundwater chemistry, TEM, and 36Cl data. *Mineralium Deposita* 43: 663–675.

Reynolds, L.J. 2000. Geology of the Olympic Dam Cu-U-Au-Ag-REE deposit. In *Hydrothermal Iron Oxide Copper-Gold and Related Deposits: A Global Perspective*, vol. 1, ed. T.M. Porter, Adelaide.

Rice, C.M., G.B. Steele, D.N. Barfod, A.J. Boyce, and M.S. Pringle. 2005. Duration of magmatic, hydrothermal, and supergene activity at Cerro Rico de Potosi, Bolivia. *Economic Geology* 100: 1647–1656.

Richards, J.P. 2002. Discussion on "Giant versus small porphyry copper deposits of Cenozoic age in northern Chile: adakitic versus normal calc-alkaline magmatism" by Oyarzun et al. (Mineralium Deposita 36: 794–798, 2001). Mineralium Deposita 37, 788–790.

Richards, J.P. 2003. Tectono-magmatic precursors for porphyry Cu-(Mo–Au) deposit formation. *Economic Geology* 98: 1515–1533.

Richards, J.P., A.J. Boyce, and M.S. Pringle. 2001. Geologic evolution of the Escondida Area, Northern Chile: A model for spatial and temporal localization of porphyry Cu mineralization. *Economic Geology* 96: 271–305.

Robb, L. 2005. *Introduction to Ore-Forming Processes*. Malden, MA: Blackwell Science.

Roedder, E. 1968. Noncolloidal origin of colloform textures in sphalerite ores. *Economic Geology* 100: 451–471.

Roedder, E. 1984. *Fluid Inclusions. Reviews in Mineralogy*, vol. 12. Mineralogical Society of America.

Rosa, C.J.P., J. McPhie, J.M.R.S. Relvas, Z. Pereira, T. Oliveira, and N. Pacheco. 2008. Facies analyses and volcanic setting of the giant Neves Corvo massive sulfide deposit, Iberian Pyrite Belt, Portugal. *Mineralium Deposita* 43: 449–466.

Rosa, C.J.P., J. McPhie, and J.M.R.S. Relvas. 2010. Type of volcanoes hosting the massive sulfide deposits of the Iberian Pyrite Belt. *Journal of Volcanology and Geothermal Research* 194: 107–126.

Rosenbaum, G., D. Giles, M. Saxon, P.G. Betts, R.F. Weinberg, and C. Duboz. 2005. Subduction of the Nazca Ridge and the Inca Plateau: Insights into the formation of ore deposits in Peru. *Earth and Planetary Science Letters* 239: 18–32.

Sáez, R., C. Moreno, F. Gonzáles, and G.R. Almodóvar. 2011. Black shales and massive sulfide deposits: Causal or casual relationships? Insights from Rammelsberg, Tharsis, and Draa Sfar. *Mineralium Deposita* 46: 585–614.

Sander, S.G., and A. Koschinsky. 2011. Metal flux from hydrothermal vents increased by organic complexation. *Nature Geoscience* 4: 145–150.

Saupé, F., and M. Arnold. 1992. Sulphur isotope geochemistry of the ores and country rocks at the Almadén mercury deposit, Ciudad Real, Spain. *Geochimica et Cosmochimica* 56: 3765–3780.

Schuiling, R.D. 2011. Troodos: A giant serpentinite diapir. *International Journal of Geosciences* 2: 98–101.

Schulz, O., F. Vavtar, and K. Dieber. 1997. Die Siderit-Erzlagerstätte Steirischer Erzberg: Eine geowissenschaftliche Studie, mit wirtschaftlicher und geschichtlicher Betrachtung. Archiv für Lagerstättenforschung der Geologischen Bundesanstalt, 65178.

Schmidt, M., R. Botz, E. Faber, M. Schmitt, J. Poggenburg, D. Garbe-Schönberg, and P. Stoffers. 2003. High-resolution methane profiles across anoxic brine-seawater boundaries in the Atlantis-II, Discovery, and Kebrit Deeps (Red Sea). *Chemical Geology* 200: 359–375.

Schwartz, M.O., S.S. Rajah, A.K. Askury, P. Putthapiban, and S. Djaswadi. 1995. The Southeast Asian tin belt. *Earth-Science Reviews* 38: 95–293.

Schwinn, G., and G. Markl. 2005. REE systematics in hydrothermal fluorite. *Chemical Geology* 216: 225–248.

Schwinn, G., T. Wagner, B. Baatartsogt, and G. Markl. 2006. Quantification of mixing processes in ore-forming hydrothermal systems by combination of stable isotope and fluid inclusion analyses. *Geochimica et Cosmochimica Acta* 70: 965–982.

Sebastian, U. 2013. *Die Geologie des Erzgebirges*. Springer Spektrum, Heidelberg.

Seifert, T. 2008. *Metallogeny and Petrogenesis of Lamprophyres in the Mid-European Variscides*. Amsterdam: IOS Press.

Seifert, T., and D. Sandmann. 2006. Mineralogy and geochemistry of indium-bearing polymetallic vein-type deposits: Implications for host minerals from the Freiberg district, Eastern Erzgebirge, Germany. *Ore Geology Reviews* 28: 1–31.

Shanks, W.C.P., and R. Thurston (ed.). 2012. Volcanogenic massive sulfide occurrence model. U.S. Geological Survey Scientific Investigations Report 2010–5070–C.

Shikazono, N. 2003. *Geochemical and Tectonic Evolution of Arc-Backarc Hydrothermal Systems: Implication for the Origin of Kuroko and Epithermal Vein-Type Mineralizations and the Global Geochemical Cycle*. Amsterdam: Elsevier.

Sillitoe, R.H. 2003. Iron oxide-copper-gold deposits: An Andean view. *Mineralium Deposita* 38: 787–812.

Sillitoe, R. 2005. Supergene oxidized and enriched porphyry copper and related deposits. Economic Geology 100th Anniversary Volume, 723–768.

Sillitoe, R. 2010. Porphyry copper systems. *Economic Geology* 105: 3–41.

Sillitoe, R.H., and J.W. Hedenquist. 2003. Linkages between volcanotectonic settings, ore-fluid compositions, and epithermal precious metal deposits. In *Volcanic, Geothermal and Ore-Forming Fluids: Rulers and Witnesses of Processes within the Earth*, ed. S.F. Simmons and I. Graham. Society of Economic Geologists Special Publication.

Skinner, B.J., D.E. White, H.J. Rose and R.E. Mays. 1967. Sulfides associated with the Salton Sea geothermal brine. Economic Geology, 316–330.

Solomon, M. 2008. Brine pool deposition for the Zn–Pb–Cu massive sulphide deposits of the Bathurst mining camp, New Brunswick, Canada. I. Comparisons with the Iberian pyrite belt. *Ore Geology Reviews* 33: 329–351.

Solomon, M., F. Tornos, R.R. Large, J.N.P. Badham, R.A. Both, and K. Zaw. 2004. Zn–Pb–Cu volcanic-hosted massive sulphide deposits: Criteria for distinguishing brine pool-type from black smoker-type sulphide deposition. *Ore Geology Reviews* 25: 259–283.

Spry, P.G., I.R. Plimer, and G.S. Teale. 2008. Did the giant Broken Hill (Australia) Zn–Pb–Ag deposit melt? *Ore Geology Reviews* 34: 223–241.

Staude, S., T. Wagner, and G. Markl. 2007. Mineralogy, mineral compositions and fluid evolution at the Wenzel hydrothermal deposit, Southern Germany: Implications for the formation of Kongsberg-Type silver deposits. *The Canadian Mineralogist* 45: 1147–1176.

Staude, S., P.D. Bons, and G. Markl. 2009. Hydrothermal vein formation by extension-driven dewatering of the middle crust: An example from SW Germany. *Earth and Planetary Science Letters* 286: 387–395.

Staude, S., T. Mordhorst, R. Neumann, W. Prebeck, and G. Markl. 2010. Compositional variation of the tennantite-tetrahedrite solid-solution series in the Schwarzwald ore district (SW Germany): The role of mineralization processes and fluid source. *Mineralogical Magazine* 74: 309–339.

Staude, S., S. Göb, K. Pfaff, F. Ströbele, W.R. Premo, and G. Markl. 2011. Deciphering fluid sources of hydrothermal systems: A combined Sr- and S-isotope study on barite (Schwarzwald, SW Germany). *Chemical Geology* 286: 1–20.

Staude, S., T. Mordhorst, S. Nau, K. Pfaff, G. Brügmann, D.E. Jacob, and G. Markl. 2012a. Hydrothermal carbonates of the Schwarzwald ore district, Southwestern Germany: Carbon source and conditions of formation using $\delta 18O$, $\delta 13C$, 87Sr/86Sr, and fluid inclusions. *The Canadian Mineralogist* 50: 1401–1434.

Staude, S., W. Werner, T. Mordhorst, K. Wemmer, D.E. Jacob, and G. Markl. 2012b. Multi-stage Ag–Bi–Co–Ni–U and Cu–Bi vein mineralization at Wittichen, Schwarzwald, SW Germany: Geological setting, ore mineralogy, and fluid evolution. *Mineralium Deposita* 47: 251–276.

Stober, I., and K. Bucher. 1999a. Origin of salinity of deep groundwater in crystalline rocks. *Terra Nova* 11: 181–185.

Stober, I., and K. Bucher. 1999b. Deep groundwater in the crystalline basement of the Black Forest region. *Applied Geochemistry* 14: 237–254.

Stoffell, B., M.S. Appold, J.J. Wilkinson, N.A. McClean, and T.E. Jeffries. 2008. Geochemistry and evolution of Mississippi Valley-Type mineralizing brines from the Tri-State and Northern Arkansas districts determined by LA-ICP-MS microanalysis of fluid inclusions. *Economic Geology* 103: 1411–1435.

Ströbele, F., S. Staude, K. Pfaff, W.R. Premo, L.H. Hildebrandt, A. Baumann, E. Pernicka, and G. Markl. 2012. Pb isotope constraints on fluid flow and mineralization processes in SW Germany. Neues Jahrbuch für Mineralogie – Abhandlungen 189, 287–309.

Thurston, P.C., J.A. Ayer, J. Goutier, and M.A. Hamilton. 2008. Depositional Gaps in Abitibi Greenstone Belt Stratigraphy: A Key to Exploration for Syngenetic Mineralization. *Economic Geology* 103: 1097–1134.

Tornos, F., C. Casquet, and J.M.R.S. Relvas. 2005. Transpressional tectonics, lower crust decoupling and intrusion of deep mafic sills: A model for the unusual metallogenesis of SW Iberia. *Ore Geology Reviews* 27: 133–163.

Tornos, F. 2006. Environment of formation and styles of volcanogenic massive sulfides: The Iberian Pyrite Belt. *Ore Geology Reviews* 28: 259–307.

Turner, R.J.W. 1992. Formation of Phanerozoic stratiform sediment hosted zinc-lead deposits: Evidence for the critical role of ocean anoxia. *Chemical Geology* 99: 165–188.

Urabe, T., and M. Kusakabe. 1990. Barite silica chimneys from the Sumisu Rift, Izu-Bonin Arc: Possible analog to hematitic chert associated with Kuroko deposits. *Earth and Planetary Science Letters* 100: 283–290.

Von Damm, K.L., L.G. Buttermore, S.E. Oosting, A.M. Bray, D. J. Fornari, M.D. Lilley, and W.C. Shanks III. 1997. Direct observation of the evolution of a seafloor "black smoker" from vapor to brine. *Earth and Planetary Science Letters* 149: 101–111.

Von Damm, K.L., M.D. Lilley, W.C. Shanks III, M. Brockington, A.M. Bray, K.M. O'Grady, E. Olson, A. Graham G. Proskurowski and the SouEPR Science Party, 2003. Extraordinary phase separation and segregation in vent fluids from the southern East Pacific Rise. Earth and Planetary Science Letters 206, 365–378.

Vry, V.H., J.J. Wilkinson, J. Seguel, and J. Millán. 2010. Multistage intrusion, brecciation, and veining at El Teniente, Chile: Evolution of a nested porphyry system. *Economic Geology* 105: 119–153.

Walenta, K. 1992. *Die Mineralien des Schwarzwaldes*. München: Christian Weise Verlag.

Weatherley, D.K., and R.W. Henley. 2013. Flash vaporization during earthquakes evidenced by gold deposits. *Nature Geoscience* 6: 294–298.

Wilcock, W.S.D., E.E.E. Hooft, D.R. Toomey, P.R. McGill, A.H. Barclay, D.S. Stakes, and T.M. Ramirez. 2009. The role of magma injection in localizing black-smoker aktivity. *Nature Geoscience* 2: 509–513.

Wilkinson, J.J., S.L. Eyre, and A.J. Boyce. 2005. Ore-forming processes in Irish-type carbonate-hosted Zn-Pb deposits: Evidence from mineralogy, chemistry, and isotopic composition of sulfides at the Lisheen mine. *Economic Geology* 100: 63–86.

Williams, A.E., and M.A. McKibben. 1989. A brine interface in the Salton Sea geothermal system, California: Fluid geochemical and isotopic characteristics. *Geochimica et Cosmochimica Acta* 53: 1905–1920.

Williams, P.J., M.D. Barton, D.A. Johnson, L. Fontboté, A. de Haller, G. Mark, N.H.S. Oliver and R. Marschnik. 2005. Iron oxide copper-gold deposits: Geology, space-time distribution, and possible modes of origin. Economic Geology 100th Anniversary Volume, 371–405.

Winckler, G., R. Kipfer, W. Aeschbach-Hertig, R. Botz, M. Schmidt, S. Schuler, and R. Bayer. 2000. Sub sea floor boiling of Red Sea Brines: New indication from noble gas data. *Geochimica et Cosmochimica Acta* 64: 1567–1575.

Yücel, M., A. Gartman, C.S. Chan, and G.W. Luther III. 2011. Hydrothermal vents as a kinetically stable source of iron-sulphide-bearing nanoparticles to the ocean. *Nature Geoscience* 4: 367–371.

Zhuang, H., J. Lu, J. Fu, and D. Liu. 1999. Two kinds oft Carlin-type gold deposite in southwestern Guizhou, China. *Chinese Science Bulletin* 44: 178–182.

Further Reading

Guilbert, J.M., and C.F. Park. 1986. *The Geology of Ore Deposits*. New York: WH Freeman.

Laznicka, P. 2010. *Giant Metallic Deposits: Future Sources of Industrial Metals*, 2nd ed. Heidelberg: Springer.

Misra, K.C. 2000. *Understanding Mineral Deposits*. Dordrecht, Niederlande: Kluwer Academic Publishers.

Neukirchen, F. 2012. *Edelsteine: Brillante Zeugen für die Erforschung der Erde*. Heidelberg: Springer Spektrum.

Okrusch, M., and S. Matthes. 2009. *Mineralogie: Eine Einführung in die spezielle Mineralogie, Petrologie und Lagerstättenkunde*, 8th ed. Heidelberg: Springer.

Pirajno, F. 2009. *Hydrothermal Processes and Mineral Sytems*. Heidelberg: Springer.

Pohl, W.L. 2011. *Economic Geology*. Chichester: Wiley-Blackwell.

Robb, L. 2005. *Introduction to Ore-Forming Processes*. Malden, MA: Blackwell Science.

Rothe, P. 2010. *Schätze der Erde*. Darmstadt: Primus Verlag.

Seidler, C. 2012. *Deutschlands verborgene Rohstoffe: Kupfer*. Hanser, München: Gold und seltene Erden.

Winter, J.D. 2001. *Igneous and Metamorphic Petrology*. New Jersey: Prentice Hall.

Deposits Formed by Sedimentation and Weathering

Weathering, transport, and sedimentation are other processes that lead to effective fractionation. Little wonder important metal deposits are found in sediments. Resources such as sand, gravel, and limestone (Chap. 7) and fossil fuels (Chap. 6) would also fit into this chapter. Changes in flow velocity in rivers lead to sorted deposits of sand and gravel. Weather-resistant minerals with a high density can be enriched to form placer deposits (Sect. 5.9). However, fine clay minerals and dissolved ions are transported into the sea. The formation of sedimentary rocks, at least on Earth, is almost always influenced by biological processes (Holland and Schidlowski 1982). This particularly applies to biogenic sediments that are built up of corals, sponges, or the shells of unicellular organisms. Sediments can comprise bird droppings (a.k.a. guano; Sect. 5.8) that are used as an important fertilizer. Moreover, organic substances such as plant residues, dead algae, and microorganisms can produce coal and crude oil. Chemical sediments are formed by precipitation from a saturated solution (e.g., evaporites; Sect. 5.7) such as salt and gypsum during the evaporation of seawater. However, living organisms can also play a role in chemical sediments. The interplay between supersaturation and microorganisms is particularly striking in the case of banded iron formations (BIFs) (Sect. 5.2).

Loose sediments are transformed into solid sedimentary rocks in a process called diagenesis by burial with further sediments. During diagenesis, substances can be precipitated from the pore water. In addition, there are of course hydrothermal deposits in sediments (already discussed in Chap. 4). The transitions between synsedimentary, diagenetic, and hydrothermal deposits are fluid. Moreover, often more than one process was involved in the formation of a given deposit making classification difficult.

Finally, weathering can also lead to the formation of deposits. In this case what remains is relevant—not the material that is removed and deposited elsewhere. Aluminum or nickel ores are known to form above suitable rocks (Sect. 5.11) during intensive chemical weathering in the tropics.

The deposits discussed in this chapter are often categorized as sedimentary, surficial, or supergene. Surficial means at and just below the Earth's surface. Sedimentary refers to deposition on the Earth's surface (sometimes including diagenetic processes). Supergene means secondary enrichment near the surface mostly by such processes as weathering, soil formation, oxidization, mobilization, and precipitation (see also Box 4.16). Residual deposits such as laterites and bauxite (Sect. 5.11) fall into the supergene category.

Allochthonous deposits are those in which material was delivered in a solid state such as placer deposits and clastic sediments (e.g., sand and gravel). Autochthonous deposits are biogenic or chemical sediments formed on the spot (e.g., by precipitation reaction or diagenetically). Residual accumulations formed during weathering are also referred to as autochthonous.

5.1 Stratiform Sediment-Hosted Copper Deposits

A range of copper deposits are found in sediment layers. Kupferschiefer ("copper shale", a regional stratigraphic unit in Central Europe, Sect. 5.1.1); the Central African copper belt (Sect. 5.1.2); and the Paleoproterozoic Kodaro-Udokan Basin in Siberia are the most important diagenetically formed deposits summarized as stratiform sediment-hosted copper (SSC) deposits. Another example is the Paradox Basin (Utah and Colorado, USA). Smaller copper deposits formed in a similar way are widespread worldwide. In addition to such diagenetic formations, hydrothermal copper ores in sandstones called red bed–type deposits (Sect. 4.13) are also grouped under this term.

© Springer Nature Switzerland AG 2020
F. Neukirchen and G. Ries, *The World of Mineral Deposits*,
https://doi.org/10.1007/978-3-030-34346-0_5

5.1.1 Kupferschiefer in Europe

Some copper deposits in Germany and Poland are bound to a Permian shale rich in organic matter (black shale) found at the base of sediments of the Zechstein Sea. This layer of copper-bearing shale is called Kupferschiefer. In most areas it is not even half a meter thick, which is why miners back in the day spoke of a seam. Mining was accordingly difficult. However, the ore zone sometimes extends into overlying and underlying rock.

The Lubin District (Poland) is one of the most important copper deposits in the world. The three active mines range from 650 m to a depth of 1200 m. The Kupferschiefer seam here is at least 2 m thick. Annual production is about 500,000 t of copper (3% of world production), more than 1000 t of silver, and almost 14,000 t of lead. Selenium, nickel, gold, platinum, and palladium are also mined here.

In German Kupferschiefer regions there are no active mines despite ore still being available. Important were Mansfeld (Box 5.1) and Sangerhausen (Saxony-Anhalt) on the eastern edge of the Harz Mountains and the Richelsdorf District (Hesse), which lies between Bad Hersfeld and Eisenach. In Lusatia (Brandenburg and Saxony) mining near Spremberg may soon be started. The copper shale here is at a depth of 800–1500 m.

There has long been controversy as to how these deposits were formed. Models ranged from synsedimentary (i.e., simultaneously with deposition of the shale; Preidl and Metzler 1984) to diagenetic formation in shale as yet unconsolidated (Sawlowicz 1989; Vaughan et al. 1989) and to subsequent (epigenetic) mineralization (Kucha and Pawlikowski 1986; Kucha and Przylowicz 1999). In recent decades it has become increasingly clear that various processes took place over a long period of time with hydrothermal solutions playing the main role during diagenesis (Oszczepalski 1999; Sun and Püttmann 2000; Bechtel et al. 2000, 2001a, b, 2002; Piestrzynski et al. 2002; Pašava et al. 2010). This section would not have been out of place in the previous chapter particularly since there are striking similarities to SEDEX deposits (Sect. 4.17) and to recent processes below the Salton Sea (Box 4.29) salt lake. However, special conditions during sedimentation would have been necessary.

Diagenesis is the consolidation of loose sediments such as sludge and sand in a solid sedimentary rock such as shale and sandstone. It begins immediately after deposition. Overburden increases as a result of burial with younger sediments (early diagenesis) and continues with further burial (late diagenesis) possibly over long periods of time. The most important processes are compaction (due to increasing pressure) and cementation (formation of new minerals in pores) both of which reduce porosity. As a result of the increasing burial of sediments diagenesis smoothly changes into low-grade metamorphism.

The Variscan Mountains had largely eroded by the end of the Permian and a large basin had formed north of them as a result of wide extension. The dry climate of the time meant that rivers mainly deposited sand that formed thick and oxidized sandstones called Rotliegendes. In places there were also large volcanoes. In the Late Permian the sea entered from the north and filled the deep part of the basin and the Zechstein Sea emerged. It stretched from today's Belgium, the Netherlands, and Eastern England via the North Sea to Denmark and large parts of Germany to Poland and Lithuania. Initially, sandstones called Weißliegendes and conglomerates at the edges of the basin were deposited. In some areas there was also limestone.

This was followed by Kupferschiefer (Fig. 5.1) that spread over an area of more than 600,000 km^2. Called black shale it is a dark-colored clay-rich rock that shows clear indications of reduced oxygen supply during deposition. The dark coloring is due to the high content of organic substances that in Kupferschiefer average out at 5%. Toward the top the carbonate content increases and the rock turns into dark gray marl.

Similar anoxic conditions prevail today in the Black Sea. Anoxic, H_2S-rich environments and sediments deposited in them are called euxinic for the Latin name of the Black Sea. It has a low-salt layer over a salt-rich and heavier water layer between which there is hardly any exchange. Due to low circulation water in the deepest part of the basin is anoxic. At the bottom of the sea a sapropel (sludge rich in organic matter) accumulates, which can contain up to 35% organic matter. The anoxic conditions make it possible for these to be preserved. Accordingly, black shale is also an important source rock for crude oil and natural gas (Chap. 6). Due to the lack of oxygen most living organisms cannot exist in this environment, but sulfate-reducing bacteria feel particularly at home here. Instead of oxygen they use sulfate from seawater as an oxidizing agent to gain energy by oxidizing organic matter or hydrogen. These bacteria excrete hydrogen sulfide (H_2S) together with water, CO_2, and organic substances. Although hydrogen sulfide provides even more hostile conditions, it leads to precipitation of pyrite and marcasite (both: FeS_2) at the same time. This happens particularly early diagenetically in sludge that has not yet or only slightly consolidated. Pyrite and marcasite mainly form tiny finely distributed mineral grains, but there are also small round aggregates called framboids. Fossils such as fish and mussels are often replaced by pyrite and marcasite. Numerous fossils have been found in Kupferschiefer (Fig. 5.2) including fish, brachiopods, and reptiles (Haubold et al. 2006). The anoxic conditions were probably local and limited to parts of the water column.

Fig. 5.1 Such a thin seam of rock just a few millimeters thick was called ore ruler. This Kupferschiefer seam is from the Reichenberg Shaft near Dens, south of Sontra (Hesse, Germany). Secondary minerals formed in the air are green and primary sulfides are yellow (*photo* © F. Neukirchen/Mineralogical collections of the TU Berlin)

Fig. 5.2 A fish (*Palaeoniscus freieslebeni*) petrified to bornite (Cu_5FeS_4) in Kupferschiefer from Mansfeld (Germany) (*photo* © F. Neukirchen/Mineralogical Collections of the TU Berlin)

Later the Zechstein Sea slowly evaporated. The increasingly concentrated seawater led to deposition of a few meters of limestone also containing organic substances. Anhydrite and thick salt layers (Sect. 5.7) followed. Then there were further cycles with evaporites. After the Permian there were coverings with further sediments.

Enrichment with copper and other metals followed during late diagenesis at temperatures between 80 and 140 °C. This probably happened over a long period of time (especially, in the Triassic and Early Jurassic) but not everywhere. In the lower Rotliegendes a large stream of salty oxidized connate water leached certain metals from the sandstone and interbedded volcanic rocks. At the southern edge of the basin (especially) and at uplifted blocks surrounded by more porous sediments the water penetrated reduced Zechstein sediments. The sudden change in the redox potential triggered a series of reactions.

In extreme cases Kupferschiefer sulfides and a good part of the organic matter disappeared. Hematite was precipitated instead. Since such oxidized reddish hematite zones are low in copper the miners called them Rote Fäule ("red rotting"). In large parts of the basin the redox front lies in sandstones below the seam of Kupferschiefer. However, in the zones with Rote Fäule it cuts through the seam of Kupferschiefer and partly

comes into contact with anhydrite of the Zechstein limestone (Fig. 5.3). Strictly speaking, the redox front is a continuous transition that is a few millimeters wide in Kupferschiefer and a few meters wide in other rocks. Sometimes economically interesting concentrations of gold (up to 10 ppm), platinum (up to 14 ppm), and palladium are present in the oxidized zone near the redox front and in the transition zone.

Box 5.1 Mansfeld Kupferschiefer

At the eastern edge of the Harz Mountains the Kupferschiefer seam crops out on the surface in a semicircle at the edge of the Mansfeld Syncline. In the center of the syncline it is located at a depth of about 1000 m. A second, somewhat smaller basin in which Kupferschiefer has also been mined is located 10 km southwest near Sangerhausen.

The Kupferschiefer seam here is broken up by a number of faults and is only 35–40 cm thick. The largest copper concentration is in the lower half of the seam. In some zones also the underlying "Sanderz" ("sand ore") and the overlying "Dachklotz" (Zechstein limestone) were mined. Copper, silver, lead, and zinc have been extracted as have by-products such as

Fig. 5.3 a Distribution of Kupferschiefer zones in Poland and depth below the surface. **b** Schematic profile of Kupferschiefer in Poland (note the different scales). Rotliegendes are terrestrial sandstones and the layers above are the lowest deposits of the Zechstein Sea. The redox front of the Rote Fäule cuts through the layers and the different zones are arranged relative to these. The highest ore grades are found in Kupferschiefer near Rote Fäule (**a** from Liedtke and Vasters 2008; **b** from Oszczepalski 1999)

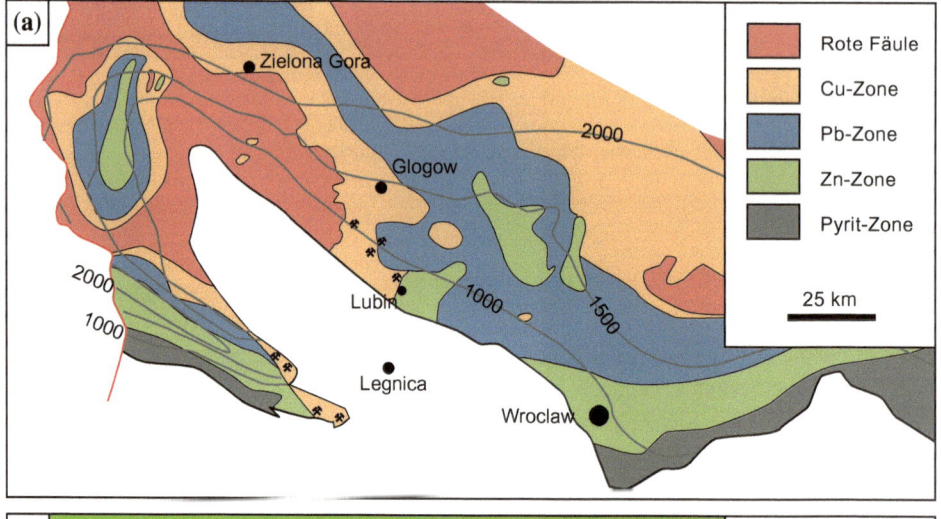

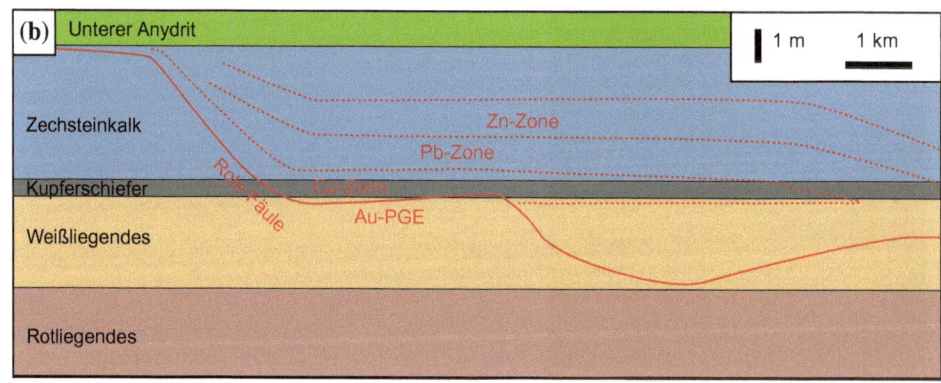

vanadium, molybdenum, cobalt, nickel, selenium, rhenium, cadmium, thallium, germanium, and gold.

Mining on copper began around the year 1200 along the outcropping layer. Later longer tunnels were avoided by sinking shafts close to each other and mining the seam only in the immediate vicinity of the shafts. In the following centuries the seam was followed downward in ever-deeper galleries leading into the interior of the syncline. Altogether more than 1000 shafts were sunk in the marginal area and small heaps piled up. In the fifteenth century the separation of silver from copper minerals was made possible by the Saiger Process.

Mansfeld also played a certain role in the Reformation. Martin Luther's father Hans Luder worked here at times as a master metallurgist. The more radical reformer Thomas Müntzer who also wanted a social revolution was at times pastor in Allstedt near Sangerhausen. He fought with the authorities and the Church and had to flee to Mülhausen (Thuringia), but was still supported by farmers and the Mansfeld miners. Later he led the rebellious peasant armies that were crushed by the princes.

Villages and pits were destroyed in the Thirty Years' War. In 1785 Germany's first steam engine became operational in Mansfeld to pump water out of the mines. Further machines followed. In the nineteenth century somewhat deeper areas were excavated at the margin of the syncline and overburden was piled up on larger flat heaps.

In the twentieth century ever-deeper shafts were added and in the center of the syncline high, pointed heaps started to pile up. Dynamite and pneumatic hammers facilitated longwall mining in the low cavity. The local mining company developed into a large enterprise that also operated a coal mine in the Ruhr area. In the days of the German Democratic Republic (GDR, a.k.a. East Germany) the Mansfeld combine continued mining operations. They were discontinued in 1990. Mining reached a depth of 995 m in the Ernst Thälmann Shaft.

Copper enrichment is greatest in the Kupferschiefer seam in the immediate vicinity of Rote Fäule in one of two locations: either where Rote Fäule cuts through the Kupferschiefer seam or where it lies immediately below it. The most important process is the replacement of pyrite by copper sulfides by such a reaction as:

$$2FeS_2 + 4Cu^+ + 3O_2 + 2H_2O \rightarrow$$
$$2Cu_2S + 2Fe^{2+} + 2SO_4^{2-} + 4H^+$$

The resulting H^+ is neutralized by carbonate. From the original pyrite content of Kupferschiefer it was determined that up to 8% Cu can be contained in the rock (Sun and Püttmann 2000; Bechtel et al. 2001b). Sometimes the ore grade is even higher when additional S^{2-} is added by thermochemical reduction of sulfate dissolved in the water. Organic matter serves not only as a reducing agent but also as a source of hydrogen for the formation of H_2S. A corresponding change in the composition of organic matter is indeed measurable. However, sulfur from organic compounds plays no role.

The dominant ore mineral in the immediate vicinity of Rote Fäule is chalcocite (Cu_2S), but farther away bornite (Cu_5FeS_4) and chalcopyrite ($CuFeS_2$) predominate. At even greater distances galena and sphalerite (lead and zinc zone) dominate. Surprisingly, silver here is found mainly in chalcocite and bornite (Oszczepalski 1999).

Mineralization is not restricted to the Kupferschiefer and reaches the overlying Zechstein limestone (especially, when it comes to lead and zinc ores). Sometimes it is also directly mineralized on contact with anhydrite. The uppermost decimeters of underlying sandstone are also partially mineralized.

In Poland the Rote Fäule covers a large area and the adjacent copper zone at Lubin is up to 30 km wide (Fig. 5.4). However, in the Mansfeld area there is a pattern in which many small zones of Rote Fäule are surrounded by a copper zone a few kilometers wide often linking directly with the next one. The lead–zinc zone is best developed in the southern and northern margin of the area.

Although epigenetic veins were later formed, they make up only a tiny part of the total ore amount. By the way, there is evidence to show that ascending deep water mixing with the connate water of the Kupferschiefer led to the precipitation of ores in hydrothermal veins in the Spessart located in, below, and above the Zechstein sediments (Wagner et al. 2010).

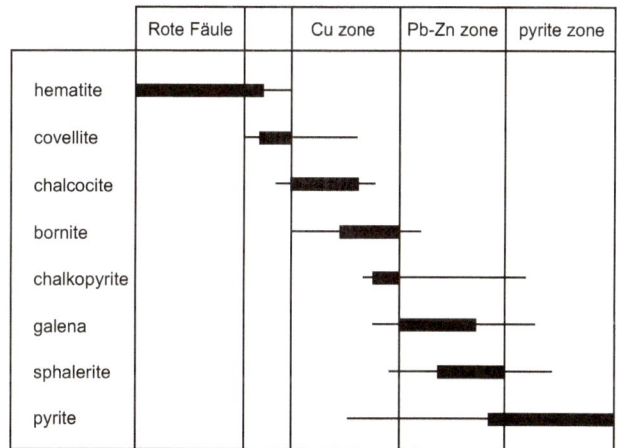

Fig. 5.4 Zoning of ore minerals in Polish Kupferschiefer (from Oszczepalski 1999)

5.1.2   Central African Copper Belt

Neoproterozoic copper–cobalt deposits in the Central African copper belt (DR Congo and Zambia) were also largely formed during diagenesis of sediments. The belt is more than 100 km wide and 700 km long and stretches from Katanga, the southern province of the Congo, almost to the southern border of Zambia. Congolese copper deposits, in particular, have a high cobalt content at the same time; in total about half the world's known cobalt reserves are located here. Some also contain nickel, uranium, thorium, silver, gold, platinum-group elements, selenium, tellurium, arsenic, molybdenum, and vanadium. World-class deposits include Kolwezi, Tenke-Fungurume, and Kamoto (Congo) and Konkola-Chililabombwe, Nchanga, Nkana, and Mufulira (Zambia).

Very different models have been proposed to explain the deposits ranging from epigenetic–hydrothermal to diagenetic and to synsedimentary. The typical zoning of ore minerals parallel to the edge of the basin speak in favor of the synsedimentary concept according to which rivers are said to have transported dissolved metals into an anoxic sea basin where they were precipitated. However, such zoning does not match fluctuations in the water level. This mismatch has led to the belief that the deposits were probably formed in several stages mainly during diagenesis (Cailteux et al. 2005; El Desouky et al. 2009). However, the main source of metals may have been the weathering of basement rocks in the surrounding area where porphyry copper deposits and other deposits were present.

Sediment deposition began with the formation of a continental rift (Roan Group sediments) in connection with the breakup of the supercontinent Rodinia. After deposition of fluviatile sandstones and conglomerates, seawater entered the lagoons and sabkhas (Box 5.11) and clastic sediments were deposited together with evaporites. In the Congo these were mainly dolomite shales (partly deposited by stromatolites; Box 5.2). These shales were rich in dolomite and some anhydrite (Mines Formation), whereas in Zambia they were mainly sandstone, silt, clay, greywacke, and some anhydrite (Musoshi Formation). The most important copper deposits are located in these layers. In Zambia clastic sediments can have one or more ore horizons. In the Congo the lower ore horizon is found in dolomite at the base of the Mines Formation, the upper ore horizon is located in dolomite-rich shale, and in places there is a third and uppermost ore horizon within carbonates. Although the lower and upper ore horizons have a combined thickness of only 15–55 m, individual deposits can be traced for hundreds of meters or several kilometers.

Sulfate-reducing bacteria living in highly saline water caused the precipitation of pyrite and the first copper sulfides (synsedimentary and early diagenetic). Some are bound to sedimentary structures such as sand ripples, erosion channels, and individual layers and are influenced by landslides. Diagenetically, they were replaced or overgrown with copper and cobalt sulfides at around 100 °C. Elevated metal content in a lagoon that evaporated and the Eh–pH gradient in the connate water could have played a role. This resulted in zoning of sulfide minerals all of which predominantly occur as tiny disseminated grains.

The continental rift developed into an ocean (Nguba Group sediments) in which platform carbonates (partly with further Cu deposits) and later deep-sea sediments were deposited. Toward the end of the Proterozoic the continents gradually came together again until they collided. The sediments were pushed together to form a fold and thrust belt (Lufilian Arc). Hot hydrothermal water flowed through the sediments and precipitated further (epigenetic) sulfides that are coarse-grained. Although partly disseminated in the rock, they are mainly found in nodules, layers, and veins.

The uranium of the copper belt was precipitated mainly during ocean formation and in a second phase during orogeny (Decrée et al. 2011). Uranium has also been found in tectonic breccias, clastic sediments, and strike-slip faults—not only copper deposits. The best known is the Cu–Co–Ni–U deposit at Shinkolobwe (Congo) from which the uranium for the atomic bombs dropped on Hiroshima and Nagasaki originated. After the Second World War Belgium accelerated uranium mining in the colony and in this way financed reconstruction of the country. During and after orogeny some epigenetic hydrothermal zinc–lead–copper deposits were formed in the platform carbonates of the ocean. Such carbonates contained several other metals (Box 4.22).

More recently (presumably intensified in the Pliocene) remobilization of copper occurred in the oxidation zone. Such copper was precipitated mainly as malachite (Fig. 5.5) in karst systems (De Putter et al. 2010). Malachite crusts or stalactites are common and exhibited in many museums. Such secondarily enriched zones are preferentially mined.

5.2   Banded Iron Formation

Banded iron formations (BIFs) are a characteristic sedimentary rock of the Precambrian. This was especially so in the period between 3.8 and 1.8 billion years ago (Archean and Paleoproterozoic). They then largely disappeared from the geological record. In the Late Proterozoic and Early Phanerozoic they are very rare. They consist of alternating layers of iron-rich bands of mostly hematite (Fe_2O_3) and magnetite (Fe_3O_4) and some chert (jasper) (i.e., microcrystalline to cryptocrystalline quartz often colored red by hematite) (Figs. 5.6 and 5.7). Individual bands are typically

Fig. 5.7 BIF from North America in the Dresden Botanical Garden. *BIF*, Banded iron formation (*photo* © André Karwath, CC-BY-SA, Wikimedia Commons)

Fig. 5.5 Malachite stalactites from L'Etoile du Congo Mine ("Star of the Congo Mine," a.k.a. Kalukuluku Mine), near Lubumbashi (DR Congo) (*photo* © Rob Lavinsky/iRocks.com, CC-BY-SA, Wikimedia Commons)

Fig. 5.6 Folded BIF of Krivoj Rog (Ukraine) with hematite (a.k.a. martite, hematite pseudomorph after magnetite), magnetite, and jasper. The ore grade is 40%. *BIF*, Banded iron formation (*photo* © F. Neukirchen/Mineralogical Collections of the TU Berlin)

0.5–3 cm thick and finely laminated internally. Less common names for BIFs are itabirite (for the Brazilian city of Itabira in the Quadrilatero Ferrifero region, Minas Gerais), jaspillite (for the mineral jasper), iron quartzite, hematite quartzite, and taconite.

BIFs are by far the most important iron ore. Current recoverable reserves are estimated at 150 billion tonnes of ore (Jorgenson 2012). Three types are usually distinguished each created at different times under different conditions

(Figs. 5.8 and 5.9). There are three prerequisites for their formation. The sea must be oxygen free thus enabling a large amount of Fe^{2+} to be dissolved in seawater, the H_2S content must be significantly lower than the Fe^{2+} input (otherwise all iron would have precipitated as sulfide; see also Sect. 5.1.1), and the iron must be precipitated by oxidation to Fe^{3+}.

Algoma-type BIFs (mostly 3.5–3 billion years old) occur mainly in Archean greenstone belts together with greywacke, submarine tuffs, and volcanic rocks. They were created at different water depths (especially, at island arcs, in back-arcs, and in rift systems). Although hydrothermal systems play a direct role (there are often volcanogenic massive sulfides in the vicinity; Sect. 4.16), BIFs were predominantly formed by oxidation at a certain distance from hot springs. Such BIFs are usually relatively small. They are mined, for example, in the Abitibi greenstone belt (see Box 4.27) in Canada. The oldest deposits of this type can be found in the Isua greenstone belt on Greenland where they have an age of about 3800 million years.

The largest iron deposits are **Superior-type** BIFs (usually 2500–1900 million years old). They emerged at a time when cyanobacteria were producing ever-increasing amounts of oxygen (Box 5.2). They formed in shelf seas, on continental slopes, or in epicontinental seas (i.e., inland seas) mostly associated with quartzites, black shales, and carbonates (Heinrich et al. 1982; Simonson 1985; Slack et al. 2007; Planavsky et al. 2009). The influence of land seems small as the lack of coarse clastic sediments shows. The deposits can be very large with an extension of tens of thousands of square kilometers and a thickness of tens or hundreds of meters. They are considered biogenic or chemical sediments because of the absence of volcanic phenomena in their vicinity. The name is derived from the BIFs around Lake Superior (Wisconsin, Michigan, Minnesota, USA, and

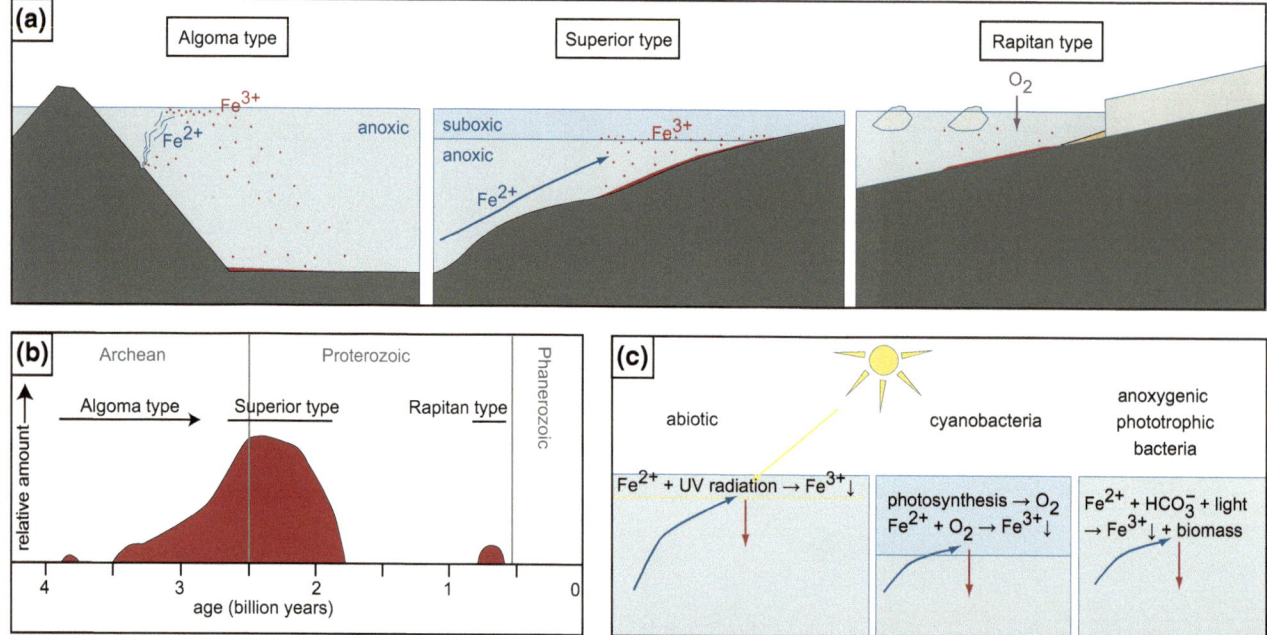

Fig. 5.8 BIFs can be divided into three types that **a** arose under different conditions and **b** at different times. The older ones are Algoma-type BIFs that originated in an anoxic ocean in direct connection with hydrothermal systems. The most important ones are Superior-type BIFs that are probably related to the development of photosynthesis that initially led to a near-surface water layer with low oxygen content. After the oceans and the atmosphere had become rich in oxygen BIFs only developed in connection with extreme ice ages during which seawater could become anoxic again for a short time (Rapitan type). **c** Three processes are mainly responsible for the oxidation of Fe(II) to Fe(III) and thus for precipitation (schematic formulae): abiotic oxidation by UV radiation, oxidation with oxygen released by photosynthetic microorganisms, and direct oxidation by certain bacteria without oxygen. *BIF*, Banded iron formation (**b** from Klein 2005)

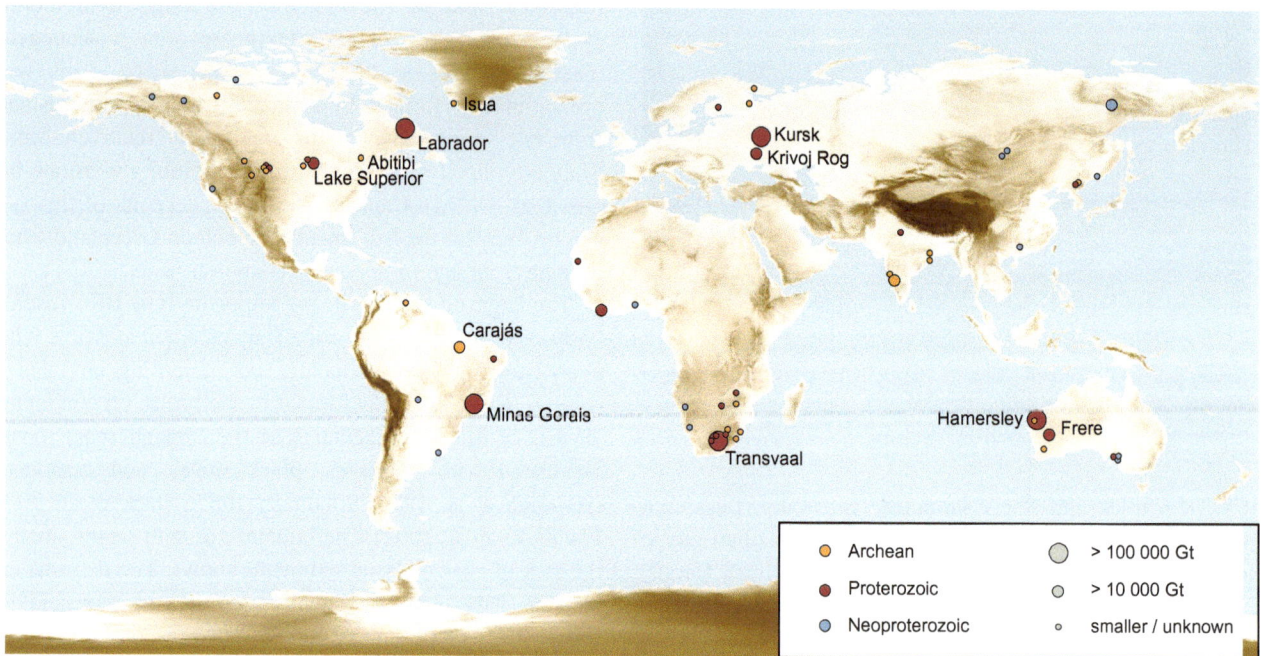

Fig. 5.9 Important BIFs by age and size. They by and large occur exclusively in the Precambrian cores of continents. *BIF*, Banded iron formation (from Bekker et al. 2010)

Ontario, Canada). The Hamersley Range (Western Australia) is the most important in the world. If you want to see a BIF in a spectacular landscape, then you should visit the gorges of Karijini national park in the Hamersley Range. Other important deposits are found in Transvaal (South Africa), Kriwoj Rog (Ukraine), Kursk (Russia; Box 5.3), Minas Gerais and Carajás (Brazil), Liberia, India (Karnataka and Orissa), and Labrador (Canada).

Rapitan-type BIFs (600–750 million years old) are the most recent and they always occur along with glacial sediments. They were created during large-scale glaciation (Snowball Earth scenario; Box 5.5). This type is economically insignificant.

The fact that these three types emerged in different periods of the Earth's history but not in more recent times is of course in need of explanation. Significant changes in the Earth's early history likely played a role here (in particular, oceans changing from an anoxic to an oxygen-rich state). However, the details are anything but clear and research over the last few decades has raised more questions than answers.

In addition to hematite and magnetite other iron minerals occur some of which can even dominate. This resulted in the concept of different facies being developed. They were originally explained by different water depths and different oxygen contents (James 1954)—now known to be a rough simplification. Carbonate facies consist of siderite (Fe^2 $^+CO_3$) or ankerite ($CaFe^{2+}(CO_3)_2$). Greenalite ((Fe^{2+}, $Mg)_6Si_4O_{10}(OH)_8$), minnesotaite ((Fe^{2+}, $Mg)_3Si_4O_{10}(OH)_2$), and stilpnomelane ($K_{0,6}(Mg, Fe^{2+}, Fe^{3+})_6Si_8Al(O, OH)_{27}\cdot2-4H_2O$) dominate in the rare silicate facies; hence the iron-rich equivalents to antigorite (serpentine) and talc (Sect. 7.10) and another iron-rich phyllosilicate (i.e. sheet silicate). Although oxide facies are the most common, they can also contain small amounts of siderite and iron silicates. However, most minerals were only created by subsequent (diagenetic and epigenetic) changes, which generally makes research more difficult. Iron (III) hydroxides, amorphous SiO_2, siderite, and greenalite are assumed to be primary minerals (Klein 2005). Sulfide facies with pyrite are probably only a subsequent modification.

In the Early Precambrian the atmosphere contained almost no oxygen. Under reducing conditions iron would have been present as Fe^{2+}, which is readily soluble. At that time the sea likely contained large quantities of dissolved iron. Most researchers believe iron came mainly from hydrothermal systems on the seabed (Horstmann et al. 2001; Klein 2005) and only a small part from weathering on land. In addition, the water was probably saturated with SiO_2 (or $Si(OH)_4$) according to which there would have been a continuous precipitation of silica gel and thus sedimentation of chert—unicellular organisms with a skeleton of opal, which are responsible for similar sediments today, did not yet exist. Several processes have been proposed that could have led to the precipitation of iron. Probably all were occurring at the same time, but which process was most important and what has changed over time are moot points.

Box 5.2 Stromatolites: early life provided oxygen in the atmosphere

The geological record coupled with observations of the atmospheres of neighboring planets suggest that the Earth's atmosphere was not always as rich in oxygen as it is today (Ries 2007, 2010). However, the early atmosphere will not have been completely free of oxygen either. The photolytic splitting of water vapor into hydrogen and oxygen by radiation from the Sun may have released small traces of oxygen (especially when comparatively light hydrogen can leave the Earth's gravitational field before it is reunited with oxygen to form water). This process also provides a good explanation for the approximately 0.13% oxygen detected in the atmosphere of Mars (Schultz 1993). Moreover, this low oxygen content was probably also necessary to protect the early phases of life (DeDuve and Hausser-Siller 1994) from the increased UV radiation of the still young Sun by forming an ozone layer. Around 4 to 3.5 billion years ago the UV radiation from the Sun was probably four to eight times stronger than that of today (Towe 1996).

However, the high oxygen content of around 21% observed in the atmosphere today is due to photosynthesis. Living beings use sunlight as an energy source to produce sugar from carbon dioxide and water. Oxygen is released as a waste product. The evolution of photosynthesizing and oxygen-producing cyanobacteria may have been one of the most significant events in the history of life on Earth. Oxygen is a potent oxidant and its massive release has fundamentally altered chemical conditions on the Earth's surface (Holland 2002). This enabled completely different ways of metabolism. None of the organisms that use these processes could have survived before the development of cyanobacteria. Determining the exact time when photosynthesis first began and oxygen production first started is associated with some problems. On the one hand, rocks that give us information about this distant time are rare and where they exist they have undergone change. On the other hand, indications are often not clear because there are also alternative processes to explain the deposits. Therefore, the main focus is on the earliest detection of stromatolites since oxygen initially remained unchanged in marine sediments due to precipitation reactions. When and how fast the oxygen content in the atmosphere increased are moot points.

Fig. 5.10 Section through a stromatolite from the Quaternary from Lake Natron (Tanzania) (*photo © F. Neukirchen*)

Stromatolites are finely laminated biogenic carbonate rocks (Fig. 5.10). Microorganisms such as cyanobacteria living in so-called algae mats precipitate substances dissolved in the water while existing sediment particles are cemented. The internal structure of very finely laminated limestones or dolomites is diverse as shown by flat layers, layers arched upward, or columns of many arched layers. The first stromatolites are known from the Warrawoona Group in Western Australia that are 3.45 billion years old (Lowe 1980); the complex stromatolites from the Swaziland Supergroup are just as old (Byerly et al. 1986). The first indications of free oxygen in seawater are seen from the same period (Hoashi et al. 2009). Stromatolites with a great variety of morphological types can be found in sediments of the Pongola Supergroup in South Africa that are about 3 billion years old. However, there are also abiotic explanatory models for stromatolite-like structures (Grotzinger and Rothman 1996) and perhaps some were deposited by anoxic bacteria. Whether the oldest stromatolite-like structures really indicate the presence of cyanobacteria is therefore not clear. The first really confirmed occurrence is in rocks 2.7 billion years old (Brocks et al. 1999; Buick 2008) and fundamental change in the atmosphere must have taken place at the latest 2.2 billion years ago as evidenced by the oldest preserved red paleosols (Beukes et al. 2002).

Ideal conditions for life already existed at that time in shallower and light-flooded areas of the oceans. The first photosynthesizing organisms such as cyanobacteria could exist here and release oxygen (Box 5.2). This oxidized Fe^{2+} to Fe^{3+} that is no longer water soluble and is precipitated in the form of iron(III) hydroxides and iron(III) oxyhydrates.

These accumulated on the bottom together with amorphous SiO_2 (chert). During diagenesis they dehydrated. Hematite and (through partial reduction) magnetite were formed. If oxygen consumption through iron oxidation did not use up the oxygen released by biological activity quickly enough, then this could lead to oxygen concentrations in the water that were toxic to bacteria and thus lead to their death (Kappler et al. 2005). The supply of bivalent iron by upwelling deep water normally ensured that oxygen released was immediately bound and could not accumulate in the seawater or the atmosphere (Kasting 1987). Accordingly, the ferrous seas buffered the oxygen produced during photosynthesis and continued to do so for a long time. Nevertheless, a suboxic layer containing some oxygen formed on the surface of the oceans and was separated from the rest of the water by a sharp chemocline. About 1.9 billion years ago, sufficient oxygen had been released to oxidize iron even in deep ocean layers. The supply of bivalent iron into the shallow seas collapsed and the formation of banded ores failed to take place.

This is the classical theory. Indeed, there is some evidence that oxygen released by photosynthesis was important (at least in Superior-type BIFs). However, the time when photosynthesis began is another moot point and begs the question of how older BIFs were created (especially, early Algoma-type BIFs).

Fe^{2+} can be oxidized by UV radiation in shallow water depths without the participation of living organisms. Since there was no ozone layer at that time radiation was particularly intense. Divalent iron ions absorb a light quantum in the UV range and thereby emit an electron. This is absorbed by H^+ and elementary hydrogen is formed. It is highly likely this process took place in the upper water layer, but it is doubted whether this can compete with the sedimentation of iron silicates and siderite. This would thus only be of minor importance (Konhauser et al. 2007).

Siderite precipitates in iron-rich water under relatively reduced conditions only if enough HCO_3^- is present. The iron silicate greenalite can form when Fe^{2+} combines with amorphous silica gel, which is known to happen relatively easily at high concentrations (Konhauser et al. 2007; Wang et al. 2009a, b).

Another possibility would be direct oxidation of iron by anoxygenic phototrophic bacteria. These bacteria form their biomass from carbon dioxide and water using light as an energy source and bivalent iron as an electron source oxidizing it to trivalent iron (Widdel et al. 1993; Ehrenreich and Widdel 1994; Konhauser et al. 2002; Weber et al. 2006; Köhler et al. 2010, 2013; Emmerich 2013). It is known that they can effectively precipitate large amounts of iron. Since conditions in the Precambrian sea would have been perfect it can be assumed that they made an important contribution.

Fig. 5.11 BIF banding is not always spectacular. This block lies in the desert of Mauritania (*photo* © Thomas Finkenbein)

However, there is still no direct evidence that they existed in the Archean.

When oxygen was being formed by photosynthesis certain aerobic bacteria might also have been important in later BIFs. Such bacteria oxidize Fe^{2+} using oxygen; hence this could lead directly to the precipitation of iron hydroxides.

This process is significantly faster and more effective than abiotic oxidation (Konhauser et al. 2002).

Another controversial topic is how the banding arises (which is not always spectacular; Figs. 5.11 and 5.12). One theory argues it is due to fluctuations in hydrothermal activity thus leading to changes in water composition. This could have a direct effect on the precipitation of amorphous silica and iron or an indirect effect via changing the concentrations of nutrients for microorganisms. Another possibility could be that strong upwelling of Fe^{2+}-rich water only occurred at certain times of the year. In other seasons the sedimentation of chert predominated because the SiO_2 concentration in the upper water layer increased due to evaporation. Another possibility is a seasonally fluctuating water temperature and thus fluctuating biological activity. For example, anoxygenic phototrophic bacteria are most active at 20–25 °C and less active at higher or lower temperatures (Posth et al. 2008). An unconventional theory (Wang et al. 2009a, b) explains fine small-scale banding as an internal, dynamically self-organizing process. This can happen when the chemical reactions taking place amplify themselves. This is the case when a product of the reaction leads to further release of the reactant or when it serves as a

Fig. 5.12 In Mauritania BIFs occur as free-standing mountains surrounded by desert sand such as Guelb BouDerga (*photo* © Thomas Finkenbein)

catalyst. This causes one reaction (precipitation of iron) or the other (precipitation of SiO_2) to be intensified alternately.

Another unconventional theory argues that BIFs were deposited in deep ocean basins by density currents (Krapez et al. 2003; Pickard et al. 2004; Lascelles 2007); hence iron was initially precipitated at hydrothermal sources. In contrast to today's black smokers (Sect. 4.15.1) the concentration of dissolved Fe^{2+} and SiO_2 when mixed with seawater is said to have been so high as to bring about colloidal suspension of hydrous iron silicates together with iron hydroxides and iron carbonates. These sedimented near the source until eventually the mud slid off and flowed as a turbidity current into the deep ocean basin. The finest particles were distributed throughout the basin. Water-containing iron silicates decayed during diagenesis and the fine laminar banding of chert and iron oxides developed. One argument for this is that interlaid sediments in a BIF in Australia were interpreted as turbidites. These are said to have slipped off the shelf edge when sea levels dropped.

However, original iron hydroxides are replaced even in early diagenesis. Hematite is formed by simple dehydration. Magnetite is formed by reducing part of Fe^{3+} back to Fe^{2+}. On the one hand, this can be achieved by certain anaerobic microorganisms and, on the other hand, by simultaneously sedimented organic substances. A significant proportion of the iron may even have been dissolved again (Konhauser et al. 2005). Oxidation of siderite or a reaction of siderite and hematite to magnetite is also conceivable. In addition, new minerals are formed. Dating of zircon from interlayered tuffs showed that the sedimentation of one BIF in the Hamersley Range (Australia) took place at the rate of 0.033 mm compact rock per year (Pickard 2002).

Box 5.3 Kursk Magnetic Anomaly

Compasses are of little use in Kursk (Russia) because of the Kursk Magnetic Anomaly (KMA). The strongest anomaly of the Earth's magnetic field is caused by the world's largest concentration of BIFs in rocks of the Precambrian craton covered by younger sediments. The BIFs were created in four different time spans. The oldest are the Oboyan Group (3.15 billion years) found in highly metamorphic gneisses and strongly transformed. The next oldest are the Mikhailovka Group found in a greenstone belt from the Middle Archean (3.15 to 2.5 billion years). BIFs of the Paleoproterozoic Kursk Group (2.5 to 2.3 billion years) are the largest and have the highest ore grade. The next largest are the slightly younger BIFs above an unconformity in the Oskol Group. Some occurrences were enriched during metamorphism and lateritic weathering to form high-grade ores (Belykh et al.

2007) mined in several large open-pit mines. Although the amount of iron available in the region is enormous, only a small part is economically recoverable.

The result is a BIF with a typical ore grade of 25–30% Fe, which is still low. All BIFs were later exposed to more or less strong metamorphism and often intensive chemical weathering near the Earth's surface. Interesting minerals can develop by metamorphism such as amphibole asbestos (Box 7.4) or the shimmering gemstone tiger's eye. Since there can also be subsequent enrichment of iron such BIFs are economically important. For example, high-grade zones in the Hamersley Range contain up to 65% Fe. Although SiO_2 is removed thus reducing the thickness of the layer and increasing the porosity, extra iron is added such that goethite or siderite replaces chert. Oxidation of magnetite to hematite via a process called martitization is typical. Whether enrichment occurred mainly during intense chemical weathering comparable to laterites (Sect. 5.11) (Morris 1985, 2002) or by hydrothermal solutions during metamorphism (Taylor et al. 2001, 2002) is highly debated.

Box 5.4 Did a meteorite impact end the formation of BIFs?

The reason the formation of BIFs came to such an abrupt end remains a mystery. The BIFs on Lake Superior (Canada and the USA) may hold the answer. The banded ores here are overlain by deposits that have now been recognized as ejecta masses of the Sudbury Impact. This impact took place about 1.85 billion years ago and left behind a crater some 200 km in diameter—the second largest crater known on Earth. This impact may be responsible not only for the richest nickel deposits on Earth (Sect. 3.3.5), but also may have something directly to do with the end of BIFs. The ocean at that time probably had a thin oxygen-containing surface layer, while the deeper water was virtually oxygen free. If an asteroid 10 km wide strikes water 1000 m deep, model calculations show that tsunamis about 1000 m high would occur at the point of impact. At a distance of 3000 km they would still be over 100 m high. These waves together with the undersea landslides that likely followed the impact and other consequences of the impact would have literally stirred up the Precambrian ocean. Since all continental masses were united in one supercontinent at that time the waves could propagate largely undisturbed across the entire ocean. The mixed ocean now contained enough oxygen even in the deeper layers to remove all divalent iron by storing it as trivalent iron in sediments preventing the transport of dissolved iron by ocean currents to distant areas where BIFs used to form (Slack and Cannon 2009).

Just over 1.8 billion years ago the sedimentation of BIFs ceased worldwide. This could be due to water of the deep sea already containing oxygen and the dissolved iron content in seawater being correspondingly low. According to the classical theory this was a gradual process. However, a meteorite impact may have been the reason (Box 5.4). But there are also indications that the oceans were still largely oxygen free and rich in iron at the time (Poulton et al. 2010). According to Poulton et al., between the oxygen-rich surface water and the anoxic and iron-rich water of the deep sea a further layer formed at medium depth that was anoxic and H_2S rich (euxinic) (see also Sect. 5.1.1). This layer occurred in a strip at least 100 km wide along coasts and caused the Fe^{2+} to precipitate as pyrite and to be deposited in shales. Thus, sulfate-reducing bacteria may well be responsible for the sudden change. A euxinic zone is formed when more than twice as much H_2S is formed as dissolved Fe^{2+} is added because then excess H_2S remains after precipitation of FeS_2. Another factor could play a role as pointed out by Wang et al. (2009b). Hydrothermal alteration of komatiite (Sect. 3.4) leaches out particularly large amounts of iron and silica on the seafloor, while alteration of basalt leads to hydrothermal fluids with significantly lower iron content. Although this ultrabasic volcanic rock was common in the Archean, later it formed less and less. Since the oceanic crust is constantly renewed by subduction and seafloor spreading this may have altered the hydrothermal systems and reduced the input of iron into the sea. In fact, after whatever it was that happened about 1.8 billion years ago the emergence of BIFs was no longer possible. Younger Rapitan-type BIFs could only emerge when the Snowball Earth scenario (Box 5.5) led to an at least partially anoxic ocean. However, in the oxidized oceans of the Phanerozoic only iron oolites (discussed subsequently) or hydrothermal iron Lahn–Dill-type deposits directly at a hot spring (Sect. 4.18) could be formed (Oftedahl 1958).

Box 5.5 Snowball Earth

In the Neoproterozoic there were apparently two major ice ages that took place about 650 and 700 million years ago. The Quaternary ice ages appear almost comfortably warm in comparison. The Earth was completely or at least largely covered with ice and looked like a giant snowball (Hoffman and Schrag 2002). Although glacial sediments of several ice ages are known from the Neoproterozoic, the Sturtian and Marinoan ice ages must have been the two most extreme glaciations. Paleomagnetic data have shown that some glacial sediments were deposited near the equator where glaciers must have reached the coasts. So-called dropstones have been found in sediments. Icebergs can drop large stones into oceans in which only fine sediments are otherwise deposited. The corresponding geographic latitude of glacial sediments was obtained in a study of detritic hematite. Since hematite is weakly magnetic the grains oriented themselves during deposition preferentially in the direction of field lines of the Earth's magnetic field. This enabled the geographical latitude to be determined statistically. This analysis was attempted at different occurrences all over the world, and wherever it was successful, tropical or subtropical latitudes were almost always found. A further special feature is that these glacier sediments are almost always directly superimposed with carbonates indicating a rapid return to a warm climate. Finally, together with the glacier sediments there were also BIFs—the first to occur after an interruption of 1.5 billion years and the last in the history of the Earth.

Although what might have brought about such a scenario is not known, it could have been that the continents were all located in tropical and subtropical latitudes. This would have increased albedo (i.e., the reflection of sunlight) and changed wind systems. It is possible that intensive tropical weathering on all continents also caused a decrease in the CO_2 content of the atmosphere. All these effects led to a cooler climate. Large glaciers formed at the continental margins close to the poles lowering the albedo even further. This obviously led relatively quickly to a self-reinforcing albedo catastrophe. The temperature

Fig. 5.13 Oolitic iron ore from Clinton, Oneida County (New York, USA) (*photo* © F. Neukirchen/Mineralogical Collections of the TU Berlin)

dropped suddenly to extreme minus temperatures and ice spread over the entire surface. This would have at least partially isolated the oceans from the atmosphere and could have shut down ocean currents. As a result anoxic zones developed in the oceans in which Fe^{2+} could accumulate in the water. This was oxidized toward the end of glaciation and formed Rapitan-type BIFs.

Defrosting a completely frozen planet is not easy because of the high albedo. However, the CO_2 content of the atmosphere probably rose to an extreme value since volcanoes continued to exist and important CO_2 sinks were shut down by the ice. This caused the temperature to rise again. As soon as the ice began to melt and the albedo reduced there was likely a rapid change to a hot greenhouse climate. It is possible that the next glaciation occurred as soon as the CO_2 content of the atmosphere had dropped again. Whether the whole Earth was covered in ice or not remains a moot point. There are sediments that suggest that the oceans were at least partially ice-free (Allen and Etienne 2008).

5.3 Iron Oolite

Iron oolites or oolitic iron ores are deposits in shallow inland seas or lagoons mostly alternating with marly clay sediments (Fig. 5.13). Other common names are **minette ore** (in Lorraine, France), **Clinton-type iron deposit** (for an occurrence in New York, USA), flaxseed ore, iron stone, and sometimes Eisenrogenstein (Germany). Although in Central Europe iron oolites from the Jurassic are widespread, the North American oolites originated in the Paleozoic. Thicknesses up to 30 m and extensions up to 150 km can occur.

Oolitic limestone is a chemical sediment typical of shallow lagoons. When water is saturated with carbonate by evaporation, it precipitates as aragonite or calcite around suspended crystallization nuclei such as fine sand or fragments of mussel shells. Particles of a certain weight sink to the bottom. Water movement caused by waves rolls them into spheres.

The name oolite means "egg stone" (from the Greek ῷόν, oon, "egg" and λίθος, lithos, "stone"). Spherical to oval grains with concentric thin coatings around a foreign body called ooids or oncoids are characteristic of such iron ores. In oolitic iron ores they consist of the minerals hematite and siderite, the iron-rich sheet silicate chamosite, and iron hydroxides (limonite). In addition to roundish ooids there are also broken and recoated ooids that together with cross-bedding indicate moving water played a part in the formation of these sedimentary structures. Oncoids form slightly differently. They are shaped more irregularly than ooids and are often larger. They are the product of biogenic precipitation of algae. Iron oolites are usually accompanied by a rich fauna of fossils that usually show iron metasomatism.

There are indications of the involvement of diagenesis in the formation of these deposits, which has led to conflicting theories about their formation. According to one theory diagenetic replacement of carbonate by iron minerals occurred in oolitic limestone (Kimberley 1980, 1989; Sorby 1857). However, some theories argue for a synsedimentary supply of ferrous solutions and chemical or biochemical precipitation (Dahanayake and Krumbein 1986). Although no recent formations of this type are known, pellets of goethite have been found in the Niger Delta at a depth of around 10 m and sediments of the iron-rich sheet silicate chamosite at greater depths (Porrenga 1967). This shows that comparable formations are possible in a synsedimentary way.

However, where do the ferrous solutions necessary for the formation of iron ores come from? Pellets of goethite found in the Niger Delta may well help in explaining this. It is reasonable to assume that the waters of slowly flowing streams in tropical-weathered landscapes with rich vegetation were rich in humic acids and organic matter flowed into lagoons. Iron can also be present as Fe^{2+} in acidic, boggy waters at positive Eh and is therefore water soluble (Fig. 5.14). Fe^{2+} can also be transported in colloidal form bound to organic matter. The pH value would increase at an estuary and the iron would flocculate and precipitate (Harder 1989). Initial enrichment in lateritic soils (Sect. 5.11) may also play a role.

The economic importance of oolitic iron ores is today low due primarily to their consistently low iron content, but also to their high content of undesirable elements such as phosphorus, aluminum, and SiO_2. However, oolitic iron ore deposits were of great importance historically. This is particularly true of the minette deposits in Luxembourg and Lorraine. With an estimated 6 billion tonnes of ore this was one of the reasons the region was contested between France and the German Empire in the 19th and twentieth centuries. The name minette ("small mine") is actually misleading since the word "small" refers to the low ore grade of 20–30%. Nevertheless, in 1919 the Lorraine Region was the second largest iron ore producer in the world with an annual output of 41 million tonnes of ore (van de Kerkhof 2002). After 1960 production fell steadily partly because of competition from higher value ore from overseas. The last mines closed in 1981 on the Luxembourg side and in 1997 in Lorraine (France). Similar but smaller occurrences can be found at Porta Westfalica (Box 5.6), Upper Franconia, and Baden-Württemberg (Germany) and in the Swiss Jura (Doggererz iron ore reserves).

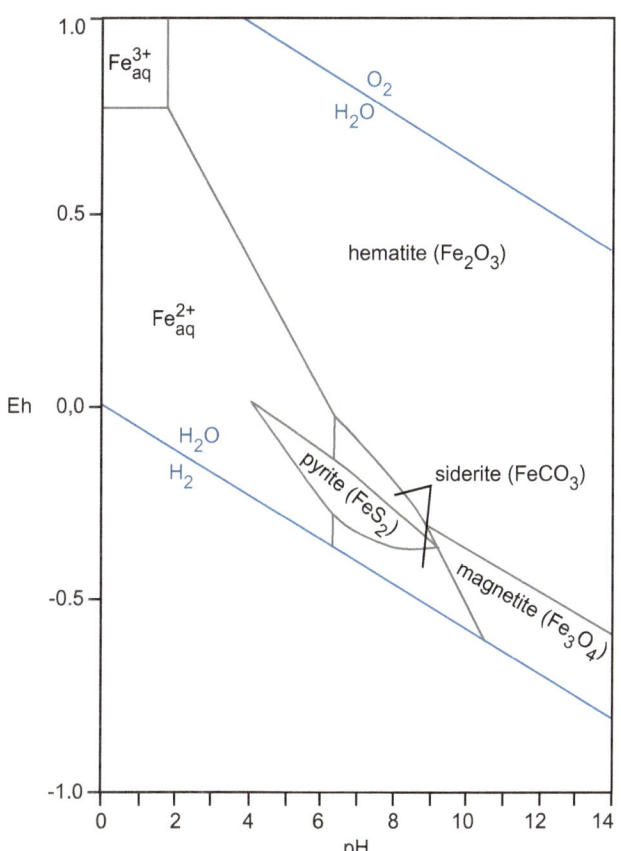

Fig. 5.15 Residual ore from the Upper Cretaceous in Peine (Lower Saxony, Germany) (*photo* © F. Neukirchen/Mineralogical Collections of the TU Berlin)

Fig. 5.14 The stability of iron minerals and dissolved iron as a function of pH and Eh (at 25 °C, 1 bar, 10^{-6} mol/L Fe, 10^{-6} mol/L S, 1 mol/L HCO_3^-). The stability fields of siderite and pyrite are significantly greater when the concentrations of HCO_3^- and S, respectively, are increased (from Garrels and Christ 1965)

Fig. 5.16 Bean ore from the Swiss Jura (*photo* © F. Neukirchen)

Box 5.6 Iron ore as raw material for road construction

On the northern slope of the Weser Mountains near Porta Westfalica (Germany) Upper Jurassic oolitic iron ore and limestone (stratigraphic unit: Korallenoolith) have been mined since 1883 and in today's Wohlverwahrt-Nammen Mine since 1935. The composition of ore in the iron-rich upper and lower seams is 13–15% Fe, 33–36% CaO, 9–11% SiO_2, and 0.18% P_2O_5. Due to its carbonate content the ore was intended as flux for Ruhr smelters and for the cement industry. Between 1934 and 1966 around 17.8 million tonnes of ore were mined from the cliff seam in the Wohlverwahrt-Nammen Mine. Maximum annual output peaked at one million tonnes during this period. Nevertheless, in 1966 it was still 473,947 tonnes corresponding to just under 5% of West German iron ore production. The mine thus made a major contribution to supplying smelters with lime flux (Barbara Rohstoffbetriebe 1991). Ever since 1987 hard rock has been extracted via so-called backfill mining for road construction. In this largely mechanized underground operation raw material is extracted by room and pillar mining (Sect. 1.11) by drilling and blasting in mining chambers about 9 m wide, 15–17 m high, and up to 600 m long. Stratification here is inclined at 18° to NNE. The extracted material is then transported by trucks to processing plants located above ground. Depending on demand 400,000–500,000 tonnes are produced annually. Since 1987 residual materials from flue gas cleaning or used sand that accumulates during sand blasting have been brought into the excavated pit as backfill. The sometimes enormous cavities may find other uses (especially, at times of energy transition). Such cavities are on show in the neighboring Kleinenbremen visitor mine. For example, an underground pumped storage hydropower station is being considered (Anonymous 2011).

Closely related to iron oolites are the **residual iron ores** of the Cretaceous in the Salzgitter area in Germany (Fig. 5.15) in which clastic goethite was washed up in its solid state. It used to be the most important German iron ore deposit. Mining ceased in 1982. There are plans to create a repository for low-level and intermediate-level radioactive waste in the Konrad Shaft. Founded in 1942, the town Salzgitter is still an important steel producer today.

5.4 Bean Ore

Bean ores are millimeter-size to centimeter-size concretions of limonite that resemble coffee beans (Fig. 5.16). They are sometimes found in caves and other karst cavities where they can form regular patches embedded in clay. Sometimes the clay has been washed out. Limonite is a mixture of different iron hydroxides such as lepidocrocite and goethite—not a mineral in itself. The irregularly shaped and concentric beans may also be hollow or contain fossils. Their formation is attributed to the solution of iron during a phase of intense tropical weathering (see also Sect. 5.11). Acidic water transports the dissolved iron into calcareous areas. Here the dissolved iron is precipitated again due to the pH value changing. Although bean ores are today negligible, up to the nineteenth century they covered a significant part of the iron demand in some regions such as southwest Germany and Switzerland.

5.5 Sedimentary Manganese Deposits

Oolitic manganese ores (Force and Cannon 1988) play a greater economic role today than corresponding iron ores. Oligocene occurrences at the Black Sea in the southern Ukraine near Nikopol (Black Sea type) merit special mention here. The manganese seam reaches a thickness of 4.5 m with a lateral extension of up to 250 km. It is mainly composed of the manganese minerals psilomelane and pyrolusite grading into rhodochrosite ($MnCO_3$) toward the basin. The deposits are located in sediments of a transgressive series (rising sea level) over weathered basement with paleosoil followed by coal-bearing lake sediments. On top lie sands (with glauconite), the manganese horizon, and finally clastic marine sediments. Similar deposits are found at Tschiaturi (Georgia) in the Caucasus and Groote Eylandt (Northern Territory, Australia).

The formation of these deposits by the introduction of manganese-containing weathering solutions into a sea basin parallels the formation of iron oolites. This poses the problem that iron had to have been separated beforehand. Manganese and iron in the Earth's crust are present at a ratio of approx. 1:50 and their geochemical behavior is very similar

under surface conditions. Mn^{2+}—unlike Mn^{3+} and Mn^{4+}—is very soluble. The stability field of Mn^{2+} as a function of Eh and pH is slightly greater than that of Fe^{2+}, which is probably the reason iron is precipitated first (Krauskopf 1957).

Manganese deposits can also form in deep seawater. Such deposits are mostly dark earthy masses called umber or wad consisting of manganese oxides. In the Black Sea this is happening today in a zone west of the Crimea. Although an ocean current brings anoxic (euxinic) deep water (see also Sect. 5.1.1) onto the shelf, manganese only precipitates when mixed with oxygen-rich water. The deep water is low in iron as pyrite has already precipitated. Manganese-rich sludge is also found in the Red Sea. Such deposits are often found in the sediments of ophiolite complexes (Box 3.7). Often they are found in the wider vicinity of volcanogenic massive sulfide deposits (Sect. 4.16) when they are then considered distal exhalative products of hot springs. Much larger occurrences formed in the Precambrian in connection with BIFs usually at some distance or somewhat higher in stratigraphy. Especially important is the Kalahari manganese field in sedimentary rocks of the Transvaal Supergroup in South Africa. Other large deposits can be found in Minas Gerais (Brazil), India, Ghana, and Gabon. They are often metamorphosed and deformed.

5.6 Manganese Nodules

These somewhat inconspicuous black–brown, potato-sized formations inspired many a fantasy in the 1970s and 1980s. About 20–40% of each nodule consists of manganese; there is also some copper (0.2–1%), nickel and cobalt; rare metal content is increased; iron makes up about 15%; and there is some aluminum. The rest is material from various deep-sea sediments such as clay and quartz (Cronan 2000). They occur in almost all oceans and at various depths, sometimes even in freshwater lakes such as Lake George in New York (Schoettle and Friedmann 1971). However, they are mainly found in the deep sea (especially, in the Pacific in a belt that stretches from Mexico and almost reaches Hawaii called the Clarion–Clipperton Fracture Zone, as well as in the Indian Ocean). Cobalt-rich crusts are known at seamounts at a water depth between 1000 and 2500 m. The nodules were discovered as early as 1868 in the Kara Sea. The Challenger Expedition 1872–1876 aboard the *HMS Challenger* discovered that such deposits are found in large parts of the oceans.

How does manganese and other metals actually get into the nodules? This question is by no means trivial considering that manganese is only found in very small quantities in seawater. However, there are some ways in which seawater can be enriched. For example, a small amount of manganese and other elements can be incorporated into the calcareous shells of many marine creatures. If the shells of these organisms sink below the carbonate compensation depth (CCD) when they die, then carbonate shells are dissolved and the elements contained in them enter the seawater. If the enriched deep water meets oxygen-rich deep water, then dissolved elements such as iron, nickel, manganese, or cobalt are oxidized and precipitate (Crerar and Barnes 1974). There are indications that microorganisms whose biofilms serve as an initial nucleus are also involved (Wang and Müller 2009; Wang et al. 2009a). At only a few millimeters per million years nodule growth is extremely slow.

Manganese nodules were long considered little more than a scientifically interesting deposit in that they were found at almost inaccessible depths. However, this changed significantly in the 1960s. Alarmed by forecasts of the Club of Rome and the oil crisis many people began to recognize the finiteness of important resources. To avoid price increases feared in the event of shortages new deposits had to be explored urgently. What could be more obvious than to extend the search for such deposits to the vast areas of the oceans? After almost 100 years the strange finds took center stage once again. In the time that ensued new plans were forged and discarded for the exploration and mining of treasures on the seabed (see also Box 5.7 and Sect. 1.13). However, falling commodity prices and the discovery of large deposits on land made expensive underwater mining increasingly unprofitable. Moreover, unresolved environmental problems placed an additional burden on plans with the result that practical technologies were often put on the back burner.

The initial euphoria about exploiting the seas was followed by a certain disillusionment and the realization that the seas are no longer a legal vacuum. Since 1994 a specially created United Nations authority called the International Seabed Authority (ISA) based in Kingston (Jamaica) and the United Nations Convention on the Law of the Sea (UNCLOS) have been monitoring the requirements that resource-hungry industrial nations have to fulfill to avoid disputes as to who actually owns mineral resources on the seabed and who is ultimately allowed to exploit them. This means that anyone wishing to engage in deep-sea mining outside nations' 200-mile exclusion zones will have to notify the competent maritime authority and apply for a license on the basis of a Deep-sea Mining Code adopted in 2000. Unfortunately, there are nations that do not sign up to the rules. UNCLOS and the Maritime Authority are strictly speaking only responsible for states that have signed the convention. Therefore, it is annoying when big global players like the United States belong to the non-signatories. ISA has already granted license areas to various partners in signatory countries. The two German claims are located in the manganese nodule belt of the Clarion–Clipperton Fracture Zone and together cover around 75,000 km^2 making them larger than the federal states of Lower Saxony and Schleswig-Holstein combined. Germany was granted sole right to mine manganese nodules there for 15 years (Lohmann and Podbregar 2012; Seidler 2012).

However, as tempting as the idea of simply picking up large quantities of such potato-shaped raw materials from the seafloor may be the great depth of the sea in which the manganese nodules are found poses some problems for metal seekers. In 1978 a consortium called Ocean Management Incorporation (OMI) led by Preussag AG brought around 800 t of manganese nodules from the Pacific Ocean to the surface to test the feasibility of using the airlift method and hydraulic vertical pumps. Both methods were shown to be suitable for underwater mining at great depths. However, mining at such great depths is costly. It was found that mining the nodules is only worthwhile if they contain certain levels of copper, cobalt, and nickel and if at least 5–10 kg of manganese nodules lie on every 1 m^2 of seabed. According to the University of Hanover (Germany) at least 5000 t per day must be extracted per ship or platform for loss-free extraction (Lohmann and Podbregar 2012).

Extracting such a large quantity of nodules requires large areas of seafloor where there are very rich manganese nodule fields. Large quantities of sediment sometimes have to be moved to gain access. This is where the problems begin. Sediment brought up to the surface along with nodules is undesirable. On the one hand, it overloads the pumps that have to bring this unwanted ballast to the sea surface. On the other hand, the manganese nodules would then have to be separated from the sediment adhering to them. The objective is to bring manganese nodules to the surface—not useless deep-sea sediments.

Such sediments pose other problems. At 3 °C, in eternal darkness, and the enormous water pressure prevailing there the cold and lightless depths of the oceans are not as hostile to life as would be expected. On the contrary, life abounds and a large number of organisms have adapted excellently to conditions prevailing there. Sedentary creatures such as sponges, anemones, bristle worms, and sea lilies are found as are mobile creatures such as squid, crabs, or sea cucumbers. All these living beings would be severely affected by the underwater mining of manganese nodules—not to mention life at higher levels in the water column. Extracted sediments would be discharged here and immense sediment plumes would extend far into the sea from the collecting vessels or platforms. Resulting chemical changes like those caused by substances dissolved in the pore water of sediments can among other things consume oxygen in the water thus leading to local oxygen depletion and the resulting migration or death of marine organisms. Fine sludge particles could also affect fish gills or the filtration apparatus of marine life. What is more, little is known about what is actually living down there or how quickly ecosystems would recover as a result of the profound change brought about by nodule mining. In an effort to protect this environment the German research association Tiefsee-Umweltschutz (TUSCH, "deep-sea environmental protection") was founded in the 1980s. A number of important experiments have been undertaken to clarify the consequences of deep-sea mining. The DISCOL Experiment (disturbance and recolonization experiment) involved the research vessel *Sonne* investigating the biocoenoses of manganese nodule fields in the Peru Basin. Bryozoans (moss animals) and other benthic organisms are even found on the nodules themselves. Once the actual state of the ecosystem had been established as far as humanly possible, an area 3.5 km² was dug up using a plough 8 m wide to simulate deep-sea mining. Apart from the disturbance of sediment an area around 20 km² was directly or indirectly affected by the activities. Although the immediate consequences were devastating, over the next seven years many of the original species migrated back into the areas changed. Because the manganese nodules were no longer on the seabed (in real mining they would be removed), however, there were no more species that were absolutely dependent on hard ground. Since resettlement of disturbed areas was primarily by adult animals and not by larvae the recovery of affected ecosystems depends strongly on the extent of the affected areas (Bluhm 2001; Borowski 2001; Thiel 2001; Thiel et al. 2001).

Box 5.7 *Glomar Explorer* and Project Azorian

Although manganese nodule fever was rampant in the late 1960s and early 1970s, there were research projects whose real goal was to uncover completely different "treasures." This includes Project Azorian (a.k. a. Project Jennifer). The CIA approached billionaire Howard Hughes who owned a company called Global Marine Inc. that operated various ships dedicated to mining marine resources. In 1971 Global Marine built the *Glomar Explorer* to search for manganese nodules off the coast of Hawaii and bring them to the surface as part of a test operation. However, it turns out this plan so beautifully suited to the times was really just a cover. The ship had a completely different mission.

A few years earlier in March 1968 the Soviet missile submarine *K129* had been lost near Hawaii. The Gulf II Class diesel–electric submarine left its base in Petropavlovsk-Kamchatsky on February 24 with a crew of 86 and had reported home several times while submerged at a shallow depth. From March onward there was a sudden radio silence. In the following weeks the Soviet Navy carried out a search along the planned course of the boat but without success. However, the activities of the Soviets were being monitored by the US Navy as a result of underwater sound stations of the Sound Surveillance System (SOSUS) previously recording an underwater explosion. The exact location of the explosion was triangulated. It was concluded that the Soviet Navy had most likely lost one of its submarines. The United States sent their own submarine the *USS Halibut* (SSGN-587) converted for underwater espionage to the region and found the wreck at a depth of 5000 m. But how could it be salvaged? Here the CIA came up with the idea of using a ship capable of extracting manganese nodules from the bottom of the sea. Equipped with the equipment necessary for salvage it succeeded in summer 1974 in retrieving parts of the submarine.

5.7 Evaporites

Evaporites (Fig. 5.17) are deposits of water-soluble minerals formed by the evaporation of water (Richter-Bernburg 1953; Warren 2010). Depending on the formation a distinction is made between marine and terrestrial deposits. They consist mainly of sulfates such as gypsum and anhydrite and of chlorides such as halite (NaCl, common salt). Salts are among the oldest and historically and culturally most important resources of humankind.

Rock salt is an important resource because of its use as table salt. In the past it was mainly used for livestock breeding (as a supplement during the breeding season), as table salt, or as a preservative. However, today it is primarily used as a raw material in the chemical industry for the production of soda ash, caustic soda, and chlorine, which are used in the production of glass, paper, aluminum, and plastics. A large part of annual production is scattered on roads in winter. In total around 20% are consumed by people and livestock, 60% are used in the chemical industry, and 13% are used for transport purposes (Pohl 2005).

The main mineral of rock salt (Fig. 5.18) is halite (NaCl). The name derives from the Ancient Greek word for salt. Although rock salt used to be an expensive commodity, today it has become a comparatively cheap raw material thus economically restricting the length of transport routes from the producer to the consumer.

Potash salts (Fig. 5.19) are mainly used in the production of fertilizer. They are also used in the chemical industry. Rubidium, magnesium, cesium, boron, and bromine are occasionally obtained as by-products. In contrast to rock salt, potash salts can only be found in a few countries in concentrations that are economical. In addition to Russia (in particular, the Kama Basin in Siberia) the main producers include Canada (in particular, the Devonian Elk Point Basin in Saskatchewan), the United States (Permian Delaware Basin in New Mexico), Germany (in particular, the Zechstein Formation), France (Rhine Rift Valley, near Mulhouse), Israel, and Jordan (Dead Sea).

Potash salt from salt domes is mined in a process termed roomwork mining by blasting (Box 5.8). When it comes to rock salt (NaCl) it is much better to pump water into the salt dome through boreholes and pump off the saturated saltwater called brine (see also Box 5.9). Often mining cavities were created, filled with water, and brine was later pumped out. Sometimes brine pours naturally from sources.

Evaporite deposits are also important in the search for fossil fuels since they are impermeable and effectively seal off oil and gas deposits. In the vicinity of salt domes traps can develop in which ascending hydrocarbons are trapped and form exploitable deposits (Sect. 6.3). Salts as impermeable rocks have a number of properties. They are able to stop liquids from entering by acting as an effective seal.

They are also water soluble making it possible to quickly and cheaply create large storage spaces for liquid and gaseous products by leaching (e.g., for the storage of natural gas). Such properties make salt rocks interesting as a possible repository for toxic or radioactive substances. The final disposal of nuclear waste in such repositories is the subject of controversial discussion. Salt can also store natural gases in pores. Various gases are encountered over and over again in salt mining posing a number of dangers to miners because of their composition and the high pressure in which the gases are held, which can even lead to explosive ingress. The crackling of gas escaping from the pores is often heard. The *Knistersalze* ("crackling salts") of the Werra Area (Germany) are well known. Their inclusions contain predominantly carbon dioxide connected with young basalt dikes. In the Wielicka District (Poland) almost pure CH_4 occurs and in districts in the Caspian Basin and in Carlsbad (New Mexico, USA) nitrogen predominantly occurs. The salts of the German Zechstein Formation often contain hydrogen sulfide.

Evaporites provide other important raw materials such as gypsum ($CaSO_4 \cdot 2H_2O$) and the anhydrous equivalent

Fig. 5.17 Salt deposits on the shore of the Dead Sea, near Wadi Mujib (Jordan). The rapidly falling water level is currently 430 m below sea level (*photo* © F. Neukirchen)

Fig. 5.18 Halite (rock salt) from Wieliczka, near Krakow (Poland) (*photo* © F. Neukirchen/Mineralogical Collections of the TU Berlin)

Table 5.1 Most important salt minerals economically (Strunz 2001)

Halite	NaCl
Sylvite (or sylvine)	KCl
Carnallite	$KMgCl3·6H_2O$
Kainite	$KMgSO_4Cl·3H_2O$
Langbeinite	$K_2Mg_2(SO_4)_3$
Polyhalite	$K_2Ca_2Mg(SO_4)_4·2H_2O$

anhydrite. Although gypsum is mainly used in construction (plaster, gypsum boards, cement), it is also used for soil improvement, in medicine, and as filler in paper. Translucent fine-grained gypsum is called alabaster. Fiber gypsum (Fig. 5.20) is common and usually found as a layer of fibrous aggregate embedded in shale or marl. Sparry plate-like slabs that can be square meters in size are also common. Gypsum or anhydrite can be formed hydrothermally (e.g., in black smokers; Sect. 4.14.1). Gypsum is also formed during the weathering of sulfides and sometimes large secondary gypsum crystals form in old mining galleries. If they are perfectly transparent, then they are called selenite. Large quantities of synthetic gypsum are produced during flue gas desulfurization and the production of phosphoric acid.

If evaporites are superimposed by other sediments, then gypsum ($CaSO_4·2H_2O$) releases its water during early diagenesis and thus converts to anhydrite ($CaSO_4$). Strontium

contained in traces in gypsum is often removed and precipitated in fissures as celestine ($SrSO_4$). When returned to the surface by geological processes, anhydrite absorbs water once again and converts into gypsum and in so doing increases in volume by 60%. Lifting damage often occurs on roads and buildings above anhydrite-containing layers such as the Gipskeuper in southwest Germany. In Staufen im Breisgau a geothermal borehole was responsible for such a transformation to take place underneath the old town. The uplift caused centimeter-sized cracks to form in the walls of historic buildings. However, gypsum is so soluble in water that karst systems such as caves and sinkholes are easily formed (e.g., on the southern edge of the Harz Mountains).

Lithium from terrestrial evaporite deposits is becoming increasingly important in the production of high-performance rechargeable batteries. Borates, soda ash (Na_2CO_3), and glauber salt (sodium sulfate) are also extracted from salt lakes (Sect. 5.7.2).

Table 5.1 lists salt minerals that are economically most important. Corresponding rocks are named for their dominant salt mineral to which the suffix "-ite" is added even if they still contain admixtures of other salt minerals (especially, potassium salts). Anyone who has ever visited a salt mine will certainly remember the vivid colors of salts on the walls of galleries. Internal deformation of the salt dome can also often be seen very well due to the coloring. Even if individual colors do not always indicate certain salt minerals, they can still give a good indication. The mineral halite (NaCl) is usually colorless, whereas potassium salts such as carnallite or polyhalite are often colored bright red by the incorporation of hematite. Potash and rock salt can also be distinguished by their taste—one of the few occasions geologists can actually use their tongue to determine the mineral content.

Box 5.8 Merkers Mine in Germany

The Merkers mining experience in Thuringia is recommended for anyone wanting to experience salt mining up close. Over the centuries more than 4600 km of tunnels were dug here to obtain the coveted potash salt.

Fig. 5.19 Red sylvinite (rock predominantly comprising sylvite (KCl)) from Bergmannssegen Mine, Lehrte, near Hanover (Germany) (*photo* © F. Neukirchen/Mineralogical Collections of the TU Berlin)

After a short introduction visitors are rapidly taken to a depth of 700 m in an elevator and get into waiting vehicles. Anyone who has ever visited a mine will perhaps be pleasantly surprised by the comfort. Although tunnels in potash mining are usually much less narrow than in coal mining, here they are as wide as a road and normal road vehicles can drive in them without problems. Even if the maximum speed underground is limited to 35 km/h, visitors get the feeling that they are traveling much faster in the tunnels. On the way through the tunnels visitors are told a lot of interesting facts about the extraction of potash salt. Then they are presented with the absolute highlight of the trip: the Crystal Grotto. Here halite crystals with an edge length of up to 1.2 m are found in a cavity discovered in 1980 and designated a national geotope by the Akademie der Geowissenschaften in 2006. The grotto is the result of Miocene volcanism in the neighboring Rhön Mountains that warmed the groundwater. Salty warm groundwater precipitated the crystals as it cooled down in the cavity.

Box 5.9 Zigong

In Zigong (Szechuan, China) people began drilling deep wells for the extraction of brine about 1000 years ago. In the nineteenth century there were so many drilling rigs in the city that old photos resemble pictures from the early oil fields of the United States. The Xinhai Well reached a depth of more than 1000 m in 1835. Not only was that an absolute record at the time, extraordinarily it was dug using a bamboo drill string driven by muscle power. The drill head of the percussion drill was the only part made of metal. Mud and cuttings were lifted using special vessels. Salt is still extracted from this well in the same way as back in the day (Fig. 5.21), but only as a demonstration of the salt museum and the drilling rig was specially reconstructed. Natural gas flows out of the wells and is used to boil down the brine to extract the salt.

The most important cations in seawater are Na^+, Ca^{2+}, Mg^{2+}, and K^+, while the most important anions are Cl^-, SO_4^{2-}, and HCO_3^-. The average salt content of the sea is around 3.5%. In other words, 35 g of dissolved salt are found in every 1000 g of seawater. Such salt comprises 27.2 g NaCl, 3.35 g $MgCl$, 2.25 $MgSO_4$, 1.127 g $CaSO_4$, 0.74 g KCl, and 0.12 g $CaCO_3$. Halite (NaCl) is by far the largest quantity. In lagoons and saltworks (Fig. 5.22) it is possible to observe the order in which salt is recovered

during evaporation. Solubility determines the order in which it takes place (Fig. 5.23). Calcium carbonates are the first to precipitate (starting from a salinity of about 6%), then gypsum (from about 15% salinity), followed by halite (from about 35% salinity). The concentration of NaCl at this point is 10 times that of seawater 90% of which has already evaporated. Potassium and magnesium salts (bitterns) are the most soluble and come last. In modern seawater they are first and foremost Mg sulfates, then mainly the potassium salt carnallite, and finally the magnesium salt bischofite $(MgCl_2 \cdot 6H_2O)$. However, the composition of seawater has changed in the course of Earth's history (Fig. 5.24). Often it was significantly richer in Ca and poorer in SO_4 than today. At those times halite was followed by carnallite and then the potash salt sylvite since there were no Mg sulfates. However the solubility of potash and magnesium salts is so good that these salts are often absent in the deposits either because they did not precipitate in the first place or because they were dissolved again at a later point in time.

The evaporation of seawater creates different rock layers on top of each other (Fig. 5.25, Table 5.2). In a bowl-shaped basin the upper zones cover a much smaller area due to the lowered water level such that layers or rather zones can lie next to each other at the same time. Although the arrangement of zones depends on the geometry of the basin, water inflows can change the salinity in certain areas of the basin enabling different zones to form at the same time.

In deep water there is often stratification of brines of different salinity only mixing seasonally or not at all. Evaporites can be precipitated from saturated brine at great depths and in shallow water. Microorganisms (algae mats) often play a role in the shore area. The addition of clay minerals results in the formation of saliferous clays (a.k.a. salt pelites) in which salt and gypsum crystals can also grow. Chevron halite forms in the water at shallow depths, is particularly rich in liquid inclusions, and has an angular

Fig. 5.20 Fiber gypsum from Morocco (*photo* © F. Neukirchen)

Fig. 5.21 In Zigong (China) brine from the historic 1000-m-deep Xinhai Well is still boiled down to salt in large pans. The fuel is natural gas that exits the borehole (*photo* © F. Neukirchen)

texture. Skeletal halite crystals sometimes grow on the surface of the brine and sink with increasing weight. Landslides and turbid currents can occur where there are steep slopes. Pelagic sediments rich in salt accumulate in deep water and coarse-grained, low-inclusion salt crystallizes at the bottom with appropriate supersaturation.

Diagenesis leads to recrystallization and reduction of the pore volume. Sulfates can release water in the process. Some minerals (in particular, potassium and magnesium salts) can be dissolved again, and they are also partly replaced by other salt minerals by adding and removing ions. Solidified low-porosity salt acts as a barrier to water because penetrating water is quickly saturated with salt and cannot dissolve any additional salt. Potassium salts are exceptions in that they can be dissolved by a saturated NaCl solution.

For salt to precipitate from seawater and terrestrial salt lakes evaporation must exceed the inflow rate or rainfall. This is particularly the case in the arid or semiarid regions of the world. Recent examples include salt lakes in California or Utah, salars of the Andes and the Dead Sea, and in marine environments the salty Kara-Bogaz-Gol Lagoon (Box 5.10) on the east coast of the Caspian Sea and sabkhas (Box 5.11)

on the coasts of the Persian Gulf east of Abu Dhabi. Figure 5.26 compares the sizes of old and recent examples.

Although additional seawater needs to be supplied into a basin to get thick marine salt deposits, enriched brine must not be able to drain off again. Ideally, a basin should allow fresh seawater to flow in near the surface via a barrier (i.e., a strip of land or a shallow seaway) capable of retaining the heavier brine. This theory was proposed as early as 1877 by Ochsenius. The Kara-Bogaz-Gol Lagoon is often given as a recent example even though it is relatively small and shallow.

Box 5.10 Kara-Bogaz-Gol Lagoon

The Kara-Bogaz-Gol Lagoon on the eastern shore of the Caspian Sea in Turkmenistan is shallow. The water level is usually about 1 m lower than the average level of the Caspian Sea. It has a water surface of about 18,400 km^2 and a water depth of only 4–7 m. Evaporation increases the salt content of water in the lagoon to 34%. Average salt content in the Caspian Sea is only 1.1–1.3%, about half that of oceans. The lagoon is separated from the Caspian Sea by two narrow

Fig. 5.22 Salt extraction from seawater in saltworks at the southern tip of La Palma, Canary Islands (Spain) (*photo* © F. Neukirchen/Blickwinkel)

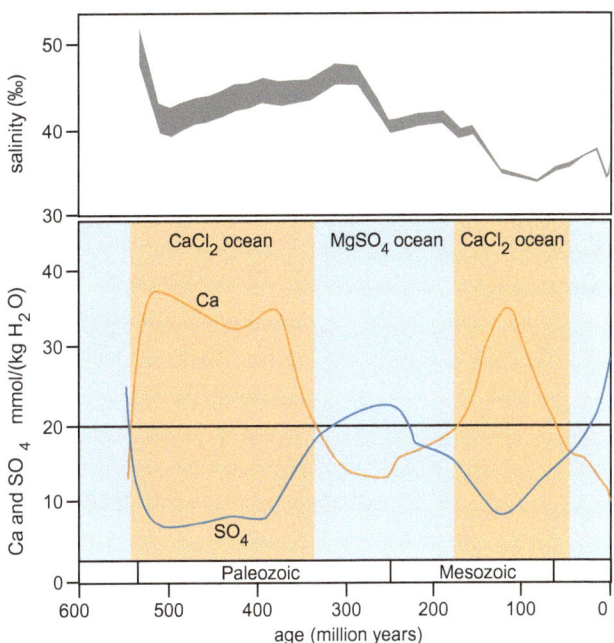

Fig. 5.24 The salinity (*top*, width according to different models) and composition (*bottom*) of seawater has changed in the course of history. Fluctuations in the concentrations of Ca and SO₄ are particularly clear. Often the oceans were significantly richer in Ca and poorer in SO₄ (CaCl₂ ocean) than today (MgSO₄ ocean). This is noticeable in the precipitation of potash and magnesium salts. Carnallite and magnesium sulfates are formed in MgSO₄ oceans, while sylvite and carnallite are formed in CaCl₂ oceans. In both cases NaCl dominates (modified from Warren 2010)

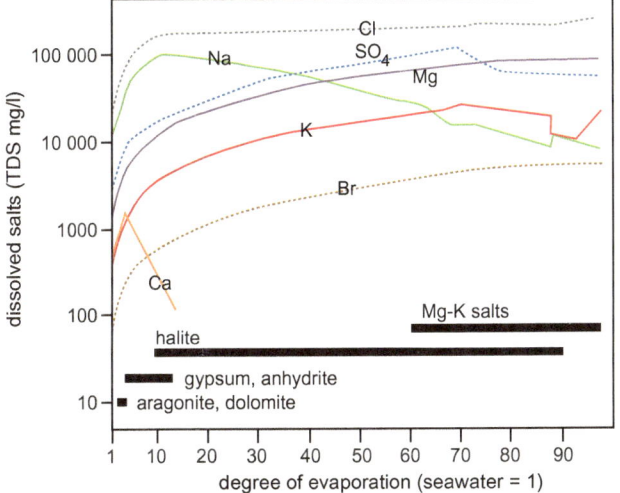

Fig. 5.23 Various minerals are recovered during the evaporation of seawater. This also changes the ratios of ions dissolved in the water. For example, NaCl-dominated brine develops into Mg–Cl–SO₄-dominated brine (from Warren 2010)

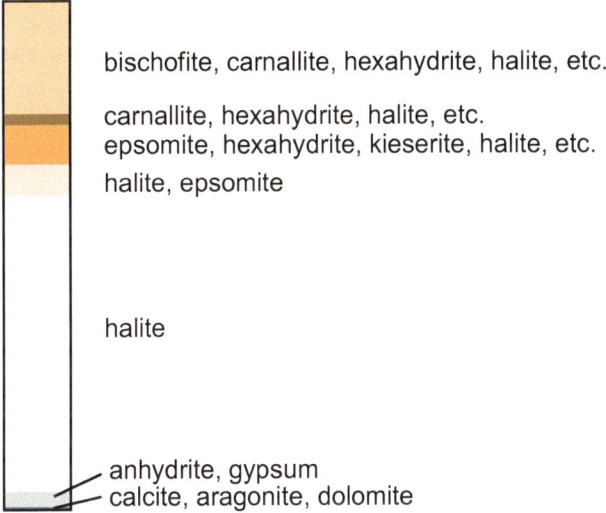

Fig. 5.25 Complete evaporation of seawater in a water column of 1000 m results in an evaporite layer only 18 m thick. Halite comprises 12 m of this layer. The upper layers are different Mg and K salts. Additional seawater usually flows in during evaporation in natural systems. This increases the overall thickness and at the same time changes the relative thicknesses of the zones and their exact composition. The uppermost zones are usually not deposited at all (from Pohl 2005)

Table 5.2 Zones (or layers) deposited successively during complete evaporation (from Valayashko 1958). Although the relative content of $MgSO_4$ and $CaCl_2$ plays a role, recent oceans are dominated by $MgSO_4$. The influx of fresh seawater changes the thickness and the exact composition of zones

Zone	Description
Gypsum anhydrite zone	Gypsum or anhydrite, calcite or aragonite Diagenetic: anhydrite, dolomite
Halite zone	Mainly halite, with gypsum, Mg-carbonate Diagenetic: anhydrite, dolomite, magnesite
Mg sulfate zone ($MgSO_4$ oceans)	Epsomite ($Mg(SO_4) \cdot 7H_2O$), halite, gypsum, polyhalite, Mg-carbonates Diagenetic: kieserite ($Mg(SO_4) \cdot H_2O$), anhydrite, magnesite
Sylvinite zone ($CaCl_2$ oceans)	Sylvite, with hexahydrite ($Mg(SO_4) \cdot 6H_2O$), polyhalite, halite Diagenetic: kainite ($KMg(SO_4)Cl \cdot 3H_2O$), langbeinite, kieserite
Carnallite zone	Carnallite, hexahydrite, halite, gypsum, polyhalite Diagenetic: kieserite, anhydrite
Bischofite zone ($MgSO_4$ oceans)	Bischofite ($MgCl_2 \cdot 6H_2O$), boron salts, carnallite, hexahydrite, halite, gypsum Diagenetic:boracite ($\alpha-(Mg,Fe)_3(B_7O_{13})Cl$), kieserite, anhydrite Since $MgCl_2$ is very soluble this zone is only present in exceptional cases

headlands and connected to it by a narrow passage (Fig. 5.27). Concerned about further lowering of the Caspian Sea a dam was built in 1980 to close the narrow passage. This not only proved completely unnecessary because the water level of the Caspian Sea had been rising since 1978, but also had many negative consequences for the lagoon and surrounding land. Cut off from inflow the lagoon dried up and turned into a salt desert. Wind-blown salt severely affected agriculture in the surrounding area and led to many environmental problems. In addition, the mining of glauber salt became impossible. Such mining is economically important for the region. Finally, the dam was removed in 1992 and the lagoon was flooded again. Due to its high salt content the lagoon would also be an ideal location for an osmosis power plant with a potential output of more than 5 GW (Dambeck 2012).

Box 5.11 Sabkha

Sabkha is the term used in the Arab world for a salt pan. Type localities can be found on the coast of the Persian Gulf in the United Arab Emirates (Fig. 5.28). Sabkhas are small bays mostly filled with siliciclastic sediments and evaporites. Al-Farraj (2005) describes their origin and development. As the sea level rises at high tide it floods coastal dune areas. Bays or lagoons separated by narrow seaways form between individual dune ridges. The narrow entrances then close at low tide and the bays flatten by wind-blown sand. The bottoms of lagoons eventually become dry at low tide and the lagoons then turn into sabkhas. While a newly formed sabkha is still flooded at high tide or during heavy rains, at an

advanced stage such events not only become increasingly rare they are also considered extraordinary events. Sabkhas are characterized by a very small relief between 10–50 cm and a very small gradient of about 1:1000 toward the coast (Butler 1969).

5.7.1 Marine Evaporites

Recently formed examples of salt deposits are tiny in comparison with the large-volume evaporites that formed in the past (Figs. 5.26 and 5.29). The Messinian salinity crisis in the Late Miocene is the youngest known example of this. Why the connection between the Mediterranean and the Atlantic (which did not exactly correspond to the current Strait of Gibraltar) was closed at that time is not yet quite clear. Candidates being discussed are uplift due to orogeny, worldwide drop in sea level (which was probably not the case at that time), and uplift due to dynamics in the Earth's mantle. According to the latter theory an island arc subduction zone moved through the Mediterranean toward the strait (rollback) causing heavy lithospheric mantle to detach beneath the edges of Africa and Spain (Duggen et al. 2003, 2005) and subsequently buoyancy raised the crust. Since evaporation was significantly higher than the amount of freshwater inflowing the water level in the Mediterranean sank and salinity increased. At some point the precipitation of gypsum and later of salt began. Although additional seawater from the Atlantic could have penetrated the basin several times, in the end all the water evaporated and in the deepest basins (i.e., more than 3000 m below normal sea level) thick salt deposits formed. At some point water again entered the basin creating the Strait of Gibraltar. The salt has since been covered by younger marine sediments.

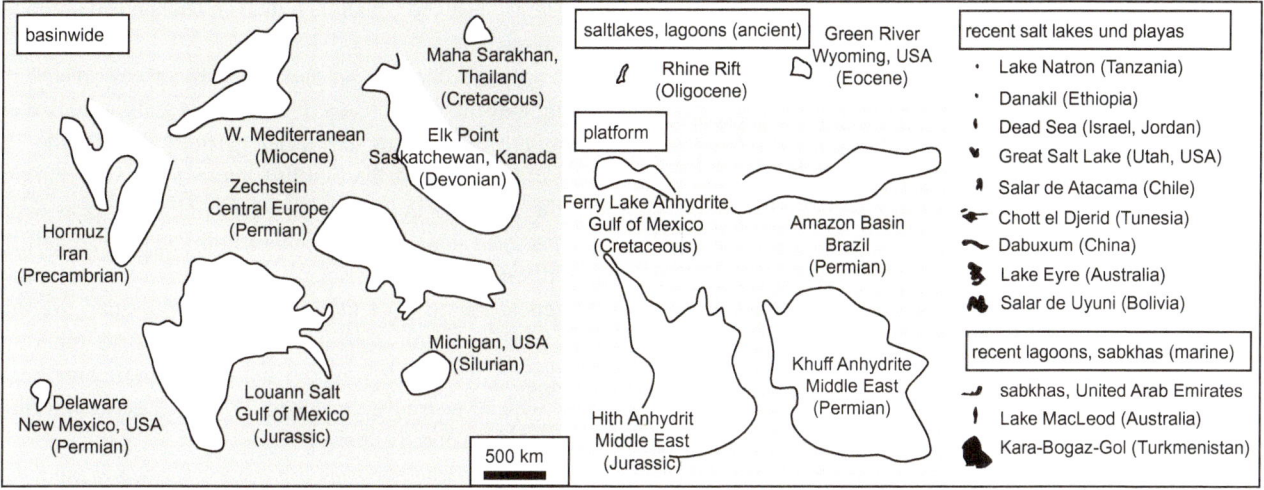

Fig. 5.26 Comparison of the size of different ancient and recent evaporites (from Warren 2010)

The Mediterranean Sea is a sea basin located between two continents that are moving toward each other. Such a constellation makes constriction of a basin relatively likely. If a continent breaks up instead, then thick salt deposits can form as well. The most recent example are thick evaporites in the northern Red Sea. These originated in the Middle Miocene when what was originally a continental rift system developed into a sea. First of all, continental crust is stretched to such an extent that some basins of the rift system are significantly deeper than sea level. Although they are not necessarily connected to each other in the beginning, seawater flows directly into some but only seeps into others.

The situation of individual basins can change over time in this way. If the climatic and hydrological conditions are right, then the basins will be filled with evaporites that will later be buried under marine sediments. While evaporites in the Red Sea originated before the onset of seafloor spreading (synrift), similar occurrences in the Atlantic only developed when seafloor spreading had already begun (postrift). This is possible because the initial opening of an ocean does not begin symmetrically and everywhere at the same time. Thus, huge salt horizons were formed in the first ocean basins that today are located in sediments of the continental slopes off the coasts of southern Brazil and Angola (Cretaceous) and in the Gulf of Mexico (Jurassic). They overlie not only stretched continental crust and sediments of the rift system but also oceanic crust. These layers have economic significance since they act as oil traps (Sect. 6.3).

Fig. 5.27 Kara-Bogaz-Gol Lagoon and part of the Caspian Sea. Water flows through a narrow passage over the headland into the lagoon (*photo* © NASA, public domain)

Fig. 5.28 Satellite image of sabkhas and lagoons near Abu Dhabi (*photo* © NASA, public domain)

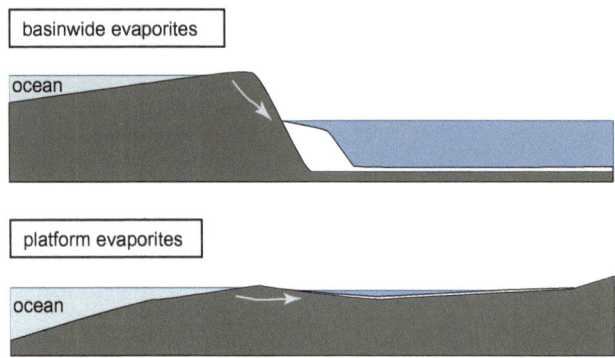

Fig. 5.29 Basinwide evaporites (mostly halite) form as a result of a constricted sea basin evaporating. The water level in the basin sinks and the salinity rises. The salinity is not the same everywhere and changes with decreasing water level. Different evaporites (*white*) are deposited in shallow water, on slopes, and at the bottom of the basin. Additional seawater can seep in. Platform evaporites form as a result of evaporation taking place in a shallow lagoon near the coast when a relatively large amount of fresh seawater can seep in during evaporation. These deposits are predominantly gypsum or anhydrite (from Warren 2010)

Sometimes large continental basins sink below sea level or are flooded when sea level rises. Incoming water forms an epicontinental sea (a.k.a. inland sea). An important example is the Zechstein Sea that covered large parts of Central Europe in the Permian (central England, North Sea, Denmark, northern Germany, Poland, Lithuania). During the time of the sea some of Europe's most important salt deposits were laid down in northern Germany. The basin was formed by extensive stretching of continental crust in the north of the already largely eroded Variscan Mountains. The sea entered from the Arctic Ocean (the Atlantic did not yet exist) into the basin via a narrow seaway. Kupferschiefer (Sect. 5.1.1) was an early deposit of this sea. This deposit was followed by evaporites some of which are more than 1000 m thick. They were created in several cycles as a result of large quantities of new seawater repeatedly flowing in. The most important cycles are the Werra Series (Zechstein 1), Staßfurt Series (Zechstein 2), Leine Series (Zechstein 3), and Aller Series (Zechstein 4). Although there are further superimposed cycles in the North Sea and in parts of northern Germany, they are only a few meters thick such as the Ohre Series (Zechstein 5), Friesland Series (Zechstein 6), and Fulda Series (Zechstein 7). Each cycle begins with a thin layer of clastic sediments such as shale or (at the edge of the basin) conglomerate. On top of them are thin carbonate rocks and then very thick layers of anhydrite and rock salt. The quantity of anhydrite and rock salt depends more or less on just how continuous the inflow of seawater was even during cycles, which led to the salinity of the Zechstein Sea fluctuating. In some cycles there were multiple changes between anhydrite and salt layers. Evaporation on several occasions was so extensive that even potash salt was deposited.

Respective deposits are not of course evenly distributed throughout the entire basin. For example, salts of the Werra Series are only found in smaller partial basins, while salts 200–750 m thick of the Staßfurt Series are spread throughout the entire basin. Salt partly formed at great water depths and partly in shallow water with turbidity currents depositing turbidite-like masses on steep slopes. Carbonates are mainly located at the edge of the basin. In places especially thick anhydrite has built reef-like ramparts close to the shore. Evaporites were later covered with thick sediments beginning with terrestrially deposited Triassic sandstone (lithostratigraphic name: Buntsandstein).

Huge evaporite horizons can also be formed in lagoons in the coastal areas of oceans (especially, when the climate was hot) (Warren 2010). This happened when sea level fluctuations were frequent and low amplitude (a few meters) over long periods of time, while sea levels were so high that parts of the continents were flooded and shallow coastal areas became more frequent. This resulted in lagoons forming on shallow coasts. Although they occasionally flooded with seawater, most of the time they were separated from the sea by a narrow land barrier. However, new seawater could seep in continuously. The influx of new seawater was often so large that salinity in the lagoon remained comparatively low despite strong evaporation. Gypsum (with interlayered carbonates and clays) was deposited and salt was present in small quantities. Such deposits are called platform evaporites as opposed to basinwide evaporites. The most important examples can be found in the Middle East in the platform sediments of the Arabian Peninsula and offshore in the Persian Gulf. Khuff anhydrite from the Permian, Hith anhydrite from the Jurassic, and corresponding layers of the Rus Formation from the Eocene are particularly noteworthy. Many of the major Arab oil and gas deposits are located under these sulfate layers (Sect. 6.4). Similar horizons can be found in Texas and New Mexico (USA), the Gulf of Mexico, and the Amazon Basin. In Germany the smaller deposits in the Gipskeuper are also worthy of mention.

A layer of salt covered by massive sediments does not retain its horizontal shape for long. Salt deforms plastically and can literally flow. Such movements are called halokinesis or salt tectonics. The overburden of younger sediments of varying thickness or density is the trigger for salt to flow from areas where the overburden pressure is high to areas where it is lower forming rolls, waves, and large salt cushions. Since the density of salt is lower than that of other rocks, movement can continue in such a way that it breaks through superimposed layers and rises as a salt diapir (salt dome) or salt wall. Different salt layers within a salt diapir are naturally wildly folded and deformed (Fig. 5.30).

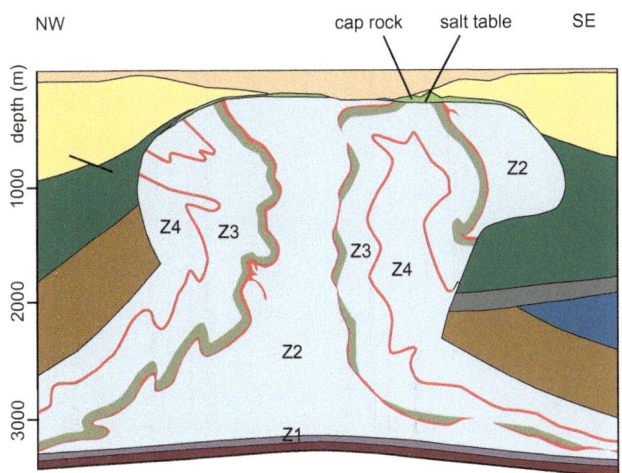

Fig. 5.30 Simplified section through the salt diapir at Gorleben (Germany) showing the most important layers of anhydrite (*olive green*), rock salt (*light blue*), and potassium salts (*red*) of the Zechstein Series 2 to 4. A cap rock with gypsum (*light green*) has formed above the salt diapir. Sediment layers of the surroundings are partly bent upward and moved at faults (modified from BGR, n.d.)

Salt flow in tectonic movements is increased during collision and extension. During orogeny thick sediment packages can be sheared off along evaporites—which are easily deformable—and moved as nappes. An important example is the Zagros Mountains (Iran). In a belt in the northeastern part of the mountain range the Arab Plate is colliding with the Eurasian Plate. Adjacent to this suture, sediments up to 14 km thick of the Arab Plate were sheared off to the southwest along the Hormuz salt layer and pushed together to form folds in a fold and thrust belt as a result of which salt accumulated in the cores of the folds in cushions up to 8 km thick. Many salt diapirs were formed in this way some of which serve as oil traps, while others have broken through to the Earth's surface (Fig. 5.31). However, since salt cushions form on normal faults during extension many salt diapirs have therefore formed in rifts, half-rifts, or along the normal faults of large basins. This also applies to the diapirs of Zechstein salt. Diapirs are usually formed in several phases during tectonic movements and by pure salt flow. If further sedimentation takes place around a diapir, then the additional load can even lead to influx of more salt into the diapir.

When a salt diapir breaks through to the surface, salt is exposed to weathering or is dissolved just below the surface by groundwater and carried away (subrosion). Less soluble components of salt such as anhydrite (or gypsum) and clay minerals are left behind, which is why the so-called cap rock (mostly gypsum) forms above an almost horizontal salt table. Sometimes it is lifted by a diapir that is still rising. For example, the Kalkberg of Bad Segeberg (Germany) consists of gypsum and anhydrite. In deserts so-called salt glaciers (or namakier) form intead with surprising similarities to real glaciers including large plateaus, downward-flowing streams, and even seracs (Talbot and Pohjola 2009).

5.7.2 Salt Lakes and Salt Pans

Salt lakes are the most important deposits of lithium, sodium carbonate (soda ash), sodium sulfate (glauber salt), and borates. Table salt and potash salt are of course also extracted. Lithium is used in the production of powerful rechargeable batteries (lithium–ion batteries) and is critically important for modern technology (Trechow 2011). This is especially the case in the production of electric cars where enormous demand is expected in the future. Lithium is obtained by the evaporation of brine, which is much easier than extracting minerals from lithium pegmatites. Soda ash is used in glass production. It can also be produced synthetically. Glauber salt is mainly used in the production of cleaning agents and in the paper industry to break down cell

Fig. 5.31 Salt diapir in a fold of the Zagros Mountains (Iran) as photographed by the Space Shuttle (*photo* © NASA)

walls. Perborates used as bleaching agents in laundry detergents are produced from borates. They are also the raw material for boron compounds such as boric acid or boron nitride.

Salt lakes form in dry regions in basins without drainage. Although some are below sea level such as Lake Eyre (Australia) and the Dead Sea, most are in mountain basins or on high plateaus in the rain shadow of mountains. For example, the Salar de Uyuni (Bolivia) is located on the Altiplano of the Andes at an altitude of 3653 m (Fig. 5.32) and the Salar de Atacama (Chile) is located between the Andes and a lower mountain range at an altitude of 2300 m (Fig. 5.33).

Although the Dead Sea is a special case with a water depth of up to 380 m, most salt lakes are relatively shallow. Some dry out seasonally—completely or partially—and leave a flat salt pan (Fig. 5.34) also called a playa. Other salt pans flood only in exceptional cases and are mainly supplied by groundwater infiltrating. Many salt lakes are surrounded by a playa of salty clay since clay minerals are deposited by streams flowing (often only seasonally) into the basin. The profile of evaporites in many basins shows repeated clay layers some of which are due to a more humid climate.

The composition of the little water that flows into these basins is very different. Moreover, the solution is not necessarily NaCl-dominated. This mainly depends on which rocks weather in the hydrological catchment area. Limestone provides Ca^{2+} and HCO_3^-, while dolomite provides Mg-rich water. Water that is Ca–Na–HCO_3^--dominated by weathering of felsic rocks, Mg–HCO_3^--dominated in the case of basic, ultrabasic rocks, and sulfide-containing rocks such as shale provides sulfate. Sometimes even older evaporites are dissolved and hydrothermal water can also make a contribution. Little wonder then that terrestrial evaporites are much more diverse in their composition than marine evaporites.

The evaporation of such waters produces brines with very different compositions (Fig. 5.35). Once the brine is saturated in a certain mineral it precipitates. The composition of the resulting brine depends on the original proportions of respective anions and cations. Usually, either the cations or the anions involved are largely removed, while the other ion (anion or cation) is still available and leads to supersaturation and precipitation of further minerals with further enrichment (Risacher et al. 2003; Warren 2010). Since some ions such as Mg^{2+} or Li^+ are adsorbed by clay minerals, this too can change the composition of the brine.

Readily soluble salts like $MgCl_2$ are hardly ever precipitated even in dry salt pans. This is even more the case with $CaCl_2$ and LiCl since brine that is still present in pores and cracks is sufficient to keep such salts in solution. Such brines are found almost exclusively in the upper 30 m of deposits because the porosity of the salt decreases to almost 0% at this depth.

Fig. 5.32 The Salar de Uyuni (Bolivia), the largest salt pan in the world, lies on the Altiplano at an altitude of 3653 m. In the rainy season it turns into a salt lake with a water depth in the decimeter range. The polygon pattern is created by the volume of water decreasing as the salt dries out (*photo* © F. Neukirchen)

Fig. 5.33 Although the Salar de Atacama (Chile) hardly ever floods, what little rainfall there is dissolves the salt and forms the spiky surface. At the edge there are small salt lakes fed by springs where flamingos sift the surface for brine shrimp and algae. The brines from this salar are one of the most important lithium resources (*photo* © F. Neukirchen)

The Salar de Atacama (Chile) is currently the most important lithium deposit accounting for almost one-third of world production. Of its area of 3200 km² the halite zone accounts for about one-third (Fig. 5.36). The marginal sulfate zone consists mainly of gypsum, carbonates, some clay, and small halite-rich patches. The thickness of evaporites deposited per year average out at only 0.1 mm. Nevertheless, the deposits (including interlayered clays) are up to 950 m thick. Although the Salar de Atacama hardly ever floods, seasonal brooks flow in at the edges and groundwater seeps in. Although salt input can be traced back to the weathering of volcanic rocks and to hydrothermal water of the Andes, older evaporites of the adjacent Cordillera de la

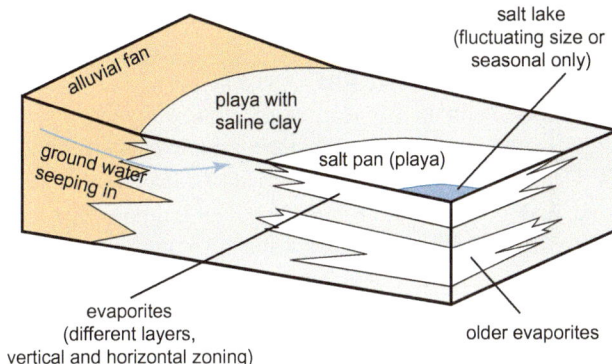

Fig. 5.34 Salt lakes are often shallow and completely or partially dry out seasonally. This is how layers of clay and evaporites (salt pan, playa) are created

Sal deposited in a salar in the Neogene and subsequently folded are dissolved and transported to the active salar.

The brine containing between 500 and 6400 ppm Li in the halite zone is pumped from wells 30 m deep and then enriched in evaporation ponds. This works thanks to the extremely dry climate of the Atacama Desert where rainfall is only 35 mm per year. This allows the possible evaporation of 3500 mm of brine per year and much more effective enrichment than comparable facilities in other parts of the world. After a year the brine is enriched to 6% lithium and transported by truck to a plant near the coast where lithium carbonate is precipitated after further purification. By-products are potassium chloride, potassium sulfate, magnesium chloride, and borate.

Lithium production is planned for the Salar de Uyuni (Bolivia). According to some estimates there is even more lithium here than in the Salar de Atacama, but the brines contain on average only 320 ppm Li and the high Mg content makes extraction more difficult. Higher grades up to over 1000 ppm Li are found in a small zone at the southeastern edge. Other important occurrences of Li-rich brine are the Salar de Hombre Muerto (Argentina) with 520 ppm; Zabuye Salt Lake (Western Tibet, China) with 500–1000 ppm; Taijinaier Lake (Qinghai, China) and Clayton Valley Playa, near Silver Peak (Nevada, USA) both with 250–300 ppm; and Great Salt Lake (Utah, USA) with 60 ppm. The occurrence of lithium in some waters of oil reservoirs should also be kept in mind since concentrations of 100–700 ppm have been found in Texas and Arkansas (Evans 2008).

Borates are almost exclusively deposited in salt lakes located in mountain regions where there is active (mainly acidic) volcanism. Hydrothermal systems can leach out easily soluble boron enriched in volcanic rocks. In some cases this happens in the basins themselves, which are filled with alternating layers of tuff and evaporites at whose faults hydrothermal solutions can rise. Evaporation produces borates such as borax (a.k.a. tincal ($Na_2B_4O_5(OH)_4 \cdot 8H_2O$)), colemanite ($CaB_3O_4(OH)_3 \cdot H_2O$), ulexite ($NaCaB_5O_6(OH)_6 \cdot 5H_2O$), and kernite (a.k.a. rasorite ($Na_2B_4O_6(OH)_2 \cdot 3H_2O$)). The solubility of these minerals is very different. When and which mineral is formed depends not only on the boron content of the brine but also on the concentration of Na and Ca.

Fig. 5.35 Schematic showing the development of brines of contrasting composition in salt lakes. When a mineral (*brown*) precipitates, the composition of the resulting brine (*blue*) depends on the original proportions of the respective anions and cations (modified from Warren 2010)

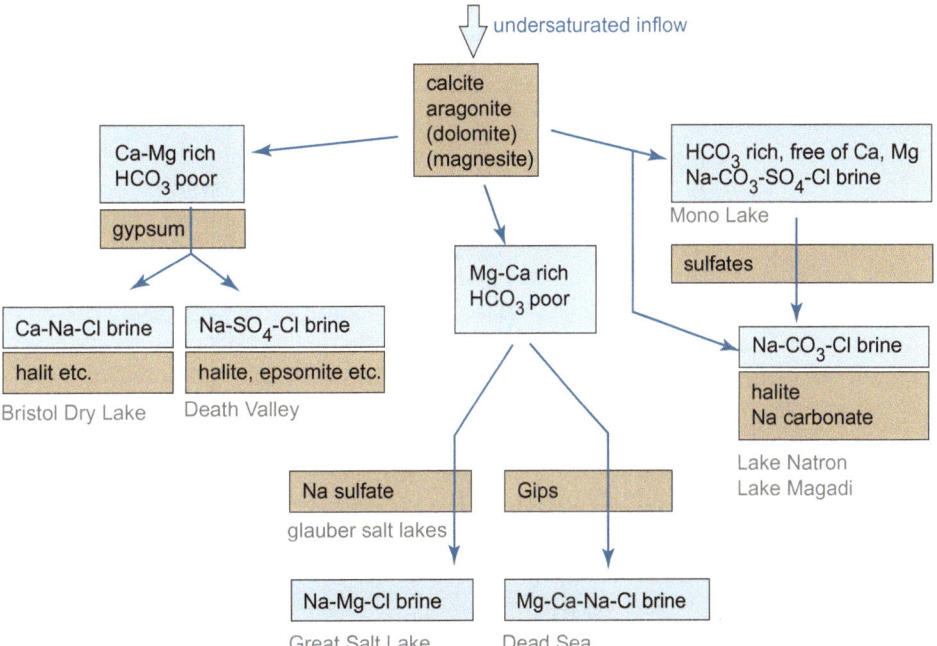

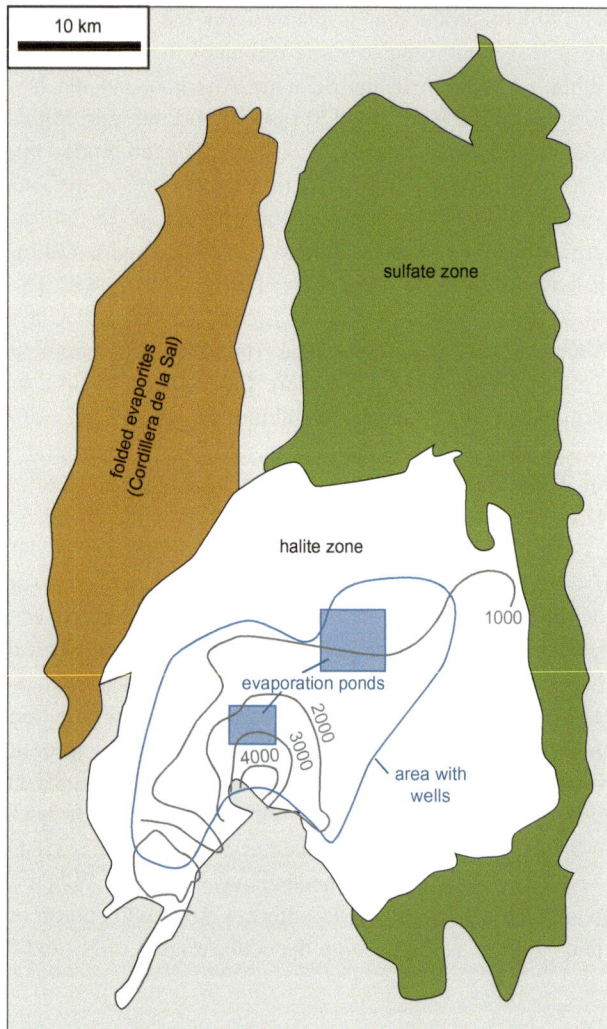

Fig. 5.36 The Salar de Atacama consists of a halite zone in the center and a marginal sulfate zone (clay, gypsum, carbonates, halite). The contour lines show the concentration of Li (ppm) in brines. Such brines are pumped from wells and enriched in evaporation basins. The adjacent Cordillera de la Sal consists of folded Neogene evaporites (from Warren 2010; Kesler et al. 2012)

Colemanite precipitates before gypsum, ulexite, and borax. It often does so simultaneously with halite. Borate layers (or halite and borate) can form on the lake bottom. Moreover, borate concretions can grow diagenetically within older evaporites and efflorescences and crusts can form at the surface of playas. Zoning of different borates can often be seen, which is either interpreted as a result of different salinity during deposition or as a diagenetic change due to groundwater infiltrating. Almost 40% of known borate reserves are located in Turkey. Other important deposits are some salars of the Andes, salt lakes of the Basin and Range Province in the western USA, and several salt lakes in China.

Sodium sulfate (glauber salt) can be precipitated from sulfate-rich, Ca-poor brines. In winter or on cold nights

when water has cooled down to 1 °C or even begins to freeze the solubility of sodium sulfate is significantly lower and mirabilite ($Na_2SO_4 \cdot 10H_2O$) crystallizes. Mirabilite often dissolves again in summer or the following day (Warren 2010). This can be observed in the countless small salt lakes in Canada, salt lakes in the Andes, and a few lakes in Africa and Asia including the Kara-Bogaz-Gol Lagoon (Box 5.10). Mirabilite is found together with epsomite, gypsum, and other minerals in the sediments of these lakes. Mirabilite can be diagenetically replaced by glauberite ($CaSO_4 \cdot Na_2SO_4$) and thernadite (Na_2SO_4). The mirabilite lakes in Canada were long believed the most important deposits, although the salt could only be extracted seasonally. Glauberite and thernadite can precipitate in warm climates if the concentration is unusually high. This was the case with the Laguna del Rey (Mexico), which is today the most important deposit of glauber salt. There is a layer of glauber salt up to 35 m thick with sodium sulfates below the gypsum-rich sediments 1–2 m thick of the active playa. The salt is obtained by leaching.

Sodium carbonates (soda ash) such as trona ($NaHCO_3 \cdot Na_2CO_3$) and nahcolite ($NaHCO_3$) are formed when some hydrogen carbonate remains after the precipitation of calcite and dolomite and the sulfate and chloride content is comparatively low. The water of these lakes is often strongly alkaline. Halite is additionally formed during further evaporation. The best known recent examples are Lake Natron and Lake Magadi in the East African Rift System in Tanzania where the weathering of alkaline volcanic rocks (and carbonatites) certainly plays a role. Obviously, there are other ways to achieve a corresponding brine. Other soda lakes can be found in the Andes, California, China, and Mongolia. Soda ash is mainly mined in Wyoming where Paleogene sediments of the Green River Formation contain 42 trona layers some of which are more than a meter thick.

Table salt from salt lakes is only significant regionally. An interesting example are the salt lakes in Tibet whose salt used to be transported by yak caravans over high passes in the Himalayas. There is no salt in the Himalayas themselves despite so much being sold as such in esoteric shops. The Dead Sea is an important producer of potash salt and bromine. NaCl, NaOH, and magnesium are also extracted from the Dead Sea.

A couple of interesting facts to complete this section are salt lakes are far from being devoid of life (Box 5.12) and the saltiest lakes on Earth are in Antarctica (Box 5.13)—not in hot deserts.

Box 5.12 Salt lakes as habitats

Even in extremely salty lakes like the Dead Sea or in the ponds at the edge of the Salar de Atacama there is life (Fig. 5.37). Life mainly consists of halophilic

Fig. 5.37 Microorganisms stain the water of Laguna Colorada (Bolivia) red. The white islands consist of borax (*photo* © F. Neukirchen)

archaea that are usually called halobacteria. They can only be found in extremely salty brines. Another extremist is the unicellular alga *Dunaliella salina*. Many of these organisms are colored red. The cyanobacterium *Spirulina* thrives in strongly alkaline lakes where flamingos feed on it. Life is more diverse in salt lakes where the salinity is lower; hence there are brine shrimp, small crabs, and sometimes even fish. On the shore of Mono Lake (California) millions of salt flies fill the air. Microorganisms, brine shrimp, and crabs are eaten by flamingos that live in many salt lakes in South America and Africa and breed on the salt pans. Many other birds find food here too.

Box 5.13 Salt in Antarctica

Although most salt deposits originated in warm and dry climates, there are examples from cold climates. When saltwater freezes the substances dissolved remain in the solution that becomes ever saltier. In Antarctica there are a number of small extremely salty lakes. Don Juan Pond with a salinity of 40% is even considered to comprise the saltiest water in the world. In addition to halite a number of other minerals are deposited here such as mirabilite, thenardite, hydrohalite, and even antarcticite ($CaCl_2 \cdot 6H_2O$) the latter of which is hardly found anywhere else due to its solubility being extremely good. The saltwater is partly derived from seawater from inclusions in the ice and from marine aerosols blown in and partly from deep groundwater. Wind absorbs moisture in the dry valleys of Antarctica causing salt to be deposited on the valley floor.

5.8 Phosphorite

Potassium, nitrogen, and phosphate are considered the most important elements for plant nutrition. For example, about 95% of phosphates extracted by mining are used in the production of fertilizers. There are very different phosphate deposits ranging from marine phosphate-rich sediments (phosphorites), guano (bird droppings) on islands such as Nauru and Christmas Island or in lakes such as Minjingu

(Tanzania) (Box 5.14), Kiruna-type deposits (Sect. 3.6) to carbonatites and phoscorites (Sect. 3.10) and their weathered products (Box 5.15).

Phosphorites are marine sedimentary rocks characterized by their high phosphate content of at least 15–20% (Fig. 5.38). The phosphate is mostly contained as cryptocrystalline fluorapatite ($Ca_5(PO_4)_3F$) or as hydroxyapatite ($Ca_5(PO_4)_3OH$). Phosphorites are often associated with carbonate rocks, shales, and sandstones. The texture can be very diverse from pelletoid or fine-grained without structure, the rocks can be hard or as soft as earth. They can contain ooids and fossils (especially, the bones and teeth of fish).

The oldest phosphorites come from the Proterozoic where they often accompany BIFs as is the case in Australia. However, they are more common in Paleozoic and Cenozoic strata. Large deposits are mined in France, Belgium, Spain, Morocco, Tunisia, China, and the United States.

Most marine phosphorite deposits were formed in shallow marine environments where the water depth was less than 200 m. They often form seams that laterally merge into non-phosphatic siderocks. The main inflow of phosphorus to the sea comes from continental weathering of phosphate-containing rocks. Rivers carry a lot of phosphate (Delaney 1998). In the ocean it is taken up by various organisms and subsequently deposited on the seafloor when they die. Bones and teeth of larger organisms such as fish can also play an important role (Baturin 2000). The deposition conditions of marine phosphorite deposits are quite diverse. They extend from the surf zone (supratidal) via tidal flats and estuaries (Baturin 2000) to deeper areas of the shelf sea. Upwelling is the upward flow of cold and nutrient-rich water from the deep sea to continental slopes. It is widely believed to have contributed to the enrichment of deposits. Upwelling leads to increased growth of phytoplankton and to increased activity of the entire oceanic food chain. Moreover, organic matter settling in these areas is today rich in phosphorus and young phosphate-rich sediments are known on many continental margins.

Pelletoid phosphates are characterized by phosphate concretions approx. 2 mm in diameter, but there are concretions several dozen centimeters in diameter. Big or small they are found on the ocean floor of continental shelves. Bioclastic phosphates (a.k.a. bone beds) consist of small fossils such as bone remnants, teeth, and coproliths (fossil excrements mainly preserved as phosphate). Fossils are sometimes subsequently cemented diagenetically with phosphate-rich cements (Baturin 2000). This process is similar to the third type in which sediments were cemented diagenetically by phosphate-rich solutions. This process is not very common and is usually associated with phosphate-rich solutions of guano deposits.

Fig. 5.38 Phosphorite from Logrosán (Spain) (*photo* © F. Neukirchen/Mineralogical Collections of the TU Berlin)

Box 5.14 Minjingu (Tanzania)

Minjingu is about 5 km east of Lake Manyara and is the most important representative of sedimentary phosphate deposits in East Africa. The phosphates are hosted in bentonite (Sect. 7.5) in phosphate layers that were presumably laid down by the sedimentation of bird guano. The exact age of the deposits is a moot point. They should likely be placed in the Upper Pliocene or the Lower Pleistocene (Schlüter 1991). The small hill Minjingu Kopje gave its name to the mine. The hill is located nearby and was once an island in a lake that was probably home to a bird colony. In addition to the remains of cormorants there are also fish remains. The deposit was discovered as a radioactive anomaly by airborne remote sensing in 1956 (van Kauwenbergh 1991). From 1983 phosphate was mined for fertilizer. Although mining was stopped in the 1990s, it was resumed in 2001. The mine was operated by the Tanzanian State Mining Company (STAMICO) in cooperation with the Finnish company Kone Oy. Approx. 20,000 t P_2O_5 per year were produced throughout this period. The ore was processed into superphosphate in Tanga, a port about 450 km away, and partly exported to Kenya. The reasons for the temporary shutdown of the mine may include the remote location of the mine, the corresponding transport routes for ore and consumables, and the lack of freshwater. Phosphate reserves at Minjingu are reported to be between 1.9 Mt and 8 Mt P_2O_5 (Atkinson and Hale 1993; van Straaten 2002).

Phosphate ore should contain at least 30% P_2O_5, have some calcium carbonate, and only small amounts of iron and aluminum for it to be economic. The phosphate-rich ore is processed after mining. It is either converted using sulfuric acid to phosphoric acid or melted to elemental phosphorus. Phosphoric acid is used to produce triple superphosphate, a calcium dihydrogen phosphate fertilizer that is readily plant available (i.e., soluble in water). Another possibility is to use phosphoric acid and ammonia to produce ammonium phosphate that is also readily plant available. Around 90% of phosphate ores extracted are consumed in agriculture. The United States is currently one of the largest producers and exporters of phosphates accounting for around 37% of P_2O_5 world exports (*USGS Minerals Yearbook* 2007). Together with China and Western Sahara (de facto Morocco) they represent the main mining countries. In the past guano deposits such as those on Christmas Island and Nauru were exploited, but today these deposits are largely depleted. Phosphates are indispensable to modern agriculture and to feeding a growing world population. Unlike many other important raw materials they cannot be replaced by other sources or other substances. Things become problematic when such important raw materials are under the control of comparatively few countries. Examples include China that has imposed 135% export taxes on phosphate exports to make exports more difficult (US deposits will largely be depleted in about 30 years) and Western Sahara is still occupied by Morocco contrary to UN resolutions. Such problems coupled with the fact that phosphate is a finite resource caused the cost to rise dramatically (up to 800%) between 2007 and 2008. Such a hike in prices resulted in a drop in demand and as a result prices fell again, but not to the previous level. It is not easy to calculate how long reserves will last in the future. For example, the USGS assumes that current stocks will last around 100 years assuming that consumption remains constant. This led Dana Cordell of the Institute for Sustainable Futures to postulate Peak Phosphorus (comparable with Peak Oil; Sect. 6.6) around 2034 (i.e., an event when maximum production is reached, the rate of further extraction drops significantly, and demand increases; Dèry and Anderson 2007; Gilbert 2009; Meier 2010). The situation may be aggravated by the increasing cultivation of crops for biofuels. On the other hand, rising commodity prices may also make low-value deposits worth exploiting that are still too expensive or too complex to extract today. A possible solution or at least relief could be the use of liquid manure or the use of phosphorus from ash from sewage sludge incineration (Meier 2010). Another possibility is the prospection of new deposits. However, rocks containing apatite such as phosphorites are not always easy to detect in the field. Airborne remote sensing is helpful because it takes advantage of the fact that phosphates can also contain uranium. To identify phosphate-containing rocks reliably they can be dabbed with nitric acid NH_4 molybdate solution. In the presence of phosphate a weak yellow coloration appears, which changes to blue when ascorbic solution is added (Germann 1981).

Box 5.15 Sukulu and Bukusu (Uganda)

The Sukulu Carbonatite Complex (Sect. 3.10) has a diameter of approx. 4 km and rises up to 180 m above the surrounding plain. It is one of five alkaline complexes located southwest of Mount Elgon (Uganda) near the Kenyan border. Their formation is closely related to the development of the East African Rift System (Barifaijo 2001). Sukulu is dated to about 40 million years (Bell and Blenkinsop 1987). It consists of ring-shaped intrusions cut by radial dikes and faults. Its composition ranges from calcite carbonatite to dolomite carbonatite to ferrocarbonatite. Common accessory minerals are apatite, magnetite, phlogopite, and pyrochlore. In the center of the ring structure divided into three valleys—northern, western, and southern—residual soils (a.k.a. residual placers) up to 60 m thick with accumulations of phosphate, magnetite, and pyrochlore have been deposited. Reedman (1984) indicates a P_2O_5 content of around 9.6% for these soils. They consist of a complex mixture of apatite, aluminum phosphates such as crandallite ($CaAl_3(PO_4)(PO_3OH)(OH)_6$), iron oxides, clay minerals, quartz and subordinated perovskite, zircon, and other weather-resistant minerals. According to van Kauwenbergh (1991) about 78% of P_2O_5 is present as apatite and about 22% as manganese or alumophosphates. Total phosphate reserves are given as 15–23 Mt (van Kauwenbergh 1991; Atkinson and Hale 1993).

The Bukusu Carbonatite (a.k.a. Busumbu Complex) is located just a few kilometers to the northeast. With a diameter of about 9–10 km it is larger than the Sukulu Carbonatite. It consists mainly of poorly exposed ultrabasic and alkaline rocks. In the 1930s large amounts of apatite were found in soils of the complex (particularly, at Busumbu Ridge in phoscorites—Sect. 3.10.1—in the south of the complex containing about 30% P_2O_5; Davies 1947). Drilling showed that the weathering zone extends to a depth of 60 m (van Kauwenbergh 1991). Phosphate is mainly present as francolite subordinated to unweathered primary fluorapatite. Francolite is a carbonate fluorapatite whose composition is variable but can be described approximately with the formula $(Ca,Mg,Sr,Na)_{10}(PO_4,SO_4,CO_3)_6F_{2-3}$. Reserves of P_2O_5 are estimated to be between 0.2 and 10 Mt (van Kauwenbergh 1991; Atkinson and Hale 1993). Phosphate was mined between 1944 and 1963.

5.9 Placer Deposits

Flowing water not only transports particles such as sand, gravel, and debris, but it also effectively sorts them by size and density. Placer deposits (a.k.a. placers) are accumulations of heavy and weather-resistant minerals formed mainly by fluviatile transport in rivers or on beaches. Placers are important deposits of gold (Fig. 5.39), tin (cassiterite), rare-earth elements (monazite, xenotime), titanium (rutile, ilmenite), zirconium (zircon, baddeleyite), tantalum (columbite–tantalite, a.k.a. coltan), various other metals, and precious stones such as diamond, sapphire, ruby, and garnet. Which heavy minerals are found depends of course on the rocks weathering in the catchment area. In addition to recent placers there are fossil placers in old sediment layers that are of great importance too.

Residual placers are a special case in that they are located directly above a primary deposit where heavy minerals were enriched as a result of light minerals washing out. They are formed by weathering—not by sedimentation. Such secondarily enriched zones of an ore deposit are often the only parts that are economically recoverable. Important examples are weathered pegmatites with residual accumulations of cassiterite and columbite–tantalite. Phosphate deposits in

Uganda (Box 5.15) are another example. Eluvial placers may be formed by soil flow on slopes and embankments below primary deposits. In loose rocks heavy minerals accumulate in small depressions called pockets.

Box 5.16 Fluviatile transport and sedimentation

Transport and sedimentation in rivers involve a complicated interplay of different processes and factors. Water flow is generally more or less turbulent as a result of which there is usually only a thin laminar layer flowing directly above the bottom. Gravel is moved by the flow by rolling and sliding along the bottom. Sand literally bounces in a process called saltation and fine particles (silt, clay) are kept in suspension. Flow velocity and turbulence clearly play a role as do properties of particles such as size and density. During floods the transport capacity of rivers is significantly increased.

The sinking velocity of spherical particles in a liquid can be used as a first approximation. This is described by Stokes' Law:

$$v = 2r^2g(\rho_p - \rho_f)/9\eta$$

where r is the radius of the particle, g the acceleration of gravity, ρ_p the density of the particle, ρ_f the density of the liquid, and η the viscosity of the liquid. The sink rate is therefore much more dependent on the radius than on the density of the particle. This also results in particles of different minerals with a certain size difference sinking equally fast in a process called hydraulic equivalence. For example, this would be the case with quartz, pyrite, and gold if the radii had a ratio of 32:2:1. It is therefore not surprising that small gold particles are often deposited together with gravel. However, this is a very gross simplification for a number of reasons such as the law does not apply in turbulent flows and particles can only sink after they have been lifted from the ground by the flow.

The uptake of particles by a flow is described by the Shields' parameter. It depends on flow velocity (more precisely, on shear stress at the bottom), the density of particles, and to a lesser extent particle size. Small low-density particles are most likely of course to initiate movement. Heavy minerals, on the other hand, tend to remain on the bottom. This is modified by the nature of the riverbed, whether it is flat or rough, and whether its particles are uniform or very different.

Although sorting mainly occurs during pickup and sinking, it also occurs during transport. This is due to the different speeds at which different particles move in suspension, by saltation, or by rolling. In a turbulent stream suspended particles are sorted according to

Fig. 5.39 Gold can be extracted from river sediments using a washing pan. Circular movement of such a pan causes water to slosh out together with light minerals leaving the heavy minerals behind (*photo* © Nate Cull, CC-BY, Wikimedia Commons)

their hydraulic equivalence by accumulating in an area where there is a certain flow velocity. In suspension where there is a high concentration of particles it becomes even more complicated because collisions between particles increasingly play a role in sorting.

The shape of particles is also important since a plate-shaped particle behaves quite differently than a spherical particle. The nature of the mineral surface and even electrostatic properties can also play a role. Solution and precipitation could be added as further mechanisms playing a role though to a lesser extent. This explains the extremely rare occurrence of unde-formed idiomorphic gold crystals in placer deposits. Such crystals would certainly not have survived a longer transport route.

By simulating transport and deposition using computers and calculating the concentrations of heavy minerals in placer deposits it has been possible to successfully calculate (at least in part) variations in ore grade for large deposits.

Fluviatile (or alluvial) placers are deposits of heavy minerals (Table 5.3) in streams and rivers. They can be active river courses or older sediments. The exact process is extremely complex and still not clearly understood even today (Box 5.16). They form in places where the flow regime changes significantly (Fig. 5.40) such as behind boulders, on sandbanks, on the convex bank (a.k.a. inner bank or slip-off slope) of a river bend, or on the edge of a basin on an alluvial fan. The interplay of sedimentation and particle pickup that ensues can lead to an accumulation of heavy minerals in the riverbed.

The highest concentration is often found above fine sediments within river sediments or directly above the underlying rock. There are two possible explanations for this. One is that heavy minerals can work their way down through the river gravel. The other is more probable in that it assumes a complete rearrangement of river gravel (e.g., by periodically recurring flood events leaving behind heavy minerals; Pohl 2005). These events usually represent the actual erosion and transport force in rivers.

Box 5.17 Nuggets

Gold nuggets have to be the best known forms in which native gold is found (Fig. 5.41). This is pri-marily due to movies and stories about the California Gold Rush. Most nuggets have a diameter of a few millimeters or a few centimeters. However, there have been some very spectacular finds such as one given the meaningful name "Welcome Stranger" found in 1869 in Moliagul (Australia) weighing about 72 kg

Table 5.3 Some heavy minerals found on placers and their density (for ease of comparison quartz has a density of 2.6 g/cm^3)

Mineral	Composition	Density
Native gold	Au	15–19 g/cm^3
Native platinum	Pt	14–19 g/cm^3
Columbite	(Fe,Mg,Mn)(Nb,Ta)$_2$O$_6$	5–8 g/cm^3
Wolframite	(Fe,Mn)WO$_4$	7.2–7.7 g/cm^3
Cassiterite	SnO$_2$	6.3–7.2 g/cm^3
Scheelite	CaWO$_4$	6.1 g/cm^3
Magnetite	Fe$_3$O$_4$	5.2 g/cm^3
Monazite	(REE,Th,Nd)PO$_4$	4.2–5.4 g/cm^3
Xenotime	(Y,Yb)PO$_4$	4.5 g/cm^3
Ilmenite	FeTiO$_3$	4–5 g/cm^3
Zircon	ZrSiO$_4$	4.6 g/cm^3
Rutile	TiO$_2$	4.2 g/cm^3
Corundum (ruby, sapphire)	Al$_2$O$_3$	3.9–4.1 g/cm^3
Spinel	MgAl$_2$O$_4$	3.6 g/cm^3
Topaz	Al$_2$SiO$_4$(F,OH)$_2$	3.5–3.6 g/cm^3
Garnet	(Mg,Fe,Ca)$_3$(Al, Fe)$_2$(SiO$_4$)$_3$	3.6–4.3 g/cm^3
Diamond	C	3.5 g/cm^3

(equivalent to 2316 troy ounces) (Anonymous 1908). Nuggets are lumps of precious metal of high purity like native gold (80–95% Au and some silver). However, there are also nuggets of other metals such as native platinum or the natural gold–silver alloy electrum. The 214-kg "gold nugget" Bernhardt Otto Holtermann found in Australia in 1972 is (usually) not considered a nugget because the gold content was comparatively low at 57 kg. Instead it is considered to be a gold-rich chunk of quartz.

This begs the question as to how a gold-rich quartz block can be turned into a gold nugget. There are a number of things that must be kept in mind here. One is that gold is very durable—so durable that it outlives the surrounding quartz during weathering. Another is that gold is usually very finely distributed in the parent rock; hence the reason gold is also finely distributed in many sedimentary deposits. To make nuggets as big as "Welcome Stranger" other factors are needed. These may well be heat and bacteria. It is striking that nug-gets are not only larger than primary gold grains but also purer. Gold from quartz veins often has a higher silver content than gold from placer deposits.

The theory that gold nuggets are produced by cold forging during fluviatile transport is no longer

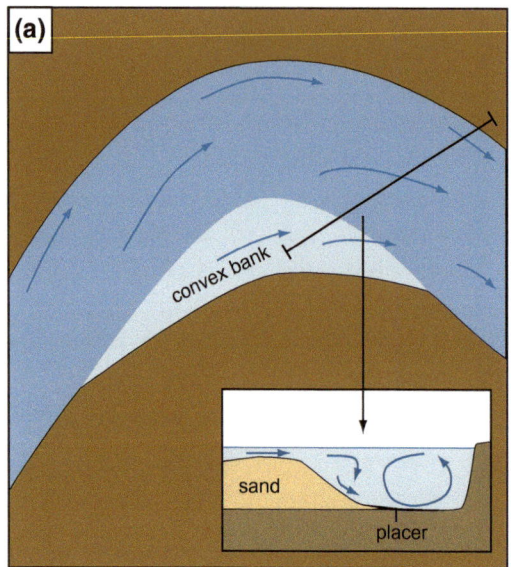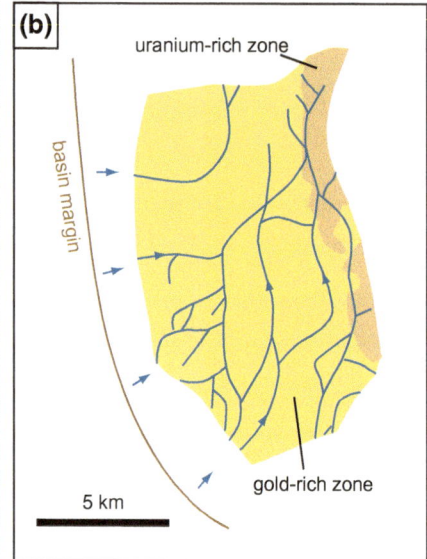

Fig. 5.40 Fluviatile placers can form when the flow changes. **a** An example is the convex bank of a river bend. **b** Much larger placers can be produced on an alluvial fan at the edge of a plain. The figure shows a fossil alluvial fan of the Welkom gold field in the Witwatersrand (South Africa), which contains two large gold-rich conglomerate layers (Basal Reef and Steynreef). The Au/U relationship changes from west (the upper part at that time) to east (see also Box 5.18) (**b** from Minter 1978)

Fig. 5.41 Gold nuggets (© Getty Images/iStockphoto)

supported. Gold can dissolve in soils (Sect. 5.11.3) and sediments even if only in small concentrations (Mann 1984; Colin et al. 1989; Bowell et al. 1993). Bacteria seem to play an important role in biofilms. Certain bacteria such as *Ralstonia metallidurans* (found in two gold mines) excrete dissolved metals because of their toxicity (Reith et al. 2006; Cabral et al. 2011). Such metals remain as a coating over existing grains. Hough et al. (2007) deny near-surface formation because they believe the healed structure of the metal can only be explained by temperatures of at least 300 °C. Accordingly, they should have been created by solution and precipitation at depth (hypogenic).

Box 5.18 Witwatersrand in South Africa

About 40% of gold ever mined comes from the Witwatersrand gold fields in South Africa (Frimmel 2002, 2008). Although the best times are over, one-third of world production is still produced here. A real gold rush began in 1886 after a large find on a farm attracted countless knights of fortune. The gold digger settlement grew into a big city in a very short time called Johannesburg.

The sediments of the Witwatersrand Supergroup (named for the mountain range of the same name) were deposited in the Archean (Robb and Meyer 1995). It used to be a wide basin in the Kaapvaal Craton probably with a large lake in the center and huge alluvial fans at the margins. Gold came from the surrounding granites and greenstone belts and was deposited in large quantities on the alluvial fans along with pitchblende (and other uranium minerals such as brannerite) and pyrite. The gold remained near the edge of the basin. Farther inward (in the lower part of alluvial fans at that time) the concentration decreased, while the content of pitchblende was higher. Later, flood basalts buried the basin probably explaining why this deposit was not eroded long ago. The whole thing was then metamorphically transformed and folded.

Most important are the layers of the Central Rand Group (2.89 to 2.76 billion years old) whose gold fields lie on the edge of the former basin in an arc 280 km long. Gold is enriched in conglomerate layers (a.k.a. reefs) several centimeters or meters thick that can be traced for dozens of kilometers. They are mined underground. Some mines reach depths of 4 km and are by far the deepest mines in the world. By-products of gold mining are osmium, iridium, other platinum-group elements, silver, and uranium.

It was long a controversial discussion whether this was really a placer deposit or whether the gold was not precipitated hydrothermally in the pores of the conglomerate. There were good arguments for the second thesis such as the gold grains being irregularly shaped —not rounded. In addition, U–Pb dating of some minerals resulted in a younger age. However, it has been possible to date the gold itself with the Re–Os isotope system. With an average age of 3.3 billion years it is clearly older than the sediments and therefore actually detrital. However, the metamorphism of placer deposits probably led to recrystallization and small-scale mobilization (Kirk et al. 2001, 2002).

The rounded pyrite grains are also older than the sediment. The fact that pyrite and pitchblende could be transported in rivers and deposited on a placer is only possible if the atmosphere at that time was by and large free of oxygen (Frimmel 2005). Oxygen was only produced in the course of the Archean by early protozoa as also reflected in changes in marine sediments (Box 5.2), but the exact temporal evolution of the Archean atmosphere is another moot point.

Another interesting point is the extremely high content of osmium in native gold from the Witwatersrand. Normal values are below 0.1 ppm, but here it ranges from 2 up to 10,350 ppm (Frimmel 2008). Due to the low solubility of osmium this cannot be explained by alteration by hydrothermal systems and there must be a magmatic origin of the Au:Os ratio. The Os:Re ratio is very similar to that found in komatiites (Kirk et al. 2002), so perhaps erosion of corresponding igneous rocks played a role.

The Witwatersrand makes up a good part of the gold present in the Earth's crust. It is reasonable to assume that there were other large gold deposits in the Archean that were later eroded. Interestingly, many large orogenic gold veins were also formed in the Archean (especially, around 2.7 billion years ago). Since the Archean was also the time when the first continents were formed and during which they became

larger mainly due to magmatism at subduction zones and hotspots, it is plausible that at this time a large part of the gold originally present in the mantle was enriched in the crust by magmatic and hydrothermal processes.

Fluviatile placers are mined for gold worldwide. Gold appears in placers in the form of tiny flakes or as nuggets (Box 5.17). During the gold rushes in California and Alaska (USA) and Victoria (Australia) young fluviatile placers were especially important. However, the most important gold province in the world is the Witwatersrand in South Africa (Box 5.18) that has fossil placers from the Archean, about 40% of all gold ever mined originates from these placers. In some gravel pits in the Rhine Rift Valley gold is still being extracted today as an additional source of income for gravel pit operators (Seidler 2012). Other examples of fluviatile placers are the historically important tin placers in the Ore Mountains (Germany and Czech Republic), placers with cassiterite and columbite–tantalite in the DR Congo, and gemstone placers in Sri Lanka and Thailand. Iron-rich placers (a.k.a. channel iron deposits) were formed in Western Australia because of the weathering of BIFs.

Many people will have seen beach placers (marine placers) for themselves during a walk on a beach in the form of extraordinarily colored sands. Currents and waves enrich heavy minerals in such sands. Black sands may contain ore minerals such as ilmenite, titanite, or cassiterite, while red sands may contain garnet or zircon. These minerals are enriched in elongated lenses or thin, strip-like layers that are repeatedly found along the coast.

Although sand is transported along coasts mainly by currents and tides, it is sorted by waves in the surf zone. Incoming waves transport material onshore where heavier grains remain on the beach preferentially while lighter grains are flushed back into the sea. Sinking speed does not play a role here, so grain size is unimportant. In addition, tides, wind, and lateral currents can have an influence as can the profile of the coast, which affects wave dynamics. It is possible that storm events play a significant part in the development of beach placers. They are preferentially formed on morphologically stable coasts where there is no strong erosion and no rapid sedimentation (Pohl 2005). Older beach placers can be flooded by a change in sea level (offshore placers) or suddenly lie on land far away from the coast (onshore placers). Modern process-oriented investigation methods of coastal formation play a major role in the modeling of beach placers (Hardisty 1990).

Beach placers of the tin belt of Southeast Asia account for the largest share of global tin production by far. They are located mainly in Thailand and Malaysia along the west coast of the Malay Peninsula and on the Indonesian Tin Islands of Belitung (a.k.a. Billiton) and Bangka (northeast of Sumatra). Offshore they are mined using dredgers, while onshore they are mainly mined using a high-pressure water jet washing out material in a small pit. Sludge is pumped off and heavy minerals are separated in a mechanical plant. In addition to cassiterite (see also Sect. 4.5) other heavy minerals are extracted as by-products including monazite and xenotime (hence the importance of Malaysia for rare-earth elements), as well as ilmenite, zircon, columbite–tantalite, wolframite, and gold.

Some beaches in India (in Kerala, Tamil Nadu, and Orissa) have high contents of monazite (Fig. 5.42). Monazite is partly mined for the extraction of rare-earth elements and thorium as nuclear fuel. Further examples can be found in South Africa and in the southeast of the United States. Known for its extremely high natural radioactivity is the monazite-rich beach of the city of Guarapari (Brazil). Zircon as a raw material for refractory materials and as zirconium ore is almost exclusively obtained from beach placers (especially, in Australia and South Africa). There are also important deposits in the United States and Brazil. The titanium ores ilmenite and rutile are also predominantly obtained from beach placers mostly together with zircon, garnet, monazite, and xenotime. The most important beaches are located on the east and west coasts of Australia, in India, Sri Lanka, South Africa, Mozambique, the United States, and Brazil.

Beach placers containing diamonds are also of great economic importance (especially, along the Diamond Coast in Namibia). Smaller diamonds of mostly very good quality can be found here in sand on the seabed and on older beach terraces spread along a length of more than 300 km. The Oyster Line discovered by the German geologist Hans Merensky (Machens 2011) is particularly famous for the exploration of diamond-bearing placer deposits and the estimation of reserves. However, the irregular distribution of diamonds and the low grades (e.g., an average of 3.6 ct/100 t in the Octha Mine on the Orange River) clearly causes problems (Van Wyk and Pienaar 1986). Geologists have a saying about this: "Diamond is like a pig—it lies where it wants to."

Eolian placers are created when wind acts as the concentrating agent by removing fine particles in deserts or on coasts. These are typically beach placers that have been further enriched by wind. There are also glacial placers in glacier sediments. Although gold can be found with a bit of luck in glacier sediments, it is not commercially exploitable. However, it has value as a hobby for amateur gold collectors (Kühne 1976, 1983; Lierl and Jans 1990). In most cases secondary enrichment by glaciofluviatile transport and subsequent processing of gravel in gravel pits plays a role here.

Fig. 5.42 Monazite sand (concentrate) of various marine and fluviatile placers (*photo* © F. Neukirchen/Mineralogical Collections of the TU Berlin)

5.10 Weathering (Introduction)

Weathering is the process that makes Earth what it is. Very few plants can live on unweathered hard rock like granite since it lacks pore spaces and nutrients. Pore spaces and nutrients are only made available by the weathering of rocks. The green cover of the Earth is carried atop a brown cover of weathering rock (soil). This layer of decaying and newly formed minerals in the course of weathering is called regolith from the Ancient Greek ῥῆγμα, *regma*, "break" and λίθος, *lithos*, "stone."

Put simply, weathering is more or less analogous to making coffee. Coffee beans must first be crushed before they make a decent coffee. Using a coffee grinder (physical weathering) pore spaces in our coffee beans are expanded beforehand and thus increase the surface area. Then water is used to extract certain easily soluble elements from the beans and remove them as a solution (chemical weathering). What remains is the less soluble remainder. Chemical weathering thus leaches out easily soluble substances and enriches substances in the soil that are difficult to dissolve. This process can form important deposits (in particular, aluminum ore (bauxite) and nickel deposits).

The minerals that make up rocks are formed under certain physical and chemical conditions. If they are exposed to changed conditions, then they usually lose their stability. In addition, rocks at depth are pressed together by the pressure and when they reach the Earth's surface through geological processes they expand. This expansion causes clefts to tear open. Moreover, there are other fissures that—depending on the geological history of the rock—are caused by tectonic, hydrothermal, or other events. Water can penetrate the rock body through these fissures. Mineral grains are dissolved on the surfaces of individual rock blocks and partially replaced by secondary minerals. Ions are carried away by water according to their solubility. In addition, Fe^{2+} is oxidized to Fe^{3+}. Chemical weathering works very effectively because soil water is slightly acidic. This is because soil water contains dissolved CO_2 (carbonic acid) and organic acids produced in the humus layer of the soil mainly by bacteria and the decomposition of dead plants. Chemical weathering is particularly effective in the tropics because the rapid growth of plants and the warm climate accelerate the formation of organic acids, the enrichment of CO_2 in the soil, and the availability of a great deal of water. This also explains why chemical weathering under soil is much faster than on a rock cliff.

The different minerals in rock are susceptible to weathering in different ways. Calcite is dissolved very easily and mobilized as Ca^{2+} and HCO_3^-. Feldspars are somewhat more durable and their composition determines the speed of weathering. Ca plagioclase (anorthite) weathers the quickest, Ca–Na plagioclase of medium composition weathers a little slower, Na plagioclase (albite) weathers even more slowly, and potassium feldspar weathers the slowest. This means that as calcium and sodium are removed the relative content of potassium increases.

Apatite in solution is important as a source of phosphorus for living beings. In ferrous minerals such as biotite and pyroxene the oxidation of iron plays an important role. Although the solubility of SiO_2 is very low, it should be kept in mind that in the long term even quartz is dissolved. Some accessory minerals such as zircon are particularly resistant to weathering. Interestingly, the weathering of igneous rocks is often such that the minerals crystallized early in unfractionated magma weather faster than those crystallizing late or in a fractionated magma. For example, olivine weathers extremely quickly, while quartz is very resistant to weathering.

This has some consequences for soils that result from weathering. One is that certain elements are removed from the rock by the weathering solution and thus are available for the formation of new minerals. Another is that stable minerals are passively enriched over time.

Weathering of a single mineral usually begins at weak zones such as internal cleavages or in the case of feldspar at exsolution lamellae (e.g., perthite texture). As weathering progresses small channels are cut into the crystal along weak zones through which the weathering solution can penetrate deeper into the mineral. Only those parts of a mineral that are relatively weathering resistant are preserved. The further the weathering cuts into the mineral, the larger its surface and the more sensitive it becomes to mechanical stress. Pore spaces are constantly expanded in the process through physical weathering when the water freezes in narrow fissures and pores (in colder climates) or when new, water-containing minerals form there (in warmer climates). The resulting pressure expands the fissures and pores and makes it easier for subsequent water to penetrate the depths of the rock and dissolve it. It is through these pore networks that freshwater loaded with oxygen, carbonic acid, and organic acids gains ever-better access to the rock and can decompose it in ever-newer and ever-deeper areas. Dissolved substances are easily removed at the same time. Newly formed minerals such as kaolinite or gibbsite are formed in the pores simply because the less mobile or non-mobile elements have to be accommodated somewhere.

Original hard rock is finally transformed into what is known as saprolite (a.k.a. rotten rock). This is largely weathered rock that can be crushed with one hand despite showing the appearance and texture of the parent rock in many places and even the same fissures and cracks. Its internal surface is much larger than that of the parent rock. Original minerals such as mica (biotite), feldspar, or amphibole are already largely weathered and replaced by newly formed clay minerals and hydroxides such as kaolinite, smectite, and gibbsite. When moist this rock is plastically deformable.

Living beings play an important role in enlarging pore spaces. The roots of plants often have to delve very deep into soil to gain access to water and locate nutrients found in the fissures of saprolite. In extreme cases this can be as much as 20 m. Symbiotic fungi support the roots of plants in this work. Roots and fungal hyphae forcibly expand small and minute fissures and fungi promote the release of nutrients by freeing up organic acids. Bacteria too are often involved in weathering.

Once the parent rock is weathered and its minerals dissolved or crushed, what remains on the surface is removed very easily by wind or water provided they are not retained by organic material and roots. The loss of material is estimated to be between 1 and 10,000 mm per year in most areas. This means that regolith has to be regenerated at least at the same rate so that plant life can flourish.

The speed at which a given rock naturally weathers depends on many factors. Granite blocks of approx. 10 cm in diameter in glacial moraines in the Sierra Nevada (California, USA) weather to saprolite only when such moraines are at least 81,000 years old. This corresponds to the so-called weathering front migrating into rock at a rate of about 0.6 m per 1000 years. Weathering to saprolite in granodiorites in southeastern Australia progresses at a rate of 4–41 m per 1000 years. Among other things this is due to the amount of rainwater available, which is 200 mm/year in the first case and 910 mm/year in the second. In addition, the weathering front usually consists of a zone into which meteoric water penetrates the rock called the vadose zone—not a flat surface that slowly penetrates downward into the parent rock. This zone is approx. 4–8 m thick in the Sierra Nevada and the distance between clefts is approx. 50 cm. Rock is weathered within this zone from all sides simultaneously starting with clefts and the smallest fissures. Granite with such clefts can take 400,000 years to weather without problems at a rate of about 0.01 m/year with a weathering zone 4 m thick. With a thicker weathering zone of 8 m the amount would double. Interestingly, these values are comparable with those of southeast Australia where saprolite is believed to form at a rate of 0.004–0.046 m/year (Dosseto et al. 2008; Graham et al. 2010).

5.11 Laterite and Bauxite

Residual deposits (a.k.a. weathering deposits) are the result of weathering. While sedimentary deposits contain substances that are removed during weathering and deposited elsewhere, residual deposits are substances that remain behind during weathering and are thus enriched such as bauxite, laterite, and residual placers (Sect. 5.9).

Laterite is the name given to a soil rich in iron and aluminum that has been formed by intensive tropical weathering. Its characteristic intense red color comes from iron oxides. The word laterite is derived from the Latin *later* meaning "brick" and refers to the use of laterite bricks as a building material. Bricks are made by cutting moist and fresh laterite into blocks and then drying them in the sun. Fe oxides make the material hard and usable (Yamaguchi n.d.). Depending on the parent rock, erosion rate, and duration of weathering, chemical weathering in a tropical climate can reach a depth of dozens or even more than a hundred meters and form laterites of corresponding thickness. Due to the great diversity of parent rocks involved, laterites can also show a variety of mineralogical and chemical compositions, thicknesses, and textures (Chowdhury et al. 1965; Aleva 1994; Tardy 1997; Dalvi et al. 2004). Weathering solutions remove soluble elements and less soluble elements are left behind. Under these conditions only the least soluble elements such as iron, aluminum, and a few extremely resistant minerals like zircon are finally left. In many laterites even quartz is completely dissolved.

Lateritic soils cover most of the humid tropics. Some authors believe that about one-third of the land surface of the world is covered by lateritic products (Tardy 1997). In addition to the local use of blocks as bricks in construction, laterite has also been used in road construction on various occasions albeit with limited success (Grace 1991). The decisive factor here is cost savings in countries that are rich in lateritic material but lack deposits of usual building materials. However, laterites are of great importance as mineral deposits. Laterites are especially important in being deposits for aluminum (bauxite), nickel, gold, rare-earth elements, and iron. Kaolin (Sect. 7.5) is also produced from laterites.

The contrasting speed at which minerals of parent rock weather creates a soil profile with horizons of different composition (Fig. 5.43). Carbonates and sulfides already leach out of rock that has hardly weathered. Although saprolite still contains mineral grains from the parent rock, they are increasingly weathered toward the top to form secondary minerals where easily soluble ions have already largely been leached out.

Laterite in the narrow sense (plasmic zone) whose composition is dominated by hardly soluble Al, Si, and Fe^{3+} is found above this zone. It consists mainly of kaolinite and iron hydroxides. Kaolinite is obtained from feldspar-rich, low-iron rocks such as leucogranite from which kaolin (a.k.

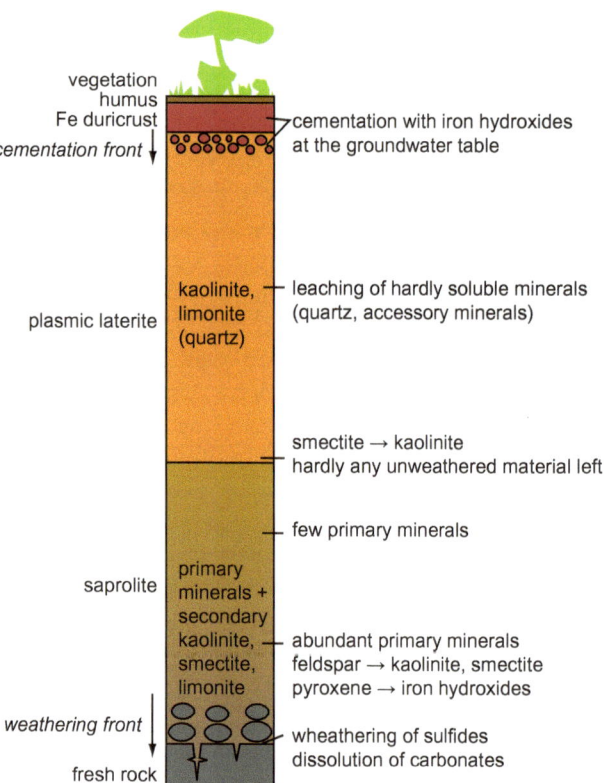

Fig. 5.43 Laterites form by intensive chemical weathering in a tropical climate. The original minerals are largely replaced by secondary minerals and soil water removes ions according to their solubility. In saprolite this process is less advanced in the plasmic laterite horizon, which is significantly depleted even of poorly soluble SiO_2 and consists almost exclusively of kaolinite and iron hydroxides. Cementation at the groundwater table forms a hard crust of iron hydroxides. The layer of humus is only thin

a. china clay) is derived. However, the iron hydroxide content is very high in laterites above iron-rich rocks such as basic igneous rocks. Nothing of the original texture of the rock can be seen here. Easily soluble ions have already been completely removed, even the SiO_2 content has been reduced, and possibly not even quartz is present. Some secondary minerals such as smectite that had formed in saprolite had decayed again. Although resistant minerals such as monazite and zircon are still present, they are very slowly being attacked. A number of metal ions are adsorbed by the surface of iron hydroxides (adsorption) and thus retained.

Cementation with iron hydroxides (ferricrete, ferricrust) or more rarely with aluminum hydroxides (alucrete) occurs at the groundwater table. Under certain circumstances there is overlying gravel of relocated laterite. The humus layer on the surface is usually very thin.

Sometimes iron-rich laterites are mined for iron ore and other times there are worthwhile enrichments of manganese. However, the deposits of aluminum, nickel, and rare-earth elements presented in the following sections are much more important.

5.11.1 Bauxite

Bauxite (Figs. 5.44 and 5.45) is the name given to aluminum ore that consists mainly of the minerals gibbsite (γ-Al (OH)$_3$), boehmite (γ-AlO(OH)), and diasporite (α-AlO (OH)). In addition, various iron oxides (such as hematite and goethite), the clay mineral kaolinite, and small amounts of titanium oxides (such as anatase) can also occur. The name bauxite derives from Les-Baux-de-Provence (France) where it was first described in 1821 by Pierre Berthier. Hoping to be able to use the reddish-brown rocks of the area as possible iron ores, to his great surprise Berthier found during analysis that his samples had no appreciable contents of any elements other than aluminum. He described his find as the "ore of Baux" and ever since aluminum-rich laterite has been called bauxite (Aleva 1994). The two types that are distinguished are laterite bauxites and karst bauxites.

Laterite bauxites are formed from very diverse, mostly feldspar-rich and iron-poor rocks such as (leuco-)granites, gneisses, and arkoses, but sometimes also above basalts, shales, and schists. To form a bauxite deposit with an Al_2O_3 content of more than 50% from parent rock with around 15% Al_2O_3, other elements including silica and iron have to be removed by weathering. For example, during bauxite formation in Los Pijiguaos (Venezuela) 61% of the mass and 77% of the volume of parent rock (granite) were lost (Meyer et al. 2002). SiO_2 and Fe^{3+} have very low solubility. The

Fig. 5.44 Bauxite (*photo* © Siimsepp/Fotolia)

solubility of Al^{3+} is extremely low in neutral water but significantly higher in strongly acidic or strongly alkaline water. At a pH below 4.5 aluminum is even more soluble than SiO_2. To remove iron and silicon from the soil and to enrich aluminum very special pH–Eh values must be maintained. The formation of aluminum hydroxides from feldspar takes place in two steps. First, the loss of SiO_2, Na^+, K^+, and Ca^{2+} leads to the formation of kaolinite. Second, loss of the remaining SiO_2 leads to the formation of gibbsite. In addition to extremely strong weathering good drainage is also required for the second step. Zones with the highest aluminum content are often located under a surficial layer rich in iron. Since average temperature and the availability of solvents (in this case water) also play a decisive role, laterite bauxites are mainly found in the tropics. Fossil laterite bauxites can therefore also serve as indicators of paleoclimate.

Australia produces more bauxite than any other country and is responsible for almost one-third of world production. China, Brazil, and India (Lee Bray 2012) are the next largest producers. Guinea, Jamaica, Guyana, and Vietnam also have large reserves. Bauxites in Europe are predominantly mined in Greece, France, and Hungary. In 2007 some 190 million tonnes of bauxite were mined. However, known reserves should be able to cover demand in the long term despite rising global demand. Around 95% of bauxites mined go into aluminum production. The rest is used for the production of abrasives, cement aggregates, special ceramics, and refractory products. Gallium is also extracted as a by-product of aluminum production.

The most important bauxite deposits in Australia include Weipa on the Cape York Peninsula, Gove in the Northern Territory, and the Darling Ranges and Kimberley in Western

Fig. 5.45 Closed bauxite mine near Otranto, Puglia (Italy) (*photo* © Loloieg, CC-BY-ND, flickr.com)

Australia. The bauxites of Weipa form bright red cliffs on the coast that are so distinctive they were described by early sailors. However, their nature was only recognized relatively late when oil was being prospected (Pohl 2005). Such bauxites evolved over arkose.

Since the formation and further development of bauxites depend very much on tropical climates and such climate regions have not shifted significantly since the Mesozoic the largest deposits of this type can be found in the tropics or subtropics. Older bauxites are comparatively rare because these rocks easily fall victim to erosion. Bauxites of the Tikhvin mining district in the Lower Carboniferous of the Moscow Basin are exceptions. They lie above sandy micaceous shales of the Upper Devonian as well as marls and dolomites of the Early Carboniferous. At the time bauxites were deposited in the Lower Carboniferous the region was on a continent drained by narrow valleys into which weathered products of Devonian shales were sedimented (Smirnov 1989). They appear as relatively narrow and

elongated lenses covered by sediments of the Lower Carboniferous and Quaternary.

Laterite bauxites differ of course from karst bauxites. Although the latter are mainly found in southern Europe in the Mediterranean region (including Les Baux), they are also known in China and on islands in the Caribbean. In most cases such deposits are limited fillings of uvalas (tublike karst depressions) or sinkholes in karst areas—not extensive deposits. The shape and thickness of deposits are controlled by the relief and altitude of the karst area. The formation process was long a moot point. According to one theory (autochthonous formation) karst bauxites are weathered products of clay in carbonate rocks (Bárdossy 1982), but this seems to apply in very few instances. The more probable theory (allochthonous formation) assumes that material from higher areas where there was lateritic weathering was supplied and sedimented in karst cavities (Valeton et al. 1987; Petrascheck 1989). It has been shown that the clay content of karst bauxites is clearly distinguishable from the clay content of underlying limestones such that bauxites cannot represent an insoluble residue of karstification. How material on a karst plateau was transported where there are no rivers became problematic for advocates of the second theory. Maybe the bauxites were sedimented after subsidence of the karst area below sea level—marine fossils such as gastropods or foraminifers were found in the bauxites of Mount Parnassus (Greece) as were limnic ostracods and plants. Maybe bauxite deposition took place in lagoons or estuaries. It is unclear whether the material was already transported as bauxite or whether shales were deposited that were subsequently exposed to intensive tropical weathering in situ. There are often indications of later deepening of karst cavities (Pohl 2005).

5.11.2 Lateritic Nickel Deposits

Laterites rich in nickel are among the most important nickel deposits since they contain 60% of all known reserves. The largest are on New Caledonia (Ries 2001), an island belonging to France in the western South Pacific about 1500 km northeast of Brisbane (Australia). In 1865 the French mining engineer Jules Garnier found the green nickel ore that would later be named garnierite (Fig. 5.46) for him. Mining began as early as 1875. At that time the nickel content of the ore was as much as 10%. Almost 100 years later (in 1974) 7 million tonnes of ore were mined containing 2.6% nickel. Such mining entailed the movement of 20 million tonnes of rock. Société Métallurgique Le Nickel (SLN) is today the largest nickel producer and is active in four areas: Thio, Kouaoua, and Poro on the east coast and Nepui on the west coast. Extracted ore is shipped to Noumea to be smelted in the Doniambo Smeltery. Ore deposits that are worth mining are close to the surface and mining regions

Fig. 5.46 Garnierite from the Camps des Sapins Mine, Thio (New Caledonia) (*photo* © Didier Descouens, CC-BY-SA, Wikimedia Commons)

as a consequence resemble terraced rice cultivation in mountainous areas rather than deep open-cast mines as is the case with porphyry copper deposits. Prior to discovery of the Sudbury Complex (Sect. 3.3.5) this was the world's most important nickel deposit. Today it contains 26% of the world's reserves and accounts for just under 7% of world production. Other important deposits of this type are found in Indonesia, the Philippines, Australia, Cuba, and Brazil.

Garnierite is a fine-grained mixture of different magnesium silicates rich in nickel such as nickel serpentine, nickel talc (willemseite), nickel smectite (pimelite), and nickel chlorite (schuchardite). There is also a quartz variety called chrysoprase colored green by nickel. Goethite can also contain up to 1.5% nickel. Since the parent rocks are peridotites that contain very little nickel this begs the question as to how nickel laterites formed. However, nickel is effectively enriched in the course of weathering and the example of New Caledonia can be used to illustrate this. The main island is about 400 km long and 40 km wide. The climate is tropical with heavy rainfall occurring mainly from January to March.

A terrane consisting mainly of basalt collided with a subduction zone at the end of the Eocene obducting a large ophiolite complex (see also Box 3.7) in a southwestern direction (Whattam 2009). The associated ultramafic rocks (peridotite: mainly harzburgite and dunite) are more or less converted into serpentinite by hydration. At about 7000 km^2 they make up one-third of the entire main island. Of the ultramafic massifs the southern one is the largest at about 5500 km^2—smaller ones can be found along the west coast. Olivine was the main nickel carrier in ultramafic rocks with 0.3% Ni followed by orthopyroxene with 0.06%.

It took several million years of tropical weathering to enrich the low nickel contents of rock to the high ore grade of the soil. Ultramafic rocks are particularly susceptible to tropical weathering. Olivine is extremely unstable under conditions found at the Earth's surface and is easily dissolved. Minerals unstable under given conditions are soon replaced by others and certain ions can go into solution. In the case of olivine weathering these are Mg^{2+} and Ni^{2+}.

The composition of olivine in the Poro area of New Caledonia can be written as $Mg_{1.82}Fe_{0.8}Ni_{0.007}(SiO_4)$. This composition is divided into individual components such as Mg olivine (forsterite), Fe olivine (fayalite), and Ni olivine for the following weathering reactions:

$$4Mg_2SiO_4(\text{forsterite}) + 10H^+$$
$$\rightarrow Mg_3Si_4O_{10}(OH)_2(\text{saponite}) + 5Mg^{2+} + 4H_2O$$

$$4Fe_2SiO_4(\text{fayalite}) + 4O_2 + 8H^+$$
$$\rightarrow Fe_2Si_4O_{10}(OH)_2(\text{nontronite}) + 6FeO(OH)(\text{goethite})$$

$$4Ni_2SiO_4(Ni\ olivine) + 10H^+$$
$$\rightarrow\ Ni_3Si_4O_{10}(OH)_2(pimelite) + 5Ni^{2+} + 4H_2O$$

Saponite, nontronite, and pimelite are clay minerals of the smectite group. Another way to obtain nickel silicates is through exchange reactions. Such reactions are typical of minerals with variable chemical compositions such as serpentine, smectite, and other silicate minerals found in laterites. An interesting reaction is the exchange of dissolved nickel for magnesium in serpentine:

$$Mg_3Si_2O_5(OH)_4 + 3Ni^{2+} \leftrightarrow Ni_3Si_2O_5(OH)_4 + 3Mg^{2+}$$

Equilibrium is found on the right-hand side of this reaction such that nickel accumulates in serpentine and magnesium in the soil solution. According to Golightly (1979) the Ni/Mg ratio in serpentine is 104 times greater than in a soil solution when the two are in equilibrium.

The different solubility of minerals ultimately leads to zoning of the laterite profile. The most easily soluble minerals are found at the base, while those with the least solubility are concentrated at the surface. Roughly speaking, the profile of a nickel laterite (Fig. 5.47) can be divided into two zones that split into further horizons (Troly et al. 1979; Guilbert and Park 1986).

Near the surface in the limonite zone the texture of parent rock is completely destroyed. Iron hydroxides predominate mostly consisting of goethite. This is the laterite horizon sensu strictu. From top to bottom it comprises (a) a hard iron crust and below it a zone with roundish iron concretions; (b) red laterite consisting of a mixture of goethite and other iron hydroxides; and (c) yellow laterite mostly consisting of fine-grained goethite (limonite).

Below the limonite zone is the saprolite zone where SiO_2 and magnesium dominate. The textures of parent rocks are still more or less recognizable. In the deeper horizons of this zone there are minerals from the original rock. The horizons are (d) earthy or soft saprolite and (e) rocky saprolite that still contains weathered fragments of the original peridotite.

Only saprolitic material is mined in New Caledonia because of the nickel silicates it contains. Although nickel contents are usually between 1.3 and 3%, in the limonite zone goethite can contain up to 1.5% Ni. However, the iron crust is depleted of nickel. The sequence can be incomplete in some deposits such as the lateritic zone of the Morro do Niquel deposit in Brazil that fell victim to erosion (Langer 1969). Apart from significant nickel enrichment the chemical composition of the saprolite zone does not differ significantly from that of the parent rock. However, drastic changes in chemism occur at the border of the limonite zone where the Ni content is significantly lower, but still slightly enriched compared with the original rock. At the same time, the SiO_2 and MgO content decreases to minimum values in the limonite zone while the Al_2O_2 and Fe_2O_3 content reaches maximum values.

The nickel content of the limonite zone is due to residual enrichment, whereas that of the saprolite zone is also based on nickel-containing soil solutions seeping down from the limonite zone. This becomes clear when considering the ratios of Fe, Al, Cr, and Ni. Elements that are not removed always show the same (or similar) element ratio in each enrichment process. This applies particularly to Fe, Al, and Cr ratios due to the low solubility of associated minerals such as iron hydroxides, aluminum hydroxides, and chromites. The ratios of Al/Fe and Cr/Fe clearly show such elements were concentrated by residual enrichment. However, the ratio of Ni/Fe in the profile deviates more strongly. It ranges from 0.04 in the initial rock to 0.25 in the rocky ore

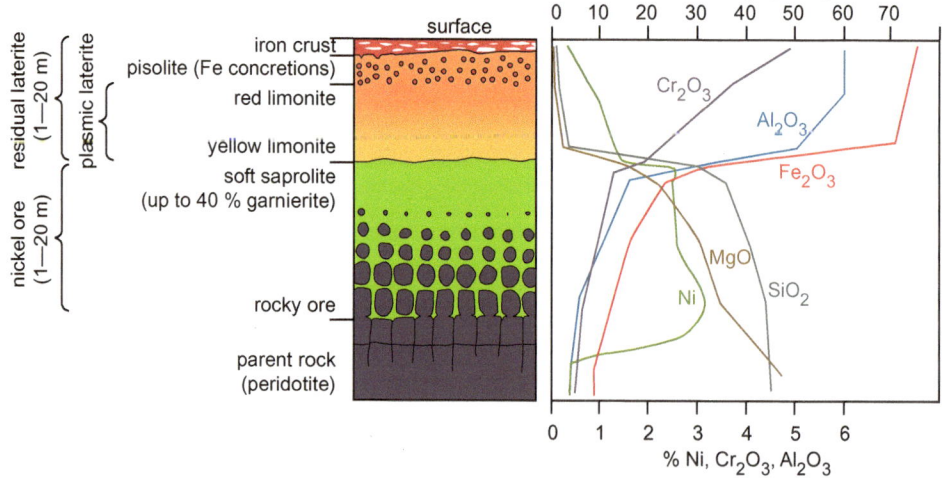

Fig. 5.47 Profile of a nickel laterite in New Caledonia. The highest concentration of nickel is found in saprolite (in nickel-rich magnesium silicates like garnierite). Although the limonite zone also has elevated nickel contents, they are not exploited in New Caledonia (from Troly et al. 1979; Guilbert and Park 1986)

and then to 0.01 in the red limonite. The transport of nickel from the limonite zone only takes place if the accumulation of nickel in this zone exceeds the ability of the dominant minerals there to incorporate nickel. Excess nickel is transported into deeper zones of the profile. A further loss of nickel from the limonite zone can be expected during mineral recrystallization. In serpentinites more or less strongly weathered in deeper zones of the profile a loss of magnesium was accompanied by the incorporation of nickel (Schellmann 1983).

Since they are easily eroded in most cases lateritic nickel deposits are very young (Miocene to Subrecent) and can be found in tropical to subtropical latitudes. However, fossil representatives are also known in areas where tropical weathering no longer takes place (Urals, Albania, Greece, Silesia). An example is the sedimentary karst nickel ores of Euboea (Greece) where residual iron ores (Sect. 5.3) are enriched at the base with concentrations up to 6% nickel. During the Cretaceous the lateritic weathering blankets of surrounding ophiolites were removed and deposited in shallow areas of the sea. Nickel was precipitated again from solutions in the underlying limestones of the Triassic and Jurassic. There are lateritic nickel deposits in Germany like those in the Saxonian Granulite Massif where Deutsche Rohstoff AG is currently endeavoring to mine the ore. It is the limonite zone that will likely be mined here.

Garnierite can be melted in an electric arc furnace and nickel can be reduced at the same time. Limonite zone ore is usually leached with sulfuric acid at high temperature and pressure.

5.11.3 Lateritic Gold Deposits

The discovery of lateritic gold ores in the 1980s was a surprise. The most important and best known representative is Boddington (Western Australia) where there are not only large quantities of bauxite but also gold-rich laterites. According to Pohl (2005) there were 60 Mt of ore holding 1.6 ppm Au. The mine started operations in 1987, but was closed on November 30, 2001 due to exhausted reserves (Louthean Publishing 2004). Since 2009 gold has been mined from the underlying rock of the Archean Yilgarn Craton that has gold-bearing hydrothermal quartz veins (Sect. 4.2). They were probably formed in two phases dated to 2.7 billion years ago and 2.612 billion years ago (McCuaig et al. 2001). The open-pit mine is likely to become Australia's largest gold mine once it has reached full capacity (Anonymous 2010, 2012). Gold accumulations in laterites are also known from many tropical countries.

The lateritic profile of Boddington consisted from top to bottom of lateritic gravel consisting of a hard 5-m-thick crust of iron hydroxides and aluminum hydroxides, 8-m-thick laterites with abundant concretions of iron hydroxides and aluminum hydroxides, and 110-m-thick clay-rich saprolites. Gold was found in the hard crust and the laterites.

Under normal surface conditions gold is largely insoluble. Nevertheless, solution textures and fine gold crystals that are occasionally discovered and very sensitive to any kind of transport prove that gold can be dissolved and transported in soil solutions (Colin et al. 1989; Santosh and Omana 1991). Gold can be dissolved in soil water in minimal concentrations in the form of complexes (Box 4.3) by means of Cl^-, I^-, Br^-, SO_4^{2-} (e.g., from the oxidation of pyrite), CO, HCN, NH_3, HCO_3^-, and organic acids (Santosh and Omana 1991; Bowell et al. 1993). The stability of complexes depends on pH and Eh such that gold is dissolved in certain horizons and precipitated in others. It is often only transported as a complex in the soil solution for a few centimeters as aurothiosulfate or as $Au(NH_3)^+$.

Many elements that accompany gold in hydrothermal systems such as silver, copper, and mercury show higher solubility when exposed to tropical weathering. They are transported much farther by the weathering solution and thus are separated from the gold. As a result gold from laterites and placer deposits is often much purer than primary rock gold. Weathering also makes this type of deposit particularly attractive since intensively weathered soils are not only easy to mine but above all cheap to mine and process in open-cast mining.

5.11.4 Lateritic Rare-Earth Element Deposits (Ion Adsorption Clay)

Rare-earth elements (REEs) can also be enriched in lateritic deposits. This type of deposit is also referred to as ion adsorption clay where REEs are adsorbed on the surface of clay minerals. These deposits are particularly important for heavy REEs that are hardly enriched during magmatic processes. In addition, they have the advantage that they can be mined and processed much more easily than other REE deposits since all that is needed is an electrolyte solution or EDTA (ethylenediaminetetraacetic acid)—the digestion of resistant minerals in acids or bases is not necessary. In China a content of 500 ppm in ion adsorption clay is already considered worthwhile.

Such deposits have so far only been mined in China. Presumably, there are similar deposits throughout the world such as the one recently discovered in Kazakhstan. Chinese occurrences are located in the south of the country in the provinces of Jiangxi, Guangdong, and Guangxi where 10,000 t of REE_2O_3 are extracted annually. The climate here is subtropical and the soils have been preserved because of the low relief. Weathering profiles are usually 15–35 m thick with REEs located in the middle and lower horizons. Light REEs are adsorbed faster and especially enriched in the middle horizon, whereas heavy REEs are enriched in the lower horizon but not to the same extent. An exception is cerium that oxidizes to Ce^{4+} and remains together with iron in the upper horizon.

REE parent rocks are igneous rocks such as granite, acidic volcanic rocks, or lamprophyres. Which REEs are most enriched in the soil depends less on the contents in parent rock than on existing accessory minerals (Bao and Zhao 2008). Igneous rocks often contain several minerals that not only have REEs in varying amounts, but also differ in their resistance to chemical weathering. Gadolinite and REE fluorocarbonates (bastnäsite, parisite) are particularly susceptible. Fergusonite, monazite, and allanite are less susceptible, while xenotime and zirkon are particularly resistant. Sometimes the most resistant minerals remain in the soil (residual placers; Sect. 5.9), while other minerals are dissolved and their REEs adsorbed by clays.

Longnan (Jiangxi) is currently the most important deposit of heavy REEs in the world. Here hydrothermally altered S-type granite that has a muscovite-rich part and a biotite-rich part is being weathered. The former part contains doverite (a fluorocarbonate containing heavy REEs), fluorite, and zircon as well as some monazite, gadolinite, xenotime, and chernovite ($YAsO_4$). The latter part contains monazite, xenotime, zircon, apatite, and fluorite. Accessories of the former part weather very easily as a result of which there is preferential release of heavy REEs.

Although high REE contents are also frequently found in lateritic soils above carbonatites (Sect. 3.10), they are above all contained in primary and secondary phosphate minerals making them less easy to process. Laterites above the Araxá Carbonatite (Brazil), for example, contain up to 13.5% REE_2O_3. One of the largest REE deposits is Mount Weld in Western Australia where carbonatite whose REEs are contained in secondary phosphates such as plumbogummite, rhabdophane, and monazite (Lottermoser 1990) is lateritically weathered. Open-pit mining recently commenced and processing and separation are slated to take place in a new factory in Malaysia.

5.12 Duricrusts

A widespread phenomenon in soils is the hard soil horizon called duricrust. Depending on the dominant cementation mineral it is called calcrete or caliche (cementation by calcite), gypcrete or gypcrust (gypsum), and salcrete (salts) or silicrete (SiO_2). Such solidified soil horizons can reach thicknesses from a few centimeters up to a few meters and are formed by vertical mass transport in weathering solutions.

Substances mobilized by weathering precipitate in humid climates as soon as chemical conditions such as pH and Eh change. Examples are ferricrusts (iron crusts) such as hard pan (in podzol soils), bog iron (Box 5.19), or ferricrusts in laterites. Although they no longer play a role as iron ore, historically they were definitely worth mining. More important today are alucretes in which the soil is cemented by aluminum hydroxides. Alucretes are mined as aluminum ore.

Box 5.19 Swords from the moors of Scandinavia

Iron-rich concretions and crusts can form in temperate climates in bogs and the soils of floodplains when they are referred to as bog iron. The decay of organic matter provides an acidic and somewhat reduced environment in which iron is soluble in water. Whenever iron comes into contact with oxygen-rich water (e.g., below peat) iron hydroxides precipitate. Bacteria may also play a role in this. In many regions such concretions were historically important from the Iron Age to the beginning of industrialization. For example, the Vikings obtained most of their iron from the moors of Scandinavia, which had the added advantage of providing the peat as fuel for smelting.

In semiarid or arid climates where evaporation is high soil solutions rise due to capillary action. Dissolved substances are concentrated and finally precipitated when water evaporates in the pore space. The main duricrusts found here are calcretes, gypcretes, and salcretes. Nodular concretions are the first to form growing and cementing the loose soil. Roots often become encrusted and plant growth in the soil is inhibited. This process is often not well understood. It not only hardens the soil and coats the roots, but also enriches salts in the topsoil and thus makes the soil extremely difficult for plant growth. Soil formations influenced by water such as duricrusts are not limited to Earth. In the Eagle Crater of Meridiani Planum the Mars rover *Opportunity* discovered rock sequences with concretions of gypsum and iron oxides that can be interpreted as duricrusts (Squyres et al. 2004).

Calcretes are duricrusts whose cementing mineral is calcite. They can form in either of the two ways mentioned above. To reiterate, either carbonate minerals dissolved in higher soil horizons precipitate at lower levels or water rising by evaporation precipitates its dissolved carbonate load near the surface. Uranium in some calcretes in arid areas has also been enriched into minable quantities. Examples are the deposits of Yelirrie (Western Australia) and Langer Heinrich

Namibia). Yelirrie was discovered in 1972 and is one of the largest uranium deposits in Australia. Various ecological concerns have so far prevented mining of the deposit. The uranium-rich horizon is located in a paleofluvial bed through which groundwater still flows today between 4 and 8 m below ground level in the lower part of the calcrete and in the upper part of underlying clay and quartz-rich sediments. Nearby there is another calcrete deposit on the edge of a salt lake. The ore mineral is carnotite ($K(UO_2)_2(VO_4)_2 \cdot 3H_2O$) cemented with calcite, iron oxides, salt, and other minerals. The weathering of granite and greenschist caused dissolved uranium and vanadium to enter sediments of the valley. This type of deposit is particularly interesting because uranium mineralization here happened under oxidized conditions— not by reduction as usual. Evaporation probably led to enrichment of uranium and vanadium, oxidation from V^{4+} to V^{5+}, and destabilization of uranyl complexes (Mann and Deutscher 1978; Carlisle 1983; Hou et al. 2007).

Literature

Al-Farraj, A. 2005. An evolutionary model for sabkha development on the north coast of the UAE. *Journal of Arid Environments* 63: 740–755.

Aleva, G.J.J. 1994. *Laterites: Concepts, geology, morphology and chemistry.* International Soil Reference and Information Centre (ISRIC).

Allen, P.A., and J.L. Etienne. 2008. Sedimentary challenge to Snowball Earth. *Nature Geoscience* 1: 817–825.

Anonymous, 1908. *The Welcome Stranger—biggest nugget known.* NZ Truth, S. 8.

Anonymous, 2010. Barnett to open Boddington Gold Mine—ABC News (Australian Broadcasting Corporation), http://www.abc.net.au/news/2010-02-03/barnett-to-open-boddington-gold-mine/320538. Accessed 2 Apr 2013.

Anonymous, 2011. *Nachnutzungskonzept Pumpspeicherkraftwerk,* 2–4. Gesteins-Perspektiven.

Anonymous, 2012. Boddington Gold Mine (BGM), Western Australia (WA)—Mining Technology. http://www.mining-technology.com/projects/boddington. Accessed 2 Apr 2013.

Atkinson, H. and M. Hale, 1993. *Phosphate production in Central and Southern Africa, 1900–1992.* Minerals Industry International, September, 22–30.

Bao, Z., and Z. Zhao. 2008. Geochemistry of mineralization with exchangeable REY in the weathering crusts of granitic rocks in South China. *Ore Geology Reviews* 33: 519–535.

Barbara Rohstoffbetriebe GmbH, 1991. Grube Wohlverwahrt-Nammen.

Bárdossy, G. 1982. *Karst bauxites. Bauxite deposits on carbonate rock.* Amsterdam, Netherlands: Elsevier.

Barifaijo, E., 2001. The petrology of the volcanic rocks of Uganda. In *GSU Newsletter, 1. Presented at the regional conference on basement geology, groundwater, mineral resources, and mining related environmental problems in Eastern Africa,* 58–59. Kampala, Uganda: Geological Society of Uganda.

Baturin, G.N. 2000. *Phosphorites on the sea floor: Origin composition and distribution.* New York: Elsevier.

Bechtel, A., Y.-N. Shieh, W.C. Elliott, S. Oszczepalski, and S. Hoernes. 2000. Mineralogy, crystallinity and stable isotopic composition of illitic clays within the Polish Zechstein basin: Implications for the genesis of Kupferschiefer mineralization. *Chemical Geology* 163: 189–205.

Bechtel, A., R. Gratzer, W. Püttmann, and S. Oszczepalski. 2001a. Variable alteration of organic matter in relation to metal zoning at the Rote Fäule front (Lubin-Sieroszowice mining district, SW Poland). *Organic Geochemistry* 32: 377–395.

Bechtel, A., Y. Sun, W. Püttmann, S. Hoernes, and J. Hoefs. 2001b. Isotopic evidence for multi-stage base metal enrichment in the Kupferschiefer from the Sangerhausen Basin, Germany. *Chemical Geology* 176: 31–49.

Bechtel, A., R. Gratzer, W. Püttmann, and S. Oszczepalski. 2002. Geochemical characteristics across the oxic/anoxic interface (Rote Fäule front) within the Kupferschiefer of the Lubin-Sieroszowice mining district (SW Poland). *Chemical Geology* 185: 9–31.

Bekker, A., J.F. Slack, N. Planavsky, B. Krapez, A. Hofmann, K.O. Konhauser, and O.J. Rouxel. 2010. Iron formation: The sedimentary product of a complex interplay among mantle, tectonic, oceanic, and biospheric processes. *Economic Geology* 105: 467–508.

Bell, K., and J. Blenkinsop. 1987. Nd and Sr isotopic compositions of East African carbonatites: Implications for mantle heterogeneity. *Geology* 15: 99–102.

Belykh, V.I., E.I. Dunai, and I.P. Lugovaya. 2007. Physicochemical formation conditions of banded iron formations and high-grade iron ores in the region of the Kursk Magnetic Anomaly: Evidence from isotopic data. *Geology of Ore Deposits* 49: 159–177.

Beukes, N.J., H. Dorland, J. Gutzmer, M. Nedachi, and H. Ohmoto. 2002. Tropical laterites, life on land, and the history of atmospheric oxygen in the Paleoproterozoic. *Geology* 30: 491–494.

BGR, n.d. Erkundungsstandort Gorleben. http://www.bgr.bund.de/DE/Themen/Endlagerung/Endlagerstandorte/Gorleben/gorleben_node.html. (Abgerufen Mai 2013).

Bluhm, H. 2001. Re-establishment of an abyssal megabenthic community after experimental physical disturbance of the seafloor. *Deep Sea Research Part II: Topical Studies in Oceanography* 48: 3841–3868.

Borowski, C. 2001. Physically disturbed deep-sea macrofauna in the Peru Basin, southeast Pacific, revisited 7 years after the experimental impact. *Deep Sea Research Part II: Topical Studies in Oceanography* 48: 3809–3839.

Bowell, R.J., R.P. Foster, and A.P. Gize. 1993. The mobility of gold in tropical rain forest soils. *Economic Geology* 88: 999–1016.

Brocks, J.J., G.A. Logan, R. Buick, and R.E. Summons. 1999. Archean molecular fossils and the early rise of eukaryotes. *Science* 285: 1033–1036.

Buick, R. 2008. When did oxygenic photosynthesis evolve? *Philosophical Transactions of the Royal Society B: Biological Sciences* 363: 2731–2743.

Butler, G.P. 1969. Modern evaporite deposition and geochemistry of coexisting brines, the sabkha, Trucial Coast, Arabian Gulf. *Journal of Petrology* 39: 70–89.

Byerly, G.R., D.R. Lower, and M.M. Walsh. 1986. Stromatolites from the 3,300–3,500-Myr Swaziland Supergroup, Barberton Mountain Land, South Africa. *Nature* 319: 489–491.

Cabral, A.R., M. Radtke, F. Munnik, B. Lehmann, U. Reinholz, H. Riesemeier, M. Tupinamba, and R. Kwitko-Ribeiro. 2011. Iodine in alluvial platinum-palladium nuggets: Evidence for biogenic precious-metal fixation. *Chemical Geology* 281: 125–132.

Cailteux, J.L.H., A.B. Kampunzu, C. Lerouge, A.K. Kaputo, and J.P. Milesi. 2005. Genesis of sediment-hosted stratiform copper-cobalt deposits, central African Copperbelt. *Journal of African Earth Sciences* 42: 134–158.

Carlisle, D. 1983. Concentration of uranium and vanadium in calcretes and gypcretes. *Geological Society, London, Special Publications* 11: 185–195.

Chowdhury, M.R., V. Venkatesh, M.A. Anandalwar and D.K. Paul, 1965. *Recent concepts on the origin of Indian laterite*. Memoirs of the Geological Survey of India A 31.

Colin, F., P. Lecomte, and B. Boulange. 1989. Dissolution features of gold particles in a lateritic profile at Dondo Mobi, Gabon. *Geoderma* 45: 241–250.

Crerar, D.A., and H. Barnes. 1974. Deposition of deep-sea manganese nodules. *Geochimica et Cosmochimica Acta* 38: 279–300.

Cronan, D.S. 2000. *Handbook of marine mineral deposits. Marine science series*. Boca Raton, FL: CRC Press.

Dahanayake, K. and W. Krumbein, 1986. *Microbial structures in oolitic iron formations*. Mineralium Deposita 21.

Dalvi, A.D., W.G. Bacon, R.C. Osborne, 2004. The past and the future of nickel laterites. In *PDAC 2004 international convention, trade show & investors exchange*, 7–10. Toronto: The Prospectors and Developers Association of Canada.

Dambeck, H. 2012. Osmosekraftwerk: Grüner Strom aus süßem Wasser—Spiegel Online. http://www.spiegel.de/wissenschaft/technik/osmosekraftwerke-liefern-oekostrom-aus-salzwasser-und-suesswasser-a-823820.html. Accessed 13 Apr 2013.

Davies, K.A. 1947. The phosphate deposits of the Eastern Province, Uganda. *Economic Geology* 42: 137–146.

Decrée, S., E. Deloule, T. De Putter, S. Dewaele, F. Mees, J. Yans, and C. Marignac. 2011. SIMS U-Pb dating of uranium mineralization in the Katanga Copperbelt: Constraints for the geodynamic context. *Ore Geology Reviews* 40: 81–89.

DeDuve, C., and I. Hausser-Siller. 1994. *Ursprung des Lebens: Präbiotische Evolution und die Entstehung der Zelle*. Heidelberg: Spektrum Akadademischer Verlag.

Delaney, M.L. 1998. Phosphorus accumulation in marine sediments and the oceanic phosphorus cycle. *Global Biogeochemical Cycles* 12: 563–572.

De Putter, T., F. Mees, S. Decrée, and S. Dewaele. 2010. Malachite, an indicator of major Pliocene Cu remobilization in a karstic environment. (Katanga, Democratic Republic of Congo). *Ore Geology Reviews* 38: 90–100.

Dèry, P. and B. Anderson, 2007. *Peak phosphorus*. Energy Bulletin.

Dosseto, A., S.P. Turner, and J. Chappell. 2008. The evolution of weathering profiles through time: New insights from uranium-series isotopes. *Earth and Planetary Science Letters* 274: 359–371.

Duggen, S., K. Hoernle, P. van den Bogaard, L. Rüpke, and J. P. Morgan. 2003. Deep roots of the Messinian salinity crisis. *Nature* 422: 602–606.

Duggen, S., K. Hoernle, P. van den Bogaard, and D. Garbe-Schönberg. 2005. Post collisional transition from subduction- to Intraplate-type magmatism in the westernmost Mediterranean: Evidence for continental-edge delamination of subcontinental lithosphere. *Journal of Petrology* 46: 1155–1201.

Ehrenreich, A., and F. Widdel. 1994. Anaerobic oxidation of ferrous iron by purple bacteria, a new type of phototrophic metabolism. *Applied Environmental Microbiology* 60: 4517–4526.

El Desouky, H.A., P. Muchez, and J. Cailteux. 2009. Two Cu-Co sulfide phases and contrasting fluid systems in the Katanga Copperbelt, Democratic Republic of Congo. *Ore Geology Reviews* 36: 315–332.

Emmerich, M., 2013. Paläontologie: Eiserne Spuren urzeitlicher Mikroben. Spektrum.de. http://www.spektrum.de/alias/palaeontologie/eiserne-spuren-urzeitlicher-mikroben/1192103. Accessed 24 Apr 2013.

Evans, R.K. 2008. *An abundance of lithium*. Santiago: World Lithium.

Force, E.R., and W.F. Cannon. 1988. Depositional model for shallow-marine manganese deposits around black shale basins. *Economic Geology* 83: 93–117.

Frimmel, H.E. 2002. Genesis of the World's largest gold deposits. *Science* 297: 1815–1817.

Frimmel, H.E. 2005. Archaean atmospheric evolution: Evidence from the Witwatersrand gold fields, South Africa. *Earth-Science Reviews* 70: 1–46.

Frimmel, H.E. 2008. Earth's continental crustal gold endowment. *Earth and Planetary Science Letters* 267: 45–55.

Garrels, R.M., and C.L. Christ. 1965. *Solutions, minerals, and equilibria*. New York: Harper & Row.

Germann, K., 1981. Phosphat-Gesteine. Lagerstätten der Steine, Erden und Industrieminerale, Vademecum. GDMB Verlag Chemie, 159–165.

Gilbert, N. 2009. The disappearing nutrient. *Nature* 461: 716–718.

Golightly, J.P. 1979. Nickeliferous laterites: a general description. In *international laterite symposium, New Orleans, Society of Mining Engineers*, 38–56. American Institute of Mining, Metallurgical, and Petroleum Engineers.

Grace, H. 1991. Investigations in Kenya and Malawi using as-dug laterite as bases for bituminous surfaced roads. *Geotechechnical and Geological Engineering* 9: 183–195.

Graham, R.C., A.M. Rossi, and K.R. Hubbert. 2010. Rock to regolith conversion: Producing hospitable substrates for terrestrial ecosystems. *GSA Today* 20: 4–9.

Grotzinger, J.P., and D.H. Rothman. 1996. An abiotic model for stromatolite morphogenesis. *Nature* 383: 423–425.

Harder, H. 1989. Mineral genesis in Ironstones: a model based upon laboratory experiments and petrographic observations. In *Phanerozoic Ironstones*, 9–18. Geological Society Special Publication.

Hardisty, J. 1990. *Beaches: Form & process: Numerical experiments with monochromatic waves on the orthogonal profile*. London, Boston: Unwin Hyman.

Haubold, H., G. Katzung, and G. Schaumberg. 2006. *Die Fossilien des Kupferschiefers: Pflanzen- und Tierwelt zu Beginn des Zechsteins; eine Erzlagerstätte und ihre Paläontologie*. Hohenwarsleben: Westarp-Wissenschaften.

Heinrich D., M. Holland and M. Schidlowski, 1982. *Mineral deposits and the evolution of the biosphere*. Berlin, Heidelberg.

Hoashi, M., D.C. Bevacqua, T. Otake, Y. Watanabe, A.H. Hickman, S. Utsunomiya, and H. Ohmoto. 2009. Primary haematite formation in an oxygenated sea 3.46 billion years ago. *Nature Geoscience* 2: 301–306.

Hoffman, P.F., and D.P. Schrag. 2002. The snowball Earth hypothesis: Testing the limits of global change. *Terra Nova* 14: 129–155.

Holland, H.D. 2002. Volcanic gases, black smokers, and the great oxidation event. *Geochimica et Cosmochimica Acta* 66: 3811–3826.

Holland, H.D., and M. Schidlowski (eds.). 1982. *Mineral deposits and the evolution of the biosphere*. Berlin: Springer.

Horstmann, U.E., D.H. Cornell, B.J. Fryer, R. Scheepers, and F. Walraven. 2001. Rare earth elements and Nd isotopic compositions in banded iron-formations of the Griqualand West Sequence, Northern Cape Province, South Africa. *Zeitschrift der Deutschen Gesellschaft für Geowissenschaften* 152: 439–465.

Hou, B., A.J. Fabris, J.L. Keeling, and M.C. Fairclough. 2007. Cenozoic palaeochannel-hosted uranium and current exploration methods, South Australia. *Mesa Journal* 46: 34–39.

Hough, R.M., C.R.M. Butt, S.M. Reddy, and M. Verrall. 2007. Gold nuggets: Supergene or hypogene? *Australian Journal of Earth Sciences* 54: 959–964.

James, H.L. 1954. Sedimentary facies of iron-formation. *Economic Geology* 49: 253–293.

Jorgenson, J.D. 2012. *World Mine Production and Reserves—Iron Ore*. USGS.

Kappler, A., C. Pasquero, K.O. Konhauser, and D.K. Newman. 2005. Deposition of banded iron formations by anoxygenic phototrophic Fe(II)-oxidizing bacteria. *Geology* 33: 865–868.

Kasting, J.F. 1987. Theoretical constraints on oxygen and carbon dioxide concentrations in the Precambrian atmosphere. *Precambrian Research* 34: 205–229.

Kesler, S.E., P.W. Gruber, P.A. Medina, G.A. Keoleian, M.P. Everson, and T.J. Wallington. 2012. Global lithium resources: Relative importance of pegmatite, brine and other deposits. *Ore Geology Reviews* 48: 55–69.

Kimberley, M.M. 1980. The Paz de Rio oolitic inland-sea iron formation. *Economic Geology* 75: 97–106.

Kimberley, M.M. 1989. Exhalative origins of iron formations. *Ore Geology Reviews* 5: 13–145.

Kirk, J., J. Ruiz, J. Chesley, S. Titley, and J. Walshe. 2001. A detrital model for the origin of gold and sulfides in the Witwatersrand basin based on Re-Os isotopes. *Geochimica et Cosmochimica Acta* 65: 2149–2159.

Kirk, J., J. Ruiz, J. Chesley, J. Walshe, and G. England. 2002. A major Archean, gold- and crust-forming event in the Kaapvaal Craton, South Africa. *Science* 297: 1856–1858.

Klein, C. 2005. Some Precambrian banded iron-formations (BIFs) from around the world: Their age, geologic setting, mineralogy, metamorphism, geochemistry, and origins. *American Mineralogist* 90: 1473–1499.

Köhler, I., K. Konhauser, and A. Kappler. 2010. Role of Microorganisms in Banded Iron Formations. In *Geomicrobiology: Molecular and environmental perspective*, ed. L.T. Barton, M. Mandl, and A. Loy. Heidelberg: Springer.

Köhler, I., K.O. Konhauser, D. Papineau, A. Bekker and A. Kappler, 2013. Biological carbon precursor to diagenetic siderite with spherical structures in iron formations. *Nature Communications* 4.

Konhauser, K.O., T. Hamade, R. Raiswell, R.C. Morris, F.G. Ferris, G. Southam, and D.E. Canfield. 2002. Could bacteria have formed the Precambrian banded iron formations? *Geology* 30: 1079–1082.

Konhauser, K.O., D.K. Newman, and A. Kappler. 2005. The potential significance of microbial Fe(III) reduction during deposition of Precambrian banded iron formations. *Geobiology* 3: 167–177.

Konhauser, K.O., L. Amskold, S.V. Lalonde, N.R. Posth, A. Kappler, and A. Anbar. 2007. Decoupling photochemical Fe(II) oxidation from shallow-water BIF deposition. *Earth and Planetary Science Letters* 258: 87–100.

Krapez, B., M.E. Barly, and A.L. Pickard. 2003. Hydrothermal and resedimented origins of the precursor sediments to banded iron formation: Sedimentological evidence from the Early Palaeoproterozoic Brockman Supersequence of Western Australia. *Sedimentology* 50: 979–1011.

Krauskopf, K.B. 1957. Separation of manganese from iron in sedimentary processes. *Geochimica et Cosmochimica Acta* 12: 61–84.

Kucha, H., and W. Przylowicz. 1999. Noble metals in organic matter and clay-organic matrices, Kupferschiefer, Poland. *Economic Geology* 94: 1137–1162.

Kucha, H., and M. Pawlikowski. 1986. Two-brine model of the genesis of strata-bound Zechstein deposits (Kupferschiefer type), Poland. *Mineralium Deposita* 21: 70–80.

Kühne, W.G. 1976. Goldtransport durch Inlandeis. Dem Andenken von Egon Erwin Kisch (1885–1948) gewidmet. *Der Aufschluss* 27: 165–169.

Kühne, W.G. 1983. Gold für uns aus der Kiesgrube. *Der Aufschluss* 34: 215–218.

Langer, E. 1969. *Die Nickellagerstätte des Morro do Niquel in Minas Gerais, Brasilien: ihr Aufschluss, ihre Bemusterung und Bewertung*. Borntraeger: Gebr.

Lascelles, D.F. 2007. Black smokers and density currents: A uniformitarian model for the genesis of banded iron-formations. *Ore Geology Reviews* 32: 381–411.

Lee Bray, E. 2012. *Bauxite and Alumina. Geological Survey*. USA: Mineral Commodity Summaries.

Liedtke, M., and J. Vasters. 2008. *Renaissance des deutschen Kupferschieferbergbaus?*, 29. Commodity Top News: Bundesamt für Geologie und Rohstoffe.

Lierl, H.-J., and W. Jans. 1990. Geschiebegold aus Schleswig-Holstein. *Geschiebekunde aktuell* 6 (47): 49–57.

Lottermoser, B.G. 1990. Rare-earth element mineralisation within the Mt. Weld carbonatite laterite. *Western Australia. Lithos* 24: 151–167.

Louthean Publishing (ed.), 2004. *The Australian mines handbook 2003/04* edition 71.

Lowe, D.R. 1980. Stromatolites 3,400-Myr old from the Archean of Western Australia. *Nature* 284: 441–443.

Machens, E. 2011. *Hans Merensky—Geologe und Mäzen: Platin*. Schweizerbart, Stuttgart: Gold und Diamanten in Afrika.

Mann, A.W. 1984. Mobility of gold and silver in lateritic weathering profiles; some observations from Western Australia. *Economic Geology* 79: 38–49.

Mann, A.W., and R.L. Deutscher. 1978. Genesis principles for the precipitation of carnotite in calcrete drainages in Western Australia. *Economic Geology* 73: 1724–1737.

McCuaig, T.C., M. Behn, H. Stein, S.G. Hagemann, N.J. McNaughton, K.F. Cassidy, D. Champion and L. Wyborn, 2001. The Boddington gold mine: a new style of Archaean Au-Cu deposit. In *Fourth International Archaean Symposium, Extended Abstracts*, 453–455.

Meier, C. 2010. Rohstoffe: Bevor der Dünger ausgeht—Spektrum.de http://www.wissenschaft-online.de/artikel/1024445%26_z=859070. Accessed 20 Mar 2013.

Meyer, F.M., U. Happel, J. Hausberg, and A. Wiechowski. 2002. The geometry and anatomy of the Los Pijiguaos bauxite deposit, Venezuela. *Ore Geology Reviews* 20: 27–54.

Minter, A.H.G. 1978. A sedimentological synthesis of placer gold, uranium and pyrite concentrations in Proterozoic Witwatersrand deposits. In *Fluvial Sedimentology*, ed. A.D. Miall, 5, 801–829. Canadian Society for Petroleum Geology, Memoir.

Morris, R.C. 1985. Genesis of iron ore in banded iron-formation by supergene and supergene-metamorphic processes—a conceptual model. In *Handbook of strata-bound and stratiform ore deposits*, ed. K.H. Wolf, 13, 73–235. Amsterdam: Elsevier.

Morris, R.C. 2002. Genesis of high-grade hematite orebodies of the Hamersley Province, Western Australia—a discussion. *Economic Geology* 97: 177–181.

Ochsenius, C. 1877. *Die Bildung der Steinsalzlager und ihrer Mutterlaugensalze unter specieller Berücksichtigung der Flötze von Douglashall in der egeln'schen Mulde*. Pfeffer, Halle: C. E. M.

Oftedahl, C. 1958. A theory of exhalative-sedimentary ores. *Geologiska Föreningen i Stockholm Förhandlingar* 80: 1–19.

Oszczepalski, S. 1999. Origin of the Kupferschiefer polymetallic mineralization in Poland. *Mineralium Deposita* 34: 599–613.

Pašava, J., S. Oszczepalski, and A. Du. 2010. Re-Os age of non-mineralized black shale from the Kupferschiefer, Poland, and implications for metal enrichment. *Mineralium Deposita* 45: 189–199.

Petrascheck, W.E. 1989. The genesis of allochthonous karst-type bauxite deposits of Southern Europe. *Mineralium Deposita* 24: 77–81.

Pickard, A.L. 2002. SHRIMP U-Pb zircon ages of tuffaceous mudrocks in the Brockman Iron Formation of the Hamersley Range, Western Australia. *Australian Journal of Earth Sciences* 49: 491–507.

Pickard, A.L., M.E. Barley, and B. Krapez. 2004. Deep-marine depositional setting of banded iron formation: Sedimentological evidence from interbedded clastic sedimentary rocks in the early Palaeoproterozoic Dales Gorge Member of Western Australia. *Sedimentary Geology* 170: 37–62.

Piestrzynski, A., J. Pieczonka, and A. Gluszek. 2002. Redbed-type gold mineralisation, Kupferschiefer, south-west Poland. *Mineralium Deposita* 37: 512–528.

Planavsky, N., O. Rouxel, A. Bekker, R. Shapiro, P. Fralick, and A. Knudsen. 2009. Iron-oxidizing microbial ecosystems thrived in late Paleoproterozoic redox-stratified oceans. *Earth and Planetary Science Letters* 286: 230–242.

Pohl, W.L. 2005. Mineralische und Energie-Rohstoffe. 5. Ed. Schweizerbart'sche Verlagsbuchhandlung, Stuttgart.

Porrenga, D. 1967. Glauconite and chamosite as depth indicators in the marine environment. *Marine Geology* 5: 495–501.

Posth, N.R., K.O. Konhauser, and A. Kappler. 2008. Alternating Si and Fe deposition caused by temperature fluctuations in Precambrian oceans. *Nature Geoscience* 10: 703–708.

Poulton, S.W., P.W. Fralick, and D.E. Canfield. 2010. Spatial variability in oceanic redox structure 1.8 billion years ago. *Nature Geoscience* 3: 486–490.

Preidl, M., and M. Metzler. 1984. The sedimentation of copper-bearing shales (Kupferschiefer) in the Sudetic foreland. *Mineralium Deposita* 19: 243–248.

Reedman, J.H. 1984. Resources of phosphate, niobium, iron, and other elements in residual soils over the Sukulu carbonatite complex, southeastern Uganda. *Economic Geology* 79: 716–724.

Reith, F., S.L. Rogers, D.C. McPhail, and D. Webb. 2006. Biomineralization of gold: Biofilms on bacterioform gold. *Science* 313: 233–236.

Richter-Bernburg, G. 1953. Über salinare Sedimentation. *Zeitschrift der Deutschen Gesellschaft für Geowissenschaften* 105: 593–645.

Ries, G. 2001. Lateritische Nickellagerstätten in Neu Kaledonien. *Der Aufschluss* 52: 79–83.

Ries, G. 2007. *Die Entwicklung der Erdatmosphäre. Der Aufschluss* 58: 217–226.

Ries, G. 2010. Die Entwicklungsgeschichte der Erdatmosphäre und ihres Sauerstoffgehaltes. *Bergbau* 61: 109–118.

Risacher, F., H. Alonso, and C. Salazar. 2003. The origin of brines and salts in Chilean salars: A hydrochemical review. *Earth-Science Reviews* 63: 249–293.

Robb, L.J., and F.M. Meyer. 1995. The Witwatersrand Basin, South Africa: Geological framework and mineralization processes. *Ore Geology Review* 10: 67–94.

Santosh, M., and P.K. Omana. 1991. Very high purity gold form lateritic weathering profiles of Nilambur, Southern India. *Geology* 19: 746–749.

Sawlowicz, Z. 1989. On the origin of copper mineralization in the Kupferschiefer: A sulphur isotope study. *Terra Nova* 1 (4): 339–343.

Schellmann, W. 1983. Geochemical principles of lateritic nickel ore formation. In *Proceedings of the international seminar of laterisation processes*, 2o, 119–135. São Paulo.

Schlüter, T. 1991. Systematik, Palökologie und Biostratonomie von Phalacrocorax kuehnaeus nov. spec., einem fossilen Kormoran (Aves: Phalacrocoracidae) aus mutmaßlich oberpliozänen Phosphoriten N Tansanias. *Berliner Geowissenschaftliche Abhandlungen. A* 134: 279–309.

Schoettle, M., and G.M. Friedmann. 1971. Fresh Water Iron-Manganese Nodules in Lake George, New York. *Geological Society of America Bulletin* 82: 101–110.

Schultz, L. 1993. *Planetologie: eine Einführung*. Basel, Boston: Birkhäuser Verlag.

Simonson, B.M. 1985. Sedimentological constraints on the origins of Precambrian iron-formations. *Geological Society of America Bulletin* 96: 244–252.

Slack, J.F., T. Grenne, A. Bekker, O.J. Rouxel, and P.A. Lindberg. 2007. Suboxic deep seawater in the late Paleoproterozoic: Evidence from hematitic chert and iron formation related to seafloor-hydrothermal sulfide deposits, central Arizona, USA. *Earth and Planetary Science Letters* 255: 243–256.

Slack, J.F., and W.F. Cannon. 2009. Extraterrestrial demise of banded iron formations 1.85 billion years ago. *Geology* 37: 1011–1014.

Smirnov, V.I. 1989. European part of the USSR. *Mineral Deposits of Europe* 4: 279–407.

Sorby, H.C. 1857. On the origin of the Cleveland Hill ironstone. *Geol. Polytechnic. Soc. West Riding Yorkshire Proc.* 3: 457–461.

Squyres, S.W., J.P. Grotzinger, R.E. Arvidson, J.F. Bell, W. Calvin, P. R. Christensen, B.C. Clark, J.A. Crisp, W.H. Farrand, K.E. Herkenhoff, J.R. Johnson, G. Klingelhöfer, A.H. Knoll, S.M. McLennan, H.Y. McSween, R.V. Morris, J.W. Rice, R. Rieder, and L.A. Soderblom. 2004. In Situ Evidence for an Ancient Aqueous Environment at Meridiani Planum, Mars. *Science* 306: 1709–1714.

Sun, Y.-Z., and W. Püttmann. 2000. The role of organic matter during copper enrichment in Kupferschiefer from the Sangerhausen basin, Germany. *Organic Geochemistry* 31: 1143–1161.

Strunz, H., 2001. Strunz mineralogical tables: chemical-structural mineral classification system. 9. Ed. Schweizerbart'sche Verlagsbuchhandlung, Stuttgart.

Talbot, C.J., and V. Pohjola. 2009. Subaerial salt extrusions in Iran as analogues of ice sheets, streams and glaciers. *Earth-Science Reviews* 97: 155–183.

Tardy, Y. 1997. *Petrology of laterites and tropical soils*. Rotterdam, Netherlands; Brookfield, VT, USA: A. A. Balkema.

Taylor, D., H.J. Dalstra, A.E. Harding, G.C. Broadbent, and M.E. Barley. 2001. Genesis of high-grade hematite orebodies of the Hamersley Province, Western Australia. *Economic Geology* 96: 837–873.

Taylor, D., H.J. Dalstra, and A.E. Harding. 2002. Genesis of high-hrade hematite orebodies of the Hamersley Province, Western Australia—a reply. *Economic Geology* 97: 179–181.

Thiel, H. 2001. Evaluation of the environmental consequences of polymetallic nodule mining based on the results of the TUSCH Research Association. *Deep Sea Research Part II: Topical Studies in Oceanography* 48: 3433–3452.

Thiel, H., G. Schriever, A. Ahnert, H. Bluhm, C. Borowski, and K. Vopel. 2001. The large-scale environmental impact experiment DISCOL—reflection and foresight. *Deep Sea Research Part II: Topical Studies in Oceanography* 48: 3869–3882.

Towe, K.M. 1996. Environmental oxygen conditions during the origin and early evolution of life. *Advances in Space Research* 18: 7–15.

Trechow, P. 2011. Lithium—ein Spannungsmacher auf Kreislaufkurs. ingenieur.de. http://www.ingenieur.de/Themen/Rohstoffe/Lithium-Spannungsmacher-Kreislaufkurs. Accessed 18 Apr 2013.

Troly, G., M. Esterle, B. Pelletier and W. Reibell, 1979. Nickel deposits in New Caledonia: some factors influencing their formation. In *Proceedings of the international symposium of lateritisation processes,* 81–119. New Orleans.

Valayashko, M.G. 1958. Die wichtigsten geochemischen Parameter für die Bildung der Kalisalzlagerstätten. *Freiburger Forschungshefte* A123: 197–233.

Valeton, I., M. Biermann, R. Reche, and F. Rosenberg. 1987. Genesis of nickel laterites and bauxites in greece during the jurassic and cretaceous, and their relation to ultrabasic parent rocks. *Ore Geology Reviews* 2: 359–404.

Van de Kerkhof, S. 2002. In: *Die Industrialisierung europäischer Montanregionen im 19. Jahrhundert*, ed. T. Pierenkemper, 225–275. Franz Steiner Verlag.

Van Straaten, P. 2002. *Rocks for crops: Agrominerals of Sub-Saharan Africa*. Nairobi, Kenya: ICAF.

van Kauwenbergh, S.J. 1991. Overview of phosphate deposits in East and Southeast Africa. *Fertilizer Research* 30: 127–150.

Van Wyk, P. and l. F. Pienaar, 1986. Diamondiferous gravels of the lower Orange River, Namaqualand, In *Mineral Deposits of Southern Africa*, 2.173–2.191. Johannesburg: Geological Society of South Africa.

Vaughan, D.J., M.A. Sweeney, G. Friedrich, R. Diedel, and C. Haranczyk. 1989. The Kupferschiefer; an overview with an appraisal of the different types of mineralization. *Economic Geology* 84: 1003–1027.

Wagner, T., M. Okrusch, S. Weyer, J. Lorenz, Y. Lahaye, H. Taubald, and R. Schmitt. 2010. The role of the Kupferschiefer in the formation of hydrothermal base metal mineralization in the Spessart ore district, Germany: Insight from detailed sulfur isotope studies. *Mineralium Deposita* 45: 217–239.

Wang, X., and W.E.G. Müller. 2009. Marine biominerals: Perspectives and challenges for polymetallic nodules and crusts. *Trends in Biotechnology* 27: 375–383.

Wang, X., H.C. Schröder, M. Wiens, U. Schloßmacher, and W.E.G. Müller. 2009a. Manganese/polymetallic nodules: Micro-structural characterization of exolithobiontic and endolithobiontic microbial biofilms by scanning electron microscopy. *Micron* 40: 350–358.

Wang, Y., H. Xu, E. Merino, and H. Konishi. 2009b. Generation of banded iron formations by internal dynamics and leaching of oceanic crust. *Nature Geoscience* 2: 781–784.

Warren, J.K. 2010. Evaporites through time: Tectonic, climatic and eustatic controls in marine and nonmarine deposits. *Earth-Science Reviews* 98: 217–268.

Weber, K.A., L.A. Achenbach, and J.D. Coates. 2006. Microorganisms pumping iron: anaerobic microbial iron oxidation and reduction. *Nature Reviews Microbiology* 4: 752–764.

Whattam, S.A. 2009. Arc-continent collisional orogenesis in the SW Pacific and the nature, source and correlation of emplaced ophiolitic nappe components. *Lithos* 113: 88–114.

Widdel, S., S. Schnell, S. Heising, A. Ehrenreich, B. Assmus, and B. Schink. 1993. Ferrous iron oxidation by anoxygenic phototrophic bacteria. *Nature* 362: 834–836.

Yamaguchi, K. E. n.d. Iron isotope compositions of Fe-oxide as a measure of water-rock interaction: An example from Precambrian tropical laterite in Botswana. Frontier Research on Earth Evolution (IFREE Report for 2003–2004).

Further Reading

Guilbert, J.M., and C.F. Park. 1986. *The geology of ore deposits*. New York: WH Freeman.

Laznicka, P. 2010. *Giant Metallic Deposits: Future sources of industrial metals*, 2nd ed. Heidelberg: Springer.

Lohmann, D., and N. Podbregar. 2012. *Im Fokus: Bodenschätze*. Springer, Heidelberg: Auf der Suche nach Rohstoffen.

Misra, K.C. 2000. *Understanding mineral deposits*. Dordrecht, Niederlande: Kluwer Academic Publishers.

Okrusch, M., and S. Matthes. 2009. *Mineralogie: Eine Einführung in die spezielle Mineralogie, Petrologie und Lagerstättenkunde*, 8th ed. Heidelberg: Springer.

Pohl, W.L. 2011. *Economic Geology*. Chichester: Wiley-Blackwell.

Robb, L. 2005. *Introduction to ore-forming processes*. Malden, Massachussetts: Blackwell Science.

Rothe, P. 2010. *Schätze der Erde*. Darmstadt: Primus Verlag.

Seidler, C. 2012. *Deutschlands verborgene Rohstoffe: Kupfer*. Hanser, München: Gold und seltene Erden.

Fossil Fuels

Petroleum, natural gas, and coal will remain the most important energy sources for some time to come (see also Box 6.1) despite all the efforts to protect the climate. Fossil fuels store the energy that living creatures of bygone times extracted from sunlight through photosynthesis. This chapter deals with the biomass that has built up and that later was converted into other compounds. Algae are the most important in the case of petroleum, while land plants are the most important in the case of coal. The formation of natural gas is more diverse in that methane can not only be produced together with petroleum and coal, but also by single-celled organisms such as anoxic methanogenic archaea. The first step on the way to a deposit is the sedimentation of organic matter without it being immediately oxidized again. Sediment must then be covered to reach the depth at which organic matter is converted into petroleum, natural gas, or coal. Up to this point development of the three is relatively similar, but as things progress the different properties of solid, liquid, or gaseous substances is important.

Coal is a carbon-rich rock. Because the content of other components decreases at greater depths, the quality of coal becomes ever higher. Methane is also released during this process. Coal can be mined if it rises close to the Earth's surface through geological processes. By contrast, oil needs to remain at a certain depth until extraction—otherwise it will be decomposed by microorganisms. There are a number of other issues such as rocks in which oil is formed have very low permeability making its extraction almost impossible; oil formation must be followed by migration into permeable rock serving as a reservoir; and oil must reach a trap sealed by impermeable rock—not flow out at the surface.

Petroleum and natural gas consist predominantly of hydrocarbons (i.e., molecules composed only of carbon and hydrogen). In typical petroleum there are thousands of different molecules made up of different numbers of carbon atoms assembled in chains with or without branches, rings, or even regular nets. Some molecules also contain nitrogen, sulfur, or oxygen. There are very different types of petroleum in each of which different molecules dominate.

Hydrocarbons that have a maximum of four carbon atoms are gaseous under standard conditions. This can be different at the depth from which they were extracted. A large amount of methane can be dissolved in petroleum under pressure. Segregation occurs only when the pressure is relieved while the petroleum is being extracted in much the same way as happens with CO_2 when a mineral water bottle is opened. The volume of oil shrinks and the separated gas can have a volume one hundred times that of the petroleum. However, a gas phase at depth can also contain heavier molecules that condense on the surface during extraction. In general, methane (CH_4) is the main component of natural gas (75–99%). The remainder are other hydrocarbons such as ethane (C_2H_6), propane (C_3H_8), butane (C_4H_{10}), and other gases like H_2S, N_2, and CO_2. Wet gas has a high content of propane, butane, and possibly larger hydrocarbons that condense to a liquid during cooling (gas condensate). Dry gas contains almost no hydrocarbons other than methane. Natural gas with a high H_2S content must be desulfurized. CO_2 can be used for the production of dry ice. Natural gas plays an increasingly important role in the energy mix of Germany and the world as fuel. Its role in avoiding carbon dioxide emissions to protect the climate will certainly increase in the near future.

Finally, there are so-called unconventional deposits of fossil fuels. These include hard-to-extract deposits such as extra-heavy crude oil, shale gas, tar sands, and oil shale (normal deposits in polar regions and the deep sea are often also counted as unconventional deposits). They are present in far greater quantities than conventional hydrocarbons. Although extraction is very complex and expensive, they are already being extracted today in ever-increasing quantities. Methane hydrates are another unconventional resource contained in permafrost soils and in marine sediments at great depths.

Several competing units of quantity are used in the case of oil and gas, and conversion depends on type and density. Barrel is the volume measure often used for petroleum in which 1 bbl = 158.987 L. BBO is the abbreviation used for billion barrels of oil. Alternately, weight is widely given in metric tonnes. Gas is measured in cubic meters (at 0 °C, 101.325 kPa) and in the United States in cubic feet (at 16 °C, 101.560 kPa). Energy content given in gigajoules (GJ) or British thermal units (Btu) is often used instead of volume. Comparisons are also used as measures such as the energy content of 1 t of oil, a unit given as tonne of oil equivalent (toe) and in Germany with coal given as hard coal unit or Steinkohleeinheit (SKE).

Box 6.1 Beginning of the Age of Oil

The oil burnt in lamps and used as a lubricant in factories during the Industrial Revolution came from whales and had nothing to do with geology. Some countries maintained large fleets of whaling ships to hunt the marine mammals. Whale oil was cooked from whale bacon and baleen was processed into corsets. Coal was the most important source of energy right up to the beginning of the twentieth century. Steam engines allowed machines to be set in motion and railways facilitated the transport of coal.

The beginning of the Age of Oil is often stated as being marked by the 25-m-deep well drilled in 1859 by Edwin Drake in Pennsylvania (USA) on behalf of a small company—the well yielded almost 10 bbl daily. In the following years the first oil boom started. It took place in the United States where innumerable

wells were drilled (Fig. 6.1) in a number of states. Distilleries processed the petroleum into lamp oil and in so doing pushed whale oil out of the market. The most active distillery operator was John D. Rockefeller who founded the first large oil company called the Standard Oil Company. Within a decade annual production in the United States rose to 5 million barrels. Despite that sounding a lot and oil actually being transported in barrels, compared with today it is of course almost insignificant. Worldwide consumption is currently 3.7 million barrels per hour.

Historically, petroleum has long been used. In East Asia and the Middle East it was used in early times as a fuel and medicine. Crude oil was extracted long before Drake in Pechelbronn, Alsace (France), in Wietze, Lüneburger Heide (Germany), in Galicia (Poland), and in many other parts of the world. In addition to oil from natural petroleum seeps it was mainly sourced from pits dug by hand. In some cases the search for salt or water led to the accidental discovery of petroleum.

The oil of Azerbaijan was described by Marco Polo in the thirteenth century. He wrote that it was inedible, burnt well, and helped as an ointment against scabies and against camel furuncles. It was also here in 1846 that a Russian engineer successfully drilled for the first time using a percussion drill, a decade before Drake. The boom of Baku followed in 1872 with the buildup of countless drilling rigs and distilleries. In the following decades about half of world production came from Azerbaijan and the other half from the United

Fig. 6.1 Drilling rigs in midway oil field (California, USA) in 1910 (*photo* © W.C. Mendenhall, USGS)

States. Ludvig and Robert Nobel, brothers of the Nobel Prize founder, happened to be in Baku when the boom began. Although they were initially looking for cheap walnut timber for rifle production, instead they founded a large oil company and invented pipelines and tankers. At the beginning of the twentieth century the bay of the Caspian Sea of the suburb Bibi Eibat was reclaimed with infill only because offshore drilling was not yet possible.

At much the same time as the light bulb was replacing oil lamps the automobile appeared and in so doing rapidly increased the demand for petroleum. Other productive oil fields were found all over the world from Sumatra, Persia, Peru, to Venezuela and Mexico. Advances in science meant oil was not only found by chance, but also through detailed mapping and analysis of geological structures. After the First World War there were important finds in the Arab world and drilling technology was developed to such an extent that wells 3000 m deep became possible. The first offshore well was drilled in 1924 from a wooden platform near Baku. However, systematic development of offshore fields only began worldwide in the 1950s.

6.1 From Peat to Coal

Box 6.2 Coke

Coal is not suitable for smelting iron because too much smoke and sulfur is released during combustion. It is therefore first converted into coke in a coking plant. Coke is a porous fuel consisting almost entirely of carbon. Low-ash medium volatile bituminous coal is particularly suitable for this purpose. It is heated to more than 1000 °C in the absence of oxygen during which it melts and volatile components that make up about one-third escape in gaseous form. Coke oven gas contains hydrogen, methane and other hydrocarbons, nitrogen, carbon monoxide, hydrogen sulfide, and other components. Sulfur and aromatic hydrocarbons are separated and the remaining gas is burned to produce energy. The melt is quenched with water and then the resulting material is broken and sieved.

Box 6.3 Coal liquefaction

Coal is available in much greater quantities than petroleum. At times of oil shortages synthetic hydrocarbons can be produced from coal and serve as synthetic

petrol, diesel, and raw materials for the chemical industry. Although this may be more energy-intensive and more expensive, such an alternative means the depletion of oil reserves does not ring the death knell of our civilization. The Fischer–Tropsch process was developed back in 1925. Although it was used on a large scale during the Second World War in Germany, it was no longer competitive afterward. In this process coal is initially gasified at more than 1000 °C and converted by reaction with air and steam to a mixture of CO and H_2 called synthesis gas. Catalysts are used to produce various higher hydrocarbons.

In addition to being burned, coal is also converted into coke (Box 6.2) or hydrocarbons (Box 6.3). It is widespread worldwide in much greater quantities than oil and gas. The United States, Russia, China, Australia, and India have the largest reserves. Coals are carbon-rich sedimentary rocks mainly derived from plant remains. They were deposited as peat in swamps or bogs. As the overburden increased this resulted in soft coal, then hard coal, and finally anthracite—coal of ever-higher quality (Figs. 6.2 and 6.3). The terms Braunkohle and Steinkohle are German for soft and hard coal, respectively. However, in English the differentiation of soft and hard coal is less common and the definition rather vague. This is due to the fact that the transition between these coal types in coalification is continuous. However, intermediate coal types do not exist in Central Europe north of the Alps. The deposits here were formed in the Carboniferous and the Neogene. Hard coal formed in the older deposits and lignite (soft coal) in the younger deposits. There are finer subdivisions of coal types and different classification systems are used in different countries (Fig. 6.4).

Coalification is the process by which peat is transformed into coal of ever-higher rank (degree of coalification or maturity). The process can be described as the metamorphism of organic matter. As the overburden increases both pressure and temperature act on the substance. Molecules change in the process. From the original plant material comprising cellulose, pectin, and lignin (Box 6.4) larger, increasingly regular structures emerge resembling the crystal lattice of graphite (Sect. 7.16) despite the many lattice defects, gaps, and impurities. Moreover, the composition changes too. To start with much water is expelled, later methane is released, and other volatile substances escape. The relative proportion of carbon increases as does the calorific value (heating value). Coalification is accompanied by a significant reduction in volume. This begins by young peat on the surface transforming into mature peat resulting in the volume being reduced by three-quarters. Based on the maturity of peat the volume of lignite can again be reduced

Fig. 6.2 Woodlike lignite from Świdnica (Poland) (*photo* © F. Neukirchen/Mineralogical Collections of the TU Berlin)

Fig. 6.3 Low-volatility bituminous coal from the Heinrich Colliery, Essen, Ruhr (Germany) (*photo* © F. Neukirchen/Mineralogical Collections of the TU Berlin)

Rank (D)		Rank (USA)	R %	C waf %	H₂O %	heating value kcal/kg (MJ/kg)
Braunkohle	Torf	peat	0.2			
	Weich-braunkohle	lignite	0.3	60	75	
					35	4000 (16.8)
	Matt-braunkohle	sub-bitumous C	0.4	71	25	5500 (23.0)
	Glanz-braunkohle	B	0.5			
		A	0.6	77	8–10	7000 (29.3)
Steinkohle	Flammkohle	high volatile bitumous C B	0.7			
	Gasflamm-kohle		0.8			
			0.9			
	Gaskohle	A	1.0			
	Fettkohle	medium volatile bitumous	1.2 / 1.4	87		8650 (36.2)
	Esskohle	low volatile bitumous	1.6 / 1.8			
	Magerkohle	semi-anthracite	2.0			
	Anthrazit	anthracite	3.0	91		8650 (36.2)
	Metaanthrazit	meta-anthracite	4.0			

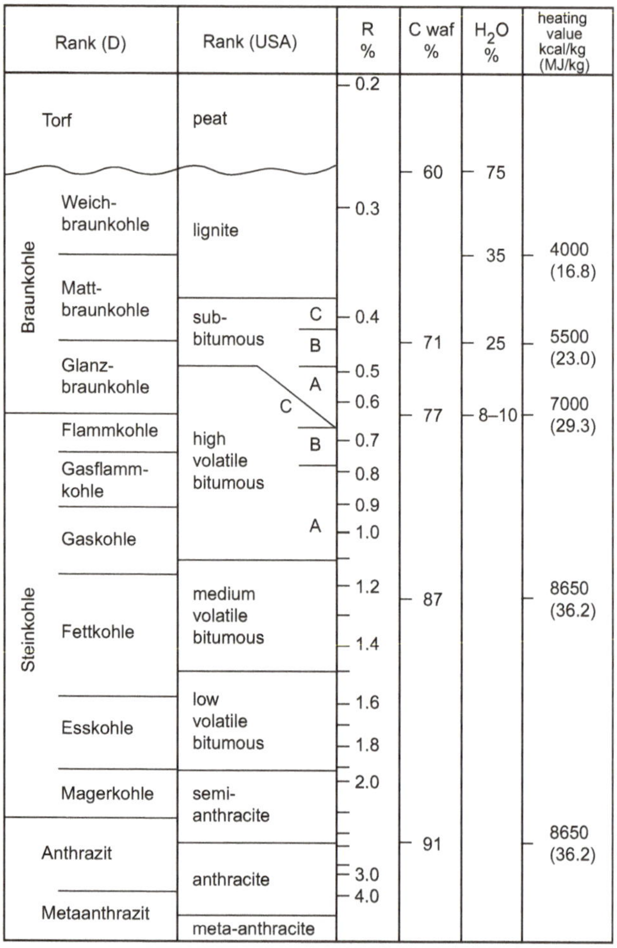

Fig. 6.4 Classification of coals in Germany and the United States. *R*, Vitrinite reflection (under oil) used to measure the degree of coalification; *C* waf, the carbon content of coal without water and ash (from Taylor et al. 1998; Pohl 2011)

to one-half and in the case of high-quality hard coal to one-tenth. A coal seam 1 m thick thus corresponds to a peat layer 10 m thick—typical of moors of today.

The most important factor in coalification is temperature, but not in the sense that a certain temperature can be assigned to a given coal rank. Also important is how long the layer was exposed to the highest temperature. In addition, hydrothermal fluids can intensify coalification (Hower and Gayer 2002) as can shearing movements. Thus, it often happens that the highest rank is not present in the deepest part of a basin. One example is the coal basin in Kentucky where hydrothermal fluids of Mississippi Valley–type (MVT) deposits (Sect. 4.12) are responsible for zones with the highest coal rank. Anthracite formed in many cases in the immediate vicinity of magmas.

Contrary to other rocks coal is composed of organic phases called macerals—not composed of minerals. There are three groups. Vitrinites (or huminites in lignite) can be traced back to woody plant material. They form the main component of most coals. They reflect light increasingly better as the degree of coalification increases. Since this is easy to measure it is therefore used for coal maturity (vitrinite reflection). Liptinites (a.k.a. exinites) are lipid-rich phases that can be traced back to spores, resins, algae, leaf skins, waxes, fats, and so on. Finally, inertinites comprise material that has already oxidized. These include fusinites in which wood fibers are still visible, but the material was oxidized before deposition in a forest fire. There are also fossil tree trunks that consist of fusinites on the outside and vitrinites on the inside.

Coal consists of dull and shiny bands that can be distinguished with the naked eye. These correspond to layers in which the aforementioned macerals are contained in different proportions. Different lithotypes are differentiated accordingly. Vitrain (bright coal) consists mainly of vitrinites. Durain (dull coal) has a high content of liptinites. Clarain (banded coal) is composed of a fine banding of alternating vitrain and durain. Fusain has a high content of fusinites.

In addition to normal humic coals formed from plant remains or peat there are the less common sapropelic coals originally deposited as sapropel (sludge rich in organic matter) and consisting mainly of liptinites including boghead coals derived from algae and cannel coals that burn like a candle and are mainly derived from spores. Sapropelic coals have a certain similarity to oil shales (Sect. 6.8) except that almost no clay minerals were deposited. A coal seam consists of areas with different coal types corresponding to deposition in different environments. Sapropelic coals typically occur in the upper part of a seam when a pond with stagnant, oxygen-poor water formed in a moor. Sometimes a seam is also covered by oil shale.

Coal also contains minor amounts of clastic material (e.g., clay minerals and sand). These remain after burning and form part of the ash. Low ash content is an important quality characteristic of coal. It is very low if the swampy environment in which it formed was not supplied with clastic material by rivers. This is particularly the case in zones far away from rivers within large swamps and in raised bogs. Low sulfur content is another factor contributing to quality. Sulfur content is large if the environment was influenced by seawater or brackish water since sulfate-reducing bacteria lead to the precipitation of sulfides in an oxygen-poor environment.

Low oxygen content during deposition is important because otherwise material will be oxidized by microorganisms or directly by oxygen. In swampy sediments such conditions are found below the water table because among other things organisms in the water quickly consume dissolved oxygen and pore water virtually stagnates. A pH value greatly reduced by organic acids restricts the activity of the microorganisms concerned. Additionally the groundwater level needs to correspond more or less to the surface all year round.

Fossils are relatively common in coal including tree stumps and roots that are still in their original position. The idea of Carboniferous and Permian tropical coal forests was developed as a result of such finds. However, peat formation was not restricted to tropical swamps comparable with the Everglades in Florida, with mangrove forests on tropical coasts, or with swampy forests on Borneo. Bogs at temperate latitudes also form peat (Figs. 6.5, 6.6 and 6.7) from which coal can form (Diessel 1992; Thomas 2012). Although plant growth and thus deposition are slower here (0.5–2.5 mm peat per year compared with up to 4 mm per year in the tropics; Volkov 2003), the oxidation of plant material is also less rapid. From time to time coal was also produced from wood that had washed up or allochthonous peat.

Deposition environments can thus be very different even within a given coal basin such as swampy shores of lakes and brackish lagoons, quiet areas between active branches of river deltas, floodplains in valleys, low-lying bogs, and raised bogs. Deposits formed at lakes are called limnic, while those formed at the coast of oceans are called paralic.

Since river courses and coastlines change over time so do depositional environments. A seam is therefore not always a layer deposited at a certain time in the entire basin but a result of temporally and spatially changing deposition conditions and environments. When sea level rises the coastline and thus peat deposition migrate inland; when sea level falls they migrate back again. This appears in a coal mine as two seams

Fig. 6.5 Moor on Tierra del Fuego (Argentina) (*photo* © F. Neukirchen/Blickwinkel)

Fig. 6.6 Peat extraction from a moor in Ireland. This is the starting material for coal formation (*photo* © Twicepix, CC-BY-SA, flickr.com)

interrupted by marine sediments (mostly shales and sandstones). Correspondingly, seams can lie at an angle to basin-wide marker horizons such as tuff layers. If peat formation is interrupted on a delta in one place because sand is temporarily deposited by a displaced river branch, then this appears as a splitting of the seam. In the marginal area near river sands the ash content of coal is often too high for economic use. Sometimes the peat has been eroded by a river and the coal seam is thin at this point or it has disappeared. Some seams have only a small extend, while others cover a large area possibly with many seam splits and varying thickness. Sediments in many coal basins form cyclically repeated alternating beds deposited in deep seawater, on coasts or by rivers with coal seams embedded in coastal and delta sediments. Such repeated series are called cyclothems and are caused by fluctuating sea levels. However, they are often not correlated across a basin since different areas of the basin sink at different rates (Süss et al. 2001, 2007) and the change in the coastline is accordingly unsystematic. In addition, there are eustatic sea level fluctuations as well. Rock layers containing coal are called coal measures.

Hard coal seams usually have thicknesses ranging from decimeters to a few meters, while lignite seams range from a few meters to dozens of meters according to the lower compression. However, very rarely there are far thicker seams (Volkov 2003) where peat formation has been stable in a single place for tens of thousands or even hundreds of thousands of years. The hard coal seam of Fushun (China) is up to 200 m thick above which lie up to 180 m of oil shale. Quang Ninh (Vietnam)—near the World Heritage Site at Ha

Long Bay—hosts the world's thickest anthracite seam at up to 60 m. The world's thickest lignite seam is in Latrobe Valley (Victoria, Australia) where in some places it is even more than 300 m thick. In the Rhenish lignite mining area near Cologne (Germany) at Garzweiler (Fig. 6.8), Hambach, and Inden the lignite seam is just over 100 m thick. This is also the case in Geiseltal (Saxony-Anhalt) and in Turov (Poland). Within these giant deposits there are actually thin interbedded layers of other sediments but of neglectable size and often the main coal seam splits into many thinner seams at the margins. Before coalification there must have been an enormous thickness of peat much greater than anything known today. Although the Philippi peat deposit in northeastern Greece in places reaches a thickness of about 200 m, more commonly thicknesses are only 5–10 m. The compaction of peat in such huge peat bogs causes the surface of bogs to lower—something the deposition of plant material has to keep up with to prevent flooding.

Basins themselves can be of different origin. Coal seams of enormous size sometimes formed on continental platforms during extensive, regular subsidence sometimes covering tens of thousands of square kilometers. Although there are usually just a few seams, they are very thick. Such deposits have very large reserves and with a few exceptions the overburden is minor making them easy to mine in open-cast mines. The downside is the coal is of low rank. Examples are lignite seams from the Jurassic in the Kansk-Achinsk Basin east of Novosibirsk (Siberia, Russia); hard coal from the Carboniferous and Permian in the Tunguska Basin, near Norilsk (Siberia, Russia); Neogene

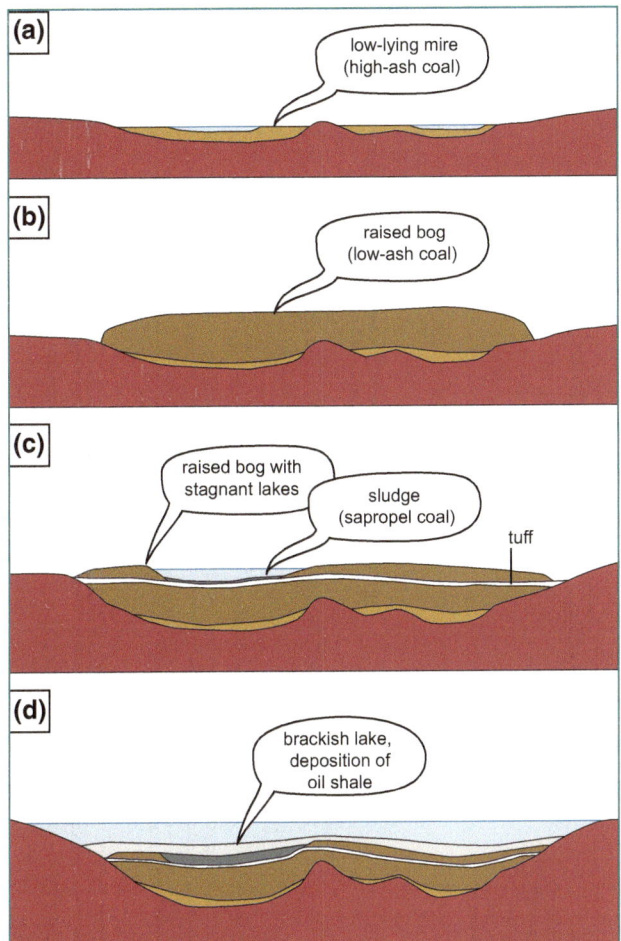

Fig. 6.7 The deposition conditions of a coal seam often changed over time. The figure shows how the Neogene lignite seam of Leoben (Austria) was deposited in a pull-apart basin. Initially, a low-lying mire (**a**) was formed whose peat still had a high content of clastic minerals. It developed into a raised mire (**b**) that was fed only by rainwater. At times a volcanic eruption caused a tuff layer to cover the surface only to be later converted into kaolinite by organic acids. Varying degrees of compaction of peat resulted in stagnant lakes (**c**) forming on the raised bog in which degraded plant material and algae remains accumulated to form a sapropel. Finally, the basin was flooded with brackish water (**d**) and an oil shale with high sulfur content was deposited (from Gruber and Sachsenhofer 2001)

(Miocene) lignite mining areas in East Germany (Central German lignite mining area between Leipzig and Braunschweig, and Lusatian mining area); and Latrobe Valley (Victoria, Australia).

Many coal deposits were formed in foreland basins of mountain ranges (comparable with the foreland basin north of the Alps). Such basins may be temporarily flooded by the sea (e.g., the Persian Gulf off the Zagros Mountains) or may contain large lakes. The Variscan orogeny in the Devonian and Carboniferous was particularly important in connection with coal. The collision between Gondwana and Laurussia and some mini continents created the supercontinent Pangaea. The result was a very wide and long mountain range that stretched across today's Europe. It was wider than today's Alps are long and at that time it was close to the equator. The Appalachians deformed again at the same time. In the Upper Carboniferous at the northern edge of the Variscan Mountains in foreland basins such as the Ruhr area, Ibbenbüren, and Aachen (Germany); Charleroi (Belgium); Upper Silesia (Poland); basins in Great Britain and Northern France; and at the western edge of the Appalachian Mountains (Pennsylvania, West Virginia, and Kentucky, USA) swampy tropical forests grew along flat coasts, deltas, and rivers. The forests were so productive that the CO_2 content of the atmosphere reduced to a very low level in the course of the Carboniferous despite initially being many times today's level. Characteristic plants were large trees belonging to Lycopodiophyta such as *Lepidodendron* and *Sigillaria* (Fig. 6.9). Their roots and trunks are common fossils. There were also tree-high giant horsetails (*Calamitaceae*) and tree ferns. Shortly thereafter other mountain formations with important coal basins followed in the Permian such as the Urals when Siberia and Pangaea collided and a mountain range similar to today's Andes in the east of Australia. In its foreland there were several large basins with coal deposits the largest being the Bowen Basin (Queensland, Australia). Foreland basins form from the deposition of rock debris derived by erosion causing further subsidence due to the weight. In this way several kilometers of sediment can be deposited in a relatively short time. Although interbedded coal seams are typically only a few meters thick, they can be very numerous (e.g., more than 100 in the Ruhr area) and extensive. Seam splits are very common. Rapid burial within a basin leads to peat being frequently converted to high-rank coal before erosion returns it to a shallower level.

Coal can also form in tectonically formed intramontane basins (within mountain ranges), rift systems, and pull-apart basins. Although there are usually just a few seams, they can be very thick. The Carboniferous Saar Basin was an intramontane basin of the Variscan Mountains with swamps and lakes. Similar deposits of the same age formed in northern Spain. Younger lignite was formed in the Eastern Alps in the Miocene during the late phase of alpine mountain formation. They are located in pull-apart basins (i.e., riftlike depressions) torn open by strike-slip faults in the Miocene (Gruber and Sachsenhofer 2001). The lignite of the Lower Rhine Basin in Germany formed in a wide continental rift partly flooded at times by the sea formed in the Miocene. Further examples in rifts are the thick coal seam of Fushun (China), the Donez Basin (Ukraine and Russia) with 130 seams from the Carboniferous in a slightly older rift, and several important districts in eastern Siberia. Finally, there are smaller coal basins that are of minor economic importance such as those found in the vicinity of salt domes and sinkholes in karst areas.

The oldest coals originated in the Proterozoic when there were no land plants. They are sapropelic coals derived from

Fig. 6.8 Bucket wheel excavator in the Garzweiler lignite open-cast mine (Germany) (*photo © Bert Kaufmann, CC-BY, flickr.com*)

algae or other microorganisms that occurred in Michigan (USA), Spitsbergen, and Greenland. Shungite (Melezhik et al. 2004) is a special example found in Karelia (Russia) consisting of coal 2 billion years old made up of small spheres. It contains fullerenes (i.e., spherical molecules of carbon that are hollow on the inside).

Although land plants appeared in the Ordovician, they only became common in the Devonian when they formed the first normal coals. Ever since coal has been continuously produced somewhere on the planet. Three periods relating to important mountain formations were particularly productive. The first period was the Upper Carboniferous to Lower Permian (already mentioned) when a lot of hard coal (Fig. 6.10) formed as a result of the Variscan orogeny—the Carboniferous is of course named for coal (carbon). In addition to the deposits in Europe, North America, and Australia there are other large ones that formed in this period in China, in southern Africa (Karoo Basin), in India, and in South America. The deposits of the supercontinent Gondwana originated in a temperate climate (especially, in the Permian). They are usually referred to as Gondwana coals. The second period was the Jurassic and Early Cretaceous and mainly concerned foreland basins of the Rocky Mountains (Laramide orogeny in the United States and Canada), Siberia, and China. The third period was the Tertiary (Paleogene, Neogene) when most of the large lignite deposits formed in the context of the Alpine orogeny as did some hard coal and anthracite deposits. Important examples are Central Europe, Australia, Indonesia, Southern Patagonia (Chile, Argentina), Colombia, Venezuela, and the west of North America.

Coal was deposited several times in some basins. One example is the huge Ordos Basin (Johnson et al. 1989; Yang et al. 2005; Yao et al. 2009) in China, which includes the northern part of Shaanxi Province north of Xian and extends to neighboring provinces (especially, Inner Mongolia). It is the largest of the many Chinese coal fields and contains natural gas and petroleum as well. In the Ordovician this was still a shelf sea at the margin of the North China Craton in which carbonates were deposited. These are problematic today for the mining industry because at times they were strongly karstified. Cave systems are to blame for sudden water intrusion that happens frequently in the mines. Other coal fields in China have to contend with the same issue (Li and Zhou 2006). The Ordos Basin came about as a result of the formation of a subduction zone to which smaller terranes attached themselves. The basin developed into a shallow coastal region with lagoons and deltas where several seams formed in the Upper Carboniferous and Lower Permian, which today contain hard coal with a high methane content. Continental collision with the South China Craton followed in the Triassic and the Qinling Mountains emerged. The basin was now in the middle of a continent. The southern part of the basin was folded and partly eroded. Further seams were deposited in the Jurassic. At times there was a large lake with a fluctuating water level and a swampy shore. In addition, large moors formed in the depressions of folds, while the folds themselves were deformed by further tectonics. Today the Ordos "basin" is largely a high plateau covered by thick loess.

Although seam gas can be extracted and used as an energy source, it is a source of danger for miners. During

Fig. 6.9 Tropical forest in the Carboniferous. Characteristic plants were large trees belonging to the Lycopodiophyta such as *Lepidodendron* (*left*) and *Sigillaria*. There were also giant horsetails belonging to the Calamitaceae (*right*) and tree ferns (*center*) (*from* © Mayers Konversationslexikon, 1885)

Fig. 6.10 Coal seam in a mine in Lancashire (UK) (*photo* © Mjtmail, CC-BY, flickr.com)

coalification methane is formed from subbituminous coal and higher coal ranks together with some propane and butane. Bituminous coal of high and medium volatility have the highest methane content. Gas has mostly escaped from coal of even higher rank and can form significant conventional natural gas deposits. Pit gas released during the mining of bituminous coal of high and medium volatility is explosive when mixed with air. It is responsible for deadly gases called firedamps; hence the need for very good ventilation. In addition, in mines coal slowly oxidizes to CO and CO_2 that accumulate in depressions and can also be fatal. In addition to methane formed during coalification microorganisms also produce biogenic methane (Thielemann et al. 2004; Fallgren et al. 2013) (especially, in former mines). The first step involves fermentative bacteria releasing acetate and other substances. These anaerobic methanogenic archaea produce methane. The methane released from active and decommissioned coal mines largely ends up in the atmosphere accounting for 7% of the world's emissions of this powerful greenhouse gas. It makes far more sense to use the gas to generate energy. The gas escaping from closed mines (coal seam methane, coal mine methane) is burned in power plants in some areas such as the Ruhr. Coalbed methane is also sometimes exploited by drilling in areas instead of mining. A seam is drilled parallel to the coalbed, water is then pumped out that was restricting the mobility of gas in the pores, and finally methane flows out of the borehole.

In the presence of oxygen coal seams can autoignite; porous hard coal is particularly susceptible. Relatively low temperatures such as those achieved by the weathering of pyrite are all that is needed. Sufficient oxygen is available at the surface at outcrops of the seam and in mines. If little oxygen is available, then incomplete combustion takes place in a fire that smolders at temperatures of a few hundred degrees. If sufficient oxygen is available, then temperatures of over 1000 °C can occur—sufficient to melt rock. Such fires are extremely difficult to control so much so that worldwide there are examples that have been burning for decades. The seam in the Burning Mountain Reserve (New South Wales, Australia) has probably been burning for 5500 years. The steaming landscape at the surface resembles a volcanic area.

Box 6.4 Composition of living organisms

The molecules that make up living organisms can largely be assigned to the following four groups. Plants consist mainly of carbohydrates (40–70%) with some such as trees also having a high lignin content. Animals are characterized by a high proportion of proteins (55–70%).

Lipids (fats, oils, waxes) are completely or largely insoluble in water. They include fatty acids, waxes,

Table 6.1 Average composition (in weight percent) of the main biomolecules compared with petroleum (Hunt 1995)

	C	H	O	S	N
Lipids	76	12	12	–	–
Proteins	53	7	22	1	17
Carbohydrates	44	6	50	–	–
Lignin	63	5	31.6	0.1	0.3
Petroleum	85	13	0.5	1	0.5

triglycerides, phospholipids, sphingolipids, and iso-prenoids such as carotene and cholesterol. Although they mainly consist of carbon and hydrogen, some also contain some oxygen, nitrogen, and phosphorus (Table 6.1).

Proteins are very large molecules composed of amino acids. They mainly consist of carbon and hydrogen, but also contain large amounts of nitrogen and oxygen. Although proteins are present in all cells, in animals the proportion is particularly high.

Carbohydrates are sugars and polymers composed of them. Sugars are hydrocarbon chains with oxygen in hydroxyl groups, aldehyde groups, and ketone groups. These chains can be combined to form branched and reticular molecules. This group includes sugar, cellulose, starch, and chitin. Plants have a very high carbohydrate content.

Lignin is a huge, widely branched polymer with many aromatic rings. It is a reinforcing element that many land plants make use of in cell walls, wood, and bark.

6.2 From Algae to Petroleum

Box 6.5 Abiotic hydrocarbons

Today almost all researchers agree that petroleum and natural gas can be traced back to biomass. However, the Soviet doctrine used to be that hydrocarbons were produced abiotically (Glasby 2006) by Fischer–Tropsch synthesis (Box 6.3). Accordingly, CO_2 and H_2 in the Earth's mantle were the starting material from which methane and higher hydrocarbons are produced catalytically supported by mineral surfaces. Such hydrocarbons would rise continuously along faults constantly refilling reservoirs. One argument was that hydrocarbons could actually be found in igneous and metamorphic rocks albeit rarely in large quantities. The idea that they migrated from sedimentary rocks

into basement rock was rejected, occurrences in rift systems and in deep shear zones were cited as arguments for abiotic formation. Moreover, thermodynamic considerations show that this reaction is quite possible. This theory was largely ignored in the West. However, at that time even the theory of a biogenic origin could not explain many questions. Too little was understood of the processes at that time.

In the West some decades later the astrophysicist Thomas Gold, a prominent advocate of abiotic origin, pointed out that methane is not rare in the solar system. The large gas planets contain a lot of methane and even some meteorites such as carbonaceous chondrites contain some methane and even traces of higher hydrocarbons. A biogenic origin can of course be excluded here. Gold therefore believed that methane had been present in large quantities in the mantle since its formation and had been continuously degassing. Larger molecules could indeed be produced by Fischer–Tropsch synthesis. Although the mantle is comparatively reduced, it is not so reduced that methane would be stable. Carbon occurs in the mantle mainly as carbonate (magnesite) and CO_2, sometimes as diamond, but only in traces as methane. A drilling program was set up in a meteorite crater in Sweden in an effort to find hydrocarbons clearly originating from the mantle. Tiny amounts of oil were indeed discovered. However, oil originating from the mantle was doubted by other researchers.

The biogenic origin is much more plausible in the case of economically interesting occurrences (Glasby 2006; Sephton and Hazen 2013) unfortunately contradicting the theory of continuous replenishment of reservoirs. The frequent association with shear zones and rift systems is more likely to be due to high geothermal gradients being advantageous for oil formation and faults forming paths for migration and forming traps. In almost all large deposits corresponding source rocks are also known (i.e., sediments with a high content of organic substances). This also applies to many occurrences in igneous rocks (Schutter 2003). For example, flood basalts contain interbedded sediments from lakes that have been dammed by lava flows.

However, this does not mean that abiotic hydrocarbons cannot exist as well. For example, fluids at some submarine hot springs contain methane and other hydrocarbons (Foustoukos and Seyfried 2004; Proskurowski et al. 2008). Obviously, the hydration of mantle peridotite to serpentinite plays an important role here. On the one hand, this produces H_2 that can react with CO_2 in Fischer–Tropsch synthesis; on the other hand, chromium-rich minerals of the mantle are suitable catalysts. Small amounts of abiotic hydrocarbons are also known from some basic and ultrabasic igneous rocks as well as from agpaitic rocks (Sect. 3.11). However, the completely different isotope composition shows that the abiotic portion of economic deposits is not even in the per mill range.

Photosynthesis-driving microorganisms such as diatoms (Fig. 6.11) and other algae, dinoflagellates, and cyanobacteria are called phytoplankton. They make by far the largest contribution to the organic material from which petroleum (North 1985; Hunt 1995; Gluyas and Swarbrick 2004; Bjørlykke 2011) is formed (Box 6.4, see also Box 6.5). In seas and lakes phytoplankton account for about 90% of total biomass. These organisms live almost exclusively in the upper 30 m of the water column where there is sufficient light to produce organic molecules and oxygen from CO_2 and H_2O. They are dependent on the presence of phosphorus, nitrogen, and iron as nutrients. They are abundant in coastal areas thanks to rivers. The upwelling of nutrient-rich water from the deep sea into shallow shelf seas can lead to particularly high organic productivity. This happens mainly on the west coasts of the tropics and subtropics as a result of the trade winds driving shallow water away from coasts thus allowing deep water to upwell. Such organisms can multiply very quickly through cell division (especially, when the nutrient level is high). Such algal blooms can be seen on satellite photos as a green coloration (Fig. 6.12).

Zooplankton occupy the next level in the food chain and account for about 10% of biomass. The most common are protozoa such as foraminifera and radiolarians, but tiny crabs (krill) and planktonic marine snails are found among them. Higher levels in the food chain occupied by crabs, fish, and whales make up only a tiny part of the biomass and are therefore completely insignificant in this context. Further contributions are made by plant remains and humic acids supplied by rivers the latter flocculating and sedimenting in estuaries.

Dead creatures sink to the bottom of the sea and almost all their biomass degrades on its way through the water column or on the bottom. This happens not only by

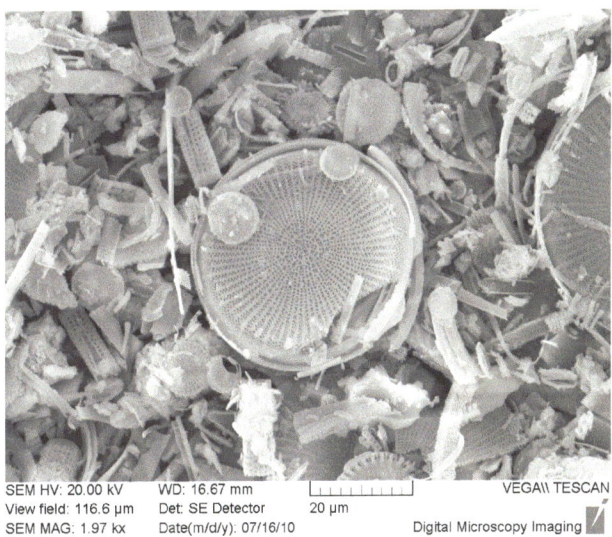

Fig. 6.11 Diatoms as seen through a scanning electron microscope. These photosynthesis-driving single-cell organisms have a skeleton of amorphous SiO_2 (opal). Due to their abundance they make an important contribution to the formation of oil (*photo* © Jasmin Stieger, CC-BY-SA, Wikimedia Commons)

oxidation with dissolved oxygen to form CO_2 and H_2O but also by microorganisms that feed on it. Living organisms burrowing through the seabed indirectly cause faster oxidation. Significant amounts of organic matter can only be preserved if water contains as little oxygen as possible.

Deep-sea basins in the Black Sea are good examples of this. The water of this sea does not circulate such that a stagnating salt-rich (and thus heavy) anoxic water layer forms at depth. Most living creatures cannot survive in it, but anoxic sulfate-reducing bacteria and archaea feel at home here. They use sulfate dissolved in water as an oxidizing agent to generate energy and in so doing oxidize organic matter or hydrogen. This produces H_2S that is toxic to most living organisms and leads to the precipitation of sulfides such as pyrite and marcasite (see also Sect. 5.1). This hostile anoxic and H_2S-rich environment is called euxinic for the Latin name of the Black Sea. The sapropel that accumulates on the bottom can contain up to 35% organic substances. After diagenesis it becomes black shale known to be a particularly fertile source rock for the formation of petroleum. Lakes in which water does not circulate regularly can become anoxic at depth and the difference in density between warm and cold water results in water stratification.

Oxygen content can become low not only in some deep-sea basins and lakes but also in the shallow water of a shelf sea when water circulation is poor and the consumption of oxygen by life is very high. Even if oxygen is present in the water its content in the pores of sediment under the seabed decreases rapidly within a few millimeters (in fine-grained sediments) or a few centimeters (in

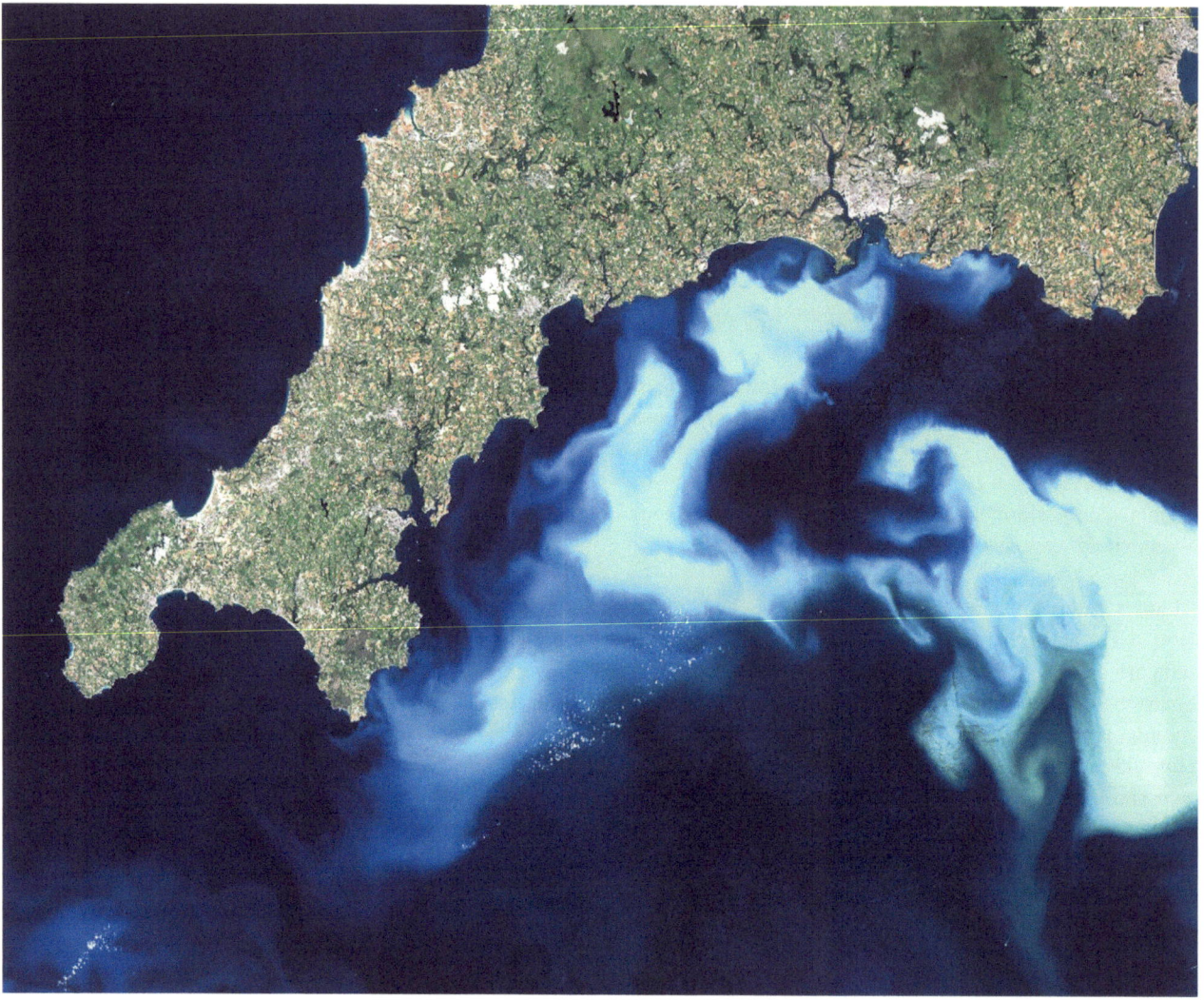

Fig. 6.12 Algal bloom off the south coast of England in a satellite photo (*photo* © NASA)

coarse-grained sediments) because microorganisms consume what little oxygen there is—oxygen that is only very slowly replaced. Covering the seabed with sediment as fine-grained as possible thus protects organic substances from oxidation.

Particularly favorable conditions for the formation of suitable source rocks prevailed during periods when a hot greenhouse climate prevailed such as the Jurassic and Cretaceous. Strong chemical weathering on land fertilized the oceans and led to rapid multiplication of plankton. At the same time the lack of glaciers in polar regions led to a reduction in the speed of ocean currents and sometimes led to entire ocean basins becoming anoxic. Rocks deposited during such anoxic events form the most important source rocks of many oil deposits. Source rocks deposited during the Jurassic and Cretaceous are responsible for almost two-thirds of oil resources (Klemme and Ulmishek 1991).

Whether the sediment deposited is a good **source rock** for petroleum depends as much on biological activity as on the rate at which substances degraded before covering. The content of organic substances in sediment is expressed as total organic carbon (TOC). In black shale this can be more than 20%, in normal shale on average 2%, in carbonate rocks on average 0.3% (but in some fine-grained carbonates several percent), and in sandstones only 0.05%. Rock with more than 0.5% is already a candidate for source rock. Important are black shales, shales, and some fine-grained carbonates.

As soon as sediment is buried under further sediments diagenesis begins. For the sediment this means compaction, pressing out of water, and cementation by newly formed minerals in the pores. Anoxic archaea living in the pores can decompose organic molecules and thereby release methane, water, and carbon dioxide. As a rule, gas simply escapes. In rare cases such as in northwestern Siberia such biogenic methane seems to have made an important contribution to large natural gas deposits as demonstrated by the very light carbon isotope signature. Organic molecules also react with each other during early diagenesis. Very large molecules that are not soluble in organic solvents are formed called **kerogen** (bitumen is the soluble part in organic solvents). Depending on which organisms have contributed to the organic matter sedimented, kerogen has different proportions of hydrogen, nitrogen, and sulfur. There are four types:

Type 1 kerogen: Formed in oceans and lakes mainly from algae and bacteria. Has a comparatively high content of lipids and very high potential for the formation of petroleum. Produces little gas. Relatively rare. Some oil shales could be mentioned.

Type 2 kerogen: Typical kerogen of anoxic marine sedimentation. Mainly formed from phytoplankton and zooplankton together with spores, pollen, and in very small amounts remains of plants and animals. Main source of oil and gas. A variant with a high sulfur content is called type 2S.

Type 3 kerogen: Formed mainly from land plants. Has a high content of lignin and cellulose. Corresponds to typical coal, but it is not necessarily about coal seams. Much more frequent are small particles finely distributed within the sediment. This releases methane making this type an important source of natural gas. Oil does not form.

Type 4 kerogen: In particular, oxidized plant material. Although rich in carbon, it contains hardly any hydrogen and therefore has no potential for the formation of petroleum or natural gas.

The compositions of kerogens and how they change with increasing maturity are usually shown in the so-called Van Krevelen diagram (Fig. 6.13) in which the H/C ratio is plotted against the O/C ratio. The diagenetic release of biogenic methane, water, and carbon dioxide and later of oil and gas reduces the contents of H and O in kerogen.

As temperature increases to about 50 °C diagenesis seamlessly merges into **catagenesis**, the main phase of oil and gas formation that corresponds to the late diagenesis of sedimentary rocks. Polymer bonds are broken during catagenesis and smaller hydrocarbon molecules split off (cracking). The temperature range in which petroleum can form is called the oil window and is relatively small between 50 and 150 °C with almost everything happening in the middle third of this range (Fig. 6.14).

Type 2 kerogen at 50 °C produces only a small amount of wet gas (methane with some propane, butane, etc.) and minimal oil. Significantly more oil is released between 90 and 120 °C. With a normal geothermal gradient this corresponds to a depth between 3 and 4 km, but in hotter or cooler basins about 1 km more or less. If the temperature continues

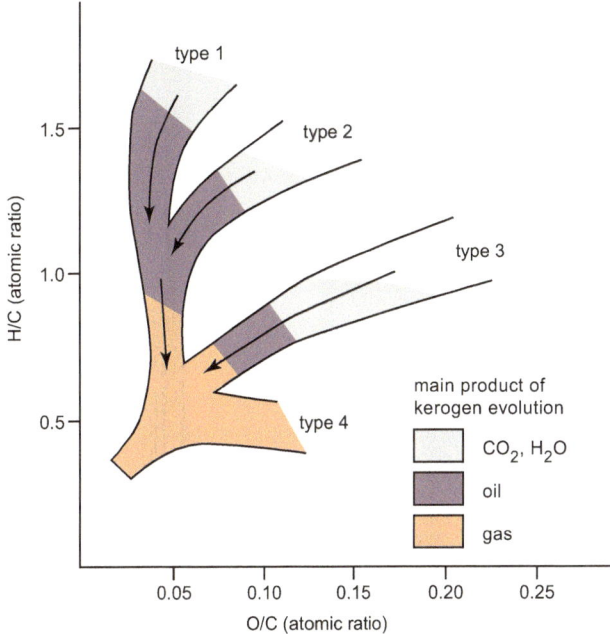

Fig. 6.13 The four types of kerogen in the so-called Van Krevelen diagram and how they change with increasing maturity. During diagenesis mainly CO_2 and H_2O (also biogenic methane) are released. During catagenesis petroleum and later natural gas are the main products. The quantities of oil and gas released cannot be read off here. However, with type 1 this is mainly oil, with type 2 it is oil and gas, with type 3 almost only gas, and with type 4 almost no oil and no gas (from North 1985)

to rise, then ever-less oil is released but significantly more gas. Even oil is cracked into smaller molecules forming light oil (Fig. 6.15) and wet gas. At more than 150 °C only gas is released in which the proportion of methane becomes ever higher and the gas drier.

Development is similar for type 1 kerogen, but the gas content is low. This is because after passing through the oil window there is only a little kerogen left. Type 3 kerogen, on the other hand, does not produce oil. Initially, a small amount of wet gas can be formed (which may not escape and then will be cracked at higher temperatures), but at higher temperatures a large amount of dry gas is released.

The release of hydrocarbons in a closed system ends as soon as organic substance hydrogen is exhausted. What remains is a carbon-rich substance called pyrobitumen from which the mineral graphite is formed during metamorphism if burial continues. However, the potential for hydrocarbons to form is much greater than the hydrogen content of the organic matter suggests since water also serves as a source of hydrogen (Seewald 2003). On the other hand, water, sulfates, and ferrous minerals oxidize part of the organic substance to CO_2. The amounts of hydrocarbons released is a question of how much time the rock has spent in the oil window. If source rock remains at the upper (cool) edge of

Fig. 6.14 Schematic representation of the quantity of hydrocarbons released by type 2 kerogen (*left*) and type 3 (*right*) with increasing temperature during burial. Apart from early biogenic methane almost only CO_2 and H_2O are released during diagenesis. Catagenesis—the cracking of large molecules—begins at about 50 °C. At a normal geothermal gradient the main part of the oil window is located at a depth of 3–4 km, whereas at greater depth only gas is released. Type 3 kerogen only produces gas. The composition of released gas and oil also changes in the downward direction where the oil becomes lighter (content of large molecules decreases) and the gas drier (content of propane and butane decreases). The curves can vary in individual cases (e.g., the time source rock spends at a certain temperature also has an effect)

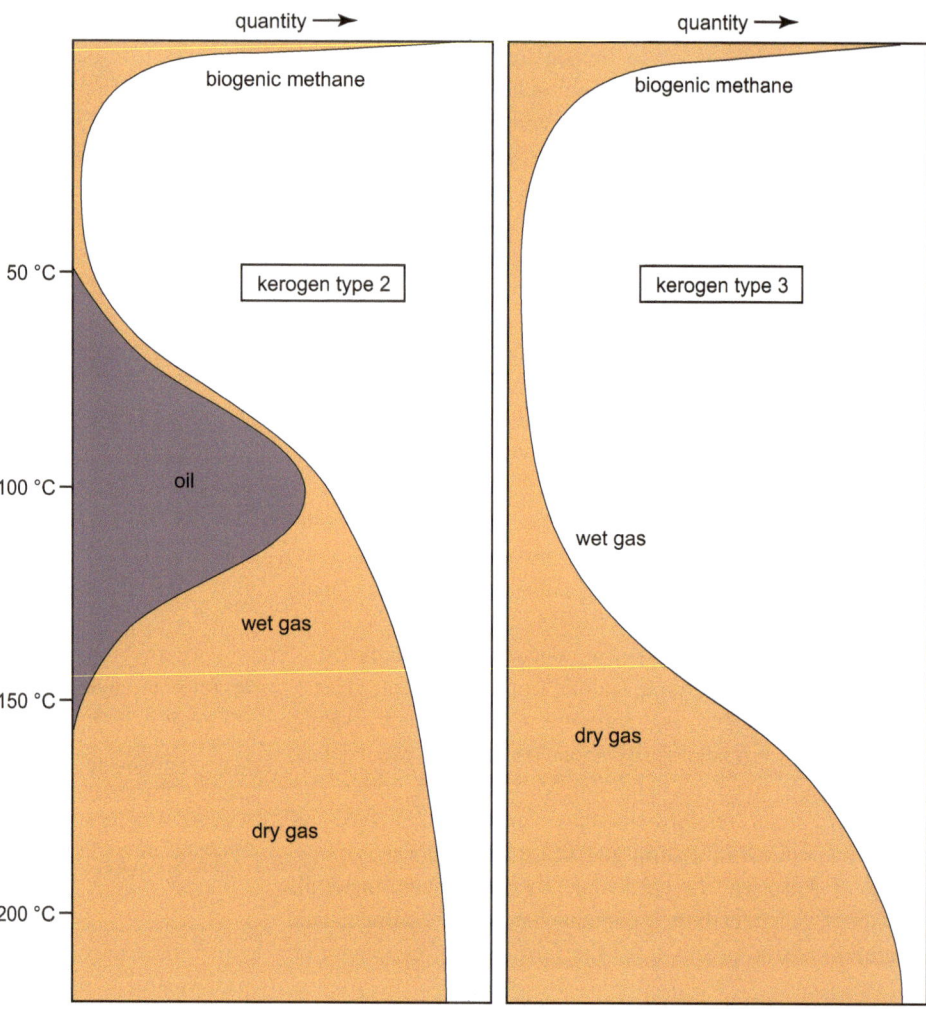

the oil window, then oil formation proceeds slowly. However, at optimal temperature a large deposit can develop within a few million years. If source rock remains at the lower edge of the oil window for a very long time a good part of the oil is cracked to gas.

At this point oil and gas have been produced, but the result is not yet a good deposit (as will be shown in a moment). After all, it is already possible to estimate whether a sediment basin contains significant amounts of hydrocarbons. Moreover, it is also known that petroleum and natural gas of diverse compositions (Box 6.6) are formed in variable proportions. It is important to know during exploration where source rocks are interbedded in the sediments of a basin and how mature they are with regard to the formation of oil and gas.

Box 6.6 Composition of petroleum

Petroleum is a mixture of various hydrocarbons and other molecules (Fig. 6.16) consisting on average of 85% C, 13% H, 1.5% S, 0.5% O, and 0.5% N.

Unsaturated hydrocarbons (with C=C double or C≡C triple bonds) do not occur in petroleum because their reactivity causes them to decompose during oil formation. Saturated hydrocarbons containing only C–C single bonds are the principal component of petroleum. Aromatic hydrocarbons also occur. They have at least one ring with delocalized double bonds (benzene ring). There are also more complex molecules. The main components of petroleum belong to one of the following groups:

Alkanes (paraffins) are saturated hydrocarbons with the formula C_nH_{2n+2}. When $n = 1$–4 they are gaseous (methane, ethane, propane, butane), when $n = 5$–15 they are liquid (pentane, hexane, heptane, octane, etc.), and when $n > 15$ they are solid (paraffin wax). In addition to straight-chain alkanes (*n*-alkanes) there are branched molecules (isoalkanes, *i*-alkanes) that have the same molecular formula but slightly different properties.

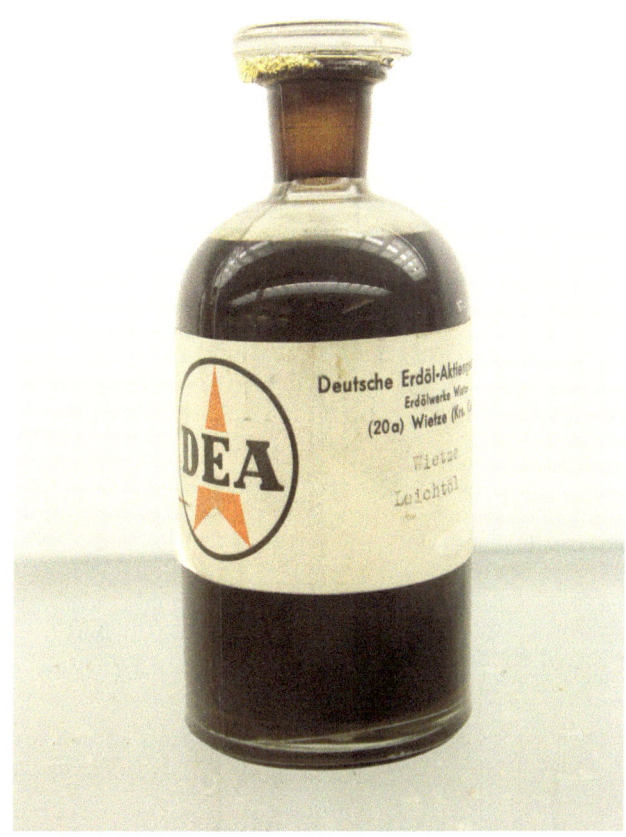

Fig. 6.15 Light crude oil from Wietze, near Celle (Germany) (*photo* © F. Neukirchen/Mineralogical Collections of the TU Berlin)

Naphthenes (cycloalkanes) are saturated cyclic hydrocarbons with the formula C_nH_{2n}. Cyclopentane (C_5H_{10}) and cyclohexane (C_6H_{12}) are particularly important components of petroleum.

Aromatics have at least one benzene ring, but they are only present in very small quantities.

Asphaltenes, **waxes, and resins** are large molecules with aromatic rings, alkane chains, and other groups. They often also contain sulfur, nitrogen, or oxygen (NSO compounds) and metals such as vanadium and nickel (the latter in particular). Among the large molecules there are compounds that are very similar to chlorophyll and regarded as a kind of molecular fossil.

Each type of oil contains thousands of different molecules. Most crude oils are a mixture of alkanes and naphthenes each accounting for about one-half with low contents of other components. Petroleum dominated by alkanes is particularly easy to process, but it is very rare. There are also oils dominated by naphthenes and asphaltenes.

The sulfur content of petroleum is very variable. In addition to sulfur-bearing organic compounds elemental sulfur and H_2S making up a few percent may also be present. Less than 0.5% S is referred to as **sweet crude oil**, while a higher content is referred to as **sour crude oil**. This is an important factor not just because SO_2 is produced during combustion and is then responsible for acid rain, but also because sulfur leads to corrosion of metals (e.g., in pipelines) and impedes the function of catalysts. Moreover, H_2S is toxic. Although desulfurization is not technically a problem, it is a cost factor.

The density of petroleum is the most common quality standard. Light crude oil has a high content of small molecules and low viscosity, whereas heavy crude oil has a high content of large molecules, high viscosity, and is difficult to process. Density is unfortunately not given in grams per cubic centimeter but in the **API gravity** unit invented by the American Petroleum Institute:

$$API \ gravity = (141.5/\rho) - 131.5$$

where ρ is relative density in relation to water at 15.6 °C (=60 °F). The density of water corresponds to 10°API. The larger the API the smaller the density. A distinction is made between light (>31.1°API), medium (22.3–31.1°API), heavy (10–22.3°API), and extra-heavy (<10°API) crude oil. The density of oil tends to decrease with increasing depth. The two types of light crude oil used as references on markets are Brent (from the North Sea) and WTI (West Texas Intermediate).

6.3   Petroleum and Natural Gas: Migration into Traps

At depth where kerogen is cracked into oil and gas the diagenesis of sediment is so advanced that it is a compact rock. Porosity is clearly reduced as a result of which pore water has already been pressed out and wafer-thin connections between pores have become considerably narrower. Source rocks are generally fine-grained rocks where diagenesis has made permeability so low that oil and gas are firmly enclosed in the pores. This is of course more true of shales than of carbonates. Extraction from these rocks is only possible using an unconventional and controversial method called fracking (Sect. 6.7).

Conventional deposits can only develop if oil and gas find a way out of impermeable source rock into permeable rock (e.g., into a sandstone). This step is called **primary**

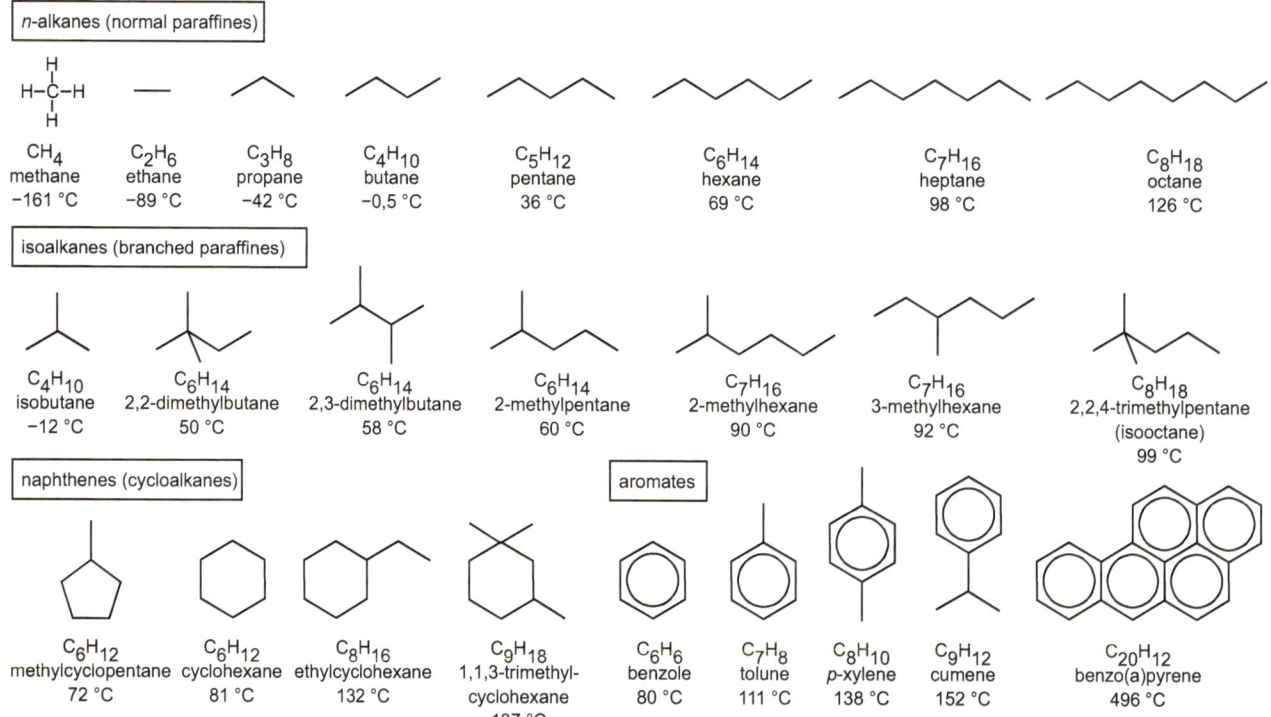

Fig. 6.16 Some organic compounds found in petroleum and natural gas and their respective boiling points. With the exception of methane: In the simplified representation, at each corner there is one C atom, the H atoms are not shown

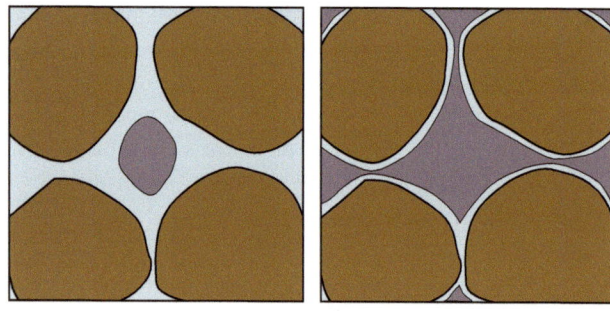

Fig. 6.17 Water normally wets mineral surfaces. If rock pores contain a lot of water (*blue*) and little oil (*violet*) (*left*), the oil cannot overcome the capillary forces even if water flows from pore to pore. If the amount of water is small (*right*), then it still wets the surfaces but the oil can form a continuous phase and thus flow easily

migration. How it works is not yet fully understood. A candidate mechanism is when small cracks appear in rock because the pressure in pores becomes too high. The lower the permeability of a rock the more difficult it becomes to expel fluids (i.e., water and later oil and gas) from ever smaller pores during compaction. Fluid pressure increases further during catagenesis due to the formation of oil and gas. Therefore, the pressure in pores of impermeable rock is usually significantly higher than in permeable sandstone. Such pressure differences have the consequence that primary migration takes the shortest route to the next permeable rock

even if this is deeper—not necessarily in an upward direction.

Another problem is that pores also contain water further restricting mobility (Box 6.7). Although the solubility of oil in water is negligible, when it comes to aromatic molecules and methane at high pressure solution of oil in water could play a role. If a pore is predominantly filled with oil such that there is a continuous oil phase from pore to pore, then the capillary force to be overcome is considerably lower. This could be the case with some source rocks. If the source rock is limestone, then organic acids and carbonic acid released during oil formation can increase permeability (Heydari and Wade 2002). Otherwise, the only explanation is diffusion. Although diffusion certainly always runs along the pressure gradient, it is however very slow. Primary migration from source rock can be a process that takes place repeatedly in phases or slowly over a long period of time. In any case much of the organic matter will remain in the rock (especially, larger, highly viscous molecules).

Box 6.7 Permeability and mobility of water and oil in the rock pores

In addition to oil (or gas) water is always present in rock pores. This can prevent the flow of oil from pore to pore, a problem that occurs in primary and secondary migration as well as in production. This effect is the main reason only part of the oil present in a deposit can be extracted. Because water normally wets the surfaces of minerals (water wet), oil is held as a droplet in the center of a pore (Fig. 6.17). If the amount of water is large, then so too is the capillary force that the oil must overcome to flow to the next pore. It is trapped in the pore while water is relatively mobile. If a pore is predominantly filled with oil such that there is a continuous oil phase from pore to pore, then the capillary force to be overcome is significantly lower and oil can flow.

Sometimes the reverse happens and oil wets mineral surfaces (oil wet) while water is held in the center. Whether this is the case depends on the contact angle between phases, which in turn depends on the composition of oil and water and the surface properties of the mineral. In this case oil and water are reversed in the illustration. Although this principally happens with extra-heavy crude oil, sometimes it is also the case with normal oil in carbonate rocks. If the amount of oil becomes too low in this case, then it will not form a continuous phase either but only a thin film on mineral surfaces. This also prevents the oil from flowing.

As soon as hydrocarbons reach permeable rock they move relatively quickly upward as far as the position of the rock layer permits. This is because their density is lower than the density of water present in the pores. This is called **secondary migration**. Here too a continuous oil phase from pore to pore is a prerequisite to effective flow—something that is easily provided in a sandstone with a large oil quantity. Oil can travel several kilometers, even hundreds of kilometers, within a permeable rock layer. If methane occurs as a separate gas phase, then the two phases may be separated. The solubility of methane in oil depends on the pressure and therefore decreases during secondary migration.

Oil rises until it escapes somewhere on the Earth's surface (Fig. 6.18) and is thus lost or until it ends up in a **trap** sealed above by impermeable rock. As soon as a trap is filled further rising oil flows past it and fills an even higher trap.

Salt, anhydrite (Sect. 5.7), and shales are the best **seals** for a trap. Although shales may leak along faults, salt flows so easily under pressure that pores and cracks are immediately closed. Hydrocarbons accumulate in a trap in the highest possible part of the permeable rock called the **reservoir**. The reservoir should have as large a pore volume as possible such that a lot of oil and gas can fit into the trap and should be as permeable as possible such that the content can flow to the well during production. Good reservoirs are sandstones, reef limestones, and limestones with lots of clefts or karst cavities (however oil-filled caves are rare exceptions). If there is a gas phase, then it forms a gas cap above the oil. Pores beneath the oil are filled with water. Of course, there are also traps that only contain gas. Strictly speaking, however, small amounts of water are also present in gas-filled and oil-filled pores—something that plays an important role in extraction.

Traps are divided into structural and stratigraphic traps (Fig. 6.19). **Structural traps** are caused by tectonic movements. Most conventional petroleum is trapped in a

Fig. 6.18 Natural oil seepage in Azerbaijan (*photo* © F. Neukirchen)

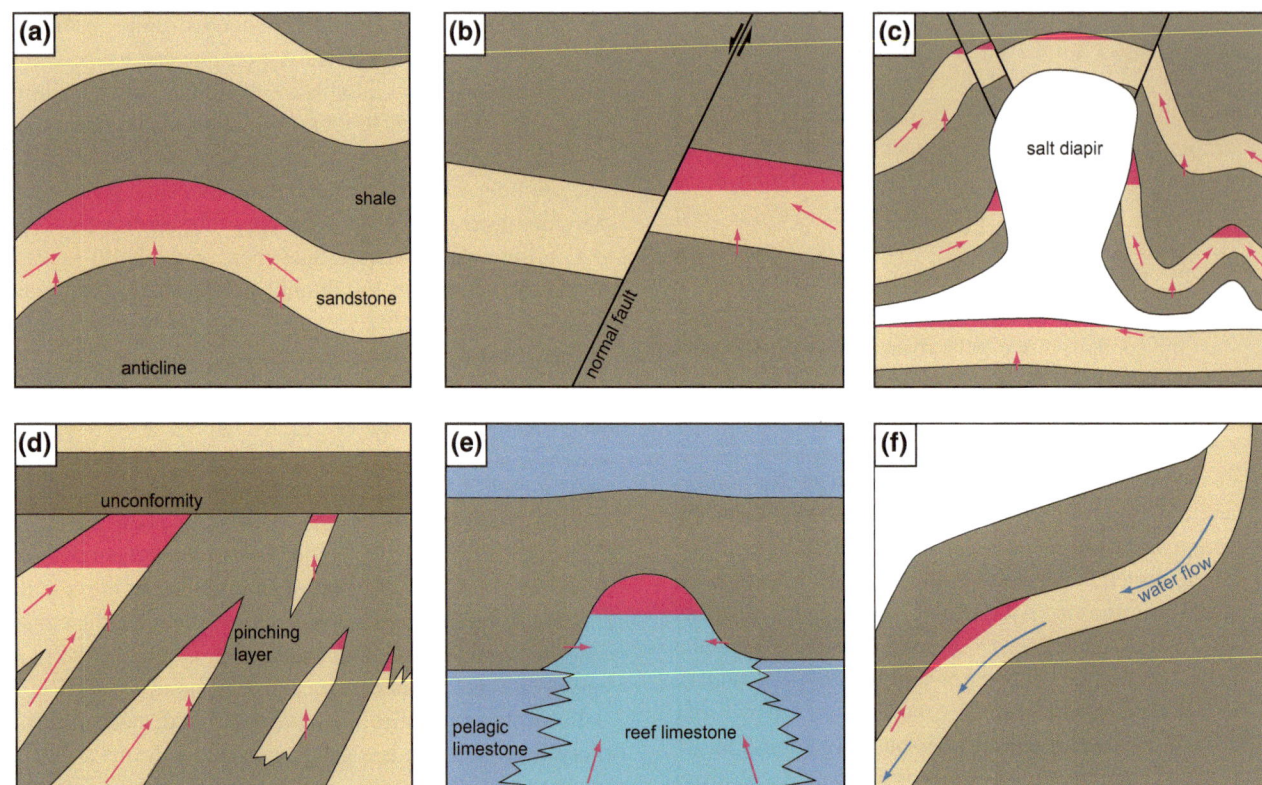

Fig. 6.19 Oil and gas traps are divided into structural traps (**a–c**), stratigraphic traps (**d**, **e**), and hydrodynamic traps (**f**). In all cases contact between permeable and impermeable layers is important. **a** By far the most important are anticlines. **b** Normal faults are frequent in sediment basins. Particularly large traps can occur in rift and horst structures. **c** When a salt diapir rises, superimposed sediments are deformed creating a variety of traps. Hydrocarbons can also accumulate below a salt horizon (in the pre-salt layer). **d** An unconformity and clastic sediments with layers pinch out. **e** A fossil reef covered with shale. **f** Hydrodynamic traps are relatively rare, in these the further rise of hydrocarbons is blocked by a stream of water. The interface between water and hydrocarbons is inclined in this case

type of fold called an anticline. For example, many anticlines with trapped oil are found in the Zagros Mountains in Iran. They were created during the collision between the Arabian Plate and Asia (the suture runs through the northeastern part of the mountain range). The sediment stack on the shelf of the Arabian Plate was thrust together creating the Zagros fold and thrust belt where sediment layers deformed into large, very regular folds. In other parts of the world anticlines are not always morphologically recognizable as mountains. There are also very rare examples of folds where oil has accumulated in an almost water-free layer in a syncline.

Faults can also create traps as long as permeable and impermeable layers are offset against each other. However, active faults are often permeable to a certain degree and allow migration to higher strata or to the surface as long as the fault is not sealed by the formation of new minerals. Traps can occur both at normal and reverse faults. Large traps of this type exist in rift and horst structures (e.g., in the North Sea in the Viking Rift between Norway and

Scotland). Deposits in the Niger Delta (Nigeria), on the other hand, are traps in a system of thrust faults.

Salt diapirs offer many possibilities for trap development. Sedimentary rocks in the surroundings of salt diapirs are broken, bent, and deformed as a result of salt rising. Anticlines, structural domes, and faults form above and around salt diapirs. Another trap is where upward-bent sediment layers contact salt. Sometimes solution cavities can form in the cap rock of salt diapirs creating an additional reservoir. Of course, hydrocarbons can also accumulate below the salt horizon (pre-salt layer) (especially, if it is slightly folded).

Stratigraphic traps form as a result of sediment deposition. For example, the permeable layer can pinch out whenever different rocks are deposited in different parts of a basin. Moreover, since sedimentation changes over time (e.g., due to fluctuations in sea level) these different rocks or facies have an interlocked profile. Some of these traps correspond to former coastlines or a sandbank, some to a former river course, others are simply the result of the thickness of strata fluctuating. Turbidity currents and other mass flows in the deep sea form sandy channels on which fine clay particles later slowly rain down—another potential trap (Janocko et al. 2013; Liu et al. 2013). Large submarine landslides lead to a similar result (Bull et al. 2009). Fossil coral reefs later covered with shales form particularly good traps because reef limestone is a good reservoir and the reef is naturally higher than limestones deposited simultaneously in deep basins or in lagoons. A good example is the Golden Lane oil field in Mexico (Wilson 1987) where reef limestones of a complete fossil atoll of the Cretaceous serve as an oil trap. Also important are unconformities (i.e., tilted and partially eroded sediments overlain with younger rather flat sediment layers). Of course, many different combinations of traps are possible in a single basin. Furthermore, since deformation can occur simultaneously during sedimentation there are combinations of stratigraphic and structural traps.

An exotic example of a trap is the Cantarell Field (Mexico), which is one of the largest in the world. The trap was created by the Chicxulub impact (Grajales-Nishimura et al. 2000). This was the meteorite impact marking the border between the Cretaceous and Paleogene believed to have been responsible for the extinction of the dinosaurs. The field is located off the Yucatán Peninsula in the Gulf of Mexico, a few hundred kilometers from the impact crater hidden in the thick sediment stack. The impact caused huge rock masses of shelf carbonates to slide off and deposit as coarse-grained breccia and in so doing formed the reservoir. The mass of fine-grained ejecta (later cemented with dolomite during diagenesis) then rained down to form the seal. This trap is not completely tight. The only reason the oil field was discovered in 1976 was because a fisherman complained that leaking oil was ruining his nets.

Other types of traps are very rare. In **hydrodynamic traps** the further rise of oil is blocked by a stream of water resulting in the interface between oil and water not being horizontal. From time to time oil has even sealed itself by forming an impermeable **asphalt layer**.

Asphalt layers are the result of subsequent change in the composition of oil via a process known as **degradation**. Such change is unfortunately bad news because alkanes and smaller molecules are the first to be degraded and thus large molecules are passively enriched. Petroleum becomes heavier and more viscous resulting in first of all heavy crude oil, then extra-heavy crude oil (heavier than water), and finally asphalt (see also Sect. 6.9). The main culprits are microorganisms that feed on petroleum in a process known as biodegradation. As soon as oil leaks at the Earth's surface oxidation by microorganisms starts. Additionally the lightest components simply evaporate and other components are being washed out by water. These are the reasons real asphalt lakes form. The best known are the La Brea Tar Pits in the middle of Los Angeles, which in the Pleistocene saw the deaths of many animals including mammoths and saber-toothed cats. There are many more asphalt lakes worldwide. The largest is Pitch Lake on the Caribbean island of Trinidad with an area of 40 ha.

However, even oil in deeper reservoirs is degraded. The discovery that anaerobic microorganisms colonize rock pores at a depth of 3 km is relatively new. However, such a "deep hot biosphere" has repeatedly been found in scientific deep drilling in oceans and continents, in the connate water of oil reservoirs, and in the deepest gold mines in South Africa. Although the density of cells is low and metabolism and cell division are extremely slow, the total biomass in deep rocks is estimated to be of a similar order of magnitude to that of all life in the oceans. Although life in oil reservoirs finds a lot of food, it is slowed down by the high temperature and a lack of other nutrients such as phosphorus. Life mainly lives at the lower boundary of the oil reservoir where it gets its nutrients from both oil and water. There are innumerable different bacteria and archaea that follow different metabolic pathways (Head et al. 2003; Aitken et al. 2004; Dolfing et al. 2008; Hallmann et al. 2008; Jones et al. 2008; Ross et al. 2010). Different substances degrade at different rates. Unbranched n-alkanes are the first to disappear followed by isoalkanes and then aromatic substances; while asphaltenes remain. In the case of extra-heavy crude oil around 50% of the original oil quantity has already been lost. Sulfate-reducing bacteria are very widespread in sulfate-rich reservoirs. They use sulfate to oxidize organic substances and release H_2S together with acetate or hydrogen carbonate. Oil is converted into methane in low-sulfate reservoirs in several steps. The first step involves syntrophic bacteria oxidizing alkanes to acetate and H_2. Although archaea combine these to CO_2 and CH_4 by means of acetoclastic methanogenesis, most of the acetate is oxidized to CO_2 and H_2 by syntrophic bacteria. However, there are other archaea involved in hydrogenotrophic methanogenesis that consume H_2 and CO_2 and release CH_4, CO_2, and H_2O. Hydrogen added by the hydrolysis of minerals or the aromatization of hydrocarbons enables almost all CO_2 to be converted to CH_4. Fermenting bacteria that also break down more resistant substances become more important once the alkanes have disappeared. Biodegradation also produces organic acids and compounds such as 25-norhopanes in addition to methane and H_2S (Bennett et al. 2006).

The lower the temperature the faster the biodegradation. Biodegradation stops at 80 °C because at this temperature microorganisms can no longer survive. Most oil reserves are more or less degraded. If this is not the case, then the oil has either only just arrived in the reservoir or the reservoir has a tectonic history in which sterilization has taken place at more than 80 °C at an appropriate depth (Wilhelms et al. 2001). Since degradation begins with migration and continues in the reservoir the timing of migration is crucial to the quality of a deposit. Incidentally, it is best when migration takes place after the traps have been created. Before turning to the subject of oil production it is worth taking a look at the region around the Persian Gulf, without doubt one of the most important petroleum provinces.

6.4 Oil from the Persian Gulf

The region around the Persian Gulf is by far the most oil rich in the world (Fig. 6.20). This is because its geological history has brought about with extraordinary timing a perfect combination of source rocks, reservoir rocks, and tightly sealed traps. Moreover, this happened three times in a row throughout the region such that three highly productive petroleum systems are stacked one on top of the other (Ahlbrandt et al. 2000; Fox and Ahlbrandt 2002; Pollastro 2003). Type 2 kerogen, which produces both oil and gas, dominates all of them. In addition, there are other oil systems in the region that are of local importance.

Today's Arabian Plate includes the Arabian Peninsula, the Persian Gulf, and a part of Iran. The tectonic suture to Eurasia runs through the northeastern part of the Zagros

Mountains called the High Zagros. Throughout the Paleozoic and Mesozoic the Arabian Plate was part of Gondwana located on the coast of the Tethys. Much of it was flooded by the sea over this long period and formed a shelf sea. With only a few interruptions continental and shallow marine sediments were deposited on the platform whose shorelines fluctuated. Sedimentation began at the end of the Precambrian when extension occurred through the region after a long phase of mountain formation. Salt horizons several kilometers thick were deposited in the basins—the best known being the Hormuz salt horizon. Salt cushions and salt diapirs rising from this horizon would later form important traps. These early rift systems already contained the first source rocks from which the important oil fields of Oman developed.

Deposition of sandstones dominated in the Paleozoic and Triassic. They were occasionally interbedded with carbonates, shales, or evaporites. The Qusaiba (locally a.k.a. Akkas, Abba, Mudawarra, and Tanf) Shale was deposited in the Early Silurian when sea levels rose at the end of the Ordovician Ice Age. This shale is buried relatively deeply in this sedimentary pile. The base of this unit called hot shale is

Fig. 6.20 The most important oil (*green*) and gas (*orange*) fields in the Gulf region

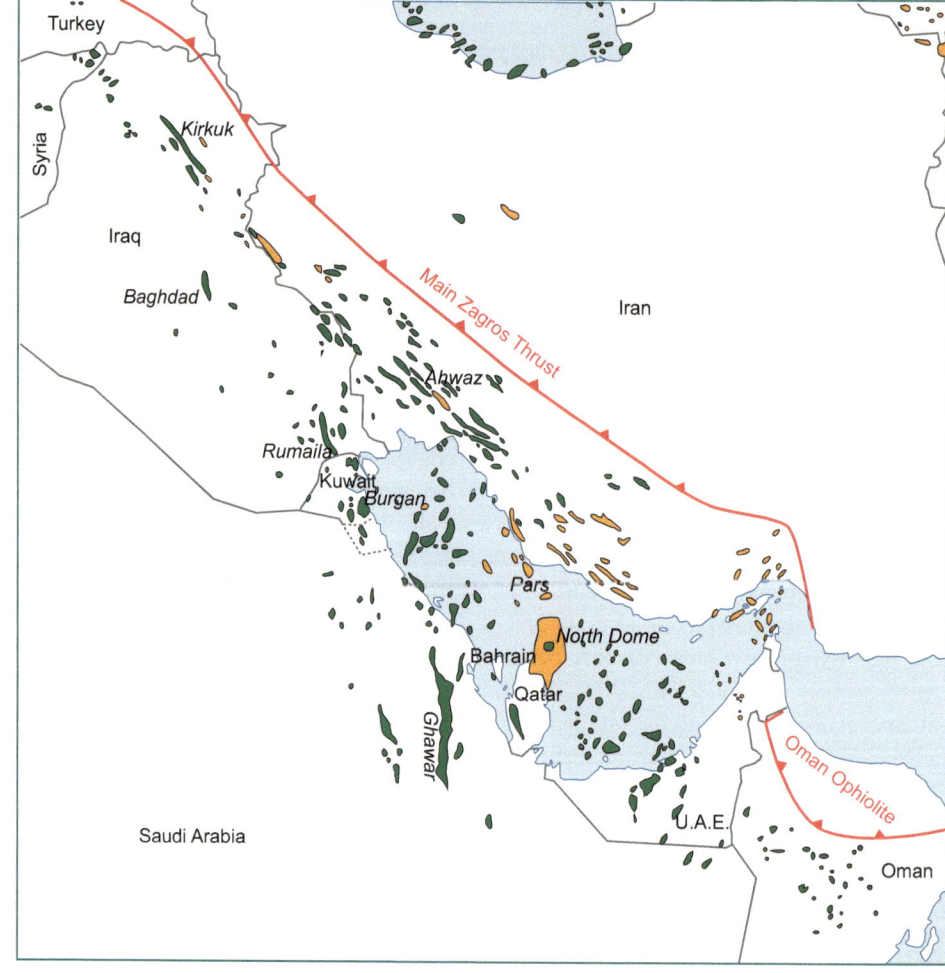

a black shale deposited under euxinic conditions. The shale is 10–65 m thick and contains several percent organic carbon (up to a maximum of 20%). Being a first-class source rock it is responsible for the deepest major petroleum system in the region.

The subsequent deposition of sandstones—known to be good reservoir rocks—was once interrupted by a phase of compression, uplift, and erosion. This was because a subduction zone had formed on the continental margin. Individual blocks of the basement were lifted and the sediment pile was arched over them to form huge folds with gently dipping limbs. The most important lifted structure was the Ghawar (Saudi Arabia) Anticline. Above the unconformity followed continental sandstones deposited by rivers and partly by wind in the Early Permian. These rocks became the most important reservoir of this system. In the Late Permian the sea level rose and the entire platform became a system of lagoons in which carbonates and anhydrite were deposited alternately in several cycles as water levels fluctuated (see also Sect. 5.7). This is the Khuff Formation that seals the petroleum system.

Extension of the back-arc of the still active subduction zone in the Triassic led to the separation of mini continent fragments that drifted through the Tethys to today form parts of Anatolia and Iran. Although the platform once again had a passive continental margin, extension had also affected the platform itself by activating faults in the basement, reinforcing large folds in the sediment stack, and bending the Khuff Formation. The traps were now complete—just in time because a little later source rock reached the oil window in the deepest basins. The oil released is of particularly high quality, very light, and low in sulfur. Almost everywhere it followed Khuff Anhydrite into the highest zones and migrated huge distances into the Ghawar Anticline and other large traps. Some oil also ended up in sandstone below the source rock and further oil remained trapped under shale horizons within the sandstones. Further traps were formed later by salt diapirs. This includes the largest gas field in the world called North Dome lying offshore Qatar where gas is trapped under Khuff Anhydrite in a domed structure above a salt diapir. The structure was probably formed during the Alpine orogeny when the source rock was already so deep that its oil was cracked into gas.

In the Jurassic the whole platform was a very shallow shelf sea in a tropical climate. Thick carbonates were deposited. Shallow water in three partial basins was at times euxinic in the Middle Jurassic as a result of low circulation. Fine-grained limestones that had a very high content of organic carbon were formed as source rocks such as the Tuwaiq Formation and Hanifa Formation (Arabian Basin and South Arabian Gulf Basin) and Sargelu Formation (Gotnia Basin, Iraq, and Kuwait). In the late Jurassic lagoons formed and alternating layers of carbonates with a porosity of 5–30% and of anhydrite were deposited. This resulted in the interbedding of very good reservoirs and very good seals (Arab Formation or Gotnia Formation). Hith Anhydrite 150 m thick finally covered everything and sealed the system completely in the whole region. This second petroleum system is the most important system in the region and responsible for some very large oil fields. Although oil formation began in the deepest basins in the Cretaceous, most of it happened in the Cenozoic. This oil is light to medium heavy and has a higher sulfur content than oil from the first system.

In the Cretaceous the coastline ran right across the platform but the position changed all the time. Sandstones were deposited on the coastal plains, whereas sandstones, shales, and carbonates were deposited on the shelf. The greenhouse climate that prevailed at the time also lead to euxinic conditions from time to time such that further source rocks (shales, carbonates) were formed—and the shales can be suitable seals. In the Paleogene a further anhydrite horizon followed called the Rus Anhydrite that once again sealed this system over a wide area. This third petroleum system is important in the Zagros Mountains and their foreland basins (Iran, Iraq, Kuwait) where source rocks reached the oil window by overthrusting and burial during mountain formation and in the United Arab Emirates they are buried under thick younger sediments. This system generated some very large oil fields including the Greater Burgan in Kuwait, which is the second largest in the world.

The Alpine orogeny commencing at the end of the Cretaceous is responsible for the traps that formed in Jurassic and Cretaceous oil systems. It began with the Oman Ophiolite being obducted in the east, the collision with Eurasia beginning in the north, and the Zagros Mountains developing. In the High Zagros nappes derived from the European continent were thrust over the edge of the Arab Plate. As a result the entire sediment stack began to move because the Hormuz salt horizon at the base formed a perfect detachment fault. The compressed sediment pile was dissected by numerous thrust faults and ramp structures and large and very regular folds were formed in the Zagros fold and thrust belt. Although these folds are much narrower than the Ghawar Anticline and have steep limbs, there is still room in the anticlines for world-class oil fields. They include most of the large fields in Iran and northern Iraq. Most of the traps in the United Arab Emirates (Alsharhan 1989) are gentle folds associated with obduction of the Oman Ophiolite.

Old structures like the Ghawar Anticline deformed once again during the Alpine orogeny. The Ghawar Anticline is by far the largest oil field in the world. It is 180 km long, 30 km wide, and according to current estimates originally contained about 100 billion barrels of recoverable oil of which less than half is left. Currently 5 million barrels per day are produced along with 57 million cubic meters of gas per day. Sediments are vaulted by basement blocks that have been lifted in several phases to form a huge anticline the limbs of which dip gently on both sides. In this way the oil released in a vast area both from the deep Paleozoic petroleum system and the Jurassic system collected here. In Saudi Arabia and Iraq there are other similar anticlines above lifted basement.

Traps associated with salt diapirs and salt cushions formed wherever thick evaporites lay at the base of sediments such as in the Zagros fold and thrust belt, the Persian Gulf, and Oman. Finally, the Arabian Plate broke away from Africa and the Red Sea came into being and with it further oil fields. The most important are in the Gulf of Suez (Egypt) where source rocks are older than the rift system and salt deposited during formation of the sea serves as a seal.

Immediately north of the Persian Gulf is the Caspian Sea. Another oil-rich region this basin was only filled in the Cenozoic, but so quickly that sediments are up to 20 km thick. Even farther north are the rich fields of the Volga-Ural Province and of Western Siberia (Russia). The greater region from the Arabian Peninsula to the north coast of Russia is sometimes referred to as the Strategic Ellipse that contains about two-thirds of all conventional oil and gas reserves.

6.5 Production of Petroleum and Natural Gas

Oil fields at the beginning of the Age of Oil were developed in a completely different way than today. Back in the day it was all about extracting the oil as easily as possible. Oil fields were often divided into innumerable claims and a real forest of drilling rigs made it possible to produce large quantities at short notice. However, this had the consequence that pressure in the reservoir decreased rapidly and most of the oil remained in the ground and was no longer recoverable. However, today a strategy is developed at great expense for each field to optimize the extraction of as much economically recoverable oil as possible. In fact, known oil reserves (Fig. 6.21) have risen significantly in recent decades as a result of the improved recovery factor—and only slightly as a result of new discoveries. Such a strategy involves deposits being explored in great detail prior to production. Continuous monitoring of the deposit using seismic methods such as 4-D seismics (Sect. 1.7) enables the strategy to be adapted to changing conditions during production.

How best to proceed depends on a number of factors such as the size, shape, and depth of the trap, the location on land (onshore) or at sea (offshore), the viscosity of the oil, and the permeability and heterogeneity of the reservoir. If permeability is very high and viscosity low, then it is often better to develop a large field with just a few production wells. In other cases many wells need to be drilled short

distances apart (Fig. 6.22). Traps are often drilled horizontally such that oil flows from a certain level into the borehole —not vertically. If there is a gas phase above the oil, then it should be the last to be extracted whenever possible because otherwise fluid pressure in the reservoir drops and a correspondingly larger proportion of oil remains unrecoverable at depth. Since it is now possible to drill several kilometers horizontally some offshore fields can be accessed by a single drilling platform (Fig. 6.23) from which many boreholes can access different points in the field. Drilling itself was described in Sect. 1.9.

When a reservoir is subject to overpressure oil automatically shoots out of the borehole during drilling. Back in the day there were occasionally violent fountains (blowouts) capable of destroying the drilling rig (Fig. 6.24). Today this is prevented by a valve system installed above the borehole during drilling (blowout preventer). Another valve system referred to as a Christmas tree is then installed for extraction the oil from which flows directly into a pipeline. In this case it is simply a case of turning on the oil tap. The valve has additional openings through which liquids can be injected or probes inserted into the borehole. As soon as the pressure has decreased or even if it has been low from the beginning, then a pump called a nodding donkey (Fig. 6.25) is installed. Such nodding donkeys are well known for moving their heads up and down while the main part of the pump is installed in the borehole.

The water to oil ratio within pores is an important factor. As described in Box 6.7 a high water content prevents the flow of oil or gas from pore to pore. This is the reason some newly developed oil and gas fields initially produce a lot of water—oil and gas are only really mobile when they form a continuous phase. However, this is also the reason production is declining when most of the oil is still in the pores. If the permeability of reservoir rock is very heterogeneous, then this problem is much greater.

Since pressure in the reservoir and therefore the force driving oil to the well both decrease only a small part of the oil available can be extracted in this way (primary recovery). Often only 5–10% is recovered, sometimes a little more. How much depends on several factors such as the viscosity of the oil and the permeability of the reservoir. Pressure drop can be partially compensated if water can flow in from an aquifer and drive the oil out or when gas expands in a gas bubble present above the oil.

Pressure drop is usually counteracted relatively early by injecting water and/or gas (N_2 or CO_2) in a process called secondary recovery. Special injection wells are required for this such that water drives oil from the sides and gas drives oil from above to the production well. Practically all the salty connate water (pumped together with oil from the well) can be disposed of in this way. Of course, this is still not enough and also seawater has to be used for flooding. By injecting acids permeability can be improved because

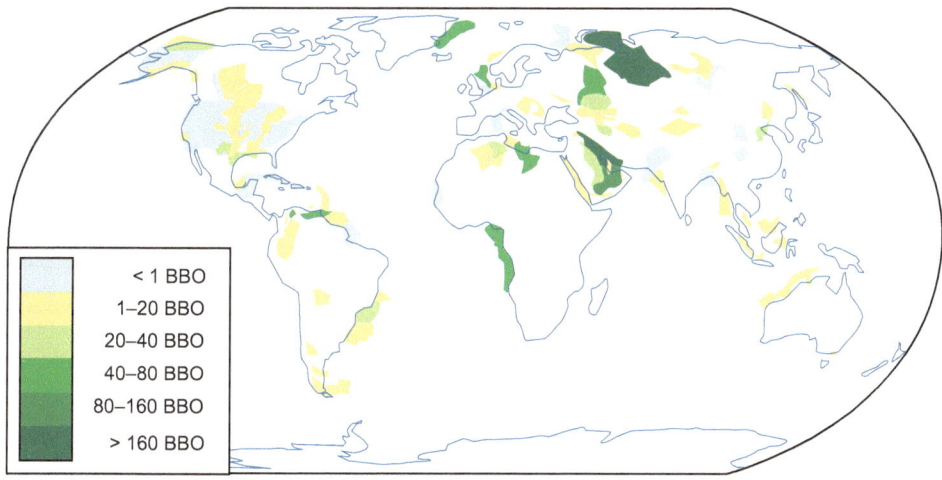

Fig. 6.21 Conventional oil reserves. Oil volume of respective basins in billion barrels of oil (BBO) (from *USGS World Petroleum Assessment 2000*)

Fig. 6.22 Small drilling rig near Baku (Azerbaijan) with (*left*) drill pipes awaiting use (*photo* © F. Neukirchen)

carbonates are dissolved. However, water preferentially flows along certain particularly permeable paths and pores remote from them are hardly affected. At some point more water than oil is pumped to the surface. Flooding with seawater also has the disadvantage that microorganisms can enter the deposit leading to increased biodegradation of the oil and increasing the sulfur content in the oil by reducing the sulfate dissolved in seawater to H_2S. In addition, sulfate from seawater can cause the precipitation of baryte ($BaSO_4$) in the pores thus reducing the permeability of reservoir rock. This is the reason sulfate is often removed from seawater before injection and biocides are added. Even using secondary production methods little more than one-third of the oil originally present in the deposit can be extracted on average, although this proportion can be significantly higher or lower depending on the circumstances.

Although individual deposits are readily depleted using such conventional methods, the lifetime of an oil field can be extended using so-called tertiary recovery methods also called enhanced oil recovery (EOR) (Thomas 2008). The extent to which technical possibilities made available today can be exploited is dictated by the oil price; depending on the circumstances an additional yield of 5–30% of the originally available oil is currently possible—the global average thus increases to around 50% of the original oil. Strictly speaking, this can no longer be considered conventional production.

One such technical possibility is to inject gas (N_2, CO_2, natural gas) or even petrol such that it dissolves in the petroleum and reduces viscosity or mobilizes heavy components thus increasing the pressure. This works quite well in homogeneous reservoirs with light to medium–heavy oil and is used very frequently in the United States. Which gas is

Fig. 6.23 A drilling platform (drilling rig) on its way to the Marlim Sul deep-sea oil field in the Campos Basin, off Brazil (*photo* © Agência Brasil, CC-BY-SA, Wikimedia Commons)

injected depends on the characteristics of the reservoir, depth, and temperature.

When it comes to heavy crude oil the injection of superheated steam via a process called steam flooding (a.k.a. huff and puff) is widely used. In this process heat reduces the viscosity of oil and oil flows more easily to the production well. For some time attempts have even been made to generate heat by igniting the reservoir, but this is not so easy to control.

Another process is polymer flooding that increases the viscosity of injected water (Han et al. 1999; Abidin et al. 2012). Water is mixed with polymers such as polyacrylamide (naturally produced from hydrocarbons) or xanthan gum (a sugar produced by the fermentation of carbohydrates; also used as a gelling agent in foods). Sometimes water is mixed with polymers and clay. Such mixtures are injected for years until about one-third or one-half of the pore volume in the reservoir is filled with it. The viscous mass more effectively drives oil to the well since it gathers up oil drops rather than simply flowing past them. Later normal water flooding can be continued. The zones that were previously most permeable and through which water flowed

preferentially are now less permeable such that other pores can now also be reached. This has been practiced on a grand scale in China for quite some time. In the Daqing Field a factory was built to produce 30,000 t of polyacrylamide per year. The optimum weight ratio of injected polymer to additionally produced oil is 1:200 in this field. Since the turn of the millennium this method has increasingly been used throughout the world. However, it does not always work; there have been pilot projects in which more polymer was injected than additional oil was produced. Polymer flooding is often combined with the injection of other chemicals. Surfactants ensure that oil and water become an emulsion. A similar effect is achieved by adding bases that react with oil forming soap at the boundary between water and oil.

Biotechnology is also used in tertiary extraction in a process called microbial enhanced oil recovery (MEOR). Its application has so far been largely limited to pilot projects. MEOR shows great potential for even higher yields. Although microorganisms are normally not welcome in oil fields (especially, sulfate-reducing bacteria whose H_2S increases the sulfur content), sometimes inhabitants of the reservoir are fed extra food such as nitrogen and molasses or special organisms are introduced (Sen 2008). The goal is to have the same effect as that of chemical flooding but possibly achieved more cost-effectively with microorganisms. Biofilms can clog the most permeable zones of a heterogeneous reservoir and thus force a flow through pores in other areas during water flooding. Some bacteria produce biosurfactants that can either be produced in the laboratory and then injected or produced in situ by bacteria introduced into the reservoir. Biopolymers can also be synthesized by bacteria in the reservoir. Microorganisms that degrade oil to methane increase the pressure in the reservoir and lead to lower viscosity by dissolving gas in the oil. Experiments are also being carried out with enzymes and genetically modified organisms.

Natural gas is transported as liquid gas (Box 6.8) in the absence of a pipeline. Petroleum is further processed in a refinery (Box 6.9).

Box 6.8 Liquid gas

Gas pipelines work very well. In the absence of gas pipelines the large volume of gas becomes a disadvantage during transport. Even if gas is compressed into compressed natural gas (CNG) the calorific value per volume is not comparable with oil. However, propane and butane can be liquefied very easily either by cooling or at normal temperature under pressure. This is called liquefied petroleum gas (LPG) that remains liquid as long as it is enclosed in a pressure vessel. LPG is therefore easily transported in tankers.

Fig. 6.24 Oil fountains (blowouts) used to occur when drilling into a pressurized reservoir. Today this is prevented by a valve system. This picture from 1916 shows a newly drilled well in the Sunset oil field, California (*photo* © I.D. Pack, USGS)

Fig. 6.25 An oil pump (nodding donkey) in Baku (Azerbaijan) (*photo* © F. Neukirchen)

The pressure does not even have to be very high. An example of liquid gas is the liquid in lighters. However, methane—the main component of natural gas— is not so easy to liquefy. It must be cooled down to −162 °C and kept almost as cold during transport. The volume drops to 1/600th, but the liquefaction of methane to liquefied natural gas (LNG) consumes much of the energy content. However, if the price of gas is high and there is no direct connection to consumers via pipelines, then it is still worthwhile. This is the reason the share of LNG in gas supply has risen significantly in recent decades. This requires very special infrastructure with liquefaction plants, loading terminals, special ships, and a corresponding terminal in the destination port where LNG is gasified again and fed into the pipeline network. Countries such as Qatar with enormous gas reserves would not be able to use their resources without this technology. This may be an answer for the European Union since it is looking for greater independence from Russia in gas supply.

It is also possible to produce higher hydrocarbons from methane in a process called gas to liquids (GTL). Such hydrocarbons are liquid and suitable as fuel, synthetic petrol, diesel, and kerosene. The first step involves methane reacting with water vapor and oxygen to form synthesis gas (CO and H_2) from which liquid hydrocarbons are synthesized in the Fischer–Tropsch process using catalysts. The process can now even be used in small plants on remote offshore platforms that previously had to burn natural gas as a result of not having a gas pipeline.

Box 6.9 Petroleum refinery

The most important process in petroleum processing is distillation (Fig. 6.26). Previously purified petroleum is separated into individual components with different boiling points. Heated to about 400 °C, petroleum enters the distillation column at the bottom. A large proportion is gaseous at this temperature and rises in the column. Toward the top the temperature decreases and smaller and smaller molecules condense and are drawn off at intermediate floors. In the lower part of the column the liquid components comprise lubricating oils (C_{26}–C_{40}), heavy distillates (C_{19}–C_{25}), and substances that become solid during cooling (molecules with >C_{40}) such as resins, asphaltenes, and heavy fuel oil from which road surfaces, insulating agents, fillers, and adhesives (in particular) are made. Between 260 and 360 °C middle distillates (C_{19}–C_{25}) condense, at 190–260 °C kerosene (C_{11}–C_{13}) condenses, and at 25–190 °C petrol (C_5–C_{10}) condenses. Smaller molecules remain gaseous (C_1–C_4). Sulfur, oxygen, and nitrogen groups are subsequently removed from intermediate products.

Diesel and heating oil are different mixtures of kerosene, light and heavy fuel oil, and various additives. Cracking is a process in which long-chain polymers are broken down into smaller molecules. The C/H ratio is reduced in the process leaving a solid substance consisting almost entirely of carbon called petroleum coke. Isomerization is the conversion of *n*-alkanes into isoalkanes using a catalyst. Organic compounds can also be combined to form many other compounds.

Fig. 6.26 Oil refinery (*photo* © Getty Images/iStockphoto)

Isoalkanes are branched alkanes that have the advantage of higher knock resistance. Knocking is the self-ignition of fuel in the cylinders of engines caused during compression by the heat present, which in the long term destroys the engine. Antiknock fuel does not autoignite. Knock resistance is determined using a test engine and indicated as the so-called octane rating relative to pure isooctane (octane number 100) and pure *n*-heptane (octane number 0). These are the numbers that appear on the dispensers of petrol stations (however, there are three different standards). In the past tetraethyl lead was simply added to petrol as an antiknock agent.

6.6 Peak Oil

The resources of the planet are of course finite. This is more so the case with petroleum than any other resource. As of 2012 some 167 billion tonnes had been consumed (worldwide cumulative production; Andruleit et al. 2012a) corresponding to about 1.2 trillion barrels. Annual production is currently just over 4 billion tonnes (four times as much as in 1960). With the economic growth of the BRIC countries (Brazil, Russia, India, China) demand is expected to continue to grow (OPEC 2012). At 168 billion tonnes the known reserves of conventional oil are pretty much the same as cumulative production to date—that is only enough for 42 years at the current rate of consumption. How long oil actually lasts cannot be predicted with certainty. However, what is certain is that oil will not run out suddenly. At some point a global production peak will be reached and then production will decline. The concept of peak oil was developed in the 1950s by the American oil geologist Marion King Hubbert. His starting point was the output of individual oil wells, which he observed rises rapidly from the time the well is found, reaches a maximum at a high level after some time, and then drops relatively quickly. From this he derived the so-called Hubbert curve that is intended to describe the development of production volume over time for entire regions with many wells or even worldwide. It looks similar to Gaussian normal distribution. Interestingly, for some regions—but by no means all regions—the curve describes the development of conventional production quite well. The exponential increase in global production from the Second World War to the first oil crisis in 1973 fitted the curve well, but since then the real curve has no longer been as regular. However, there is no compelling argument to believe that real development must be symmetrical and that the drop after peak oil will be as rapid as the increase before peak oil. The theory of peak oil is surrounded by numerous

scenarios of doom ranging from world wars and famines to the complete collapse of human civilization as production declines. Although that can be dismissed as exaggerated panic mongering, nevertheless there is a need for action. The sooner a switch to alternative technologies takes place the better.

The timing of the global production peak is impossible to predict. In 1989 it was proclaimed that the peak had been reached (Campbell 1989), which as is well known did not happen. Even today the interpretation of data by experts varies widely. The pessimists say the point has long since been passed; others say several decades are left before it is reached. In 2009 the German Geological Survey (BGR) spoke of 2035 being the production peak, although the follow-up study is much more optimistic (Cramer and Andruleit 2009; Andruleit et al. 2012a). The biggest optimists (such as Maugeri 2004) believe that the peak is far from being near. Perhaps the Age of Oil will end when not all oil has been consumed much as the Age of Coal in the nineteenth century ended before coal ran out.

There are several reasons for caution in this discussion. On the one hand, in recent decades improvements such as monitoring with 4-D seismics and so-called tertiary recovery methods have led to a significant increase in the recoverable share of conventional oil fields. The resulting expansion of resources has been significantly greater than the discovery of new oil fields. Of course, there are limits to this development since it will certainly never be possible to recover 100%. However, even a small improvement can result in a significant amount of oil yield. In addition, early estimations of the oil quantity of a field are often too low and fields then "grow" with further exploration. On the other hand, originally only conventional oil was considered. However, unconventional sources such as tar sands in Canada and shale oil in the United States already account for a considerable proportion of production and the share is rising. This is where enormous additional resources come into play that put conventional oil in the shade. Although recent forecasts take unconventional oil into account, different estimates of recoverable quantities range widely.

A further argument put forward by some pessimists is that new finds reached a peak in the 1960s and then became scarce. Now it is argued that this is because almost all large fields are already known. This is where another false assumption comes into play—namely, that large oil fields are the first to be found. However, a small, near-surface field is often easier to find and cheaper to develop, while some large fields are difficult to find and expensive to develop. However, the lull in new discoveries could be argued as being due to the price of oil being relatively low in the 1990s and correspondingly little being invested in exploration. Since the price has risen significantly since the turn of the millennium exploration has intensified leading to the discovery of several large fields. There are still sedimentary basins that have not yet been sufficiently investigated—even if they are becoming ever fewer. There are of course the polar regions. The USGS believes that Iraq has the greatest potential worldwide for conventional oil fields that have not yet been discovered because hardly any investments have been made in further exploration as a result of the size of known oil fields.

However, many new fields discovered in the last decade are in the deep sea. Accessing them involves not only enormous technological effort but also very high development costs (Anonymous 2010, 2011). Wealth as Arab countries know it is not likely to come from such fields. A good example of a deep-sea find is the Lula Field discovered offshore Brazil in 2006. It is one of the most important new discoveries and could cover the world's needs for three months. It has 2000 m of water and 5000 m of rock above it and is located in a trap under a thick salt horizon. At the time of discovery there was only one platform in the world producing oil in similar water depths—a decade earlier the world record was half as deep. Moreover, drilling through salt under correspondingly high pressure is not easy and can only be done thanks to new technology termed measurement while drilling at an acceptable risk. New platforms working here cost many times more than shallow-water platforms and running costs are higher. Production from the Lula Field began in 2013. When it comes to depth the record holder since 2010 has been Perdido Spar in the Gulf of Mexico, which operates at a water depth of 2438 m and has almost reached the outer edge of the continental slope. Actual deep-sea basins have little sediment cover; hence no oil can be expected there. People are of course working at their technical limits in the deep sea as evidenced by accidents such as the Deepwater Horizon accident (Box 6.10). The enormous cost of developing such fields blurs the line between conventional and unconventional oil wells. It is safe to assume that further large finds will be made at similar water depths in the coming decades, perhaps even in polar regions. However, the time will come when there will be a real shortage of conventional oil.

Global production maximum should be reached at some point; otherwise, the curve will be much more irregular than predicted by early peak oil theorists. However, peak cheap oil was probably passed a long time ago. This will likely lead to renewable energy sources becoming more competitive (and using them is also beneficial for the climate). The number of fields easily exploitable and still waiting to be discovered is likely to be small. Profit margins will continue to rise for countries that still have reserves of cheap oil. Saudi Arabia and Iraq still have large oil fields in their backyards that are yet to be developed because fields currently producing already cover desired production quotas.

The higher the proportion of oil produced at high cost the more likely the price of oil will rise even if enormous resources are still available—irrespective of whether sophisticated tertiary production is employed or new normal fields in the deep sea, polar regions, and the remotest parts of Siberia are found or unconventional sources such as tar sands or shale oil are used. That does not mean the price of oil cannot fall. If capacity is expanded too quickly, then the price of oil will fall below the production costs of unconventional oil with corresponding consequences for the companies involved. Global production depends not only on available resources, but also on consumption and the price producers can achieve.

Box 6.10 Deepwater Horizon

The disaster at the Deepwater Horizon drilling platform (Fig. 6.27) in the Gulf of Mexico, 66 km off the coast of Louisiana, was the worst oil disaster in history. On the evening of April 20, 2010 a methane blowout occurred and high-pressure natural gas escaped and ignited explosively on the platform. For a little over a day the platform was engulfed in flames before it sank. Eleven workers were killed. The well was in the process of being completed and experiencing considerable problems in concreting casings in the deepest part of the well partly due to the use of unsuitable cement. Pressure displays were also misinterpreted and necessary tests were omitted. Finally, the blowout preventer installed on the seabed failed. The platform's engines suffocated in the gas cloud causing the platform to drift. Deficiencies in the security system did the rest.

When the platform sank it was by no means the end of the story. Oil and gas poured at high pressure from several leaks on the seabed at a depth of 1400 m. According to later estimates the initial amount of oil discharged was 62,000 bbl per day. At first, an unsuccessful attempt was made to close the blowout preventer using diving robots. Then a steel dome was put over the leak to guide the oil through a pipe, again unsuccessfully. Although this method works in shallow water, it does not in the deep sea because methane hydrates form (Sect. 6.10) clogging the pipe. The oil spill was so strong that it was not possible to inject heavy drilling fluid through the blowout preventer (top kill). All that could be done for the time being was to simply insert a pipe into the leak and intercept at least some of the oil, while an attempt was made to get the leak under control by sawing off pipes and replacing caps. Two drilling ships began to drill so-called relief wells that would link up with the well

Fig. 6.27 Deepwater Horizon in flames on April 21, 2010 (*photo* © US Coast Guard/USGS)

some 5500 m under the seabed with the aim of closing it with bentonite and cement (bottom kill). On the 85th day they finally managed to put a 75-t cap on the leak and close it temporarily. In the interim 4.9 million barrels of oil had escaped not counting the portion captured. To make matters worse a few days later the ships had to leave the disaster zone because of a storm. Only when the storm had passed could the well be finally plugged first from above by pumping in cement and bentonite, which was possible because the oil pressure had decreased, and then also from below via relief wells.

Some of the oil floating on the sea surface was flared off and some collected by special ships. Although the oil spill was enormous and reached coastal regions on both sides of the Mississippi Delta, fortunately a smaller area was affected than had been feared. Among the victims were 151 dolphins, hundreds of turtles, and many birds. The oxygen content of water below the oil dropped. Damage in the deep sea seemed to be even greater as a result of much of the oil sedimenting on the seabed and killing organisms living there and in deep water (Schrope 2011). The reason for sedimentation was probably the depth of the water with large drops of oil taking four hours to ascend through the water column and small ones much longer. There was clearly enough

time for microorganisms to consume the light components of oil making the oil heavier and allowing it to sink again.

6.7 Fracking: Shale Gas and Tight Oil

Shale gas and shale oil are hydrocarbons firmly trapped in the pores of shales. Although they are incapable of leaving the source rock, they can be recovered by fracking. Confusingly, the term shale oil is sometimes used as a synonym for synthetic crude oil derived from oil shale (see next section). Tight gas and tight oil are hydrocarbons trapped in reservoir rocks whose permeability is very low.

Arguably the most controversial method of producing hydrocarbons is fracking—short for hydraulic fracturing. Put simply, liquid is pressed into a borehole at very high pressure with the aim of opening up small cracks in the rock and thereby interconnecting small hydrocarbon bubbles trapped in the rock and connecting them to the borehole. This makes it possible to develop reserves that cannot be extracted using conventional methods because the permeability of the rocks concerned such as shale or slate is very low. When conventional deposits are developed it might be enough to drill into reservoir rock and the substance hoped for reaches the

Fig. 6.28 Before fracking commences a borehole between about 1000 and 1500 m long is drilled horizontally or rather parallel to bedding in the corresponding layer. The borehole is lined with a casing. Small explosive charges create holes in the casing. Hydraulic pressure builds up and creates cracks in the rock. This allows gas trapped in rock pores to flow into the borehole. Sand prevents the cracks from closing again. The fluid used for fracking contains many chemicals that must not get into drinking water

pipeline by itself (possibly with a little help). However, in the case of unconventional deposits as large an area as possible of the rock of interest must be affected by drilling. This is achieved by drilling horizontally or rather parallel to bedding of the corresponding layer (Figs. 6.28 and 6.29). This means that more sophisticated drilling technology is needed in unconventional deposits. It must be possible for staff not only to steer the drill head accordingly, but also to fully understand where it is at the moment, underground. It is here that the old miner's problem is encountered. You do not really know what the shovel—or rather the drill head in this case—will unearth. Although seismics can clearly be used to locate layers, it has an accuracy of around 0.1% leading to inaccuracies in the meter range at a depth of 1–5 km. Various methods can help such as measuring rock density at the drill head using gamma quanta from a caesium-137 source, taking advantage of (salt)water-bearing layers being better electrical conductors than hydrocarbon-bearing ones, and determining the presence of

oil or gas and the thickness of the layer by measuring the specific resistance of the rock and combining it with porosity already measured. Measuring the polarization of electro-magnetic waves shows the possible electrical anisotropy of the rock and gives an estimate of the direction in which the fluids will preferentially move. Other methods used in the search for oil and gas are nuclear magnetic resonance, direct hydraulic measurements, and extraction of rock samples.

The next problem to be addressed is the low permeability of rock. If a hole is simply drilled into low-permeable rock, then only the rock directly affected by the drilling is able to release its trapped substances unlike the case with high-permeable reservoir rock. However, if rock is interspersed with fine cracks, then a much larger quantity of trapped substances can be recovered (Fig. 6.30). This can only be done if the pressure of the liquid pressed into the borehole exceeds geological stress acting on the rocks. Water is normally used as the fracking fluid. However, water alone is not sufficient because more is required than simply creating cracks in the rock. They must also be kept open long enough to be able to recover trapped substances. The weight of overlying rocks at least 1000 m thick ensures that the cracks just formed close again as soon as trapped substances are recovered.

To keep such cracks open various chemical additives are added to the water all of which are more or less harmful to the environment and the exact composition of which is usually regarded as an industrial secret by the relevant companies. The most unproblematic additive is sand that keeps the cracks open once they have been created. Other substances such as butyl diglycol are intended to increase the carrying capacity of water for sand. Acids are used to dissolve rock and thus help create cracks. Polyacrylamides reduce friction between rock and the hydrocarbons to be extracted. Other substances such as isopropanol increase the viscosity of fluids, while yet others are designed to prevent corrosion or depositions. Biocides are also found in fracking fluids. People must not come into contact with many of the substances used unless they are protected. Release of such substances into the environment is prohibited by law. Use of these substances, the environmental hazards they pose, and the risk for earthquakes make fracking the subject of very intense and very emotional debates.

If the method is so complex and risky, why is it used? The world's population has been highly dependent on the availability of fossil fuels such that known and easily accessible fossil hydrocarbon sources on Earth are more or less exhausted. To meet still growing demand for cheap energy, production companies must not only move to ever-more remote regions, but also develop ever-more complex production techniques to develop sources that are ever-more difficult to produce. This is where new unconventional sources come into play such as shale gas and shale oil. Since it is becoming increasingly difficult to find

Fig. 6.29 Drilling tower for a shale gas well in the Fayetteville Shale of the Arkoma Basin (Arkansas, USA) (*photo* © Bill Cunningham, USGS)

Fig. 6.30 Used perforating gun. Small holes are made in the casing by explosives where pressure is then built up by injecting fracking liquid into the borehole (*photo* © Bill Cunningham, USGS)

conventional deposits, exploration geologists are focusing on source rocks from which hydrocarbons have not yet escaped. The situation is analogous to that of a human being fed from a basket full of apples and now hungry only to find that it is almost empty. Observing apples high in the canopy of a tree nearby what is needed is a ladder—and the equivalent of this ladder is fracking.

Developing reservoirs for gas or hydrothermal energy production by means of deliberately produced microfractures is not a new idea. Hydraulic fracturing was used experimentally for the first time in 1947 and first used commercially in 1949 (Montgomery and Smith 2010). Tight gas has been extracted by means of hydraulic fracturing for around 35 years in the Cloppenburg region of Germany (Ewen et al. 2011).

No other means of extracting fossil fuels has been subjected to so much criticism as hydraulic fracturing. Apart from the risk for earthquakes the main concern of many people is the use of additives and where they end up. Popular movies such as *Gasland* by filmmaker Josh Fox from 2010 have contributed to this. The movie shows dramatic images including scenes from Colorado in which water coming out of a tap is ignited with a lighter. Unfortunately, the movie fails to point out that burning springs exist in nature where combustible natural gas is dissolved in water and released when pressure is relieved such as in West Virginia and Kentucky (USA) and in Azerbaijan. Well before the start of fracking activity in the region concerned the Colorado Division of Water determined in 1976 that the water there sometimes contains combustible gases (Watts 2011; Ecomides 2011; Lingenhöhl 2011). The movie reached a wide audience and certainly fueled critical debate about unconventional gas production not just in Germany but worldwide.

Problems with hydraulic fracturing are primarily caused by additives in fracking fluids some of which (as mentioned above) should not come into free contact with the environment as they are toxic and sometimes carcinogenic. Normally, that should be avoidable, but there are a few problems. Although deposits may be separated from aquifers by large rock packages, the drill and later production pipes first have of course to be driven through the upper layers of such packages before the oil and gas stored there can be brought to the surface. The first critical point is therefore sealing the borehole against aquifers in upper geological levels. Moreover, fracking does not stop at a single drilling. To exploit a deposit in an economically viable way this method requires a large number of boreholes because the effect only works in the immediate vicinity of boreholes.

Another problem is that large volumes of water that can range anywhere between 4.5 million and 19 million litres are needed for fracking. Moreover, even more water is needed when wells need to be refracked, which is often the case (Andrews 2010; Abdalla and Drohan 2010). The provision of such large quantities of water can lead to considerable problems and conflicts of interest (especially, in areas where there are chronic water shortages such as the Karoo in South Africa; Urbina 2011). After fracking any water mixed with additives must be pumped out of the borehole, transported away, or deposited where, of course, it can come into contact with the environment.

What is the real risk of groundwater contamination during fracking? In principle, any contact between fracking fluids used and groundwater should be avoided. Despite gas-bearing strata being located well away from aquifers (separated usually by hundreds of meters or more than 1000 m of rock including impermeable strata) and the borehole being equipped with appropriate casings in groundwater-carrying areas, experience in the United States has shown that accidents can always occur. If the tightness of the well is not guaranteed, then gas from the deposit may well get into groundwater as a study of the Marcellus Shale showed (Tollefson 2013; Jackson et al. 2013). This is particularly true of fracking that took place in the 1990s and early 2000s when a large number of companies without appropriate risk and quality management participated in the development of unconventional deposits—a serious problem according to a number of studies (Lingenhöhl 2011; Jackson et al. 2013). However, Ritva Westendorf-Lahouse, a spokeswoman for ExxonMobil, points to over 300 accident-free fracks in about 50 years (Nestler 2013).

Another problem is concern that it is fracking itself (i.e., violent breaking up or shattering of rock) that creates corresponding paths for the contamination of groundwater in the first place. Although this is comparatively unlikely since the length of cracks produced depends among other things on energy introduced into the rock, it cannot be ruled out. The hydraulic pressure necessary to generate kilometer-long fissures cannot be generated with current techniques. Fracks are preceded by so-called minifracks or datafracks in which important rock-mechanical parameters for the modeling of crack propagation are determined. The longest upward-directed fracks ever recorded caused by fracking were 588 m in the Barnett Shale and 536 m in the Marcellus Shale (both in the United States). There are natural hydraulic fractures that extend upward more than 1000 m (Davies et al. 2012). Such long vertical fissures are actually not desirable because crack propagation should only take place within a gas-bearing target formation whenever possible. The extent to which fissures created by fracking can be linked to natural fissures in rock is a moot point. The BGR believes that no pressure would built up for fracking in this case (BGR 2011). Whether or not this view is too optimistic in view of our lack of understanding of underground fault zones is yet another moot point.

The BGR considers overlying salts of the Zechstein as a hydrological and mechanical barrier (Kosinowski et al. 2012) when it comes to tight gas in the sandstones of the Rotliegend in Germany. The BGR also considers clay deposits of the Jurassic, Lower Cretaceous, and Tertiary like the Rupelton of the Paleogene as comparable barriers (Kosinowski et al. 2012) when it comes to stratigraphically higher unconventional natural gas deposits in the Posidonia Shale (Jurassic) and the Wealden (Cretaceous).

Sometimes there is a further natural obstacle to the migration of pollutants present in fracking fluids reaching groundwater. The groundwater in northern Germany, for example, is vertically subdivided into a freshwater horizon near the surface and a brine horizon below. Due to the difference in density between freshwater and saltwater there is no significant mixing. This could well inhibit the vertical spread of fracking fluids and deep connate water since both have a higher salinity and thus higher density than usable near-surface groundwater. In this case they would only spread in saltwater (Kosinowski et al. 2012).

After fracking what happens to the fluids contaminated? They not only include the recovered fracking fluids, but also normal drilling water and possibly contaminated brines extracted from the depth. However, the last two also occur in conventional wells and are therefore not a special feature of hydraulic fracturing. Water reflux can also contain dissolved substances such as benzenes or radioactive salts. Since water is often injected into deep rock layers reprocessing may be a way to separate hazardous substances from the water and reuse the rest. However, many municipal wastewater treatment plants are ill equipped to handle such large quantities of water and its content of substances, which can lead to problems as examples in Pennsylvania show (Levy 2011). Special mobile processing plants may be able to help here. When there is no solution it is possible to do without water and change to other fracking fluids. A possible alternative would be propane (Fischer 2013).

The main concern of many people is the possibility of earthquakes being triggered by fracking. In principle, fracking is little more than artificially triggered microquakes. Although they are normally so weak that they cannot be felt on the Earth's surface without technical aids, sometimes there can be stronger shocks believed to be directly related to fracking. Two events in spring 2011 in Blackpool (UK) with magnitudes of 2.3 (April 1, 2011) and 1.5 (May 27, 2011) were related to fracking and caused English authorities to require participating companies to observe seismic activity in their areas very closely (Department of Energy and Climate Change 2012; Vukomanovic 2011). Similar events have taken place elsewhere including Germany such as the magnitude 3.0 earthquake in the Neuenkirchen-Tewel natural gas field on February 13, 2012. Since no fracking had been carried out in this gas field for the previous two years in this case it would be quite possible that the quake had nothing to do with fracking. Conventional gas and oil production can also significantly increase the seismicity of an area if the deposit is emptied and stress conditions in the rock change as a result. In the Neuenkirchen-Tewel gas field there had already been an earthquake of magnitude 4.5 in 2004 (Schrammar 2012). Although earthquakes of magnitudes 3–4 are perceptible to humans, they usually do not cause any damage. Experts believe that stronger earthquakes are not caused by fracking (Arbeitskreis gesellschaftliche Akteure 2011). Sometimes earthquakes attributed to fracking can be traced back to the injection of used water, as was the case in Ohio (Anonymous 2012). Studies of the Barnett Shale in the United States have shown that between November 2009 and September 2011 about 67 shocks of magnitude 1.5 or more occurred. Epicenters in 24 of them could be located very precisely and turned out to be within a radius of 3.2 km of one of the boreholes. Of the 161 drill holes in a study area approx. 90% showed no seismicity (Frohlich 2012). Drill holes in rock layers that are already subject to increased geological stress are particularly at risk. Energy can then easily be released by human activities. Here, too, there is certainly still a great need for research.

The current boom in unconventional natural gas and oil production enabled the United States (Fig. 6.31) to drastically reduce its dependence on energy imports; in recent years it has been able to supply up to 83% of its energy needs from its own sources. As a result US gas prices fell—in some cases very dramatically. Europe indirectly benefited from this because the US need to import liquefied petroleum gas from Qatar fell. As a result previously dominant suppliers such as Russian Gazprom had to offer their customers discounts (Tenbrock and Vorholz 2013).

Europe (Germany, in particular) is heavily dependent on imports of fossil fuels. The BGR estimates that there are around 1.3 trillion cubic meters of recoverable shale gas reserves (Andruleit et al. 2012b). Other estimates predict volumes between 0.7 trillion and 2.3 trillion cubic meters (Nestler 2013). These values are significantly higher than the quantities of conventional natural gas resources of 0.15 trillion cubic meters or natural gas reserves of 0.146 trillion cubic meters.

While America is wallowing in the luxury of having gas reserves for the next 100 years, dark clouds are gathering on the horizon clouding the view a little. However, it is not environmental concerns that are playing a role here. Due to countless drillings in one of the productive formations—the Marcellus Shale—more precise data are now available and there are indications that around 42% less shale gas can be produced here than originally planned, although it might turn out to be possibly only 10% less (Tenbrock and Vorholz 2013). In addition, only a few fields in the United States provide the bulk of the unconventional gas and oil produced. When it comes to shale gas 88% comes from 6 of the country's 30 fields. The situation for shale oil is even more striking with just 2 of 21 fields accounting for 81% of production (Hughes 2013).

Fig. 6.31 Map of the United States showing the distribution of hydrocarbon-containing shales from which gas and oil are currently being extracted by fracking. Some basins contain several corresponding layers. The most important formations are marked in *red*. Shale oil comes almost exclusively from Bakken and Eagle Ford Shale (from EIA 2011)

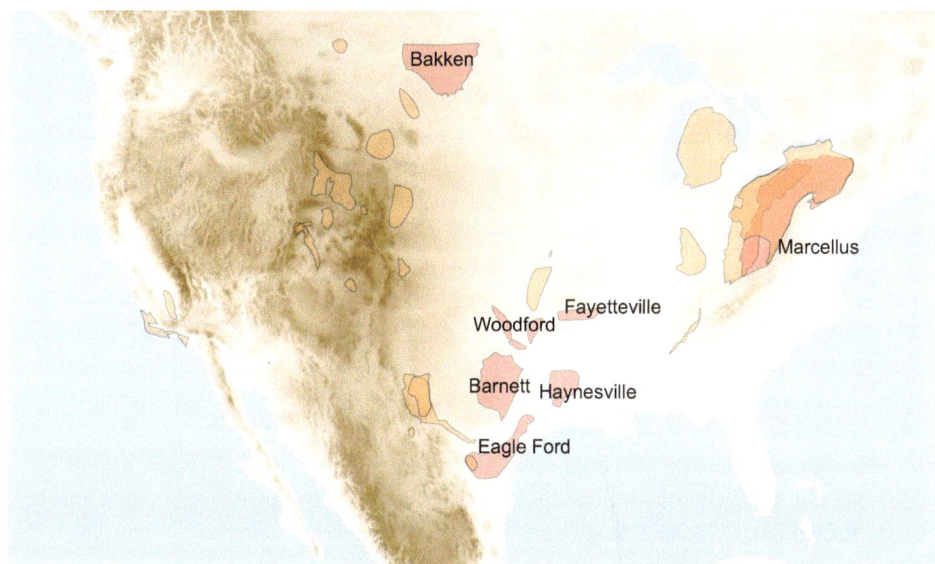

Some of the trouble surrounding estimates can be explained by problems in funding. Gas prices in the United States have fallen sharply to around USD4 per million British thermal units (1 Btu is the thermal energy needed to heat a British pound of water by one degree Fahrenheit). However, a profit would only be possible at around $8 as a result of the production costs incurred (Gärtner 2013; Tenbrock and Vorholz 2013).

Shale gas wells are rapidly exhausted. They typically produce 80–95% less after 3 years than at the outset. Often the flow rate falls by 30–50% within a year. Therefore, a large number of additional wells constantly have to be drilled to keep overall output constant. In the Haynesville gas field (Arkansas, Louisiana, and Texas, USA) this amounts to more than 800 wells per year each at a cost of approx. USD9 million just to keep production at the 2012 level (Hughes 2013). For example, around USD42 billion needed to drill 7200 wells a year is offset by only USD33 billion in revenue from gas sales (Hughes 2013). This coupled with questionable reserves puts production companies under strong pressure some of which are already comparing the boom in fracking with the real estate bubble of 2008 (Gärtner 2013; Tenbrock and Vorholz 2013). The US Energy Information Administration (EIA) does not believe there will be a major boom in unconventional gas and oil production in Europe (Uken 2013).

The subject of fracking will certainly occupy minds for some time to come. The temptation to tap into previously untapped resources will increase as conventional resources run out. This question will become ever more urgent with every fossil fuel price round. Although suspected quantities are considerable, they remain finite. The EIA estimates the worldwide amount of recoverable shale gas at around 207 trillion cubic meters (EIA 2013). In contrast, the BGR estimates there are 173.7 trillion cubic meters of unconventional gas resources worldwide (Andruleit et al. 2012b). Countries such as the United States, China, Argentina, Algeria, South Africa, Mexico, Australia, and Russia have the largest resources. The order varies according to different authors due to still incomplete data in many areas of the world. With at least 1.3 trillion cubic meters Germany is roughly 20th in the rankings, while Poland and France, each with more than 5 trillion cubic meters, have more potential as does the Ukraine. These figures should be compared with consumption and with total gas resources and reserves. Global consumption in 2010 was around 3.2 trillion cubic meters. Global gas resources (including shale gas and tight gas) were about 531 trillion cubic meters, while reserves were about 192 trillion cubic meters (Andruleit et al. 2011).

When it comes to risks it should not be forgotten that the conventional recovery of fossil fuels is also fraught with them. Earthquakes reaching magnitude 4 have been caused by coal mining. The earthquake in Saarwellingen (Germany) in 2008 is a very good example of this (Anonymous 2008). Coal mining can also have a far-reaching impact on the groundwater level of a region. The pumping of water from pits lowers the groundwater level. The weathering of pyrite leads to the formation of sulfuric acid and the release of heavy metals. These substances can cause significant pollution when the groundwater level rises again. Open-cast mines and dumps not only consume large tracts of land but also are blots on the landscape. Even conventional oil and gas production has its problems. Although the environmental risks associated with oil production are obvious, smaller earthquakes can also be triggered by conventional oil and gas production. Possible aftershocks of natural gas production had already been registered (particularly, in the Netherlands and Lower Saxony) before the introduction of fracking or in areas where it is not used (Müller 2013; Mix 2012). This is simply due to the fact that stress can build up and discharge when something is removed from the Earth. This can also lead to subsidence. Pumping groundwater out of tectonically active regions is sufficient for this to happen as shown by an event in southern Spain (dapd 2012).

Many of the dangers associated with fracking are therefore not new when it comes to the mining and extraction of fossil fuels. They are certainly not unique to fracking technology. By following the recommendations of the BGR and the already very high environmental standards in Germany the risk of this technology should be manageable (Ewen et al. 2011). What is important is to avoid tectonically active regions, areas where drinking water collects, medicinal spring protection areas, and areas with potential permeable migration paths irrespective of whether they are natural or artificially created (e.g., by mining activities). At the end of the day it makes more sense to use technology wisely and subject it to home-grown safety precautions than to import gas produced under dubious conditions (especially, when there is a risk of pollution and hazards). Whether fracking is then competitive with conventional gas is a different matter.

6.8 Oil Shale

Even more difficult is the use of source rocks that have not yet reached the oil window (i.e., temperature range in which oil forms) and whose organic substance is still present as kerogen. Although they contain no oil and are not necessarily a shale, they are called oil shales. However, they release gas and synthetic crude oil when heated to about 500 °C via pyrolysis. It is also possible to burn them directly.

Oil shales are clayey–marly sediments or inferior sapropelic coals with a kerogen content between 4 and 50% that is relatively often type 1 kerogen. Theoretically, 40–600 L oil per tonne of rock can be extracted from them without taking into account the high energy requirement of heating. In fact, only a fraction is recoverable. Although this is rarely economical, it could of course change if oil prices continue to rise. Oil shales were deposited in large lakes such as the Green River Formation in the United States in the Eocene, in shallow shelf seas such as kukersite in Estonia in the Ordovician, at the edge of swamps in coal basins such as in Fushun (China), and in small quantities in crater lakes of volcanoes and in maar craters such as Messel, near Darmstadt (Germany).

Messel is one of the most famous fossil sites in the world. Countless animals from the Eocene have been found here. They are so well preserved that the soft parts of mammals and the colors of insect wings can still be seen. Oil shale was mined as fuel in the past; hence many first-class fossils have certainly been burned.

Open-cast mining is the most efficient means of extraction and direct combustion in a power plant is the most energy-efficient way of using oil shale. Residues also form a good raw material for cement production. Although 90% of Estonia's electricity demand is covered by oil shale, it is of no importance to any other country. The Dotternhausen cement plant in Germany uses oil shale internally to generate electricity and then produces cement with fired oil shale.

Oil shale is heated in a retort to extract synthetic crude oil. Estonia and China are the only two nations to do this on a grand scale. In China thick layers of oil shale overlying a thick coal seam are mined in large open-pit coal mines. Exploration is being carried out in some other countries in view of the high price of oil and consideration is being given to starting up or expanding capacities.

Various in situ methods have been developed in which oil shale is heated underground for years either electrically or by hot gases. Oil can then be extracted by fracking even where open-pit mining is not possible. Pilot plants have shown that this is possible. Although the energy produced is greater than the energy used, it would only pay off if oil prices were significantly higher. However, it is not exactly environmentally friendly.

There are three formations in the United States that each have a theoretical oil content corresponding to the global amount of conventional oil resources. They comprise Cretaceous oil shales in the Piceance Basin (Colorado), Eocene oil shales in the Green River Formation (Colorado, Wyoming, Utah), and Paleogene oil shales in the Uintah Basin (Utah). Whether significant quantities of oil will ever be extracted is impossible to predict. Further occurrences can be found in Russia, Brazil, DR Congo, Morocco, Jordan, and many other countries.

6.9 Tar Sand and Heavy Crude Oil

Tar sands (a.k.a. oil sands) in the eastern foreland of the Rocky Mountains in Alberta currently make Canada the sixth largest oil producer in the world. The Athabasca tar sands are by far the world's largest known deposit of crude bitumen. There are several smaller ones in the area such as Peace River and Cold Lake. Together they cover an area of 141,000 km^2. Even if only 10% of these sands are considered recoverable, they equate to 174.5 billion barrels—almost double the amount of Ghawar, the world's largest known conventional oil field.

Deposits of heavy crude oil in the Orinoco Belt in Venezuela are of a similar magnitude. Although the total quantity is somewhat smaller, the recoverable share is assumed to be much higher. Estimates of the recoverable quantity have repeatedly been revised upward. In 2009 the USGS (2009) quoted an incredible 513 billion barrels—almost twice Saudi Arabia's accurately known oil resources. The Orinoco Belt covers an area 300 km long and up to 100 km wide that follows the course of the river that gave it its name. Strictly speaking, it is predominantly extra-heavy crude oil (i.e., heavier than water). Heavy crude oil transitions to and from bitumen (asphalt, tar) along a continuous spectrum in which the differentiation of terms is not clearly defined.

Smaller but still enormous deposits of heavy crude oil and bitumen are found worldwide such as in California, the Andes, the Middle East, the Caspian Basin, the Republic of Congo (Brazzaville), Eastern Siberia, and China. Reserves are significantly larger than those of conventional oil even taking cumulative production to date into account. However, such unconventional oils have the disadvantage that they are difficult to extract, transport, and process. Not only do they have a very high density, they are also extremely viscous. The spectrum ranges from honeylike to solid asphalt. That is because they are made up almost entirely of very large molecules. In addition, the content of heavy metals such as nickel and vanadium and the content of sulfur, nitrogen, and oxygen are significantly increased. However, hydrogen content is significantly lower than in normal petroleum.

As already described in detail in Sect. 6.3 such substances are produced by degradation of normal petroleum (in particular, by microorganisms in a process termed biodegradation) because oil migrated into a shallow reservoir instead of remaining in a deep trap. The largest deposits of heavy crude oil and tar sands are found on the outer edges of foreland basins of mountain ranges (Head et al. 2003; Jacome et al. 2003; Hein 2006). The basin with the Athabasca tar sands formed during the Laramide orogeny (i.e., the formation of the Rocky Mountains in the late

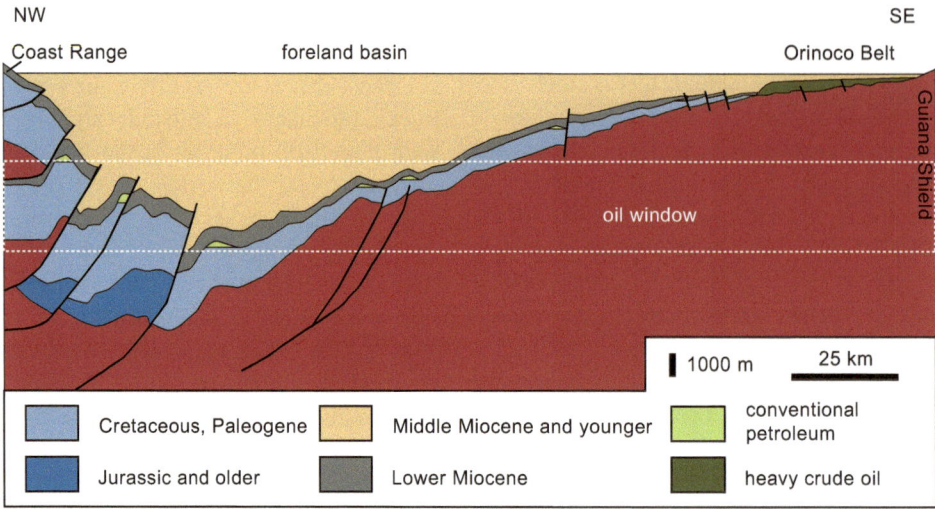

NW   SE
Coast Range   foreland basin   Orinoco Belt

oil window

1000 m   25 km

Cretaceous, Paleogene   Middle Miocene and younger   conventional petroleum

Jurassic and older   Lower Miocene   heavy crude oil

Fig. 6.32 Schematic section through the foreland basin of the Venezuelan Coast Range (result of the collision between the South American Plate and the Caribbean Plate) to the marginal Orinoco Belt (Venezuela). Although oil released by source rocks (Cretaceous and older) in the deep area of the basin has only partly remained in conventional traps, a large part has migrated to the outer edge of the basin into loose sandstones and degraded there to heavy crude oil (from Jacome et al. 2003)

Cretaceous and Paleogene), whereas the Orinoco Belt is at the outer edge of the foreland basin of the Venezuelan Coast Range (virtually the northeastern end of the Andes running along the Caribbean coast). Source rock in the deepest part of the foreland basin reached the oil window during mountain formation allowing oil to migrate under an impermeable horizon to the other end of the basin (Fig. 6.32) and end up in still unconsolidated sand at shallow depth (in Venezuela a smaller part remained in conventional traps on the way).

If the overburden of the relevant layer is small, then it is easiest to extract tar sand via open-cast mining. This is the case on the northern edge of the Athabasca Basin where tar sand is overlain only by peaty soil and soft sediments. Where once a pristine boreal coniferous forest had grown now large open-cast mines have been established (Johnson 2012; Fig. 6.33). Tar sand is loaded by excavators onto dump trucks that transport the coarse material to a crusher. This material is then made into a slurry by adding water. Sand, bitumen, and water are separated in large basins and the sand is then sedimented in tailing lakes. On average, 1 t of oil sand produces only half a barrel of oil.

Farther south in the Athabasca Basin the cover is too thick. Nevertheless, a method has been developed to allow production by drilling. This involves the drilling of two boreholes one above the other that run horizontally for hundreds of meters in the respective layer separated by a few meters. Superheated steam is then injected through the upper borehole heating the bitumen to such an extent that it becomes less viscous and flows to the lower borehole by gravity. This results in a hot emulsion of condensed water and bitumen that can be pumped to the surface. Water is reused to generate steam after separation. The procedure is called steam-assisted gravity drainage (SAGD). Although the yield (up to 60%) is very high, the material requirements and energy requirements for steam generation are likewise high. Two to three barrels of water must be vaporized per barrel of recovered bitumen.

Some heavy crude oil in the Orinoco Belt is easier to extract. Its viscosity is not quite so high even in normal conditions, and the reservoir is so deep that it is already warm. Horizontal boreholes are drilled into which heavy crude oil flows by itself. However, special pumps are necessary to lift the viscous mass and yield amounts to only 15% of the available oil (the abovementioned USGS estimate assumes more expensive methods and a yield of 45%). South of the Athabasca tar sands Canada also has a belt of heavy crude oil where a drilling method called cold heavy oil production with sand (CHOPS) has become established that does not attempt to prevent sand from entering the borehole using a filter (as is normally the case), but instead deliberately pumps a mixture of heavy crude oil and sand to the surface with special pumps.

Although heavy crude oil and bitumen cannot be pumped through a pipeline or processed in a normal refinery, it is possible to dilute the viscous mass with naphtha or gas condensate and then send it through a pipeline. The solvent can then be distilled off at the other end and sent back through a second pipeline.

The next step involves upgrading bitumen to synthetic crude oil by cracking large molecules—something that is even more expensive with bitumen than with heavy crude oil and consumes a lot of energy. There are two possible ways of doing this. Coking involves some bitumen being cracked into smaller hydrocarbon molecules, while a solid substance

Fig. 6.33 Extraction of Athabasca tar sands at Fort McMurray (Alberta, Canada) (*photo* © dan_prat/iStockphoto)

called petroleum coke remains. This consists almost exclusively of carbon and in this case has a high sulfur content. The other possibility is thermal cracking in the presence of hydrogen. Although this is more expensive, the yield of liquid hydrocarbons is much higher. Hydrogen is previously produced from methane. The sulfur-rich synthetic oil still has to be desulfurized, which produces enormous amounts of elemental sulfur—more than could be sold. The final result is synthetic crude oil that can compete with the highest quality crude oils. The mining and upgrading of Canadian tar sands are not exactly environmentally friendly, and both are also quite expensive. The running costs are about USD70 per barrel, so it is only worth it if the oil price is very high.

6.10   Methane Hydrates

Although the idea of flammable ice (Fig. 6.34) sounds crazy, methane hydrate is exactly that. It is a solid substance consisting of water and methane that is only stable under cold temperatures and high pressure. It has the approximate composition $CH_4 \cdot 5.75 H_2O$. The reason ice floats is that it has a lower density than water as a result of water molecules

Fig. 6.34 Methane hydrate on fire (*photo* © J. Pinkston and L. Stern, USGS)

being dipoles arranged in the ice to form a very wide-meshed crystal lattice that is held together by hydrogen bonds. A methane molecule that is not involved in the structure with bonds fits into the large cages in between (Sloan and Koh 2007). Such structures are also called clathrates from the Latin *clatratus* meaning "trapped in a cage." Other gases can also form clathrates such as gas hydrates with ethane, propane, carbon dioxide, or nitrogen. Three different structures with cages of different sizes are known (I, II, and H). Gas hydrates occurring in nature usually contain almost exclusively methane with a cubic crystal lattice (structure I). If 1 m^3 of methane hydrate is thawed, then 164 m^3 gaseous methane and 0.8 m^3 liquid water are obtained.

Methane hydrate (Fig. 6.35) is stable in the sediments of oceans worldwide as long as the water column above is at least 300–500 m (depending on temperature). The stability

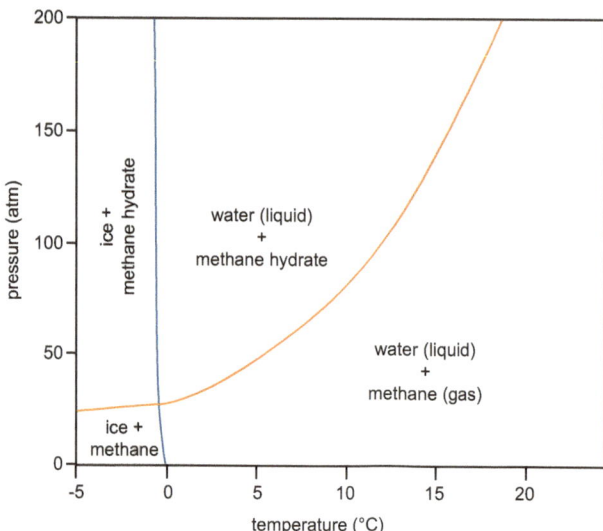

Fig. 6.35 Phase diagram of the H$_2$O–CH$_4$ system in which methane hydrates are stable above the *orange* curve

zone is limited downward by temperature. At a water depth of 500 m the stability zone relates to the uppermost sediment layer. With increasing water depth the stability zone becomes thicker such that at 2000 m it is several hundred meters (depending on the geothermal gradient). Accordingly, large quantities of methane hydrates are contained in deep areas of continental slopes that drop into the deep sea, but of course only where there is methane. Although they are mainly found in rock pores such as unsolidified sand, they are also found in cracks (Cook et al. 2008). Methane hydrates also form in shallow water in polar regions and even on continents in permafrost in a zone 300–500 m thick that extends from the lower part of frozen soil down into sediments. Large deposits are found mainly in northern Siberia, northern Canada, and northern Alaska. However, 99% of all methane hydrates are likely to be located on continental slopes. The quantities are enormous but estimates vary by several orders of magnitude. What is certain is that far more carbon is bound here than in all other coal, oil, and gas deposits combined. However, only a tiny part is potentially recoverable. Although carbon isotopes show that it is almost always biogenic methane released by anoxic methanogenic microorganisms living in the sediment (Kvenvolden 1995), sometimes it is natural gas rising from the depth or a mixture instead.

Methane hydrates have been discussed as a potential energy source for some time now and several countries have tried to exploit them experimentally (Ruppel et al. 2008). However, even with continental deposits this is costly (Ruppel 2011a) and currently not economic. Methane hydrate can be thawed at a well by injecting steam to increase the temperature or by pumping to decrease the pore

pressure. However, this is very energy consuming and time consuming since thawing is strongly endothermic and the methane released increases the pressure again. Moreover, it is not easy to ensure that methane ends up in the pipe and not in the atmosphere. An interesting new method is to pump in CO$_2$ to replace methane in clathrates without thawing them. In this way methane could be extracted and carbon dioxide disposed of at the same time.

Methane hydrates are usually perceived as a problem instead of an energy source. This also applies to conventional oil drilling in the deep sea where attempts are made not to drill through sediments that have a high content of methane hydrates. It is feared that drilling or subsequent production could lead to local thawing and that the gas would damage the drill pipe or casing or would be released (Hovland and Gudmestad 2001). It is also suspected that a massive leak of methane in the form of many small bubbles or a single giant bubble could sink entire ships or drilling platforms (May and Monaghan 2003). This could be the reason ships mysteriously disappear from time to time such as in the Bermuda Triangle in the western Atlantic. Methane hydrates are also known to be a problem in pipelines in cold regions because they form at low temperatures when the gas is not completely free of water and then clog the pipe.

Discussion about methane hydrates in the context of climate change is much more extensive. Since methane is 25 times more effective as a greenhouse gas than carbon dioxide the thawing of methane hydrates could significantly speed up climate change. It is generally assumed that once the tipping point is passed a self-reinforcing process with increasing warming would occur. Although the occurrence of methane in permafrost would be particularly problematic here, in the case of the deep sea it is relatively probable that almost all degassed methane would be oxidized to carbon dioxide by living organisms during ascent in the water (Ruppel 2011b)—carbon dioxide is less effective as a greenhouse gas. The result would be a more acidic and oxygen-poor sea that would limit the living conditions of many living creatures and the oceans would absorb CO$_2$ more slowly from the atmosphere.

Methane hydrates may play a role in large submarine landslides on continental slopes that in turn can trigger tsunamis. The best known example is the huge Storragaa Landslide in the Atlantic off Norway that triggered a huge tsunami in the North Atlantic 8000 years ago. Although what caused the landslide remains controversial (Meinert et al. 2005; Brown et al. 2006), it is conceivable that the slope might have become unstable due to thawing of methane hydrates or that the slope slipped along a gas-rich layer at the base of the stability zone of methane hydrates. Maybe the thawing of methane hydrates occurred as a result of pressure relief during slipping off intensifying it. Other researchers believe that methane hydrates played a minor role at most and consider an earthquake to be the trigger. This discussion is also relevant to extraction of methane hydrates. There are critical voices warning that submarine landslides could be triggered.

Literature

Abdalla, C., and J. Drohan. 2010. Water withdrawals for development of Marcellus Shale Gas in Pennsylvania. Penn State Cooperative Extension.

Abidin, A.Z., T. Puspasari, and W.A. Nugroho. 2012. Polymers for enhanced oil recovery technology. *Procedia Chemistry* 4: 11–16.

Ahlbrandt, T.S., R.M. Pollastro, T.R. Klett, C.J. Schenk, S.J. Lindquist and J.E. Fo., 2000. Region 2 assessment summary—Middle East and North Africa. In U.S. Geological Survey World petroleum assessment 2000—Description and results. U.S. Geological Survey Digital Data Series 60.

Aitken, C.M., D.M. Jones, and S.R. Larter. 2004. Anaerobic hydrocarbon biodegradation in deep subsurface oil reservoirs. *Nature* 431: 291–294.

Alsharhan, A.S. 1989. Petroleum geology of the United Arab Emirates. *Journal of Petroleum Geology* 12: 253–288.

Andrews, A. 2010. Unconventional Gas Shales: Development, Technology, and Policy Issues. Diane Publishing.

Andruleit, H., H.G. Babies, J. Meßner, S. Rehder, M. Schauer, and S. Schmidt. 2011. Reserven, Ressourcen und Verfügbarkeit von Energierohstoffen 2011. Deutsche Rohstoffagentur (DERA) in der Bundesanstalt für Geologie und Rohstoffe.

Andruleit, H., H.G. Babies, A. Bahr, J. Kus, J. Meßner, and M. Schauer. 2012a. Energiestudie 2012. Reserven, Ressourcen und Verfügbarkeit von Energierohstoffen. Deutsche Rohstoffagentur in der Bundesanstalt für Geowissenschaften und Rohstoffe.

Andruleit, H., A. Bahr, C. Bönnemann, J. Erbacher, D. Franke, J. P. Gerling, N. Gestermann, T. Himmelsbach, M. Kosinowski, S. Krug, R. Pierau, T. Pletsch, U. Rogalla, S. Schlömer and NiKo-Projekt-Team. 2012b. Abschätzung des Erdgaspotenzials aus dichten Tongesteinen (Schiefergas) in Deutschland. Bundesanstalt für Geowissenschaften und Rohstoffe.

Anonymous, 2008. Saarland: Kohleabbau löst Erdbeben mit Stärke 4,0 aus. Spiegel Online, February 24, 2008. http://www.spiegel.de/panorama/saarland-kohleabbau-loest-erdbeben-mit-staerke-4-0-aus-a-537345.html.

Anonymous. 2010. Plumbing the depths. The Economist, 6. März 2010.

Anonymous. 2011. In deep waters. Brazil's offshore oil. The Economist, 3. February 2011.

Anonymous, 2012. Ohio quakes probably triggered by waste disposal well, say seismologists. Lamont-Doherty Earth Observatory. June 01, 2012. http://www.ldeo.columbia.edu/news-events/seismologists-link-ohio-earthquakes-waste-disposal-wells.

Arbeitskreis gesellschaftliche Akteure. 2011. Fracking und seismische Ereignisse – Erdbeben und Fracking. Fracking, Erdgassuche in Deutschland. September 12, 2011. http://dialog-erdgasundfrac.de/fracking-und-seismische-ereignisse-erdbeben-und-fracking.

Bennett, B., M. Fustic, P. Farrimond, H. Huang, and S.R. Larter. 2006. 25-Norhopanes: Formation during biodegradation of petroleum in the subsurface. *Organic Geochemistry* 37: 787–797.

BGR. 2011. Faktenblatt „Fracking".

Bjørlykke, K. 2011a. *Petroleum Geoscience*. Heidelberg: Springer.

Brown, H.E., W.S. Holbrook, M.J. Hornbach, and J. Nealon. 2006. Slide structure and role of gas hydrate at the northern boundary of the Storegga Slide, offshore Norway. *Marine Geology* 220: 179–186.

Bull, S., J. Cartwright, and M. Huuse. 2009. A review of kinematic indicators from mass-transport complexes using 3D seismic data. *Marine and Petroleum Geology* 26: 1132–1151.

Campbell, C. 1989. Oil Price Leap in the Early Nineties. Noroil, Kingston-upon-Thames.

Cook, A.E., D. Goldberg, and R.L. Kleinberg. 2008. Fracture-controlled gas hydrate systems in the northern Gulf of Mexico. *Marine and Petroleum Geology* 25: 932–941.

Cramer, B., and H. Andruleit (ed.). 2009. Energierohstoffe 2009. Reserven, Ressourcen und Verfügbarkeit. Bundesanstalt für Geowissenschaften und Rohstoffe (BGR).

dapd. 2012. Wasserentnahme war schuld an Erdbeben in Spanien. http://de.nachrichten.yahoo.com/wasserentnahme-war-schuld-erdbeben-spanien-061419737.html.

Davies, R.J., S.A. Mathias, J. Moss, S. Hustoft, and L. Newport. 2012. Hydraulic fractures: How far can they go? *Marine and Petroleum Geology* 37: 1–6.

Department of Energy and Climate Change. 2012. The unconventional hydrocarbon resources of Britain's onshore basins – shale gas.

Diessel, C.F.K. 1992a. *Coal-Bearing Depositional Systems*. Heidelberg: Springer.

Dolfing, J., S.R. Larter, and I.M. Head. 2008. Thermodynamic constraints on methanogenic crude oil biodegradation. *The ISME Journal* 2: 442–452.

Ecomides, M. 2011. Don't be swayed by faucets on fire and other anti-fracking propaganda. Forbes. March 07, 2011. http://www.forbes.com/sites/greatspeculations/2011/03/07/dont-be-swayed-by-faucets-on-fire-and-other-anti-fracking-propaganda/.

EIA. 2011. Lower 48 states shale plays (Karte). http://www.eia.gov/oil_gas/rpd/shale_gas.jpg (abgerufen 10.06.2013).

EIA. 2013. Analysis & Projections. Technically Recoverable Shale Oil and Shale Gas Resources: An Assessment of 137 Shale Formations in 41 Countries Outside the United States. Accessed June 18, 2013. http://www.eia.gov/analysis/studies/worldshalegas/.

Ewen, C., D. Borchardt, S. Richter, and R. Hammerbacher. 2011. Risikostudie Fracking-Sicherheit und Umweltverträglichkeit der Fracking-Technologie für die Erdgasgewinnung aus unkonventionellen Quellen (Übersichtsfassung), 2011.

Fallgren, P.H., S. Jin, C. Zeng, Z. Ren, A. Lu and P.J.S. Colberg. 2013. Comparison of coal rank for enhanced biogenic natural gas production. *International Journal of Coal Geology* (Published online).

Fischer, L. 2013. Fracking ohne Wasser - Die umweltfreundliche Option? Fischblog. Accessed April 14, 2013. http://www.scilogs.de/wblogs/blog/fischblog/technik/2013-01-14/fracking-ohne-wasser-die-umweltfreundliche-option.

Foustoukos, D.I., and W.E. Seyfried Jr. 2004. Hydrocarbons in hydrothermal vent fluids: The role of chromium-bearing catalysts. *Science* 304: 1002–1005.

Fox, J.E., and T.S. Ahlbrandt. 2002. Petroleum geology and total petroleum systems of the Widyan Basin and Interior Platform of Saudi Arabia and Iraq. U.S. Geological Survey Bulletin 2202–E.

Frohlich, C. 2012. Two-year survey comparing earthquake activity and injection-well locations in the Barnett Shale, Texas. *Proceedings of the National Academy of Sciences* (August 6). https://doi.org/10.1073/pnas.1207728109. http://www.pnas.org/content/early/2012/07/30/1207728109.abstract.

Gärtner, M. 2013. Fracking: Amerikas Schiefergas-Boom droht jähes Ende. Manager Magazin. May 13, 2013. http://www.manager-magazin.de/unternehmen/artikel/0,2828,899442,00.html.

Glasby, G.P. 2006. Abiogenic origin of hydrocarbons: An historical overview. *Resource Geology* 56: 85–98.

Gluyas, J., and R. Swarbrick. 2004a. *Petroleum Geoscience*. Malden, MA: Blackwell Publishing.

Grajales-Nishimura, J.M., E. Cedillo-Pardo, C. Rosales-Domínguez, D. J. Morán-Zenteno, W. Alvarez, P. Claeys, J. Ruíz-Morales, J. García-Hernández, P. Padilla-Avila, and A. Sánchez Ríos. 2000. Chicxulub impact: The origin of reservoir and seal facies in the southeastern Mexico oil fields. *Geology* 28: 307–310.

Gruber, W., and R.F. Sachsenhofer. 2001. Coal deposition in the Noric Depression (Eastern Alps): Raised and low-lying mires in Miocene pull-apart basins. *International Journal of Coal Geology* 48: 89–114.

Hallmann, C., L. Schwark, and K. Grice. 2008. Community dynamics of anaerobic bacteria in deep petroleum reservoirs. *Nature Geoscience* 1: 588–591.

Han, D.K., C.Z. Yang, Z.Q. Zhang, Z.H. Lou, and Y.I. Chang. 1999. Recent development of enhanced oil recovery in China. *Journal of Petroleum Science and Engineering* 22: 181–188.

Head, I.M., D.M. Jones, and S.R. Larter. 2003. Biological activity in the deep subsurface and the origin of heavy oil. *Nature* 426: 344–352.

Hein, F.J. 2006. Heavy oil and oil (tar) sands in North America: An overview & summary of contributions. *Natural Resources Research* 15: 67–84.

Heydari, E., and W.J. Wade. 2002. Massive recrystallization of low-Mg calcite at high temperatures in hydrocarbon source rocks: Implications for organic acids as factors in diagenesis. *AAPG Bulletin* 86: 1285–1303.

Hovland, M. and O.T. Gudmestad. 2001. Potential influence of gas hydrates on seabed installations. In *Natural Gas Hydrates; Occurrence, Distribution and Detection*, vol. 124, ed. C.K. Paull and W. Dillon, 307–315. American Geophysical Union Monograph.

Hower, J.C., and R.A. Gayer. 2002. Mechanisms of coal metamorphism: Case studies from Paleozoic coalfields. *International Journal of Coal Geology* 50: 215–245.

Hughes, J.D. 2013. Rohstoffe: Schiefergas im Realitätstest. Spektrum. de. Accessed march 07, 2013. http://www.spektrum.de/alias/ rohstoffe/schiefergas-im-realitaetstest/1185968.

Hunt, J.M. 1995a. *Petroleum Geochemistry and Geology*. New York: Freeman & Co.

Jackson, R.B., A. Vengosh, T.H. Darrah, N.R. Warner, A. Down, R. J. Poreda, S.G. Osborn, K. Zhao, and J.D. Karr. 2013. Increased stray gas abundance in a subset of drinking water wells near Marcellus shale gas extraction. *Proceedings of the National Academy of Sciences*. https://doi.org/10.1073/pnas.1221635110, http://www. pnas.org/content/early/2013/06/19/1221635110.abstract.

Jacome, M.I., N. Kusznir, F. Audemard, and S. Flint. 2003. Tectonostratigraphic evolution of the Maturin foreland basin—Eastern Venezuela. In *The Circum-Gulf of Mexico and the Caribbean— Hydrocarbon Habitats, Basin Formation, and Plate Tectonics*, vol. 79, ed. C. Bartolini, R.T. Buffler and J. Blickwede, 735–749. American Association of Petroleum Geologists Memoir.

Janocko, M., W. Nemec, S. Henriksen, and M. Warchol. 2013. The diversity of deep-water sinuous channel belts and slope valley-fill complexes. *Marine and Petroleum Geology* 41: 7–34.

Johnson, R. 2012. Canadian Oil Sands Flyover. Business Insider, http:// www.businessinsider.com/canadian-oil-sands-flyover-2012-5.

Johnson, E.A., S. Liu, and Y. Zhang. 1989. Depositional environments and tectonic controls on the coal-bearing Lower to Middle Jurassic Yan'an Formation, southern Ordos Basin, China. *Geology* 17: 1123–1126.

Jones, D.M., I.M. Head, N.D. Gray, J.J. Adams, A.K. Rowan, C.M. Aitken, B. Bennett, H. Huang, A. Brown, B.F.J. Bowler, T. Oldenburg, M. Erdmann, and S.R. Larter. 2008. Crude-oil biodegradation via methanogenesis in subsurface petroleum reservoirs. *Nature* 451: 176–181.

Klemme, H.D., and G.F. Ulmishek. 1991. Effective petroleum source rocks of the World: Stratigraphic distribution and controlling depositional factors. *AAPG Bulletin* 75: 1809–1851.

Kosinowski, M., U. Berner, D. Franke, C. Gaedicke, P. Gerling, T. Himmelsbach, R. Jatho, et al. 2012. Stellungnahme der Bundesanstalt für Geowissenschaften und Rohstoffe zum Gutachten des Umweltbundesamtes (UBA) „Umweltauswirkungen von Fracking bei der Aufsuchung und Gewinnung von Erdgas aus unkonventionellen Lagerstätten – Risikobewertung, Handlungsempfehlungen und Evaluierung bestehender rechtlicher Regelungen und Verwaltungsstrukturen". UFOPLAN - NR. 3711 23 299. Bundesanstalt für Geowissenschaften und Rohstoffe. http://www.bgr.bund.de/DE/ Themen/Energie/Downloads/BGR-Stellungnahme-UBA2012.pdf? __blob=publicationFile&v=3.

Kvenvolden, K.A. 1995. A review of the geochemistry of methan in natural gas hydrate. *Organic Geochemistry* 23: 997–1008.

Levy, M. 2011. Fracking wastewater disposal process to be altered in Pennsylvania. Huffington Post, April 19, 2011. http://www. huffingtonpost.com/2011/04/20/fracking-wastewater-disposal- pennsylvania_n_851441.html.

Li, G., and W. Zhou. 2006. Impact of karst water on coal mining in North China. *Environmental Geology* 49: 449–457.

Lingenhöhl, D. 2011. Schiefergas: Das Risiko ist beherrschbar. Spektrum.de. http://www.spektrum.de/alias/schiefergas/das-risiko- ist-beherrschbar/1123445.

Liu, L., T. Zhang, X. Zhao, S. Wu, J. Hu, X. Wang, and Y. Zhang. 2013. Sedimentary architecture models of deepwater turbidite channel systems in the Niger Delta continental slope, West Africa. *Petroleum Science* 10: 139–148.

Maugeri, L. 2004. Oil: Never cry wolf—Why the petroleum age is far from over. *Science* 304: 1114–1115.

May, D.A., and J.J. Monaghan. 2003. Can a single bubble sink a ship? *American Journal of Physics* 71: 842–849.

Melezhik, V.A., M.M. Filippov, and A.E. Romanshkin. 2004. A giant Palaeoproterozoic deposit of shungite in NW Russia: Genesis and practical applications. *Ore Geology Reviews* 24: 135–154.

Meinert, J., M. Vanneste, S. Bünz, K. Andreassen, H. Haflidason, and H.P. Sejrup. 2005. Ocean warming and gas hydrate stability on the mid-Norwegian margin at the Storegga Slide. *Marine and Petroleum Geology* 22: 233–244.

Mix, M. 2012. Erdgasbohrungen lösen Erdbeben aus. August 08, 2012. http://www.kreiszeitung.de/lokales/verden/langwedel/ erdgasbohrungen-loesen-erdbeben-2656288.html.

Montgomery, C.T., and M. Smith. 2010. Hydraulic fracturing: The past, present and future. *Journal of Petroleum Technology* 62: 26.

Müller, T. 2013. Niederlande: Wenn die Erde täglich bebt. Zeit online, February 18, 2013. http://www.zeit.de/wirtschaft/2013-02/ niederlande-erdgas-foerderung.

Nestler, R. 2013. Energie: Deutschland und das Schiefergas. Spektrum. de. Accessed January 07, 2013. http://www.spektrum.de/alias/ energie/deutschland-und-das-schiefergas/1179965.

North, F.K. 1985. *Petroleum Geology*. Allen & Unwin.

OPEC. 2012. World Oil Outlook. https://www.opec.org/opec_web/ static_files_project/media/downloads/publications/WOO2012.pdf.

Pollastro, R.M. 2003. Total petroleum systems of the Paleozoic and Jurassic, greater Ghawar uplift and adjoining provinces of central Saudi Arabia and northern Arabian-Persian Gulf. U. S. Geological Survey Bulletin 2202-H.

Proskurowski, G., M.D. Lilley, J.S. Seewald, G.L. Früh-Green, E. J. Olson, J.E. Lupton, S.P. Sylva, and D.S. Kelley. 2008. Abiogenic hydrocarbon production at Lost City Hydrothermal Field. *Science* 319: 604–607.

Ross, A.S., P. Farrimond, M. Erdmann, and S.R. Larter. 2010. Geochemical compositional gradients in a mixed oil reservoir indicative of ongoing biodegradation. *Organic Geochemistry* 41: 307–320.

Ruppel, C. 2011a. Methane hydrates and the future of natural gas. MITEI Natural Gas Report, Supplementary Paper 4, The Future of natural gas. MIT energy initiative study.

Ruppel, C. 2011b. Methane hydrates and contemporary climate change. *Nature Education Knowledge* 3: 29.

Ruppel, C., R. Boswell, and E. Jones. 2008. Scientific results from Gulf of Mexico Gas Hydrates Joint Industry Project Leg 1 drilling: Introduction and overview. *Marine and Petroleum Geology* 25: 819–829.

Schrammar, S. 2012. Erdgasförderung als Erdbeben-Auslöser? Deutschlandfunk, February 16, 2012. http://www.dradio.de/dlf/sendungen/umwelt/1679187/.

Schrope, M. 2011. Deep wounds. *Nature* 472: 152–154.

Schutter, S.R. 2003. Hydrocarbon occurrence and exploration in and around igneous rocks. *Geological Society London, Special Publication* 214: 7–33.

Seewald, J.S. 2003. Organic-inorganic interactions in petroleum-producing sedimentary basins. *Nature* 426: 327–333.

Sen, R. 2008. Biotechnology in petroleum recovery: The microbial EOR. *Progress in Energy and Combustion Science* 34: 714–724.

Sephton, M.A., and R.M. Hazen. 2013. On the origins of deep hydrocarbons. *Reviews in Mineralogy and Geochemistry* 75: 449–465.

Sloan, E.D., and C.A. Koh. 2007. *Clathrate Hydrates of Natural Gases*, 3 ed. CRC Press, Taylor & Francis Group.

Süss, M.P., A. Schäfer, and G. Drozdzewiski. 2001. Discussion. A sequence stratigraphic model for the Lower Coal Measures (Upper Carboniferous) of the Ruhr district, north-west Germany. *Sedimentology* 48: 1171–1186.

Süss, M.P., G. Drozdzewski, and A. Schäfer. 2007. Sedimentary environment dynamics and the formation of coal in the Pennsylvanian Variscan foreland in the Ruhr Basin (Germany, Western Europe). *International Journal of Coal Geology* 69: 267–287.

Taylor, G.H., M. Teichmüller, A. Davis, C.F.K. Diessel, R. Littke, and P. Robert. 1998a. *Organic Petrology*. Berlin and Stuttgart: Gebrüder Borntraeger.

Tenbrock, C., and F. Vorholz. 2013. Fracking: Amerika im Gasrausch. Zeit Online, February 07, 2013. http://www.zeit.de/2013/07/Fracking-USA-Erdgas-Umwelt.

Thielemann, T., B. Cramer, and A. Schippers. 2004. Coalbed methane in the Ruhr Basin, Germany: A renewable energy resource? *Organic Geochemistry* 35: 1537–1549.

Thomas, L. 2012a. *Coal Geology*, 2nd ed. Oxford: Wiley-Blackwell.

Thomas, S.. 2008. Enhanced Oil Recovery—An overview. Oil & Gas Science and Technology – Revue d'IFP Energies Nouvelles 63, 9–19.

Tollefson, J. 2013. Gas drilling taints groundwater. *Nature* 498: 415–416.

Uken, M. 2013. Internationale Energieagentur: Europas vergebliche Hoffnung auf Fracking. Zeit Online, November 06, 2013. http://www.zeit.de/wirtschaft/2013-06/internationale-energieagentur-fracking-europa.

Urbina, I. 2011. South African farmers see threat from fracking. New York Times, December 30, 2011. http://www.nytimes.com/2011/12/31/world/south-african-farmers-see-threat-from-fracking.html?_r=0.

USGS. 2000. U.S. Geological Survey World petroleum assessment 2000.

USGS. 2009. An estimate of recoverable heavy oil resources of the Orinoco Oil Belt, Venezuela. Fact Sheet 2009–3028.

Volkov, V.N. 2003. Phenomenon of the formation of very thick coal beds. *Lithology and Mineral Resources* 38: 223–232.

Vukomanovic, O. 2011. UK firm says shale fracking caused earthquakes. Reuters, February 11, 2011. http://www.reuters.com/article/2011/11/02/us-gas-fracking-idUSTRE7A160020111102.

Watts, A. 2011. The Gasland movie: A Fracking shame—Director pulls video to hide inconvenient truths. Watts up with that? April 06, 2011. http://wattsupwiththat.com/2011/06/04/the-gasland-movie-a-fracking-shame-director-pulls-video-to-hide-inconvenient-truths/.

Wilhelms, A., S.R. Larter, I. Head, P. Farrimond, R. di-Primio, and C. Zwach. 2001. Biodegradation of oil in uplifted basins prevented by deep-burial sterilization. *Nature* 411: 1034–1037.

Wilson, H.H. 1987. The structural evolution of the golden lane, Tampico Embayment, Mexico. *Journal of Petroleum Geology* 10: 5–40.

Yang, Y., W. Li, and L. Ma. 2005. Tectonic and stratigraphic controls of hydrocarbon systems in the Ordos basin: A multicycle cratonic basin in central China. *AAPG Bulletin* 89: 255–269.

Yao, Y., D. Liu, D. Tang, S. Tang, Y. Che, and W. Huang. 2009. Preliminary evaluation of the coalbed methane production potential and its geological controls in the Weibei Coalfield, Southeastern Ordos Basin, China. *International Journal of Coal Geology* 78: 1–15.

Further Reading

Bjørlykke, K. 2011b. *Petroleum Geoscience*. Heidelberg: Springer.

Diessel, C.F.K. 1992b. *Coal-Bearing Depositional Systems*. Heidelberg: Springer.

Gluyas, J., and R. Swarbrick. 2004b. *Petroleum Geoscience*. Malden, MA: Blackwell Publishing.

Hunt, J.M. 1995b. *Petroleum Geochemistry and Geology*. New York: Freeman & Co.

Lohmann, D., and N. Podbregar. 2012. *Im Fokus: Bodenschätze*. Springer, Heidelberg: Auf der Suche nach Rohstoffen.

North, F.K. 1985. Petroleum Geology. Allen & Unwin.

Pohl, W.L. 2011. *Economic Geology*. Chichester: Wiley-Blackwell.

Rothe, P. 2010. *Schätze der Erde*. Darmstadt: Primus Verlag.

Seidler, C. 2012. *Deutschlands verborgene Rohstoffe: Kupfer*. Hanser, München: Gold und seltene Erden.

Taylor, G.H., M. Teichmüller, A. Davis, C.F.K. Diessel, R. Littke, and P. Robert. 1998b. *Organic Petrology*. Berlin and Stuttgart: Gebrüder Borntraeger.

Thomas, L. 2012b. *Coal Geology*, 2nd ed. Oxford: Wiley-Blackwell.

Industrial Minerals and Rocks

When it comes to mineral deposits very few people think of sand and gravel despite such commodities being consumed in far greater quantities than metals. Raw materials such as natural stone, lime, clay, kaolin, gravel, and sand are referred to as industrial minerals and rocks. In Germany 600 million tonnes of these are extracted each year from quarries, dredging lakes, sand pits, and gravel pits. They are mainly used in construction as cement, concrete, brick, or natural stone. They also serve as raw materials for ceramics and glass and some other applications. A number of minerals are used in engineering due to their special properties (industrial minerals). Salt, gypsum, and phosphate would comfortably fit into this chapter, but they have already been discussed in Chap. 5.

7.1 Sand, Gravel, and Natural Stones

When rock erodes it is crushed and rounded during transport in rivers and deposited as gravel. Continued disaggregation and weathering in rivers or on beaches results in the formation of sand consisting of ever-more resistant minerals with increasing distance from the eroded area. Although the weathering of granite, gneiss, and sandstone results primarily in quartz, feldspars are also present when transport distances are short. The weathering of basalt predominantly results in pyroxene. Moreover, limestone sand, dolomite sand, gypsum sand, and sand that consists mainly of heavy minerals (Sect. 5.9) such as monazite, rutile, zircon, garnet, and magnetite are also present.

Sand and gravel are mainly used as aggregates in concrete and in road construction. Annual worldwide demand is about 40 billion tonnes. Mining takes place in gravel pits, quarry pits, on beaches, and the seabed. The exact composition of sand used for construction is less important than grain size and shape. Countries with abundant desert sand such as Saudi Arabia and the United Arab Emirates often have to import sand for their skyscrapers. Global demand is

so high that mining can lead to strange stories. Small islands in the Maldives and Indonesia were completely removed. Sometimes sand is even stolen: in Jamaica an entire hotel beach disappeared overnight without a trace in 2008—thieves transporting around 500 truckloads undetected.

Pure quartz sand is an important raw material for glass and ceramics, whereas regular quartz sand is used in metal foundries. Sand blasting is mainly carried out using sands of corundum, garnet, or other minerals since quartz dust is known to trigger silicosis.

Many rocks are split into blocks or sawn into slabs. They serve as masonry stones, façade decoration, floor slabs, paving stones, or gravestones. In addition to properties such as hardness, resistance to weathering, and processing behavior, aesthetics often plays an important role in selection of the right natural stone for the project under consideration. Intrusive rocks such as granite, diorite, and gabbro, as well as volcanic rocks such as basalt, tuff, or porphyry are popular as are metamorphic rocks such as marble and serpentinite and sediments such as sandstone, limestone and dolomite. Slate is easily split into slabs and is used among other things as roofing slate.

Fissures, the schistosity of rocks, and the cleavage of minerals are exploited when splitting rocks with chisels or wedges. Hard stones are sawn with a cutting disk in which diamonds are embedded. Wire saws are sometimes used in which a steel wire covered with hard materials is pulled over the rock and slowly cuts into it.

7.2 Lime, Marl, and Dolomite

Limestone is used directly as a building material and consumed in large quantities in industry too. Most applications require it to be burned to quicklime (CaO) at a temperature of at least 900 °C and if necessary converted with water to slaked lime ($Ca(OH)_2$) in an exothermic reaction in which a lot of heat is generated. Cement (Box 7.1) accounts for only a small part of consumption. Added quicklime is used to form slag in blast furnaces. In flue gas desulfurization a lime suspension is sprayed through exhaust gases leading to the formation of gypsum ($CaSO_4 \cdot 2H_2O$). In sugar production a suspension of slaked lime (a.k.a. milk of lime) is used to precipitate undesirable constituents. Acidic soils can be improved by liming with quicklime or limestone flour. Ground limestone or calcium carbonate formed from slaked lime by absorption of CO_2 serves as filler in paper. Limestone also serves as a raw material for glass production.

Box 7.1 Cement, concrete, and mortar

Cement is the binder in mortar and concrete. Concrete involves mixing cement powder with water, gravel, and additives; mortar involves mixing cement powder with water, sand, and additives. A chemical reaction hardens the material in a process called setting in which very fine-grained needlelike minerals with a feltlike texture are formed.

The simplest variant is lime mortar in which burnt lime (CaO) (a.k.a. quicklime or caustic lime) is produced by burning (calcinating) ground limestone usually in a rotary kiln:

$$CaCO_3 \rightarrow CaO + CO_2$$

When the powder is mixed with water an exothermic reaction occurs in which the oxide reacts to form a hydroxide ($Ca(OH)_2$), slaked lime, or hydrated lime. This in turn absorbs CO_2 from the air during setting and thus reacts back to $CaCO_3$. It only works when exposed to air.

By contrast, Portland cement can also harden in water (hydraulic cement). Together with marl or limestone other additives such as clay, sand, and iron ore are added to the rotary kiln and fired at about 1400 °C to form a clinker containing not only CaO but also phases such as Ca_3SiO_5, Ca_2SiO_4, $Ca_3Al_2O_6$, and $Ca_2(Al,Fe)_2O_5$. The clinker is mixed with gypsum and ground to a powder. Different chemical reactions take place when hardening with water and different water-containing calcium silicates are formed besides $Ca(OH)_2$. Cement with special physical and chemical properties can be produced by adding other substances.

Some tuffs or ground pumice can also be used as cement. Fine fragments of volcanic glass begin to crystallize into finely felted minerals in the presence of water.

Limestone is a sedimentary rock consisting mainly of calcite ($CaCO_3$) and varying amounts of quartz, clay minerals, organic substances, pyrite, or iron hydroxides. Limestone from reefs formed by the skeletons of corals and sponges is very pure. Limestones from lagoons often also contain dolomite, while banked limestones from the deep sea often also contain clay minerals. Moreover, limestone sinter (travertine) deposited at some springs can be used as can crusts in soils (calcrete), carbonatites (Sect. 3.10), and marble. Marble is formed by metamorphic transformation of limestone the result of which is mainly increased grain size.

Marl is a rock consisting mainly of calcite and clay minerals. Although it is mainly used for cement production, some compositions can even be burned directly into cement or sintered into cement clinker without further additives.

The rock dolomite consists mainly of the mineral of the same name—$CaMg(CO_2)_3$. The classical idea was that dolomite formed when limestone absorbed Mg^{2+} from water and dissolved Ca^{2+} at the same time in a process called dolomitization. It was further believed that seawater, connate water during diagenesis, or hydrothermal water was responsible for this. However, such a reaction does not occur under normal conditions. We now know that sulfate-reducing bacteria in very salty water lead directly to the deposition of dolomite (McKenzie and Vasconcelos 2009; Krause et al. 2012). There are examples from sabkhas (Box 5.11) in the United Arab Emirates, salt lakes in Australia, and coastal lagoons in Brazil. In addition to direct precipitation of dolomite in so-called algal mats the chemical conditions brought about by bacteria (supersaturation of dolomite, undersaturation of calcite) also lead to dolomitization a few centimeters deeper.

Dolomite is used as an aggregate in blast furnaces and in concrete. It is also used in the liming of soils, in glass production, as a scouring agent, filter material, and in the production of refractory materials.

7.3 Tuff, Pumice, Perlite, Pozzulan, and Trass

Acidic and intermediate magmas often have a very high content of fluids. In volcanic eruptions degassing causes the magma to foam driving material out of the vent at high speed and fragmenting it into fine material. In the ash cloud rising above the volcano the melt cools down and solidifies into glass. This so-called ash consists of fine-grained fragments of glass. They are often Y-shaped since they are fragments of former walls separating gas bubbles in the foam. Such deposits are called tuff in which glass can crystallize over time into a fine-grained solid rock with feltlike minerals. Although pumice (Fig. 7.1) comprises larger pieces of solidified magma foam that rain down near the crater, fine material can be transported very far by winds. Ignimbrites are tuffs with ash and pumice deposited by pyroclastic flows.

Fig. 7.1 Disused pumice quarry at Wingertsberg, near Mendig, Eifel (Germany) (*photo* © F. Neukirchen)

Solidified tuffs make popular bricks. Uncompacted tuffs are mainly used as aggregates in concrete to reduce weight and achieve effective thermal insulation. Coarsely ground pumice or pumice sand is used to produce lightweight building blocks that can be bonded with cement. In horticulture pumice is often mixed into the soil to increase pore volume. Stonewashed jeans are worked with pumice.

Pozzulan and trass are names given to certain tuffs that are mixed with cement to produce hydraulic cement that hardens particularly well under water. This is because SiO_2 can easily be dissolved from volcanic glass (and from some zeolites formed during the weathering of glass) and calcium silicates can therefore easily form. In addition, newly formed mineral phases are rich in aluminum making this cement much more durable. This can clearly be seen in sunken port facilities dating from Roman times whose cement survived 2000 years better than many a modern quay made of Portland cement survived 100 years (Jackson et al. 2013).

Perlite is a glass-rich rock that forms by hydration of obsidian or acidic lavas. It mostly occurs as a small deposit in the form of a lava dome. When perlite is heated, water escapes and a bubble-laden material is formed used to produce lightweight concrete with good thermal insulation.

7.4 Feldspar, Quartz, and Mica

Feldspar, quartz, and mica are rock-forming minerals and the main components of many rocks. They are also important raw materials. Although they are very common, they can only be used as raw materials if they can be obtained in sufficient purity.

Quartz (SiO_2) is used as a raw material for glass (Box 7.2), ceramics, silicon, silicon alloys, silicon carbide, ferrosilicon for steel production, silica gel, silicones, silanes, and amorphous fumed silica (anticaking agent, filler in plastics and paints, abrasives in toothpaste). Quartz is piezoelectric and can be excited by electromagnetic fields to vibrate. It is used as a digital pendulum in clocks, radio transmitters, and radio receivers. However, only flawless and non-twinned crystals are suitable for this purpose. Quartz crystals are therefore grown hydrothermally in autoclaves using natural quartz as raw material. Synthetic quartz crystals are also used as prisms and lenses. Quartz as raw material should be as free of inclusions as possible. Quartz sand, quartzite (a metamorphic rock formed by the transformation of sandstone and consisting predominantly of quartz), and

Table 7.1 The most common micas and their formula (they may contain other elements)

Muscovite	$KAl_2(AlSi_3O_{10})(OH)_2$
Biotite	$K(Fe,Mg)_3(AlSi_3O_{10})(OH)_2$
Phlogopite (Mg biotite)	$K\dot{M}g_3(AlSi_3O_{10})(OH)_2$

quartz from hydrothermal veins (which is particularly pure; Sect. 4.1) and from pegmatites (Sect. 3.8) are extracted.

Potassium feldspar is a solid solution i.e. series of mixtures between alkali feldspar (orthoclase, microcline, $KAlSi_3O_8$) and albite ($NaAlSi_3O_8$). Plagioclase is a solid solution between albite and anorthite ($CaAl_2Si_2O_8$).

Potassium feldspar has a low melting point and is therefore used as flux for the production of ceramics and glass. It is mainly obtained from pegmatites where it partly forms very large crystals and in so doing simplifies the production of very pure concentrate. With the help of flotation, magnetic separators, and suchlike feldspar can also be extracted from granites, syenites, nepheline syenites, and other rocks. Both potassium feldspar and plagioclase are used as fillers. Anorthosites (Sect. 3.5) are also used.

Mica are phyllosilicates (a.k.a. sheet silicates) that occur in the form of leaves or books that can be split into thin, flexible panes. Like a sandwich their crystal lattice arrangement consists of two layers with silicon and aluminum oxide tetrahedrons and an intermediate octahedron layer with either aluminum (e.g., muscovite) or iron and magnesium (e.g., biotite) (see Table 7.1 for compositions). Since the iron content of biotite interferes with many technical applications, phlogopite (the Mg endmember of biotite) and muscovite are of particular economic importance. Phlogopite is much more heat resistant than muscovite. In pegmatites, mica can form plates several meters thick that can easily be split into thin slices. In Russia these were used as window panes. The word muscovite derives from "Moscow glass." Although muscovite is mainly extracted from pegmatites, phlogopite occurs in metasomatic reaction zones between acidic and basic rocks and magmatically in alkaline rocks (Sect. 3.9). Ground mica is also extracted from other mica-rich rocks. Micas are good electrical and thermal insulators; hence their use in electrical engineering. However, they have been replaced by ceramics in many applications. Ground mica can be used as an insulating material, an additive in drilling fluids, as a glossy pigment in paints and cosmetics, and as filler in plaster and plastics.

Box 7.2 Glass

Glass is an amorphous substance usually produced by rapid supercooling of a melt when there is no time for crystallization. Although glasses do not have a melting point, they soften over a longer temperature range. Even at normal temperature they behave like a highly viscous liquid flowing imperceptibly slowly. Very different compositions can be quenched to glass as long as the melt is cooled down quickly enough. Most types of glass are silicate glass consisting mainly of SiO_2. Silicon oxide tetrahedrons are arranged in a chaotic 3D network in which other ions are scattered.

There are also natural silicate glasses. Acidic volcanic rocks mostly solidify in a glassy way as obsidian, pumice, or ash. Basic magmas can have a glassy groundmass when cooled down quickly. When quenched with water basic magmas can have a glassy surface, but this glass begins to crystallize relatively quickly into very fine-grained crystals.

Artificial glass is almost always soda-lime glass containing SiO_2 and some Na_2O, K_2O, CaO, and Al_2O_3. Other substances are added to achieve color (mainly metal oxides) or to improve its properties. A high proportion of PbO leads to glasses with strong light refraction and dispersion such as lead glass, flint glass, and strass. Crystal glass is a high-quality glass with a high content of other metal oxides (although it is not crystalline). Borosilicate glass with a high B_2O_3 content used in the manufacture of test tubes is particularly resistant to chemicals. K_2O and ZnO lower the melting point, Al_2O_3 increases the breaking strength, and so on.

Glass is made from a mixture of quartz sand (SiO_2), sodium carbonate (soda or soda ash (Na_2CO_3) found in salt lakes (Sect. 5.7.2) or synthetically from NaCl), potassium carbonate (potash (K_2CO_3) made from KCl (potassium salt)), feldspar, and lime (possibly together with waste glass and the abovementioned additives).

Other types of glass are also produced. Quartz glass consists only of SiO_2 and has a very high melting point. Chalcogenide glasses do not contain oxides or silicates; instead they have sulfides, tellurides, or selenides. For example, germanium–arsenic–tellurium glass is particularly permeable to infrared and therefore used in infrared optics. Fluoride glasses consist of F^- and metals such as Zr, La, Nd, Pr, Ba, Na, and Al; they are used for special optical applications. Moreover, there are amorphous metal alloys and some plastics that are called glass.

Irrespective of whether they are called glasses, sapphire glass (scratch resistant) and fluorite glass (for lenses with low chromatic aberration) are disks sawn from crystals—not glasses.

7.5 Clay and Kaolin

Clay minerals are a group of phyllosilicates (a.k.a. sheet silicates) usually occurring in the form of tiny (<2 μm) flakes. They are an important component of soils, deep-sea mud, shales, and hydrothermal alteration zones. They are water-containing or hydroxyl-containing silicates with a high aluminum content. Soils or loose masses with a high proportion of clay minerals are called clay.

The crystal lattice arrangement of clay minerals comprises different layers. The tetrahedron layers consist of interconnected silicon oxide tetrahedrons (Si is partly replaced by Al). The fourth tip of each tetrahedron is connected to the next layer in which Al^{3+}, Fe^{3+}, Fe^{2+}, Mg^{2+}, and Mn^{2+} sit on an octahedral position (i.e., between 6 oxygen or more precisely O^{2-} or $(OH)^{-}$). There may also be wide interlayer spaces in which H_2O and all kinds of ions including larger ones such as Na^+ and K^+ fit. Depending on how these layers are arranged clay minerals are categorized as 1:1 or 2:1 (i.e., 1 or 2 tetrahedral layers per 1 octahedral layer) (Fig. 7.2). In addition, there are clay minerals with alternating 1:1 and 2:1 parts.

There are countless clay minerals some of which are very similar to each other (Table 7.2). They can only be determined by X-ray diffraction. They are suitable to varying degrees as raw materials for ceramics (Box 7.3) or for other technical applications. In general, clay minerals are formed during the weathering or hydrothermal alteration of feldspar in which alkalis and possibly also some SiO_2 are removed, while immobile aluminum (in particular) remains. Which clay minerals are produced depends not only on the source material, but also on the pH, temperature, and how far weathering or alteration has progressed.

Kaolinite is a white 1:1 clay mineral that particularly forms in acidic environments during the advanced weathering of feldspar. Alkalis and iron have to be removed by water as does some of the SiO_2 for kaolinite to form. For example, potassium feldspar:

2 $KAlSi_3O_8$ (potassium feldspar) + 3 H_2O → $Al_2Si_2O_5(OH)_4$ (kaolinite) + 4 SiO_2 (hardly soluble) + 2 K^+ + 2 $(OH)^-$

Dickite and nacrite are very similar minerals. Kaolin (a.k.a. china clay; Fig. 7.3) is a soft rock consisting mainly of these two minerals together with some quartz and possibly other minerals such as muscovite, illite, anatase, goethite, and hematite. Allophanes (amorphous gels) and halloysite are intermediate reaction products that may also be present. The name kaolin derives from the Gaoling (high mountain) Mine in the province of Jiangxi (China) where the "white earth"—as it is sometimes called—was found.

Kaolin particularly forms during soil formation in humid tropical climates above feldspar-rich, iron-poor parent rocks

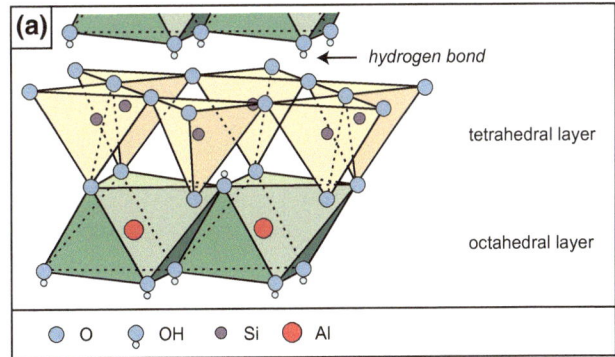

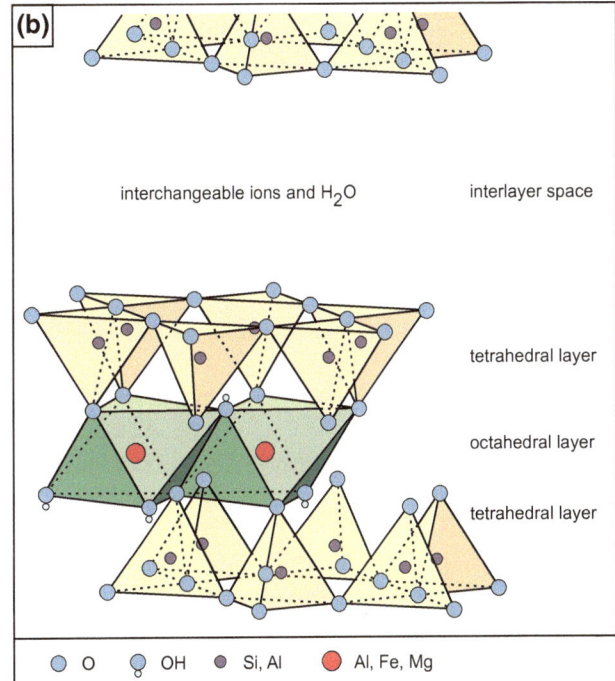

Fig. 7.2 Structure of clay minerals: **a** kaolinite and **b** smectite. With 1:1 clay minerals such as kaolinite packages of one tetrahedron layer and one octahedron layer are repeated and held together by hydrogen bonds. With 2:1 clay minerals such as smectite packages are composed of tetrahedron–octahedron–tetrahedron layers. Water and ions are found in the interlayer space between the packages both of which can be exchanged

such as granite, rhyolite, and arkose; hence it is subject to a certain form of lateritic weathering (Sect. 5.11). This also applies to occurrences in Germany originating in the hot

Table 7.2 The most important clay minerals and their approximate composition

Kaolinite	$Al_2Si_2O_5(OH)_4$
Montmorillonite	$Na_{0.33}(Al,Mg,Fe)_2(Si_4O_{10}) \cdot nH_2O$
Illite	$K_{0.65}Al_2(Si_{3.35}Al_{0.65}O_{10})(OH_2)$
Vermiculite	$Mg_{2.35}(Mg,Fe^{3+},Al)(Si,Al)_4O_{10}(OH_2) \cdot 4H_2O$

Fig. 7.3 Kaolin (china clay) from Dölau, near Halle (Germany) (*photo* © F. Neukirchen/Mineralogical Collections of the TU Berlin)

climate of the Late Cretaceous and Paleogene such as Upper Lusatia around Königswartha-Caminau on granodiorites, in Kemmlitz (near Oschatz, northwest Saxony) and Salzmünde (near Halle) on porphyry, and in Hirschau-Schnaittenbach (Upper Palatinate) on arkose. There are also some occurrences in Thuringia and the Westerwald. Moreover, some shales (especially, when accompanied by illite) can weather to kaolin. Kaolin particularly forms in connection with high-sulfidation epithermal systems (Sect. 4.3) and with greisen (Sect. 4.6) in hydrothermal alteration zones.

Kaolin is principally known for its role in the production of porcelain; little surprise then that the traditional centers of porcelain production such as Meissen, Selb, or Kahla are located near kaolin deposits. Kaolin is also used for a number of other products. Around 50% of the kaolin produced in Germany is used as a coating for paper and around 35% is used as filler (see Box 7.5) in the paper and rubber industry. Kaolin is also used in cosmetics (e.g., as a base for the production of powders). Kaolin can be added to foods as a separating agent, bleaching agent, or carrier. It cannot be absorbed by the intestine and is therefore harmless. It is authorized in the European Union as a food additive and given the number E559. Global production in 2003 amounted to around 45.6 million tonnes. The world's known recoverable reserves are estimated at around 14.2 billion tonnes, which should theoretically last for 300 years if production remains unchanged. Unwanted components are separated by sieving and slurrying.

Montmorillonite and other minerals of the smectite group are 2:1 clay minerals characterized by the fact that ions in interlayer spaces can be exchanged thus making an important contribution to the fertility of soils. They can also absorb and store water by swelling. The weathering of tuffs can lead to the formation of clays consisting mainly of smectites together with some quartz, opal, feldspar, and volcanic glass. Very similar clays may also be formed by sedimentation in the sea or lakes. Such material is called bentonite whose many properties make it suitable for a variety of applications. Bentonite has the useful property of swelling up when slurrying; hence its use in filling cavities. Although bentonite sludge is liquefied abruptly by vibrations, at rest it is comparatively solid and mechanically stable—a property called thixotropic. It is also a good ion exchanger. Bentonite is used as a drilling fluid (Sect. 1.9); a lubricant in tunnel excavation; a support for building pits; a seal for waste deposits; a binder in ore pellets and foundry sands; a raw material for ceramics; a filler in paints; a clarifier in the production of wine, fruit juices, cocoa butter, and edible oils; a degreaser of wool; a sealant and ion exchanger in ponds and aquariums; an additive for improving sandy soils; and an absorbent in cat litter.

Illite is the most common clay mineral in shales and an important component in many soils. It occurs during weathering or hydrothermal alteration of feldspar under less extreme conditions than is the case with kaolinite. It can also be formed from montmorillonite by absorption of Ka^+ (e.g., by fertilization). Its structure and composition are similar to muscovite (mica) into which it transforms during metamorphism.

Clay generally consists of various clay minerals and contains other fine-grained minerals such as quartz, feldspar, mica, some iron hydroxides, pyrite, and possibly carbonate. This is the reason some clays are more suitable for ceramics than others—suitability can be improved by adding further minerals. Pottery clays must not contain coarse minerals or rock fragments. They must be fine-grained, easily malleable, and react in the right way when fired in the kiln. No cracks must form during drying. The requirements for bricks are lower since the clay may be coarse and contain higher contents of iron minerals, carbonate, and so on. Since clays are not permeable to water they are also used in hydraulic engineering.

Vermiculite is a special clay mineral in that it expands 5 to even 50 times its volume when heated to more than 850 °C. Its leaflike crystals take on the form of worms; hence its name. Expanded it has a very low density that at the same time is strongly absorbing. It is used as an insulator against heat and sound, as aggregate in concrete and screed, as packaging material, as an absorbent of liquids (e.g., oil in accidents, cat litter), and as a soil additive in horticulture. Vermiculite is formed during the weathering or hydrothermal alteration of micas such as biotite and phlogopite. Although it is also present in many soils, it is mainly extracted from weathered or altered mica-rich rocks in alkaline rock complexes (Sect. 3.9) such as Phalaborwa (Box 3.15) and Kovdor (Sect. 3.10.1). In contrast to other clay minerals its crystals are often several millimeters or even centimeters in size; hence it is sometimes counted as mica.

Box 7.3 Ceramics

Inorganic non-metallic materials that are easily formed at normal temperature (when moist) and become solid at high heat (firing, sintering) are called ceramics. Depending on the mixing ratio and type of ingredients, grain size (coarse or fine ceramics), and firing temperature it is possible to produce polycrystalline materials of different porosity and with very different properties such as hardness, fracture toughness, density, thermal expansion, and thermal conductivity. Ceramics could be described as synthetic fine-grained rocks consisting of tiny crystals and possibly also some glass (Chiang et al. 1997; Richerson 2006; Routschka and Wuthnow 2007).

Classical ceramics were invented more than 24,000 years ago. They are silicate ceramics produced from mixtures of clay minerals, quartz, and potassium feldspar (Fig. 7.4) possibly with additional ingredients such as Al_2O_3 or $CaCO_3$. Although some natural clays are already suitable, other minerals may have to be added if necessary. Such ceramics include porcelain, stoneware, earthenware, and architectural ceramics (bricks). Porcelain is a white, dense, pore-free ceramic consisting mainly of the aluminosilicate mullite and a glass phase. The starting material is a mixture of kaolinite, quartz, and potassium feldspar. Stoneware is the name given to dense, pore-free, glassy (sintered) ceramics produced from clay minerals, quartz, and potassium feldspar. Earthenware differs from this in having high porosity. The aluminosilicates andalusite, kyanite, and sillimanite are also used for special ceramics. During firing in the kiln they are converted into mullite plus SiO_2.

Other silicate ceramics used in technical applications are made from soapstone together with clay, feldspar, magnesite, and possibly other ingredients such as steatite, cordierite, and forsterite.

A number of oxides can also be used as ceramic materials. Densely sintered Al_2O_3 (corundum) is the most important. It is characterized by high hardness, strength, temperature resistance, wear resistance, and excellent electrical insulation. It is used in medical

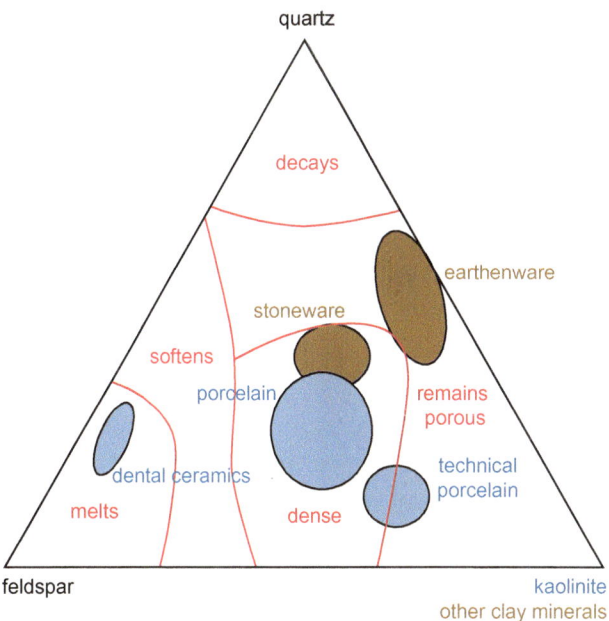

Fig. 7.4 Classical ceramics are produced from a mixture of clay minerals, quartz, and feldspar. Different mixing ratios lead to materials with different properties (from Gauckler 2005)

implants, in machine parts that move such as pump pistons, as an abrasive, and as an insulator in spark plugs. By adding some ZrO_2 it forms its own phase in the microstructure, toughness is further improved (the result is called zironia toughened alumina, ZTA). Zirconium oxide (ZrO_2) is another high-performance ceramic. Cubic high-temperature modification normally converts into tetragonal modification during cooling, which can be prevented by doping with MgO, CaO, or Y_2O_3 (cubically stabilized zirconium oxide). Accordingly, there are zirconium oxide ceramics with different properties such as polycrystalline tetragonal zirconium (TZP) oxide, partially stabilized zirconium (PSZ) oxide, and fully stabilized zirconium oxide. Further examples are oxide ceramics with BeO, MgO, or TiO_2.

A number of other oxide ceramics are characterized by special electrical and magnetic properties. They include electrical insulators, dielectrics, semiconductors, electrical conductors, and superconductors. Although ceramics made of Fe_2O_3 and possibly other metal oxides are not electrically conductive, they can be magnetized (ferrimagnetic). They are referred to as ferrites and depending on their properties are used as coil cores in electromagnets or as permanent magnets. Some important ceramics consist of phases in the perovskite structure. For example, $BaTiO_3$ is an excellent capacitor, mixtures of $PbZrO_3$ and $PbTiO_3$ are piezoelectric ceramics, and electrooptical ceramics such as $(Pb,La)(Zr,Ti)O_3$, $LiNbO_3$, and $LiTaO_3$ change their refractive index when induced by electrical or magnetic fields.

Finally, there are ceramics made of carbides such as silicon carbide (SiC), boron carbide (B_4C), tungsten carbide (WC) and nitrides such as silicon nitride (Si_3N_4), aluminum nitride (AlN), and boron nitride (BN) that are grouped together as non-oxide ceramics. They are characterized by extreme hardness and resistance.

Ceramics are often classified according to their intended use such as tableware, sanitary ceramics, refractory ceramics, tiles, ceramic abrasives, and technical ceramics. Technical ceramics include high-performance ceramics, structural ceramics, engineering ceramics, electroceramics, and (for the medical field) bioceramics.

Shaping usually takes place when clay is in a moist state by pressing, extruding, casting, free modeling, or on the classic potter's wheel. Thin coatings can be produced by chemical vapor deposition (CVD) and other special processes. Additives such as glue, wax, gelatin, or paraffin oil are sometimes added to make shaping possible in the first place; they then evaporate during firing. Other ingredients such as styrofoam balls, coal dust, or sawdust can increase porosity.

The green compact formed is first dried to expel pore water and then fired or sintered at high temperature. Chemical reactions occur that are comparable with contact metamorphism (in particular, aqueous minerals like clay minerals are replaced by anhydrous phases like mullite). Other phases such as Al_2O_3 recrystallize. In both cases a new structure is formed that is more solid and has a smaller pore volume. The volume of the workpiece also decreases slightly. At certain compositions and high temperatures a minor component (e.g., feldspar) even melts. The melt can close pores and form a thin film between grain boundaries. As it cools it solidifies into glass that may later crystallize into fine minerals. Melt formation can be facilitated by the addition of further substances (fluxes). Glass makes porcelain impermeable to water and gives it luster. Dental ceramics have a particularly high glass content. Refractory ceramics must not contain any ingredients that have a low melting point. If they do have a glass phase, then they have been produced by sintering at a very high temperature.

Ceramics are often glazed when they are coated with a thin film of glass that forms a smooth, waterproof surface that can be decorated by coloring. This is done by applying mineral flour—consisting predominantly of quartz together with fluxes such as potassium feldspar, calcite, boron, or lead compounds as well as temperature-insensitive color pigments—to the surface of the prefired (usually at low temperature) piece. It is then fired at a high temperature.

7.6   Aluminium Silicates

The aluminium silicates andalusite, kyanite, and sillimanite have the composition Al_2SiO_5. They are often summarized as sillimanite minerals. They are used in the production of special ceramics and refractory materials when they are converted during firing into mullite and SiO_2. They particularly form during the metamorphism of shales (i.e., in metapelites) and other aluminum-rich rocks such as andalusite in low-grade metamorphism (low-grade contact or regional metamorphism), sillimanite at high temperature (contact metamorphism), and kyanite at high pressure. Although these schists are only suitable for extraction if the content of the respective mineral is high enough, this is only the case in smaller lenses or thin layers. Metasomatic or hydrothermal depletion of SiO_2 can also lead to the formation of aluminosilicates (e.g., when acidic igneous rocks contact ultramafic rocks and in epithermal alteration zones). In addition, they accumulate together with other heavy minerals in placer deposits.

Contact metamorphic metapelites in the Bushveld Complex (Sect. 3.3.3) in South Africa (Transvaal Super Group) and contact metamorphic metapelite in Glomel (France) are the principal andalusite deposits.

Kyanite is mainly obtained from placers. Lenticular or layered kyanite-rich portions in quartzites in the Appalachian Mountains (Virginia, North Carolina, South Carolina, USA) are the most important primary deposits. They are highly metamorphosed former sediments and volcanic rocks of an island arc. There are smaller occurrences in the Alps and many other regions.

Sillimanite is obtained almost exclusively from placers. Although primary occurrences are rarely mined, there are examples in hornfels (contact metamorphic), slates, and gneisses. Pegmatites that have lost SiO_2 through metasomatism should also be mentioned.

7.7 Wollastonite

The structure and composition of wollastonite ($CaSiO_3$) are similar to those of pyroxenes. The mineral forms white needles or fibers that are mainly used as fillers in ceramics and plastics. Some can also be used as a harmless substitute for asbestos. Small amounts of wollastonite may be present in alkaline rocks. However, only metamorphic or metasomatic occurrences in marbles and skarns are economically important (Sect. 4.9). It is formed at high temperature by the reaction:

$$CaCO_3 + SiO_2 \rightarrow CaSiO_3 + CO_2$$

SiO_2 is either supplied by hydrothermal solutions or already contained in the limestone as quartz sand. The reaction quickly comes to a standstill if CO_2 cannot escape; hence it must be an open system. Although many skarns contain wollastonite, worthwhile occurrences are nevertheless relatively rare. Some garnet is mined and sold as a by-product. Wollastonite is also produced synthetically.

7.8 Garnet

Garnet comprises a group of minerals that includes several common rock forming minerals. Red pyrope ($Mg_3Al_2(SiO_4)_3$) is an important mineral in the Earth's mantle (garnet peridotite) and in the high-pressure metamorphic rock eclogite. Dark-red almadine ($Fe_3Al_3(SiO_4)_3$) is typical of garnet mica schist. Grossular ($Ca_3Al_3(SiO_4)_3$) is typical of skarns. Spessartine, uwarowite, andradite, and melanite should also be mentioned. High-quality garnet crystals are popular gemstones. Garnet sand mainly obtained from beach placers is used as an inexpensive abrasive and for sandblasting.

7.9 Olivine (Forsterite)

Forsterite (Mg_2SiO_4) is the magnesium endmember of the olivine group (Fig. 7.5). It is a rock-forming mineral in the Earth's mantle and in basic and ultrabasic igneous rocks in which a small part of Mg^{2+} is always replaced by Fe^{2+}. Due to its high melting point magnesium-rich olivine is processed into refractory materials. An important requirement is that it contains no serpentine since olivine transforms into this mineral through hydration even along cracks. In addition, olivine is often mixed into iron ore pellets to influence slag formation. Precious stone quality crystals are traded as peridot.

Olivine is extracted from ultramafic rocks called dunites that consist almost exclusively of olivine. They may be pieces of lithospheric mantle in an ophiolite complex

Fig. 7.5 Magnesium-rich olivine in mantle rocks is green. This piece (*left*) comes from the Åheim olivine mine in Norway. Olivine from igneous rocks (*right*, from Oldoinyo Lengai, Tanzania) usually contains a little more iron and has a lower melting point (*photo* © F. Neukirchen)

(Box 3.7). Dunite remains when large amounts of basaltic magma are melted from normal peridotite of the mantle, which also contains clinopyroxene and orthopyroxene. Dunites can also be formed as crystal cumulates at the base of magma chambers holding basic or ultrabasic magmas.

7.10 Magnesite, Talc, and Soapstone

Climbers know ground magnesite ($MgCO_3$) as chalk. It is also used as a food additive (e.g., it prevents grains of salt from clumping together when moist). Magnesite is also the most important raw material in the production of magnesium. The mineral is mainly fired in large quantities to form magnesia (MgO) that is used as a mineral fertilizer, filler in paper and plastics, and as an adsorbent. Sorel cement used in industrial flooring is a mixture of MgO and $MgCl_2$.

Many refractory materials (Routschka and Wuthnow 2007) such as bricks for lining blast furnaces also contain MgO. The powder can be sintered at temperatures around 2000 °C to fire bricks called magnesite brick. It can also be sintered when mixed with chromite (magnesite chrome brick), spinel, zirconium oxide, and other materials. In some refractory materials (e.g., chrome brick) sintered magnesia is only a minor component and serves as cement.

Talc (Fig. 7.6) ($Mg_3Si_4O_{10}(OH)_2$) is the softest of all minerals and can even be scratched with a fingernail. This phyllosilicate is a lubricant and at the same time is water repellent. Talc is traded under the name talcum powder. It is used as filler in paper, plastics, medicines, and paints; as a food additive; and as a powder.

Soapstone is a very soft rock consisting mainly of talc and magnesite. The ease with which it can be worked has made it

Fig. 7.6 Piece of a talc vein 2–3 cm wide within soapstone (*upper left* part of the specimen) from Norway. Both are the result of metamorphism of serpentinite by interaction with fluids (*photo* © F. Neukirchen)

popular with people working in the arts and crafts. It can also be used as a refractory material. Larger quantities are needed for the production of technical ceramics. Mixing soapstone with a little clay and feldspar results in a very strong ceramic called steatite that is mainly used as an insulator in electrical components. If the clay content is higher, then cordierite ceramics are obtained that can withstand strong temperature fluctuations and from which catalysts can be built. If the magnesium content is increased instead, then refractory forsterite ceramics are obtained. The problem with soapstone is that it often contains fine needles of chrysotile (fibrous serpentine, serpentine asbestos; see Box 7.4) that must not under any circumstances end up in the lung.

Talc and magnesite can form in ultramafic rocks such as peridotite and serpentinite (especially, in ophiolite complexes; see Box 3.7) and in dolomite by interaction with fluids. At shallower depths peridotite—the main constituent of the Earth's mantle—takes up water and converts to serpentinite—rock consisting predominantly of the mineral serpentine. Although the serpentine modification antigorite is very common, at low temperature chrysotile (fibrous serpentine) is stable and forms instead mainly along cracks. Sometime a rock comprising olivine and talc is formed instead as a result of metamorphism in a temperature range of approx. 550–650 °C and depending on pressure (Bucher and Grapes 2011). Almost pure talc can be formed when fluids add SiO_2 or remove Mg^{2+}.

When peridotite (or serpentinite) comes into contact with fluids containing CO_2, magnesite can form. Although magnesite is also found in small quantities in the mantle, it is much more frequent in the mantle rocks of ophiolite complexes, which are often more or less strongly transformed into magnesite along cracks. Serpentinites react so easily with CO_2 that even a low-CO_2 fluid is sufficient.

Soapstone is formed during high-grade metamorphism (amphibolite facies) of peridotite or serpentinite in the presence of a fluid containing CO_2.

The most important talc deposits are metasomatic formations in dolomite in which SiO_2 and possibly also Mg^{2+} have been added by fluids and Ca^{2+} and CO_2 have been removed. There is a large deposit of talc in Trimouns in the Pyrenees (France) where a layer of dolomite 5 km long has almost completely transformed into talc.

Large lenticular magnesite deposits can also occur in dolomite layers. Their formation is probably similar to dolomitization (Sect. 7.2), but further advanced, in which here too a sedimentary–diagenetic or a subsequent metasomatic transformation is conceivable.

Box 7.4   Asbestos: From a sought-after raw material to a hazardous pollutant

It is hard to imagine today that asbestos was once a raw material in high demand. Asbestos is the name given to fibrous crystals that are so thin and flexible that they can be processed into textiles. In ancient times asbestos was used to braid wicks for oil lamps and later on to protect firefighters from fires. Insulation material made from asbestos and roofs covered with cement asbestos were once popular. Although asbestos is very heat resistant, a good fire retardant, resistant to chemicals, heat insulating, has a very high tensile strength, and is inexpensive, it is exceedingly dangerous to health. If fibers are inhaled, then they can settle in the lungs like small needles and cause asbestosis or lung cancer. As a consequence asbestos is little used today and buildings incorporating asbestos are extensively renovated—if not demolished. In many countries its use has been banned.

The mineral chrysotile (a.k.a. white asbestos, serpentine asbestos; Fig. 7.7) was mostly used. It is a modification of serpentine ($Mg_3Si_2O_5(OH)_4$) that is stable at low temperature and a phyllosilicate whose crystal lattice is rolled into small needles. The mineral is relatively abundant in the serpentinites of ophiolite complexes (especially, in fissures).

Several minerals of the amphibole group also occur in the form of asbestos fibers. Fibrous riebeckite (a.k.a. blue asbestos, crocidolite) and grunerite (a.k.a. brown asbestos, amosite) occur in southern Africa in metamorphic banded iron formations (Sect. 5.2). Tremolite asbestos and actinolite asbestos are other metamorphic formations. The fiber size of amphibole asbestos makes these varieties even more harmful than chrysotile asbestos.

Fig. 7.7 Chrysotile (fibrous serpentine) from Zululand (South Africa) (*photo* © F. Neukirchen/Mineralogical Collections of the TU Berlin)

7.11 Corundum

Corundum (Al_2O_3) is the second hardest of all minerals. High-quality crystals are expensive gemstones (Neukirchen 2012) such as ruby (red) and sapphire (blue, green, yellow). However, much more common are coarse masses of corundum, quartz, and iron minerals that are referred to as emery. They were formed by metamorphism (Feenstra and Wunder 2002) of bauxite (Sect. 5.11.1) (e.g., on Naxos). Corundum can also be produced synthetically from bauxite as an intermediate step in aluminum production (Bayer process). Larger crystals can be grown using the Verneuil method.

Corundum is used as a grinding and polishing agent, for sandblasting (since unlike quartz dust it does not cause silicosis), and in cutoff wheels (i.e., cutting disks) and other tools. It is used as aggregate for hard concrete and ceramics, as well as in the production of refractory materials and special ceramics. Colorless synthetic sapphire is used in the manufacture of scratch-resistant window panes. Titanium-doped sapphire is an important solid state laser.

7.12 Diamond

Elemental carbon in a cubic crystal lattice in the form of diamond is the hardest mineral. Diamond is not only an important gemstone, but is also used in large quantities as a grinding and polishing agent and in cutting tools. Further applications are windows or coatings designed to resist extreme conditions (space travel, reactors in the chemical industry) and in the manufacture of optical instruments. Every year about 30 t of rough diamonds are mined one-fifth of which is cut into gemstones. In addition, about 100 t of synthetic diamonds are manufactured per year. Natural diamond is formed mainly by redox reactions in the Earth's mantle and crystals then reach the surface through special magmas such as kimberlite. For more information about diamonds and kimberlites see Neukirchen (2012).

7.13 Diatomite (Diatomaceous Earth)

Diatoms are very common single-cell organisms whose cell walls are made of amorphous SiO_2 (opal). There are many different species with different shapes such as wheels, rods, boats, and barrels (see also Sect. 6.2). They reproduce by cell division in which the cell wall divides into two halves—the other half of the cell wall is newly formed by both diatoms. Most species live in water both sea and fresh. Diatoms can proliferate rapidly in lakes and sea basins if the water has a very high SiO_2 content. The reasons for this may be hydrothermal (as is the case in Myvatn, a lake in Iceland), tuffs in the bedrock, blown in by wind, or rock weathering. Diatomite occurrences in northern Germany formed in lakes of interglacials (short warm periods between the ice ages) that filled gullies excavated by glaciers. Diatoms sink to the bottom and form a chalky rock when they die. Such rock has many pores, low density, is absorbent, and heat and acid resistant. It is used in thermal insulation; as a water filter; as filler (Box 7.5) for plastics, rubber, paints, asphalt, and powder; and as a carrier for pesticides. It serves as an absorber in cat litter and for oil after accidents. It is also used as a mild polishing agent. Diatomite can be white, gray, greenish, or black depending on the content of organic substances, pyrite, and other minerals. It is slurried during processing to separate unwanted components and then dried in an oven. For some purposes it is fired at 800 °C (calcined) to get rid of organic components.

Box 7.5 Fillers
Fillers are additives used in the manufacture of paper, plastics, paints, and foodstuffs. They have little effect on the main properties of the product. Fillers can consist of mineral powder, wood flour, glucose, glass

fibers, and many other substances. They have a number of uses such as increasing the volume of a substance, as carriers of active substances in tablets, and making material easier to process (e.g., anticaking agent in a powder). Moreover, fillers are also used to specifically improve certain properties such as mica powder in concrete to improve thermal insulation; paper becomes white, opaque, less absorbent, and gets a smooth surface by adding kaolin, gypsum, chalk lime, or baryte; plastics can be made stiffer, more heat or abrasion resistant, or electrically conductive by such additives; and adjusting the shade of paints by diluting pigments or to improve consistency of the suspension. Mineral fillers are important raw materials that are consumed by industry in large quantities.

7.14 Fluorite and Baryte

Fluorite (Fig. 7.8a) and baryte (Fig. 7.8b) are among the most frequent gangue minerals in hydrothermal deposits. Gangue is the term miners used to describe the material accompanying ore at that time believed to be worthless. It has now become a sought-after raw material and many a former silver mine is now mining gangue.

Fluorite (CaF_2) is used as a flux in the smelting of metals (especially, in steel production) where it significantly lowers the melting point. Fluorite is the raw material used in the production of fluorine, hydrofluoric acid (HF),

hydrofluorocarbons, and fluorides such as cryolite, ammonium fluoride, and sodium fluoride. Cryolite serves as a flux in aluminum production (Sect. 1.16), whereas ammonium fluoride and sodium fluoride are used in the prevention of dental caries. This works because some of the $(OH)^-$ in the enamel of teeth, which consists largely of the mineral apatite with the composition $Ca_5(PO_4)_3(OH)$, is replaced by F^- (fluorapatite). This makes the enamel more resistant to acids produced by bacteria. Fluorite crystals or synthetic fluorides are ground into special lenses used in the assembly of high-quality objectives with low chromatic aberration. Fluorite is a major component in some Mississippi Valley type (MVT) ore deposits (Sect. 4.12), metasomatic deposits, and hydrothermal veins (Sect. 4.1).

Baryte ($BaSO_4$) has a very high density of 4.5 g/cm^3. This is exploited in deep drilling (Sect. 1.9) (e.g., for petroleum). When used as an additive in drilling mud it increases the density and thus the pressure of mud against rock thus stabilizing the borehole. Baryte is used in the manufacture of a special concrete that is particularly heavy and less radiolucent. Baryte is also used in the production of white pigment. In photographic paper (baryta paper) and plastics it serves as a filler. Baryte is formed in large quantities at hot springs on the seabed. The most important baryte deposits are sedimentary exhalative (SEDEX) deposits (Sect. 4.17), volcanogenic massive sulfide (VMS) deposits (Sect. 4.16), and large baryte deposits without sulfides that were formed analogously at less hot springs such as white smokers (Sect. 4.15.1). Further occurrences are MVT deposits (Sect. 4.12), hydrothermal veins (Sect. 4.1), evaporites (Sect. 5.7), and baryte sinter in terrestrial sediments and karst systems.

(a)

(b)

Fig. 7.8 **a** Fluorite and **b** baryte from the Clara Mine in Oberwolfach, Black Forest (Germany) (*photos* © F. Neukirchen, Markl Collection/Tübingen)

7.15 Zeolites

The zeolite mineral group includes more than 70 natural minerals (Table 7.3). There are also hundreds of synthetic zeolites that precipitate from saturated solutions or are produced from other zeolites by chemical reactions. Although their composition and structure resemble those of feldspar, the structure of silicon and alumina tetrahedrons in the crystal lattice is so wide that water fits into micropores in between (Fig. 7.9). This can be expelled by heating and then absorbed again without changing the crystal lattice. As a result of some other molecules fitting into the micropores some zeolites are used as molecular sieves. This is because they are permeable to certain gases or liquids (e.g., a specific gas can be filtered from a gas stream). Important applications include ion exchangers, absorbers (oil or petrol in accidents, cat litter), desiccants, nitrogen binders, water treatment, aquarium filters, and the binding of radioactive substances. The absorption heat can also be used for heating or indirectly for cooling.

Synthetic zeolites are used in large quantities in detergents where they decalcify water. Although sodium phosphate used to be employed for this purpose, it led to overfertilization of water bodies. Certain synthetic zeolites serve as catalysts (e.g., to crack hydrocarbons). The reaction takes place within micropores the active centers of which are metals and lattice points introduced into the lattice. They can release H^+ and thus act as an acid.

Zeolites are very common in nature. Beautiful crystals can be found on fissures and in gas bubbles in volcanic rocks. Zeolites are also formed during the decomposition of volcanic glasses. Tiny crystals are formed in sediments during diagenesis or very low-grade metamorphism (zeolite facies). Sediments of salt lakes (Sect. 5.7.2) may also contain zeolites. However, only rocks with a very high zeolite content are worth mining. They are formed mainly by hydrothermal alteration of volcanic glass in tuffs (Sect. 7.3) by alkaline water (pH 8–10 at 50–300 °C). Since normal hydrothermal systems on volcanoes (epithermal systems; Sect. 4.3) are strongly acidic this must be external water. One possibility is an ignimbrite   (tuff deposited by

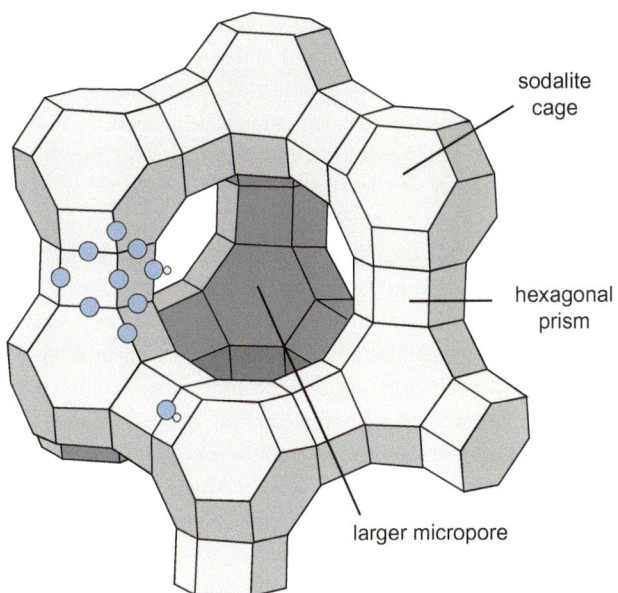

Fig. 7.9 Structure of zeolite Y (faujasite). Silicon or aluminum are located at the intersections of lines and oxygen or $(OH)^-$ is located in between (indicated in *blue*). Sodium is found in the hexagonal prisms, along with water in the sodalite cages, and completely hydrated along with further water in the larger micropores (modified from © Hermann Luyken, public domain, Wikimedia Commons)

pyroclastic flows) that has flowed into an alkaline salt lake. Another is the alteration of tuff by hydrothermal water entering from carbonate rocks.

7.16 Graphite

The stable modification of carbon is graphite (Fig. 7.10) under normal conditions, the structure involves each carbon atom forming covalent bonds with three neighboring atoms resulting in a 2D lattice with hexagonal meshes. The fourth electron of the outermost shell is relatively mobile between the layers. This together with the cohesion between layers having very low binding energy result in this mineral having

Table 7.3 Natural zeolites most important economically

Phillipsite	$KCa(Al_3Si_5O_{16})\cdot6\ H_2O$
Chabazite	$Ca(Al_2Si_4O_{12})\cdot6\ H_2O$
Erionite	$NaK_2MgCa_{1.5}(Al_8Si_{28}O_{72})\cdot28\ H_2O$
Ferrierite	$(Na,K)Mg_2Ca_{0.5}(Al_6Si_{30}O_{72})\cdot20\ H_2O$
Clinoptilolite	$(Na,K)_6(Al_6Si_{30}O_{72})\cdot20\ H_2O$
Mordenite	$Na_3KCa_2(Al_8Si_{40}O_{96})\cdot28\ H_2O$

Frequent but economically insignificant examples of this mineral group are natrolite, stilbite, heulandite, and analcime. Synthetic zeolites are usually named by appending letters (zeolite A, etc.)

Fig. 7.10 Graphite from Mataraka Plumbago Mine (Sri Lanka) (*photo* © F. Neukirchen/Mineralogical Collections of the TU Berlin)

two important properties: it is a good electrical conductor and is so soft that it is even used as a lubricant. In addition, it is resistant to acids, stable at extremely high temperatures (as long as it does not burn), and a good heat conductor.

Grain size plays a role in the application of graphite in addition to purity (Kwiecinska and Petersen 2004) such that high-quality flake graphite can be distinguished from inexpensive so-called amorphous graphite, which is however very fine-grained powder—not amorphous. During processing, graphite can be separated from other minerals by flotation. Graphite can also be produced synthetically by heating coke or petroleum coke to more than 3000 °C. The Acheson Process in which a mixture of coke and clay is electrically heated around a graphite rod is another way of producing graphite synthetically. At very high temperatures SiC is formed and converted to graphite by the evaporation of silicon at temperatures of more than 4150 °C.

Graphite is used in metallurgy for casting molds and crucibles, for carburizing steel, and as a reducing additive in blast furnaces. It serves as an electrode in batteries, arc furnaces, and in fused salt electrolysis in aluminum production. Carbon brushes in electric motors and generators ensure electrical contact between moving parts. Slide bearings and graphite seals are self-lubricating. Plastics can be made electrically conductive with graphite as filler. The chemical industry uses it for corrosion-resistant reactors. Graphite-moderated nuclear power plants have been built (an inglorious example being Chernobyl). An everyday example of graphite is the pencil (which also contains clay to increase hardness).

Three different graphite deposits can be distinguished. The most important are occurrences in highly metamorphic rocks (mostly amphibolite or granulite facies) in which the organic matter of former sediments has been converted to graphite. Almost pure graphite is produced during the high-grade metamorphism of coal seams; it is the final step in the coalification of plant remains via peat, soft coal, hard coal, and anthracite. Corresponding metamorphism of other sediments results in graphite-containing shale, gneiss, quartzite, and so on. Exact formation conditions depend not only on pressure and temperature, but also on the fluid phase during metamorphism and the already existing structure of organic substances (long-chain polymers have to be modified more strongly than ring structures). Shearing movements play an important supporting role. Accordingly, graphite formation begins with very fine-grained graphite with many lattice defects (upper greenschist facies), while stronger metamorphism produces larger crystals. Although there are corresponding deposits in Europe (e.g., in the Alps and Norway), by far the largest known reserves are in China.

Similarly, organic substances such as coal, kerogen, oil, and bitumen can also be converted to graphite during contact metamorphism. However, only small crystals are formed. An important deposit of this type is found near La Colorada, Sonora (Mexico).

Finally, there are epigenetic graphite deposits in veins and shear zones where graphite was precipitated from a fluid mostly by reduction of CO or CO_2 and sometimes from CH_4. The fluid can come from magmas, from metamorphic reactions, from the Earth's mantle, or it can be mobilized organic carbon. Precipitation takes place by cooling, fluid mixing, or changing oxygen fugacity (e.g., in a reaction with pyrrhotite or other iron minerals). Possible reactions are, for example:

$$2CO \rightarrow C + CO_2$$
$$CH_4 \rightarrow C + 2H_2$$
$$CO_2 + 2H_2 \rightarrow C + H_2O$$

The best known epigenetic graphite veins are found in Sri Lanka where carbon isotopes indicate the magmatic origin of the fluid (Touzain et al. 2010). Epigenetic graphite deposits also tend to develop under metamorphic conditions because they promote the nucleation and growth of larger crystals (Pasteris 1999), whereas graphite formed from a cooling fluid at shallow depth is very fine grained. Graphite is also found in meteorites (especially, in carbonaceous chondrites) and in some igneous rocks.

7.17   Sulfur

There are two different processes leading to the formation of elemental sulfur in nature. The first involves gases escaping from volcanoes containing not only water and CO_2 but also H_2S and SO_2 in different proportions. The reaction:

$$4H_2S + 2SO_2 \rightarrow 6S + 4H_2O$$

separates sulfur directly as a solid from cooling gas in a process called sublimation in which SO_2 can be partly formed by oxidation of H_2S. This happens above all at H_2S-rich gas outlets called solfataras and to a lesser extent at H_2S-rich fumaroles. Although filigree crystals sometimes form directly in crevices (Fig. 7.11), more common are earthy yellow masses and impregnations in rock pores. Due to the low melting point of sulfur, droplets or small streams of molten sulfur can also be seen at hot-gas outlets solidifying into solid sulfur.

Sulfur has been mined worldwide at a number of volcanoes such as Fossa Volcano on Vulcano (Italy). A small sulfur mine at the summit of the 6176-m-high Aucanquilcha Volcano (Chile) used to be the highest mine in the world. Especially spectacular are the solfataras at Kawah Ijen (Java, Indonesia). Mining here involved pipes being driven into the

Fig. 7.11 Sulfur crystals at a solfatara at the crater rim of Fossa Volcano on Vulcano island (Italy) (*photo* © F. Neukirchen)

solfataras enabling molten sulfur to drip through the pipes and solidify into massive blocks dragged out of the crater by workers in baskets (Fig. 7.12). So much SO_2 is introduced into the crater lake of Kawah Ijen that it is full of sulfuric acid with a pH of 0.1.

The second process is of greater economic importance and involves sulfur being indirectly formed by sulfate-reducing bacteria. Bacteria find conditions in the cap rocks of salt diapirs particularly suitable (Fig. 7.13) when natural gas or petroleum flows into them from surrounding sediments. Sulfate-reducing bacteria use the reduction of sulfate dissolved in water combined with simultaneous oxidation of organic substances (or hydrogen) to generate energy. This produces HS^- and CO_2 or HCO_3^-, for example, through the reaction:

$$SO_4^{2-} + CH_4 \rightarrow HS^- + HCO_3^- + H_2O$$

Hydrogen sulfide then oxidizes to sulfur by reaction with sulfate, by oxygen-rich water, or by the action of other bacteria. Hydrogen carbonate combines with Ca^{2+} to calcite. In this way lentils or thin layers of sulfur are formed in the sulfates of the cap rock, while some of the sulfates are replaced by calcite. Important deposits of this type can be found near Caltanisetta in Sicily (Italy), on the northern edge of the Carpathian Mountains (especially, in Poland), in Iraq, and along the Gulf Coast in Mexico and the United States. However, they are currently only mined in Poland. Underground deposits can be mined using the Frasch Process in which three tubes lying one inside the other are inserted into a borehole. Hot water at about 150 °C is

Fig. 7.12 Sulfur vapors and sulfur deposits at Kawah Ijen (Indonesia) (*photo* © F. Neukirchen)

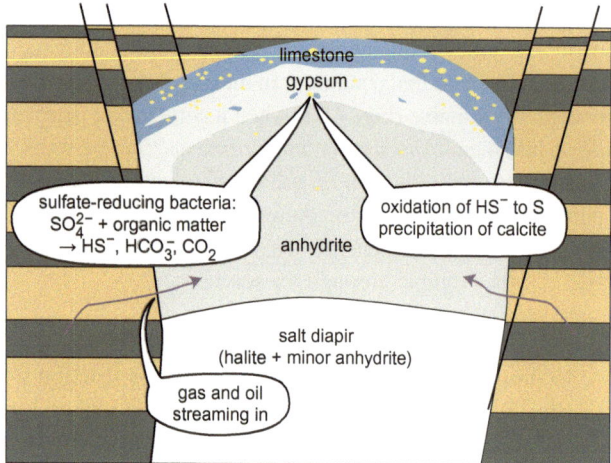

Fig. 7.13 Important deposits of elemental sulfur are formed by the action of sulfate-reducing bacteria in the cap rocks (mostly gypsum) of salt diapirs. The bacteria consume sulfate dissolved in water (gypsum, anhydrite) and organic substances (natural gas, petroleum) and excrete HS^-, which is then oxidized to sulfur. Hydrogen carbonate, another metabolic product, leads to the precipitation of calcite. Sulfur (*yellow*) forms smaller lentils and deposits in the cap rocks that have been partially converted to limestone

injected through one under high pressure. The sulfur melts and is brought to the surface by compressed air through the conveying pipe.

However, the mining of elemental sulfur accounts for only a tiny fraction of sulfur production. The desulfurization of natural gas containing H_2S (sour gas) and petroleum produces large quantities. The mining of tar sands makes Canada one of the main sulfur exporters. Therefore other important producers are countries that produce natural gas and oil as well as locations where there are large refineries. In addition, sulfuric acid is produced in large quantities as a by-product of sulfide smelting.

The most important application of elemental sulfur is the vulcanization of rubber. However, about 90% of the sulfur is processed into sulfuric acid known as the workhorse of the chemical industry. The production of mineral fertilizers consumes about 60% of this amount. The process involves the insoluble phosphate apatite reacting with sulfuric acid to form gypsum and phosphoric acid, which in turn is converted to other phosphates. Ammonium sulfate fertilizer is produced by sulfuric acid reacting with ammonia. In addition to phosphoric acid other acids are produced from sulfuric acid and corresponding minerals such as hydrochloric acid by reaction with halite (HCl, table salt), hydrofluoric acid by reaction with fluorite, nitric acid by reaction with nitrates (especially, Chile saltpeter ($NaNO_3$)). A mixture of sulfuric and nitric acid is the basis for the production of explosives such as dynamite. Sulfuric acid is also used for the production

of surfactants and organic dyes, for the leaching or digestion of some ores, and as electrolyte in car batteries.

Literature

Bucher, K., and R. Grapes. 2011. *Petrogenesis of Metamorphic Rocks*, 8th ed. Heidelberg: Springer.

Chiang, Y.-M., D. Birnie, and D. Kingery. 1997a. *Physical Ceramics: Principles for Ceramic Science and Engineering*. New York: Wiley.

Feenstra, A., and B. Wunder. 2002. Dehydration of diasporite to corundite in nature and experiment. *Geology* 30: 119–122.

Gauckler, L.J. 2005. *Materialwisenschaft I: Keramik*. ETH Zürich: Vorlesungsskript.

Jackson, M.D., S.R. Chae, R. Taylor, P. Li, C. Meral, S.R. Mulcahy, A. M. Emwas, J. Moon, S. Yoon, G. Vola, H.-R. Wenk, and P.J.M. Monteiro. 2013. Unlocking the secrets of Al-tobermorite in Roman seawater concrete. American Mineralogist (in Press).

Krause, S., V. Liebetrau, S. Gorb, M. Sánchez-Román, J.A. McKenzie, and T. Treude. 2012. Microbial nucleation of Mg-rich dolomite in exopolymeric substances under anoxic modern seawater salinity: New insight into an old enigma. *Geology* 40: 587–590.

Kwiecinska, B., and H.I. Petersen. 2004. Graphite, semi-graphite, natural coke, and natural char classification—ICCP system. *International Journal of Coal Geology* 57: 99–116.

McKenzie, J.A., and C. Vasconcelos. 2009. Dolomite Mountains and the origin of the dolomite rock of which they mainly consist: Historical developments and new perspectives. *Sedimentology* 56: 205–219.

Neukirchen, F. 2012a. *Edelsteine: Brillante Zeugen für die Erforschung der Erde*. Heidelberg: Springer Spektrum.

Pasteris, J.D. 1999. Causes of the uniformly high crystallinity of graphite in large epigenetic deposits. *Journal of Metamorphic Geology* 17: 779–787.

Richerson, D.. 2006. Modern Ceramic Engineering. Properties, Processing and Use in Design. Oxford: Taylor & Francis.

Routschka, G., and H. Wuthnow (eds.). 2007. *Taschenbuch Feuerfeste Werkstoffe. Aufbau, Eigenschaften Prüfung*, 4th ed. Essen: Vulkan-Verlag.

Touzain, P., N. Balasooriya, K. Bandaranayake, and C. Descolas-Gros. 2010. Vein graphite from the Bogala and Kahatagaha-Kolongaha-mines, Sri Lanka: A possible origin. *Canadian Mineralogist* 48: 1373–1384.

Further Reading

Chiang, Y.-M., D. Birnie, and D. Kingery. 1997b. *Physical Ceramics: Principles for Ceramic Science and Engineering*. New York: Wiley.

Neukirchen, F. 2012b. *Edelsteine: Brillante Zeugen für die Erforschung der Erde*. Heidelberg: Springer Spektrum.

Okrusch, M., and S. Matthes. 2009. *Mineralogie: Eine Einführung in die spezielle Mineralogie, Petrologie und Lagerstättenkunde*, 8th ed. Heidelberg: Springer.

Patton, T.C. 1973. *Pigment Handbook*. New York: Wiley.

Pohl, W.L. 2011. *Economic Geology*. Chichester: Wiley-Blackwell.

Rothe, P. 2010. *Schätze der Erde*. Darmstadt: Primus Verlag.

Seidler, C. 2012. *Deutschlands verborgene Rohstoffe: Kupfer*. Hanser, München: Gold und seltene Erden.

US Geological Survey. Mineral Yearbooks. http://minerals.usgs.gov/minerals/pubs/myb.html.

Glossary

A-type granite Anorogenic granite (or similar plutonic rock) formed during extension in continental rifts or during post-orogenic collapse. Typically alkaline rock.

Accessory minerals Minerals that make up less than 1% of a rock.

Acidic (magma) Melt or igneous rock with >66% SiO_2.

Aggregate Batch of many crystals of a single mineral.

Agpaitic rock Peralkaline nepheline syenite with complex sodium zirconium/titanium/SEE silicates such as eudialyte, aenigmatite, etc., due to an extremely high content of sodium and rare elements.

Alkaline rock Silica-undersatured igneous rock rich in alkalis ($Na_2O + K_2O$), e.g. nephelinite, phonolite, nepheline syenite and many more rocks of different composition. Corresponding melts are referred to as alkaline magmas.

Alloy Metallic material consisting of several elements.

Alteration Transformation of a rock by interaction with a (hydrothermal) fluid.

Amalgam Alloy of mercury and gold or mercury and silver. Liquid at high mercury content.

Andesite Intermediate volcanic rock with a medium composition between basalt and rhyolite. Typical for subduction zones.

Anorthosite Igneous rock (plutonic rock) consisting of >90% plagioclase.

Argillic alteration Alteration by acidic hydrothermal fluids near the Earth's surface with a resulting rock of predominantly clay minerals. Typical for high-sulfidation epithermal veins and at the Earth's surface over porphyry copper deposits.

Ash Combustion residue.

Ash (magma) Fine magma fragments extruded by gas-rich, explosive volcanic eruptions. Ash deposits are called tuff.

Asthenosphere Soft part of the upper mantle, below the lithosphere, may contain small amounts of melt (basaltic magma).

Atmophile Volatile elements (according to Goldschmidt) such as hydrogen, carbon, nitrogen, noble gases.

Back-arc In the hinterland of a subduction zone, the pull of the submerging plate causes extension, which can result in the formation of new oceanic crust.

Barrel Volume measure used for crude oil, 1 bbl = 158.987 l.

Barren rock Economically unusable side rock of a deposit. Also sterile rock.

Basalt "Basic" volcanic rock formed by melting the Earth's mantle. Consists mainly of pyroxene and plagioclase with little olivine or quartz. Classified as tholeiite basalt, alkali basalt, alkali olivine basalt etc. depending on composition. The corresponding plutonic rock is gabbro.

Basement Older igneous and metamorphic rocks, i.e. continental crust without sediments.

F. Neukirchen and G. Ries, *The World of Mineral Deposits*, https://doi.org/10.1007/978-3-030-34346-0

Basic (magma) Melt or igneous rock with SiO_2 content between 45 and 52%.

Batholith A gigantic elongate body composed of countless plutons of granites, gabbros and so on. They are formed below the volcanoes of the subduction zones.

Bauxite Aluminium ore consisting mainly of gibbsite, boehmite and kaolinite.

BBO Billion barrels of oil.

Bentonite Swellable clay (mostly montmorillonite) that is frequently used in construction engineering and drilling as drilling fluid.

Biomining Oxidation of sulfide ores and leaching of certain elements using microorganisms. Also: bioleaching, biooxidation.

Black shale Shale with a high content of organic substances.

Black smokers Hot springs (300–400 °C) with chimney-shaped structures on the seabed with precipitation of sulfides, sulfate and amorphous SiO_2. See also VMS.

Blast furnace Iron and steelmaking furnace in which ore is reduced to liquid pig iron at high temperature.

Blueschist Metamorphic rock formed during the transformation of basalt under high pressure.

Breccia Rock composed of angular rock fragments, either formed by sedimentation with a very short transport path, or by landslides, or by breaking in situ through tectonic movements, magmatism or hydrothermal fluids whose pressure exceeds the rock pressure. The fine-grained matrix consists of hydrothermally precipitated minerals, rock flour or solidified magma.

Brent A light type of petroleum from the North Sea whose price is used as a basis for other types of crude oil on the stock exchange. See also WTI.

Brine Highly saline aqueous solution.

Calc-alkaline magma Typical fractionation trend in the magmas of the subduction zones at continental margins, from basalt via andesite to rhyolite.

Calcite Frequent carbonate mineral $CaCO_3$. Limestone consists mainly of calcite.

Caldera Crater-shaped collapse structure on a volcano caused by the collapse of the underlying magma chamber.

Carbonate Mineral with $(CO_3)^{2-}$ as anion complex, e.g. calcite and dolomite. Also short for carbonate rock.

Carbonate platform Shallow sea area with very fast carbonate sedimentation, which can keep up with the slow sinking due to its own weight, e.g. Bahamas, Great Barrier Reef.

Carbonatite Ingeous carbonate rock.

Cast iron Iron with a high carbon content (approx. 2–4%), is brittle and not forgeable.

Catathermal Previously used for hydrothermal systems above 300 °C, now used for hydrothermal systems at great depths, also called hypothermal. See also epithermal, mesothermal.

Cementation zone In connection with the formation of a surficial oxidation zone in a sulfide deposit, metals such as copper and silver may accumulate just below the groundwater level, resulting in secondary sulfides.

Chalcophile "Sulfur-loving elements" (according to Goldschmidt), for example copper, lead, zinc, silver.

Chemical weathering Weathering due to dissolving processes. Very intense in humid-warm climate below soil.

Chert Fine-grained, sedimentary rock consisting of silica. Mostly derived from opal shells of protozoa, from which quartz is formed during diagenesis. Also exhalative at submarine hot springs.

Chloritization Alteration to a chlorite-rich rock, see also sericitic alteration.

Cleft Crack in the rock, often caused by pressure relief during ascent.

Coal Rock with very high carbon content.

Colloform Fine-grained cauliflower- or kidney-shaped aggregates with a shiny surface. Also: botryoidal.

Coltan The minerals of the solid solution series columbite-tantalite. Most important tantalum ore.

Compatible The ion is enriched in the crystals during the fractionation between magma and rock (both during melting and crystallization) and depleted in the melt.

Complex (chemistry) A compound consisting of a central particle (e.g. metal ion) and one or more attached ligands (e.g. Cl^-). Complexes allow a significantly higher solubility of the corresponding metal in a fluid.

Complex (igneous rocks) A body of rock composed of several different plutonic rocks.

Conglomerate Sedimentary rock formed from gravel.

Contact metamorphism Metamorphism caused by magma at high temperature, e.g. by emplacement of a granite pluton into sedimentary rocks.

Converter Special large crucible in metallurgical plants, e.g. for refining pig iron.

Country rock Surrounding rock.

Craton The oldest (Precambrian) cores of the continents.

Critical point If pressure and temperature exceed the critical point (374 °C and 221 bar in pure water), the phase boundary between liquid and gas (boiling point) disappears in favor of a continuous transition.

Crust Uppermost shell of the Earth, oceanic (approx. 5 km thick) and continental crust (approx. 35 km thick, under mountains up to double) have a different composition.

Crystal Solid, homogeneous, anisotropic body. In a crystal, the atoms are regularly arranged on fixed lattice sites. If ideally shaped it has crystal faces whose arrangement, shape and symmetry is derived from the crystal lattice. However, crystals can also be irregularly shaped as part of a rock. Almost all minerals form crystals.

Deposit Occurrence of ores, or petroleum, natural gas or other commodities.

Diagenesis Consolidation of loose material to a sedimentary rock. With slight burial, a cement of quartz, calcite or hematite, for example, is precipitated from the circulating water, which cements sand grains or pebbles. In addition, the pore volume is reduced by compaction.

Diapir Rising finger-shaped rock body, for example in the mantle or as a salt dome.

Dike Fissure in which magma rises or rather rose and cooled to an igneous rock.

Diopside Mineral of the pyroxene group, $CaMg(Si_2O_6)$.

Disseminated ore Rock speckled with ore minerals. The ore minerals are either formed syngenetically (together with the rock) or epigenetically (later) through mineralization in the rock pores.

Dolomite Carbonate mineral, $CaMg(CO_3)_2$, or sedimentary rock composed of this mineral.

Drill core Cylindrical rock sample taken from a borehole with a special drill head.

Electrum Alloy of silver and gold occurring naturally.

Epigenetic The ores are younger than the host rock. Opposite of syngenetic.

Epithermal Previously for low temperature hydrothermal systems (<200 °C), today for shallow depth hydrothermal systems. See also mesothermal, catathermal.

Euxinic Water with low oxygen content and high content of H_2S, such as in the deep sea of the Black Sea.

Evaporite Sediment left behind by evaporation of water, such as salt or gypsum.

Exhalative "Exhaled" from sources on the seabed. See also VMS and SEDEX.

Exploration Search for unknown deposits and more detailed investigation of potential deposits.

Facies Properties of sedimentary rocks attributable to the sedimentary environment (e.g. lagoon and reef). Metamorphic facies: Metamorphic grade, given relative to the transformation products of basalt at corresponding pressure and temperature (greenschist facies, amphibolite facies, granulite facies, blueschist facies, eclogite facies).

Fault Planar fractures with displacement of tectonic blocks. See also normal fault, reverse fault, strike-slip fault, thrust fault.

Feldspar Group of silicate minerals, plagioclase and potassium feldspar are important rock-forming minerals.

Feldspathoid Group of silicate minerals similar to feldspar, but containing less SiO_2. In silica-undersatured igneous rocks.

Felsic The (bright) minerals quartz, feldspar and feldspathoids, or a rock consisting mainly of them. See also mafic.

Fenitization Metasomatic transformation of a rock in the vicinity of carbonatites or alkaline rocks, with a strong enrichment of alkalis.

Fluid Liquids, gases and supercritical fluids. In particular, mixtures of water, dissolved salts and gases such as CO_2 and H_2S.

Footwall Underlying stratum.

Fracking Hydraulic Fracturing. Hydraulic generation of cracks in an impermeable rock, allowing the extraction of crude oil and natural gas tightly entrapped in pores.

Fractional crystallization Change in the composition of a magma during crystallization. Leads e.g. from basalt ("basic") via andesite to rhyolite ("acidic"). Most important process of fractionation in magmatic systems.

Fractionation Enrichment or depletion of different elements in different processes.

Gabbro Plutonic rock with the same composition as basalt.

Gangue Minerals of a (hydrothermal) deposit without metal content, especially quartz, fluorite, baryte and carbonates.

Garnet Group of silicate minerals, including almandine and pyrope.

Gas hydrate Ice-like substance consisting of H_2O and gases such as methane.

Glass "Frozen rock melt", in contrast to crystals there is no systematic arrangement of ions in glass.

Gossan Surficial part of the oxidation zone of sulfide deposits in which water-soluble metals were dissolved and removed. Insoluble metals remain, especially iron oxidized to Fe^{3+}.

Granite Coarse-grained igneous rock of feldspar, quartz and mica, cooled to a pluton at depth. The equivalent of the "acidic" volcanic rock rhyolite. Formed by fractional crystallization of basaltic magma or by partial melting of continental crust.

Gravimetry Measurement of variations in the Earth's gravitational field.

Graywacke Sedimentary rock consisting of small rock fragments, sand and clay minerals. Deposition in the sea near coastal mountains.

Greenstone belt Precambrian formation predominantly consisting of greenschist (metamorphically transformed basalt).

Greisen Metasomatic rock that consists almost exclusively of quartz and is sometimes found above granite. May contain mica, topaz, tourmaline and the ores wolframite and cassiterite. Formed by converting the solid granite by reactive solutions released from the last residual melt of the granite magma.

Hangingwall Overlying stratum.

Hardness The scratch hardness of a mineral, i.e. the resistance to scratching with a sharp-edged material, is an important distinguishing feature. As a rule, hardness is given on a relative scale compared to ten specific minerals (Mohs scale of hardness). The absolute differences in hardness between these reference minerals are very different.

Heavy metal Arbitrary term for certain metals which, depending on the definition, either have a high density or a high atomic weight or are particularly toxic.

High Field Strength Elements (HFSE) Elements with a high ionic charge in relation to the ionic radius. For example uranium, zirconium, niobium, rare-earth elements.

Hot spot Volcanism caused by a mantle diapir (e.g. Hawaii), independent of plate boundaries.

Hydrometallurgy Leaching of certain metals from ores using suitable chemicals such as acids or cyanides and subsequent processing by solvent extraction, ion exchangers, precipitation reactions, electrolysis and so on.

Hydrothermal Crystallization from hot aqueous solution.

Hypothermal Also catathermal, previously used for hydrothermal systems above 300 °C, today used for hydrothermal systems at great depths. See also epithermal, mesothermal.

I-type granite Granite produced by fractional crystallization of basaltic magma or by partial melting of gabbro.

Igneous rock Synonym for magmatic rock. Rock formed by cooling magma (molten rock). Either plutonic or volcanic rock.

Ignimbrite Volcanic rock deposited from pyroclastic flows (pumice and ash flows).

IOCG Iron oxide copper gold deposit.

Incompatible The ion is enriched in the melt during the fractionation between magma and rock (both during melting and crystallization).

Intermediate (magma) Melt or igneous rock with SiO_2 content between 52 and 66%.

Island arc A chain of volcanic islands above a subduction zone.

Karst Terrain morphology formed by the dissolution of rocks (especially limestone, but also gypsum, salt), including underground systems such as caves.

Kerogen Organic substance that cannot be dissolved in organic solvents. Intermediate product in the formation of crude oil and natural gas.

Large ion lithophile Elements with a large ionic radius, such as potassium, rubidium, strontium.

Laterite Red soft rock formed by intensive tropical weathering.

Lava Magma that flows on the Earth's surface.

Light metal Metals or alloys of low density.

LIL Large ion lithophile, ions with large radius.

Limestone Sedimentary rock consisting mainly of calcite. Either deposited by living organisms such as corals, algae, bacteria, etc. or by chemical precipitation.

Liquidus Temperature at which the last crystals disappear during fusion and the rock is completely melted (or when the magma cools down, the first crystals form). See also Solidus.

Lithophile "Rock-loving elements" (according to Goldschmidt), especially elements found in silicate minerals of the crust and the mantle, such as aluminum, sodium, potassium, but also rare-earth elements, uranium, tantalum.

Lithosphere Earth's crust and the uppermost, rigid part of the mantle. Below is the soft asthenosphere.

Mafic All rock-forming minerals except quartz, feldspar and feldspathoid. Mainly dark minerals. Or rock with a high proportion of mafic minerals. See also felsic.

Magma Molten rock, including crystals and dissolved gases. See also igneous rock.

Magma chamber Sub-volcanic body of molten rock that can cool to a plutonic rock or rise further.

Magnetics Measurement of anomalies of the Earth's magnetic field.

Mantle Middle shell of the Earth, between the Earth's crust and core. Consists of peridotite. The uppermost part behaves rigidly (lithospheric mantle), while the area below is ductile (asthenosphere). In the lower mantle the minerals of the peridotite are transformed into other minerals which are only stable under extremely high pressure.

Mantle diapir Hot mantle material rising finger- or mushroom-shaped in the mantle. Responsible for hot spots.

Marble Metamorphic carbonate rock.

Massive sulfide Massive ore body consisting almost exclusively of sulfides.

Maturity The maturity (depending on temperature and time) of a source rock and its kerogen with regard to the formation of petroleum and natural gas or the maturity of coal formation.

Mesothermal Previously for hydrothermal systems of medium temperature (200–300 °C), today for hydrothermal systems at medium depth (a few kilometers). See also epithermal, catathermal.

Metal Elements or alloys in which metallic bonds predominate, i.e. the outermost electrons of each atom are freely mobile. Metals have a high electrical conductivity, a high thermal conductivity, are ductile and have a metallic luster.

Metalloids Elements which are both in the periodic table and in their properties between metals and non-metals. Metalloids are semiconductors.

Metallurgy Science of metal extraction and processing

Melilitite Strongly silica-undersaturated volcanic rock rich in alkalis. Contains in addition to the mineral melilite, $(Ca, Na)_2(Mg,Al)Si_2O_7$, mostly mafic minerals, possibly also nepheline and glass.

Metaluminous rock Igneous rock with $Na_2O + K_2O < Al_2O_3 < Na_2O + K_2O + CaO$.

Metamorphic rock Rock formed by transformation (in solid state) from other rocks by increasing temperature and pressure, for example mica schist, greenschist, gneiss, marble, eclogite.

Metasomatism Change of a rock by mass exchange with an aqueous solution or a magma.

Meteoric water Rainwater and groundwater formed from it.

Methane CH_4, most important component in natural gas.

Methane hydrate Ice-like substance consisting of H_2O and CH_4. Methane is trapped in the cages of the ice crystal lattice.

Mica Important sheet silicate minerals, especially biotite and muscovite. Common in granite, gneiss and mica schist.

Mid-ocean ridge Constructive plate boundary. Mountains in the ocean, at whose central trench new oceanic crust is formed.

Migmatite Partially melted rock (transition metamorphic/igneous rocks). Often looks like gneiss with light streaks.

Mineral Lifeless, homogeneous and natural solid of the Earth or other celestial bodies. Apart from a few exceptions, minerals are inorganic compounds (or elements) and form crystals.

Moho Boundary between the Earth's crust and mantle. Short for Mohorovičić discontinuity.

Monazite Rare-earth element phosphate.

MVT Mississippi Valley type. Important type of hydrothermal lead-zinc deposits.

Mylonite Rock plastically deformed at a fault or shear zone by strong movement.

Nappe Sheetlike tectonic block in orogenic belts bounded by shallow-dipping thrust faults. Also: thrust sheet.

Native metal Metals occurring in elementary form in nature, for example native gold, platinum, silver, copper.

Natural gas Mainly methane, plus propane, butane, H_2S, CO_2 and other gases. "Wet gas" has a higher proportion of propane and butane, "dry gas" is almost only methane. "Sour gas" has a high sulfur content.

Nephelinite Strongly silica-undersaturated volcanic rock with nepheline and other mainly alkali-rich minerals, free of feldspar.

Normal fault A fault in an extensional setting in which the hangingwall (the block above the fault) moves downward relative to the footwall block.

Oil sand Synonym for tar sand.

Oil shale Kerogen-rich rock that emits synthetic crude oil and gas when heated.

Olivine Silicate mineral, $MgSiO_4$. Important in mantle rocks and basalt. Gemstone under the name peridot.

Ophiolite Oceanic lithosphere emplaced in continental crust, consisting of mantle rock (peridotite or serpentinite), gabbro, basalt dikes, pillow lavas and sediments.

Ore A mineral or rock that can be mined for economic reasons, usually to extract metals.

Ore grade Content of the economically interesting metal in the ore, expressed as a percentage or g/t.

Orogeny Mountain building at convergent plate margins by continental collision or accretion of terranes on a subduction zone. E.g. Alpine Orogeny, Variscan Orogeny. The term mostly refers to tectonic movements and deformation and not to the resulting uplift itself.

Oxidation zone Surficial zone of sulfide deposits with oxidization of sulfides. Certain metals such as copper and silver are dissolved and precipitated somewhat deeper. This occurs either in the form of so-called "oxidic" ores (oxides, carbonates, sulfates, etc.) or in the cementation zone below the groundwater table in the form of secondary sulfides. Iron remains in the leached zone (see gossan).

Oxygen fugacity Oxygen concentration, thermodynamically corrected taking into account the free energy of different phases.

Paragenesis The association of jointly formed minerals that were therefore in equilibrium with each other.

Pegmatite Igneous rock with huge (centimeter to several meters large) crystals. Occurs together with granite and similar plutonic rocks and has a similar composition to these. Sometimes contains rare metals or gemstones.

Peralkaline rock Igneous rock with $Al_2O_3 < Na_2O + K_2O$.

Peraluminous rock Igneous rock with $Al_2O_3 > Na_2O + K_2O + CaO$.

Peridotite Rock of the Earth's mantle, consisting of olivine, pyroxene (diopside and enstatite) and garnet or spinel. Converts to serpentinite by hydration.

Petroleum Mixture of hydrocarbons that is liquid under standard conditions.

PGE See platinum group elements.

Phenocrysts Larger crystals in a fine-grained volcanic rock.

Phoscorite Igneous rock of magnetite, apatite and olivine, which sometimes occurs together with carbonatite in alkaline rock complexes.

Phosphorite Phosphorus-rich sedimentary rock.

Phyllic alteration See sericitic alteration.

Physical weathering Mechanical rock fragmentation, e.g. by frost shattering.

Pig iron Non-forgeable iron produced in a blast furnace with a very high carbon content (similar to cast iron), which is converted into steel by reducing the carbon content (refining).

Pillow lava Pillow-shaped basalt lava formed under water by quenching the surface.

Placer deposit Secondary deposit of high-density, weather-resistant minerals deposited in a river bed or on a beach. Especially gold, platinum, cassiterite, monazite, zircon.

Plagioclase Important rock-forming silicate mineral of the feldspar group with solid solution of $Na(AlSi_3O_8)$ and $Ca(Al_2Si_2O_8)$.

Plate (lithospheric plate) Includes continental and oceanic crust and the rigid part of the mantle (i.e. the lithosphere). The plates "float" and move on the easily deformable asthenosphere.

Plate tectonics Theory of the mobility of the lithospheric plates.

Platinum group elements (PGE) The very similar metals osmium, iridium, platinum, ruthenium, rhodium and palladium.

Pluton Larger body of igneous rock (e.g. granite or gabbro) cooled at depth. Also intrusion.

Plutonic rock Igneous rock that cooled slowly at depth in a pluton and is therefore coarse-grained (Also intrusive rock).

Pneumatolytic Obsolete term for hydrothermal formations at extremely high temperatures (>400 °C) at which pressure fluctuations were considered decisive. These included greisen and some skarns.

Podiform Lenticular.

Porphyry Volcanic rock with large phenocrysts in a fine-grained matrix.

Porphyry copper deposit Important hydrothermal deposit type.

Potassium feldspar Important rock-forming silicate mineral of the feldspar group, $(Na,K)(AlSi_3O_8)$.

ppb Parts per billion. 1 ppb = 0.0000001%.

ppm Parts per million. 1 ppm = 0.0001%.

Precambrian Geological age in which the first primitive creatures appeared: From the formation of the Earth (about 4.6 billion years ago) to the beginning of the Cambrian (542 million years ago first mass occurrence of shellfish). Divided into Hadean (no rocks preserved), Archean and Proterozoic.

Precious metal Difficult to oxidize metals especially gold, silver and platinum group elements.

Propylitic alteration Slight alteration at relatively low temperature especially by heated groundwater (only minor amounts of magmatic fluid). Similar to greenschist metamorphism. In the vicinity of porphyry copper deposits and hydrothermal veins.

Prospecting Search for unknown deposits.

Pumice "Frozen magma foam", is extruded during explosive volcanic eruptions.

Pyrometallurgy Processing of ores or metals in a furnace. This includes, for example, the reduction of oxides with carbon monoxide and the roasting of sulfides.

Pyroxene Group of silicate minerals (chain silicates), including diopside, augite and enstatite.

Quartz Important rock-forming mineral, SiO_2. Perfect crystals are called rock crystal.

Rare-earth elements (REE) Group of elements comprising lanthan and the lanthanides (in the periodic table cerium to lutetium), as well as yttrium and scandium.

Reef A term mainly used in South Africa for thin ore layers in hard rock, especially if they contain platinum or gold.

Refining (metals) Purification of impure metals.

Refining (oil) Separation of crude oil into different components, in particular by distillation.

Remote sensing Obtaining geophysical data using satellites, aircraft or helicopters.

Reserves Secured deposits that can be mined profitably for the given market situation and technology.

Reservoir Porous rock into which petroleum and natural gas have flowed.

Resources Deposits including currently unprofitable or only suspected deposits.

Reverse fault A fault in a compressional setting at which the hangingwall (the block above the fault) moves upward relative to the footwall block. See also thrust fault.

Rhyolite Volcanic rock with high SiO_2 content ("acidic"). Formed by fractional crystallization from basaltic magma or by melting continental crust. Often as rock glass (obsidian) or finely fragmented as ash.

Rift valley Elongated depression caused by extension, which is limited on both sides by normal faults.

Roasting "Burning" of sulfide ore under oxidizing conditions. Converts sulfides to oxides and allows further smelting.

Rock A natural mixture of minerals.

Rotary drilling Method used for deep drilling with rotating drill pipes.

S-type granite Granite formed by melting sediments or corresponding metamorphic rocks.

Seam A sedimentary stratiform deposit, especially coal.

Seam gas Methane contained in coal.

SEDEX Sedimentary exhalative deposit. Important type of deposit formed at hot springs on the seabed.

Sediment Rock deposited on the Earth's surface (excluding volcanic rocks). There are clastic sediments (sandstone, shale, conglomerate), biogenic sediments (almost all limestones) and sediments formed by chemical precipitation (salt, gypsum, some limestones).

Semiconductors Elements or compounds with electrical properties that differ from both electrical conductors and insulators.

Sericitic alteration Alteration of a rock to predominantly sericite (very fine-grained muscovite) and chlorite, occurs in many types of deposits. Also called phyllic alteration. Flowing transition to chloritization.

Serpentinite Rock consisting predominantly of the minerals of the serpentine group. Formed from peridotite (the mantle rock) by absorption of water.

Siderophile "Iron-loving elements" (according to Goldschmidt) such as manganese, cobalt, nickel, gold, platinum.

Silicate Minerals with (SiO_4)-tetrahedron as anion complex. These include almost all important rock-forming minerals.

Silicification Alteration with enrichment of SiO_2 by precipitation of additional quartz, chalcedony or opal or by leaching of other substances. Especially near the surface above porphyry copper deposits and in high-sulfidation epithermal systems.

Shale Fine-grained clastic sedimentary rock, with high content of clay minerals.

Shale gas Gas firmly trapped in the pores of shales, can be obtained by fracking.

Shale oil Crude oil firmly trapped in the pores of shales can be obtained by fracking. Sometimes also for synthetic crude oil obtained from oil shale.

Shear zone Fault with ductile deformation.

Shelf Shallow sea above continental crust at a continental margin.

Schist Metamorphic rocks that break into slabs because they consist of aligned leaf-shaped minerals. Notably mica schist formed by the metamorphic transformation of shales.

Skarn Metasomatic silicate rock formed by reaction of a carbonate rock with hydrothermal solutions.

Slag Melting residue from metal smelting.

Slate Low-grade metamorphic rocks formed from shale.

Smelter Plant for extracting metals from ores (smelting). This is done either in a special furnace (pyrometallurgical), by solution and precipitation (hydrometallurgical) or by electrolysis in a solution or melt (electrometallurgical).

Smelting Metal extraction from ores and subsequent further processing.

Solidus Temperature at which the first melt is formed when a rock is heated (or the last melt residues solidify when magma cools). See also Liquidus.

Source rock Rock with a high content of organic substances from which crude oil and natural gas can form.

Steel Iron alloy with a carbon content of 0.01% to a maximum of 2%, may contain added steel refiners such as manganese, chromium, vanadium.

Sterile rock Economically unusable side rock of a deposit. Also barren rock.

Stockwork Ore-containing fissure network (also stringer zone).

Stratabound A deposit that is limited to a specific rock layer (e.g. a certain layer of limestone, sandstone, shale or tuff). The deposit itself can be stratiform, but also irregularly shaped or discordant.

Stratiform Layered deposit parallel to the rock layers (concordant), e.g. coal seam, Kupferschiefer, SEDEX.

Streak color When a mineral is scratched over a ceramic board, a line of characteristic color is created. Important distinctive feature.

Strike-slip fault A fault where two tectonic blocks glide past each other laterally (also called transform fault in the case of plate boundaries).

Stringer zone Ore-containing fissure network (Also stockwork zone).

Subduction zone Plate boundary at which oceanic crust submerges under a continent (e.g. Andes) or another oceanic plate (e.g. Mariana arc) and sinks into the mantle.

Sulfide Mineral with S^{2-} as anion. Many ore minerals are sulfides, e.g. chalcopyrite, sphalerite, galena.

Sulfosalts Generic term for minerals with $(AsS_3)^{3-}$ or $(SbS_3)^{3-}$. Also referred to as "complex sulfides".

Superconductor Material whose electrical resistance drops below the "transition temperature" to 0. The transition temperature is usually close to absolute zero (-273.15 °C) but significantly higher for high-temperature superconductors (these are still icy temperatures, but can be achieved by cooling with e.g. liquid nitrogen).

Supercritical See critical point.

Supergene Secondary enrichment near the surface mostly by processes like weathering, soil formation, oxidization, mobilization and precipitation. E.g. enrichment in the oxidation zone of sulfide deposits.

Surficial At or just below the Earth's surface.

Suture Seam between two continents after continental collision. Often contains ophiolites.

Syngenetic Ores and side rocks formed together. Opposite to epigenetic.

Tailings Muddy waste from ore processing which is sedimented in a sludge pond.

Tar sand Sandstone with a high content of asphalt and extra-heavy crude oils resulting from the degradation of crude oil.

Tectonics Deformation and displacement of rocks caused by stress (directed pressure) and the resulting brittle or ductile deformation.

Telethermal Obsolete term for hydrothermal formations below 100 °C.

Terrane Crustal blocks like mini-continents of different size (e.g. Madagascar, Seychelles), but also seamount chains, island arcs, submarine basalt plateaus. Can be welded to the edge of a continent at subduction zones.

Tethys Ocean vanished by the Alpine orogeny (in the Alps, the Taurus and Zagros and the Himalayas).

Texture Geometry of the mineral grains in the rock.

Tholeiite Silica-oversatured basalt, which contains orthopyroxene and possibly quartz in contrast to silica-undersaturated alkali olivine basalt. Typical for magmatism at mid-ocean ridges. Fractionation of tholeiite initially leads to a sharp increase in the iron content (tholeiitic trend).

Thrust fault Flat dipping fault typical for continental collision with overthrusting of tectonic nappes.

Tight oil Petroleum trapped in a reservoir rock with very low permeability, can be extracted by fracking.

Transition metals Metals in the d-block of the periodic table (atomic numbers 21–30, 39–48, 57–80, 89–112). Almost all metals belong to this group.

Tuff Loose or subsequently solidified deposits of volcanic ash. The material either moved down the volcanic cone as a pyroclastic flow or rained down from the eruption cloud during an ash eruption.

Turbidite Avalanche-like submarine turbidity current or its deposits.

Ultrabasic Melt or magmatite with <45% SiO_2.

Ultramafic Rock consisting of more than 90% mafic minerals.

Unconformity Boundary where sediments lie at another angle above tilted and partly eroded older rocks.

Vein A fissure that has been completely or partially filled by hydrothermal minerals.

VHMS Volcanic-hosted massive sulfide. Corresponds to volcanogenic massive sulfide deposits (VMS), but without genetic interpretation.

Volcanic rock Extruded igneous rock deposited on the Earth's surface, such as basalt, andesite or rhyolite (Also: extrusive rock). Can be a hard rock (e.g. a lava flow), rock glass (obsidian) or loose ash deposits (tuff), which in turn harden over time.

Volcanogenic massive sulfide deposit (VMS) An important type of deposit formed by hot springs on the seabed.

White smokers Vent shaped springs on the seabed that deposit mainly anhydrite or barite. They are similar to black smokers, but slightly cooler (250–300 °C).

Wrought iron Iron that contains almost no carbon. Elastic, easy to forge, but relatively soft.

WTI West Texas Intermediate, a light type of petroleum whose price is used to determine the price of other types of crude oil on the stock exchange. See also Brent.

Zeolite Group of silicate minerals (framework silicates) whose wide-meshed crystal lattice contains exchangeable ions and water.

Index

A

Aachen (Germany), 289
Abiotic hydrocarbons, 292
Abitibi Greenstone Belt (Canada), 114, 181, 213
Absorber, 330, 335, 337
Abyssal pegmatite, 120
Acanthite, 63, 166, 184
Accessory, 68, 77, 118, 130
Access ramp, 28
Accretionary orogen, 161, 182
Acheson process, 338
Acidic magma, 81, 89, 326
Acidithiobacillus ferrooxidans, 11, 36
Acidithiobacillus thiooxidans, 36
Acid mine drainage, 34
Actinolite, 192, 334
Adsorption, 271
Advanced SpaceborneThermal Emission
    and Reflectionradiometer (ASTER), 18
Aegirine, 118, 124, 132
Aenigmatite, 130, 136
Aeolian placer, 268
Aerial image, 17
Aeromagnetics, 21
Agpaitic, 118
Agpaitic rock, 130
Airborne geophysics, 20
Aircraft, 17, 19
Airgun, 23
Aitik (Sweden), 189
Alabaster, 250
Alaska-Urals type, 102
Albania, 213
Albite, 328
Alemão (Brazil), 189
Algae, 293
Algea mat, 240
Algeria, 316
Algoma type, 237
Aljustrel (Portugal), 216
Alkali feldspar, 328
Alkaline manganese battery, 54
Alkaline rocks, 84, 123
Alkane, 296
Allanite, 68, 71, 119
Allard Lake anorthosite complex, 116
Allchar (Macedonia), 196
Aller series (Zechstein), 256

Alligator River District (Australia), 202
Allochthonous, 231, 287
Allophane, 329
Alluvial fan, 265, 266
Alluvial placer, 265
Almadén (Spain), 166
Almadine, 333
Alpine orogeny, 199, 303
Alpine type, 196
Altenberg (Germany), 38, 156, 188
Alteration zone, 19
Alucrete, 271, 276
Aluminium, 65, 271
    smelter, 40
Aluminium bronze, 59
Aluminosilicates, 332
Alunite, 175, 180
Alvikite, 124
Amalgam, 75
Amalgam process, 36, 63, 75
Amazonite, 119
Amblygonite, 72, 119, 122
Ammonium phosphate, 263
Amorphous graphite, 338
Amosite, 334
Amphibole asbestos, 334
Anatase, 66
Andacollo (Chile), 178
Andalusite, 331, 332
Andesite, 83
Andradite, 192
Anglesite, 60, 181
Anhydrite, 248, 299, 303
Ankerite, 194, 239
Annaberg (Germany), 154
Annabergite, 75
Anode sludge, 40, 63, 74
Anorthite, 328
Anorthosite, 83, 95, 115
Anoxic, 222, 232, 246, 293
Anoxygenic phototrophic bacteria, 240
Antamina (Peru), 194
Antartica, 96
Anthracite, 285
Anticline, 300
Antigorite, 334
Antimony, 75, 195
    critical resources, 12

© Springer Nature Switzerland AG 2020
F. Neukirchen and G. Ries, *The World of Mineral Deposits*,
https://doi.org/10.1007/978-3-030-34346-0